AF840993

# Springer Texts in Statistics

*Series Editors*
G. Casella
S. Fienberg
I. Olkin

For other titles published in this series, go to
www.springer.com/series/417

Robert W. Keener

# Theoretical Statistics

## Topics for a Core Course

Springer

Robert W. Keener
Department of Statistics
University of Michigan
Ann Arbor, MI 48109-1092
USA



ISSN 1431-875X
ISBN 978-1-4614-2670-7 ISBN 978-0-387-93839-4 (eBook)
DOI 10.1007/978-0-387-93839-4
Springer New York Dordrecht Heidelberg London



Printed on acid-free paper

Springer is part of Springer Science+Business Media (www.springer.com)

To Michael, Herman, Carl, and Billy

# Preface

This book evolved from my notes for a three-semester sequence of core courses on theoretical statistics for doctoral students at the University of Michigan. When I first started teaching these courses, I used *Theory of Point Estimation* and *Testing Statistical Hypotheses* by Lehmann as texts, classic books that have certainly influenced my writings.

To appreciate this book students will need a background in advanced calculus, linear algebra, probability, and some analysis. Some of this material is reviewed in the appendices. And, although the content on statistics is reasonably self-contained, prior knowledge of theoretical and applied statistics will be essential for most readers.

In teaching core courses, my philosophy has been to try to expose students to as many of the central theoretical ideas and topics in the discipline as possible. Given the growth of statistics in recent years, such exposition can only be achieved in three semesters by sacrificing depth. Although basic material presented in early chapters of the book is covered carefully, many of the later chapters provide brief introductions to areas that could take a full semester to develop in detail.

The role of measure theory in advanced statistics courses deserves careful consideration. Although few students will need great expertise in probability and measure, all should graduate conversant enough with the basics to read and understand research papers in major statistics journals, at least in their areas of specialization. Many, if not most, of these papers will be written using the language of measure theory, if not all of its substance. As a practical matter, to prepare for thesis research many students will want to begin studying advanced methods as soon as possible, often before they have finished a course on measure and probability. In this book I follow an approach that makes such study possible. Chapter 1 introduces probability and measure theory, stating many of the results used most regularly in statistics. Although this material cannot replace an honest graduate course on probability, it gives most students the background and tools they need to read and understand most theoretical derivations in statistics. As we use this material in the rest of the book, I avoid esoteric mathematical details unless they are central to a proper understanding of issues at hand. In addition to the intrinsic value of concepts from measure theory, there are several other advantages to this approach. First, results in the book can be stated precisely and at their proper level of generality, and most of the proofs presented are essentially rigorous.[1] In addition, the use of material from probability, measure theory, and analysis in a statistical context will help students appreciate its value and will motivate some to study and learn probability at a deeper level. Although this approach is a challenge for some students, and may make some statistical issues a bit harder to understand and appreciate, the advantages outweigh these concerns.

As a caveat I should mention that some sections and chapters, mainly later in the book, are more technical than most and may not be accessible without a sufficient background in mathematics. This seems unavoidable to me; the topics considered cannot be covered properly otherwise.

Conditioning arguments are used extensively in the book. To keep the derivations as intuitive and accessible as possible, the presentation is based on (regular) conditional distributions to avoid conditioning on $\sigma$-fields.[2] As long as the conditioning information can be viewed as a random vector, conditional distributions exist and this approach entails no loss of generality. Conditional distributions are introduced in Chapter 1, with the conditioning variable discrete, and the law of total probability or *smoothing* is demonstrated in this case. A more general treatment of conditioning is deferred to Chapter 6. But I mention in Chapter 1 that smoothing identities are completely general, and use these identities in Chapter 6 to motivate the technical definition of conditional distributions.

With advances in technology for sharing and collecting information, large data sets are now common. Large sample methods have increasing value in statistics and receive significant attention in this book. With large amounts of data, statisticians will often seek the flexibility of a semi- or nonparametric model, models in which some parameters are viewed as smooth functions. At a technical and practical level, there is considerable value in viewing functions as points in some space. This notion is developed in various ways in this text. The discussion of asymptotic normality for the maximum likelihood estimator is structured around a weak law of large numbers for random functions, an approach easily extended to cover estimating equations and robustness. Weak compactness arguments are used to study optimal testing. Finally, there is an introduction to Hilbert space theory, used to study a spline approach to nonparametric regression. Modern statisticians need some knowledge of functional analysis. To help students meet the challenge of learning this material, the presentation here builds intuition by noting similarities between infinite-

[1] A reader with a good background in probability should have little trouble filling in any missing technical details.

[2] Filtrations and conditioning on $\sigma$-fields are mentioned in Chapter 20 on sequential analysis.

dimensional and finite-dimensional spaces, and provides motivation by linking the mathematical results to significant statistical applications.

If you are a professor using this book as a text, please note that results from Chapters 1 through 4 and Sections 6.1 and 6.2 are used extensively in the rest of the book, and any unfamiliar material on these pages should be covered with care. But much of the rest of the book can be resequenced or omitted to suit your preferences. Chapters 7 and 15 on Bayesian methods should be covered in order, as should Chapters 12, 13, and 17 on hypothesis testing. Chapter 11 on empirical Bayes estimation uses results from Chapter 7, and Chapter 14 on the general linear model uses results on testing from Chapter 12. Results on large sample theory from Chapters 8 and (to a lesser extent) 9 are used in Chapters 15 through 20. As I mentioned earlier, results in some chapters and sections[3] are more mathematically challenging; depending on the maturity of your students you may want to omit or cover this material superficially, possibly without proofs or derivations. For these chapters and sections and others, title footnotes indicate whether the material is optional and how the results will be used later.

Finally, a few words of appreciation are in order. To Michael Woodroofe, Herman Chernoff, and Carl Bender, who have had such an impact on my personal development as a mathematician, probabilist, and statistician; to friends, family, colleagues, and the Department for support and encouragement; and to past students, reviewers, and editors for a wealth of useful suggestions. This manuscript was typeset using LaTeXand figures were produced using MATLAB. Finally, a special thanks to future students; the notion that this book will help some of you has kept me believing it to be a worthwhile project.

Ann Arbor, Michigan
June 24, 2010

ROBERT KEENER

[3] My list would include Sections 6.4, 9.1, 9.9, 12.5, 12.6, and 12.7; and Chapters 13 and 16.

# Notation

*Absolute Continuity*, $P \ll \mu$: The measure $P$ is absolutely continuous with respect to (or $P$ is dominated by) the measure $\mu$. See page 7.

*Convergence in Distribution*: $Y_n \Rightarrow Y$. See page 131.

*Convergence in Probability*: $Y_n \xrightarrow{p} Y$. See page 129.

*Cumulants*: $\kappa_{r_1,\dots,r_s}$. See page 30.

*Derivatives*: If $h$ is a differentiable function from some subset of $\mathbb{R}^m$ into $\mathbb{R}^m$, then $Dh(x)$ is a matrix of partial derivatives with $[Dh(x)]_{ij} = \partial h_i(x)/\partial x_j$.

*Floor and Ceiling*: For $x \in \mathbb{R}$, the floor of $x$, denoted $\lfloor x \rfloor$, is the is the largest integer $y$ with $y \le x$. The ceiling $\lceil x \rceil$ of $x$ is the smallest integer $y \ge x$.

*Inner Product*: $\langle x, y \rangle$. See page 374.

*Inverse Functions*: If $f$ is a function on $D$ with range $R = f(D)$, then $f^{-1}$, mapping $2^R \to 2^D$, is defined by $f^{-1}(B) = \{x \in D : f(x) \in B\}$. If $f$ is one-to-one, the inverse function $f^{\leftarrow}$ is defined so that $f^{\leftarrow}(y) = x$ when $y = f(x)$.

*Maximum and Minimum*: $x \wedge y \stackrel{\text{def}}{=} \min\{x, y\}$ and $x \vee y \stackrel{\text{def}}{=} \max\{x, y\}$.

*Norms*: For $x \in \mathbb{R}^p$, $\|x\|$ is the usual Euclidean norm. For functions, $\|f\|_\infty = \sup|f|$, and $\|f\|_2 = \left[\int f\,d\mu\right]^{1/2}$. For points $x$ in an inner product space, $\|x\| = \langle x, x\rangle^{1/2}$.

*Point Mass*: $\delta_c$ is a probability measure that all of its mass to the point $c$, so $\delta(\{c\}) = 1$.

*Scales of Magnitude*: $O(\cdot)$, $O_p(\cdot)$, $o(\cdot)$, and $o_p(\cdot)$. See page 141.

*Set Notation*: The complement of a set $A$ is denoted $A^c$. For two sets $A$ and $B$, $AB$ or $A \cap B$ denotes the intersection, $A \cup B$ denotes the union, and $A - B \stackrel{\text{def}}{=} AB^c$ will denotes the set difference. Infinite unions and intersections of sets $A_1, A_2, \dots$ are denoted

$$\bigcup_{i=1}^{\infty} A_i \stackrel{\text{def}}{=} \{x : x \in A_i, \forall i\}$$

and

$$\bigcap_{i=1}^{\infty} A_i \stackrel{\text{def}}{=} \{x : x \in A_i, \exists i\}.$$

*Stochastic Transition Kernel*: $Q$ is a stochastic transition kernel if $Q_x(\cdot)$ is a probability measure for all $x$ and $Q_x(B)$ is a measurable function of $x$ for every Borel set $B$.

*Topology*: For a set $S$, $\overline{S}$ is the closure, $S^o$ the interior, and $\partial S = \overline{S} - S^o$ is the boundary. See page 432.

*Transpose*: The transpose of a vector or matrix $x$ is denoted $x'$.

# Contents

# 1

# Probability and Measure

Much of the theory of statistical inference can be appreciated without a detailed understanding of probability or measure theory. This book does not treat these topics with rigor. But some basic knowledge of them is quite useful. Much of the literature in statistics uses measure theory and is inaccessible to anyone unfamiliar with the basic notation. Also, the notation of measure theory allows one to merge results for discrete and continuous random variables. In addition, the notation can handle interesting and important applications involving censoring or truncation in which a random variable of interest is neither discrete nor continuous. Finally, the language of measure theory is necessary for stating many results correctly. In the sequel, measure-theoretic details are generally downplayed or ignored in proofs, but the presentation is detailed enough that anyone with a good background in probability should be able to fill in any missing details.

In this chapter measure theory and probability are introduced, and several of the most useful results are stated without proof.

## 1.1 Measures

A measure $\mu$ on a set $\mathcal{X}$ assigns a nonnegative value $\mu(A)$ to many subsets $A$ of $\mathcal{X}$. Here are two examples.

*Example 1.1.* If $\mathcal{X}$ is countable, let

$$\mu(A) = \#A = \text{number of points in } A.$$

This $\mu$ is called *counting measure* on $\mathcal{X}$.

*Example 1.2.* Let $\mathcal{X} = \mathbb{R}^n$ and define

$$\mu(A) = \int \cdots \int_A dx_1 \cdots dx_n.$$

R.W. Keener, *Theoretical Statistics: Topics for a Core Course*, Springer Texts in Statistics,
DOI 10.1007/978-0-387-93839-4_1, 

With $n = 1$, 2, or 3, $\mu(A)$ is called the length, area, or volume of $A$, respectively. In general, this measure $\mu$ is called *Lebesgue measure* on $\mathbb{R}^n$. Actually, for some sets $A$ it may not be clear how one should evaluate the integral "defining" $\mu(A)$, and, as we show, the theory of measure is fundamentally linked to basic questions about integration.

The measures in these examples differ from one another in an interesting way. Counting measure assigns mass to individual points, $\mu(\{x\}) = 1$ for $x \in \mathcal{X}$, but the Lebesgue measure of any isolated point is zero, $\mu(\{x\}) = 1$. In general, if $\mu(\{x\}) > 0$, then $x$ is called an *atom* of the measure with mass $\mu(\{x\}) > 0$.

It is often impossible to assign measures to all subsets $A$ of $\mathcal{X}$. Instead, the domain[1] of a measure $\mu$ will be a $\sigma$-field.

**Definition 1.3.** *A collection $\mathcal{A}$ of subsets of a set $\mathcal{X}$ is a* $\sigma$-field *(or $\sigma$-algebra) if*

1. *$\mathcal{X} \in \mathcal{A}$ and $\emptyset \in \mathcal{A}$.*
2. *If $A \in \mathcal{A}$, then $A^c = \mathcal{X} - A \in \mathcal{A}$.*
3. *If $A_1, A_2, \ldots \in \mathcal{A}$, then $\bigcup_{i=1}^{\infty} A_i \in \mathcal{A}$.*

The following definition gives the basic properties that must be satisfied for a set function $\mu$ to be called a measure. These properties should be intuitive for Examples 1.1 and 1.2.

**Definition 1.4.** *A function $\mu$ on a $\sigma$-field $\mathcal{A}$ of $\mathcal{X}$ is a* measure *if*

1. *For every $A \in \mathcal{A}$, $0 \le \mu(A) \le \infty$; that is, $\mu : \mathcal{A} \to [0, \infty]$.*
2. *If $A_1, A_2, \ldots$ are disjoint elements of $\mathcal{A}$ ($A_i \cap A_j = \emptyset$ for all $i \ne j$), then*

$$\mu\left(\bigcup_{i=1}^{\infty} A_i\right) = \sum_{i=1}^{\infty} \mu(A_i).$$

One interesting and useful consequence of the second part of this definition is that if measurable sets $B_n$, $n \ge 1$, are increasing ($B_1 \subset B_2 \subset \cdots$), with union $B = \bigcup_{n=1}^{\infty} B_n$, called the limit of the sequence, then

$$\mu(B) = \lim_{n\to\infty} \mu(B_n). \tag{1.1}$$

This can be viewed as a continuity property of measures.

For notation, if $\mathcal{A}$ is a $\sigma$-field of subsets of $\mathcal{X}$, the pair $(\mathcal{X}, \mathcal{A})$ is called a *measurable space*, and if $\mu$ is a measure on $\mathcal{A}$, the triple $(\mathcal{X}, \mathcal{A}, \mu)$ is called a *measure space*.

A measure $\mu$ is *finite* if $\mu(\mathcal{X}) < \infty$ and *$\sigma$-finite* if there exist sets $A_1, A_2, \ldots$ in $\mathcal{A}$ with $\mu(A_i) < \infty$ for all $i = 1, 2, \ldots$ and $\bigcup_{i=1}^{\infty} A_i = \mathcal{X}$. All measures considered in this book are $\sigma$-finite.

---

[1] See Appendix A.1 for basic information and language about functions.

A measure $\mu$ is called a *probability measure* if $\mu(\mathcal{X}) = 1$, and then the triple $(\mathcal{X}, \mathcal{A}, \mu)$ is called a *probability space.* For probability (or other finite) measures, something analogous to (1.1) holds for decreasing sets. If measurable sets $B_1 \supset B_2 \supset \cdots$ have intersection $B = \bigcap_{n=1}^{\infty} B_n$, then

$$\mu(B) = \lim_{n\to\infty} \mu(B_n). \tag{1.2}$$

*Example 1.1, continued.* Counting measure given by $\mu(A) = \#A$ can be defined for any subset $A \subset \mathcal{X}$, so in this example, the $\sigma$-field $\mathcal{A}$ is the collection of all subsets of $\mathcal{X}$. This $\sigma$-field is called the *power set* of $\mathcal{X}$, denoted $\mathcal{A} = 2^{\mathcal{X}}$.

*Example 1.2, continued.* The Lebesgue measure of a set $A$ can be defined, at least implicitly, for any set $A$ in a $\sigma$-field $\mathcal{A}$ called the Borel sets of $\mathbb{R}^n$. Formally, $\mathcal{A}$ is the smallest $\sigma$-field that contains all "rectangles"

$$(a_1, b_1) \times \cdots \times (a_n, b_n) = \{x \in \mathbb{R}^n : a_i < x_i < b_i, i = 1, \dots, n\}.$$

Although there are many subsets of $\mathbb{R}^n$ that are not Borel, none of these sets can be written explicitly.

## 1.2 Integration

The goal of this section is to properly define integrals of "nice" functions $f$ against a measure $\mu$. The integral written as $\int f\, d\mu$ or as $\int f(x)\, d\mu(x)$ when the variable of integration is needed. To motivate later developments, let us begin by stating what integration is for counting and Lebesgue measure.

*Example 1.5.* If $\mu$ is counting measure on $\mathcal{X}$, then the integral of $f$ against $\mu$ is

$$\int f\, d\mu = \sum_{x\in\mathcal{X}} f(x).$$

*Example 1.6.* If $\mu$ is Lebesgue measure on $\mathbb{R}^n$, then the integral of $f$ against $\mu$ is

$$\int f\, d\mu = \int \cdots \int f(x_1, \dots, x_n)\, dx_1 \dots dx_n.$$

It is convenient to view $x$ as the vector $(x_1, \dots, x_n)'$ and write this integral against Lebesgue measure as $\int \cdots \int f(x)\, dx$ or $\int f(x)\, dx$.

The modern definition of integration given here is less constructive than the definition offered in most basic calculus courses. The construction is driven by basic properties that integrals should satisfy and proceeds arguing that for "nice" functions $f$ these properties force a unique value for $\int f\, d\mu$. A key regularity property for the integrand is that it is "measurable" according to the following definition.

**Definition 1.7.** *If $(\mathcal{X}, \mathcal{A})$ is a measurable space and $f$ is a real-valued function on $\mathcal{X}$, then $f$ is* measurable *if*

$$f^{-1}(B) \stackrel{\text{def}}{=} \{x \in \mathcal{X} : f(x) \in B\} \in \mathcal{A}$$

*for every Borel set $B$.*

Although there are many functions that are not measurable, they cannot be stated explicitly. Continuous and piecewise continuous functions are measurable. A more interesting example is the function $f : \mathbb{R} \to \mathbb{R}$ with $f(x) = 1$ when $x$ is an irrational number in $(0, 1)$, and $f(x) = 0$ otherwise. With the Riemann notion of integration used in basic calculus courses, for this function $f$, $\int f(x)\,dx$ is not defined. The more general methods presented here give the natural answer, $\int f(x)\,dx = 1$. In the sequel, functions of interest are generally presumed to be measurable.

The indicator function $1_A$ of a set $A$ is defined as

$$1_A(x) = I\{x \in A\} = \begin{cases} 1, & x \in A; \\ 0, & x \notin A. \end{cases}$$

Here are the basic properties for integrals.

1. For any set $A$ in $\mathcal{A}$, $\int 1_A\,d\mu = \mu(A)$.
2. If $f$ and $g$ are nonnegative measurable functions, and if $a$ and $b$ are positive constants,
$$\int (af + bg)\,d\mu = a \int f\,d\mu + b \int g\,d\mu. \tag{1.3}$$
3. If $f_1 \le f_2 \le \cdots$ are nonnegative measurable functions, and if $f(x) = \lim_{n\to\infty} f_n(x)$, then
$$\int f\,d\mu = \lim_{n\to\infty} \int f_n\,d\mu.$$

The first property provides the link between $\int f\,d\mu$ and the measure $\mu$, the second property is linearity, and the third property is useful for taking limits of integrals.

Using the first two properties, if $a_1, \ldots, a_m$ are positive constants, and if $A_1, \ldots, A_m$ are sets in $\mathcal{A}$, then

$$\int \left( \sum_{i=1}^m a_i 1_{A_i} \right) d\mu = \sum_{i=1}^m a_i \mu(A_i).$$

Functions of this form are called *simple*. Figure 1.1 shows the graph of the simple function $1_{(1/2,\pi)} + 2 1_{(1,2)}$. The following result asserts that nonnegative measurable functions can be approximated by simple functions.

**Theorem 1.8.** *If $f$ is nonnegative and measurable, then there exist nonnegative simple functions $f_1 \le f_2 \le \cdots$ with $f = \lim_{n\to\infty} f_n$.*

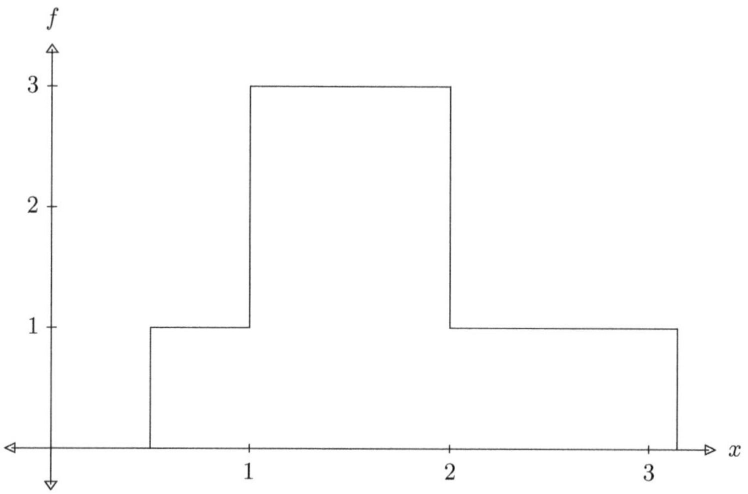


**Fig. 1.1.** The simple function $1_{(1/2,\pi)} + 21_{(1,2)}$.

This result along with the third basic property for integrals allows us to integrate any nonnegative measurable function $f$, at least in principle. The answer is unique; different choices for the increasing sequence of simple functions give the same answer. To integrate a general measurable function $f$, introduce the positive and negative parts

$$f^+(x) = \max\{f(x), 0\} \text{ and } f^-(x) = -\min\{f(x), 0\}.$$

Then $f^+$ and $f^-$ are both nonnegative and measurable, and $f = f^+ - f^-$. The integral of $f$ should generally be the difference between the integral of $f^+$ and the integral of $f^-$. This difference is ambiguous only when the integrals of $f^+$ and $f^-$ are both infinite. So, if either $\int f^+\, d\mu < \infty$ or $\int f^-\, d\mu < \infty$, we define

$$\int f\, d\mu = \int f^+\, d\mu - \int f^-\, d\mu.$$

With this definition the linearity in (1.3) holds unless the right-hand side is formally $\infty - \infty$. Note also that because $|f| = f^+ + f^-$, this definition gives a finite value for $\int f\, d\mu$ if and only if

$$\int f^+\, d\mu + \int f^-\, d\mu = \int |f|\, d\mu < \infty.$$

When $\int |f|\, d\mu < \infty$, $f$ is called *integrable*.

## 1.3 Events, Probabilities, and Random Variables

Let $P$ be a probability measure on a measurable space $(\mathcal{E}, \mathcal{B})$, so $(\mathcal{E}, \mathcal{B}, P)$ is a probability space. Sets $B \in \mathcal{B}$ are called *events*, points $e \in \mathcal{E}$ are called *outcomes*, and $P(B)$ is called the *probability* of $B$.

A measurable function $X : \mathcal{E} \to \mathbb{R}$ is called a *random variable*. The probability measure $P_X$ defined by

$$P_X(A) = P\bigl(\{e \in \mathcal{E} : X(e) \in A\}\bigr) \overset{\text{def}}{=} P(X \in A)$$

for Borel sets $A$ is called the *distribution* of $X$. The notation

$$X \sim Q$$

is used to indicate that $X$ has distribution $Q$; that is, $P_X = Q$. The *cumulative distribution function* of $X$ is defined by

$$F_X(x) = P(X \le x) = P\bigl(\{e \in \mathcal{E} : X(e) \le x\}\bigr) = P_X\bigl((-\infty, x]\bigr),$$

for $x \in \mathbb{R}$.

## 1.4 Null Sets

Let $\mu$ be a measure on $(\mathcal{X}, \mathcal{A})$. A set $N$ is called *null* (or null with respect to $\mu$) if

$$\mu(N) = 0.$$

If a statement holds for $x \in \mathcal{X} - N$ with $N$ null, the statement is said to hold *almost everywhere* (a.e.) or a.e. $\mu$. For instance, $f = 0$ a.e. $\mu$ if and only if $\mu\bigl(\{x \in \mathcal{X} : f(x) \ne 0\}\bigr) = 0$.

There is an alternative language for similar ideas when $\mu$ is a probability measure. Suppose some statement holds if and only if $x \in B$. Then the statement holds (a.e. $\mu$) if and only if $\mu(B^c) = 0$ if and only if $\mu(B) = 1$. This can be expressed by saying "the statement holds *with probability one*."

The values of a function on a null set cannot affect its integral. With this in mind, here are a few useful facts about integration that are fairly easy to appreciate:

1. If $f = 0$ (a.e. $\mu$), then $\int f\,d\mu = 0$.
2. If $f \ge 0$ and $\int f\,d\mu = 0$, then $f = 0$ (a.e. $\mu$).
3. If $f = g$ (a.e. $\mu$), then $\int f\,d\mu = \int g\,d\mu$ whenever either one of the integrals exists.
4. If $\int 1_{(c,x)} f\,d\mu = 0$ for all $x > c$, then $f(x) = 0$ for a.e. $x > c$. The constant $c$ here can be $-\infty$.

As a consequence of 2, if $f$ and $g$ are integrable and $f > g$, then $\int f\,d\mu > \int g\,d\mu$ (unless $\mu$ is identically zero).

## 1.5 Densities

Densities play a basic role in statistics. In many situations the most convenient way to specify the distribution of a random vector $X$ is to give its density. Also, densities give likelihood functions used to compute Bayes estimators or maximum likelihood estimators. The density for a measure exists whenever it is absolutely continuous with respect to another measure according to the following definition.

**Definition 1.9.** *Let $P$ and $\mu$ be measures on a $\sigma$-field $\mathcal{A}$ of $\mathcal{X}$. Then $P$ is called* absolutely continuous *with respect to $\mu$, written $P \ll \mu$, if $P(A) = 0$ whenever $\mu(A) = 0$.*

**Theorem 1.10 (Radon–Nikodym).** *If a finite measure $P$ is absolutely continuous with respect to a $\sigma$-finite measure $\mu$, then there exists a nonnegative measurable function $f$ such that*

$$P(A) = \int_A f\,d\mu \stackrel{\text{def}}{=} \int f 1_A\,d\mu.$$

The function $f$ in this theorem is called the Radon–Nikodym derivative of $P$ with respect to $\mu$, or the *density* of $P$ with respect to $\mu$, denoted

$$f = \frac{dP}{d\mu}.$$

By the third fact about integration and null sets in the previous section, the density $f$ may not be unique, but if $f_0$ and $f_1$ are both densities, then $f_0 = f_1$ (a.e. $\mu$). If $X \sim P_X$ and $P_X$ is absolutely continuous with respect to $\mu$ with density $p = dP_X/d\mu$, it is convenient to say that $X$ *has density* $p$ with respect to $\mu$.

*Example 1.11. Absolutely Continuous Random Variables.* If a random variable $X$ has density $p$ with respect to Lebesgue measure on $\mathbb{R}$, then $X$ or its distribution $P_X$ is called *absolutely continuous* with density $p$. Then, from the Radon–Nikodym theorem,

$$F_X(x) = P(X \le x) = P_X\big((-\infty, x]\big) = \int_{-\infty}^{x} p(u)\,du.$$

Using the fundamental theorem of calculus, $p$ can generally be found from the cumulative distribution function $F_X$ by differentiation, $p(x) = F_X'(x)$.

*Example 1.12. Discrete Random Variables.* Let $\mathcal{X}_0$ be a countable subset of $\mathbb{R}$. The measure $\mu$ defined by

$$\mu(B) = \#(\mathcal{X}_0 \cap B)$$

for Borel sets $B$ is (also) called *counting measure on* $\mathcal{X}_0$. As in Example 1.5,

$$\int f\,d\mu = \sum_{x\in\mathcal{X}_0} f(x).$$

Suppose $X$ is a random variable and that

$$P(X \in \mathcal{X}_0) = P_X(\mathcal{X}_0) = 1.$$

Then $X$ is called a *discrete random variable.* Suppose $N$ is a null set for $\mu$, so $\mu(N) = 0$. From the definition of $\mu$, $\#(N \cap \mathcal{X}_0) = 0$, which means that $N \cap \mathcal{X}_0 = \emptyset$ and so $N \subset \mathcal{X}_0^c$. Then $P_X(N) = P(X \in N) \le P(X \in \mathcal{X}_0^c) = 1 - P(X \in \mathcal{X}_0) = 0$. Thus $N$ must also be a null set for $P_X$, and this shows that $P_X$ is absolutely continuous with respect to $\mu$. The density $p$ of $P_X$ with respect to $\mu$ satisfies

$$P(X \in A) = P_X(A) = \int_A p\,d\mu = \sum_{x\in\mathcal{X}_0} p(x)1_A(x).$$

In particular, if $A = \{y\}$ with $y \in \mathcal{X}_0$, then $X \in A$ if and only if $X = y$, and so

$$P(X = y) = \sum_{x\in\mathcal{X}_0} p(x)1_{\{y\}}(x) = p(y).$$

This density $p$ is called the *mass function* for $X$. Note that because $\mathcal{X}_0^c$ is a null set, the density $p(y)$ can be defined arbitrarily when $y \notin \mathcal{X}_0$. The natural convention is to take $p(y) = 0$ for $y \notin \mathcal{X}_0$, for then $p(y) = P(Y = y)$ for all $y$.

## 1.6 Expectation

If $X$ is a random variable on a probability space $(\mathcal{E}, \mathcal{B}, P)$, then the *expectation* or *expected value* of $X$ is defined as

$$EX = \int X\,dP. \tag{1.4}$$

This formula is rarely used. Instead, if $X \sim P_X$ it can be shown that

$$EX = \int x\,dP_X(x).$$

Also, if $Y = f(X)$, then

$$EY = Ef(X) = \int f\,dP_X. \tag{1.5}$$

Integration against $P_X$ in these two formulas is often accomplished using densities. If $P_X$ has density $p$ with respect to $\mu$, then

$$\int f\,dP_X = \int fp\,d\mu. \tag{1.6}$$

This identity allows formal substitution of $p\,d\mu$ for $dP_X$, which makes the derivative notation $p = dP_X/d\mu$ seem natural. Together these results can all be viewed as change of variable results. Proofs of these results are based on the methods used to define integrals. It is easy to show that (1.5) and (1.6) hold when $f$ is an indicator function. By linearity they must then hold for positive simple functions, and then a limiting argument shows that they hold for general measurable $f$, at least when the integrals exist. Specializing these results to absolutely continuous and discrete random variables we have the following important examples.

*Example 1.13.* If $X$ is an absolutely continuous random variable with density $p$, then

$$EX = \int x\,dP_X(x) = \int xp(x)\,dx$$

and

$$Ef(X) = \int f(x)p(x)\,dx. \tag{1.7}$$

*Example 1.14.* If $X$ is discrete with $P(X \in \mathcal{X}_0) = 1$ for a countable set $\mathcal{X}_0$, if $\mu$ is counting measure on $\mathcal{X}_0$, and if $p$ is the mass function given by $p(x) = P(X = x)$, then

$$EX = \int x\,dP_X(x) = \int xp(x)\,d\mu(x) = \sum_{x\in\mathcal{X}_0} xp(x)$$

and

$$Ef(X) = \sum_{x\in\mathcal{X}_0} f(x)p(x). \tag{1.8}$$

Expectation is a linear operation. If $X$ and $Y$ are random variables and $a$ and $b$ are nonzero constants, then

$$E(aX + bY) = aEX + bEY, \tag{1.9}$$

provided $EX$ and $EY$ both exist and the right-hand side is not $\infty - \infty$. This follows easily from the definition of expectation (1.4), because integration is linear (1.3). Another important property of expectation is that if $X$ and $Y$ have finite expectations and $X < Y$ (a.e. $P$), then $EX < EY$. Also, using linearity and the second fact about integration in Section 1.4, if $X \le Y$ (a.e. $P$) and both have finite expectations, $EX \le EY$ with equality only if $X = Y$ (a.e. $P$).

The *variance* of a random variable $X$ with finite expectation is defined as

$$\mathrm{Var}(X) = E(X - EX)^2.$$

If $X$ is absolutely continuous with density $p$, by (1.7)

$$\mathrm{Var}(X) = \int (x - EX)^2 p(x)\,dx,$$

and if $X$ is discrete with mass function $p$, by (1.8)

$$\mathrm{Var}(X) = \sum_{x \in \mathcal{X}_0} (x - EX)^2 p(x).$$

Using (1.9),

$$\mathrm{Var}(X) = E\bigl(X^2 - 2XEX + (EX)^2\bigr) = EX^2 - (EX)^2,$$

a result that is often convenient for explicit calculation.

The covariance between two random variables $X$ and $Y$ with finite expectations is defined as

$$\mathrm{Cov}(X, Y) = E(X - EX)(Y - EY), \tag{1.10}$$

whenever the expectation exists. Note that $\mathrm{Cov}(X, X) = \mathrm{Var}(X)$. Using (1.9),

$$\begin{aligned}\mathrm{Cov}(X, Y) &= E\bigl(XY - XEY - YEX + (EX)(EY)\bigr)\\ &= EXY - (EX)(EY). \end{aligned}\tag{1.11}$$

The covariance between two variables might be viewed as a measure of the linear association between the two variables. But because covariances are influenced by the measurement scale, a more natural measure is the *correlation*, defined using the covariance as

$$\mathrm{Cor}(X, Y) = \frac{\mathrm{Cov}(X, Y)}{\bigl[\mathrm{Var}(X)\mathrm{Var}(Y)\bigr]^{1/2}}.$$

Correlations always lie in $[-1, 1]$, with values $\pm 1$ arising when there is a perfect linear relation between the two variables.[2]

## 1.7 Random Vectors

If $X_1, \dots, X_n$ are random variables, then the function $X : \mathcal{E} \to \mathbb{R}^n$ defined by

[2] This follows from the covariance inequality (4.11).

$$X(e) = \begin{pmatrix} X_1(e) \\ \vdots \\ X_n(e) \end{pmatrix}, \qquad e \in \mathcal{E},$$

is called a *random vector.*[3] Much of the notation and many of the results presented in this chapter for random variables extend naturally and directly to random vectors. For instance, the distribution $P_X$ of $X$ is defined by

$$P_X(B) = P(X \in B) \stackrel{\text{def}}{=} P\bigl(\{e \in \mathcal{E} : X(e) \in B\}\bigr)$$

for Borel sets $B \in \mathbb{R}^n$, and notation $X \sim P_X$ means that $X$ has distribution $P_X$. The random vector $X$ or its distribution $P_X$ is called *absolutely continuous* with density $p$ if $P_X$ is absolutely continuous with respect to Lebesgue measure on $\mathbb{R}^n$. In this case

$$P(X \in B) = \int \underset{B}{\cdots} \int p(x)\, dx.$$

The random vector $X$ is *discrete* if $P(X \in \mathcal{X}_0) = 1$ for some countable set $\mathcal{X}_0 \subset \mathbb{R}^n$. If $p(x) = P(X = x)$, then $P_X$ has density $p$ with respect to counting measure on $\mathcal{X}_0$ and

$$P(X \in B) = \sum_{x \in \mathcal{X}_0 \cap B} p(x).$$

The expectation of a random vector $X$ is the vector of expectations,

$$EX = \begin{pmatrix} EX_1 \\ \vdots \\ EX_n \end{pmatrix}.$$

If $T : \mathbb{R}^n \to \mathbb{R}$ is a measurable function, then $T(X)$ is a random variable, and, as in (1.5),

$$ET(X) = \int T\, dP_X$$

whenever the expectation or integral exists. If $P_X$ has a density $p$ with respect to a dominating measure $\mu$, this integral can be expressed as $\int Tp\, d\mu$, which becomes

$$\sum_{x \in \mathcal{X}_0} T(x)p(x) \quad \text{or} \quad \int \cdots \int T(x)p(x)\, dx$$

in the discrete and absolutely continuous cases with $\mu$ counting or Lebesgue measure, respectively.

---

[3] Equivalently, the vector-valued function $X$ is measurable: $X^{-1}(B) \in \mathcal{E}$ for every Borel set $B \in \mathbb{R}^n$.

## 1.8 Covariance Matrices

A matrix $W$ is called a *random matrix* if the entries $W_{ij}$ are random variables. If $W$ is a random matrix, then $EW$ is the matrix of expectations of the entries,

$$(EW)_{ij} = EW_{ij}.$$

If $v$ is a constant vector, $A$, $B$, and $C$ are constant matrices, $X$ is a random vector, and $W$ is a random matrix, then

$$E[v + AX] = v + AEX \tag{1.12}$$

and

$$E[A + BWC] = A + B(EW)C. \tag{1.13}$$

These identities follow easily from basic properties of expectation because $(v + AX)_i = v_i + \sum_j A_{ij}X_j$ and $(A + BWC)_{ij} = A_{ij} + \sum_k \sum_l B_{ik}W_{kl}C_{lj}$.

The *covariance* of a random vector $X$ is the matrix of covariances of the variables in $X$; that is,

$$\bigl[\mathrm{Cov}(X)\bigr]_{ij} = \mathrm{Cov}(X_i, X_j).$$

If $\mu = EX$ and $(X - \mu)'$ denotes the transpose of $X - \mu$, a (random) row vector, then

$$\mathrm{Cov}(X_i, X_j) = E(X_i - \mu_i)(X_j - \mu_j) = E\bigl[(X - \mu)(X - \mu)'\bigr]_{ij},$$

and so

$$\mathrm{Cov}(X) = E(X - \mu)(X - \mu)'. \tag{1.14}$$

Similarly, using (1.11) or (1.13),

$$\mathrm{Cov}(X) = EXX' - \mu\mu'.$$

To find covariances after an affine transformation, because the transpose of a product of two matrices (or vectors) is the product of the transposed matrices in reverse order, using (1.14), (1.12), and (1.13), if $v$ is a constant vector, $A$ is a constant matrix, and $X$ is a random vector, then

$$\begin{aligned}\mathrm{Cov}(v + AX) &= E(v + AX - v - A\mu)(v + AX - v - A\mu)' \\ &= EA(X - \mu)(X - \mu)'A' \\ &= A\bigl[E(X - \mu)(X - \mu)'\bigr]A' \\ &= A\mathrm{Cov}(X)A'. \end{aligned} \tag{1.15}$$

## 1.9 Product Measures and Independence

Let $(\mathcal{X}, \mathcal{A}, \mu)$ and $(\mathcal{Y}, \mathcal{B}, \nu)$ be measure spaces. Then there exists a unique measure $\mu \times \nu$, called the *product measure*, on $(\mathcal{X} \times \mathcal{Y}, \mathcal{A} \vee \mathcal{B})$ such that

$$(\mu \times \nu)(A \times B) = \mu(A)\nu(B),$$

for all $A \in \mathcal{A}$ and all $B \in \mathcal{B}$. The $\sigma$-field $\mathcal{A} \vee \mathcal{B}$ is defined formally as the smallest $\sigma$-field containing all sets $A \times B$ with $A \in \mathcal{A}$ and $B \in \mathcal{B}$.

*Example 1.15.* If $\mu$ and $\nu$ are Lebesgue measures on $\mathbb{R}^n$ and $\mathbb{R}^m$, respectively, then $\mu \times \nu$ is Lebesgue measure on $\mathbb{R}^{n+m}$.

*Example 1.16.* If $\mu$ and $\nu$ are counting measures on countable sets $\mathcal{X}_0$ and $\mathcal{Y}_0$, then $\mu \times \nu$ is counting measure on $\mathcal{X}_0 \times \mathcal{Y}_0$.

The following result shows that integration against the product measure $\mu \times \nu$ can be accomplished by iterated integration against $\mu$ and $\nu$, in either order.

**Theorem 1.17 (Fubini).** *If $f \geq 0$, then*

$$\begin{aligned}\int f\,d(\mu \times \nu) &= \int \left[\int f(x,y)\,d\nu(y)\right] d\mu(x) \\ &= \int \left[\int f(x,y)\,d\mu(x)\right] d\nu(y).\end{aligned}$$

*Dropping the restriction $f \geq 0$, if $\int |f|\,d(\mu \times \nu) < \infty$ then these equations hold.*

Taking $f = 1_S$, this result gives a way to compute $(\mu \times \nu)(S)$ when $S$ is not the Cartesian product of sets in $\mathcal{A}$ and $\mathcal{B}$.

**Definition 1.18 (Independence).** *Two random vectors, $X \in \mathbb{R}^n$ and $Y \in \mathbb{R}^m$ are* independent *if*

$$P(X \in A, Y \in B) = P(X \in A)P(Y \in B), \tag{1.16}$$

*for all Borel sets $A$ and $B$.*

If $Z = \binom{X}{Y}$, then $Z \in A \times B$ if and only if $X \in A$ and $Y \in B$, and (1.16) can be expressed in terms of the distributions of $X$, $Y$, and $Z$ as

$$P_Z(A \times B) = P_X(A)P_Y(B).$$

This shows that the distribution of $Z$ is the product measure,

$$P_Z = P_X \times P_Y.$$

The density of $Z$ is also given by the product of the densities of $X$ and $Y$. This can be shown using Fubini's theorem and (1.6) to change variables of integration. Specifically, suppose $P_X$ has density $p_X$ with respect to $\mu$ and $P_Y$ has density $p_Y$ with respect to $\nu$. Then

$$\begin{aligned} P(Z \in S) &= \int 1_S \, d(P_X \times P_Y) \\ &= \int \left[ \int 1_S(x, y) \, dP_X(x) \right] dP_Y(y) \\ &= \int \left[ \int 1_S(x, y) p_X(x) \, d\mu(x) \right] p_Y(y) \, d\nu(y) \\ &= \int 1_S(x, y) p_X(x) p_Y(y) \, d(\mu \times \nu)(x, y). \end{aligned}$$

This shows that $P_Z$ has density $p_X(x)p_Y(y)$ with respect to $\mu \times \nu$. In applications, $\mu$ and $\nu$ will generally be counting or Lebesgue measure. Note that the level of generality here covers mixed cases in which one of the random vectors is discrete and the other is absolutely continuous.

Whenever $Z = \binom{X}{Y}$, $P_Z$ is called the *joint distribution* of $X$ and $Y$, and a density for $P_Z$ is called the *joint density* of $X$ and $Y$. So when $X$ and $Y$ are independent with densities $p_X$ and $p_Y$, their joint density is $p_X(x)p_Y(y)$.

These ideas extend easily to collections of several random vectors. If $Z$ is formed from random vectors $X_1, \dots, X_n$, then a density or distribution for $Z$ is called a joint density or joint distribution, respectively, for $X_1, \dots, X_n$. The vectors $X_1, \dots, X_n$ are *independent* if

$$P(X_1 \in B_1, \dots, X_n \in B_n) = P(X_1 \in B_1) \times \cdots \times P(X_n \in B_n)$$

for any Borel sets $B_1, \dots, B_n$. Then $P_Z = P_{X_1} \times \cdots \times P_{X_n}$, where this product is the unique measure $\mu$ satisfying

$$\mu(B_1 \times \cdots \times B_n) = P_{X_1}(B_1) \times \cdots \times P_{X_n}(B_n).$$

The following proposition shows that functions of independent variables are independent.

**Proposition 1.19.** *If $X_1$, ..., $X_n$ are independent random vectors, and if $f_1$, ..., $f_n$ are measurable functions, then $f_1(X_1)$, ..., $f_n(X_n)$ are independent.*

If $X_i$ has density $p_{X_i}$ with respect to $\mu_i$, $i = 1, \dots, n$, then $X_1, \dots, X_n$ have joint density $p$ given by

$$p(x_1, \dots, x_n) = p_{X_1}(x_1) \times \cdots \times p_{X_n}(x_n)$$

with respect to $\mu = \mu_1 \times \cdots \times \mu_n$. If $X_1, \dots, X_n$ are independent, and they all have the same distribution, $X_i \sim Q$, $i = 1, \dots, n$, then $X_1, \dots, X_n$ are called *independent and identically distributed* (i.i.d.), and the collection of variables is called a *random sample* from $Q$.

## 1.10 Conditional Distributions

Suppose $X$ and $Y$ are random vectors. If $X$ is observed and we learn that $X = x$, then $P_Y$ should no longer be viewed as giving appropriate probabilities for $Y$. Rather, we should modify $P_Y$ taking account of the new information that $X = x$. When $X$ is discrete this can be accomplished using the standard formula for conditional probabilities of events: Let $p_X(x) = P(X = x)$, the mass function for $X$; take $\mathcal{X}_0 = \{x : p_X(x) > 0\}$, the set of possible values for $X$; and define

$$Q_x(B) = P(Y \in B|X = x) = \frac{P(Y \in B, X = x)}{P(X = x)} \tag{1.17}$$

for Borel sets $B$ and $x \in \mathcal{X}_0$. Then for any $x \in \mathcal{X}_0$, it is easy to show that $Q_x$ is a probability measure, called the *conditional distribution* for $Y$ given $X = x$.

Formally, conditional probabilities should be *stochastic transition kernels.* These are defined as functions $Q : \mathcal{X} \times \mathcal{B} \to [0, 1]$ satisfying two properties. First, for $x \in \mathcal{X}$, $Q_x(\cdot)$ should be a probability measure on $\mathcal{B}$; and second, for any $B \in \mathcal{B}$, $Q_x(B)$ should be a measurable function of $x$.

For completeness, we should also define $Q_x(B)$ above when $x \notin \mathcal{X}_0$. How this is done does not really matter; taking $Q_x$ to be some fixed probability measure for $x \notin \mathcal{X}_0$ would be one simple possibility.

Conditional distributions also exist when $X$ is not discrete, but the definition is technical and is deferred to Chapter 6. However, the most important results in this section hold whether $X$ is discrete or not. In particular, if $X$ and $Y$ are independent and $X$ is discrete, by (1.17) $Q_x$ equals $P_Y$, regardless of the value of $x \in \mathcal{X}_0$. This fact remains true in general and is the basis for a host of interesting and useful calculations.

Integration against a conditional distributions gives a conditional expectation. Specifically, the conditional expectation of $f(X, Y)$ given $X = x$ is defined as

$$E\big[f(X,Y)\,\big|\,X = x\big] = \int f(x,y)\,dQ_x(y). \tag{1.18}$$

Suppose $X$ and $Y$ are both discrete with $Y$ taking values in a countable set $\mathcal{Y}_0$ and $X$ taking values in $\mathcal{X}_0$ as defined above. Then $Z = \binom{X}{Y}$ takes values in the countable set $\mathcal{X}_0 \times \mathcal{Y}_0$ and is discrete with mass function $p_Z(z) = P(Z = z) = P(X = x, Y = y)$, where $z = \binom{x}{y}$. By (1.17), $Q_x(\mathcal{Y}_0) = 1$ and so $Q_x$ is discrete with mass function $q_x$ given by

$$q_x(y) = Q_x\big(\{y\}\big) = P(Y = y|X = x) = \frac{P(Y = y, X = x)}{P(X = x)} \tag{1.19}$$

for $x \in \mathcal{X}_0$. Then the conditional expectation in (1.18) can be calculated as a sum,

$$H(x) = E\big[f(X,Y)\,\big|\,X = x\big] = \sum_{y \in \mathcal{Y}_0} f(x,y)q_x(y).$$

For regularity, suppose $E|f(X,Y)| < \infty$. Noting from (1.19) that $P(X = x, Y = y) = q_x(y)p_X(x)$, the expectation of $f(X,Y)$ can be written as

$$\begin{aligned} Ef(X,Y) &= \sum_{\binom{x}{y} \in \mathcal{X}_0 \times \mathcal{Y}_0} f(x,y)P(X = x, Y = y) \\ &= \sum_{x \in \mathcal{X}_0} \sum_{y \in \mathcal{Y}_0} f(x,y)q_x(y)p_X(x) \\ &= \sum_{x \in \mathcal{X}_0} H(x)p_X(x) \\ &= EH(X). \end{aligned}$$

This is a fundamental result in conditioning, called the *law of total probability*, *the tower property*, or *smoothing*. In fact, smoothing identities are so basic that they form the basis for general definitions of conditional probability and expectation when $X$ is not discrete. The random variable $H(X)$ obtained evaluating $H$ from (1.18) at $X$ is denoted

$$H(X) = E\bigl[f(X,Y)\bigm|X\bigr].$$

With this convenient notation the smoothing identity is just

$$Ef(X,Y) = EE\bigl[f(X,Y)\bigm|X\bigr].$$

In particular, when $f(X,Y) = Y$ this becomes

$$EY = EE(Y|X).$$

When $Y = 1_B$, the indicator of an event $B$, $EY = P(B)$ and this identity becomes

$$P(B) = EP(B|X),$$

where $P(B|X) \stackrel{\text{def}}{=} E(1_B|X)$. Finally, these identities also hold when the initial expectation or probability is conditional. Specifically,[4]

$$E(Y|X) = E\bigl[E(Y|X,W)\bigm|X\bigr] \tag{1.20}$$

and

$$P(B|X) = E\bigl[P(B|X,Y)\bigm|X\bigr].$$

[4] See Problem 1.46.

## 1.11 Problems[5]

*1. Prove (1.1). Hint: Define $A_1 = B_1$ and $A_n = B_n - B_{n-1}$ for $n \geq 2$. Show that the $A_n$ are disjoint and use the countable additivity property of measure. Note: By definition, $\sum_{n=1}^{\infty} c_n = \lim_{N\to\infty} \sum_{n=1}^{N} c_n$.

2. For a set $B \subset \mathbb{N} = \{1, 2, \dots\}$, define

$$\mu(B) = \lim_{n\to\infty} \frac{\#\big[B \cap \{1, \dots, n\}\big]}{n},$$

when the limit exists, and let $\mathcal{A}$ denote the collection of all such sets.

   a) Find $\mu(E)$, $\mu(O)$, and $\mu(S)$, where $E = \{2, 4, \dots\}$, all even numbers, $O = \{1, 3, \dots\}$, all odd numbers, and $S = \{1, 4, 9, \dots)$, all perfect squares.
   b) If $A$ and $B$ are disjoint sets in $\mathcal{A}$, show that $\mu(A \cup B) = \mu(A) + \mu(B)$.
   c) Is $\mu$ a measure? Explain your answer.

3. Suppose $\mu$ is a measure on the Borel sets of $(0, \infty)$ and that $\mu\big((x, 2x]\big) = \sqrt{x}$, for all $x > 0$. Find $\mu\big((0, 1]\big)$.

4. Let $\mathcal{X} = \{1, 2, 3, 4\}$. Find the smallest $\sigma$-field $\mathcal{A}$ of subsets of $\mathcal{X}$ that contains the sets $\{1\}$ and $\{1, 2, 3\}$.

5. *Truncation.* Let $\mu$ be a measure on $(\mathcal{X}, \mathcal{A})$ and let $A$ be a set in $\mathcal{A}$. Define $\nu$ on $\mathcal{A}$ by

$$\nu(B) = \mu(A \cap B), \qquad B \in \mathcal{A}.$$

   Show that $\nu$ is a measure on $(\mathcal{X}, \mathcal{A})$.

6. Suppose $\mathcal{A}$ and $\mathcal{B}$ are $\sigma$-fields on the same sample space $\mathcal{X}$. Show that the intersection $\mathcal{A} \cap \mathcal{B}$ is also a $\sigma$-field on $\mathcal{X}$.

7. Let $\mathcal{X}$ denote the rational numbers in $(0, 1)$, and let $\mathcal{A}$ be all subsets of $\mathcal{X}$, $\mathcal{A} = 2^{\mathcal{X}}$. Let $\mu$ be a real-valued function on $\mathcal{A}$ satisfying

$$\mu\big[(a, b) \cap \mathcal{X}\big] = b - a, \qquad \text{for all } a < b,\ a \in \mathcal{X},\ b \in \mathcal{X}.$$

   Show that $\mu$ cannot be a measure.

*8. Prove *Boole's inequality*: For any events $B_1, B_2, \dots,$

$$P\left(\bigcup_{i\geq 1} B_i\right) \leq \sum_{i\geq 1} P(B_i).$$

   Hint: One approach would be to establish the result for finite collections by induction, then it extend to countable collections using (1.1). Another idea is to use Fubini's theorem, noting that if $B$ is the union of the events, $1_B \leq \sum 1_{B_i}$.

[5] Solutions to starred problems in each chapter are given at the back of the book.

9. *Cantor set.* The Cantor set can be defined recursively. Start with the closed unit interval $[0,1]$ and form $K_1$ by removing the open middle third, so

$$K_1 = [0.1/3] \cup [2/3, 1].$$

Next, form $K_2$ by removing the two open middle thirds from the intervals in $K_1$, so

$$K_2 = [0, 1/9] \cup [2/9, 1/3] \cup [2/3, 7/9] \cup [8/9, 1].$$

Continue removing middle thirds to form $K_2, K_3, \dots$. The Cantor set $K$ is the limit or intersection of these sets,

$$K = \bigcap_{n=1}^{\infty} K_n.$$

Show that $K$ is a Borel set and find its length or Lebesgue measure. Remark: $K$ and $[0,1]$ have the same cardinality.

*10. Let $\mu$ and $\nu$ be measures on $(\mathcal{E}, \mathcal{B})$.
   a) Show that the sum $\eta$ defined by $\eta(B) = \mu(B) + \nu(B)$ is also a measure.
   b) If $f$ is a nonnegative measurable function, show that

$$\int f\,d\eta = \int f\,d\mu + \int f\,d\nu.$$

   Hint: First show that this result holds for nonnegative simple functions.

*11. Suppose $f$ is the simple function $1_{(1/2,\pi]} + 21_{(1,2]}$, and let $\mu$ be a measure on $\mathbb{R}$ with $\mu\{(0, a^2]\} = a$, $a > 0$. Evaluate $\int f\,d\mu$.

*12. Suppose that $\mu\{(0, a)\} = a^2$ for $a > 0$ and that $f$ is defined by

$$f(x) = \begin{cases} 0, & x \le 0; \\ 1, & 0 < x < 2; \\ \pi, & 2 \le x < 5; \\ 0, & x \ge 5. \end{cases}$$

Compute $\int f\,d\mu$.

*13. Define the function $f$ by

$$f(x) = \begin{cases} x, & 0 \le x \le 1; \\ 0, & \text{otherwise}. \end{cases}$$

Find simple functions $f_1 \le f_2 \le \cdots$ increasing to $f$ (i.e., $f(x) = \lim_{n\to\infty} f_n(x)$ for all $x \in \mathbb{R}$). Let $\mu$ be Lebesgue measure on $\mathbb{R}$. Using our formal definition of an integral and the fact that $\mu\bigl((a,b]\bigr) = b - a$ whenever $b > a$ (this might be used to formally define Lebesgue measure), show that $\int f\,d\mu = 1/2$.

14. Suppose $\mu$ is a measure on $\mathbb{R}$ with $\mu([0,a]) = e^a$, $a \ge 0$. Evaluate $\int (1_{(0,2)} + 2 1_{[1,3]})\,d\mu$.
15. Suppose $\mu$ is a measure on subsets of $\mathbb{N} = \{1,2,\dots\}$ and that

$$\mu(\{n, n+1, \dots\}) = \frac{n}{2^n}, \qquad n = 1,2,\dots.$$

Evaluate $\int x\,d\mu(x)$.

*16. Define $F(a-) = \lim_{x \uparrow a} F(x)$. Then, if $F$ is nondecreasing, $F(a-) = \lim_{n\to\infty} F(a - 1/n)$. Use (1.1) to show that if a random variable $X$ has cumulative distribution function $F_X$,

$$P(X < a) = F_X(a-).$$

Also, show that

$$P(X = a) = F_X(a) - F_X(a-).$$

*17. Suppose $X$ is a geometric random variable with mass function

$$p(x) = P(X = x) = \theta(1-\theta)^x, \qquad x = 0,1,\dots,$$

where $\theta \in (0,1)$ is a constant. Find the probability that $X$ is even.

*18. Let $X$ be a function mapping $\mathcal{E}$ into $\mathbb{R}$. Recall that if $B$ is a subset of $\mathbb{R}$, then $X^{-1}(B) = \{e \in \mathcal{E} : X(e) \in B\}$. Use this definition to prove that

$$X^{-1}(A \cap B) = X^{-1}(A) \cap X^{-1}(B),$$

$$X^{-1}(A \cup B) = X^{-1}(A) \cup X^{-1}(B),$$

and

$$X^{-1}\left(\bigcup_{i=0}^{\infty} A_i\right) = \bigcup_{i=0}^{\infty} X^{-1}(A_i).$$

*19. Let $P$ be a probability measure on $(\mathcal{E}, \mathcal{B})$, and let $X$ be a random variable. Show that the distribution $P_X$ of $X$ defined by $P_X(B) = P(X \in B) = P\big(X^{-1}(B)\big)$ is a measure (on the Borel sets of $\mathbb{R}$).

20. Suppose $X$ is a Poisson random variable with mass function

$$p(x) = P(X = x) = \frac{\lambda^x e^{-\lambda}}{x!}, \qquad x = 0,1,\dots,$$

where $\lambda > 0$ is a constant. Find the probability that $X$ is even.

*21. Let $X$ have a uniform distribution on $(0,1)$; that is, $X$ is absolutely continuous with density $p$ defined by

$$p(x) = \begin{cases} 1, & x \in (0,1); \\ 0, & \text{otherwise.} \end{cases}$$

Let $Y_1$ and $Y_2$ denote the first two digits of $X$ when $X$ is written as a binary decimal (so $Y_1 = 0$ if $X \in (0,1/2)$ for instance). Find $P(Y_1 = i, Y_2 = j)$, $i = 0$ or $1$, $j = 0$ or $1$.

*22. Let $\mathcal{E} = (0, 1)$, let $\mathcal{B}$ be the Borel subsets of $\mathcal{E}$, and let $P(A)$ be the length of $A$ for $A \in \mathcal{B}$. ($P$ would be called the *uniform probability measure* on $(0, 1)$.) Define the random variable $X$ by

$$X(e) = \min\{e, 1/2\}.$$

Let $\mu$ be the sum of Lebesgue measure on $\mathbb{R}$ and counting measure on $\mathcal{X}_0 = \{1/2\}$. Show that the distribution $P_X$ of $X$ is absolutely continuous with respect to $\mu$ and find the density of $P_X$.

*23. The standard normal distribution $N(0, 1)$ has density $\phi$ given by

$$\phi(x) = \frac{e^{-x^2/2}}{\sqrt{2\pi}}, \qquad x \in \mathbb{R},$$

with respect to Lebesgue measure $\lambda$ on $\mathbb{R}$. The corresponding cumulative distribution function is $\Phi$, so

$$\Phi(x) = \int_{-\infty}^{x} \phi(z)\, dz$$

for $x \in \mathbb{R}$. Suppose that $X \sim N(0, 1)$ and that the random variable $Y$ equals $X$ when $|X| < 1$ and is 0 otherwise. Let $P_Y$ denote the distribution of $Y$ and let $\mu$ be counting measure on $\{0\}$. Find the density of $P_Y$ with respect to $\lambda + \mu$.

*24. Let $\mu$ be a $\sigma$-finite measure on a measurable space $(X, \mathcal{B})$. Show that $\mu$ is absolutely continuous with respect to some probability measure $P$. Hint: You can use the fact that if $\mu_1, \mu_2, \dots$ are probability measures and $c_1, c_2, \dots$ are nonnegative constants, then $\sum c_i \mu_i$ is a measure. (The proof for Problem 1.10 extends easily to this case.) The measures $\mu_i$ you will want to consider are truncations of $\mu$ to sets $A_i$ covering $X$ with $\mu(A_i) < \infty$, given by $\mu_i(B) = \mu(B \cap A_i)$. With the constants $c_i$ chosen properly, $\sum c_i \mu_i$ will be a probability measure.

*25. The monotone convergence theorem states that if $0 \le f_1 \le f_2 \cdots$ are measurable functions and $f = \lim f_n$, then $\int f\, d\mu = \lim \int f_n\, d\mu$. Use this result to prove the following assertions.

a) Show that if $X \sim P_X$ is a random variable on $(\mathcal{E}, \mathcal{B}, P)$ and $f$ is a nonnegative measurable function, then

$$\int f\big(X(e)\big)\, dP(e) = \int f(x)\, dP_X(x).$$

Hint: Try it first with $f$ an indicator function. For the general case, let $f_n$ be a sequence of simple functions increasing to $f$.

b) Suppose that $P_X$ has density $p$ with respect to $\mu$, and let $f$ be a nonnegative measurable function. Show that

$$\int f\, dP_X = \int fp\, d\mu.$$

*26. *The gamma distribution.*

a) The gamma function is defined for $\alpha > 0$ by

$$\Gamma(\alpha) = \int_0^\infty x^{\alpha-1} e^{-x}\,dx.$$

Use integration by parts to show that $\Gamma(x+1) = x\Gamma(x)$. Show that $\Gamma(x+1) = x!$ for $x = 0, 1, \dots$.

b) Show that the function

$$p(x) = \begin{cases} \frac{1}{\Gamma(\alpha)\beta^\alpha} x^{\alpha-1} e^{-x/\beta}, & x > 0; \\ 0, & \text{otherwise}, \end{cases}$$

is a (Lebesgue) probability density when $\alpha > 0$ and $\beta > 0$. This density is called the gamma density with parameters $\alpha$ and $\beta$. The corresponding probability distribution is denoted $\Gamma(\alpha, \beta)$.

c) Show that if $X \sim \Gamma(\alpha, \beta)$, then $EX^r = \beta^r \Gamma(\alpha + r)/\Gamma(\alpha)$. Use this formula to find the mean and variance of $X$.

*27. Suppose $X$ has a uniform distribution on $(0, 1)$. Find the mean and covariance matrix of the random vector $\binom{X}{X^2}$.

*28. If $X \sim N(0, 1)$, find the mean and covariance matrix of the random vector $\binom{X}{I\{X>c\}}$.

29. Let $X$ be a random vector in $\mathbb{R}^n$ with $EX_i^2 < \infty$, $i = 1, \dots, n$, and let $A = EXX'$. Show that $A$ is nonnegative definite: $v'Av \ge 0$ for all $v \in \mathbb{R}^n$.

30. Let $W$ be absolutely continuous with density

$$p(x) = \begin{cases} \lambda e^{-\lambda x}, & x > 0; \\ 0, & \text{otherwise}, \end{cases}$$

where $\lambda > 0$ (the exponential density with failure rate $\lambda$), and define $X = \lfloor W \rfloor$ and $Y = W - X$. Here $\lfloor \cdot \rfloor$ is the *floor* or *greatest integer* function: $\lfloor x \rfloor$ is the greatest integer less than or equal to $x$.

a) Find $P(X = k)$, $k \ge 0$, the mass function for $X$.
b) Find $P(Y \le y | X = k)$, $y \in (0, 1)$. What is the cumulative distribution function for $Y$?
c) Find $EY$ and $\mathrm{Var}(Y)$.
d) Compute $EW$. Use linearity and your answer to (c) to find $EX$.
e) Find the covariance matrix for the random vector $\binom{Y}{W}$.

31. Let $X$ be an absolutely continuous random variable with density

$$p(x) = \begin{cases} 2x, & x \in (0, 1); \\ 0, & \text{otherwise}. \end{cases}$$

a) Find the mean and variance of $X$.
b) Find $E\sin(X)$.

c) Let $Y = I\{X > 1/2\}$. Find $\mathrm{Cov}(X, Y)$.

*32. Suppose $E|X| < \infty$ and let

$$h(t) = \frac{1 - E\cos(tX)}{t^2}.$$

Use Fubini's theorem to find $\int_0^\infty h(t)\,dt$. Hint:

$$\int_0^\infty \big(1 - \cos(u)\big)u^{-2}\,du = \frac{\pi}{2}.$$

*33. Suppose $X$ is absolutely continuous with density $p_X(x) = xe^{-x}$, $x > 0$ and $p_X(x) = 0$, $x \le 0$. Define $c_n = E(1 + X)^{-n}$. Use Fubini's theorem to evaluate $\sum_{n=1}^\infty c_n$.

34. Let $Z$ have a standard normal distribution, introduced in Problem 1.23.

a) For $n = 1, 2, \dots$, show that

$$EZ^{2n} = (2n - 1)!! \stackrel{\text{def}}{=} (2n - 1) \times (2n - 3) \times \cdots \times 1.$$

Hint: Use an inductive argument based on an integration by parts identity or formulas for the gamma function.

b) Use the identity in (a) and Fubini's theorem to evaluate

$$\sum_{n=1}^\infty \frac{(2n - 1)!!}{3^n n!}.$$

35. Prove Proposition 1.19.

*36. Suppose $X$ and $Y$ are independent random variables, and let $F_X$ and $F_Y$ denote their cumulative distribution functions.

a) Use smoothing to show that the cumulative distribution function of $S = X + Y$ is

$$F_S(s) = P(X + Y \le s) = EF_X(s - Y). \tag{1.21}$$

b) If $X$ and $Y$ are independent and $Y$ is almost surely positive, use smoothing to show that the cumulative distribution function of $W = XY$ is $F_W(w) = EF_X(w/Y)$ for $w > 0$.

*37. Differentiating (1.21) with respect to $s$ one can show that if $X$ is absolutely continuous with density $p_X$, then $S = X+Y$ is absolutely continuous with density

$$p_S(s) = Ep_X(s - Y)$$

for $s \in \mathbb{R}$. Use this formula to show that if $X$ and $Y$ are independent with $X \sim \Gamma(\alpha, 1)$ and $Y \sim \Gamma(\beta, 1)$, then $X + Y \sim \Gamma(\alpha + \beta, 1)$.

*38. Let $Q_\lambda$ denote the exponential distribution with failure rate $\lambda$, given in Problem 1.30. Let $X$ be a discrete random variable taking values in $\{1, \dots, n\}$ with mass function

$$P(X = k) = \frac{2k}{n(n+1)}, \qquad k = 1, \dots, n,$$

and assume that the conditional distribution of $Y$ given $X = x$ is exponential with failure rate $x$,

$$Y|X = x \sim Q_x.$$

a) Find $E[Y|X]$.
b) Use smoothing to compute $EY$.

*39. Let $X$ be a discrete random variable uniformly distributed on $\{1, \dots, n\}$, so $P(X = k) = 1/n$, $k = 1, \dots, n$, and assume that the conditional distribution of $Y$ given $X = x$ is exponential with failure rate $x$.
a) For $y > 0$ find $P[Y > y|X]$.
b) Use smoothing to compute $P(Y > y)$.
c) Determine the density of $Y$.

40. Let $X$ and $Y$ be independent absolutely continuous random variables, $X$ with density $p_X(x) = e^{-x}$, $x > 0$, $p_X(x) = 0$, $x < 0$, and $Y$ uniformly distributed on $(0, 1)$. Let $V = X/(X + Y)$.
a) Find $P(V > c|Y = y)$ for $c \in (0, 1)$.
b) Use smoothing to compute $P(V > c)$.
c) What is the density of $V$?

41. Suppose that $X$ has the standard exponential distribution with density $p_X(x) = e^{-x}$, $x \geq 0$, $p_X(x) = 0$, $x < 0$; that $Y$ has a (discrete) uniform distribution on $\{1, \dots, n\}$; and that $X$ and $Y$ are independent.
a) Find the joint density of $X$ and $Y$. Use it to compute $P(X+Y > 3/2)$.
b) Find the covariance matrix for $Z = \binom{X}{X+Y}$.
c) Find $E\bigl[\exp\bigl(XY/(1+Y)\bigr) \bigm| X\bigr]$.
d) Use smoothing to compute $E\exp\bigl(XY/(1+Y)\bigr)$.

42. Two measures $\mu$ and $\nu$ on $(\mathcal{X}, \mathcal{A})$ are called (mutually) *singular* if $\mu(A) = \nu(A^c) = 0$ for some $A \in \mathcal{A}$. For instance, Lebesgue measure and counting measure on some countable subset $\mathcal{X}_0$ of $\mathbb{R}$ are singular (take $A = \mathcal{X}_0$). Let $\mu$ and $\nu$ be singular measures on the Borel sets of $\mathbb{R}$, and let $Q_0$ and $Q_1$ be probability measures absolutely continuous with respect to $\mu$ and $\nu$, respectively, with densities

$$q_0 = \frac{dQ_0}{d\mu} \text{ and } q_1 = \frac{dQ_1}{d\nu}.$$

Let $X$ have a Bernoulli distribution with success probability $p$, and assume that

$$Y|X = 0 \sim Q_0 \text{ and } Y|X = 1 \sim Q_1.$$

a) Use the result in Problem 1.10 to show that $Q_1$ has density $q_1 1_A$ with respect to $\mu + \nu$, where $\mu(A) = \nu(A^c) = 0$.
b) Use smoothing to derive a formula for $P(Y \in B)$ involving $Q_0$, $Q_1$, and $p$.

c) Find a density for $Y$ with respect to $\mu + \nu$.

43. Let $X$ and $Y$ be independent random variables with $X$ uniformly distributed on $(0, 1)$ and $Y$ uniformly distributed on $\{1, \dots, n\}$. Define $W = Ye^{XY}$.
    a) Find $E[W|Y = y]$.
    b) Use smoothing to compute $EW$.
44. The standard exponential distribution is absolutely continuous with density $p(x) = 1_{(0,\infty)}(x)e^{-x}$. Let $X$ and $Y$ be independent random variables, both from this distribution, and let $Z = X/Y$.
    a) For $z > 0$, find $P(Z \le z|Y = y)$.
    b) Use smoothing and the result in part (a) to compute

$$P(Z \le z), \qquad z > 0.$$

    c) Find the covariance between $Y$ and $I\{Z \le z\}$.
45. Show that $E[f(X)Y|X] = f(X)E(Y|X)$.
46. If $E|Y| < \infty$ and $f$ is a bounded function, then by smoothing,

$$E[f(X)Y] = E\big[f(X)E(Y|X)\big].$$

By (1.20) we should then have

$$E[f(X)Y] = E\Big[f(X)E\big[E(Y|X, W)\,\big|\,X\big]\Big].$$

Use a smoothing argument to verify that this equation holds, demonstrating that (1.20) works in this case.

# 2

# Exponential Families

Inferential statistics is the science of learning from data. Data are typically viewed as random variables or vectors, but in contrast to our discussion of probability, distributions for these variables are generally unknown. In applications, it is often reasonable to assume that distributions come from a suitable class of distributions. In this chapter we introduce classes of distributions called *exponential families*. Examples include the binomial, Poisson, normal, exponential, geometric, and other distributions in regular use. From a theoretical perspective, exponential families are quite regular. In addition, moments for these distributions can often be computed easily using the differential identities in Section 2.4.

## 2.1 Densities and Parameters

Let $\mu$ be a measure on $\mathbb{R}^n$, let $h : \mathbb{R}^n \to \mathbb{R}$ be a nonnegative function, and let $T_1, \ldots, T_s$ be measurable functions from $\mathbb{R}^n$ to $\mathbb{R}$. For $\eta \in \mathbb{R}^s$, define

$$A(\eta) = \log \int \exp\left[\sum_{i=1}^{s} \eta_i T_i(x)\right] h(x)\, d\mu(x). \tag{2.1}$$

Whenever $A(\eta) < \infty$, the function $p_\eta$ given by

$$p_\eta(x) = \exp\left[\sum_{i=1}^{s} \eta_i T_i(x) - A(\eta)\right] h(x), \qquad x \in \mathbb{R}^n, \tag{2.2}$$

integrates to one; that is, $\int p_\eta \, d\mu = 1$. So, this construction gives a family of probability densities indexed by $\eta$. The set

$$\Xi = \{\eta : A(\eta) < \infty\}$$

is called the *natural parameter space*, and the family of densities $\{p_\eta : \eta \in \Xi\}$ is called an *s-parameter exponential family* in *canonical form*.

R.W. Keener, *Theoretical Statistics: Topics for a Core Course*, Springer Texts in Statistics, DOI 10.1007/978-0-387-93839-4_2, © Springer Science+Business Media, LLC 2010

*Example 2.1.* Suppose $\mu$ is Lebesgue measure on $\mathbb{R}$, $h = 1_{(0,\infty)}$, $s = 1$, and $T_1(x) = x$. Then

$$\begin{aligned} A(\eta) &= \log \int_0^\infty e^{\eta x}\,dx \\ &= \begin{cases} \log(-1/\eta), & \eta < 0; \\ \infty, & \eta \ge 0. \end{cases} \end{aligned}$$

Thus, $p_\eta(x) = \exp[\eta x - \log(-1/\eta)]1_{(0,\infty)}(x)$ is a density for $\eta \in \Xi = (-\infty, 0)$. In form, these are the exponential densities, which are usually parameterized by the mean or failure rate instead of the canonical parameter $\eta$ here.

To allow other parameterizations for an exponential family of densities, let $\eta$ be a function from some space $\Omega$ into $\Xi$ and define

$$p_\theta(x) = \exp\left[\sum_{i=1}^s \eta_i(\theta)T_i(x) - B(\theta)\right] h(x)$$

for $\theta \in \Omega$, $x \in \mathbb{R}^n$, where $B(\theta) = A\bigl(\eta(\theta)\bigr)$. Families $\{p_\theta : \theta \in \Omega\}$ of this form are called *s-parameter exponential families.*

*Example 2.2.* The normal distribution $N(\mu, \sigma^2)$ has density

$$\begin{aligned} p_\theta(x) &= \frac{1}{\sqrt{2\pi\sigma^2}} e^{-(x-\mu)^2/(2\sigma^2)} \\ &= \frac{1}{\sqrt{2\pi}} \exp\left[\frac{\mu}{\sigma^2}x - \frac{1}{2\sigma^2}x^2 - \left(\frac{\mu^2}{2\sigma^2} + \log\sigma\right)\right], \end{aligned}$$

where $\theta = (\mu, \sigma^2)$. This is a two-parameter exponential family with $T_1(x) = x$, $T_2(x) = x^2$, $\eta_1(\theta) = \mu/\sigma^2$, $\eta_2(\theta) = -1/(2\sigma^2)$, $B(\theta) = \mu^2/(2\sigma^2) + \log\sigma$, and $h(x) = 1/\sqrt{2\pi}$.

*Example 2.3.* If $X_1, \dots, X_n$ is a random sample from $N(\mu, \sigma^2)$, then their joint density is

$$\begin{aligned} p_\theta(x_1, \dots, x_n) &= \prod_{i=1}^n \left[\frac{1}{\sqrt{2\pi\sigma^2}} e^{-(x_i-\mu)^2/(2\sigma^2)}\right] \\ &= \frac{1}{(2\pi)^{n/2}} \exp\left[\frac{\mu}{\sigma^2}\sum_{i=1}^n x_i - \frac{1}{2\sigma^2}\sum_{i=1}^n x_i^2 - n\left(\frac{\mu^2}{2\sigma^2} + \log\sigma\right)\right]. \end{aligned}$$

These densities also form a two-parameter exponential family with $T_1(x) = \sum_{i=1}^n x_i$, $T_2(x) = \sum_{i=1}^n x_i^2$, $\eta_1(\theta) = \mu/\sigma^2$, $\eta_2(\theta) = -1/(2\sigma^2)$, $B(\theta) = n\bigl[\mu^2/(2\sigma^2) + \log\sigma\bigr]$, and $h(x) = 1/(2\pi)^{n/2}$.

The similarity between these two examples is not accidental. If $X_1, \dots, X_n$ is a random sample with common marginal density

$$\exp\left[\sum_{i=1}^{s} \eta_i(\theta) T_i(x) - B(\theta)\right] h(x),$$

then their joint density is

$$\exp\left[\sum_{i=1}^{s} \eta_i(\theta) \left(\sum_{j=1}^{n} T_i(x_j)\right) - nB(\theta)\right] \prod_{j=1}^{n} h(x_j), \tag{2.3}$$

which is an $s$-parameter exponential family with the same functions $\eta_1, \dots, \eta_s$, and with

$$\tilde{T}_i(x) = \sum_{j=1}^{n} T_i(x_j), \qquad \tilde{B}(\theta) = nB(\theta), \qquad \tilde{h}(x) = \prod_{i=1}^{n} h(x_i),$$

where the tilde is used to indicate that the function is for the family of joint densities.

## 2.2 Differential Identities

In canonical exponential families it is possible to relate moments and cumulants for the statistics $T_1, \dots, T_s$ to derivatives of $A$. The following theorem plays a central role.

**Theorem 2.4.** *Let $\Xi_f$ be the set of values for $\eta \in \mathbb{R}^s$ where*

$$\int |f(x)| \exp\left[\sum_{i=1}^{s} \eta_i T_i(x)\right] h(x)\, d\mu(x) < \infty.$$

*Then the function*

$$g(\eta) = \int f(x) \exp\left[\sum_{i=1}^{s} \eta_i T_i(x)\right] h(x)\, d\mu(x)$$

*is continuous and has continuous partial derivatives of all orders for $\eta \in \Xi_f^\circ$ (the interior of $\Xi_f$). Furthermore, these derivatives can be computed by differentiation under the integral sign.*

A proof of this result is given in Brown (1986), a monograph on exponential families with statistical applications. Although the proof is omitted here, key ideas from it are of independent interest and are presented in the next section. As an application of this result, if $f = 1$, then $\Xi_f = \Xi$, and, by (2.1),

$$g(\eta) = e^{A(\eta)} = \int \exp\left[\sum_{i=1}^{s} \eta_i T_i(x)\right] h(x)\, d\mu(x).$$

Differentiating this expression with respect to $\eta_j$, which can be done under the integral if $\eta \in \Xi^\circ$, gives

$$\begin{aligned} e^{A(\eta)} \frac{\partial A(\eta)}{\partial \eta_j} &= \int \frac{\partial}{\partial \eta_j} \exp\left[\sum_{i=1}^{s} \eta_i T_i(x)\right] h(x)\, d\mu(x) \\ &= \int T_j(x) \exp\left[\sum_{i=1}^{s} \eta_i T_i(x)\right] h(x)\, d\mu(x). \end{aligned}$$

Using the definition (2.2) of $p_\eta$, division by $e^{A(\eta)}$ gives

$$\frac{\partial A(\eta)}{\partial \eta_j} = \int T_j(x) p_\eta(x)\, d\mu(x).$$

This shows that if data $X$ has density $p_\eta$ with respect to $\mu$, then

$$E_\eta T_j(X) = \frac{\partial A(\eta)}{\partial \eta_j} \tag{2.4}$$

for any $\eta \in \Xi^\circ$.

## 2.3 Dominated Convergence

When $s = 1$, (2.4) is obtained differentiating the identity

$$e^{A(\eta)} = \int e^{\eta T(x)} h(x)\, d\mu(x),$$

passing the derivative inside the integral. To understand why this should work, suppose the integral is finite for $\eta \in [-2\epsilon, 2\epsilon]$ and consider taking the derivative at $\eta = 0$. If the function is differentiable at zero, the derivative will be the following limit:

$$\begin{aligned} \lim_{n\to\infty} \frac{e^{A(\epsilon/n)} - e^{A(0)}}{\epsilon/n} &= \lim_{n\to\infty} \int \frac{e^{\epsilon T(x)/n} - 1}{\epsilon/n} h(x)\, d\mu(x) \\ &= \lim_{n\to\infty} \int f_n(x)\, d\mu(x), \end{aligned}$$

where

$$f_n(x) = \frac{e^{\epsilon T(x)/n} - 1}{\epsilon/n} h(x). \tag{2.5}$$

As $n \to \infty$, $f_n(x) \to f(x) \overset{\text{def}}{=} T(x)h(x)$. So the desired result follows provided

$$\int f_n \, d\mu \to \int f \, d\mu$$

as $n \to \infty$. This seems natural, but is not automatic (see Example 2.6). The following basic result gives a sufficient condition for this sort of convergence.

**Theorem 2.5 (Dominated Convergence).** *Let $f_n$, $n \geq 1$ be a sequence of functions with $|f_n| \leq g$ (a.e. $\mu$) for all $n \geq 1$. If $\int g \, d\mu < \infty$ and $\lim_{n\to\infty} f_n(x) = f(x)$ for a.e. $x$ under $\mu$, then*

$$\int f_n \, d\mu \to \int f \, d\mu$$

*as $n \to \infty$.*

*Example 2.6.* To appreciate the need for a "dominating" function $g$ in this theorem, suppose $\mu$ is Lebesgue measure on $\mathbb{R}$, define $f_n = 1_{(n,n+1)}$, $n \geq 1$, and take $f = 0$. Then $f_n(x) \to f(x)$ as $n \to \infty$, for all $x$. But $\int f_n \, d\mu = 1$, for all $n \geq 1$, and these values do not converge to $\int f \, d\mu = 0$.

To apply dominated convergence in our original example with $f_n$ given by (2.5), the following bounds are useful:

$$|e^t - 1| \leq |t| e^{|t|}, \qquad t \in \mathbb{R},$$

and

$$|t| \leq e^{|t|}, \qquad t \in \mathbb{R}.$$

Using these,

$$\begin{aligned} \left| \frac{e^{\epsilon T(x)/n} - 1}{\epsilon/n} \right| &\leq \frac{|\epsilon T(x)/n|}{\epsilon/n} e^{|\epsilon T(x)/n|} \\ &\leq \frac{1}{\epsilon} |\epsilon T(x)| e^{|\epsilon T(x)|} \leq \frac{1}{\epsilon} e^{|2\epsilon T(x)|} \leq \frac{1}{\epsilon} \left( e^{2\epsilon T(x)} + e^{-2\epsilon T(x)} \right). \end{aligned}$$

The left-hand side of this bound multiplied by $h(x)$ is $|f_n(x)|$, so

$$|f_n(x)| \leq \frac{1}{\epsilon} \left( e^{2\epsilon T(x)} + e^{-2\epsilon T(x)} \right) h(x) \stackrel{\text{def}}{=} g(x).$$

The dominating function $g$ has a finite integral because

$$\int e^{\pm 2\epsilon T(x)} h(x) \, d\mu(x) = e^{A(\pm 2\epsilon)} < \infty.$$

So, by dominated convergence $\int f_n \, d\mu \to \int f \, d\mu$ as $n \to \infty$, as desired.

## 2.4 Moments, Cumulants, and Generating Functions

Let $T = (T_1, \dots, T_s)'$ be a random vector in $\mathbb{R}^s$. Note that the dot product of $T$ with a constant $u \in \mathbb{R}^s$ is $u \cdot T = u'T$. The *moment generating function* of $T$ is defined as

$$M_T(u) = Ee^{u_1 T_1 + \cdots + u_s T_s} = Ee^{u \cdot T}, \qquad u \in \mathbb{R}^s,$$

and the *cumulant generating function* is

$$K_T(u) = \log M_T(u).$$

According to the following lemma, the moment generating function $M_X$ determines the distribution of $X$, at least if it is finite in some open interval.

**Lemma 2.7.** *If the moment generating functions $M_X(u)$ and $M_Y(u)$ for two random vectors $X$ and $Y$ are finite and agree for $u$ in some set with a nonempty interior, then $X$ and $Y$ have the same distribution, $P_X = P_Y$.*

Expectations of products of powers of $T_1, \dots, T_s$ are called *moments* of $T$, denoted

$$\alpha_{r_1,\dots,r_s} = E[T_1^{r_1} \times \cdots \times T_s^{r_s}].$$

The following result shows that these moments can generally be found by differentiating $M_T$ at $u = 0$. The proof is omitted, but is similar to the proof of Theorem 2.4. Here, dominated convergence would be used to justify differentiation under an expectation.

**Theorem 2.8.** *If $M_T$ is finite in some neighborhood of the origin, then $M_T$ has continuous derivatives of all orders at the origin, and*

$$\alpha_{r_1,\dots,r_s} = \frac{\partial^{r_1}}{\partial u_1^{r_1}} \cdots \frac{\partial^{r_s}}{\partial u_s^{r_s}} M_T(u) \bigg|_{u=0}.$$

The corresponding derivatives of $K_T$ are called *cumulants*, denoted

$$\kappa_{r_1,\dots,r_s} = \frac{\partial^{r_1}}{\partial u_1^{r_1}} \cdots \frac{\partial^{r_s}}{\partial u_s^{r_s}} K_T(u) \bigg|_{u=0}.$$

When $s = 1$, $K_T' = M_T'/M_T$ and $K_T'' = [M_T M_T'' - (M_T')^2]/M_T^2$. At $u = 0$, these equations give

$$\kappa_1 = ET \text{ and } \kappa_2 = ET^2 - (ET)^2 = \mathrm{Var}(T).$$

Generating functions can be quite useful in the study of sums of independent random vectors. As a preliminary to this investigation, the following lemma shows that in regular situations, the expectation of a product of independent variables is the product of the expectations.

**Lemma 2.9.** *Suppose $X$ and $Y$ are independent random variables. If $X$ and $Y$ are both positive, or if $E|X|$ and $E|Y|$ are both finite, then*

$$EXY = EX \times EY.$$

*Proof.* Viewing $|XY|$ as a function $g$ of $Z = \binom{X}{Y} \sim P_X \times P_Y$, by Fubini's theorem,

$$E|XY| = \int g\, d(P_X \times P_Y) = \int \left( \int |x||y|\, dP_X(x) \right) dP_Y(y).$$

The inner integral is $|y|E|X|$, and the outer integral then gives $E|X| \times E|Y|$, so $E|XY| = E|X| \times E|Y|$. This proves the lemma if $X$ and $Y$ are both positive, because then $X = |X|$ and $Y = |Y|$. If $E|X| < \infty$ and $E|Y| < \infty$, then $E|XY| < \infty$, so the same steps omitting absolute values prove the lemma.

□

By iteration, this lemma extends easily to products of several independent variables.

Suppose $T = Y_1 + \cdots + Y_n$, where $Y_1, \ldots, Y_n$ are independent random vectors in $\mathbb{R}^s$. Then by Proposition 1.19, the random variables $e^{u \cdot Y_1}, \ldots, e^{u \cdot Y_n}$ are independent, and

$$M_T(u) = Ee^{u \cdot T} = E[e^{u \cdot Y_1} \times \cdots \times e^{u \cdot Y_n}] = M_{Y_1}(u) \times \cdots \times M_{Y_n}(u).$$

Taking logarithms,

$$K_T(u) = K_{Y_1}(u) + \cdots + K_{Y_n}(u).$$

Derivatives at the origin give cumulants, and thus cumulants for the sum $T$ will equal the sum of the corresponding cumulants of $Y_1, \ldots, Y_n$. This is a well-known result for the mean and variance.

If $X$ has density from a canonical exponential family (2.2), and if $T = T(X)$, then $T$ has moment generating function

$$\begin{aligned} E_\eta e^{u \cdot T(X)} &= \int e^{u \cdot T(x)} e^{\eta \cdot T(x) - A(\eta)} h(x)\, d\mu(x) \\ &= e^{A(u+\eta) - A(\eta)} \int e^{(u+\eta) \cdot T(x) - A(u+\eta)} h(x)\, d\mu(x), \end{aligned}$$

provided $u + \eta \in \Xi$. The final integrand is $p_{u+\eta}$, which integrates to one. So, the moment generating function is $e^{A(u+\eta)-A(\eta)}$, and the cumulant generating function is

$$K_T(u) = A(u + \eta) - A(\eta).$$

Taking derivatives, the cumulants for $T$ are

$$\kappa_{r_1, \ldots, r_s} = \frac{\partial^{r_1}}{\partial \eta_1^{r_1}} \cdots \frac{\partial^{r_s}}{\partial \eta_s^{r_1}} A(\eta).$$

*Example 2.10.* If $X$ has the Poisson distribution with mean $\lambda$, then

$$P(X=x)=\frac{\lambda^x e^{-\lambda}}{x!}=\frac{1}{x!}e^{x\log\lambda-\lambda},\qquad x=0,1,\dots.$$

The mass functions for $X$ form an exponential family, but the family is not in canonical form. The canonical parameter here is $\eta=\log\lambda$. The mass function expressed using $\eta$ is

$$P(X=x)=\frac{1}{x!}\exp[\eta x-e^\eta],\qquad x=0,1,\dots,$$

and so $A(\eta)=e^\eta$. Taking derivatives, all of the cumulants of $T=X$ are $e^\eta=\lambda$.

*Example 2.11.* The class of normal densities formed by varying $\mu$ with $\sigma^2$ fixed can be written as

$$p_\mu(x)=\exp\left[\frac{\mu x}{\sigma^2}-\frac{\mu^2}{2\sigma^2}\right]\frac{e^{-x^2/(2\sigma^2)}}{\sqrt{2\pi\sigma^2}}.$$

These densities form an exponential family with $T(x)=x$, canonical parameter $\eta=\mu/\sigma^2$, and $A(\eta)=\sigma^2\eta^2/2$. The first two cumulants are $\kappa_1=A'(\eta)=\sigma^2\eta=\mu$ and $\kappa_2=A''(\eta)=\sigma^2$. Because $A$ is quadratic, all higher-order cumulants, $\kappa_3,\kappa_4,\dots$, are zero.

To calculate moments from cumulants when $s=1$, repeatedly differentiate the identity $M=e^K$. This gives $M'=K'e^K$, $M''=(K''+K'^2)e^K$, $M'''=(K'''+3K'K''+K'^3)e^K$, and $M''''=(K''''+3K''^2+4K'K'''+6K'^2K''+K'^4)e^K$. At zero, these equations give

$$ET=\kappa_1,\qquad ET^2=\kappa_2+\kappa_1^2,\qquad ET^3=\kappa_3+3\kappa_1\kappa_2+\kappa_1^3,$$

and

$$ET^4=\kappa_4+3\kappa_2^2+4\kappa_1\kappa_3+6\kappa_1^2\kappa_2+\kappa_1^4.$$

For instance, if $X\sim\text{Poisson}(\lambda)$, $EX=\lambda$, $EX^2=\lambda+\lambda^2$, $EX^3=\lambda+3\lambda^2+\lambda^3$, and $EX^4=\lambda+7\lambda^2+6\lambda^3+\lambda^4$, and if $X\sim N(\mu,\sigma^2)$, $EX^3=3\mu\sigma^2+\mu^3$ and $EX^4=3\sigma^4+6\mu^2\sigma^2+\mu^4$.

The expressions above expressing moments as functions of cumulants can be solved to express cumulants as functions of moments. The algebra is easier if the variables are centered. Note that for $c\in\mathbb{R}^s$,

$$M_{T+c}(u)=Ee^{u\cdot(T+c)}=e^{u\cdot c}Ee^{u\cdot T}=e^{u\cdot c}M_T(u),$$

and so $K_{T+c}(u)=u\cdot c+K_T(u)$. Taking derivatives, it is clear that the constant $c$ only affects first-order cumulants. So with $s=1$, if $j\ge 2$, the $j$th cumulant $\kappa_j$ for $T$ will be the same as the $j$th cumulant for $T-ET$. The equations above then give

$$\kappa_3 = E(T - ET)^3$$

and

$$E(T - ET)^4 = \kappa_4 + 3\kappa_2^2,$$

and so

$$\kappa_4 = E(T - ET)^4 - 3\text{Var}^2(T).$$

In higher dimensions, the first-order cumulants are the means of $T_1, \dots, T_s$, and second-order cumulants are covariances between these variables. Formulas for mixed cumulants in higher dimensions become quite complicated as the order increases.

## 2.5 Problems[1]

*1. Consider independent Bernoulli trials with success probability $p$ and let $X$ be the number of failures before the first success. Then $P(X = x) = p(1 - p)^x$, for $x = 0, 1, \dots$, and $X$ has the geometric distribution with parameter $p$, introduced in Problem 1.17.
   a) Show that the geometric distributions form an exponential family.
   b) Write the densities for the family in canonical form, identifying the canonical parameter $\eta$, and the function $A(\eta)$.
   c) Find the mean of the geometric distribution using a differential identity.
   d) Suppose $X_1, \dots, X_n$ are i.i.d. from a geometric distribution. Show that the joint distributions form an exponential family, and find the mean and variance of $T$.

*2. Determine the canonical parameter space $\Xi$, and find densities for the one-parameter exponential family with $\mu$ Lebesgue measure on $\mathbb{R}^2$, $h(x, y) = \exp\bigl[-(x^2 + y^2)/2\bigr]/(2\pi)$, and $T(x, y) = xy$.

3. Suppose that $X_1, \dots, X_n$ are independent random variables and that for $i = 1, \dots, n$, $X_i$ has a Poisson distribution with mean $\lambda_i = \exp(\alpha + \beta t_i)$, where $t_1, \dots, t_n$ are observed constants and $\alpha$ and $\beta$ are unknown parameters. Show that the joint distributions for $X_1, \dots, X_n$ form a two-parameter exponential family and identify the statistics $T_1$ and $T_2$.

*4. Find the natural parameter space $\Xi$ and densities $p_\eta$ for a canonical one-parameter exponential family with $\mu$ Lebesgue measure on $\mathbb{R}$, $T_1(x) = \log x$, and $h(x) = (1 - x)^2$, $x \in (0, 1)$, and $h(x) = 0$, $x \notin (0, 1)$.

*5. Find the natural parameter space $\Xi$ and densities $p_\eta$ for a canonical one-parameter exponential family with $\mu$ Lebesgue measure on $\mathbb{R}$, $T_1(x) = -x$, and $h(x) = e^{-2\sqrt{x}}/\sqrt{x}$, $x > 0$, and $h(x) = 0$, $x \le 0$. (Hint: After a change of variables, relevant integrals will look like integrals against a normal density. You should be able to express the answer using $\Phi$, the standard normal cumulative distribution function.) Also, determine the mean and variance for a variable $X$ with this density.

[1] Solutions to the starred problems are given at the back of the book.

*6. Find the natural parameter space $\Xi$ and densities $p_\eta$ for a canonical two-parameter exponential family with $\mu$ counting measure on $\{0,1,2\}$, $T_1(x)=x$, $T_2(x)=x^2$, and $h(x)=1$ for $x\in\{0,1,2\}$.

*7. Suppose $X_1,\dots,X_n$ are independent geometric variables with $p_i$ the success probability for $X_i$. Suppose these success probabilities are related to a sequence of "independent" variables $t_1,\dots,t_n$, viewed as known constants, through

$$p_i = 1-\exp(\alpha+\beta t_i), \qquad i=1,\dots,n.$$

Show that the joint densities for $X_1,\dots,X_n$ form a two-parameter exponential family, and identify the statistics $T_1$ and $T_2$.

*8. Assume that $X_1,\dots,X_n$ are independent random variables with $X_i\sim N(\alpha+\beta t_i,1)$, where $t_1,\dots,t_n$ are observed constants and $\alpha$ and $\beta$ are unknown parameters. Show that the joint distributions for $X_1,\dots,X_n$ form a two-parameter exponential family, and identify the statistics $T_1$ and $T_2$.

*9. Suppose that $X_1,\dots,X_n$ are independent Bernoulli variables (a random variable is Bernoulli if it only takes on values 0 and 1) with

$$P(X_i=1)=\frac{\exp(\alpha+\beta t_i)}{1+\exp(\alpha+\beta t_i)}.$$

Show that the joint distributions for $X_1,\dots,X_n$ form a two-parameter exponential family, and identify the statistics $T_1$ and $T_2$.

10. Suppose a researcher is interested in how the variance of a response $Y$ depends on an independent variable $x$. Natural models might be those in which $Y_1,\dots,Y_n$ are independent mean zero normal variables with the variance of $Y_i$ some function of a linear function of $x_i$:

$$\operatorname{Var}(Y_i)=g(\theta_1+\theta_2 x_i).$$

Suggest a form for the function $g$ such that the joint distributions for the $Y_i$, as the parameters $\theta$ vary, form a two-parameter exponential family.

11. Find the natural parameter space $\Xi$ and densities $p_\eta$ for a canonical one-parameter exponential family with $\mu$ Lebesgue measure on $\mathbb{R}$, $T_1(x)=x$, and $h(x)=\sin x$, $x\in(0,\pi)$, and $h(x)=0$, $x\notin(0,\pi)$.

12. *Truncation.* Let $\{p_\theta:\theta\in\Omega\}$ be an exponential family of densities with respect to some measure $\mu$, where

$$p_\theta(x)=h(x)\exp\left[\sum_{i=1}^s \eta_i(\theta)T_i(x)-B(\theta)\right].$$

In some situations, a potential observation $X$ with density $p_\theta$ can only be observed if it happens to lie in some region $S$. For regularity, assume that $\Lambda(\theta)\stackrel{\text{def}}{=}P_\theta(X\in S)>0$. In this case, the appropriate distribution for the observed variable $Y$ is given by

$$P_\theta(Y \in B) = P_\theta(X \in B|X \in S), \qquad B \in \mathcal{B}.$$

This distribution for $Y$ is called the *truncation* of the distribution for $X$ to the set $S$.

a) Show that $Y$ has a density with respect to $\mu$, giving a formula for its density $q_\theta$.
b) Show that the densities $q_\theta$, $\theta \in \Omega$, form an exponential family.

14. Find densities $p_\eta$ for a canonical one-parameter exponential family if $\mu$ is counting measure on $\mathcal{X}_0 = \{-1, 0, 1\}^3$, $h$ is identically one, and $T(x)$ is the median of $x_1$, $x_2$, and $x_3$.

*15. For an exponential family in canonical form, $ET_j = \partial A(\eta)/\partial \eta_j$. This can be written in vector form as $ET = \nabla A(\eta)$. Derive an analogous differential formula for $E_\theta T$ for an $s$-parameter exponential family that is not in canonical form. Assume that $\Omega$ has dimension $s$. Hint: Differentiation under the integral sign should give a system of linear equations. Write these equations in matrix form.

16. Find the natural parameter space $\Xi$ and densities $p_\eta$ for a canonical one-parameter exponential family with $\mu$ counting measure on $\{1, 2, \ldots\}$, $h(x) = x^2$, and $T(x) = -x$. Also, determine the mean and variance for a random variable $X$ with this density. Hint: Consider what Theorem 2.4 has to say about derivatives of $\sum_{x=1}^\infty e^{-\eta x}$.

*17. Let $\mu$ denote counting measure on $\{1, 2, \ldots\}$. One common definition for $\sum_{k=1}^\infty f(k)$ is $\lim_{n\to\infty} \sum_{k=1}^n f(k)$, and another definition is $\int f\, d\mu$.

a) Use the dominated convergence theorem to show that the two definitions give the same answer when $\int |f|\, d\mu < \infty$. Hint: Find functions $f_n$, $n = 1, 2, \ldots$, so that $\sum_{k=1}^n f(k) = \int f_n\, d\mu$.
b) Use the monotone convergence theorem, given in Problem 1.25, to show the definitions agree if $f(k) \geq 0$ for all $k = 1, 2, \ldots$.
c) Suppose $\lim_{n\to\infty} f(n) = 0$ and that $\int f^+\, d\mu = \int f^-\, d\mu = \infty$ (so that $\int f\, d\mu$ is undefined). Let $K$ be an arbitrary constant. Show that the list $f(1), f(2), \ldots$ can be rearranged to form a new list $g(1), g(2), \ldots$ so that

$$\lim_{n\to\infty} \sum_{k=1}^n g(k) = K.$$

18. Let $\lambda$ be Lebesgue measure on $(0, \infty)$. The "Riemann" definition of $\int_0^\infty f(x)\, dx$ for a continuous function $f$ is

$$\lim_{c\to\infty} \int_0^c f(x)\, dx,$$

when the limit exists. Another definition is $\int f\, d\lambda$. Use the dominated convergence theorem to show that these definitions agree when $f$ is integrable, $\int |f|\, d\lambda < \infty$. Hint: Let $c_n$ be a sequence of constants with $c_n \to \infty$, and find functions $f_n$ such that $\int f_n(x)\, dx = \int_0^{c_n} f(x)\, dx$.

*19. Let $p_n$, $n = 1, 2, \ldots$, and $p$ be probability densities with respect to a measure $\mu$, and let $P_n$, $n = 1, 2, \ldots$, and $P$ be the corresponding probability measures.

a) Show that if $p_n(x) \to p(x)$ as $n \to \infty$, then $\int |p_n - p|\, d\mu \to 0$. Hint: First use the fact that $\int (p_n - p)\, d\mu = 0$ to argue that $\int |p_n - p|\, d\mu = 2\int (p - p_n)^+\, d\mu$. Then use dominated convergence.

b) Show that $|P_n(A) - P(A)| \le \int |p_n - p|\, d\mu$. Hint: Use indicators and the bound $|\int f\, d\mu| \le \int |f|\, d\mu$.

Remark: Distributions $P_n$, $n \ge 1$, are said to *converge strongly* to $P$ if $\sup_A |P_n(A) - P(A)| \to 0$. The two parts above show that pointwise convergence of $p_n$ to $p$ implies strong convergence. This was discovered by Scheffé.

20. Let $h$ be a bounded differentiable function on $[0, \infty)$, vanishing at zero, $h(0) = 0$.

a) Show that

$$\int_0^\infty |h(1/x^2)|\, dx < \infty.$$

Hint: Because $h$ is differentiable at 0, $h(x)/x \to h'(0)$ as $x \downarrow 0$, and $|h(x)| \le cx$ for $x$ sufficiently small.

b) If $Z$ has a standard normal distribution, $Z \sim N(0, 1)$, find

$$\lim_{n\to\infty} nEh\bigl(1/(n^2Z^2)\bigr).$$

Hint: Be careful with your argument: the answer should not be zero.

21. Let $\mu$ be counting measure on $\{1, 2, \ldots\}$, and let $f_n = c_n 1_{\{n\}}$, $n = 1, 2, \ldots$, for some constants $c_1, c_2, \ldots$.

a) Find $f(x) = \lim_{n\to\infty} f_n(x)$ for $x = 1, 2, \ldots$.

b) Show that these functions $f_n$ can be dominated by an integrable function; that is, there exists $g$ with $\int g\, d\mu < \infty$ and $|f_n| \le g$, $n = 1, 2, \ldots$, if and only if

$$\sum_{n=1}^\infty |c_n| < \infty.$$

c) Find constants $c_1, c_2, \ldots$ that provide an example of functions $f_n$ that cannot be dominated by an integrable function, so the assumption of the dominated convergence theorem fails, but $\int f_n\, d\mu \to \int f\, d\mu$.

*22. Suppose $X$ is absolutely continuous with density

$$p_\theta(x) = \begin{cases} \dfrac{e^{-(x-\theta)^2/2}}{\sqrt{2\pi}\Phi(\theta)}, & x > 0; \\ 0, & \text{otherwise.} \end{cases}$$

Find the moment generating function of $X$. Compute the mean and variance of $X$.

*23. Suppose $Z \sim N(0,1)$. Find the first four cumulants of $Z^2$. Hint: Consider the exponential family $N(0,\sigma^2)$.

*24. Find the first four cumulants of $T = XY$ when $X$ and $Y$ are independent standard normal variates.

*25. Find the third and fourth cumulants of the geometric distribution.

*26. Find the third cumulant and third moment of the binomial distribution with $n$ trials and success probability $p$.

*27. Let $T$ be a random vector in $\mathbb{R}^2$.

a) Express $\kappa_{2,1}$ as a function of the moments of $T$.

b) Assume $ET_1 = ET_2 = 0$ and give an expression for $\kappa_{2,2}$ in terms of moments of $T$.

*28. Suppose $X \sim \Gamma(\alpha, 1/\lambda)$, with density

$$\frac{\lambda^\alpha x^{\alpha-1} e^{-\lambda x}}{\Gamma(\alpha)}, \qquad x > 0.$$

Find the cumulants of $T = (X, \log X)$ of order 3 or less. The answer will involve $\psi(\alpha) = d \log \Gamma(\alpha)/d\alpha = \Gamma'(\alpha)/\Gamma(\alpha)$.

29. Let $X_1, \dots, X_n$ be independent random variables, and let $\alpha_i$ and $t_i$, $i = 1, \dots, n$, be known constants. Suppose $X_i \sim \Gamma(\alpha_i, 1/\lambda_i)$ with $\lambda_i = \theta_1 + \theta_2 t_i$, $i = 1, \dots, n$, where $\theta_1$ and $\theta_2$ are unknown parameters. Show that the joint distributions form a two-parameter exponential family. Identify the statistic $T$ and give its mean and covariance matrix. (Similar models arise in "parameter design" experiments used to study the effects of various factors on process variation.)

30. In independent Bernoulli trials with success probability $p$, the variable $X$ counting the number of failures before the $m$th success has a negative binomial distribution with mass function

$$P(X = x) = \binom{m+x-1}{m-1} p^m (1-p)^x, \qquad x = 0, 1, \dots.$$

Find the moment generating function of $X$, along with the first three moments and first three cumulants of $X$.

31. An estimator $\hat{\theta}$ is called *unbiased* for a parameter $\theta$ if $E\hat{\theta} = \theta$. If $X_1, \dots, X_n$ are i.i.d., then the sample moment

$$\hat{\alpha}_r = \frac{1}{n} \sum_{i=1}^n X_i^r$$

is an unbiased estimator of $\alpha_r = EX_i^r$. Unbiased estimators for cumulants are called *K-statistics*. They are a bit harder to identify than unbiased estimators for moments, because cumulants depend on powers of moments. For example, $\kappa_2 = \alpha_2 - \alpha_1^2$.

a) One natural estimator for $\alpha_1^2$ is $\overline{X}^2 = \hat{\alpha}_1^2$. Find the expected value of this estimator. When is it biased?

b) Show that

$$\binom{n}{2}^{-1} \sum_{1 \le i < j \le n} X_i X_j$$

is unbiased for $\alpha_1^2$. Give a formula relating this estimator to sample moments $\hat{\alpha}_1$ and $\hat{\alpha}_2$.

c) Give an unbiased estimator for $\kappa_2 = \text{Var}(X_i)$ based on sample moments $\hat{\alpha}_1$ and $\hat{\alpha}_2$.

d) Find the expected value of $\hat{\alpha}_1^3$, showing that it is usually biased for $\alpha_1^3$. Relate the unbiased estimator

$$\binom{n}{3}^{-1} \sum_{1 \le i < j < k \le n} X_i X_j X_k$$

to sample moments.

e) Find an unbiased estimator for $\kappa_3$ based on the first three sample moments, $\hat{\alpha}_1$, $\hat{\alpha}_2$, and $\hat{\alpha}_3$.

32. By Taylor expansion, for $\theta \in (0, 1)$,

$$-\log(1 - \theta) = \sum_{x=1}^{\infty} \frac{\theta^x}{x}.$$

From this, for $\theta \in (0, 1)$,

$$p_\theta(x) = \frac{\theta^x}{-x \log(1 - \theta)}, \qquad x = 1, 2, \ldots,$$

is the mass function for a probability distribution, called the *log series distribution*. Let $X$ be a discrete random variable with this distribution. Find the mean and variance of $X$.

33. A discrete random variable $X$ on $\{0, 1, \ldots\}$ has a *power series distribution* if its mass function has form

$$P(X = x) = \frac{a(x)\theta^x}{C(\theta)}, \qquad x = 0, 1, 2, \ldots.$$

Derive formulas for the mean and variance of this distribution involving derivatives of $C$.

# 3

# Risk, Sufficiency, Completeness, and Ancillarity

The initial section of this chapter develops a basic framework for inference. Later sections concern the notion of sufficiency that arises when data can be summarized without any loss of information.

## 3.1 Models, Estimators, and Risk Functions

Inferential statistics can be viewed as the science or art of learning about an unknown parameter $\theta$ from data $X$. For most applications, $X$ will be a random vector. The parameter $\theta$ may be a single constant, but more commonly takes values in some subset of $\mathbb{R}^p$. The parameter $\theta$ and data $X$ are related through a model in which the distribution of $X$ is determined by $\theta$. The distribution when the parameter is $\theta$ is denoted $P_\theta$, and we write

$$X \sim P_\theta.$$

Formally, a model should be a mapping $\theta \rightsquigarrow P_\theta$, but more commonly, a model is written as the set of distributions for $X$, $\mathcal{P} = \{P_\theta : \theta \in \Omega\}$, where the parameter space $\Omega$ is the set of all possible values for $\theta$.

A statistic is a function of the data $X$ and can be viewed as providing partial information about the data when the function is many-to-one. A typical example, when $X$ is a random vector in $\mathbb{R}^n$, would be the sample average

$$\delta(X) = \overline{X} = \frac{X_1 + \cdots + X_n}{n}.$$

In a broad sense there are two major categories for inference: estimation and hypothesis testing. For now we focus on estimation, in which the goal is to find statistic $\delta$ so that $\delta(X)$ is close to $g(\theta)$. Then $\delta$ or $\delta(X)$ is called an estimator of $g(\theta)$. The case $g(\theta) = \theta$ is allowed, and, when $\theta$ is a vector, $g(\theta) = \theta_k$ may be a fairly typical situation of interest. Hypothesis testing is introduced and studied in Chapter 12.

R.W. Keener, *Theoretical Statistics: Topics for a Core Course*, Springer Texts in Statistics,
DOI 10.1007/978-0-387-93839-4_3, © Springer Science+Business Media, LLC 2010

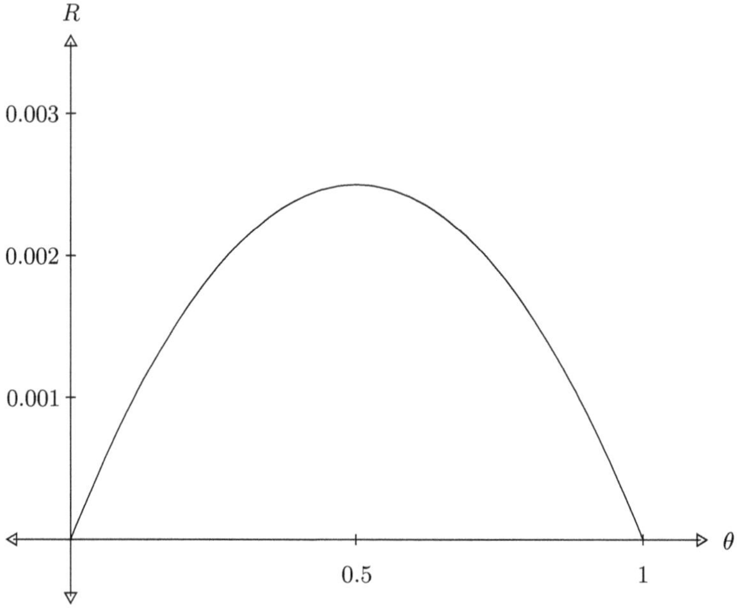


**Fig. 3.1.** Binomial risk function: $R(\theta, \delta)$.

*Example 3.1.* For a coin toss, the chance of heads is very close to $1/2$. Suppose instead we stand a coin on its edge, balancing it with a finger on top, and spin it by flicking it with a different finger. If this is done 100 times, with the trials independent and a common chance $\theta$ of heads on each spin, then the total number of heads $X$ should have a binomial distribution. Thus

$$X \sim \text{Binomial}(100, \theta).$$

Viewing $X$ as our data and taking

$$P_\theta = \text{Binomial}(100, \theta), \qquad \theta \in [0, 1] = \Omega,$$

our model $\mathcal{P} = \{P_\theta : \theta \in \Omega\}$ is the set of binomial distributions with 100 trials.

In this example, a natural estimator of $\theta$ is $\delta(X) = X/100$, the proportion of heads in 100 spins. In the sequel we study the performance of estimators, trying to decide when an estimator, such as $\delta$ here, is good or optimal in some sense. An adequate answer to these questions must involve criteria that judge the performance of estimators. One standard approach to making such judgments is called *decision theory.* For estimation, this approach begins with a loss function $L$ chosen so that $L(\theta, d)$ is the loss associated with estimating $g(\theta)$ by a value $d$. It is natural to assume $L\bigl(\theta, g(\theta)\bigr) = 0$, so that there is no loss for the correct answer, and $L(\theta, d) \geq 0$, for all $\theta$ and $d$. Because $X$ is

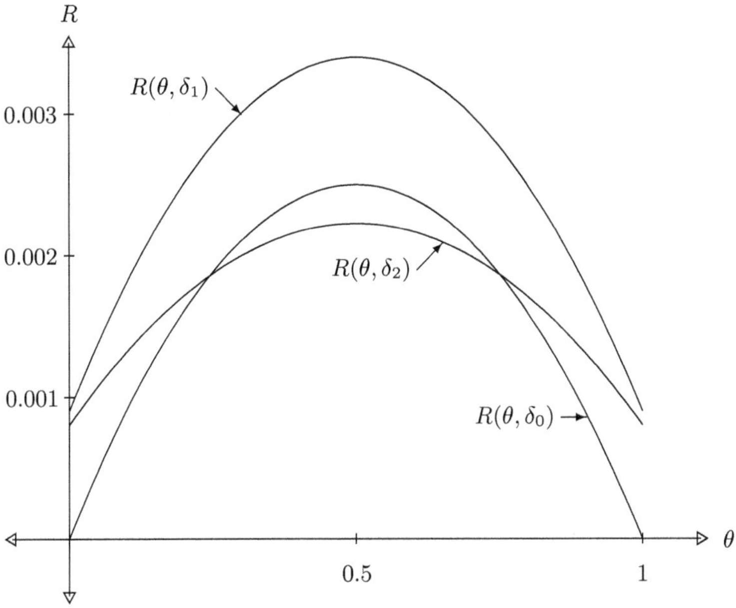


**Fig. 3.2.** Risks for $\delta_0$, $\delta_1$, and $\delta_2$.

random, $L\big(\theta, \delta(X)\big)$ is random and can be large if we are unlucky, even if $\delta$ is an excellent estimator. Accordingly, an estimator $\delta$ is judged by its average loss or *risk function* $R$, defined as

$$R(\theta, \delta) = E_\theta L\big(\theta, \delta(X)\big).$$

Here $E_\theta$ denotes expectation when $X \sim P_\theta$.

*Example 3.1, continued.* Suppose $X \sim \text{Binomial}(100, \theta)$, $\delta(X) = X/100$, $g(\theta) = \theta$, and the loss function is given by $L(\theta, d) = (\theta - d)^2$, called squared error loss. Then the risk function for $\delta$ is

$$R(\theta, \delta) = E_\theta(\theta - X/100)^2 = \frac{\theta(1-\theta)}{100}, \qquad \theta \in [0, 1].$$

A plot of this risk function is given in Figure 3.1.

A fundamental problem arises when one compares estimators using risk functions: if the risk functions for two estimators cross, there is no clear decision which estimator is best. For instance, in our binomial example, if $\delta_0(X)$ is the original estimator $X/100$, $\delta_1(X) = (X+3)/100$, and $\delta_2(X) = (X+3)/106$, then $R(\theta, \delta_0) = \theta(1-\theta)/100$, $R(\theta, \delta_1) = \big(9 + 100\theta(1-\theta)\big)/100^2$, and $R(\theta, \delta_2) = (9 - 8\theta)(1 + 8\theta)/106^2$. These functions are plotted together in Figure 3.2. Looking at the graph, $\delta_0$ and $\delta_2$ are both better than $\delta_1$, but the comparison between $\delta_0$ and $\delta_2$ is ambiguous. When $\theta$ is near $1/2$, $\delta_2$ is the preferable estimator, but if $\theta$ is near 0 or 1, $\delta_0$ is preferable. If $\theta$ were known, we could choose between $\delta_0$ and $\delta_2$. However, if $\theta$ were known, there would be no need to estimate its value.

## 3.2 Sufficient Statistics

Suppose $X$ and $Y$ are independent with common Lebesgue density

$$f_\theta(x) = \begin{cases} \theta e^{-\theta x}, & x \geq 0; \\ 0, & x < 0, \end{cases}$$

and let $U$ be independent of $X$ and $Y$ and uniformly distributed on $(0,1)$. Take $T = X + Y$, and define

$$\tilde{X} = UT \text{ and } \tilde{Y} = (1-U)T.$$

Let us find the joint density of $\tilde{X}$ and $\tilde{Y}$. The density of $T$ is needed, and this can be found by smoothing. Because $X$ and $Y$ are independent,[1]

$$\begin{aligned} P(T \leq t | Y = y) &= P(X + Y \leq t | Y = y) \\ &= E[I\{X + Y \leq t\} | Y = y] \\ &= \int I\{x + y \leq t\} dP_X(x) \\ &= F_X(t - y). \end{aligned}$$

So $P(T \leq t|Y) = F_X(t - Y)$ and

$$F_T(t) = P(T \leq t) = EF_X(t - Y).$$

This formula holds generally. Specializing to our specific problem, $F_X(t - Y)$ is $1 - e^{-\theta(t-Y)}$ on $Y < t$ and is zero on $Y \geq t$. Writing the expected value of this variable as an integral against the density of $Y$, for $t \geq 0$,

$$F_T(t) = \int_0^t \bigl(1 - e^{-\theta(t-y)}\bigr)\theta e^{-\theta y}\, dy = 1 - e^{-\theta t} - t\theta e^{-\theta t}.$$

Taking derivatives, $T$ has density

$$p_T(t) = F_T'(t) = t\theta^2 e^{-\theta t}, \qquad t \geq 0,$$

with $p_T(t) = 0$ for $t < 0$. Because $T$ and $U$ are independent, they have joint density

$$p_\theta(t, u) = \begin{cases} t\theta^2 e^{-\theta t}, & t \geq 0,\ u \in (0,1); \\ 0, & \text{otherwise.} \end{cases}$$

[1] The distribution of $T$ naturally depends on $\theta$, but for convenience this dependence is suppressed in the notation here.

From this,

$$P\left(\begin{pmatrix}\tilde X\\ \tilde Y\end{pmatrix}\in B\right)=\iint 1_B\bigl(tu,t(1-u)\bigr)p_\theta(t,u)\,du\,dt.$$

Changing variables to $x=ut$, $du=dx/t$ in the inner integral, and reversing the order of integration using Fubini's theorem,

$$P\left(\begin{pmatrix}\tilde X\\ \tilde Y\end{pmatrix}\in B\right)=\iint 1_B(x,t-x)t^{-1}p_\theta(t,x/t)\,dt\,dx.$$

Now a change of variables to $y=t-x$ in the inner integral gives

$$P\left(\begin{pmatrix}\tilde X\\ \tilde Y\end{pmatrix}\in B\right)=\iint 1_B(x,y)(x+y)^{-1}p_\theta\left(x+y,\frac{x}{x+y}\right)dy\,dx.$$

Thus $\tilde X$ and $\tilde Y$ have joint density

$$\frac{p_\theta\left(x+y,\frac{x}{x+y}\right)}{x+y}=\begin{cases}\theta^2e^{-\theta(x+y)}, & x\ge 0,\ y\ge 0;\\ 0, & \text{otherwise.}\end{cases}$$

This density is the same as the joint density of $X$ and $Y$, and so this calculation shows that the joint distribution of $\tilde X$ and $\tilde Y$ is the same as the joint distribution of $X$ and $Y$. Considered as data that provide information about $\theta$, the pair $(\tilde X,\tilde Y)$ should be just as informative as $(X,Y)$. But $(\tilde X,\tilde Y)$ can be computed from $T=X+Y$ and $U$. Because the distribution of $U$ does not depend on $\theta$, it could be generated numerically on a computer or obtained from a table of random numbers. Thus $T$ by itself also provides as much information about $\theta$ as the pair $(X,Y)$ because we could construct fake data $(\tilde X,\tilde Y)$ equivalent to $(X,Y)$ using any convenient variable $U$ that is uniformly distributed on $(0,1)$. The sum $T=X+Y$ is called a *sufficient statistic.* This construction of fake data works because the conditional distribution $Q_t$ for $X$ and $Y$ given $T=t$, given explicitly by[2]

$$Q_t(B)=P_\theta\left[\begin{pmatrix}X\\ Y\end{pmatrix}\in B\,\middle|\,T=t\right]=P\left[\begin{pmatrix}Ut\\ (1-U)t\end{pmatrix}\in B\right],$$

does not depend on $\theta$. This motivates the following definition in a general setting.

**Definition 3.2.** *Suppose $X$ has distribution from a family $\mathcal{P}=\{P_\theta:\theta\in\Omega\}$. Then $T=T(X)$ is a* sufficient statistic *for $\mathcal{P}$ (or for $X$, or for $\theta$) if for every $t$ and $\theta$, the conditional distribution of $X$ under $P_\theta$ given $T=t$ does not depend on $\theta$.*

[2] See Chapter 6 for a proper treatment of conditional distributions.

Suppose $T$ is sufficient, and let

$$Q_t(B) = P_\theta(X \in B|T = t).$$

Then $P_\theta(X \in B|T) = Q_T(B)$ and, by smoothing,

$$P_\theta(X \in B) = E_\theta P_\theta(X \in B|T) = E_\theta Q_T(B).$$

Suppose we use a random number generator to construct "fake" data $\tilde{X}$ from $T$ taking $\tilde{X} \sim Q_t$ when $T = t$. Then

$$\tilde{X}|T = t \sim Q_t$$

and by smoothing

$$P_\theta(\tilde{X} \in B) = E_\theta P_\theta(\tilde{X} \in B|T) = E_\theta Q_T(B). \tag{3.1}$$

So $X$ and $\tilde{X}$ have the same distribution.

**Theorem 3.3.** *Suppose $X$ has distribution from a family $\mathcal{P} = \{P_\theta : \theta \in \Omega\}$ and that $T = T(X)$ is sufficient. Then for any estimator $\delta(X)$ of $g(\theta)$ there exists a randomized estimator based on $T$ that has the same risk function as $\delta(X)$.*

*Proof.* A randomized estimator is one that can be constructed from $T$ with auxiliary random number generation. Inasmuch as $\tilde{X}$ can be constructed from $T$ by random number generation, $\delta(\tilde{X})$ is a randomized estimator, and its risk is the same as the risk of $\delta(X)$ because $X$ and $\tilde{X}$ both have the same distribution $P_\theta$. □

A similar notion of sufficiency, due to Blackwell (1951), can be used to compare experiments. It is natural here to identify an experiment with the model that gives distributions for data from the experiment. Let $\tilde{\mathcal{P}} = \{\tilde{P}_\theta : \theta \in \Omega\}$ and $\mathcal{P} = \{P_\theta : \theta \in \Omega\}$ be models for two experiments. As before, the notion is that $\tilde{\mathcal{P}}$ is sufficient if fake data can be created using an observation from $\tilde{\mathcal{P}}$ and external randomization, with the distributions for the fake data, as $\theta$ varies, the same as distributions for real data from the other experiment.

**Definition 3.4.** *Model $\tilde{\mathcal{P}}$ is* sufficient *for $\mathcal{P}$ if there is a stochastic transition kernel $Q$ such that*

$$P_\theta(B) = \int Q_t(B)\, d\tilde{P}_\theta(t)$$

*for every Borel set $B$ and all $\theta \in \Omega$.*

The argument that this supports randomization to construct fake data is the same as (3.1). Suppose $\tilde{\mathcal{P}}$ is sufficient for $\mathcal{P}$, and let $T \sim \tilde{P}_\theta$ be data from the sufficient experiment. When $T = t$, the fake data $\tilde{X}$ is obtained by sampling from $Q_t$, so that

$$P(\tilde{X} \in B|T = t) = Q_t(B).$$

Then, by smoothing,

$$P(\tilde{X} \in B) = EP(\tilde{X} \in B|T) = EQ_T(B) = \int Q_t(B)\,d\tilde{P}_\theta(t) = P_\theta(B).$$

This shows that $\tilde{X} \sim P_\theta$, so the distributions for $\tilde{X}$, as desired, are the same as distributions for real data from the other experiment $\mathcal{P}$, regardless of the value of $\theta \in \Omega$.

## 3.3 Factorization Theorem

From the definition of sufficiency in the last section it is easy to understand the sense in which a sufficient statistic $T(X)$ conveys all of the information about $\theta$ from data $X$, at least when the model is correct. But the definition is less useful for finding a sufficient statistic, or trying to determine whether a specific statistic is sufficient. When distributions in the family are specified by densities, sufficiency can be checked using the factorization theorem simply by looking at the form of the densities.

**Definition 3.5.** *A family of distributions $\mathcal{P} = \{P_\theta : \theta \in \Omega\}$ is* dominated *if there exists a measure $\mu$ with $P_\theta$ absolutely continuous with respect to $\mu$, for all $\theta \in \Omega$.*

**Theorem 3.6 (Factorization Theorem).** *Let $\mathcal{P} = \{P_\theta : \theta \in \Omega\}$ be a family of distributions dominated by $\mu$. A necessary and sufficient condition for a statistic $T$ to be sufficient is that there exist functions $g_\theta \geq 0$ and $h \geq 0$ such that the densities $p_\theta$ for the family satisfy*

$$p_\theta(x) = g_\theta\bigl(T(x)\bigr)h(x), \quad \text{for a.e. } x \text{ under } \mu.$$

A proof of this result is given in Section 6.4. It depends in part on a proper definition of conditional distributions.

*Example 3.7.* Suppose $X_1, \dots, X_n$ are i.i.d. absolutely continuous variables with common marginal density

$$f_\theta(x) = \begin{cases} (\theta+1)x^\theta, & x \in (0,1); \\ 0, & \text{otherwise}, \end{cases}$$

for $\theta > -1$. Then their joint density $p_\theta$ is

$$p_\theta(x) = \prod_{i=1}^n f_\theta(x_i) = \prod_{i=1}^n (\theta+1)x_i^\theta = (\theta+1)^n \left(\prod_{i=1}^n x_i\right)^\theta, \qquad x \in (0,1)^n,$$

with $p_\theta(x) = 0$ if $x \notin (0,1)^n$. Taking $g_\theta(t) = (\theta+1)^n t^\theta$ and $h = 1_{(0,1)^n}$, from the factorization theorem, $T = \prod_{i=1}^n X_i$ is sufficient.

When $\{x : p_\theta(x) > 0\}$ depends on $\theta$, care is needed to ensure that formulas for $p_\theta$ used in the factorization theorem work for all $x$. To accomplish this, indicator functions are often used.

*Example 3.8.* Suppose $X_1, \dots, X_n$ are a random sample from the uniform distribution on $(\theta, \theta + 1)$. Then the common marginal density is

$$f_\theta(x) = \begin{cases} 1, & x \in (\theta, \theta + 1); \\ 0, & \text{otherwise.} \end{cases}$$

So $f_\theta = 1_{(\theta,\theta+1)}$. The joint density is

$$p_\theta(x) = \prod_{i=1}^{n} 1_{(\theta,\theta+1)}(x_i),$$

which equals one if and only if $\max_i x_i < \theta + 1$ and $\min_i x_i > \theta$. So

$$p_\theta(x) = 1_{(\theta,\infty)}(\min_i x_i) 1_{(-\infty,\theta+1)}(\max_i x_i).$$

By the factorization theorem, $T = (\min_i X_i, \max_i X_i)$ is sufficient.

## 3.4 Minimal Sufficiency

If $T$ is sufficient for a family of distributions $\mathcal{P}$, and if $T = f(\tilde{T})$, then $\tilde{T}$ is also sufficient. This follows easily from the factorization theorem when the family is dominated. This suggests the following definition.

**Definition 3.9.** *A statistic $T$ is* minimal sufficient *if $T$ is sufficient, and for every sufficient statistic $\tilde{T}$ there exists a function $f$ such that $T = f(\tilde{T})$ (a.e. $\mathcal{P}$). Here (a.e. $\mathcal{P}$) means that the set where equality fails is a null set for every $P \in \mathcal{P}$.*

*Example 3.10.* If $X_1, \dots, X_{2n}$ are i.i.d. from $N(\theta, 1)$, $\theta \in \mathbb{R}$, then

$$\tilde{T} = \begin{pmatrix} \sum_{i=1}^{n} X_i \\ \sum_{i=n+1}^{2n} X_i \end{pmatrix}$$

is sufficient but not minimal. It can be shown that $T = \sum_{i=1}^{2n} X_i$ is minimal sufficient here,[3] and $T = f(\tilde{T})$ if we take $f(t) = t_1 + t_2$.

If $\mathcal{P}$ is a dominated family, then the density $p_\theta(X)$, viewed as a function of $\theta$, is called the *likelihood function.* By the factorization theorem, any sufficient statistic must provide enough information to graph the shape of the likelihood, where two functions are defined to have the same shape if they are proportional. The next result shows that a statistic is minimal sufficient if there is a one-to-one relation between the statistic and the likelihood shape.

[3] This follows from Example 3.12 below.

**Theorem 3.11.** *Suppose $\mathcal{P} = \{P_\theta : \theta \in \Omega\}$ is a dominated family with densities $p_\theta(x) = g_\theta\bigl(T(x)\bigr)h(x)$. If $p_\theta(x) \propto_\theta p_\theta(y)$ implies $T(x) = T(y)$, then $T$ is minimal sufficient.*[4]

*Proof.* A proper proof of this result unfortunately involves measure-theoretic niceties, but here is the basic idea. Suppose $\tilde{T}$ is sufficient. Then $p_\theta(x) = \tilde{g}_\theta\bigl(\tilde{T}(x)\bigr)\tilde{h}(x)$ (a.e. $\mu$). Assume this equation holds for all $x$. If $T$ is not a function of $\tilde{T}$, then there must be two data sets $x$ and $y$ that give the same value for $\tilde{T}$, $\tilde{T}(x) = \tilde{T}(y)$, but different values for $T$, $T(x) \neq T(y)$. But then

$$p_\theta(x) = \tilde{g}_\theta\bigl(\tilde{T}(x)\bigr)\tilde{h}(x) \propto_\theta \tilde{g}_\theta\bigl(\tilde{T}(y)\bigr)\tilde{h}(y) = p_\theta(y),$$

and from the condition on $T$ in the theorem, $T(x)$ must equal $T(y)$. Thus $T$ is a function of $\tilde{T}$. Because $\tilde{T}$ was an arbitrary sufficient statistic, $T$ is minimal. □

Although a proper development takes more work, this result in essence says that the shape of the likelihood is minimal sufficient, and so a minimal sufficient "statistic" exists for dominated families.[5] When this result is used, if the implication only fails on a null set (for the family), $T$ will still be minimal sufficient. In particular, if the implication holds unless $p_\theta(x)$ and $p_\theta(y)$ are identically zero as $\theta$ varies, then $T$ will be minimal sufficient.

*Example 3.12.* Suppose $\mathcal{P}$ is an $s$-parameter exponential family with densities

$$p_\theta(x) = e^{\eta(\theta)\cdot T(x) - B(\theta)}h(x),$$

for $\theta \in \Omega$. By the factorization theorem, $T$ is sufficient. Suppose $p_\theta(x) \propto_\theta p_\theta(y)$. Then

$$e^{\eta(\theta)\cdot T(x)} \propto_\theta e^{\eta(\theta)\cdot T(y)},$$

which implies that

$$\eta(\theta)\cdot T(x) = \eta(\theta)\cdot T(y) + c,$$

where the constant $c$ may depend on $x$ and $y$, but is independent of $\theta$. If $\theta_0$ and $\theta_1$ are any two points in $\Omega$,

$$\bigl[\eta(\theta_0) - \eta(\theta_1)\bigr]\cdot T(x) = \bigl[\eta(\theta_0) - \eta(\theta_1)\bigr]\cdot T(y)$$

and

[4] The notation "$\propto_\theta$" here means that the two expressions are proportional when viewed as functions of $\theta$. So $p_\theta(x) \propto_\theta p_\theta(y)$ here would mean that there is a "proportionality constant" $c$ that may depend on $x$ and $y$, so $c = c(x,y)$, such that $p_\theta(x) = c(x,y)p_\theta(y)$, for all $\theta \in \Omega$.

[5] At a technical level this may fail without a bit of regularity. Minimal sufficient $\sigma$-fields must exist in this setting, but there may be no minimal sufficient statistic if $\mathcal{P}$ is not separable (under total variation norm). For discussion and counterexamples, see Bahadur (1954) and Landers and Rogge (1972).

$$\big[\eta(\theta_0) - \eta(\theta_1)\big] \cdot \big[T(x) - T(y)\big] = 0.$$

This shows that $T(x) - T(y)$ is orthogonal to every vector in

$$\eta(\Omega) \ominus \eta(\Omega) \stackrel{\text{def}}{=} \{\eta(\theta_0) - \eta(\theta_1) : \theta_0 \in \Omega, \theta_1 \in \Omega\},$$

and so it must lie in the orthogonal complement of the linear span[6] of $\eta(\Omega) \ominus \eta(\Omega)$. In particular, if the linear span of $\eta(\Omega) \ominus \eta(\Omega)$ is all of $\mathbb{R}^s$, then $T(x)$ must equal $T(y)$. So, in this case, $T$ will be minimal sufficient.

*Example 3.13.* Suppose $X_1, \dots, X_n$ are i.i.d. absolutely continuous variables with common marginal density

$$f_\theta(x) = \frac{1}{2} e^{-|x-\theta|}.$$

Then the joint density is

$$p_\theta(x) = \frac{1}{2^n} \exp\biggl\{-\sum_{i=1}^n |x_i - \theta|\biggr\}.$$

The variables $X_{(1)} \le X_{(2)} \le \cdots \le X_{(n)}$ found by listing $X_1, \dots, X_n$ in increasing order are called the *order statistics.* By the factorization theorem, $T = (X_{(1)}, \dots, X_{(n)})'$, is sufficient. Suppose $p_\theta(x) \propto_\theta p_\theta(y)$. Then the difference between $\sum_{i=1}^n |x_i - \theta|$ and $\sum_{i=1}^n |y_i - \theta|$ is constant in $\theta$. Both of these functions are piecewise linear functions of $\theta$ with a slope that increases by two at each order statistic. The difference can only be constant in $\theta$ if $x$ and $y$ have the same order statistics. Thus the order statistics are minimal sufficient for this family of distributions.

## 3.5 Completeness

Completeness is a technical condition that strengthens sufficiency in a useful fashion.

**Definition 3.14.** *A statistic $T$ is* complete *for a family $\mathcal{P} = \{P_\theta : \theta \in \Omega\}$ if*

$$E_\theta f(T) = c, \quad \textit{for all } \theta,$$

*implies $f(T) = c$ (a.e. $\mathcal{P}$).*

*Remark 3.15.* Replacing $f$ by $f - c$, the constant $c$ in this definition could be taken to be zero.

[6] See Appendix A.3 for a review of vector spaces and the geometry of $\mathbb{R}^n$.

*Example 3.16.* Suppose $X_1, \dots, X_n$ are i.i.d. from a uniform distribution on $(0, \theta)$. Using indicator functions, the joint density is $I\{\min x_i > 0\} I\{\max x_i < \theta\}/\theta^n$, and so $T = \max\{X_1, \dots, X_n\}$ is sufficient by the factorization theorem (Theorem 3.6). By independence, for $t \in (0, \theta)$,

$$\begin{aligned} P_\theta(T \le t) &= P_\theta(X_1 \le t, \dots, X_n \le t) \\ &= P_\theta(X_1 \le t) \times \cdots \times P_\theta(X_n \le t) = (t/\theta)^n. \end{aligned}$$

Differentiating this expression, $T$ has density $nt^{n-1}/\theta^n$, $t \in (0, \theta)$. Suppose $E_\theta f(T) = c$ for all $\theta > 0$; then

$$E_\theta\bigl[f(T) - c\bigr] = \frac{n}{\theta^n} \int_0^\theta \bigl[f(t) - c\bigr] t^{n-1}\, dt = 0.$$

From this (using fact 4 about integration in Section 1.4) $\bigl[f(t) - c\bigr]t^{n-1} = 0$ for a.e. $t > 0$. So $f(T) = c$ (a.e. $\mathcal{P}$), and $T$ is complete.

**Theorem 3.17.** *If $T$ is complete and sufficient, then $T$ is minimal sufficient.*

*Proof.* Let $\tilde{T}$ be a minimal sufficient statistic, and assume $T$ and $\tilde{T}$ are both bounded random variables. Then $\tilde{T} = f(T)$. Define $g(\tilde{T}) = E_\theta[T|\tilde{T}]$, noting that this function is independent of $\theta$ because $\tilde{T}$ is sufficient. By smoothing, $E_\theta g(\tilde{T}) = E_\theta T$, and so $E_\theta\bigl[T - g(\tilde{T})\bigr] = 0$, for all $\theta$. But $T - g(\tilde{T}) = T - g\bigl(f(T)\bigr)$, a function of $T$, and so by completeness, $T = g(\tilde{T})$ (a.e. $\mathcal{P}$). This establishes a one-to-one relationship between $T$ and $\tilde{T}$. From the definition of minimal sufficiency, $T$ must also be minimal sufficient.

For the general case, first note that sufficiency and completeness are both preserved by one-to-one transformations, so two statistics can be considered equivalent if they are related by a one-to-one (bimeasurable) function. But there are one-to-one bimeasurable functions from $\mathbb{R}^n$ to $\mathbb{R}$, and so any random vector is equivalent to a single random variable.[7] Using this, if $T$ and $\tilde{T}$ are random vectors, the result follows easily from the one-dimensional case transforming both of them to equivalent random variables. ☐

**Definition 3.18.** *An exponential family with densities $p_\theta(x) = \exp\bigl\{\eta(\theta) \cdot T(x) - B(\theta)\bigr\}h(x)$, $\theta \in \Omega$, is said to be of* full rank *if the interior of $\eta(\Omega)$ is not empty and if $T_1, \dots, T_s$ do not satisfy a linear constraint of the form $v \cdot T = c$ (a.e. $\mu$).*

If $\Omega \subset \mathbb{R}^s$ and $\eta$ is continuous and one-to-one (injective), and the interior of $\Omega$ is nonempty, then the interior of $\eta(\Omega)$ cannot be empty. This follows from the "invariance of domain" theorem of Brouwer (1912).

If the interior of $\eta(\Omega)$ is not empty, then the linear span of $\eta(\Omega) \ominus \eta(\Omega)$ will be all of $\mathbb{R}^s$, and, by Example 3.12, $T$ will be minimal sufficient. The following result shows that in this case $T$ is also complete.

[7] For instance, the function $g : \mathbb{R}^2 \to \mathbb{R}$ that alternates the decimal digits of its arguments, and thus, for instance, $g(12.34\dots, 567.89\dots) = 506172.8394\dots$, is one-to-one and bimeasurable.

**Theorem 3.19.** *In an exponential family of full rank, $T$ is complete.*

**Definition 3.20.** *A statistic $V$ is called* ancillary *if its distribution does not depend on $\theta$. So, $V$ by itself provides no information about $\theta$.*

An ancillary statistic $V$, by itself, provides no useful information about $\theta$. But in some situations $V$ can be a function of a minimal sufficient statistic $T$. For instance, in Example 3.13 differences $X_{(i)} - X_{(j)}$ between order statistics are ancillary. But they are functions of the minimal sufficient $T$ and are relevant to inference. The following result of Basu shows that when $T$ is complete it will contain no ancillary information. See Basu (1955, 1958) or Lehmann (1981) for further discussion.

**Theorem 3.21 (Basu).** *If $T$ is complete and sufficient for $\mathcal{P} = \{P_\theta : \theta \in \Omega\}$, and if $V$ is ancillary, then $T$ and $V$ are independent under $P_\theta$ for any $\theta \in \Omega$.*

*Proof.* Define $q_A(t) = P_\theta(V \in A|T = t)$, so that $q_A(T) = P_\theta(V \in A|T)$, and define $p_A = P_\theta(V \in A)$. By sufficiency and ancillarity, neither $p_A$ nor $q_A(t)$ depend on $\theta$. Also, by smoothing,

$$p_A = P_\theta(V \in A) = E_\theta P_\theta(V \in A|T) = E_\theta q_A(T),$$

and so, by completeness, $q_A(T) = p_A$ (a.e. $\mathcal{P}$). By smoothing,

$$\begin{aligned} P_\theta(T \in B, V \in A) &= E_\theta 1_B(T) 1_A(V) \\ &= E_\theta E_\theta\big(1_B(T)1_A(V) \bigm| T\big) \\ &= E_\theta 1_B(T) E_\theta\big(1_A(V) \bigm| T\big) \\ &= E_\theta 1_B(T) q_A(T) \\ &= E_\theta 1_B(T) p_A \\ &= P_\theta(T \in B) P_\theta(V \in A). \end{aligned}$$

Here $A$ and $B$ are arbitrary Borel sets, and so $T$ and $V$ are independent. □

*Example 3.22.* Suppose $X_1, \dots, X_n$ are i.i.d. from $N(\mu, \sigma^2)$, and take $\mathcal{P} = \mathcal{P}_\sigma = \{N(\mu, \sigma^2)^n : \mu \in \mathbb{R}\}$. (Thus $\mathcal{P}_\sigma$ is the family of all normal distributions with standard deviation the fixed value $\sigma$.) With $\overline{x} = (x_1 + \dots + x_n)/n$, the joint density can be written as

$$\frac{1}{(2\pi\sigma^2)^{n/2}} \exp\left[\frac{n\mu}{\sigma^2}\overline{x} - \frac{n\mu^2}{2\sigma^2} - \frac{1}{2\sigma^2}\sum_{i=1}^n x_i^2\right].$$

These densities for $\mathcal{P}_\sigma$ form a full rank exponential family, and so the average $\overline{X} = (X_1 + \dots + X_n)/n$ is a complete sufficient statistic for $\mathcal{P}_\sigma$. Define

$$S^2 = \frac{1}{n-1}\sum_{i=1}^n (X_i - \overline{X})^2,$$

called the sample variance. For the family $\mathcal{P}_\sigma$, $S^2$ is ancillary. To see this, let $Y_i = X_i - \mu$, $i = 1, \dots, n$. Because

$$\begin{aligned} P_\mu(Y_i \le y) = P_\mu(X_i \le y + \mu) &= \int_{-\infty}^{y+\mu} \exp\left[-\frac{1}{2\sigma^2}(x-\mu)^2\right] \frac{dx}{\sqrt{2\pi\sigma^2}} \\ &= \int_{-\infty}^{y} \exp\left[-\frac{1}{2\sigma^2}u^2\right] \frac{du}{\sqrt{2\pi\sigma^2}}, \end{aligned}$$

and the integrand is the density for $N(0, \sigma^2)$, $Y_i \sim N(0, \sigma^2)$. Then $Y_1, \dots, Y_n$ are i.i.d. from $N(0, \sigma^2)$. Because $\overline{Y} = (Y_1 + \cdots + Y_n)/n = \overline{X} - \mu$, $X_i - \overline{X} = Y_i - \overline{Y}$, $i = 1, \dots, n$, and

$$S^2 = \frac{1}{n-1} \sum_{i=1}^{n} (Y_i - \overline{Y})^2.$$

Because the joint distribution of $Y_1, \dots, Y_n$ depends on $\sigma$ but not $\mu$, $S^2$ is ancillary for $\mathcal{P}_\sigma$. Hence, by Basu's theorem, $\overline{X}$ and $S^2$ are independent.[8]

## 3.6 Convex Loss and the Rao–Blackwell Theorem

**Definition 3.23.** *A real-valued function $f$ on a convex set $\mathcal{C}$ in $\mathbb{R}^p$ is called* convex *if, for any $x \ne y$ in $\mathcal{C}$ and any $\gamma \in (0, 1)$,*

$$f\big[\gamma x + (1-\gamma)y\big] \le \gamma f(x) + (1-\gamma) f(y). \tag{3.2}$$

*The function $f$ is* strictly convex *if* (3.2) *holds with strict inequality.*

Geometrically, $f$ is strictly convex if the graph of $f$ for values between $x$ and $y$ lies below the chord joining $\big(x, f(x)\big)$ and $\big(y, f(y)\big)$, illustrated in Figure 3.3. If $p = 1$ and $f''$ exists and is nonnegative on $\mathcal{C}$, the $f$ is convex. The next result is the supporting hyperplane theorem in one dimension.

**Theorem 3.24.** *If $f$ is a convex function on an open interval $\mathcal{C}$, and if $t$ is an arbitrary point in $\mathcal{C}$, then there exists a constant $c = c_t$ such that*

$$f(t) + c(x - t) \le f(x), \qquad \forall x \in \mathcal{C}.$$

*If $f$ is strictly convex, then $f(t) + c(x - t) < f(x)$ for all $x \in \mathcal{C}$, $x \ne t$.*

The left-hand side of this inequality is a line through $\big(t, f(t)\big)$. So, this result says that we can always find a line below the graph of $f$ touching the graph of $f$ at $t$. This is illustrated in Figure 3.4.

[8] The independence established here plays an important role when distribution theory for this example is considered in more detail in Section 4.3. Independence can also be established using spherical symmetry of the multivariate normal distribution, an approach developed in a more general setting in Chapter 14.

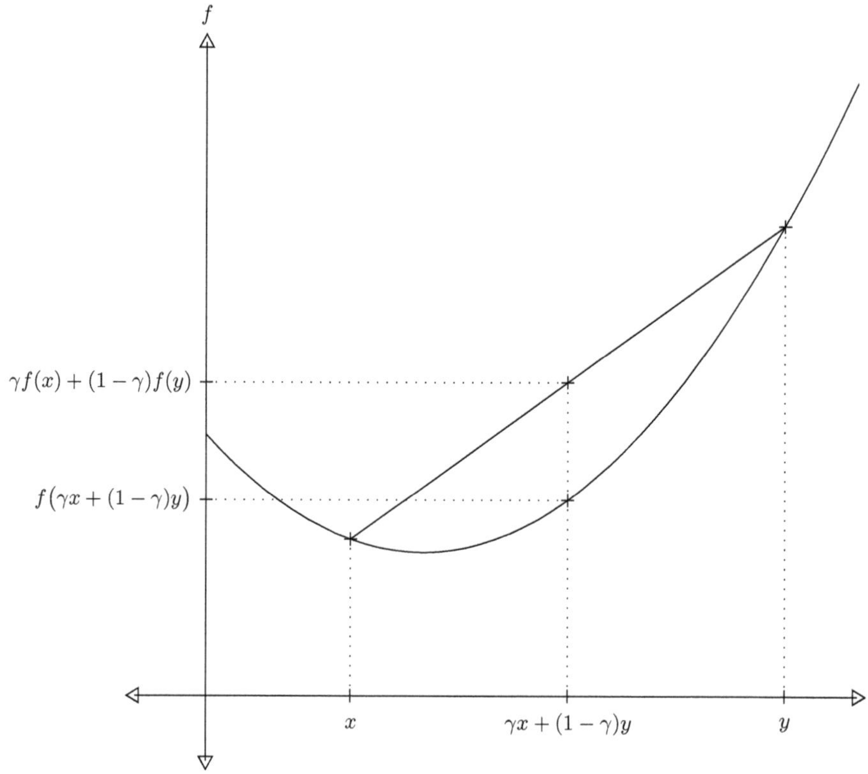


**Fig. 3.3.** A convex function.

**Theorem 3.25 (Jensen's Inequality).** *If $\mathcal{C}$ is an open interval, $f$ is a convex function on $\mathcal{C}$, $P(X \in \mathcal{C}) = 1$, and $EX$ is finite, then*

$$f(EX) \le Ef(X).$$

*If $f$ is strictly convex, the inequality is strict unless $X$ is almost surely constant.*

*Proof.* By Theorem 3.24 with $t = EX$, for some constant $c$,

$$f(EX) + c(x - EX) \le f(x), \qquad \forall x \in \mathcal{C},$$

and so

$$f(EX) + c(X - EX) \le f(X), \qquad \text{(a.e. } P\text{)}.$$

The first assertion of the theorem follows taking expectations. If $f$ is strictly convex this bound will be strict on $X \neq EX$. The second assertion of the theorem then follows using fact 2 from Section 1.4. □

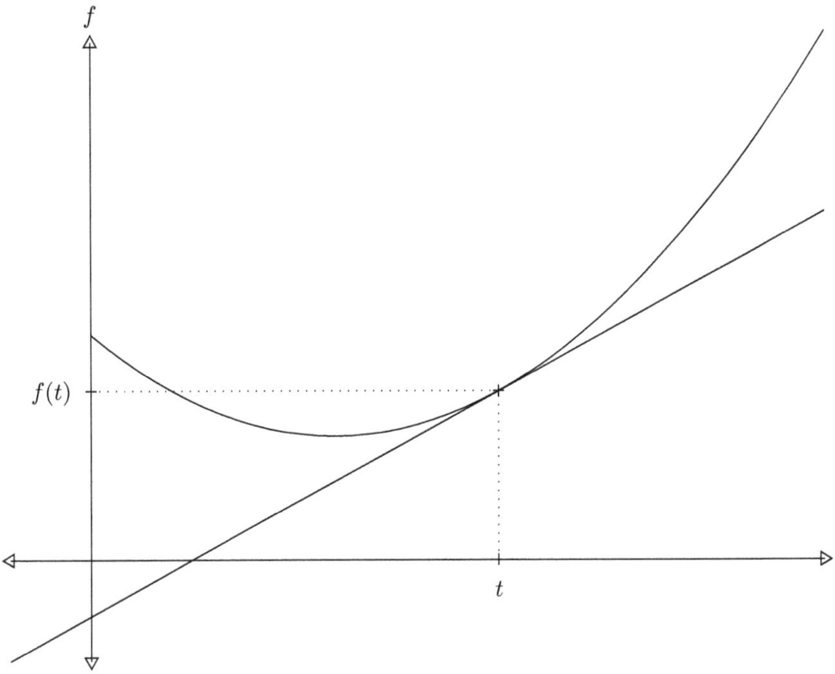


**Fig. 3.4.** Convex function with support line.

*Remark 3.26.* Jensen's inequality also holds in higher dimensions, with $X$ a random vector.

*Example 3.27.* The functions $1/x$ and $-\log x$ are strictly convex on $(0,\infty)$. If $X>0$, then $1/EX \le E[1/X]$ and $\log EX \ge E\log(X)$. These inequalities are strict unless $X$ is constant.

If $\delta(X)$ is an estimator of $g(\theta)$, then the risk of $\delta$ for a loss function $L(\theta,d)$ is $R(\delta,\theta) = E_\theta L\bigl(\theta,\delta(X)\bigr)$. Suppose $T$ is a sufficient statistic. By Theorem 3.3 there is a randomized estimator based on $T$ with the same risk as $\delta$. The following result shows that for convex loss functions there is generally a non-randomized estimator based on $T$ that has smaller risk than $\delta$.

**Theorem 3.28 (Rao–Blackwell).** *Let $T$ be a sufficient statistic for $\mathcal{P} = \{P_\theta : \theta\in\Omega\}$, let $\delta$ be an estimator of $g(\theta)$, and define $\eta(T) = E[\delta(X)|T]$. If $\theta\in\Omega$, $R(\theta,\delta)<\infty$, and $L(\theta,\cdot)$ is convex, then*

$$R(\theta,\eta) \le R(\theta,\delta).$$

*Furthermore, if $L(\theta,\cdot)$ is strictly convex, the inequality will be strict unless $\delta(X)=\eta(T)$ (a.e. $P_\theta$).*

*Proof.* Jensen's inequality with expectations against the conditional distribution of $\delta(X)$ given $T$ gives

$$L(\theta, \eta(T)) \le E_\theta\big[L(\theta, \delta(X)) \mid T\big].$$

Taking expectations, $R(\theta, \eta) \le R(\theta, \delta)$. The assertion about strict inequality follows after a bit of work from the second assertion in Jensen's inequality. □

This result shows that with convex loss functions the only estimators worth considering, at least if estimators are judged solely by their risk, are functions of $T$ but not $X$. It can also be used to show that any randomized estimator is worse than a corresponding nonrandomized estimator. Using the probability integral transformation,[9] any randomized estimator can be viewed as a function of $X$ and $U$, where $X$ and $U$ are independent and the distribution of $U$ does not depend on $\theta$. But if $X$ and $U$ are both considered as data, then $X$ is sufficient, and with convex loss the risk of a randomized estimator $\delta(X, U)$ will be worse than the risk of the estimator $E\big(\delta(X, U) \bigm| X\big)$, which is based solely on $X$.

## 3.7 Problems[10]

1. An estimator $\delta$ is called *inadmissible* if there is a competing estimator $\tilde{\delta}$ with a better risk function, that is, if $R(\theta, \tilde{\delta}^*) \le R(\theta, \delta)$, for all $\theta \in \Omega$, and $R(\theta, \tilde{\delta}^*) < R(\theta, \delta)$, for some $\theta \in \Omega$. If there is no estimator with a better risk function, $\delta$ is *admissible.* Consider estimating success probability $\theta \in [0, 1]$ from data $X \sim \text{Binomial}(n, \theta)$ under squared error loss. Define $\delta_{a,b}$ by

$$\delta_{a,b}(X) = a\frac{X}{n} + (1 - a)b,$$

which might be called a linear estimator, because it is a linear function of $X$.

   a) Find the variance and bias of $\delta_{a,b}$. (The bias of an arbitrary estimator $\delta$ of $\theta$ is defined as $b(\theta, \delta) = E_\theta\delta(X) - \theta$.)
   b) If $a > 1$, show that $\delta_{a,b}$ is inadmissible by finding a competing linear estimator with better risk. Hint: The risk of an arbitrary estimator $\delta$ under squared error loss is $\text{Var}_\theta\big(\delta(X)\big) + b^2(\theta, \delta)$. Find an unbiased estimator with smaller variance.
   c) If $b > 1$ or $b < 0$, and $a \in [0, 1)$, show that $\delta_{a,b}$ is inadmissible by finding a competing linear estimator with better risk. Hint: Find an estimator with the same variance but better bias.

[9] If the inverse for a possibly discontinuous cumulative distribution function $F$ is defined as $F^{\Leftarrow}(t) = \sup\{x : F(x) \le t\}$, and if $U$ is uniformly distributed on $(0, 1)$, then the random variable $F^{\Leftarrow}(U)$ has cumulative distribution function $F$.

[10] Solutions to the starred problems are given at the back of the book.

d) If $a < 0$, find a linear estimator with better risk than $\delta_{a,b}$.

*2. Suppose data $X_1, \dots, X_n$ are independent with

$$P_\theta(X_i \le x) = x^{t_i\theta}, \qquad x \in (0,1),$$

where $\theta > 0$ is the unknown parameter, and $t_1, \dots, t_n$ are known positive constants. Find a one-dimensional sufficient statistic $T$.

*3. An object with weight $\theta$ is weighed on scales with different precision. The data $X_1, \dots, X_n$ are independent, with $X_i \sim N(\theta, \sigma_i^2)$, $i = 1, \dots, n$, with the standard deviations $\sigma_1, \dots, \sigma_n$ known constants. Use sufficiency to suggest a weighted average of $X_1, \dots, X_n$ to estimate $\theta$. (A weighted average would have form $\sum_{i=1}^n w_i X_i$, where the $w_i$ are positive and sum to one.)

*4. Let $X_1, \dots, X_n$ be a random sample from an arbitrary discrete distribution $P$ on $\{1,2,3\}$. Find a two-dimensional sufficient statistic.

5. For $\theta \in \Omega = (0,1)$, let $\tilde{P}_\theta$ denote a discrete distribution with mass function

$$\tilde{p}_\theta(t) = (1+t)\theta^2(1-\theta)^t, \qquad t = 0, 1, \dots,$$

and let $P_\theta$ denote the binomial distribution with two trials and success probability $\theta$. Show that the model $\tilde{\mathcal{P}} = \{\tilde{P}_\theta : \theta \in \Omega\}$ is sufficient for the binomial model $\mathcal{P} = \{P_\theta : \theta \in \Omega\}$. Identify the stochastic transition $Q$ by giving the mass functions

$$q_t(x) = Q_t(\{x\}), \qquad x = 0, 1, 2,$$

for $t = 0, 1, \dots$.

*6. The beta distribution with parameters $\alpha > 0$ and $\beta > 0$ has density

$$f_{\alpha,\beta}(x) = \begin{cases} \dfrac{\Gamma(\alpha+\beta)}{\Gamma(\alpha)\Gamma(\beta)} x^{\alpha-1}(1-x)^{\beta-1}, & x \in (0,1); \\ 0, & \text{otherwise.} \end{cases}$$

Suppose $X_1, \dots, X_n$ are i.i.d. from a beta distribution.

a) Determine a minimal sufficient statistic (for the family of joint distributions) if $\alpha$ and $\beta$ vary freely.

b) Determine a minimal sufficient statistic if $\alpha = 2\beta$.

c) Determine a minimal sufficient statistic if $\alpha = \beta^2$.

*7. *Logistic regression.* Let $X_1, \dots, X_n$ be independent Bernoulli variables, with $p_i = P(X_i = 1)$, $i = 1, \dots, n$. Let $t_1, \dots, t_n$ be a sequence of known constants that are related to the $p_i$ via

$$\log \frac{p_i}{1-p_i} = \alpha + \beta t_i,$$

where $\alpha$ and $\beta$ are unknown parameters. Determine a minimal sufficient statistic for the family of joint distributions.

*8. The multinomial distribution, derived later in Section 5.3, is a discrete distribution with mass function

$$\frac{n!}{x_1! \times \cdots \times x_s!} p_1^{x_1} \times \cdots \times p_s^{x_s},$$

where $x_0, \dots, x_s$ are nonnegative integers summing to $n$, where $p_1, \dots, p_s$ are nonnegative probabilities summing to one, and $n$ is the sample size. Let $N_{11}, N_{12}, N_{21}, N_{22}$ have a multinomial distribution with $n$ trials and success probabilities $p_{11}, p_{12}, p_{21}, p_{22}$. (A common model for a two-by-two contingency table.)
   a) Give a minimal sufficient statistic if the success probabilities vary freely over the unit simplex in $\mathbb{R}^4$. (The unit simplex in $\mathbb{R}^p$ is the set of all vectors with nonnegative entries summing to one.)
   b) Give a minimal sufficient statistic if the success probabilities are constrained so that $p_{11}p_{22} = p_{12}p_{21}$.

*9. Let $f$ be a positive integrable function on $(0, \infty)$. Define

$$c(\theta) = 1 \Big/ \int_\theta^\infty f(x)\,dx,$$

and take $p_\theta(x) = c(\theta)f(x)$ for $x > \theta$, and $p_\theta(x) = 0$ for $x \le \theta$. Let $X_1, \dots, X_n$ be i.i.d. with common density $p_\theta$.
   a) Show that $M = \min\{X_1, \dots, X_n\}$ is sufficient.
   b) Show that $M$ is minimal sufficient.

*10. Suppose $X_1, \dots, X_n$ are i.i.d. with common density $f_\theta(x) = (1 + \theta x)/2$, $|x| < 1$; $f_\theta(x) = 0$, otherwise, where $\theta \in [-1, 1]$ is an unknown parameter. Show that the order statistics are minimal sufficient. (Hint: A polynomial of degree $n$ is uniquely determined by its value on a grid of $n+1$ points.)

11. Consider a two-sample problem in which $X_1, \dots, X_n$ is a random sample from $N(\mu, \sigma_x^2)$ and $Y_1, \dots, Y_m$ is an independent random sample from $N(\mu, \sigma_y^2)$. Let $P_\theta$ denote the joint distribution of these $n + m$ variables, with $\theta = (\mu, \sigma_x^2, \sigma_y^2)$. Find a minimal sufficient statistic for this family of distributions.

12. Let $Z_1$ and $Z_2$ be independent standard normal random variables with common density $\phi(x) = \exp(-x^2/2)/\sqrt{2\pi}$, and suppose $X$ and $Y$ are related to these variables by

$$X = Z_1 \text{ and } Y = (X + Z_2)\theta,$$

where $\theta > 0$ is an unknown parameter. (This might be viewed as a regression model in which the independent variable is measured with error.)
   a) Find the joint density for $X$ and $Y$.
   b) Suppose our data $(X_1, Y_1), \dots, (X_n, Y_n)$ are i.i.d. random vectors with common distribution that of $X$ and $Y$ in part (a),

$$\begin{pmatrix} X_i \\ Y_i \end{pmatrix} \sim \begin{pmatrix} X \\ Y \end{pmatrix}, \qquad i = 1, \dots, n.$$

Find a minimal sufficient statistic.

13. Let $X_1, \dots, X_n$ be independent Poisson variables with $\lambda_i = EX_i$, $i = 1, \dots, n$. Let $t_1, \dots, t_n$ be a sequence of known constants related to the $\lambda_i$ by

$$\log \lambda_i = \alpha + \beta t_i, \qquad i = 1, \dots, n,$$

where $\alpha$ and $\beta$ are unknown parameters. Find a minimal sufficient statistic for the family of joint distributions.

14. Let $X_1, \dots, X_n$ be i.i.d. from a discrete distribution $Q$ on $\{1, 2, 3\}$. Let $p_i = Q(\{i\}) = P(X_j = i)$, $i = 1, 2, 3$, and assume we know that $p_1 = 1/3$, but have no additional knowledge of $Q$. Define

$$N_i = \#\{j \le n : X_j = i\}.$$

   a) Show that $T = (N_1, N_2)$ is sufficient.
   b) Is $T$ minimal sufficient? If so, explain why. If not, find a minimal sufficient statistic.

15. Use completeness for the family $N(\theta, 1)$, $\theta \in \mathbb{R}$ to find an essentially unique solution $f$ of the following integral equation:

$$\int f(x) e^{\theta x}\, dx = \sqrt{2\pi} e^{\theta^2/2}, \qquad \theta \in \mathbb{R}.$$

*16. Let $X_1, \dots, X_n$ be a random sample from an absolutely continuous distribution with density

$$f_\theta(x) = \begin{cases} 2x/\theta^2, & x \in (0, \theta); \\ 0, & \text{otherwise.} \end{cases}$$

   a) Find a one-dimensional sufficient statistic $T$.
   b) Determine the density of $T$.
   c) Show directly that $T$ is complete.

*17. Let $X, X_1, X_2, \dots$ be i.i.d. from an exponential distribution with failure rate $\lambda$ (introduced in Problem 1.30).
   a) Find the density of $Y = \lambda X$.
   b) Let $\overline{X} = (X_1 + \cdots + X_n)/n$. Show that $\overline{X}$ and $(X_1^2 + \cdots + X_n^2)/\overline{X}^2$ are independent.

18. Let $X_1, \dots, X_n$ be independent, with $X_i \sim N(t_i\theta, 1)$, where $t_1, \dots, t_n$ are a sequence of known constants (not all zero).
   a) Show that the least squares estimator $\hat{\theta} = \sum_{i=1}^n t_i X_i / \sum_{i=1}^n t_i^2$ is complete sufficient for the family of joint distributions.
   b) Use Basu's theorem to show that $\hat{\theta}$ and $\sum_{i=1}^n (X_i - t_i\hat{\theta})^2$ are independent.

19. Let $X$ and $Y$ be independent Poisson variables, $X$ with mean $\theta$, and $Y$ with mean $\theta^2$, $\theta \in (0, \infty)$.
   a) Find a minimal sufficient statistic for the family of joint distributions.

b) Is your minimal sufficient statistic complete? Explain.

20. Let $Z_1, \dots, Z_n$ be i.i.d. standard normal variates, and let $Z$ be the random vector formed from these variables. Use Basu's theorem to show that $\|Z\|$ and $Z_1/\|Z\|$ are independent.
21. Let $X_1, \dots, X_n$ be i.i.d. from the uniform distribution on $(0, 1)$, and let $M = \max\{X_1, \dots, X_n\}$. Show that $X_1/M$ and $M$ are independent.
22. Let $(X_1, Y_1), \dots, (X_n, Y_n)$ be i.i.d. and absolutely continuous with common density

$$f_\theta(x, y) = \begin{cases} 2/\theta^2, & x > 0, y > 0, x + y < \theta; \\ 0, & \text{otherwise.} \end{cases}$$

(This is the density for a uniform distribution on the region inside a triangle in $\mathbb{R}^2$.)

a) Find a minimal sufficient statistic for the family of joint distributions.
b) Find the density for your minimal sufficient statistic.
c) Is the minimal sufficient statistic complete?

23. Suppose $X$ has a geometric distribution with success probability $\theta \in (0, 1)$, $Y$ has a geometric distribution with success probability $2\theta - \theta^2$, and $X$ and $Y$ are independent. Find a minimal sufficient statistic $T$ for the family of joint distributions. Is $T$ complete?
24. Let data $X$ and $Y$ be independent variables with

$$X \sim \text{Binomial}(n, \theta) \text{ and } Y \sim \text{Binomial}(n, \theta^2),$$

with $\theta \in (0, 1)$ an unknown parameter.

a) Find a minimal sufficient statistic.
b) Is the minimal sufficient statistic complete? If it is, explain why; if it is not, find a nontrivial function $g$ such that $E_\theta g(T) = 0$ for all $\theta$.

25. Let $X_1, \dots, X_n$ be i.i.d. absolutely continuous random variables with common density

$$f_\theta(x) = \begin{cases} \theta e^{-\theta x}, & x > 0; \\ 0, & x \le 0, \end{cases}$$

where $\theta > 0$ is an unknown parameter.

a) Find the density of $\theta X_i$.
b) Let $X_{(1)} \le \cdots \le X_{(n)}$ be the order statistics and $\overline{X} = (X_1 + \cdots + X_n)/n$ the sample average. Show that $\overline{X}$ and $X_{(1)}/X_{(n)}$ are independent.

26. Two teams play a series of games, stopping as soon as one of the teams has three wins. Assume the games are independent and that the chance the first team wins is an unknown parameter $\theta \in (0, 1)$. Let $X$ denote the number of games the first team wins, and $Y$ the number of games the other team wins.

a) Find the joint mass function of $X$ and $Y$.

b) If our data are $X$ and $Y$, find a minimal sufficient statistic.
c) Is the minimal sufficient statistic in part (b) complete? Explain your reasoning.

27. Let $X_1, \dots, X_n$ be i.i.d. from a uniform distribution on $(-\theta, \theta)$, where $\theta > 0$ is an unknown parameter.
a) Find a minimal sufficient statistic $T$.
b) Define
$$V = \frac{\overline{X}}{\max_i X_i - \min_i X_i},$$
where $\overline{X} = (X_1 + \cdots + X_n)/n$, the sample average. Show that $T$ and $V$ are independent.

28. Show that if $f$ is defined and bounded on $(-\infty, \infty)$, then $f$ cannot be convex (unless it is constant).

*29. Find a function on $(0, \infty)$ that is bounded and strictly convex.

*30. Use convexity to show that the canonical parameter space $\Xi$ of a one-parameter exponential family must be an interval. Specifically, show that if $\eta_0 < \eta < \eta_1$, and if $\eta_0$ and $\eta_1$ both lie in $\Xi$, then $\eta$ must lie in $\Xi$.

*31. Let $f$ and $g$ be positive probability densities on $\mathbb{R}$. Use Jensen's inequality to show that
$$\int \log\left(\frac{f(x)}{g(x)}\right) f(x)\,dx > 0,$$
unless $f = g$ a.e. (If $f = g$, the integral equals zero.) This integral is called the *Kullback–Leibler information.*

32. The geometric mean of a list of positive constants $x_1, \dots, x_n$ is
$$\tilde{x} = \left(x_1 \times \cdots \times x_n\right)^{1/n},$$
and the arithmetic mean is the average $\overline{x} = (x_1 + \cdots + x_n)/n$. Show that $\tilde{x} \le \overline{x}$.

# 4

# Unbiased Estimation

Example 3.1 shows that a clean comparison between two estimators is not always possible: if their risk functions cross, one estimator will be preferable for $\theta$ in some subset of the parameter space $\Omega$, and the other will be preferable in a different subset of $\Omega$. In some cases this problem will not arise if both estimators are unbiased. We may then be able to identify a best unbiased estimator. These ideas and limitations of the theory are discussed in Sections 4.1 and 4.2. Sections 4.3 and 4.4 concern distribution theory and unbiased estimation for the normal one-sample problem in which data are i.i.d. from a normal distribution. Sections 4.5 and 4.6 introduce Fisher information and derive lower bounds for the variance of unbiased estimators.

## 4.1 Minimum Variance Unbiased Estimators

An estimator $\delta$ is called *unbiased* for $g(\theta)$ if

$$E_\theta \delta(X) = g(\theta), \qquad \forall \theta \in \Omega. \tag{4.1}$$

If an unbiased estimator exists, $g$ is called *U-estimable.*

*Example 4.1.* Suppose $X$ has a uniform distribution on $(0, \theta)$. Then $\delta$ is unbiased if

$$\int_0^\theta \delta(x)\theta^{-1}\,dx = g(\theta), \qquad \forall \theta > 0,$$

or if

$$\int_0^\theta \delta(x)\,dx = \theta g(\theta), \qquad \forall \theta > 0. \tag{4.2}$$

So $g$ cannot be U-estimable unless $\theta g(\theta) \to 0$ as $\theta \downarrow 0$. If $g'$ exists, then differentiating (4.2), by the fundamental theorem of calculus,

$$\delta(x) = \frac{d}{dx}\big(xg(x)\big) = g(x) + xg'(x).$$

R.W. Keener, *Theoretical Statistics: Topics for a Core Course*, Springer Texts in Statistics,
DOI 10.1007/978-0-387-93839-4_4, 

For instance, if $g(\theta) = \theta$, $\delta(X) = 2X$.

*Example 4.2.* If $X$ has the binomial distribution with $n$ trials and success probability $\theta$, and if $g(\theta) = \sin\theta$, then $\delta$ will be unbiased if

$$\sum_{k=0}^{n} \delta(k)\binom{n}{k}\theta^k(1-\theta)^{n-k} = \sin\theta, \qquad \forall\theta \in (0,1).$$

The left hand side of this equation is a polynomial in $\theta$ with degree at most $n$. The sine function cannot be written as a polynomial, therefore $\sin\theta$ is not U-estimable.

With squared error loss, $L(\theta, d) = \bigl(d - g(\theta)\bigr)^2$, the risk of an unbiased estimator $\delta$ is

$$R(\theta, \delta) = E_\theta\bigl(\delta(X) - g(\theta)\bigr)^2 = \mathrm{Var}_\theta\bigl(\delta(X)\bigr),$$

and so the goal is to minimize the variance.

**Definition 4.3.** *An unbiased estimator $\delta$ is* uniformly minimum variance unbiased (UMVU) *if*

$$\mathrm{Var}_\theta(\delta) \le \mathrm{Var}_\theta(\delta^*), \qquad \forall\theta \in \Omega,$$

*for any competing unbiased estimator $\delta^*$.*

In a general setting there is no reason to suspect that there will be a UMVU estimator. However, if the family has a complete sufficient statistic, a UMVU will exist, at least when $g$ is U-estimable.

**Theorem 4.4.** *Suppose $g$ is U-estimable and $T$ is complete sufficient. Then there is an essentially unique unbiased estimator based on $T$ that is UMVU.*

*Proof.* Let $\delta = \delta(X)$ be any unbiased estimator and define

$$\eta(T) = E[\delta|T],$$

as in the Rao–Blackwell theorem (Theorem 3.28). By smoothing,

$$g(\theta) = E_\theta\delta = E_\theta E_\theta[\delta|T] = E_\theta\eta(T),$$

and thus $\eta(T)$ is unbiased. Suppose $\eta^*(T)$ is also unbiased. Then

$$E_\theta\bigl[\eta(T) - \eta^*(T)\bigr] = 0, \qquad \forall\theta \in \Omega,$$

and by completeness, $\eta(T) - \eta^*(T) = 0$ (a.e. $\mathcal{P}$). This shows that the estimator $\eta(T)$ is essentially unique; any other unbiased estimator based on $T$ will equal $\eta(T)$ except on a $\mathcal{P}$-null set. The estimator $\eta(T)$ has minimum variance by the Rao–Blackwell theorem with squared error loss. Specifically, if $\delta^*$ is any unbiased estimator, then $\eta^*(T) = E_\theta(\delta^*|T)$ is unbiased by the calculation above. With squared error loss, the risk of $\delta^*$ or $\eta^*(T)$ is the variance, and so

$$\mathrm{Var}_\theta(\delta^*) \ge \mathrm{Var}_\theta\bigl(\eta^*(T)\bigr) = \mathrm{Var}_\theta\bigl(\eta(T)\bigr), \qquad \forall\theta \in \Omega.$$

Thus, $\eta(T)$ is UMVU. □

From the uniqueness assertion in this theorem, if $T$ is complete sufficient and $\eta(T)$ is unbiased, $\eta(T)$ must be UMVU. Viewing (4.1) as an equation for $\delta$, any solution of the form $\delta = \eta(T)$ will be UMVU. This approach provides one strategy to find these estimators.

*Example 4.5.* Let $X_1, \dots, X_n$ be i.i.d. from the uniform distribution on $(0, \theta)$. From Example 3.16, $T = \max\{X_1, \dots, X_n\}$ is complete and sufficient for the family of joint distributions. Suppose $\eta(T)$ is unbiased for $g(\theta)$. Then

$$\int_0^\theta \eta(t) \frac{nt^{n-1}}{\theta^n}\, dt = g(\theta), \qquad \theta > 0,$$

which implies

$$n \int_0^\theta t^{n-1} \eta(t)\, dt = \theta^n g(\theta), \qquad \theta > 0.$$

If $g$ is differentiable and $\theta^n g(\theta) \to 0$ as $\theta \downarrow 0$, then differentiation with respect to $\theta$ gives

$$n\theta^{n-1}\eta(\theta) = \frac{d}{d\theta}\bigl(\theta^n g(\theta)\bigr),$$

and so

$$\eta(t) = \frac{1}{nt^{n-1}} \frac{d}{dt}\bigl(t^n g(t)\bigr) = g(t) + \frac{tg'(t)}{n}, \qquad t > 0.$$

When $g$ is a constant $c$ this argument shows the $\eta(T)$ must also equal $c$, and so $T$ is complete.

When $g$ is the identity function, $g(\theta) = \theta$, $\eta(t) = (n + 1)t/n$. Thus, $(n + 1)T/n$ is the UMVU of $\theta$. Another unbiased estimator is $\delta = 2\overline{X}$. By the theory we have developed, $\eta(T)$ must have smaller variance than $\delta$. In this example the comparison can be done explicitly. Since

$$E_\theta T^2 = \int_0^\theta t^2 \frac{nt^{n-1}}{\theta^n}\, dt = \frac{n}{n+2}\theta^2$$

and

$$E_\theta \eta^2(T) = \left(\frac{n+1}{n}\right)^2 E_\theta T^2 = \frac{(n+1)^2}{n(n+2)}\theta^2,$$

we have

$$\operatorname{Var}_\theta\bigl(\eta(T)\bigr) = E_\theta \eta^2(T) - \bigl(E_\theta \eta(t)\bigr)^2 = \frac{(n+1)^2}{n(n+2)}\theta^2 - \theta^2 = \frac{\theta^2}{n(n+2)}.$$

When $n = 1$, $T = 2X_1$, and so this formula implies that $\operatorname{Var}_\theta(2X_i) = \theta^2/3$. Because $\delta$ is an average of these variables,

$$\operatorname{Var}_\theta(\delta) = \frac{\theta^2}{3n}.$$

The ratio of the variance of $\eta(T)$ to the variance of $\delta$ is

$$\frac{\mathrm{Var}_\theta\big(\eta(T)\big)}{\mathrm{Var}_\theta(\delta)} = \frac{3}{n+2}.$$

As $n \to \infty$ this ratio tends to zero, and so $\eta(T)$ is much more accurate than $\delta$ when $n$ is large.

The proof of Theorem 4.4 also suggests another way to find UMVU estimators. If $\delta$ is an arbitrary unbiased estimator, then $\eta(T) = E[\delta|T]$ will be UMVU. So if any unbiased estimator can be identified, the UMVU can be obtained by computing its conditional expectation.

*Example 4.6.* Let $X_1, \dots, X_n$ be i.i.d. Bernoulli variables with $P_\theta(X_i = 1) = \theta = 1 - P_\theta(X_i = 0)$, $i = 1, \dots, n$. The marginal mass function can be written as $\theta^x(1-\theta)^{1-x}$, $x = 0$ or 1, and so the joint mass function is

$$\prod_{i=1}^n \theta^{x_i}(1-\theta)^{1-x_i} = \theta^{T(x)}(1-\theta)^{n-T(x)},$$

where $T(x) = x_1 + \cdots + x_n$. These joint mass functions form an exponential family with

$$T = T(X) = X_1 + \cdots + X_n \sim \mathrm{Binomial}(\theta, n)$$

as a complete sufficient statistic. Consider unbiased estimation of $g(\theta) = \theta^2$. One unbiased estimator is $\delta = X_1X_2$. The UMVU estimator must be $\eta(T) = E_\theta[X_1X_2|T] = P_\theta(X_1 = X_2 = 1|T)$. Because

$$\begin{aligned} P_\theta(X_1 = X_2 = 1, T = t) &= P_\theta\left(X_1 = X_2 = 1, \sum_{i=3}^n X_i = t-2\right) \\ &= \theta^2\binom{n-2}{t-2}\theta^{t-2}(1-\theta)^{n-t}, \end{aligned}$$

$$\begin{aligned} P_\theta(X_1 = X_2 = 1|T = t) &= \frac{P_\theta(X_1 = X_2 = 1, T = t)}{P_\theta(T = t)} \\ &= \frac{\theta^2\binom{n-2}{t-2}\theta^{t-2}(1-\theta)^{n-t}}{\binom{n}{t}\theta^t(1-\theta)^{n-t}} = \frac{t}{n}\frac{t-1}{n-1}. \end{aligned}$$

So $T(T-1)/(n^2-n)$ is the UMVU estimator of $\theta^2$.

## 4.2 Second Thoughts About Bias

Although the approach developed in the previous section often provides reasonable estimators, the premise that one should only consider unbiased estimators is suspect. Estimators with considerable bias may not be worth considering, but estimators with small bias may be quite reasonable.

*Example 4.1, continued.* As before, $X_1, \ldots, X_n$ are i.i.d. from the uniform distribution on $(0, \theta)$, and $T = \max\{X_1, \ldots, X_n\}$ is complete sufficient. The UMVU estimator $(n+1)T/n$ is a multiple of $T$, but is it the best multiple of $T$? To address this question, let us calculate the risk of $\delta_a = aT$ under squared error loss. From our prior calculations,

$$E_\theta T = \frac{n\theta}{n+1} \text{ and } E_\theta T^2 = \frac{n\theta^2}{n+2}.$$

So the risk of $\delta_a$ is

$$\begin{aligned} R(\theta, \delta_a) &= E_\theta (aT - \theta)^2 \\ &= a^2 E_\theta T^2 - 2a\theta E_\theta T + \theta^2 \\ &= \theta^2 \left( \frac{n}{n+2} a^2 - \frac{2n}{n+1} a + 1 \right). \end{aligned}$$

This is a quadratic function of $a$ minimized when $a = (n+2)/(n+1)$. With this choice for $a$, $R(\theta, \delta_a) = \theta^2/(n+1)^2$, slightly smaller than the risk $\theta^2/(n^2+2n)$ for the UMVU estimator. With squared error loss, the risk of an arbitrary estimator $\delta$ can be written as

$$R(\theta, \delta) = E_\theta \big(\delta - g(\theta)\big)^2 = \text{Var}_\theta(\delta) + b^2(\theta, \delta),$$

where $b(\theta, \delta) = E_\theta \delta - g(\theta)$ is the bias of $\delta$. In this example, the biased estimator $\delta_a$ has smaller variance than the UMVU estimator, which more than compensates for a small amount of additional risk due to the bias. Possibilities for this kind of trade-off between bias and variance arise fairly often in statistics. In nonparametric curve estimation these trade-offs often play a key role. (See Section 18.1.)

*Example 4.7.* Suppose $X$ has mass function

$$P_\theta(X = x) = \frac{\theta^x e^{-\theta}}{x!(1 - e^{-\theta})}, \qquad x = 1, 2, \ldots.$$

This is the density for a Poisson distribution truncated to $\{1, 2, \ldots\}$. The mass functions for $X$ form an exponential family and $X$ is complete sufficient. Consider estimating $g(\theta) = e^{-\theta}$ (the proportion lost through truncation). If $\delta(X)$ is unbiased, then

$$e^{-\theta} = \sum_{k=1}^{\infty} \frac{\delta(k)\theta^k e^{-\theta}}{k!(1 - e^{-\theta})}, \qquad \theta > 0,$$

and so

$$\sum_{k=1}^{\infty} \frac{\delta(k)}{k!} \theta^k = 1 - e^{-\theta} = \sum_{k=1}^{\infty} \frac{(-1)^{k+1}}{k!} \theta^k, \qquad \theta > 0.$$

These power series will agree if and only if they have equal coefficients for $\theta^k$. Hence $\delta(k)$ must be $(-1)^{k+1}$, and the UMVU estimator is $(-1)^{X+1}$, which is 1 when $X$ is odd and $-1$ when $X$ is even! In this example the only unbiased estimator is absurd.

## 4.3 Normal One-Sample Problem—Distribution Theory

In this section distributional results related to sampling from a normal distribution are derived. To begin, here are a few useful properties about normal variables. Let $X \sim N(\mu, \sigma^2)$ and take $Z = (X - \mu)/\sigma$.

1. The distribution of $Z$ is standard normal, $Z \sim N(0, 1)$. More generally, if $a$ and $b$ are constants, $aX + b \sim N(a\mu + b, a^2\sigma^2)$.

   *Proof.* The cumulative distribution function of $Z$ is

$$P(Z \le z) = P\left(\frac{X-\mu}{\sigma} \le z\right) = P(X \le \mu + z\sigma)$$
$$= \int_{-\infty}^{\mu+z\sigma} \frac{\exp\bigl[-(x-\mu)^2/(2\sigma^2)\bigr]}{\sqrt{2\pi\sigma^2}}\,dx = \int_{-\infty}^{z} \frac{e^{-u^2/2}}{\sqrt{2\pi}}\,du.$$

   Taking a derivative with respect to $z$, $Z$ has density $e^{-z^2/2}/\sqrt{2\pi}$ and so $Z \sim N(0, 1)$. The second assertion can be established in a similar fashion. □

2. The moment generating function of $Z$ is $M_Z(u) = e^{u^2/2}$, $u \in \mathbb{R}$.

   *Proof.* Completing the square,

$$M_Z(u) = Ee^{uZ} = \int e^{uz}\frac{e^{-z^2/2}}{\sqrt{2\pi}}\,dz = e^{u^2/2}\int \frac{e^{-(z-u)^2/2}}{\sqrt{2\pi}}\,dz.$$

   The integrand here is the density for $N(u, 1)$, which integrates to one, and the result follows. □

3. The moment generating function of $X$ is

$$M_X(u) = e^{u\mu + u^2\sigma^2/2}, \qquad u \in \mathbb{R}.$$

   *Proof.*

$$M_X(u) = Ee^{uX} = Ee^{u(\mu+\sigma Z)} = e^{u\mu}Ee^{u\sigma Z} = e^{u\mu}M_Z(u\sigma) = e^{u\mu+u^2\sigma^2/2}.$$

□

4. If $X_1 \sim N(\mu_1, \sigma_1^2)$ and $X_2 \sim N(\mu_2, \sigma_2^2)$ are independent, then $X_1 + X_2 \sim N(\mu_1 + \mu_2, \sigma_1^2 + \sigma_2^2)$.

   *Proof.*

$$\begin{aligned} M_{X_1+X_2}(u) = M_{X_1}(u)M_{X_2}(u) &= e^{u\mu_1+u^2\sigma_1^2/2}e^{u\mu_2+u^2\sigma_2^2/2} \\ &= e^{u(\mu_1+\mu_2)+u^2(\sigma_1^2+\sigma_2^2)/2}, \qquad u \in \mathbb{R}, \end{aligned}$$

   which is the moment generating function for $N(\mu, \sigma^2)$ with $\mu = \mu_1 + \mu_2$ and $\sigma^2 = \sigma_1^2 + \sigma_2^2$. So the assertion follows by Lemma 2.7. □

Let $X_1, \ldots, X_n$ be a random sample from $N(\mu, \sigma^2)$. By Example 2.3, the joint densities parameterized by $\theta = (\mu, \sigma^2)$ form a two-parameter full rank exponential family with complete sufficient statistic $T = \left(\sum_{i=1}^n X_i, \sum_{i=1}^n X_i^2\right)$. It is often more convenient to work with statistics

$$\overline{X} = \frac{X_1 + \cdots + X_n}{n} \quad \text{and} \quad S^2 = \frac{1}{n-1}\sum_{i=1}^n (X_i - \overline{X})^2,$$

called the sample mean and variance. Using the identity $\sum_{i=1}^n (X_i - \overline{X})^2 = \sum_{i=1}^n X_i^2 - n\overline{X}^2$, we have

$$\overline{X} = \frac{T_1}{n} \quad \text{and} \quad S^2 = \frac{T_2 - T_1^2/n}{n-1}, \tag{4.3}$$

or

$$T_1 = n\overline{X} \quad \text{and} \quad T_2 = (n-1)S^2 + n\overline{X}^2. \tag{4.4}$$

This establishes a one-to-one relationship between $T$ and $(\overline{X}, S^2)$. One-to-one relationships preserve sufficiency and completeness, and so $(\overline{X}, S^2)$ is also a complete sufficient statistic.

Iterating Property 4, $X_1 + \cdots + X_n \sim N(n\mu, n\sigma^2)$. Dividing by $n$, by Property 1, $\overline{X} \sim N(\mu, \sigma^2/n)$. We know from Example 3.22 that $\overline{X}$ and $S^2$ are independent, but the derivation to find the marginal distribution of $S^2$ is a bit more involved.

The gamma distribution, introduced in Problem 1.26, with parameters $\alpha > 0$ and $\beta > 0$, denoted $\Gamma(\alpha, \beta)$, has density

$$f_{\alpha,\beta}(x) = \begin{cases} \dfrac{x^{\alpha-1}e^{-x/\beta}}{\beta^\alpha \Gamma(\alpha)}, & x > 0; \\ 0, & \text{otherwise}, \end{cases} \tag{4.5}$$

where $\Gamma(\cdot)$ is the gamma function defined as

$$\Gamma(\alpha) = \int_0^\infty x^{\alpha-1}e^{-x}\,dx, \qquad \Re(\alpha) > 0.$$

Useful properties of the gamma function include

$$\Gamma(\alpha+1)=\alpha\Gamma(\alpha), \qquad \Re(\alpha)>0$$

(which follows after integration by parts from the definition), $\Gamma(n+1)=n!$, $n=1,2,\ldots$, and $\Gamma(1/2)=\sqrt{\pi}$. It is not hard to show that if $X\sim\Gamma(\alpha,1)$, then $\beta X\sim\Gamma(\alpha,\beta)$. For this reason, $\beta$ is called a *scale parameter*, and $\alpha$ is called the *shape parameter* for the distribution.

If $X\sim\Gamma(\alpha,\beta)$, then, for $u<1/\beta$,

$$\begin{aligned}M_X(u)=Ee^{uX}&=\int_0^\infty e^{ux}\frac{x^{\alpha-1}e^{-x/\beta}}{\beta^\alpha\Gamma(\alpha)}\,dx\\&=\frac{1}{(1-u\beta)^\alpha}\frac{1}{\Gamma(\alpha)}\int_0^\infty y^{\alpha-1}e^{-y}\,dy=\frac{1}{(1-u\beta)^\alpha},\end{aligned}$$

where the change of variables $y=(1-u\beta)x/\beta$ gives the third equality. From this, if $X\sim\Gamma(\alpha_x,\beta)$ and $Y\sim\Gamma(\alpha_y,\beta)$ are independent, then

$$M_{X+Y}(u)=M_X(u)M_Y(u)=\frac{1}{(1-u\beta)^{\alpha_x+\alpha_y}}.$$

This is the moment generating function for $\Gamma(\alpha_x+\alpha_y,\beta)$, and so

$$X+Y\sim\Gamma(\alpha_x+\alpha_y,\beta). \tag{4.6}$$

The chi-square distributions are special cases of the gamma distribution, generally defined as sums of independent squared standard normal variables. If $Z\sim N(0,1)$, then

$$M_{Z^2}(u)=\int e^{uz^2}\frac{1}{\sqrt{2\pi}}e^{-z^2/2}\,dz=\frac{1}{\sqrt{1-2u}}\int\frac{1}{\sqrt{2\pi}}e^{-x^2/2}\,dx=\frac{1}{\sqrt{1-2u}},$$

where the change of variables $x=z\sqrt{1-2u}$ gives the second equality. The distribution of $Z^2$ is called the chi-square distribution on one degree of freedom, denoted $\chi_1^2$. But the moment generating function for $Z^2$ just computed is the moment generating function for $\Gamma(1/2,2)$. So $\chi_1^2=\Gamma(1/2,2)$.

**Definition 4.8.** *The* chi-square distribution *on $p$ degrees of freedom, $\chi_p^2$, is the distribution of the sum $Z_1^2+\cdots+Z_p^2$ when $Z_1,\ldots,Z_p$ are i.i.d. from $N(0,1)$.*

Repeated use of (4.6) shows that

$$\chi_p^2=\Gamma(p/2,2),$$

which has moment generating function

$$\frac{1}{(1-2u)^{p/2}}, \qquad u<1/2. \tag{4.7}$$

Returning to sampling from a normal distribution, let $X_1, \dots, X_n$ be i.i.d. from $N(\mu, \sigma^2)$, and define $Z_i = (X_i - \mu)/\sigma$, so that $Z_1, \dots, Z_n$ are i.i.d. from $N(0, 1)$. Then

$$\overline{Z} = \frac{1}{n}\sum_{i=1}^{n} \frac{X_i - \mu}{\sigma} = \frac{\sum_{i=1}^{n} X_i - n\mu}{n\sigma} = \frac{\overline{X} - \mu}{\sigma}.$$

Note that $\sqrt{n}\,\overline{Z} \sim N(0, 1)$, and so $n\overline{Z}^2 \sim \chi_1^2$. Next,

$$\begin{aligned} V \stackrel{\text{def}}{=} \frac{(n-1)S^2}{\sigma^2} &= \sum_{i=1}^{n} \left(\frac{X_i - \overline{X}}{\sigma}\right)^2 \\ &= \sum_{i=1}^{n} \left(\frac{X_i - \mu}{\sigma} - \frac{\overline{X} - \mu}{\sigma}\right)^2 \\ &= \sum_{i=1}^{n} (Z_i - \overline{Z})^2. \end{aligned}$$

Expanding the square,

$$V = \sum_{i=1}^{n} (Z_i - \overline{Z})^2 = \sum_{i=1}^{n} (Z_i^2 - 2Z_i\overline{Z} + \overline{Z}) = \sum_{i=1}^{n} Z_i^2 - n\overline{Z}^2,$$

and thus

$$V + n\overline{Z}^2 = \sum_{i=1}^{n} Z_i^2 \sim \chi_n^2. \tag{4.8}$$

By Basu's theorem (see Example 3.22), $\overline{X}$ and $S^2$ are independent. Because $n\overline{Z}^2$ is a function of $\overline{X}$, and $V$ is a function of $S^2$, $V$ and $n\overline{Z}^2$ are independent. Using this independence and formula (4.7) for the moment generating function for $\chi_n^2$, (4.8) implies

$$M_V(u) M_{n\overline{Z}^2}(u) = \frac{1}{(1 - 2u)^{n/2}}.$$

But $n\overline{Z}^2 \sim \chi_1^2$ with moment generating function $1/\sqrt{1 - 2u}$, and thus

$$M_V(u) = \frac{1}{(1 - 2u)^{(n-1)/2}}. \tag{4.9}$$

This is the moment generating function for $\chi_{n-1}^2$, and thus

$$V = \frac{(n-1)S^2}{\sigma^2} \sim \chi_{n-1}^2.$$

This along with $\overline{X} \sim N(\mu, \sigma^2/n)$ implicitly determines the joint distribution of $\overline{X}$ and $S^2$, because these two variables are independent.

## 4.4 Normal One-Sample Problem—Estimation

Results from the last section lead directly to a variety of UMVU estimates. First note that for $n+r>1$,

$$\begin{aligned} ES^r &= E\left[\frac{\sigma^r}{(n-1)^{r/2}}V^{r/2}\right] \\ &= \frac{\sigma^r}{(n-1)^{r/2}}\int_0^\infty \frac{x^{(r+n-3)/2}e^{-x/2}}{2^{(n-1)/2}\Gamma\bigl[(n-1)/2\bigr]}\,dx \\ &= \frac{\sigma^r 2^{r/2}\Gamma\bigl[(r+n-1)/2\bigr]}{(n-1)^{r/2}\Gamma\bigl[(n-1)/2\bigr]}. \end{aligned} \tag{4.10}$$

From this,

$$\frac{(n-1)^{r/2}\Gamma\bigl[(n-1)/2\bigr]}{2^{r/2}\Gamma\bigl[(r+n-1)/2\bigr]}S^r$$

is an unbiased estimate of $\sigma^r$. This estimate is UMVU because it is a function of the complete sufficient statistic $(\overline{X}, S^2)$. In particular, when $r=2$, $S^2$ is UMVU for $\sigma^2$. Note that the UMVU estimate for $\sigma$ is not $S$, although $S$ is a common and natural choice in practice. By Stirling's formula

$$\frac{(n-1)^{r/2}\Gamma\bigl[(n-1)/2\bigr]}{2^{r/2}\Gamma\bigl[(r+n-1)/2\bigr]} = 1 - \frac{r(r-2)}{4n} + O(1/n^2),$$

as $n\to\infty$.[1] For large $n$, the bias of $S^r$ as an estimate of $\sigma^r$ will be slight.

Because $E\overline{X}=\mu$, $\overline{X}$ is the UMVU estimator of $\mu$. However, $\overline{X}^2$ is a biased estimator of $\mu^2$ as

$$E\overline{X}^2 = (E\overline{X})^2 + \operatorname{Var}(\overline{X}) = \mu^2 + \sigma^2/n.$$

The bias can be removed by subtracting an unbiased estimate of $\sigma^2/n$. Doing this, $\overline{X}^2 - S^2/n$ is UMVU for $\mu^2$.

The parameter $\mu/\sigma$ might be interpreted as a signal-to-noise ratio. The unbiased estimate of $\sigma^{-1}$ given above only depends on $S^2$ and is independent of $\overline{X}$, the unbiased estimate of $\mu$. Multiplying these estimates together,

$$\frac{\overline{X}\sqrt{2}\,\Gamma\bigl[(n-1)/2\bigr]}{S\sqrt{n-1}\,\Gamma\bigl[(n-2)/2\bigr]}$$

is UMVU for $\mu/\sigma$.

The $p$th quantile for $N(\mu,\sigma^2)$ is a value $x$ such that $P(X_i\le x)=p$. If $\Phi$ is the cumulative distribution function for $N(0,1)$, then as $Z_i=(X_i-\mu)/\sigma\sim N(0,1)$,

---

[1] Here $O(1/n^2)$ represents a remainder bounded in magnitude by some multiple of $1/n^2$. See Section 8.6.

$$P(X_i \le x) = P\left(Z_i \le \frac{x-\mu}{\sigma}\right) = \Phi\left(\frac{x-\mu}{\sigma}\right).$$

This equals $p$ if $(x-\mu)/\sigma = \Phi^{\leftarrow}(p)$, and so the $p$th quantile of $N(\mu, \sigma^2)$ is

$$x = \mu + \sigma\Phi^{\leftarrow}(p).$$

The UMVU estimate of this quantile is

$$\overline{X} + \frac{\sqrt{n-1}\Gamma\left[(n-1)/2\right]}{\sqrt{2}\Gamma(n/2)} S\Phi^{\leftarrow}(p).$$

## 4.5 Variance Bounds and Information

From (1.10) and (1.11), the covariance between two random variables $X$ and $Y$ is

$$\operatorname{Cov}(X,Y) = E(X-EX)(Y-EY) = EXY - (EX)(EY).$$

In particular, if either mean, $EX$ or $EY$, is zero, $\operatorname{Cov}(X,Y) = EXY$. Letting $\sigma_X = \sqrt{\operatorname{Var}(X)}$ and $\sigma_Y = \sqrt{\operatorname{Var}(Y)}$, then because

$$E\left[(X-EX)\sigma_Y \pm (Y-EY)\sigma_X\right]^2 = 2\sigma_X\sigma_Y\left(\sigma_X\sigma_Y \pm \operatorname{Cov}(X,Y)\right) \ge 0,$$

we have the bound

$$|\operatorname{Cov}(X,Y)| \le \sigma_X\sigma_Y \quad \text{or} \quad \operatorname{Cov}^2(X,Y) \le \operatorname{Var}(X)\operatorname{Var}(Y), \tag{4.11}$$

called the *covariance inequality.*

Using the covariance inequality, if $\delta$ is an unbiased estimator of $g(\theta)$ and $\Psi$ is an arbitrary random variable, then

$$\operatorname{Var}_\theta(\delta) \ge \frac{\operatorname{Cov}_\theta^2(\delta,\psi)}{\operatorname{Var}_\theta(\psi)}. \tag{4.12}$$

The right hand side of this inequality involves $\delta$, so this seems rather useless as a bound for the variance of $\delta$. To make headway we need to choose $\psi$ cleverly, so that $\operatorname{Cov}_\theta(\delta,\psi)$ is the same for all $\delta$ that are unbiased for $g(\theta)$.

Let $\mathcal{P} = \{P_\theta : \theta \in \Omega\}$ be a dominated family with densities $p_\theta$, $\theta \in \Omega \subset \mathbb{R}$. As a starting point, $E_{\theta+\Delta}\delta - E_\theta\delta$ gives the same value $g(\theta+\Delta) - g(\theta)$ for any unbiased $\delta$. Here $\Delta$ must be chosen so that $\theta + \Delta \in \Omega$. Next, we write $E_{\theta+\Delta}\delta - E_\theta\delta$ as a covariance under $P_\theta$. To do this we first express $E_{\theta+\Delta}\delta$ as an expectation under $P_\theta$, which is accomplished by introducing a likelihood ratio. This step of the argument involves a key assumption that $p_{\theta+\Delta}(x) = 0$ whenever $p_\theta(x) = 0$. Define $L(x) = p_{\theta+\Delta}(x)/p_\theta(x)$ when $p_\theta(x) > 0$, and $L(x) = 1$, otherwise. (This function $L$ is called a likelihood ratio.) From the assumption,

$$L(x)p_\theta(x) = \frac{p_{\theta+\Delta}(x)}{p_\theta(x)} p_\theta(x) = p_{\theta+\Delta}(x), \qquad \text{a.e. } x,$$

and so, for any function $h$ integrable under $P_{\theta+\Delta}$,

$$E_{\theta+\Delta} h(X) = \int h p_{\theta+\Delta}\, d\mu = \int h L p_\theta\, d\mu = E_\theta L(X) h(X).$$

Taking $h = 1$, $E_\theta L = 1$; and taking $h = \delta$, $E_{\theta+\Delta}\delta = E_\theta L\delta$. So if we define

$$\psi(X) = L(X) - 1,$$

then $E_\theta \psi = 0$ and

$$E_{\theta+\Delta}\delta - E_\theta\delta = E_\theta L\delta - E_\theta\delta = E_\theta\psi\delta = \mathrm{Cov}_\theta(\delta, \psi).$$

Thus

$$\mathrm{Cov}_\theta(\delta, \psi) = g(\theta + \Delta) - g(\theta)$$

for any unbiased estimator $\delta$. With this choice for $\psi$, (4.12) gives

$$\mathrm{Var}_\theta(\delta) \ge \frac{\left[g(\theta+\Delta) - g(\theta)\right]^2}{\mathrm{Var}_\theta(\psi)} = \frac{\left[g(\theta+\Delta) - g(\theta)\right]^2}{E_\theta \left(\dfrac{p_{\theta+\Delta}(X)}{p_\theta(X)} - 1\right)^2}, \tag{4.13}$$

called the Hammersley–Chapman–Robbins inequality.

Under suitable regularity, the dominated convergence theorem can be used to show that the lower bound in (4.13), which can be written as

$$\frac{\left[\dfrac{g(\theta+\Delta) - g(\theta)}{\Delta}\right]^2}{E_\theta \left(\dfrac{\left[p_{\theta+\Delta}(X) - p_\theta(X)\right]/\Delta}{p_\theta(X)}\right)^2},$$

converges to

$$\frac{\left[g'(\theta)\right]^2}{E_\theta \left(\dfrac{\partial p_\theta(X)/\partial\theta}{p_\theta(X)}\right)^2}$$

as $\Delta \to 0$. The denominator here is called *Fisher information*, denoted $I(\theta)$, and given by

$$I(\theta) = E_\theta \left(\frac{\partial \log p_\theta(X)}{\partial\theta}\right)^2. \tag{4.14}$$

With enough regularity to interchange integration and differentiation,

$$0 = \frac{\partial}{\partial\theta}1 = \frac{\partial}{\partial\theta}\int p_\theta(x)\,d\mu(x) = \int \frac{\partial}{\partial\theta}p_\theta(x)\,d\mu(x)$$
$$= \int \frac{\partial \log p_\theta(x)}{\partial\theta}p_\theta(x)\,d\mu(x) = E_\theta\frac{\partial \log p_\theta(X)}{\partial\theta},$$

and so

$$I(\theta) = \operatorname{Var}_\theta\left(\frac{\partial \log p_\theta(X)}{\partial\theta}\right). \tag{4.15}$$

If we can pass two partial derivatives with respect to $\theta$ inside the integral $\int p_\theta\,d\mu = 1$, then

$$\int \frac{\partial^2 p_\theta(x)}{\partial\theta^2}\,d\mu(x) = E_\theta\left[\frac{\partial^2 p_\theta(X)/\partial\theta^2}{p_\theta(X)}\right] = 0.$$

From this, inasmuch as

$$\frac{\partial^2 \log p_\theta(X)}{\partial\theta^2} = \frac{\partial^2 p_\theta(X)/\partial\theta^2}{p_\theta(X)} - \left(\frac{\partial \log p_\theta(X)}{\partial\theta}\right)^2,$$
$$I(\theta) = -E_\theta\frac{\partial^2 \log p_\theta(X)}{\partial\theta^2}. \tag{4.16}$$

For calculations, this formula is often more convenient than (4.14).

A lower bound based on Fisher information can be derived in much the same way as the Hammersley–Chapman–Robbins inequality, but tends to involve differentiation under an integral sign. Let $\delta$ have mean $g(\theta) = E_\theta\delta$ and take $\psi = \partial \log p_\theta/\partial\theta$. With sufficient regularity,

$$g'(\theta) = \frac{\partial}{\partial\theta}\int \delta p_\theta\,d\mu = \int \delta\frac{\partial}{\partial\theta}p_\theta\,d\mu = \int \delta\psi p_\theta\,d\mu,$$

or

$$g'(\theta) = E_\theta\delta\psi. \tag{4.17}$$

In a given application, this might be established using dominated convergence. If $\delta$ is identically one, then $g(\theta) = 1$, $g'(\theta) = 0$, and we anticipate $E_\theta\psi = 0$. Then (4.17) shows that $\operatorname{Cov}_\theta(\delta, \psi) = g'(\theta)$. Using this in (4.12) we have the following result.

**Theorem 4.9.** *Let $\mathcal{P} = \{P_\theta : \theta \in \Omega\}$ be a dominated family with $\Omega$ an open set in $\mathbb{R}$ and densities $p_\theta$ differentiable with respect to $\theta$. If $E_\theta\psi = 0$, $E_\theta\delta^2 < \infty$, and (4.17) hold for all $\theta \in \Omega$, then*

$$\operatorname{Var}_\theta(\delta) \ge \frac{\left[g'(\theta)\right]^2}{I(\theta)}, \qquad \theta \in \Omega.$$

This result is called the Cramér–Rao, or information, bound. The regularity condition (4.17) is troublesome. It involves the estimator $\delta$, thus the theorem leaves open the possibility that some estimators, not satisfying (4.17), may have variance below the stated bound. This has been addressed in various ways. Under very weak conditions Woodroofe and Simons (1983) show that the bound holds for any estimator $\delta$ for almost all $\theta$. Other authors impose more restrictive conditions on the model $\mathcal{P}$, but show that the bound holds for all $\delta$ at all $\theta \in \Omega$.

Suppose $\mathcal{P} = \{P_\theta : \theta \in \Omega\}$ is a dominated family with densities $p_\theta$ and Fisher information $I$. If $h$ is a one-to-one function from $\Xi$ to $\Omega$, then the family $\mathcal{P}$ can be reparameterized as $\tilde{\mathcal{P}} = \{Q_\xi : \xi \in \Xi\}$ with the identification $Q_\xi = P_{h(\xi)}$. Then $Q_\xi$ has density $q_\xi = p_{h(\xi)}$. Letting $\theta = h(\xi)$, by the chain rule, Fisher information $\tilde{I}$ for the reparameterized family $\tilde{\mathcal{P}}$ is given by

$$\begin{aligned}\tilde{I}(\xi) &= \tilde{E}_\xi \left(\frac{\partial \log q_\xi(X)}{\partial \xi}\right)^2 = \tilde{E}_\xi \left(\frac{\partial \log p_{h(\xi)}(X)}{\partial \xi}\right)^2 \\ &= \bigl[h'(\xi)\bigr]^2 E_\theta \left(\frac{\partial \log p_\theta(X)}{\partial \theta}\right)^2 = \bigl[h'(\xi)\bigr]^2 I(\theta). \qquad (4.18)\end{aligned}$$

*Example 4.10. Exponential Families.* Let $\mathcal{P}$ be a one-parameter exponential family in canonical form with densities $p_\eta$ given by

$$p_\eta(x) = \exp\bigl[\eta T(x) - A(\eta)\bigr] h(x).$$

Then

$$\frac{\partial \log p_\eta(X)}{\partial \eta} = T - A'(\eta),$$

and so by (4.15),

$$I(\eta) = \mathrm{Var}_\eta\bigl(T - A'(\eta)\bigr) = \mathrm{Var}_\eta(T) = A''(\eta).$$

Because

$$\frac{\partial^2 \log p_\eta(X)}{\partial \eta^2} = -A''(\eta),$$

this formula for $I(\eta)$ also follows immediately from (4.16). If the family is parameterized instead by $\mu = A'(\eta) = E_\eta T$, then by (4.18)

$$A''(\eta) = I(\mu)\bigl[A''(\eta)\bigr]^2,$$

and so, because $A''(\eta) = \mathrm{Var}(T)$,

$$I(\mu) = \frac{1}{\mathrm{Var}_\mu T}.$$

Note that because $T$ is UMVU for $\mu$, the lower bound $\mathrm{Var}_\mu(\delta) \ge 1/I(\mu)$ for an unbiased estimator $\delta$ of $\mu$ is sharp in this example.

*Example 4.11. Location Families.* Suppose $\epsilon$ is an absolutely continuous random variable with density $f$. The family of distributions $\mathcal{P} = \{P_\theta : \theta \in \mathbb{R}\}$ with $P_\theta$ the distribution of $\theta + \epsilon$ is called a *location family*. Using a change of variables $x = \theta + e$,

$$\begin{aligned}\int g(x)\,dP_\theta(x) &= E_\theta g(X) = Eg(\theta+\epsilon)\\ &= \int g(\theta+e)f(e)\,de = \int g(x)f(x-\theta)\,dx,\end{aligned}$$

and so $P_\theta$ has density $p_\theta(x) = f(x-\theta)$. Fisher information for this family is given by

$$\begin{aligned}I(\theta) = E_\theta\left(\frac{\partial \log f(X-\theta)}{\partial\theta}\right)^2 &= E_\theta\left(-\frac{f'(X-\theta)}{f(X-\theta)}\right)^2\\ &= E\left(\frac{f'(\epsilon)}{f(\epsilon)}\right)^2 = \int \frac{\big[f'(x)\big]^2}{f(x)}\,dx.\end{aligned}$$

So for location families, $I(\theta)$ is constant and does not vary with $\theta$.

If two (or more) independent vectors are observed, then the total Fisher information is the sum of the Fisher information provided by the individual observations. To see this, suppose $X$ and $Y$ are independent, and that $X$ has density $p_\theta$ and $Y$ has density $q_\theta$ (dominating measures for the distributions of $X$ and $Y$ can be different). Then by (4.15), the Fisher information observing $X$ is

$$I_X(\theta) = \mathrm{Var}_\theta\left(\frac{\partial \log p_\theta(X)}{\partial\theta}\right),$$

and the Fisher information observing $Y$ is

$$I_Y(\theta) = \mathrm{Var}_\theta\left(\frac{\partial \log q_\theta(Y)}{\partial\theta}\right).$$

As $X$ and $Y$ are independent, their joint density is $p_\theta(x)q_\theta(y)$, and Fisher information observing both vectors $X$ and $Y$ is

$$\begin{aligned}I_{X,Y}(\theta) &= \mathrm{Var}_\theta\left(\frac{\partial \log\big[p_\theta(X)q_\theta(Y)\big]}{\partial\theta}\right)\\ &= \mathrm{Var}_\theta\left(\frac{\partial \log p_\theta(X)}{\partial\theta} + \frac{\partial \log q_\theta(Y)}{\partial\theta}\right)\\ &= \mathrm{Var}_\theta\left(\frac{\partial \log p_\theta(X)}{\partial\theta}\right) + \mathrm{Var}_\theta\left(\frac{\partial \log q_\theta(Y)}{\partial\theta}\right)\\ &= I_X(\theta) + I_Y(\theta).\end{aligned}$$

Iterating this, the Fisher information for a random sample of $n$ observations will be $nI(\theta)$ if $I(\theta)$ denotes the Fisher information for a single observation.

## 4.6 Variance Bounds in Higher Dimensions

When the parameter $\theta$ takes values in $\mathbb{R}^s$, Fisher information will be a matrix, defined in regular cases by

$$\begin{aligned} I(\theta)_{i,j} &= E_\theta\left[\frac{\partial \log p_\theta(X)}{\partial\theta_i}\frac{\partial \log p_\theta(X)}{\partial\theta_j}\right] \\ &= \operatorname{Cov}_\theta\left(\frac{\partial \log p_\theta(X)}{\partial\theta_i}, \frac{\partial \log p_\theta(X)}{\partial\theta_j}\right) \\ &= -E_\theta\left[\frac{\partial^2 \log p_\theta(X)}{\partial\theta_i\partial\theta_j}\right]. \end{aligned}$$

The first two lines here are equal because

$$E_\theta \nabla_\theta \log p_\theta(X) = 0,$$

and, as before, the third formula requires extra regularity necessary to pass a second derivative inside an integral. Using matrix notation,

$$\begin{aligned} I(\theta) &= E_\theta\big(\nabla_\theta \log p_\theta(X)\big)\big(\nabla_\theta \log p_\theta(X)\big)' \\ &= \operatorname{Cov}_\theta\big(\nabla_\theta \log p_\theta(X)\big) = -E_\theta \nabla_\theta^2 \log p_\theta(X), \end{aligned}$$

where $\nabla_\theta$ is the gradient with respect to $\theta$, $\nabla_\theta^2$ is the Hessian matrix of second order derivatives, and prime denotes transpose. The lower bound for the variance of an unbiased estimator $\delta$ of $g(\theta)$, where $g : \Omega \to \mathbb{R}$, is

$$\operatorname{Var}_\theta(\delta) \geq \nabla g(\theta)' I^{-1}(\theta) \nabla g(\theta).$$

*Example 4.12. Exponential Families.* If $\mathcal{P}$ is an $s$-parameter exponential family in canonical form with densities

$$p_\eta(x) = \exp\big[\eta \cdot T(x) - A(\eta)\big] h(x),$$

then

$$\frac{\partial^2 \log p_\eta(X)}{\partial\eta_i\partial\eta_j} = -\frac{\partial^2 A(\eta)}{\partial\eta_i\partial\eta_j}.$$

Thus

$$I(\eta)_{i,j} = \frac{\partial^2 A(\eta)}{\partial\eta_i\partial\eta_j}.$$

This can be written more succinctly as

$$I(\eta) = \nabla^2 A(\eta).$$

The final formula in this section is a multivariate extension of (4.18). As before, let $\mathcal{P} = \{P_\theta : \theta \in \Omega\}$ be a dominated family with densities $p_\theta$ and Fisher information $I$, but now $\Omega$ is a subset of $\mathbb{R}^s$. Let $h$ be a differentiable one-to-one function from $\Xi$ to $\Omega$ and introduce the family $\tilde{\mathcal{P}} = \{Q_\xi : \xi \in \Xi\}$ with $Q_\xi = P_{h(\xi)}$. The density for $Q_\xi$ is $q_\xi = p_{h(\xi)}$, and by the chain rule,

$$\frac{\partial \log q_\xi(X)}{\partial \xi_i} = \sum_j \frac{\partial \log p_\theta(X)}{\partial \theta_j} \frac{\partial h_j(\xi)}{\partial \xi_i},$$

where $\theta = h(\xi)$. If $Dh$ represents the matrix of partial derivatives of $h$ given by

$$[Dh(\xi)]_{i,j} = \frac{\partial h_i(\xi)}{\partial \xi_j},$$

then $\partial \log q_\xi(X)/\partial \xi_i$ is the $i$th entry of $[Dh(\xi)]' \nabla_\theta \log p_\theta(X)$. So

$$\nabla_\xi \log q_\xi(X) = [Dh(\xi)]' \nabla_\theta \log p_\theta(X)$$

and

$$\begin{aligned}
\tilde{I}(\xi) &= \tilde{E}_\xi [\nabla_\xi \log q_\xi(X)] [\nabla_\xi \log q_\xi(X)]' \\
&= E_\theta [Dh(\xi)]' [\nabla_\theta \log p_\theta(X)] [\nabla_\theta \log p_\theta(X)]' [Dh(\xi)] \\
&= [Dh(\xi)]' I(\theta) [Dh(\xi)].
\end{aligned}$$

## 4.7 Problems[2]

*1. Let $X_1, \dots, X_m$ and $Y_1, \dots, Y_n$ be independent variables with the $X_i$ a random sample from an exponential distribution with failure rate $\lambda_x$, and the $Y_j$ a random sample from an exponential distribution with failure rate $\lambda_y$.
   a) Determine the UMVU estimator of $\lambda_x/\lambda_y$.
   b) Under squared error loss, find the best estimator of $\lambda_x/\lambda_y$ of form $\delta = c\overline{Y}/\overline{X}$.
   c) Find the UMVU estimator of $e^{-\lambda_x} = P(X_1 > 1)$.

*2. Let $X_1, \dots, X_n$ be a random sample from $N(\mu_x, \sigma^2)$, and let $Y_1, \dots, Y_m$ be an independent random sample from $N(\mu_y, 2\sigma^2)$, with $\mu_x$, $\mu_y$, and $\sigma^2$ all unknown parameters.
   a) Find a complete sufficient statistic.
   b) Determine the UMVU estimator of $\sigma^2$. Hint: Find a linear combination $L$ of $S_x^2 = \sum_{i=1}^n (X_i - \overline{X})^2/(n-1)$ and $S_y^2 = \sum_{j=1}^m (Y_j - \overline{Y})^2/(m-1)$ so that $(\overline{X}, \overline{Y}, L)$ is complete sufficient.

[2] Solutions to the starred problems are given at the back of the book.

c) Find a UMVU estimator of $(\mu_x - \mu_y)^2$.

d) Suppose we know the $\mu_y = 3\mu_x$. What is the UMVU estimator of $\mu_x$?

*3. Let $X_1, \dots, X_n$ be a random sample from the Poisson distribution with mean $\lambda$. Find the UMVU for $\cos\lambda$. (Hint: For Taylor expansion, the identity $\cos\lambda = (e^{i\lambda} + e^{-i\lambda})/2$ may be useful.)

*4. Let $X_1, \dots, X_n$ be independent normal variables, each with unit variance, and with $EX_i = \alpha t_i + \beta t_i^2$, $i = 1, \dots, n$, where $\alpha$ and $\beta$ are unknown parameters and $t_1, \dots, t_n$ are known constants. Find UMVU estimators of $\alpha$ and $\beta$.

*5. Let $X_1, \dots, X_n$ be i.i.d. from some distribution $Q_\theta$, and let $\overline{X} = (X_1 + \cdots + X_n)/n$ be the sample average.

a) Show that $S^2 = \sum(X_i - \overline{X})^2/(n-1)$ is unbiased for $\sigma^2 = \sigma^2(\theta) = \mathrm{Var}_\theta(X_i)$.

b) If $Q_\theta$ is the Bernoulli distribution with success probability $\theta$, show that $S^2$ from (a) is UMVU.

c) If $Q_\theta$ is the exponential distribution with failure rate $\theta$, find the UMVU estimator of $\sigma^2 = 1/\theta^2$. Give a formula for $E_\theta[X_i^2|\overline{X} = c]$ in this case.

*6. Suppose $\delta$ is a UMVU estimator of $g(\theta)$; $U$ is an unbiased estimator of zero, $E_\theta U = 0$, $\theta \in \Omega$; and that $\delta$ and $U$ both have finite variances for all $\theta \in \Omega$. Show that $U$ and $\delta$ are uncorrelated, $E_\theta U\delta = 0$, $\theta \in \Omega$.

*7. Suppose $\delta_1$ is a UMVU estimator of $g_1(\theta)$, $\delta_2$ is UMVU estimator of $g_2(\theta)$, and that $\delta_1$ and $\delta_2$ both have finite variance for all $\theta$. Show that $\delta_1 + \delta_2$ is UMVU for $g_1(\theta) + g_2(\theta)$. Hint: Use the result in the previous problem.

*8. Let $X_1, \dots, X_n$ be i.i.d. absolutely continuous variables with common density $f_\theta$, $\theta > 0$, given by

$$f_\theta(x) = \begin{cases} \theta/x^2, & x > \theta; \\ 0, & x \le \theta. \end{cases}$$

Find the UMVU estimator for $g(\theta)$ if $g(\theta)/\theta^n \to 0$ as $\theta \to \infty$ and $g$ is differentiable.

9. Let $X$ be a single observation from a Poisson distribution with mean $\lambda$. Determine the UMVU estimator for

$$e^{-2\lambda} = \left[P_\lambda(X=0)\right]^2.$$

*10. Suppose $X$ is an exponential variable with density $p_\theta(x) = \theta e^{-\theta x}$, $x > 0$; $p_\theta(x) = 0$, otherwise. Find the UMVU estimator for $1/(1+\theta)$.

*11. Let $X_1, \dots, X_3$ be i.i.d. geometric variables with common mass function $f_\theta(x) = P_\theta(X_i = x) = \theta(1-\theta)^x$, $x = 0, 1, \dots$. Find the UMVU estimator of $\theta^2$.

*12. Let $X$ be a single observation, absolutely continuous with density

$$p_\theta(x) = \begin{cases} \frac{1}{2}(1+\theta x), & |x| < 1; \\ 0, & |x| \geq 1. \end{cases}$$

Here $\theta \in [-1, 1]$ is an unknown parameter.

a) Find a constant $a$ so that $aX$ is unbiased for $\theta$.

b) Show that $b = E_\theta|X|$ is independent of $\theta$.

c) Let $\theta_0$ be a fixed parameter value in $[-1, 1]$. Determine the constant $c = c_{\theta_0}$ that minimizes the variance of the unbiased estimator $aX + c(|X| - b)$ when $\theta = \theta_0$. Is $aX$ uniformly minimum variance unbiased?

13. Let $X_1, \ldots, X_m$ be i.i.d. from a Poisson distribution with parameter $\lambda_x$ and let $Y_1, \ldots, Y_n$ be i.i.d. from a Poisson distribution with parameter $\lambda_y$, with all $n + m$ variables independent.

a) Find the UMVU of $(\lambda_x - \lambda_y)^2$.

b) Give a formula for the chance $X_i$ is odd, and find the UMVU estimator of this parameter.

14. Let $X_1, \ldots, X_n$ be i.i.d. from an arbitrary discrete distribution on $\{0, 1, 2\}$. Let $T_1 = X_1 + \cdots + X_n$ and $T_2 = X_1^2 + \cdots + X_n^2$.

a) Show that $T = (T_1, T_2)$ is complete sufficient.

b) Let $\mu = EX_i$. Find the UMVU of $\mu^3$.

15. Let $X_1, \ldots, X_n$ be i.i.d. absolutely continuous random variables with common marginal density $f_\theta$ given by

$$f_\theta(x) = \begin{cases} e^{\theta - x}, & x \geq \theta; \\ 0, & x < \theta. \end{cases}$$

Find UMVU estimators for $\theta$ and $\theta^2$.

16. Let $X_1$ and $X_2$ be independent discrete random variables with common mass function

$$P(X_i = x) = -\frac{\theta^x}{x \log(1-\theta)}, \qquad x = 1, 2, \ldots,$$

where $\theta \in (0, 1)$.

a) Find the mean and variance of $X_1$.

b) Find the UMVU of $\theta / \log(1 - \theta)$.

17. Let $X_1, \ldots, X_n$ be i.i.d. absolutely continuous variables with common density $f_\theta$, $\theta \in \mathbb{R}$, given by

$$f_\theta(x) = \begin{cases} \dfrac{\phi(x)}{\Phi(\theta)}, & x < \theta; \\ 0, & x \geq \theta. \end{cases}$$

(This is the density for the standard normal distribution truncated above at $\theta$.)

a) Derive a formula for the UMVU for $g(\theta)$. (Assume $g$ is differentiable and behaves reasonably as $\theta \to \pm\infty$.)

b) If $n = 3$ and the observed data are $-2.3$, $-1.2$, and 0, what is the estimate for $\theta^2$?

18. Let $X_1$ and $X_2$ be i.i.d. discrete variables with common mass function

$$f_\theta(x) = P_\theta(X_i = x) = (x+1)\theta^2(1-\theta)^x, \qquad x = 0, 1, \dots,$$

where $\theta \in (0,1)$.
a) Compute $E\big[1/(X_i+1)\big]$.
b) Find the mass function for $X_1 + X_2$.
c) Use conditioning to find the UMVU for $\theta$.

19. Let $X$ have a binomial distribution with $n$ trials and success probability $\theta \in (0,1)$. If $m \le n$, find the UMVU estimator of $\theta^m$.

20. Let $X_1, \dots, X_n$ be i.i.d. and absolutely continuous with common marginal density $f_\theta$ given by

$$f_\theta(x) = \begin{cases} 2x/\theta^2, & 0 < x < \theta; \\ 0, & \text{otherwise}, \end{cases}$$

where $\theta > 0$ is an unknown parameter. Find the UMVU estimator of $g(\theta)$ if $g$ is differentiable and $\theta^{2n} g(\theta) \to 0$ as $\theta \downarrow 0$.

21. Let $X_1, \dots, X_n$ be i.i.d. from an exponential distribution with density

$$f_\theta(x) = \begin{cases} \theta e^{-\theta x}, & x > 0; \\ 0, & \text{otherwise}. \end{cases}$$

Find UMVU estimators for $\theta$ and $\theta^2$.

22. Suppose $X_1, \dots, X_n$ are independent with $X_j \sim N(0, j\theta^2)$, $j = 1, \dots, n$. Find the UMVU estimator of $\theta$.

23. For $\theta > 0$, let

$$\Delta_\theta = \{(x,y) \in \mathbb{R}^2 : x > 0, y > 0, x + y < \theta\},$$

the interior of a triangle. Let $(X_1, Y_1)$, ..., $(X_n, Y_n)$ be i.i.d. from the uniform distribution on $\Delta_\theta$, so their common density is $2 1_{\Delta_\theta}/\theta^2$.
a) Find a complete sufficient statistic $T$.
b) Find the UMVU estimators of $\theta$ and $\cos\theta$.

*24. In the normal one-sample problem, the statistic $t = \sqrt{n}\,\overline{X}/S$ has the noncentral $t$-distribution on $n-1$ degrees of freedom and noncentrality parameter $\delta = \sqrt{n}\mu/\sigma$. Use our results on distribution theory for the one-sample problem to find the mean and variance of $t$.

25. Let $X_1, \dots, X_n$ be i.i.d. from $N(\mu, \sigma^2)$, with $\mu$ and $\sigma$ both unknown.
a) Find the UMVU estimator of $\mu^3$.
b) Find the UMVU estimator of $\mu^2/\sigma^2$.

26. Let $X_1, \dots, X_n$ be independent with

$$X_i \sim N(m_i\mu, m_i\sigma^2), \qquad i = 1, \dots, n,$$

where $m_1, \dots, m_n$ are known constants and $\mu$ and $\sigma$ are unknown parameters. (These sort of data would arise if i.i.d. variables from $N(\mu, \sigma^2)$ were divided into groups, with the $i$th group having $m_i$ observations, and the observed data are totals for the $n$ groups.)

a) Find UMVU estimators for $\mu$ and $\sigma^2$.
b) Show that the estimators in part (a) are independent.

27. Let $Z_1$ and $Z_2$ be independent standard normal random variables. Find $E\sqrt{|Z_1/Z_2|}$.

*28. Let $X_1, \dots, X_n$ be i.i.d. from the uniform distribution on $(0, \theta)$.
   a) Use the Hammersley–Chapman–Robbins inequality to find a lower bound for the variance of an unbiased estimator of $\theta$. This bound will depend on $\Delta$. Note that $\Delta$ cannot vary freely but must lie in a suitable set.
   b) In principle, the best lower bound can be found taking the supremum over $\Delta$. This calculation cannot be done explicitly, but an approximation is possible. Suppose $\Delta = -c\theta/n$. Show that the lower bound for the variance can be written as $\theta^2 g_n(c)/n^2$. Determine $g(c) = \lim_{n\to\infty} g_n(c)$.
   c) Find the value $c_0$ that maximizes $g(c)$ over $c \in (0,1)$ and give an approximate lower bound for the variance of $\delta$. (The value $c_0$ cannot be found explicitly, but you should be able to come up with a numerical value.)

29. Determine the Fisher information $I(\theta)$ for the density $f_\theta(x) = (1+\theta x)/2$, $x \in (-1,1)$, $f_\theta(x) = 0$, $x \notin (-1,1)$.

*30. Suppose $X_1, \dots, X_n$ are independent with $X_i \sim N(\alpha+\beta t_i, 1)$, $i = 1, \dots, n$, where $t_1, \dots, t_n$ are known constants and $\alpha$, $\beta$ are unknown parameters.
   a) Find the Fisher information matrix $I(\alpha, \beta)$.
   b) Give a lower bound for the variance of an unbiased estimator of $\alpha$.
   c) Suppose we know the value of $\beta$. Give a lower bound for the variance of an unbiased estimator of $\alpha$ in this case.
   d) Compare the estimators in parts (b) and (c). When are the bounds the same? If the bounds are different, which is larger?
   e) Give a lower bound for the variance of an unbiased estimator of the product $\alpha\beta$.

*31. Find the Fisher information for the Cauchy location family with densities $p_\theta$ given by

$$p_\theta(x) = \frac{1}{\pi\big[(x-\theta)^2 + 1\big]}.$$

Also, what is the Fisher information for $\theta^3$?

*32. Suppose $X$ has a Poisson distribution with mean $\theta^2$, so the parameter $\theta$ is the square root of the usual parameter $\lambda = EX$. Show that the Fisher information $I(\theta)$ is constant.

*33. Consider the exponential distribution with failure rate $\lambda$. Find a function $h$ defining a new parameter $\theta = h(\lambda)$ so that Fisher information $I(\theta)$ is constant.

*34. Consider an autoregressive model in which $X_1 \sim N\big(\theta, \sigma^2/(1-\rho^2)\big)$ and the conditional distribution of $X_{j+1}$ given $X_1 = x_1, \dots, X_j = x_j$, is $N\big(\theta + \rho(x_j - \theta), \sigma^2\big)$, $j = 1, \dots, n-1$.
   a) Find the Fisher information matrix, $I(\theta, \sigma)$.
   b) Give a lower bound for the variance of an unbiased estimator of $\theta$.
   c) Show that the sample average $\overline{X} = (X_1 + \cdots + X_n)/n$ is an unbiased estimator of $\theta$, compute its variance, and compare its variance with the lower bound. Hint: Define $\epsilon_j = X_j - \theta$ and $\eta_{j+1} = \epsilon_{j+1} - \epsilon_j$. Use smoothing to argue that $\eta_2, \dots, \eta_n$ are i.i.d. $N(0, \sigma^2)$ and are independent of $\epsilon_1$. Similarly, $X_i$ is independent of $\eta_{i+1}, \eta_{i+2}, \dots$. Use these facts to find first $\mathrm{Var}(X_2) = \mathrm{Var}(\epsilon_2)$, then $\mathrm{Var}(X_3)$, $\mathrm{Var}(X_4)$, .... Finally, find $\mathrm{Cov}(X_{i+1}, X_i)$, $\mathrm{nCov}(X_{i+2}, X_i)$, and so on.

35. Consider the binomial distribution with $n$ trials and success probability $p$. Find a function $h$ defining a new parameter $\theta = h(p)$ so that Fisher information $I(\theta)$ is constant.

36. Let $X_1, \dots, X_n$ be i.i.d. with common density $f_\theta(x) = e^{\theta - x}$, $x > \theta$, $f_\theta(x) = 0$, otherwise.
   a) Find lower bounds for the variance of an unbiased estimator of $\theta$ using the Hammersley–Chapman–Robbins inequality. These bounds will depend on the choice of $\Delta$.
   b) What choice of $\Delta$ gives the best (largest) lower bound?

37. Suppose $X$ has a Poisson distribution with mean $\lambda$, and that given $X = n$, $Y$ is Poisson with mean $n\theta$.
   a) Find the Fisher information matrix.
   b) Derive a formula for $\mu_Y = EY$.
   c) Find a lower bound for the variance of an unbiased estimator of $\mu_Y$.
   d) Compare the bound in part (c) with the variance of $Y$.

38. Suppose $X$ has a geometric distribution with parameter $\theta$, so $P(X = x) = \theta(1-\theta)^x$, $x = 0, 1, \dots$, and that given $X = n$, $Y$ is binomial with $x$ trials and success probability $p$.
   a) Find the Fisher information matrix.
   b) Give a lower bound for the variance of an unbiased estimator of $\mu_Y = EY$. Compare the lower bound with $\mathrm{Var}(Y)$.

39. Let $X$ have a "triangular" shaped density given by

$$f_\theta(x) = \begin{cases} 2(\theta - x)/\theta^2, & x \in (0, \theta); \\ 0, & \text{otherwise.} \end{cases}$$

   a) Use the Hammersley–Chapman–Robbins inequality to derive lower bounds for the variance of an unbiased estimator of $\theta$ based on a single observation $X$. These bounds will depend on the choice of $\Delta$.

b) What is in fact the smallest possible variance for an unbiased estimator $\delta(X)$ of $\theta$? Compare this value with the lower bounds in part (a).

40. Let $X$ have a geometric distribution with success probability $\theta$, so $P_\theta(X = x) = \theta(1-\theta)^x$, $x = 0, 1, \ldots$. What is the smallest possible variance for an unbiased estimator $\delta(X)$ of $\theta$? Compare this variance with the Cramér–Rao lower bound in Theorem 4.9.

41. Let $X_1, \ldots, X_n$ be i.i.d. random variables (angles) from the von Mises distribution with Lebesgue density

$$p_\theta(x) = \begin{cases} \dfrac{\exp\{\theta_1 \sin x + \theta_2 \cos x\}}{2\pi I_0(\|\theta\|)}, & x \in (0, 2\pi); \\ 0, & \text{otherwise.} \end{cases}$$

Here $\|\theta\|$ denotes the Euclidean length of $\theta$, $\|\theta\| = (\theta_1^2 + \theta_2^2)^{1/2}$, and the function $I_0$ is a modified Bessel function.

a) Find the Fisher information matrix, expressed using $I_0$ and its derivatives.

b) Give a lower bound for the variance of an unbiased estimator of $\|\theta\|$.

42. Let $\theta = (\alpha, \lambda)$ and let $P_\theta$ denote the gamma distribution with shape parameter $\alpha$ and scale $1/\lambda$. So $P_\theta$ has density

$$p_\theta(x) = \begin{cases} \dfrac{\lambda^\alpha x^{\alpha-1} e^{-x\lambda}}{\Gamma(\alpha)}, & x > 0; \\ 0, & \text{otherwise.} \end{cases}$$

a) Find the Fisher information matrix $I(\theta)$, expressed using the "psi" function $\psi \overset{\text{def}}{=} \Gamma'/\Gamma$ and its derivatives.

b) What is the Cramér–Rao lower bound for the variance of an unbiased estimator of $\alpha + \lambda$?

c) Find the mean $\mu$ and variance $\sigma^2$ for $P_\theta$. Show that there is a one-to-one correspondence between $\theta$ and $(\mu, \sigma^2)$.

d) Find the Fisher information matrix if the family of gamma distributions is parameterized by $(\mu, \sigma^2)$, instead of $\theta$.

# 5

# Curved Exponential Families

Curved exponential families may arise when the parameters of an exponential family satisfy constraints. For these families the minimal sufficient statistic may not be complete, and UMVU estimation may not be possible. Curved exponential families arise naturally with data from sequential experiments, considered in Section 5.2, and Section 5.3 considers applications to contingency table analysis.

## 5.1 Constrained Families

Let $\mathcal{P} = \{P_\eta : \eta \in \Xi\}$ be a full rank $s$-parameter canonical exponential family with complete sufficient statistic $T$. Consider a submodel $\mathcal{P}_0$ parameterized by $\theta \in \Omega$ with $\tilde{\eta}(\theta)$ the value for the canonical parameter associated with $\theta$. So

$$\mathcal{P}_0 = \{P_{\tilde{\eta}(\theta)} : \theta \in \Omega\}.$$

Often $\tilde{\eta} : \Omega \to \Xi$ is one-to-one and onto. In this case $\mathcal{P}_0 = \mathcal{P}$ and the choice of parameter, $\theta$ or $\eta$, is dictated primarily by convenience. Curved exponential families may arise when $\mathcal{P}_0$ is a strict subset of $\mathcal{P}$, generally with $\Omega \subset \mathbb{R}^r$ and $r < s$. Here are two possibilities.

1. Points $\eta$ in the range of $\tilde{\eta}$, $\tilde{\eta}(\Omega) = \{\tilde{\eta}(\theta) : \theta \in \Omega\}$, satisfy a nontrivial linear constraint. In this case, $\mathcal{P}_0$ will be a $q$-parameter exponential family for some $q < s$. The statistic $T$ will still be sufficient, but will not be minimal sufficient.
2. The points $\eta$ in $\tilde{\eta}(\Omega)$ do not satisfy a linear constraint. In this case, $\mathcal{P}_0$ is called a *curved exponential family*. Here $T$ will be minimal sufficient (see Example 3.12), but may not be complete.

*Example 5.1.* Joint distributions for a sample from $N(\mu, \sigma^2)$ form a two-parameter exponential family with canonical parameter

R.W. Keener, *Theoretical Statistics: Topics for a Core Course*, Springer Texts in Statistics,
DOI 10.1007/978-0-387-93839-4_5, © Springer Science+Business Media, LLC 2010

$$\eta = \left(\frac{\mu}{\sigma^2}, -\frac{1}{2\sigma^2}\right)$$

and complete sufficient statistic

$$T = \left(\sum_{i=1}^n X_i, \sum_{i=1}^n X_i^2\right).$$

(See Example 2.3.) If $\mu$ and $\sigma^2$ are equal and we let $\theta$ denote the common value, then our subfamily will consist of joint distributions for a sample from $N(\theta, \theta)$, $\theta > 0$. Then

$$\tilde{\eta}(\theta) = \left(1, -\frac{1}{2\theta}\right),$$

and the range of $\tilde{\eta}$ is the half-line indicated in Figure 5.1. Because points $\eta$ in $\tilde{\eta}(\Omega)$ satisfy the linear constraint $\eta_1 = 1$, the subfamily should be exponential with less than two parameters. This is easy to check; the joint densities form a full rank one-parameter exponential family with $\sum_{i=1}^n X_i^2$ as the canonical complete sufficient statistic.

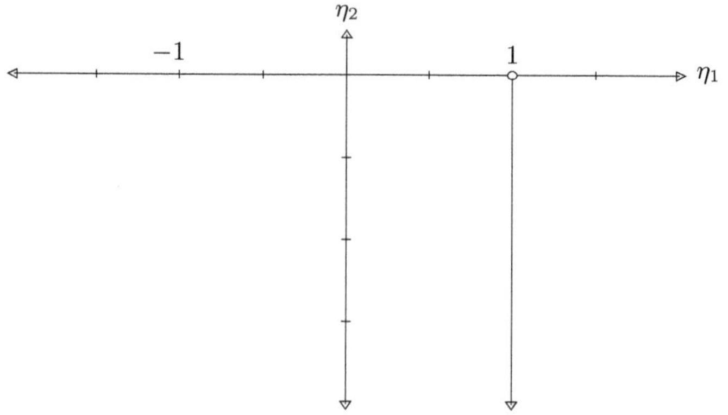


**Fig. 5.1.** Range of $\tilde{\eta}(\theta) = \big(1, -1/(2\theta)\big)$.

Suppose instead $\sigma = |\mu|$, so the subfamily will be joint distributions for a sample from $N(\theta, \theta^2)$, $\theta \in \mathbb{R}$. In this case

$$\tilde{\eta}(\theta) = \left(\frac{1}{\theta}, -\frac{1}{2\theta^2}\right).$$

Now the range space $\tilde{\eta}(\Omega)$ is the parabola in Figure 5.2. Points in this range space do not satisfy a linear constraint, so in this case we have a curved exponential family and $T$ is minimal sufficient. Because

$$E_\theta T_1^2 = (E_\theta T_1)^2 + \mathrm{Var}_\theta(T_1) = n^2\theta^2 + n\theta^2,$$

and

$$E_\theta T_2 = nE_\theta X_i^2 = n\big((E_\theta X_i)^2 + \mathrm{Var}_\theta(X_i)\big) = 2n\theta^2,$$

we have

$$E_\theta\big(2T_1^2 - (n+1)T_2\big) = 0, \qquad \theta \in \mathbb{R}.$$

Thus $g(T) = 2T_1^2-(n+1)T_2$ has zero mean regardless the value of $\theta$. Inasmuch as $g(T)$ is not zero (unless $n = 1$), $T$ is not complete.

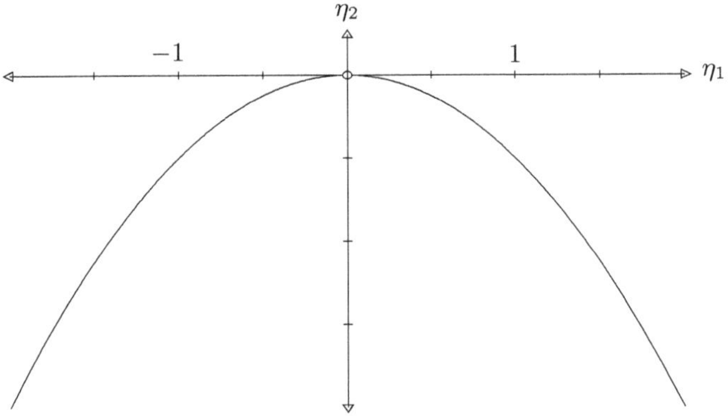


**Fig. 5.2.** Range of $\tilde{\eta}(\theta) = \big(1/\theta, -1/(2\theta^2)\big)$.

*Example 5.2.* If our data consist of two independent random samples, $X_1, \ldots, X_m$ from $N(\mu_x, \sigma_x^2)$ and $Y_1, \ldots, Y_n$ from $N(\mu_y, \sigma_y^2)$, then the joint distributions form a four-parameter exponential family indexed by $\theta = (\mu_x, \mu_y, \sigma_x^2, \sigma_y^2)$. A canonical sufficient statistic for the family is

$$T = \left(\sum_{i=1}^m X_i, \sum_{j=1}^n Y_j, \sum_{i=1}^m X_i^2, \sum_{j=1}^n Y_j^2\right),$$

and the canonical parameter is

$$\eta = \left(\frac{\mu_x}{\sigma_x^2}, \frac{\mu_y}{\sigma_y^2}, -\frac{1}{2\sigma_x^2}, -\frac{1}{2\sigma_y^2}\right).$$

By (4.3) and (4.4), an equivalent statistic would be $(\overline{X}, \overline{Y}, S_x^2, S_y^2)$, where $S_x^2 = \sum_{i=1}^m (X_i - \overline{X})^2/(m-1)$ and $S_y^2 = \sum_{i=1}^n (Y_i - \overline{Y})^2/(n-1)$. Results from Section 4.3 provide UMVU estimates for $\mu_x$, $\mu_y$, $\sigma_x^r$, $\sigma_y^r$, etc.

If the variances for the two samples agree, $\sigma_x^2 = \sigma_y^2 = \sigma^2$, then $\eta$ satisfies the linear constraint $\eta_3 = \eta_4$. In this case the joint distributions form a three-parameter exponential family with complete sufficient statistic $(T_1, T_2, T_3 + T_4)$. An equivalent sufficient statistic here is $(\overline{X}, \overline{Y}, S_p^2)$, where

$$S_p^2 = \frac{\sum_{i=1}^m (X_i - \overline{X})^2 + \sum_{i=1}^n (Y_i - \overline{Y})^2}{n+m-2} = \frac{(m-1)S_x^2 + (n-1)S_y^2}{n+m-2},$$

called the *pooled sample variance.* Again the equivalence follows easily from (4.3) and (4.4). Also, because $S_x^2$ and $S_y^2$ are independent, from the definition of the chi-square distribution and (4.9),

$$\frac{(n+m-2)S_p^2}{\sigma^2} \sim \chi^2_{n+m-2}.$$

Again, results from Section 4.3 provide UMVU estimates for various parameters of interest.

Another subfamily arises if the means for the two samples are the same, $\mu_x = \mu_y$. In this case the joint distributions form a curved exponential family, and $T$ or $(\overline{X}, \overline{Y}, S_x^2, S_y^2)$ are minimal sufficient. In this case these statistics are not complete because $E(\overline{X} - \overline{Y}) = 0$ for all distributions in the subfamily.

## 5.2 Sequential Experiments

The protocol for an experiment is *sequential* if the data are observed as they are collected, and the information from the observations influences how the experiment is performed. For instance, the decision whether to terminate a study at some stage or continue collecting more data might be based on prior observations. Or, in allocation problems sampling from two or more populations, the choice of population sampled at a given stage could depend on prior data.

There are two major reasons why a sequential experiment might be preferred over a classical experiment. A sequential experiment may be more efficient. Here efficiency gains might be quantified as a reduction in decision theoretic risk, with costs for running the experiment added to the usual loss function. There are also situations in which certain objectives can only be met with a sequential experiment. Here is one example.

*Example 5.3. Estimating a Population Size.* Consider a lake (or some other population) with $M$ fish. Here $M$ is considered an unknown parameter, and the goal of the experiment is to estimate $M$. Data to estimate $M$ are obtained from a "capture–recapture" experiment. This experiment has two phases. First, $k$ fish are sampled from the lake and tagged so they can be identified. These fish are then returned to the lake. At the second stage, fish are sampled at random from the lake. Note that during this phase a sampled fish is tagged with probability $\theta = k/M$. (Actually, there is an assumption here that at the second stage tagged and untagged fish are equally likely to be captured; this premise seems suspect for real fish.) In terms of $\theta$,

$$M = k/\theta,$$

and so the inferential goal is basically to estimate $1/\theta$. Information from the second stage of this experiment can be coded using Bernoulli variables $X_1, \dots, X_N$, where $N$ denotes the sample size, and $X_i$ is one if the $i$th fish is tagged or zero if the $i$th fish is not tagged.

In mathematical terms we have a situation in which potential data $X_1, X_2, \dots$ are i.i.d. Bernoulli variables with success probability $\theta$. If the sample size is fixed, $N = n$, then our data have joint density

$$\prod_{i=1}^{n} \theta^{x_i}(1-\theta)^{1-x_i} = \theta^{T(x)}(1-\theta)^{n-T(x)},$$

where $T(x) = x_1 + \cdots + x_n$. These densities form an exponential family with $T$ as a sufficient statistic. Because $T$ has a binomial distribution with mean $n\theta$, $T/n$ is unbiased for $\theta$ and hence UMVU. But there can be no unbiased estimate of $1/\theta$ because

$$E_\theta \delta(T) \le \max_{0 \le k \le n} \delta(k),$$

which is less than $1/\theta$ once $\theta$ is sufficiently small. Note that if $\theta$ is much smaller than $1/n$, then $T$ will be zero with probability close to one. The real problem here is that when $T = 0$ we cannot infer much about the relative size of $\theta$ from our data.

Inverse binomial sampling avoids the problem just noted by continued sampling until $m$ of the $X_i$ equal one. The number of observations $N$ is now a random variable. Also, this is a sequential experiment because the decision to stop sampling is based on observed data.

Intuitively, data from inverse binomial sampling would be the list

$$X = (X_1, \dots, X_N).$$

There is a bit of a technical problem here: this list is not a random vector because the number of entries $N$ is random. The most natural way around this trouble involves a more advanced notion of "data" in which the information from an experiment is viewed as the $\sigma$-field of events that can be resolved from the experiment. Here this $\sigma$-field would include events such as $\{T = k\}$ or $\{N = 7\}$, but would preclude events such as $\{X_{N+2} = 0\}$. See Chapter 20 for a discussion of this approach. Fortunately, in this example we can avoid these technical issues in the following fashion. Let $Y_1$ be the number of zeros in the list $X$ before the first one, and let $Y_i$ be the number of zeros between the $(i-1)$st and $i$th one, $i = 2, \dots, m$. Note that the list $X$ can be recovered from $Y = (Y_1, \dots, Y_m)$. If, for instance, $Y = (2, 0, 1)$, then $X$ must be $(0, 0, 1, 1, 0, 1)$. The variables $Y_1, \dots, Y_m$ are i.i.d. with

$$\begin{aligned} P_\theta(Y_i = y) &= P(X_1 = 0, \dots, X_y = 0, X_{y+1} = 1) \\ &= (1-\theta)^y \theta \\ &= \exp\bigl(y \log(1-\theta) + \log \theta\bigr), \qquad y = 0, 1, \dots. \end{aligned}$$

This is the mass function for the geometric distribution. It is a one-parameter exponential family with canonical parameter $\eta = \log(1-\theta)$ and

$$A(\eta) = -\log\theta = -\log(1-e^\eta).$$

Thus

$$E_\theta Y_i = A'(\eta) = \frac{e^\eta}{1-e^\eta} = \frac{1-\theta}{\theta}.$$

The family of joint distributions of $Y_1,\dots,Y_m$ has $T=\sum_{i=1}^m Y_i = N-m$ as a complete sufficient statistic. The statistic $T$ counts the number of failures before the $m$th success and has the negative binomial distribution with mass function

$$P(T=t) = \binom{m+t-1}{m-1}\theta^m(1-\theta)^t, \qquad t=0,1,\dots.$$

Inasmuch as

$$E_\theta T = mE_\theta Y_i = \frac{m}{\theta} - m,$$

$$\frac{T+m}{m} = \frac{N}{m}$$

is UMVU for $1/\theta$.

The following result gives densities for a sequential experiment in which data $X_1, X_2, \dots$ are observed until a stopping time $N$. This stopping time is allowed to depend on the data, but clairvoyance is prohibited. Formally, this is accomplished by insisting that

$$\{N=n\} = \bigl\{(X_1,\dots,X_n)\in A_n\bigr\}, \qquad n=1,2,\dots,$$

for some sequence of sets $A_1, A_2, \dots$.

**Theorem 5.4.** *Suppose $X_1, X_2, \dots$ are i.i.d. with common marginal density $f_\theta$, $\theta\in\Omega$. If $P_\theta(N<\infty)=1$ for all $\theta\in\Omega$, then the total data, viewed informally*[1] *as $(N, X_1,\dots,X_N)$, have joint density*

$$\prod_{i=1}^n f_\theta(x_i). \tag{5.1}$$

*When $f_\theta$ comes from an exponential family, so that*

$$f_\theta(x) = e^{\eta(\theta)\cdot T(x) - B(\theta)}h(x),$$

[1] One way to be more precise is to view the information from the observed data as a $\sigma$-field. This approach is developed in Section 20.2, and Theorem 20.6 (Wald's fundamental identity) from this section is the mathematical basis for the theorem here.

*then the joint density is*

$$\exp\left[\eta(\theta)\cdot\sum_{i=1}^{n}T(x_i)-nB(\theta)\right]\prod_{i=1}^{n}h(x_i). \tag{5.2}$$

*These densities form an exponential family with canonical parameters* $\eta_1(\theta)$, $\dots$, $\eta_s(\theta)$, *and* $-B(\theta)$, *and sufficient statistic* $\left(\sum_{i=1}^{N}T(X_i), N\right)$.

By (5.1), the likelihood for the sequential experiment is the same as the likelihood that would be used ignoring the optional stopping and treating $N$ as a fixed constant. In contrast, distributional properties of standard estimators are generally influenced by optional stopping. For instance, the sample average $(X_1+\cdots+X_N)/N$ is generally a biased estimator of $E_\theta X_1$. (See Problems 5.10 and 5.12 for examples.)

The exponential family (5.2) has an extra canonical parameter $-B(\theta)$, therefore sequential experiments usually lead to curved exponential families. The inverse binomial example is unusual in this regard, basically because the experiment is conducted so that $\sum_{i=1}^{N}X_i$ must be the fixed constant $m$.

## 5.3 Multinomial Distribution and Contingency Tables

The multinomial distribution is a generalization of the binomial distribution arising from $n$ independent trials with outcomes in a finite set, $\{a_0,\dots,a_s\}$ say. Define vectors

$$e_0=\begin{pmatrix}1\\0\\0\\\vdots\\0\\0\end{pmatrix},\quad e_1=\begin{pmatrix}0\\1\\0\\\vdots\\0\\0\end{pmatrix},\dots,\quad e_s=\begin{pmatrix}0\\0\\0\\\vdots\\0\\1\end{pmatrix}$$

in $\mathbb{R}^{s+1}$, and take $Y_i=e_j$ if trial $i$ has outcome $a_j$, $i=1,\dots,n$, $j=0,\dots,s$. Then $Y_1,\dots,Y_n$ are i.i.d. If $p_j$, $j=0,\dots,s$, is the chance of outcome $a_j$, then $P(Y_i=e_j)=p_j$. If we define

$$X=\begin{pmatrix}X_0\\X_1\\\vdots\\X_s\end{pmatrix}=\sum_{i=1}^{n}Y_i,$$

then $X_j$ counts the number of trials with $a_j$ as the outcome. By independence, the joint mass function of $Y_1,\dots,Y_n$ will be an $n$-fold product of success probabilities $p_0, \dots, p_s$. The number of times that $p_j$ arises in this product will be the number of trials with outcome $a_j$, and so

$$P(Y_1 = y_1, \dots, Y_n = y_n) = \prod_{j=0}^{s} p_j^{x_j} = \exp\left[\sum_{j=0}^{s} x_j \log p_j\right],$$

where $x = x(y) = y_1 + \cdots + y_n$. Thus the joint distribution for $Y_1, \dots, Y_n$ form an $(s+1)$-parameter exponential family with canonical sufficient statistic $X$. But this family is not of full rank because $X_0 + \cdots + X_s = n$. Taking advantage of this constraint

$$P(Y_1 = y_1, \dots, Y_n = y_n) = \exp\left[\sum_{i=1}^{s} x_i \log(p_i/p_0) + n \log p_0\right],$$

which is a full rank $s$-parameter exponential family with complete sufficient statistic $(X_1, \dots, X_s)$. There is a one-to-one correspondence between this statistic and $X$, therefore $X$ is also complete sufficient.

The distribution of $X$ can be obtained from the distribution of $Y$ as

$$P(X = x) = \sum_{(y_1,\dots,y_n):\sum_{i=1}^{n} y_i = x} P(Y = y). \tag{5.3}$$

The probabilities in this sum all equal $\prod_{j=1}^{s} p_j^{x_j}$, and so this common value must be multiplied by the number of ways the $y_j$ can sum to $x$. This is equal to the number of ways of partitioning the set of trials $\{1, \dots, n\}$ into $s+1$ sets, the first with $x_0$ elements, the next with $x_1$ elements, and so on. This count is a *multinomial coefficient* given by

$$\binom{n}{x_0, \dots, x_s} = \frac{n!}{x_0! \times \cdots \times x_s!}.$$

This formula can be derived recursively. There are $\binom{n}{x_0}$ ways to choose the first set, then $\binom{n-x_0}{x_1}$ ways to choose the second set, and so on. The product of these binomial coefficients simplifies to the stated result. Using a multinomial coefficient to evaluate the sum in (5.3),

$$P(X_0 = x_0, \dots, X_s = x_s) = \binom{n}{x_0, \dots, x_s} p_0^{x_0} \times \cdots \times p_s^{x_s},$$

provided $x_0, \dots, x_s$ are nonnegative integers summing to $n$. This is the mass function for the *multinomial distribution*, and we write

$$X \sim \text{Multinomial}(p_0, \dots, p_s; n).$$

The marginal distribution of $X_j$, because $X_j$ counts the number of trials with $a_j$ as an outcome, is binomial with success probability $p_j$. Because $X$ is complete sufficient, $X_j/n$ is UMVU for $p_j$. Unbiased estimation of the product $p_jp_k$ of two different success probabilities is more interesting as $X_j$ and $X_k$ are dependent. One unbiased estimator $\delta$ is the indicator that $Y_1$ is $a_j$ and $Y_2$ is $a_k$. The chance that $X = x$ given $\delta = 1$ is a multinomial probability for $n-2$ trials with outcome $a_j$ occurring $x_j - 1$ times and outcome $a_k$ occurring $x_k - 1$ times. Therefore

$$\begin{aligned} P(\delta = 1|X = x) &= \frac{P(\delta = 1)P(X = x|\delta = 1)}{P(X = x)} \\ &= \frac{p_jp_k\binom{n-2}{x_0,\dots,x_j-1,\dots,x_k-1,\dots,x_s}p_0^{x_0}\times\cdots\times p_s^{x_s}/(p_jp_k)}{\binom{n}{x_0,\dots,x_s}p_0^{x_0}\times\cdots\times p_s^{x_s}} \\ &= \frac{x_jx_k}{n(n-1)}. \end{aligned}$$

Thus $X_jX_k/(n^2 - n)$ is UMVU for $p_jp_k$, $j \neq k$.

In applications, the success probabilities $p_0, \dots, p_s$ often satisfy additional constraints. In some cases this will lead to a full rank exponential family with fewer parameters, and in other cases it will lead to a curved exponential family. Here are two examples of the former possibility.

*Example 5.5. Two-Way Contingency Tables.* Consider a situation with $n$ independent trials, but now for each trial two characteristics are observed: Characteristic $A$ with possibilities $A_1, \dots, A_I$, and Characteristic $B$ with possibilities $B_1, \dots, B_J$. Let $N_{ij}$ denote the number of trials in which the combination $A_iB_j$ is observed, and let $p_{ij}$ denote the chance of $A_iB_j$ on any given trial. Then

$$N = (N_{11}, N_{12}, \dots, N_{IJ}) \sim \text{Multinomial}(p_{11}, p_{12}, \dots, p_{IJ}; n).$$

These data and the sums

$$N_{i+} = \sum_{j=1}^{J} N_{ij}, \qquad i = 1, \dots, I,$$

and

$$N_{+j} = \sum_{i=1}^{I} N_{ij}, \qquad j = 1, \dots, J,$$

are often presented in a contingency table with the following form:

| | $B_1$ | $\cdots$ | $B_J$ | Total |
|---|---|---|---|---|
| $A_1$ | $N_{11}$ | $\cdots$ | $N_{1J}$ | $N_{1+}$ |
| $\vdots$ | $\vdots$ | | $\vdots$ | $\vdots$ |
| $A_I$ | $N_{I1}$ | $\cdots$ | $N_{IJ}$ | $N_{I+}$ |
| Total | $N_{+1}$ | $\cdots$ | $N_{+J}$ | $n$ |

If characteristics $A$ and $B$ are independent, then

$$p_{ij} = p_{i+}p_{+j}, \qquad i = 1, \dots, I, \quad j = 1, \dots, J,$$

where $p_{i+} = \sum_{j=1}^{J} p_{ij}$ is the chance of $A_i$, and $p_{+j} = \sum_{i=1}^{I} p_{ij}$ is the chance of $B_j$. With independence, the mass function of $N$ can be written as

$$\binom{n}{n_{11}, \dots, n_{IJ}} \prod_{i=1}^{I} \prod_{j=1}^{J} (p_{i+}p_{+j})^{n_{ij}}.$$

Letting $n_{i+} = \sum_{j=1}^{J} n_{ij}$, $i = 1, \dots, I$, and $n_{+j} = \sum_{i=1}^{I} n_{ij}$, $j = 1, \dots, J$,

$$\prod_{i=1}^{I} \prod_{j=1}^{J} p_{i+}^{n_{ij}} = \prod_{i=1}^{I} p_{i+}^{n_{i+}} \text{ and } \prod_{i=1}^{I} \prod_{j=1}^{J} p_{+j}^{n_{ij}} = \prod_{j=1}^{J} p_{+j}^{n_{+j}}.$$

So the mass function of $N$ can be written

$$\binom{n}{n_{11}, \dots, n_{IJ}} \prod_{i=1}^{I} p_{i+}^{n_{i+}} \prod_{j=1}^{J} p_{+j}^{n_{+j}}.$$

Using the constraints $\sum_{i=1}^{I} n_{i+} = \sum_{j=1}^{J} n_{+j} = n$ and $\sum_{i=1}^{I} p_{i+} = \sum_{j=1}^{J} p_{+j} = 1$, this mass function equals

$$\binom{n}{n_{11}, \dots, n_{IJ}} \exp\Biggl[\sum_{i=2}^{I} n_{i+} \log\left(\frac{p_{i+}}{p_{1+}}\right) + \sum_{j=2}^{J} n_{+j} \log\left(\frac{p_{+j}}{p_{+1}}\right) + n \log(p_{1+}p_{+1})\Biggr].$$

These mass functions form a full rank $(I+J-2)$-parameter exponential family with canonical sufficient statistic

$$(N_{2+}, \dots, N_{I+}, N_{+2}, \dots, N_{+J}).$$

The equivalent statistic

$$(N_{1+}, \dots, N_{I+}, N_{+1}, \dots, N_{+J})$$

is also complete sufficient. In this model, $N_{i+} \sim \text{Binomial}(n, p_{i+})$ and $N_{+j} \sim \text{Binomial}(n, p_{+j})$ are independent. So $\hat{p}_{i+} = N_{i+}/n$ and $\hat{p}_{+j} = N_{+j}/n$ are UMVU estimates of $p_{i+}$ and $p_{+j}$, respectively, and $\hat{p}_{i+}\hat{p}_{+j}$ is the UMVU estimate of $p_{ij} = p_{i+}p_{+j}$.

*Example 5.6. Tables with Conditional Independence.* Suppose now that three characteristics, $A$, $B$, and $C$, are observed for each trial, with $N_{ijk}$ the number of trials that result in combination $A_iB_jC_k$, and $p_{ijk}$ the chance of this combination. Situations frequently arise in which it seems that characteristics $A$ and $B$ should be unrelated, but they are not independent because both are influenced by the third characteristic $C$. An appropriate model may be that $A$ and $B$ are conditionally independent given $C$. This leads naturally to the following constraints on the cell probabilities:

$$p_{ijk} = p_{++k} \frac{p_{i+k}}{p_{++k}} \frac{p_{+jk}}{p_{++k}}, \qquad i = 1, \dots, I, \; j = 1, \dots, J, \; k = 1, \dots, K,$$

where a "+" as a subscript indicates that the values for that subscript should be summed. Calculations similar to those for the previous example show that the mass functions with these constraints form a full rank $\bigl(K(I+J-1)-1\bigr)$-parameter exponential family with sufficient statistics $N_{++k}$, $k = 1, \dots, K$, $N_{i+k}$, $i = 1, \dots, I$, $k = 1, \dots, K$, and $N_{+jk}$, $j = 1, \dots, J$, $k = 1, \dots, K$.

## 5.4 Problems[2]

*1. Suppose $X$ has a binomial distribution with $m$ trials and success probability $\theta$, $Y$ has a binomial distribution with $n$ trials and success probability $\theta^2$, and $X$ and $Y$ are independent.
   a) Find a minimal sufficient statistic $T$.
   b) Show that $T$ is not complete, providing a nontrivial function $f$ with $E_\theta f(T) = 0$.

*2. Let $X$ and $Y$ be independent Bernoulli variables with $P(X = 1) = p$ and $P(Y = 1) = h(p)$ for some known function $h$.
   a) Show that the family of joint distributions is a curved exponential family unless

$$h(p) = \frac{1}{1 + \exp\left\{a + b \log \frac{p}{1-p}\right\}}$$

   for some constants $a$ and $b$.
   b) Give two functions $h$, one where $(X, Y)$ is minimal but not complete, and one where $(X, Y)$ is minimal and complete.

3. Let $X$ and $Y$ be independent Poisson variables.
   a) Suppose $X$ has mean $\lambda$, and $Y$ has mean $\lambda^2$. Do the joint mass functions form a curved two-parameter exponential family or a full rank one-parameter exponential family?
   b) Suppose instead $X$ has mean $\lambda$, and $Y$ has mean $2\lambda$. Do the joint mass functions form a curved two-parameter exponential family, or a full rank one-parameter exponential family?

[2] Solutions to the starred problems are given at the back of the book.

4. Consider the two-sample problem with $X_1, \dots, X_m$ i.i.d. from $N(\mu_x, \sigma_x^2)$ and $Y_1, \dots, Y_n$ i.i.d. from $N(\mu_y, \sigma_y^2)$, and all $n+m$ variables mutually independent.
   a) Find the UMVU estimator for the ratio of variances, $\sigma_x^2/\sigma_y^2$.
   b) If the two variances are equal, $\sigma_x^2 = \sigma_y^2 = \sigma$, find the UMVU estimator of the normalized difference in means $(\mu_x - \mu_y)/\sigma$.
5. Consider the two-sample problem with $X_1, \dots, X_m$ i.i.d. from $N(\mu_x, \sigma^2)$ and $Y_1, \dots, Y_n$ i.i.d. from $N(\mu_y, \sigma^2)$, and all $n+m$ variables mutually independent. Fix $\alpha \in (0,1)$ and define a parameter $q$ so that $P(X_i > Y_i + q) = \alpha$. Find the UMVU of $q$.
*6. Two teams A and B play a series of games, stopping as soon as one of the teams has 4 wins. Assume that game outcomes are independent and that on any given game team A has a fixed chance $\theta$ of winning. Let $X$ and $Y$ denote the number of games won by the first and second team, respectively.
   a) Find the joint mass function for $X$ and $Y$. Show that as $\theta$ varies these mass functions form a curved exponential family.
   b) Show that $T = (X, Y)$ is complete.
   c) Find a UMVU estimator of $\theta$.
*7. Consider a sequential experiment in which observations are i.i.d. from a Poisson distribution with mean $\lambda$. If the first observation $X$ is zero, the experiment stops, and if $X > 0$, a second observation $Y$ is observed. Let $T = 0$ if $X = 0$, and let $T = 1 + X + Y$ if $X > 0$.
   a) Find the mass function for $T$.
   b) Show that $T$ is minimal sufficient.
   c) Does this experiment give a curved two-parameter exponential family or full rank one-parameter exponential family?
   d) Is $T$ a complete sufficient statistic? Hint: Write $e^{\lambda} E_{\lambda} g(T)$ as a power series in $\lambda$ and derive equations for $g$ setting coefficients for $\lambda^x$ to zero.
8. Potential observations $(X_1, Y_1)$, $(X_2, Y_2)$, ... in a sequential experiment are i.i.d. The marginal distribution of $X_i$ is Poisson with parameter $\lambda$, the marginal distribution of $Y_i$ is Bernoulli with success probability $1/2$, and $X_i$ and $Y_i$ are independent. Suppose we continue observation, stopping the first time that $Y_i = 1$, so that the sample size is

$$N = \inf\{i : Y_i = 1\}.$$

   a) Show that the joint densities form an exponential family, and identify a minimal sufficient statistic. Is the family curved?
   b) Find two different unbiased estimators of $\lambda$, both functions of the minimal sufficient statistic. Is the minimal sufficient statistic complete?
9. Consider an experiment observing independent Bernoulli trials with unknown success probability $\theta \in (0,1)$. Suppose we observe trial outcomes until there are two successes in a row.
   a) Find a minimal sufficient statistic.

b) Give a formula for the mass function of the minimal sufficient statistic.
c) Is the minimal sufficient statistic complete? If it is, explain why, and if it is not, find a nontrivial function with constant expectation.

10. Consider a sequential experiment in which $X_1$ and $X_2$ are independent exponential variables with failure rate $\lambda$. If $X_1 < 1$, sampling stops after the first observation; if not, the second variable $X_2$ is also sampled. So $N = 1$ if $X_1 < 1$ and $N = 2$ if $X_1 \geq 1$.
   a) Do the densities for this experiment form a curved two-parameter exponential family or a one-parameter exponential family?
   b) Find $E\overline{X}$, and compare this expectation with the mean $1/\lambda$ of the exponential distribution.
11. Suppose independent Bernoulli trials are performed until the number of successes and number of failures differ by 2. Let $X$ denote the number of successes, $Y$ the number of failures (so $|X - Y| = 2$), and $\theta$ the chance of success.
   a) Find the joint mass function for $X$ and $Y$. Show that these mass functions form a curved exponential family with $T = (X, Y)$.
   b) Show that $T$ is complete.
   c) Find the UMVU estimator for $\theta$.
   d) Find $P(X > Y)$.
12. Consider a sequential experiment in which the potential observations $X_1, X_2, \ldots$ are i.i.d. from a geometric distribution with success probability $\theta \in (0, 1)$, so

$$P(X_i = x) = \theta(1 - \theta)^x, \qquad x = 0, 1, \ldots.$$

The sampling rule calls for a single observation ($N = 1$) if $X_1 = 0$, and two observations ($N = 2$) if $X_1 \geq 1$. Define

$$T = \sum_{i=1}^{N} X_i.$$

   a) Do the densities for this experiment form a curved two-parameter exponential family or a one-parameter exponential family?
   b) Show that $T$ is minimal sufficient.
   c) Find the mass function for $T$.
   d) Is $T$ complete? Explain why or find a function $g$ such that $g(T)$ has constant expectation.
   e) Find $E\overline{X}$. Is $\overline{X}$ an unbiased estimator of $EX_1$?
   f) Find the UMVU estimator of $EX_1$.

*13. Consider a single two-way contingency table and define $R = N_{11}+N_{12}$ (the first row sum), $C = N_{11}+N_{21}$ (the first column sum), and $D = N_{11}+N_{22}$ (the sum of the diagonal entries).
   a) Show that the joint mass function can be written as a full rank three-parameter exponential family with $T = (R, C, D)$ as the canonical sufficient statistic.

b) Relate the canonical parameter associated with $D$ to the "cross-product ratio" $\alpha$ defined as $\alpha = p_{11}p_{22}/(p_{12}p_{21})$.

c) Suppose we observe $m$ independent two-by-two contingency tables. Let $n_i$, $i = 1, \dots, m$, denote the trials for table $i$. Assume that cell probabilities for the tables may differ, but that the cross-product ratios for all $m$ tables are all the same. Show that the joint mass functions form a full rank exponential family. Express the sufficient statistic as a function of the variables $R_1, \dots, R_m$, $C_1, \dots, C_m$, and $D_1, \dots, D_m$.

14. Consider a two-way contingency table with a multinomial distribution for the counts $N_{ij}$ and with $I = J$. If the probabilities are symmetric, $p_{ij} = p_{ji}$, do the mass functions form a curved exponential family, or a full rank exponential family? With this constraint, identify a minimal sufficient statistic. Also, if possible, give UMVU estimators for the $p_{ij}$.

15. Let $(N_{11k}, N_{12k}, N_{21k}, N_{22k})$, $k = 1, \dots, n$, be independent two-by-two contingency tables. The $k$th table has a multinomial distribution with $m$ trials and success probabilities

$$\left(\frac{1+\theta_k}{4}, \frac{1-\theta_k}{4}, \frac{1-\theta_k}{4}, \frac{1+\theta_k}{4}\right).$$

Note that $\theta_k$ can be viewed as a measure of dependence in table $k$. (If $\theta_k = 0$ there is independence in table $k$.) Consider a model in which

$$\log\left[\frac{1+\theta_k}{1-\theta_k}\right] = \alpha + \beta x_k, \qquad k = 1, \dots, m,$$

where $\alpha$ and $\beta$ are unknown parameters, and $x_1, \dots, x_n$ are known constants. Show that the joint densities form an exponential family and identify a minimal sufficient statistic. Is this statistic complete?

*16. For an $I \times J$ contingency table with independence, the UMVU estimator of $p_{ij}$ is $\hat{p}_{i+}\hat{p}_{+j} = N_{i+}N_{+j}/n^2$.

a) Determine the variance of this estimator, $\mathrm{Var}(\hat{p}_{i+}\hat{p}_{+j})$.

b) Find the UMVU estimator of the variance in (a).

17. In some applications the total count in a contingency table would most naturally be viewed as a random variable. In these cases, a Poisson model might be more natural than the multinomial model in the text.

a) Let $X_1, \dots, X_p$ be independent Poisson variables, and let $\lambda_i$ denote the mean of $X_i$. Show that $T = X_1 + \dots + X_p$ has a Poisson distribution, and find

$$P(X_1 = x_1, \dots, X_p = x_p | T = n),$$

the conditional mass function for $X$ given $T = n$.

b) Consider a model for a two-by-two contingency table in which entries $N_{11}, \dots, N_{22}$ are independent Poisson variables, and let $\lambda_{ij}$ denote the mean of $N_{ij}$. With the constraint $\lambda_{11}\lambda_{22} = \lambda_{12}\lambda_{21}$, do the joint mass functions for these counts form a curved four-parameter exponential family or a three-parameter exponential family?

18. Consider a two-way contingency table with a multinomial distribution for the counts $N_{ij}$ with $I = J$. Assume that the cell probabilities $p_{ij}$ are constrained to have the same marginal values,

$$p_{i+} = p_{+i}, \qquad i = 1, \dots, I.$$

a) If $I = 2$, find a minimal sufficient statistic $T$. Is $T$ complete?
b) Find a minimal sufficient statistic $T$ when $I = 3$. Is this statistic complete?
c) Suppose we add an additional constraint that the characteristics are independent, so

$$p_{ij} = p_{i+}p_{+j}, \qquad i = 1, \dots, I, \quad j = 1, \dots, I.$$

Give a minimal sufficient statistic when $I = 2$, and determine whether it is complete.

# 6

# Conditional Distributions

Building on Section 1.10, this chapter provides a more thorough and proper treatment of conditioning. Section 6.4 gives a proof of the factorization theorem (Theorem 3.6).

## 6.1 Joint and Marginal Densities

Let $X$ be a random vector in $\mathbb{R}^m$, let $Y$ be a random vector in $\mathbb{R}^n$, and let $Z = (X, Y)$ in $\mathbb{R}^{m+n}$. Suppose $P_Z$ has density $p_Z$ with respect to $\mu \times \nu$ where $\mu$ and $\nu$ are measures on $\mathbb{R}^m$ and $\mathbb{R}^n$. This density $p_Z$ is called the *joint density* of $X$ and $Y$. Then

$$P(Z \in B) = \iint 1_B(x,y) p_Z(x,y)\, d\mu(x)\, d\nu(y).$$

By Fubini's theorem, the order of integration here can be reversed. To compute $P(X \in A)$ from this formula, note that $X \in A$ if and only if $Z \in A \times \mathbb{R}^n$. Then because $1_{A\times\mathbb{R}^n}(x,y) = 1_A(x)$,

$$\begin{aligned} P(X \in A) = P(Z \in A \times \mathbb{R}^n) &= \iint 1_A(x) p_Z(x,y)\, d\nu(y)\, d\mu(x) \\ &= \int_A \left\{ \int p_Z(x,y)\, d\nu(y) \right\} d\mu(x). \end{aligned}$$

From this, $X$ has density

$$p_X(x) = \int p_Z(x,y)\, d\nu(y) \tag{6.1}$$

with respect to $\mu$. This density $p_X$ is called the *marginal density* of $X$. Similarly, $Y$ has density

$$p_Y(y) = \int p_Z(x,y)\, d\mu(x),$$

called the marginal density of $Y$.

R.W. Keener, *Theoretical Statistics: Topics for a Core Course*, Springer Texts in Statistics,
DOI 10.1007/978-0-387-93839-4_6, © Springer Science+Business Media, LLC 2010

*Example 6.1.* Suppose $\mu$ is counting measure on $\{0,1,\dots,k\}$ and $\nu$ is Lebesgue measure on $\mathbb{R}$. Define

$$p_Z(x,y) = \begin{cases} \binom{k}{x} y^x(1-y)^{k-x}, & x = 0,1,\dots,k,\ y \in (0,1); \\ 0, & \text{otherwise.} \end{cases}$$

By (6.1), $X$ has density

$$p_X(x) = \int_0^1 \binom{k}{x} y^x(1-y)^{k-x}\,dy = \frac{1}{k+1}, \qquad x = 0,1,\dots,k.$$

(The identity $\int_0^1 u^{\alpha-1}(1-u)^{\beta-1}\,du = \Gamma(\alpha)\Gamma(\beta)/\Gamma(\alpha+\beta)$ is used to evaluate the integral.) This is the density for the uniform distribution on $\{0,1,\dots,k\}$. To find the marginal density of $Y$ we integrate the joint density against $\mu$. For $y \in (0,1)$,

$$p_Y(y) = \int p_Z(x,y)\,d\mu(x) = \sum_{x=0}^{k} \binom{k}{x} y^x(1-y)^{k-x} = 1;$$

and if $y \notin (0,1)$, $p_Y(y) = 0$. Therefore $Y$ is uniformly distributed on $(0,1)$.

## 6.2 Conditional Distributions

Let $X$ and $Y$ be random vectors. The definition of the conditional distribution $Q_x$ of $Y$ given $X = x$ is related to our fundamental smoothing identity. Specifically, if $E|f(X,Y)| < \infty$, we should have

$$Ef(X,Y) = EE[f(X,Y)|X], \tag{6.2}$$

with $E[f(X,Y)|X] = H(X)$ and

$$H(x) = E[f(X,Y)|X=x] = \int f(x,y)\,dQ_x(y).$$

Written out, (6.2) becomes

$$Ef(X,Y) = \int H(x)\,dP_X(x) = \iint f(x,y)\,dQ_x(y)\,dP_X(x). \tag{6.3}$$

The formal definition requires that (6.3) holds when $f$ is an indicator of $A \times B$. Then (6.2) or (6.3) will hold for general measurable $f$ provided $E|f(X,Y)| < \infty$.

**Definition 6.2.** *The function $Q$ is a* conditional distribution *for $Y$ given $X$, written*

$$Y|X = x \sim Q_x,$$

*if*

1. *$Q_x(\cdot)$ is a probability measure for all $x$,*
2. *$Q_x(B)$ is a measurable function of $x$ for any Borel set $B$, and*
3. *for any Borel sets $A$ and $B$,*

$$P(X \in A, Y \in B) = \int_A Q_x(B)\,dP_X(x).$$

When $X$ and $Y$ are random vectors, conditional distributions will always exist.[1] Conditional probabilities can be defined in more general settings, but assignments so that $Q_x(\cdot)$ is a probability measure may not be possible.

The stated definition of conditional distributions is not constructive. In the setting of Section 6.1 in which $X$ and $Y$ have joint density $p_Z$ with respect to $\mu \times \nu$, conditional distributions can be obtained explicitly using the following result.

**Theorem 6.3.** *Suppose $X$ and $Y$ are random vectors with joint density $p_Z$ with respect to $\mu \times \nu$. Let $p_X$ be the marginal density for $X$ given in (6.1), and let $E = \{x : p_X(x) > 0\}$. For $x \in E$, define*

$$p_{Y|X}(y|x) = \frac{p_Z(x, y)}{p_X(x)}, \tag{6.4}$$

*and let $Q_x$ be the probability measure with density $p_{Y|X}(\cdot|x)$ with respect to $\nu$. When $x \notin E$, define $p_{Y|X}(y|x) = p_0(y)$, where $p_0$ is the density for an arbitrary fixed probability distribution $P_0$, and let $Q_x = P_0$. Then $Q$ is a conditional distribution for $Y$ given $X$.*

*Proof.* Part one of the definition is apparent, and part two follows from measurability of $p_Z$. It is convenient to establish (6.3) directly; part three of the definition then follows immediately. First note that $P(X \in E) = 1$, and without loss of generality we can assume that $p_Z(x, y) = 0$ whenever $x \notin E$. (If not, just change $p_Z(x, y)$ to $p_Z(x, y)1_E(x)$—these functions agree for a.e. $(x, y)$ $(\mu \times \nu)$, and either can serve as the joint density.) Then $p_Z(x, y) = p_X(x)p_{Y|X}(y|x)$ for all $x$ and $y$, and the right-hand side of (6.3) equals

$$\iint f(x, y)p_{Y|X}(y|x)\,d\nu(y)\,p_X(x)\,d\mu(x)$$
$$= \iint f(x, y)p_Z(x, y)\,d\nu(y)\,d\mu(x) = Ef(X, Y).$$

□

When $X$ and $Y$ are independent, $p_Z(x, y) = p_X(x)p_Y(y)$, and so

[1] When $X$ is a random variable, this is given as Theorem 33.3 of Billingsley (1995). See Chapter 5 of Rao (2005) for more general cases.

$$p_{Y|X}(y|x) = \frac{p_X(x)p_Y(y)}{p_X(x)} = p_Y(y),$$

for $x \in E$. So the conditional and marginal distributions for $Y$ are the same (for a.e. $x$).

*Example 6.1, continued.* Because $p_Y(y) = 1$, $y \in (0, 1)$,

$$p_{X|Y}(x|y) = p_Z(x, y) = \binom{k}{x} y^x (1-y)^{k-x}, \qquad x = 0, 1, \dots, k.$$

As a function of $x$ with $y$ fixed, this is the mass function for the binomial distribution with success probability $y$ and $k$ trials. So

$$X|Y = y \sim \text{Binomial}(k, y). \tag{6.5}$$

Similarly, recalling that $p_X(x) = 1/(k+1)$,

$$\begin{aligned} p_{Y|X}(y|x) &= \frac{p_Z(x,y)}{p_X(x)} = (k+1)\binom{k}{x} y^x (1-y)^{k-x} \\ &= \frac{\Gamma(k+2)}{\Gamma(x+1)\Gamma(k-x+1)} y^{x+1-1}(1-y)^{k-x+1-1}, \qquad y \in (0,1). \end{aligned}$$

This is the density for the beta distribution, and so

$$Y|X = x \sim \text{Beta}(x+1, k-x+1).$$

To illustrate how smoothing might be used to calculate expectations in this example, as the binomial distribution in (6.5) has mean $ky$,

$$E[X|Y] = kY.$$

So, by smoothing,

$$EX = EE[X|Y] = kEY = k\int_0^1 y\,dy = \frac{k}{2}.$$

Summation against the mass function for $X$ gives the same answer:

$$EX = \sum_{x=0}^{k} \frac{x}{k+1} = \frac{k}{2}.$$

To compute $EX^2$ using smoothing, because the binomial distribution in (6.5) has second moment $ky(1-y) + k^2y^2$,

$$\begin{aligned} EX^2 &= EE[X^2|Y] = E\big[kY(1-Y) + k^2Y^2\big] \\ &= \int_0^1 \big(ky(1-y) + k^2y^2\big)\,dy = \frac{k(1+2k)}{6}. \end{aligned}$$

Summation against the mass function for $X$ gives

$$EX^2 = \sum_{x=0}^{k} \frac{x^2}{k+1},$$

so these calculations show indirectly that

$$\sum_{x=0}^{k} x^2 = \frac{k(k+1)(2k+1)}{6}.$$

(This can also be proved by induction.)

## 6.3 Building Models

To develop realistic models for two or more random vectors, it is often convenient to specify a joint density, using (6.4), as

$$p_Z(x,y) = p_X(x)p_{Y|X}(y|x).$$

The thought process using this equation would involve first choosing a marginal distribution for $X$ and then combining this marginal distribution with a suitable distribution for $Y$ if $X$ were known. This equation can be extended to several vectors. If $p(x_k|x_1,\dots,x_{k-1})$ denotes the conditional density of $X_k$ given $X_1 = x_1, \dots, X_{k-1} = x_{k-1}$, then the joint density of $X_1, \dots, X_n$ is

$$p_{X_1}(x_1)p(x_2|x_1)\cdots p(x_n|x_1,\dots,x_{n-1}). \tag{6.6}$$

*Example 6.4. Models for Time Series.* Statistical applications in which variables are observed over time are widespread and diverse. Examples include prices of stocks, measurements of parts from a production process, or growth curve data specifying size or dimension of a person or organism over time. In most of these applications it is natural to suspect that the observations will be dependent. For instance, if $X_k$ is the log of a stock price, a model with

$$X_k|X_1 = x_1, \dots, X_{k-1} = x_{k-1} \sim N(x_{k-1} + \mu, \sigma^2)$$

may be natural. If $X_1 \sim N(x_0 + \mu, \sigma^2)$, then by (6.6), $X_1, \dots, X_n$ will have joint density

$$\prod_{k=1}^{n} \frac{1}{\sqrt{2\pi\sigma^2}} \exp\left[-\frac{(x_k - x_{k-1} - \mu)^2}{2\sigma^2}\right].$$

Differences $X_k - X_{k-1}$ here are i.i.d. from $N(\mu, \sigma^2)$, and the model here for the joint distribution is called a *random walk.*

Another model for variables that behave in a more stationary fashion over time might have

$$X_k|X_1 = x_1, \dots, X_{k-1} = x_{k-1} \sim N(\rho x_{k-1}, \sigma^2),$$

where $|\rho| < 1$. If $X_1 \sim N(\rho x_0, \sigma^2)$, then by (6.6) the joint density is

$$\prod_{k=1}^{n} \frac{1}{\sqrt{2\pi\sigma^2}} \exp\left[-\frac{(x_k - \rho x_{k-1})^2}{2\sigma^2}\right].$$

This is called an *autoregressive* model.

*Example 6.5. A Simple Model for Epidemics.* For any degree of realism, statistical models for epidemics must allow substantial dependence over time, and conditioning arguments can be quite useful in attempts to incorporate this dependence in a natural fashion. To illustrate, let us develop a simple model based on suspect assumptions. Improvements with more realistic assumptions should give practical and useful models

Let $N$ denote the size of the population of interest, and let $X_i$ denote the number of infected individuals in the population at stage $i$. Assume that once someone is infected, they stay infected. Also, assume that the chance an infected individual infects a noninfected individual during the time interval between two stages is $p = 1 - q$ and that all chances for infection are independent. Then the chance a noninfected person stays noninfected during the time interval between stages $k$ and $k+1$, given $X_k = x_k$ (and other information about the past), is $q^{x_k}$, and so the number of people newly infected during this time interval, $X_{k+1} - X_k$, will have a binomial distribution. Specifically

$$X_{k+1} - X_k | X_1 = x_1, \dots, X_k = x_k \sim \text{Binomial}(N - x_k, 1 - q^{x_k}).$$

This leads to conditional densities (mass functions)

$$p(x_{k+1}|x_1, \dots, x_k) = \binom{N - x_k}{x_{k+1} - x_k} \left(1 - q^{x_k}\right)^{x_{k+1} - x_k} \left(q^{x_k}\right)^{N - x_{k+1}}.$$

The product of these gives the joint mass function for $X_1, \dots, X_n$.

## 6.4 Proof of the Factorization Theorem[2]

To prove the factorization theorem (Theorem 3.6) we need to work directly from the definition of conditional distributions, for in most cases $T$ and $X$ will not have a joint density with respect to any product measure. To begin, suppose $P_\theta$, $\theta \in \Omega$, has density

$$p_\theta(x) = g_\theta\big(T(x)\big)h(x) \tag{6.7}$$

[2] This section is optional; the proof is fairly technical.

with respect to $\mu$. Modifying $h$, we can assume without loss of generality that $\mu$ is a probability measure equivalent[3] to the family $\mathcal{P} = \{P_\theta : \theta \in \Omega\}$. Let $E^*$ and $P^*$ denote expectation and probability when $X \sim \mu$; let $G^*$ and $G_\theta$ denote marginal distributions for $T(X)$ when $X \sim \mu$ and $X \sim P_\theta$; and let $Q$ be the conditional distribution for $X$ given $T$ when $X \sim \mu$. To find densities for $T$,

$$\begin{aligned} E_\theta f(T) &= \int f\bigl(T(x)\bigr) g_\theta\bigl(T(x)\bigr) h(x)\, d\mu(x) \\ &= E^* f(T) g_\theta(T) h(X) \\ &= \iint f(t) g_\theta(t) h(x)\, dQ_t(x)\, dG^*(t) \\ &= \int f(t) g_\theta(t) w(t)\, dG^*(t), \end{aligned}$$

where

$$w(t) = \int h(x)\, dQ_t(x).$$

If $f$ is an indicator function, this shows that $G_\theta$ has density $g_\theta(t)w(t)$ with respect to $G^*$. Next, define $\tilde{Q}_t$ to have density $h/w(t)$ with respect to $Q_t$, so that

$$\tilde{Q}_t(B) = \int_B \frac{h(x)}{w(t)}\, dQ_t(x).$$

(On the null set $w(t) = 0$, $\tilde{Q}_t$ can be an arbitrary probability measure.) Then

$$\begin{aligned} E_\theta f(X,T) &= E^* f(X,T) g_\theta(T) h(X) \\ &= \iint f(x,t) g_\theta(t) h(x)\, dQ_t(x)\, dG^*(t) \\ &= \iint f(x,t) \frac{h(x)}{w(t)}\, dQ_t(x) g_\theta(t) w(t)\, dG^*(t) \\ &= \iint f(x,t)\, d\tilde{Q}_t(x)\, dG_\theta(t). \end{aligned}$$

By (6.3) this shows that $\tilde{Q}$ is a conditional distribution for $X$ given $T$ under $P_\theta$. Because $\tilde{Q}$ does not depend on $\theta$, $T$ is sufficient.

Before considering the converse—that if $T$ is sufficient the densities $p_\theta$ must have form (6.7)—we should discuss mixture distributions. Given a marginal probability distribution $G^*$ and a conditional distribution $Q$, we can define a *mixture distribution* $\hat{P}$ by

$$\hat{P}(B) = \int Q_t(B)\, dG^*(t) = \iint 1_B(x)\, dQ_t(x)\, dG^*(t).$$

[3] "Equivalence" here means that $\mu(N) = 0$ if and only if $P_\theta(N)$, $\forall \theta \in \Omega$. The assertion here is based on a result that any dominated family is equivalent to the mixture of some countable subfamily.

Then for integrable $f$,

$$\int f\,d\hat{P} = \iint f(x)\,dQ_t(x)\,dG^*(t).$$

(By linearity, this must hold for simple functions $f$, and the general case follows taking simple functions converging to $f$.)

Suppose now that $T$ is sufficient, with $Q$ the conditional distribution for $X$ given $T$. Let $g_\theta$ be the $G^*$ density of $T$ when $X \sim P_\theta$. (This density will exist, for if $G^*(N) = 0$, $\mu(N_0) = 0$ where $N_0 = T^{-1}(N)$, and so $G_\theta(N) = P_\theta(T \in N) = P_\theta(X \in N_0) = \int_{N_0} p_\theta\,d\mu = 0$.) Then

$$\begin{aligned} P_\theta(X \in B) &= E_\theta P_\theta(X \in B|T) \\ &= E_\theta Q_T(B) \\ &= \int Q_t(B) g_\theta(t)\,dG^*(t) \\ &= \iint 1_B(x)\,dQ_t(x)\,g_\theta(t)\,dG^*(t) \\ &= \iint 1_B(x) g_\theta\bigl(T(x)\bigr)\,dQ_t(x)\,dG^*(t) \\ &= \int_B g_\theta\bigl(T(x)\bigr)\,d\hat{P}(x). \end{aligned}$$

This shows that $P_\theta$ has density $g_\theta\bigl(T(\cdot)\bigr)$ with respect to $\hat{P}$.

The mixture distribution $\hat{P}$ is absolutely continuous with respect to $\mu$. To see this, suppose $\mu(N) = 0$. Then $P_\theta(N) = \int Q_t(N)\,dG_\theta(t) = 0$, which implies $G_\theta(\tilde{N}) = 0$, where $\tilde{N} = \{t : Q_t(N) > 0\}$. Because $\mu$ is equivalent to $\mathcal{P}$ and $G_\theta(\tilde{N}) = P_\theta(T \in \tilde{N}) = 0$, $\forall \theta \in \Omega$, $P^*(T \in \tilde{N}) = G^*(\tilde{N}) = 0$. Thus $Q_t(N) = 0$ (a.e. $G^*$) and so $\hat{P}(N) = \int Q_t(N)\,dG^*(t) = 0$. Taking $h = d\hat{P}/dP^*$, $P_\theta$ has density $g_\theta\bigl(T(x)\bigr)h(x)$ with respect to $P^*$.

## 6.5 Problems[4]

1. *The beta distribution.*
   a) Let $X$ and $Y$ be independent random variables with $X \sim \Gamma(\alpha, 1)$ and $Y \sim \Gamma(\beta, 1)$. Define new random variables $U = X + Y$ and $V = X/(X + Y)$. Find the joint density of $U$ and $V$. Hint: If $p$ is the joint density of $X$ and $Y$, then

$$\begin{aligned} P\{(U, V) \in B\} &= P\left\{\left(X + Y, \frac{X}{X+Y}\right) \in B\right\} \\ &= \int 1_B\left(x + y, \frac{x}{x+y}\right) p(x, y)\,dx\,dy. \end{aligned}$$

[4] Solutions to the starred problems are given at the back of the book.

Next, change variables to write this integral as an integral against $u = x+y$ and $v = x/(x+y)$. The change of variables can be accomplished either using Jacobians or writing the double integral as an iterated integral and using ordinary calculus.

b) Find the marginal density for $V$. Use the fact that this density integrates to one to compute $\int_0^1 x^{\alpha-1}(1-x)^{\beta-1}dx$. This density for $V$ is called the beta density with parameters $\alpha$ and $\beta$. The corresponding distribution is denoted Beta$(\alpha,\beta)$.

c) Compute the mean and variance of the beta distribution.

d) Find the marginal density for $U$.

*2. Let $X$ and $Y$ be independent random variables with cumulative distribution functions $F_X$ and $F_Y$.

a) Assuming $Y$ is continuous, use smoothing to derive a formula expressing the cumulative distribution function of $X^2Y^2$ as the expected value of a suitable function of $X$. Also, if $Y$ is absolutely continuous, give a formula for the density.

b) Suppose $X$ and $Y$ are both exponential with the same failure rate $\lambda$. Find the density of $X-Y$.

*3. Suppose that $X$ and $Y$ are independent and positive. Use a smoothing argument to show that if $x \in (0,1)$, then

$$P\left(\frac{X}{X+Y} \le x\right) = EF_X\left(\frac{xY}{1-x}\right), \tag{6.8}$$

where $F_X$ is the cumulative distribution function of $X$.

*4. Differentiating (6.8), if $X$ is absolutely continuous with density $p_X$, then $V = X/(X+Y)$ is absolutely continuous with density

$$p_V(x) = E\left[\frac{Y}{(1-x)^2}p_X\left(\frac{xY}{1-x}\right)\right], \qquad x \in (0,1).$$

Use this formula to derive the beta distribution introduced in Problem 6.1, showing that if $X$ and $Y$ are independent with $X \sim \Gamma(\alpha,1)$ and $Y \sim \Gamma(\beta,1)$, then $V = X/(X+Y)$ has density

$$p_V(x) = \frac{\Gamma(\alpha+\beta)}{\Gamma(\alpha)\Gamma(\beta)}x^{\alpha-1}(1-x)^{\beta-1}$$

for $x \in (0,1)$.

*5. Let $X$ and $Y$ be absolutely continuous with joint density

$$p(x,y) = \begin{cases} 2, & 0 < x < y < 1; \\ 0, & \text{otherwise.} \end{cases}$$

a) Find the marginal density of $X$ and the marginal density of $Y$.

b) Find the conditional density of $Y$ given $X = x$.

c) Find $E[Y|X]$.
d) Find $EXY$ by integration against the joint density of $X$ and $Y$.
e) Find $EXY$ by smoothing, using the conditional expectation you found in part (c).

*6. Let $\mu$ be Lebesgue measure on $\mathbb{R}$ and let $\nu$ be counting measure on $\{0,1,\dots\}^2$. Suppose the joint density of $X$ and $Y$ with respect to $\mu \times \nu$ is given by

$$p(x, y_1, y_2) = x^2(1-x)^{y_1+y_2}$$

for $x \in (0,1)$, $y_1 = 0,1,2,\dots$, and $y_2 = 0,1,2,\dots$.

a) Find the marginal density of $X$.
b) Find the conditional density of $X$ given $Y = y$ (i.e., given $Y_1 = y_1$ and $Y_2 = y_2$).
c) Find $E[X|Y]$ and $E[X^2|Y]$. Hint: The formula

$$\int_0^1 x^{\alpha-1}(1-x)^{\beta-1}\,dx = \frac{\Gamma(\alpha)\Gamma(\beta)}{\Gamma(\alpha+\beta)}$$

may be useful.
d) Find $E\big[1/(4+Y_1+Y_2)\big]$. Hint: Find $EX$ using the density in part (a) and find an expression for $EX$ using smoothing and the conditional expectation in part (c).

7. Let $X$ and $Y$ be random variables with joint Lebesgue density

$$p(x,y) = \begin{cases} 2y^2 e^{-xy}, & x > 0,\ y \in (0,1); \\ 0, & \text{otherwise.} \end{cases}$$

a) Find the marginal density for $Y$.
b) Find the conditional density for $X$ given $Y = y$.
c) Find $P(X > 1|Y = y)$, $E[X|Y = y]$, and $E[X^2|Y = y]$.

8. Suppose $X$ has the standard exponential distribution with marginal density $e^{-x}$, $x > 0$, and that

$$P(Y = y|X = x) = \frac{x^y e^{-x}}{y!}, \qquad y = 0,1,\dots.$$

a) Find the joint density for $X$ and $Y$. Identify the dominating measure.
b) Find the marginal density of $Y$.
c) Find the conditional density of $X$ given $Y = y$.
d) Find $EY$ using the marginal density in part (b).
e) As the conditional distribution of $Y$ given $X = x$ is Poisson with parameter $x$, $E(Y|X = x) = x$. Use this to find $EY$ by smoothing.

9. Suppose that $X$ is uniformly distributed on the interval $(0,1)$ and that

$$P(Y = y|X = x) = (1-x)x^y, \qquad y = 0,1,\dots.$$

a) Find the joint density for $X$ and $Y$. What is the measure for integrals against this density?

b) Find the marginal density of $Y$.
c) Find the conditional density of $X$ given $Y = y$.
d) Find $E[X|Y = y]$. Find $P(X < 1/2|Y = 0)$ and $P(X < 1/2|Y = 1)$.

10. Suppose $X$ and $Y$ are independent, both uniformly distributed on $(0, 1)$. Let $M = \max\{X, Y\}$ and $Z = \min\{X, Y\}$.
 a) Find the conditional distribution of $Z$ given $M = m$.
 b) Suppose instead that $X$ and $Y$ are independent but uniformly distributed on the finite set $\{1, \dots, k\}$. Give the conditional distribution of $Z$ given $M = j$.

*11. Suppose $X$ and $Y$ are independent, both absolutely continuous with common density $f$. Let $M = \max\{X, Y\}$ and $Z = \min\{X, Y\}$. Determine the conditional distribution for the pair $(X, Y)$ given $(M, Z)$.

*12. Let $X$ and $Y$ be independent exponential variables with failure rate $\lambda$, so the common marginal density is $\lambda e^{-\lambda x}$, $x > 0$. Let $T = X + Y$. Give a formula expressing $E[f(X, Y)|T = t]$ as a one-dimensional integral. Hint: Review the initial example on sufficiency in Section 3.2.

13. Suppose $X$ and $Y$ are absolutely continuous with joint density

$$\frac{\exp\left[-\frac{x^2-2\rho xy+y^2}{2(1-\rho^2)}\right]}{2\pi\sqrt{1-\rho^2}}.$$

This is a bivariate normal density with

$$EX = EY = 0, \qquad \mathrm{Var}(X) = \mathrm{Var}(Y) = 1,$$

and

$$\mathrm{Cor}(X, Y) = \rho.$$

Determine the conditional distribution of $Y$ given $X$. (Naturally, the answer should depend on the correlation coefficient $\rho$.) Use smoothing to find the covariance between $X^2$ and $Y^2$.

*14. Let $X$ and $Y$ be absolutely continuous with density $p(x, y) = e^{-x}$, if $0 < y < x$; $p(x, y) = 0$, otherwise.
 a) Find the marginal densities of $X$ and $Y$.
 b) Compute $EY$ and $EY^2$ integrating against the marginal density of $Y$.
 c) Find the conditional density of $Y$ given $X = x$, and use it to compute $E[Y|X]$ and $E[Y^2|X]$.
 d) Find the expectations of $E[Y|X]$ and $E[Y^2|X]$ integrating against the marginal density of $X$.

15. Suppose $X$ has a Poisson distribution with mean $\lambda$ and that given $X = x$, $Y$ has a binomial distribution with $x$ trials and success probability $p$. (If $X = 0$, $Y = 0$.)
 a) Find the marginal distribution of $Y$.
 b) Find the conditional distribution of $X$ given $Y$.
 c) Find $E[Y^2|X]$.
 d) Compute $EY^2$ by smoothing, using the result in part (c).

e) Compute $EY^2$ integrating against the marginal distribution from part (a).
f) Find $E[X|Y]$ and use this to compute $EX$ by smoothing.

16. Let $X = (X_1, X_2)$ be an absolutely continuous random vector in $\mathbb{R}^2$ with density $f$, and let $T = X_1 + X_2$.
a) Find the joint density for $X_1$ and $T$.
b) Give a formula for the density of $T$.
c) Give a formula for the conditional density of $X_1$ given $T = t$.
d) Give a formula for $E\bigl[g(X) \bigm| T = t\bigr]$. Hint: View $g(X)$ as a function of $T$ and $X_1$ and use the conditional density you found in part (c).
e) Suppose $X_1$ and $X_2$ are i.i.d. standard normal. Then $X_1 - X_2 \sim N(0, 2)$ and $T \sim N(0, 2)$. Find $P\bigl(|X_1 - X_2| < 1 \bigm| T\bigr)$ using your formula from part (d). Integrate this against the density for $T$ to show that smoothing gives the correct answer.

17. Let $X_1, \dots, X_n$ be jointly distributed Bernoulli variables with mass function

$$P(X_1 = x_1, \dots, X_n = x_n) = \frac{s_n!(n - s_n)!}{(n+1)!},$$

where $s_n = x_1 + \cdots + x_n$.
a) Find the joint mass function for $X_1, \dots, X_{n-1}$. (Your answer should simplify.)
b) Find the joint mass function for $X_1, \dots, X_k$ for any $k < n$.
c) Find $P(X_{k+1} = 1 | X_1 = x_1, \dots, X_k = x_k)$, $k < n$.
d) Let $S_n = X_1 + \cdots + X_n$. Find

$$P(X_1 = x_1, \dots, X_n = x_n | S_n = s).$$

e) Let $Y_k = (1 + S_k)/(k + 2)$. For $k < n$, find

$$E(Y_{k+1} | X_1 = x_1, \dots, X_k = x_k),$$

expressing your answer as a function of $Y_k$. Use smoothing to relate $EY_{k+1}$ to $EY_k$. Find $EY_k$ and $ES_k$.

18. Suppose $X \sim N(0, 1)$ and $Y|X = x \sim N(x, 1)$.
a) Find the mean and variance of $Y$.
b) Find the conditional distribution of $X$ given $Y = y$.

19. Let $X$ be absolutely continuous with a positive continuous density $f$ and cumulative distribution function $F$. Take $Y = X^2$.
a) Find the cumulative distribution function and the density for $Y$.
b) For $y > 0$, $y \neq x^2$, find

$$\lim_{\epsilon \downarrow 0} P\bigl[X \le x \bigm| Y \in (y - \epsilon, y + \epsilon)\bigr].$$

c) The limit in part (b) should agree with the cumulative distribution function for a discrete probability measure $Q_y$. Find the mass function for this discrete distribution.

d) Show that $Q$ is a conditional distribution for $X$ given $Y$. Specifically, show that it satisfies the conditions in Definition 6.2.

20. Suppose $X$ and $Y$ are conditionally independent given $W = w$ with

$$X|W = w \sim N(aw, 1) \text{ and } Y|W = w \sim N(bw, 1).$$

Use smoothing to derive formulas relating $EX$, $EY$, $\mathrm{Var}(X)$, $\mathrm{Var}(Y)$, and $\mathrm{Cov}(X, Y)$ to moments of $W$ and the constants $a$ and $b$.

21. Suppose $Y \sim N(\nu, \tau^2)$ and that given $Y = y$, $X_1, \dots, X_n$ are i.i.d. from $N(y, \sigma^2)$. Taking $\overline{x} = (x_1 + \cdots + x_n)/n$, show that the conditional distribution of $Y$ given $X_1 = x_1, \dots, X_n = x_n$ is normal with

$$E(Y|X_1 = x_1, \dots, X_n = x_n) = \frac{\nu/\tau^2 + n\overline{x}/\sigma^2}{1/\tau^2 + n/\sigma^2}$$

and

$$\mathrm{Var}(Y|X_1 = x_1, \dots, X_n = x_n) = \frac{1}{1/\tau^2 + n/\sigma^2}.$$

Remark: If *precision* is defined as the reciprocal of the variance, these formulas state that the precision of the conditional distribution is the sum of the precisions of the $X_i$ and $Y$, and the mean of the conditional distribution is an average of the $X_i$ and $\nu$, weighted by the precisions of the variables.

22. A building has a single elevator. Times between stops on the first floor are presumed to follow an exponential distribution with failure rate $\theta$. In a time interval of duration $t$, the number of people who arrive to ride the elevator has a Poisson distribution with mean $\lambda t$.

a) Suggest a joint density for the time $T$ between elevator stops and the number of people $X$ that board when it arrives.

b) Find the marginal mass function for $X$.

c) Find $EX^2$.

d) Let $\lambda > 0$ and $\theta > 0$ be unknown parameters, and suppose we observe data $X_1, \dots, X_n$ that are i.i.d. with the marginal mass function of $X$ in part (b). Suggest an estimator for the ratio $\theta/\lambda$ based on the average $\overline{X}$. With these data, if $n$ is large will we be able to estimate $\lambda$ accurately?

# 7

# Bayesian Estimation

As mentioned in Section 3.1, a comparison of two estimators from their risk functions will be inconclusive whenever the graphs of these functions cross. This difficulty will not arise if the performance of an estimator is measured with a single number. In a Bayesian approach to inference the performance of an estimator $\delta$ is judged by a weighted average of the risk function, specifically by

$$\int R(\theta, \delta)\, d\Lambda(\theta), \tag{7.1}$$

where $\Lambda$ is a specified probability measure on the parameter space $\Omega$.

## 7.1 Bayesian Models and the Main Result

The weighted average (7.1) arises as expected loss using $\delta$ in a *Bayesian probability model* in which both the unknown parameter and data are viewed as random. For notation, $\Theta$ is the random parameter with $\theta$ a possible value for $\Theta$. In the Bayesian model,

$$\Theta \sim \Lambda,$$

with $\Lambda$ called the *prior* distribution because it represents probabilities before data are observed, and $P_\theta$ is the conditional distribution of $X$ given $\Theta = \theta$, that is,

$$X|\Theta = \theta \sim P_\theta.$$

Then

$$E\big[L(\Theta, \delta(X)) \mid \Theta = \theta\big] = \int L\big(\theta, \delta(x)\big)\, dP_\theta(x) = R(\theta, \delta),$$

and by smoothing,

$$EL(\Theta, \delta) = EE\big[L(\Theta, \delta) \mid \Theta\big] = ER(\Theta, \delta) = \int R(\theta, \delta)\, d\Lambda(\theta).$$

R.W. Keener, *Theoretical Statistics: Topics for a Core Course*, Springer Texts in Statistics,
DOI 10.1007/978-0-387-93839-4_7, 

In Bayesian estimation, the choice of the prior distribution $\Lambda$ is critical. In some situations $\Theta$ may be random in the usual frequentist sense with a random process producing the current parameter $\Theta$ and other random parameters in the past and future. Then $\Lambda$ would be selected from prior experience with the random process. For instance, the parameter $\Theta$ may be the zip code on a letter estimated using pixel data from an automatic scanner. The prior distribution here should just reflect chances for various zip codes. More commonly, the parameter $\Theta$ cannot be viewed as random in a frequentist sense. The general view in these cases would be that the prior $\Lambda$ should reflect the researchers' informed subjective opinion about chances for various values of $\Theta$. Both of these ideas regarding selection of $\Lambda$ may need to be tempered with a bit of pragmatism. Calculations necessary to compute estimators may be much easier if the prior distribution has a convenient form.

An estimator that minimizes (7.1) is called *Bayes*. Lacking information from data $X$, the best estimate is just the constant minimizing $EL(\Theta, d)$ over allowed values of $d$. The following result shows that a Bayes estimator can be found in a similar fashion. The key difference is that one should now minimize the conditional expected loss given the data, that is $E\bigl[L(\Theta, d) \bigm| X = x\bigr]$. This conditional expected loss is called the *posterior risk* and would be computed integrating against the conditional distribution for $\Theta$ given $X = x$, called the *posterior distribution* of $\Theta$.

**Theorem 7.1.** *Suppose $\Theta \sim \Lambda$, $X|\Theta = \theta \sim P_\theta$, and $L(\theta, d) \ge 0$ for all $\theta \in \Omega$ and all $d$. If*

*a) $EL(\Theta, \delta_0) < \infty$ for some $\delta_0$,*

*and*

*b) for a.e. $x$ there exists a value $\delta_\Lambda(x)$ minimizing*

$$E\bigl[L(\Theta, d) \bigm| X = x\bigr]$$

*with respect to $d$,*

*then $\delta_\Lambda$ is a Bayes estimator.*

*Proof.* Let $\delta$ be an arbitrary estimator. Then for a.e. $x$,

$$\begin{aligned} E\bigl[L(\Theta, \delta(X)) \bigm| X = x\bigr] &= E\bigl[L(\Theta, \delta(x)) \bigm| X = x\bigr] \\ &\ge E\bigl[L(\Theta, \delta_\Lambda(x)) \bigm| X = x\bigr] \\ &= E\bigl[L(\Theta, \delta_\Lambda(X)) \bigm| X = x\bigr], \end{aligned}$$

and so

$$E\bigl[L(\Theta, \delta(X)) \bigm| X\bigr] \ge E\bigl[L(\Theta, \delta_\Lambda(X)) \bigm| X\bigr],$$

almost surely. Taking expectations, by smoothing,

$$EL\bigl(\Theta, \delta(X)\bigr) \ge EL\bigl(\Theta, \delta_\Lambda(X)\bigr).$$

Thus $\delta_\Lambda$ is Bayes. ☐

## 7.2 Examples

*Example 7.2. Weighted Squared Error Loss.* Suppose

$$L(\theta, d) = w(\theta)\bigl(d - g(\theta)\bigr)^2.$$

By Theorem 7.1, $\delta_\Lambda(x)$ should minimize

$$\begin{aligned} E\bigl[w(\Theta)\bigl(d - g(\Theta)\bigr)^2 \bigm| X = x\bigr] = d^2 E\bigl[w(\Theta) \bigm| X = x\bigr] \\ - 2dE\bigl[w(\Theta)g(\Theta) \bigm| X = x\bigr] \\ + E\bigl[w(\Theta)g^2(\Theta) \bigm| X = x\bigr]. \end{aligned}$$

This is a quadratic function of $d$, minimized when the derivative

$$2dE\bigl[w(\Theta) \bigm| X = x\bigr] - 2E\bigl[w(\Theta)g(\Theta) \bigm| X = x\bigr]$$

equals zero. Thus

$$\delta_\Lambda(x) = \frac{E\bigl[w(\Theta)g(\Theta) \bigm| X = x\bigr]}{E\bigl[w(\Theta) \bigm| X = x\bigr]}. \tag{7.2}$$

If the weight function $w$ is identically one, then

$$\delta_\Lambda(X) = E\bigl[g(\Theta) \bigm| X\bigr],$$

called the *posterior mean* of $g(\Theta)$.

If $\mathcal{P}$ is a dominated family with $p_\theta$ the density for $P_\theta$, and if $\Lambda$ is absolutely continuous with Lebesgue density $\lambda$, then the joint density of $X$ and $\Theta$ is

$$p_\theta(x)\lambda(\theta).$$

By (6.1), the marginal density of $X$ is

$$q(x) = \int p_\theta(x)\lambda(\theta)\,d\theta,$$

and by (6.4), the conditional density of $\Theta$ given $X = x$ is

$$\lambda(\theta|x) = \frac{p_\theta(x)\lambda(\theta)}{q(x)}.$$

Using this, (7.2) becomes

$$\delta_\Lambda(x) = \frac{\int w(\theta)g(\theta)p_\theta(x)\lambda(\theta)\,d\theta}{\int w(\theta)p_\theta(x)\lambda(\theta)\,d\theta}.$$

(The factor $1/q(x)$ common to both the numerator and denominator cancels.)

*Example 7.3. Binomial.* When $P_\theta$ is the binomial distribution with $n$ trials and success probability $\theta$, the beta distribution $\mathrm{Beta}(\alpha, \beta)$ with density

$$\lambda(\theta) = \frac{\Gamma(\alpha+\beta)}{\Gamma(\alpha)\Gamma(\beta)}\theta^{\alpha-1}(1-\theta)^{\beta-1}, \qquad \theta \in (0,1),$$

is a common choice for the prior distribution of $\Theta$. (For a derivation of this density, see Problem 6.1.) The beta density integrates to one, therefore

$$\int_0^1 \theta^{\alpha-1}(1-\theta)^{\beta-1}\,d\theta = \frac{\Gamma(\alpha)\Gamma(\beta)}{\Gamma(\alpha+\beta)}. \tag{7.3}$$

Using this,

$$\begin{aligned} E\Theta &= \frac{\Gamma(\alpha+\beta)}{\Gamma(\alpha)\Gamma(\beta)}\int_0^1 \theta^{1+\alpha-1}(1-\theta)^{\beta-1}\,d\theta \\ &= \frac{\Gamma(\alpha+\beta)}{\Gamma(\alpha)\Gamma(\beta)}\frac{\Gamma(\alpha+1)\Gamma(\beta)}{\Gamma(\alpha+\beta+1)} \\ &= \frac{\alpha}{\alpha+\beta}. \end{aligned}$$

The marginal density of $X$ in the Bayesian model is

$$\begin{aligned} q(x) &= \int p_\theta(x)\lambda(\theta)\,d\theta \\ &= \int_0^1 \binom{n}{x}\frac{\Gamma(\alpha+\beta)}{\Gamma(\alpha)\Gamma(\beta)}\theta^{x+\alpha-1}(1-\theta)^{n-x+\beta-1}\,d\theta \\ &= \binom{n}{x}\frac{\Gamma(\alpha+\beta)}{\Gamma(\alpha)\Gamma(\beta)}\frac{\Gamma(x+\alpha)\Gamma(n-x+\beta)}{\Gamma(n+\alpha+\beta)}. \end{aligned}$$

This is the mass function for the beta-binomial distribution, sometimes used in a non-Bayesian setting to model variables that exhibit more variation than would be anticipated from a binomial model. Dividing $p_\theta(x)\lambda(\theta)$ by this mass function $q(x)$,

$$\lambda(\theta|x) = \frac{\Gamma(n+\alpha+\beta)}{\Gamma(\alpha+x)\Gamma(\beta+n-x)}\theta^{x+\alpha-1}(1-\theta)^{n-x+\beta-1}, \qquad \theta \in (0,1).$$

This shows that

$$\Theta|X = x \sim \mathrm{Beta}(x+\alpha, n-x+\beta).$$

The updating necessary to find the posterior distribution from the prior and the observed data is particularly simple here; just increment $\alpha$ by the number of successes observed, and increment $\beta$ by the number of failures. Prior distributions that ensure a posterior distribution from the same class are called *conjugate*. See Problem 7.4 for a class of examples. With squared error loss,

$$\delta_\Lambda(X) = E[\Theta|X] = \frac{X+\alpha}{n+\alpha+\beta}.$$

Straightforward algebra gives

$$\delta_\Lambda(X) = \left[\frac{n}{n+\alpha+\beta}\right]\frac{X}{n} + \left[1 - \frac{n}{n+\alpha+\beta}\right]\frac{\alpha}{\alpha+\beta},$$

which shows that the Bayes estimator here is a weighted average of the UMVU estimator $X/n$ and the prior mean $E\Theta = \alpha/(\alpha+\beta)$.

*Example 7.4. Negative Binomial.* From a sequence of Bernoulli trials with success probability $\theta$, let $X$ be the number of failures before the second success. Then

$$p_\theta(x) = P_\theta(X = x) = (x+1)\theta^2(1-\theta)^x, \qquad x = 0, 1, \dots.$$

Consider estimation of $g(\Theta) = 1/\Theta$ for a Bayesian model in which $\Theta$ is uniformly distributed on $(0,1)$. Then

$$\lambda(\theta|x) \propto_\theta p_\theta(x)\lambda(\theta) \propto_\theta \theta^2(1-\theta)^x.$$

This is proportional to the density for Beta$(3, x+1)$, and so

$$\Theta|X = x \sim \text{Beta}(3, x+1).$$

The posterior mean is

$$\begin{aligned}\delta_0(x) = E[\Theta^{-1}|X = x] &= \frac{\Gamma(x+4)}{\Gamma(3)\Gamma(x+1)}\int_0^1 \theta(1-\theta)^x\,d\theta \\ &= \frac{\Gamma(x+4)\Gamma(2)\Gamma(x+1)}{\Gamma(3)\Gamma(x+1)\Gamma(x+3)} = \frac{x+3}{2}.\end{aligned}$$

Recalling from Example 5.3 that the UMVU estimator of $1/\theta$ for this model is

$$\delta_1(x) = \frac{x+2}{2},$$

we have the curious result that

$$\delta_0(X) = \delta_1(X) + \frac{1}{2}.$$

So the estimator $\delta_0$ has constant bias $b(\theta, \delta_0) = E_\theta\delta_0(X) - 1/\theta = 1/2$. With squared error loss, the risk of any estimator is its variance plus the square of its bias. Because $\delta_0$ and $\delta_1$ differ by a constant they have the same variance, and so

$$R(\theta, \delta_0) = \text{Var}_\theta(\delta_0) + 1/4 = \text{Var}_\theta(\delta_1) + 1/4 = R(\theta, \delta_1) + 1/4.$$

Thus the UMVU estimator $\delta_1$ has uniformly smaller risk than $\delta_0$! An estimator is called *inadmissible* if a competing estimator has a better[1] risk function. And an inadmissible estimators is generally not Bayes, because an estimator with a better risk function usually has smaller integrated risk. See Theorems 11.6 and 11.7. Trouble arises in this innocuous example because condition (a) in Theorem 7.1 fails. When this happens, any estimator will minimize (7.1), and Bayesian calculations may lead to a poor estimator.

## 7.3 Utility Theory[2]

In much of this book there is a basic presumption that risk or expected loss should be used to compare and judge estimators. This may be reasonably intuitive, but there is an important philosophical question regarding why expectation should play such a central role. *Utility theory* provides motivation for this approach, showing that if preferences between probability distributions obey a few basic axioms, then one distribution will be preferred over another if and only if its expected utility is greater. The treatment of utility theory here is a bit sketchy. For more details see Chapter 7 of DeGroot (1970).

Let $R$ be a set of rewards. These rewards could be numerical or monetary, but more ethereal settings in which a reward might be some degree of fame or happiness could also be envisioned. Let $\mathcal{R}$ denote all probability distributions on $(R, \mathcal{F})$, where $\mathcal{F}$ is some $\sigma$-field. The distributions $P \in \mathcal{R}$ are called "lotteries." The idea here is that if you play some lottery $P \in \mathcal{R}$ you will receive a random reward in $R$ according to the distribution $P$. Let "$\precsim$" indicate preferences among lotteries in $\mathcal{R}$. Formally, $\precsim$ should be a *complete ordering* of $\mathcal{R}$; that is, it should satisfy these conditions:

1. If $P_1$ and $P_2$ are lotteries in $\mathcal{R}$, then either $P_1 \prec P_2$, $P_2 \prec P_1$, or $P_1 \simeq P_2$. (Here $P_1 \simeq P_2$ means that $P_1 \precsim P_2$ and $P_2 \precsim P_1$; and $P_1 \prec P_2$ means that $P_1 \precsim P_2$, but $P_1 \not\simeq P_2$.)
2. If $P_1$, $P_2$, and $P_3$ are lotteries in $\mathcal{R}$ with $P_1 \precsim P_2$ and $P_2 \precsim P_3$, then $P_1 \precsim P_3$.

It is also convenient to identify a reward $r \in R$ with the degenerate probability distribution in $\mathcal{R}$ that assigns unit mass to $\{r\}$. (To ensure this is possible, the $\sigma$-field $\mathcal{F}$ must contain all singletons $\{r\}$, $r \in R$.) We can then define reward intervals

$$[r_1, r_2] = \{r \in R : r_1 \precsim r \precsim r_2\}.$$

A lottery $P \in \mathcal{R}$ is called *bounded* if $P([r_1, r_2]) = 1$ for some rewards $r_1$ and $r_2$ in $R$. Let $\mathcal{R}_B$ denote the collection of all bounded lotteries in $\mathcal{R}$.

---

[1] See Section 11.3 for a formal definition.

[2] This section covers optional material not used in later chapters.

**Definition 7.5.** *A (measurable) function $U : R \to \mathbb{R}$ is a* utility function *for* $\precsim$ *if*

$$P_1 \precsim P_2 \text{ if and only if } E_{P_1}U \le E_{P_2}U,$$

*whenever the expectations exist. (Here $E_P U = \int U\,dP$.)*

The following example shows that utility functions may or may not exist.

*Example 7.6.* Suppose $R$ contains two rewards, $a$ and $b$, and let $P_\theta$ be the lottery that gives reward $a$ with probability $\theta$. Suppose $b \prec a$. Then if $\precsim$ has a utility function $U$, $U(a)$ is larger than $U(b)$. Inasmuch as the expected utility of $P_\theta$ is

$$\int U(r)\,dP_\theta(r) = \theta U(a) + (1-\theta)U(b) = \theta\big[U(a) - U(b)\big] + U(b),$$

the expected utility of $P_\theta$ increases as $\theta$ increases. Hence $P_{\theta_1} \precsim P_{\theta_2}$ if and only if $\theta_1 \le \theta_2$. Similarly, if $a \simeq b$, then $U(a) = U(b)$ and all lotteries are equivalent under $\precsim$. But preferences between lotteries do not have to behave in this fashion. For instance, if someone views rewards $a$ and $b$ as comparable, but finds pleasure in the excitement of not knowing the reward they will receive, a preference relation in which $P_{\theta_1} \precsim P_{\theta_2}$ if and only if $|\theta_1 - 1/2| \ge |\theta_2 - 1/2|$ may be appropriate. For this preference relation there is no utility function.

Under axioms given below, utility functions will exist. The language makes extensive use of pairwise mixtures of distributions. If $P_1$ and $P_2$ are lotteries and $\alpha \in (0,1)$, then the mixture $\alpha P_1 + (1-\alpha)P_2$ can be viewed (by smoothing) as a lottery that draws from $P_1$ with probability $\alpha$ and draws from $P_2$ with probability $1-\alpha$. In particular, because we associate rewards with degenerate lotteries, $\alpha r_1 + (1-\alpha)r_2$ represents a lottery that gives reward $r_1$ with probability $\alpha$ and reward $r_2$ with probability $1-\alpha$.

A1) If $P_1$, $P_2$, and $P$ are bounded lotteries in $\mathcal{R}_B$, and if $\alpha \in (0,1)$, then $P_1 \prec P_2$ if and only if $\alpha P_1 + (1-\alpha)P \prec \alpha P_2 + (1-\alpha)P$.

It is also easy to see that $P_1 \precsim P_2$ if and only if $\alpha P_1 + (1-\alpha)P \precsim \alpha P_2 + (1-\alpha)P$. As a further consequence, if $P_1 \precsim Q_1$ and $P_2 \precsim Q_2$, all in $\mathcal{R}_B$, and $\alpha \in (0,1)$, then

$$\alpha P_1 + (1-\alpha)P_2 \precsim \alpha Q_1 + (1-\alpha)P_2 \precsim \alpha Q_1 + (1-\alpha)Q_2.$$

If $P_1 \simeq Q_1$ and $P_2 \simeq Q_2$, again all in $\mathcal{R}_B$, the reverse inequalities also hold, and

$$\alpha P_1 + (1-\alpha)P_2 \simeq \alpha Q_1 + (1-\alpha)Q_2. \tag{7.4}$$

As a final consequence of this axiom, if $r_1 \prec r_2$ are rewards in $R$, and if $\alpha$ and $\beta$ are constants in $[0,1]$, then

$$\alpha r_2 + (1-\alpha)r_1 \prec \beta r_2 + (1-\beta)r_1 \text{ if and only if } \alpha < \beta. \tag{7.5}$$

A2) If $P_1 \prec P \prec P_2$ are bounded lotteries, then there exist constants $\alpha$ and $\beta$ in $(0,1)$ such that $P \prec \alpha P_2 + (1-\alpha)P_1$ and $P \succ \beta P_2 + (1-\beta)P_1$.

The following result follows from this axiom, and is used shortly to construct a candidate utility function.

**Theorem 7.7.** *If $r_1 \precsim r \precsim r_2$ are rewards in $R$, then there exists a unique value $\nu \in [0,1]$ such that*

$$r \simeq \nu r_2 + (1-\nu)r_1.$$

*Proof.* Consider $S = \{\alpha \in [0,1] : r \prec \alpha r_2 + (1-\alpha)r_1\}$, an interval by (7.5), and let $\nu$ be the lower endpoint of $S$, $\nu = \inf S$. If $\nu r_2 + (1-\nu)r_1 \prec r$ then $\nu < 1$, and by the second axiom

$$r \succ \beta r_2 + (1-\beta)\bigl(\nu r_2 + (1-\nu)r_1\bigr) = \bigl(\nu + \beta(1-\nu)\bigr)r_2 + \bigl(1-\nu-\beta(1-\nu)\bigr)r_1$$

for some $\beta \in (0,1)$. This would imply that $\nu$ is not the lower endpoint of $S$. But if $\nu r_2 + (1-\nu)r_1 \succ r$, then $\nu > 0$, and by the second axiom

$$r \prec \alpha r_1 + (1-\alpha)\bigl(\nu r_2 + (1-\nu)r_1\bigr) = (1-\alpha)\nu r_2 + \bigl(1-(1-\alpha)\nu\bigr)r_1,$$

for some $\alpha \in (0,1)$, again contradicting $\nu = \inf S$. Thus $r \simeq \nu r_2 + (1-\nu)r_1$. Uniqueness follows from similar considerations. □

Let $s_0 \prec s_1$ be fixed rewards in $R$. Utility functions, if they exist, are not unique, for if $U$ is a utility, and if $a$ and $b$ are constants with $b > 0$, then $a + bU$ is also a utility function. From this, if a utility function exists, there will be a utility function with $U(s_0) = 0$ and $U(s_1) = 1$. The construction below gives this utility function.

Suppose $r \in [s_0, s_1]$. Then by Theorem 7.7,

$$r \simeq \nu s_1 + (1-\nu)s_0,$$

for some $\nu \in [0,1]$. If a utility function exists, then the expected utilities for the two lotteries in this equation must agree, which means that we must have

$$U(r) = \nu.$$

If instead $r \precsim s_0$, then by Theorem 7.7,

$$s_0 \simeq \nu s_1 + (1-\nu)r,$$

for some $\nu \in (0,1)$. Equating expected utilities, $0 = \nu + (1-\nu)U(r)$, and so we need

$$U(r) = -\frac{\nu}{1-\nu}.$$

Finally, if $s_1 \prec r$, then by Theorem 7.7,

$$s_1 \simeq \nu r + (1 - \nu)s_0,$$

and equating expected utilities we must have

$$U(r) = \frac{1}{\nu}.$$

The following technical axiom is needed to ensure that this function $U$ is measurable.

A3) For any $r_1$, $r_2$, and $r_3$ in $R$, and any $\alpha$ and $\beta$ in $[0, 1]$,

$$\{r \in R : \alpha r + (1 - \alpha)r_1 \precsim \beta r_2 + (1 - \beta)r_3\} \in \mathcal{F}.$$

Let $P$ be a bounded lottery, so that $P\{[r_1, r_2]\} = 1$ for some $r_1$ and $r_2$ in $R$. The final axiom concerns a two-stage lottery in which the first stage is $P$, and the second stage trades in $P$ for an equivalent mixture of $r_1$ and $r_2$. To be specific, define a function $\alpha : [r_1, r_2] \to [0, 1]$ using Theorem 7.7 so that

$$r \simeq \alpha(r)r_2 + \bigl(1 - \alpha(r)\bigr)r_1.$$

From the construction of $U$ it can be shown that

$$\alpha(r) = \frac{U(r) - U(r_1)}{U(r_2) - U(r_1)}. \tag{7.6}$$

For instance, if $s_0 \precsim r_1 \precsim r \precsim r_2 \precsim s_1$, from the construction of $U$,

$$r_1 \simeq U(r_1)s_1 + \bigl(1 - U(r_1)\bigr)s_0, \qquad r_2 \simeq U(r_2)s_1 + \bigl(1 - U(r_2)\bigr)s_0,$$

and, using (7.4),

$$\begin{aligned} &\alpha r_2 + (1 - \alpha)r_1 \\ &\quad \simeq \alpha\bigl[U(r_2)s_1 + \bigl(1 - U(r_2)\bigr)s_0\bigr] + (1 - \alpha)\bigl[U(r_1)s_1 + \bigl(1 - U(r_1)\bigr)s_0\bigr] \\ &\quad = \bigl[\alpha U(r_2) + (1 - \alpha)U(r_1)\bigr]s_1 + \bigl[1 - \alpha U(r_2) - (1 - \alpha)U(r_1)\bigr]s_0. \end{aligned}$$

Because $r \simeq U(r)s_1 + \bigl(1 - U(r)\bigr)s_0$, $r \simeq \alpha r_2 + (1 - \alpha)r_1$ when

$$U(r) = \alpha U(r_2) + (1 - \alpha)U(r_1).$$

Solving for $\alpha$ we obtain (7.6).

In the two-stage lottery, if the reward for the first stage, sampled from $P$, is $r$, then the second stage is $\alpha(r)r_2 + \bigl(1 - \alpha(r)\bigr)r_1$. Conditioning on the outcome of the first stage, this two-stage lottery gives reward $r_2$ with probability

$$\beta = \int \alpha(r)\, dP(r).$$

Otherwise, the two-stage lottery gives reward $r_1$. The final axiom asserts that under $\precsim$ this two-stage lottery is equivalent to $P$.

A4) $P \simeq \beta r_2 + (1-\beta) r_1$.

Based on the stated axioms, the final result of this section shows that the function $U$ constructed above is a utility function for bounded lotteries.

**Theorem 7.8.** *If axioms A1 through A4 hold, and $P_1$ and $P_2$ are bounded lotteries, then*

$$P_1 \precsim P_2 \text{ if and only if } E_{P_1}U \le E_{P_2}U.$$

*Proof.* Choose $r_1$ and $r_2$ so that $P_1\{[r_1, r_2]\}$ and $P_2\{[r_1, r_2]\}$ both equal one. By the fourth axiom and (7.6),

$$P_1 \simeq \left[\frac{E_{P_1}U - U(r_1)}{U(r_2) - U(r_1)}\right] r_2 + \left[\frac{U(r_2) - E_{P_1}U}{U(r_2) - U(r_1)}\right] r_1$$

and

$$P_2 \simeq \left[\frac{E_{P_2}U - U(r_1)}{U(r_2) - U(r_1)}\right] r_2 + \left[\frac{U(r_2) - E_{P_2}U}{U(r_2) - U(r_1)}\right] r_1.$$

By (7.5), $P_1 \precsim P_2$ if and only if

$$\frac{E_{P_1}U - U(r_1)}{U(r_2) - U(r_1)} \le \frac{E_{P_2}U - U(r_1)}{U(r_2) - U(r_1)},$$

which happens if and only if $E_{P_1}U \le E_{P_2}U$. □

## 7.4 Problems[3]

*1. Consider a Bayesian model in which the prior distribution for $\Theta$ is exponential with failure rate $\eta$, so that $\lambda(\theta) = \eta e^{-\eta\theta}$, $\theta > 0$. Given $\Theta = \theta$, the data $X_1, \dots, X_n$ are i.i.d. from the Poisson distribution with mean $\theta$. Determine the Bayes estimator for $\Theta$ if the loss function is $L(\theta, d) = \theta^p (d-\theta)^2$, with $p$ a fixed positive constant.

*2. Consider a Bayesian model in which the prior distribution for $\Theta$ is absolutely continuous with density $\lambda(\theta) = 1/(1+\theta)^2$, $\theta > 0$. Given $\Theta = \theta$, our datum is a single variable $X$ uniformly distributed on $(0, \theta)$. Give an equation to find the Bayes estimate $\delta_\Lambda(X)$ of $\Theta$ if the loss function is $L(\theta, d) = |d - \theta|$. Determine $P(\delta_\Lambda(X) < \Theta | X = x)$, explicitly.

*3. In a Bayesian approach to simple linear regression, suppose the intercept $\Theta_1$ and slope $\Theta_2$ of the regression line are *a priori* independent with $\Theta_1 \sim N(0, \tau_1^2)$ and $\Theta_2 \sim N(0, \tau_2^2)$. Given $\Theta_1 = \theta_1$ and $\Theta_2 = \theta_2$, data $Y_1, \dots, Y_n$ are independent with $Y_i \sim N(\theta_1 + \theta_2 x_i, \sigma^2)$, where the variance $\sigma^2$ is known, and $x_1, \dots, x_n$ are constants summing to zero, $x_1 + \cdots + x_n = 0$. Find the Bayes estimates of $\Theta_1$ and $\Theta_2$ under squared error loss.

---

[3] Solutions to the starred problems are given at the back of the book.

*4. *Conjugate prior distributions.* Let $\mathcal{P} = \{P_\theta, \theta \in \Omega\}$ be a one-parameter canonical exponential family with densities $p_\theta$ given by

$$p_\theta(x) = h(x)e^{\theta T(x) - A(\theta)}.$$

Here $\Omega$ is an interval. Let $\Lambda = \Lambda_{\alpha,\beta}$ be an absolutely continuous prior distribution with density

$$\lambda(\theta) = \begin{cases} \exp\{\alpha\theta - \beta A(\theta) - B(\alpha, \beta)\}, & \theta \in \Omega; \\ 0, & \text{otherwise,} \end{cases}$$

where

$$B(\alpha, \beta) = \log \int_\Omega \exp\{\alpha\theta - \beta A(\theta)\}\, d\theta.$$

These densities $\Lambda_{\alpha,\beta}$ form a canonical two-parameter exponential family. Let $\Xi = \{(\alpha, \beta) : B(\alpha, \beta) < \infty\}$ be the canonical parameter space. Assume for regularity that $\lambda(\theta) \to 0$ as $\theta$ approaches either end of the interval $\Omega$, regardless the value of $(\alpha, \beta) \in \Xi$.

a) With the stated regularity, $\int_\Omega \lambda'(\theta)\, d\theta = 0$. Use this to give an explicit formula for $EA'(\Theta)$ when $\Theta \sim \Lambda_{\alpha,\beta}$. (The answer should be a simple function of $\alpha$ and $\beta$.)

b) Consider a Bayesian model in which $\Theta \sim \Lambda_{\alpha,\beta}$ and given $\Theta = \theta$, $X_1, \dots, X_n$ are i.i.d. with common density $P_\theta$ from the exponential family $\mathcal{P}$. Determine the Bayes estimate of $A'(\Theta)$ under squared error loss. Show that this estimate is a weighted average of $EA'(\Theta)$ and the average $\overline{T} = \big[T(X_1) + \cdots + T(X_n)\big]/n$.

c) Demonstrate the ideas in parts (a) and (b) when $P_\theta$ is the exponential distribution with failure rate $\theta$ and mean $1/\theta$: identify the prior distributions $\Lambda_{\alpha,\beta}$, and give an explicit formula for the Bayes estimator of the mean $1/\theta$.

5. Consider an autoregressive model in which $X_1 \sim N\big(\theta, \sigma^2/(1 - \rho^2)\big)$ and the conditional distribution of $X_{j+1}$ given $X_1 = x_1, \dots, X_j = x_j$ is $N\big(\theta + \rho(x_j - \theta), \sigma^2\big)$, $j = 1, \dots, n - 1$. Suppose the values for $\rho$ and $\sigma$ are fixed constants, and consider Bayesian estimation with $\Theta \sim N(0, \tau^2)$. Find Bayes estimates for $\Theta$ and $\Theta^2$ under squared error loss.

*6. Consider a Bayesian model in which the random parameter $\Theta$ has a Bernoulli prior distribution with success probability $1/2$, so $P(\Theta = 0) = P(\Theta = 1) = 1/2$. Given $\Theta = 0$, data $X$ has density $f_0$, and given $\Theta = 1$, $X$ has density $f_1$.

a) Find the Bayes estimate of $\Theta$ under squared error loss.

b) Find the Bayes estimate of $\Theta$ if $L(\theta, d) = I\{\theta \neq d\}$ (called *zero-one loss*).

*7. Consider Bayesian estimation in which the parameter $\Theta$ has a standard exponential distribution, so $\lambda(\theta) = e^{-\theta}$, $\theta > 0$, and given $\Theta = \theta$, $X_1, \dots, X_n$ are i.i.d. from an exponential distribution with failure rate $\theta$. Determine the Bayes estimator of $\Theta$ if the loss function is $L(\theta, d) = (d - \theta)^2/d$.

8. Consider a Bayesian model in which the prior distribution for $\Theta$ is standard exponential and the density for $X$ given $\Theta = \theta$ is

$$p_\theta(x) = \begin{cases} e^{\theta - x}, & x > \theta; \\ 0, & \text{otherwise.} \end{cases}$$

   a) Find the marginal density for $X$ and $EX$ in the Bayesian model.
   b) Find the Bayes estimator for $\Theta$ under squared error loss. (Assume $X > 0$.)
9. Suppose $\Theta \sim \Lambda$ and $X|\Theta = \theta \sim P_\theta$, and let $f$ be a nonnegative measurable function. Use smoothing to write $Ef(\Theta, X)$ as an iterated integral. (This calculation shows that specification of a Bayesian model in this fashion determines the joint distribution of $X$ and $\Theta$.)
10. Suppose we observe two independent observations, $(X_1, Y_1)$ and $(X_2, Y_2)$ from an absolutely continuous bivariate distribution with density

$$\frac{\sqrt{1-\theta^2}}{2\pi} \exp\left[-\frac{1}{2}(x^2 + y^2 - 2\theta xy)\right].$$

    Find the Bayes estimate for $\Theta$ under squared error loss if the prior distribution is uniform on $(-1, 1)$.
11. Consider a Bayesian model in which the prior distribution for $\Theta$ is uniform on $(0, 1)$ and given $\Theta = \theta$, $X_i$, $i \geq 1$, are i.i.d. Bernoulli with success probability $\theta$. Find

$$P(X_{n+1} = 1 | X_1, \dots, X_n).$$

12. *Bayesian prediction.*
    a) Let $X$ and $Y$ be jointly distributed, with $X$ a random variable and $Y$ a random vector. Suppose we are interested in predicting $X$ from $Y$. The efficacy of a predictor $f(Y)$ might be measured using the expected squared error, $E\big(X - f(Y)\big)^2$. Use a smoothing argument to find the function $f$ minimizing this quantity.
    b) Consider a Bayesian model in which $\Theta$ is a random parameter, and given $\Theta = \theta$, random variables $X_1, \dots, X_{n+1}$ are i.i.d. from a distribution $P_\theta$ with mean $\mu(\theta)$. With squared error loss, the best estimator of $\mu(\theta)$ based on $X_1, \dots, X_n$ is

$$\hat{\mu} = E\big[\mu(\Theta) \bigm| X_1, \dots, X_n\big].$$

    Show that $\hat{\mu}$ is also the best predictor for $X_{n+1}$ based on $Y = (X_1, \dots, X_n)$. You can assume that $\Theta$ is absolutely continuous, and that the family $\mathcal{P} = \{P_\theta : \theta \in \Omega\}$ is dominated with densities $p_\theta$, $\theta \in \Omega$.
13. Consider a Bayesian model in which $\Theta$ is absolutely continuous with density

$$\lambda(\theta) = \begin{cases} \dfrac{e^{-1/\theta}}{\theta^2}, & \theta > 0; \\ 0, & \text{otherwise,} \end{cases}$$

and given $\Theta = \theta$, $X_1, \dots, X_n$ are i.i.d. $N(0, \theta)$. Find the Bayes estimator for $\Theta$ under squared error loss.

14. Consider a Bayesian model in which given $\Theta = \theta$, $X_1, \dots, X_n$ are i.i.d. from a Bernoulli distribution with mean $\theta$.
    a) Let $\bigl(\pi(1), \dots, \pi(n)\bigr)$ be a permutation of $(1, \dots, n)$. Show that

    $$\bigl(X_{\pi(1)}, \dots, X_{\pi(n)}\bigr) \text{ and } \bigl(X_1, \dots, X_n\bigr)$$

    have the same distribution. When this holds the variables involved are said to be *exchangeable*.
    b) Show that $\operatorname{Cov}(X_i, X_j) \geq 0$. When will this covariance be zero?
15. Consider a Bayesian model in which $\Theta$ is absolutely continuous with density

$$\lambda(\theta) = \begin{cases} \dfrac{4\theta^3}{(1+\theta)^5}, & \theta > 0; \\ 0, & \text{otherwise,} \end{cases}$$

and given $\Theta = \theta > 0$, data $X$ and $Y$ are absolutely continuous with density

$$p_\theta(x, y) = \begin{cases} 1/\theta^3, & |x| < \theta y < \theta^2; \\ 0, & \text{otherwise,} \end{cases}$$

Find the Bayes estimator of $\Theta$ under squared error loss.

*16. (For fun) Let $X$ and $Y$ be independent Cauchy variables with location $\theta$.
    a) Show that $X$ and the average $A = (X + Y)/2$ have the same distribution.
    b) Show that $P_\theta(|A - \theta| < |X - \theta|) > 1/2$, so that $A$ is more likely to be closer to $\theta$ that $X$. (Hint: Graph the region in the plane where the event in question occurs.)

# 8

# Large-Sample Theory

To this point, most of the statistical results in this book concern properties that hold in some exact sense. An estimator is either sufficient or not, unbiased or not, Bayes or not. If exact properties are impractical or not available, statisticians often rely on approximations. This chapter gives several of the most basic results from probability theory used to derive approximations. Several notions of convergence for random variables and vectors are introduced, and various limit theorems are presented. These results are used in this chapter and later to study and compare the performance of various estimators in large samples.

## 8.1 Convergence in Probability

Our first notion of convergence holds if the variables involved are close to their limit with high probability.

**Definition 8.1.** *A sequence of random variables $Y_n$* converges in probability *to a random variable $Y$ as $n \to \infty$, written*

$$Y_n \xrightarrow{p} Y,$$

*if for every $\epsilon > 0$,*

$$P\big(|Y_n - Y| \geq \epsilon\big) \to 0$$

*as $n \to \infty$.*

**Theorem 8.2 (Chebyshev's Inequality).** *For any random variable $X$ and any constant $a > 0$,*

$$P\big(|X| \geq a\big) \leq \frac{EX^2}{a^2}.$$

R.W. Keener, *Theoretical Statistics: Topics for a Core Course*, Springer Texts in Statistics, DOI 10.1007/978-0-387-93839-4_8, © Springer Science+Business Media, LLC 2010

*Proof.* Regardless of the value of $X$,

$$I\{|X| \geq a\} \leq X^2/a^2.$$

The result follows by taking expectations. $\square$

**Proposition 8.3.** *If* $E(Y_n - Y)^2 \to 0$ *as* $n \to \infty$, *then* $Y_n \xrightarrow{p} Y$.

*Proof.* By Chebyshev's inequality, for any $\epsilon > 0$,

$$P\bigl(|Y_n - Y| \geq \epsilon\bigr) \leq \frac{E(Y_n - Y)^2}{\epsilon^2} \to 0. \qquad \square$$

*Example 8.4.* Suppose $X_1, X_2, \dots$ are i.i.d., with mean $\mu$ and variance $\sigma^2$, and let $\overline{X}_n = (X_1 + \cdots + X_n)/n$. Then

$$E(\overline{X}_n - \mu)^2 = \mathrm{Var}(\overline{X}_n) = \sigma^2/n \to 0,$$

and so $\overline{X}_n \xrightarrow{p} \mu$ as $n \to \infty$. In fact, $\overline{X}_n \xrightarrow{p} \mu$ even when $\sigma^2 = \infty$, provided $E|X_i| < \infty$. This result is called the *weak law of large numbers.*

**Proposition 8.5.** *If* $f$ *is continuous at* $c$ *and if* $Y_n \xrightarrow{p} c$, *then* $f(Y_n) \xrightarrow{p} f(c)$.

*Proof.* Because $f$ is continuous at $c$, given any $\epsilon > 0$, there exists $\delta_\epsilon > 0$ such that $|f(y) - f(c)| < \epsilon$ whenever $|y - c| < \delta_\epsilon$. Thus

$$P\bigl(|Y_n - c| < \delta_\epsilon\bigr) \leq P\bigl(|f(Y_n) - f(c)| < \epsilon\bigr),$$

which implies

$$P\bigl(|f(Y_n) - f(c)| \geq \epsilon\bigr) \leq P\bigl(|Y_n - c| \geq \delta_\epsilon\bigr) \to 0. \qquad \square$$

In statistics there is a family of distributions of interest, indexed by a parameter $\theta \in \Omega$, and the symbol $\xrightarrow{P_\theta}$ is used to denote convergence in probability with $P_\theta$ as the underlying probability measure.

**Definition 8.6.** *A sequence of estimators* $\delta_n$, $n \geq 1$, *is* consistent *for* $g(\theta)$ *if for any* $\theta \in \Omega$,

$$\delta_n \xrightarrow{P_\theta} g(\theta)$$

*as* $n \to \infty$.

If $R(\theta, \delta_n) = E_\theta\bigl(\delta_n - g(\theta)\bigr)^2$ is the mean squared error, or risk, of $\delta_n$ under squared error loss, then by Proposition 8.3, $\delta_n$ will be consistent if $R(\theta, \delta_n) \to 0$ as $n \to \infty$, for any $\theta \in \Omega$. Letting $b_n(\theta) = E_\theta\delta_n - g(\theta)$, called the *bias* of $\delta_n$,

$$R(\theta, \delta_n) = b_n^2(\theta) + \mathrm{Var}_\theta(\delta_n),$$

and so sufficient conditions for consistency are that $b_n(\theta) \to 0$ and $\mathrm{Var}_\theta(\delta_n) \to 0$ as $n \to \infty$, for all $\theta \in \Omega$.

Convergence in probability extends directly to higher dimensions. If $Y$, $Y_1$, $Y_2, \dots$ are random vectors in $\mathbb{R}^p$, then $Y_n$ converges in probability to $Y$, written $Y_n \xrightarrow{p} Y$, if for every $\epsilon > 0$, $P(\|Y_n - Y\| > \epsilon) \to 0$ as $n \to \infty$. Equivalently, $Y_n \xrightarrow{p} Y$ if $[Y_n]_i \xrightarrow{p} [Y]_i$ as $n \to \infty$ for $i = 1, \dots, p$. Proposition 8.5 also holds as stated, with the same proof, for random vectors $Y_n$ and $c \in \mathbb{R}^p$, with $f$ a vector-valued function, $f : \mathbb{R}^p \to \mathbb{R}^q$. For instance, since addition and multiplication are continuous functions from $\mathbb{R}^2 \to \mathbb{R}$, if $X_n \xrightarrow{p} a$ and $Y_n \xrightarrow{p} b$ as $n \to \infty$, then

$$X_n + Y_n \xrightarrow{p} a + b \text{ and } X_n Y_n \xrightarrow{p} ab, \tag{8.1}$$

as $n \to \infty$.

## 8.2 Convergence in Distribution

If a sequence of estimators $\delta_n$ is consistent for $g(\theta)$, then the distribution of the error $\delta_n - g(\theta)$ must concentrate around zero as $n$ increases. But convergence in probability will not tell us how rapidly this concentration occurs or the shape of the error distribution after suitable magnification. For this, the following notion of convergence in distribution is more appropriate.

**Definition 8.7.** *A sequence of random variables $Y_n$, $n \ge 1$, with cumulative distribution functions $H_n$,* converges in distribution *(or law) to a random variable $Y$ with cumulative distribution function $H$ if*

$$H_n(y) \to H(y)$$

*as $n \to \infty$ whenever $H$ is continuous at $y$. For notation we write $Y_n \Rightarrow Y$ or $Y_n \Rightarrow P_Y$.*

One aspect of this definition that may seem puzzling at first is that pointwise convergence of the cumulative distribution functions only has to hold at continuity points of $H$. Here is a simple example that should make this seem more natural.

*Example 8.8.* Suppose $Y_n = 1/n$, a degenerate random variable, and that $Y$ is always zero. Then

$$H_n(y) = P(Y_n \le y) = I\{1/n \le y\}.$$

If $y > 0$, then $H_n(y) = I\{1/n \le y\} \to 1$ as $n \to \infty$, for eventually $1/n$ will be less than $y$. If $y \le 0$, then $H_n(y) = I\{1/n \le y\} = 0$ for all $n$, and so $H_n(y) \to 0$ as $n \to \infty$. Because $H(y) = P(Y \le y) = I\{0 \le y\}$, comparisons with the limits just obtained show that $H_n(y) \to H(y)$ if $y \ne 0$. But $H_n(0) = 0 \to 0 \ne 1 = H(0)$. In this example, $Y_n \Rightarrow Y$, but the cumulative distribution functions $H_n(y)$ do not converge to $H(y)$ when $y = 0$, a discontinuity point of $H$.

**Theorem 8.9.** *Convergence in distribution, $Y_n \Rightarrow Y$, holds if and only if $Ef(Y_n) \to Ef(Y)$ for all bounded continuous functions $f$.*

*Remark 8.10.* The convergence of expectations in this theorem is often taken as the definition for convergence in distribution. One advantage of this as a definition is that it generalizes easily to random vectors. Extensions to more abstract objects, such as random functions, are even possible.

**Corollary 8.11.** *If $g$ is a continuous function and $Y_n \Rightarrow Y$, then*

$$g(Y_n) \Rightarrow g(Y).$$

*Proof.* If $f$ is bounded and continuous, then $f \circ g$ is also bounded and continuous. Since $Y_n \Rightarrow Y$,

$$Ef\bigl(g(Y_n)\bigr) \to Ef\bigl(g(Y)\bigr).$$

Because $f$ is arbitrary, this shows that the second half of Theorem 8.9 holds for the induced sequence $g(Y_n)$ and $g(Y)$. So by the equivalence, $g(Y_n) \Rightarrow g(Y)$. ☐

For convergence in distribution, the central limit theorem is our most basic tool. For a derivation and proof, see Appendix A.7, or any standard text on probability.

**Theorem 8.12 (Central Limit Theorem).** *Suppose $X_1, X_2, \dots$ are i.i.d. with common mean $\mu$ and variance $\sigma^2$. Take $\overline{X}_n = (X_1 + \dots + X_n)/n$. Then*

$$\sqrt{n}(\overline{X}_n - \mu) \Rightarrow N(0, \sigma^2).$$

As an application of this result, let $H_n$ denote the cumulative distribution function of $\sqrt{n}(\overline{X}_n - \mu)$ and note that

$$\begin{aligned} P(\mu - a/\sqrt{n} < \overline{X}_n \le \mu + a/\sqrt{n}) &= P\bigl(-a < \sqrt{n}(\overline{X}_n - \mu) \le a\bigr) \\ &= H_n(a) - H_n(-a) \\ &\to \Phi(a/\sigma) - \Phi(-a/\sigma). \end{aligned}$$

This information about the distribution of $\overline{X}_n$ from the central limit theorem is more detailed than information from the weak law of large numbers, that $\overline{X}_n \xrightarrow{p} \mu$.

The central limit theorem is certainly one of the most useful and celebrated results in probability and statistics, and it has been extended in numerous ways. Theorems 9.27 and 9.40 provide extensions to averages of i.i.d. random vectors and martingales, respectively. Other extensions concern situations in which the summands are independent but from different distributions or weakly dependent in a suitable sense. In addition, some random variables will be approximately normal because their difference from a variable in one of these central limit theorems converges to zero, an approach used repeatedly later in this book. Results bounding the error in the central limit theorem have also been derived. With the assumptions of Theorem 8.12, the Berry–Esséen theorem, given as Theorem 16.5.1 of Feller (1971), states that

$$\bigl|P\bigl(\sqrt{n}(\overline{X}_n - \mu) \le x\bigr) - \Phi(x/\sigma)\bigr| \le \frac{3E|X_1 - \mu|^3}{\sigma^3\sqrt{n}}. \tag{8.2}$$

The next result begins to develop a calculus for convergence of random variables combining convergence in distribution with convergence in probability.

**Theorem 8.13.** *If $Y_n \Rightarrow Y$, $A_n \xrightarrow{p} a$, and $B_n \xrightarrow{p} b$, then*

$$A_n + B_n Y_n \Rightarrow a + bY.$$

The central limit theorem stated only provides direct information about distributions of averages. Many estimators in statistics are not exactly averages, but can be related to averages in some fashion. In some of these cases, clever use of the central limit theorem still provides a limit theorem for an estimator's distribution. A first possibility would be for variables that are smooth functions of an average and can be written as $f(\overline{X}_n)$. The Taylor approximation

$$f(\overline{X}_n) \approx f(\mu) + f'(\mu)(\overline{X}_n - \mu)$$

with the central limit theorem motivates the following proposition.

**Proposition 8.14 (Delta Method).** *With the assumptions in the central limit theorem, if $f$ is differentiable at $\mu$, then*

$$\sqrt{n}\bigl(f(\overline{X}_n) - f(\mu)\bigr) \Rightarrow N\bigl(0, [f'(\mu)]^2\sigma^2\bigr).$$

*Proof.* For convenience, let us assume that $f$ has a continuous derivative[1] and write

$$f(\overline{X}_n) = f(\mu) + f'(\mu_n)(\overline{X}_n - \mu),$$

where $\mu_n$ is an intermediate point lying between $\overline{X}_n$ and $\mu$. Since $|\mu_n - \mu| \le |\overline{X}_n - \mu|$ and $\overline{X}_n \xrightarrow{p} \mu$, $\mu_n \xrightarrow{p} \mu$, and since $f'$ is continuous, $f'(\mu_n) \xrightarrow{p} f'(\mu)$ by Proposition 8.5. If $Z \sim N(0, \sigma^2)$, then $\sqrt{n}(\overline{X}_n - \mu) \Rightarrow Z \sim N(0, \sigma^2)$ by the central limit theorem. Thus by Theorem 8.13,

$$\sqrt{n}\bigl(f(\overline{X}_n) - f(\mu)\bigr) = f'(\mu_n)\bigl[\sqrt{n}(\overline{X}_n - \mu)\bigr] \Rightarrow f'(\mu)Z \sim N\bigl(0, [f'(\mu)]^2\sigma^2\bigr).$$

This use of Taylor's theorem to approximate distributions is called the *delta method.* □

---

[1] A proof under the stated condition takes a bit more care; one approach is given in the discussion following Proposition 8.24.

By Theorem 8.9, if $X_n \Rightarrow X$ and $f$ is bounded and continuous, $Ef(X_n) \to Ef(X)$. If $f$ is continuous but unbounded, convergence of $Ef(X_n)$ may fail. The theorem below shows that convergence will hold if the variables are uniformly integrable according to the following definition.

**Definition 8.15.** *Random variables $X_n$, $n \ge 1$, are* uniformly integrable *if*

$$\sup_{n\ge 1} E\bigl[|X_n| I\{|X_n| \ge t\}\bigr] \to 0,$$

*as $t \to \infty$.*

Because $E|X_n| \le t + E\bigl[|X_n|I\{|X_n| \ge t\}\bigr]$, if $\sup_{n\ge 1} E\bigl[|X_n|I\{|X_n| \ge t\}\bigr]$ is finite for some $t$, $\sup_n E|X_n| < \infty$. Thus uniform integrability implies $\sup_n E|X_n| < \infty$. But the converse can fail. If $Y_n \sim \text{Bernoulli}(1/n)$ and $X_n = nY_n$, then $E|X_n| = 1$ for all $n$, but the variables $X_n$, $n \ge 1$, are not uniformly integrable.

**Theorem 8.16.** *If $X_n \Rightarrow X$, then $E|X| \le \liminf E|X_n|$. If $X_n$, $n \ge 1$, are uniformly integrable and $X_n \Rightarrow X$, then $EX_n \to EX$. If $X$ and $X_n$, $n \ge 1$, are nonnegative and integrable with $X_n \Rightarrow X$ and $EX_n \to EX$, then $X_n$, $n \ge 1$, are uniformly integrable.*

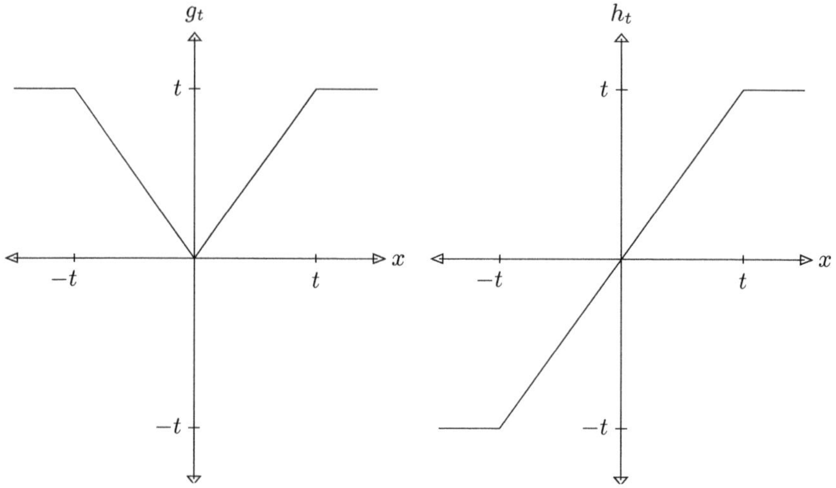


**Fig. 8.1.** Functions $g_t$ and $h_t$.

*Proof.* For $t > 0$, define functions[2] $g_t(x) = |x| \wedge t$ and $h_t(x) = -t \vee (x \wedge t)$, pictured in Figure 8.1. These functions are bounded and continuous, and so if

[2] Here $x \wedge y \overset{\text{def}}{=} \min\{x, y\}$ and $x \vee y \overset{\text{def}}{=} \max\{x, y\}$.

$X_n \Rightarrow X$, $Eg_t(X_n) \to Eg_t(X)$ and $Eh_t(X_n) \to Eh_t(X)$. For the first assertion in the theorem,

$$\liminf E|X_n| \geq \liminf E|X_n| \wedge t = E|X| \wedge t,$$

and the right-hand side increases to $E|X|$ as $t \to \infty$ by monotone convergence (Problem 1.25).

For the second assertion, by uniform integrability and the first result, $E|X| < \infty$. Since $|X_n - h_t(X)| \leq |X_n| I\{|X_n| \geq t\}$,

$$\begin{aligned}\limsup |EX_n - EX| &\leq \limsup |Eh_t(X_n) - Eh_t(X)| + E\,|X - h_t(X)| \\ &\quad + \sup E\,|X_n - h_t(X_n)| \\ &\leq E\,|X - h_t(X)| + \sup E\left[|X_n| I\{|X_n| \geq t\}\right],\end{aligned}$$

which decreases to zero as $t \to \infty$.

For the final assertion, since the variables are nonnegative with $EX_n \to EX$ and $Eg_t(X_n) \to Eg_t(X)$, then for any $t > 0$,

$$E(X_n - t)^+ = EX_n - Eg_t(X_n) \to EX - Eg_t(X) = E(X - t)^+.$$

Using this, since $xI\{x \geq 2t\} \leq 2(x - \alpha)^+$, $x > 0$,

$$\limsup E\big[X_n I\{X_n \geq 2t\}\big] \leq \limsup 2E(X_n - t)^+ = 2E(X - t)^+.$$

By dominated convergence $E(X - t)^+ \to 0$ as $t \to \infty$. Thus

$$\lim_{t\to\infty} \limsup E\big[|X_n| I\{X_n \geq 2t\}\big].$$

Uniform integrability follows fairly easily from this (see Problem 8.9). ☐

## 8.3 Maximum Likelihood Estimation

Suppose data vector $X$ has density $p_\theta$. This density, evaluated at $X$ and viewed as a function of $\theta$, $L(\theta) = p_\theta(X)$, is called the *likelihood function*, and the value $\hat{\theta} = \hat{\theta}(X)$ maximizing $L(\cdot)$ is called the *maximum likelihood estimator* of $\theta$. The maximum likelihood estimator of $g(\theta)$ is defined[3] to be $g(\hat{\theta})$. For explicit calculation it is often convenient to maximize the log-likelihood function, $l(\theta) = \log L(\theta)$. instead of $L(\cdot)$.

*Example 8.17.* Suppose the density for our data $X$ comes from a canonical one-parameter exponential family with density

$$p_\eta(x) = \exp\big\{\eta T(x) - A(\eta)\big\}h(x).$$

[3] It is not hard to check that this definition remains consistent if different parameters are used to specify the model.

Then the maximum likelihood estimator $\hat{\eta}$ of $\eta$ maximizes

$$l(\eta) = \log p_\eta(X) = \eta T - A(\eta) + \log h(X).$$

Because $l''(\eta) = -A''(\eta) = -\mathrm{Var}_\eta(T) < 0$, the $\hat{\eta}$ is typically[4] the unique solution of

$$0 = l'(\eta) = T - A'(\eta).$$

Letting $\psi$ denote the inverse function of $A'$,

$$\hat{\eta} = \psi(T).$$

If our data are a random sample $X_1, \dots, X_n$ with common density $p_\eta$, then the joint density is $\prod_{i=1}^n p_\eta(x_i)$ and the log-likelihood is

$$l(\eta) = \eta \sum_{i=1}^n T(X_i) - nA(\eta) + \log \prod_{i=1}^n h(X_i).$$

The maximum likelihood estimator $\hat{\eta}$ solves

$$0 = l'(\eta) = \sum_{i=1}^n T(X_i) - nA'(\eta),$$

and so

$$\hat{\eta} = \psi(\overline{T}), \quad \text{where } \overline{T} = \frac{1}{n}\sum_{i=1}^n T(X_i).$$

It is interesting to note that the maximum likelihood estimator for the mean of $T$, $E_\eta T(X_i) = A'(\eta)$, is

$$A'(\hat{\eta}) = A'\big(\psi(\overline{T})\big) = \overline{T}.$$

The maximum likelihood estimator here is an unbiased function of the complete sufficient statistic; therefore, it is also UMVU. But in general maximum likelihood estimators may have some bias.

Since the maximum likelihood estimator in this example is a function of an average of i.i.d. variables, its asymptotic distribution can be determined using the delta method, Proposition 8.14. By the implicit function theorem, $\psi$ has derivative $(1/A'') \circ \psi$. This derivative evaluated at $A'(\eta) = E_\eta T(X_i)$ is

$$\frac{1}{A''\big[\psi\big(A'(\eta)\big)\big]} = \frac{1}{A''(\eta)}.$$

Because $\mathrm{Var}_\eta\big(T(X_i)\big) = A''(\eta)$, by Proposition 8.14

[4] Examples are possible in which $l(\cdot)$ is strictly increasing or strictly decreasing. The equation here holds whenever $T \in \eta'(\Xi)$.

$$\sqrt{n}(\hat{\eta} - \eta) \Rightarrow N\big(0, 1/A''(\eta)\big). \tag{8.3}$$

Note that since the Fisher information from each observation is $A''(\eta)$, by the Cramér–Rao lower bound, if an estimator $\tilde{\eta}$ is unbiased for $\eta$, then

$$\mathrm{Var}_\eta\big[\sqrt{n}(\tilde{\eta} - \eta)\big] = n\mathrm{Var}_\eta(\tilde{\eta}) \geq \frac{1}{A''(\eta)}.$$

So (8.3) can be interpreted as showing that $\hat{\eta}$ achieves the Cramér–Rao lower bound in an asymptotic sense. For this reason, $\hat{\eta}$ is considered asymptotically efficient. A rigorous treatment of asymptotic efficiency is delicate and technical; a few of the main developments are given in Section 16.6.

## 8.4 Medians and Percentiles

Let $X_1, \ldots, X_n$ be random variables. These variables, arranged in increasing order, $X_{(1)} \leq X_{(2)} \leq \cdots \leq X_{(n)}$, are called *order statistics*. The first order statistic $X_{(1)}$ is the smallest value, $X_{(1)} = \min\{X_1, \ldots, X_n\}$, and the last order statistic $X_{(n)}$ is the largest value, $X_{(n)} = \max\{X_1, \ldots, X_n\}$. The *median* is the middle order statistic when $n$ is odd, or (by convention) the average of the two middle order statistics when $n$ is even:

$$\tilde{X} = \begin{cases} X_{(m)}, & n = 2m - 1; \\ \frac{1}{2}(X_{(m)} + X_{(m+1)}), & n = 2m. \end{cases}$$

The median $\tilde{X}$ and mean $\overline{X}$ are commonly used to describe the center or overall location of the variables $X_1, \ldots, X_n$. One possible advantage for the median is that it will not be influenced by a few extreme values. For instance, if the data are $(1, 2, 3, 4, 5)$, then both $\tilde{X}$ and $\overline{X}$ are 3. But if the data are $(1, 2, 3, 4, 500)$, $\tilde{X}$ is still 3, but $\overline{X} = 102$. If we view them as estimators, it is also natural to want to compare the error distributions of $\overline{X}$ and $\tilde{X}$. For a random sample, the error distribution of $\overline{X}$ can be approximated using the central limit theorem. In what follows, we derive an analogous result for $\tilde{X}$.

Assume now that $X_1, X_2, \ldots$ are i.i.d. with common cumulative distribution function $F$, and let $\tilde{X}_n$ be the median of the first $n$ observations. For regularity, assume that $F$ has a unique median $\theta$, so $F(\theta) = 1/2$, and that $F'(\theta)$ exists and is finite and positive. Let us try to approximate

$$P\big(\sqrt{n}(\tilde{X}_n - \theta) \leq a\big) = P(\tilde{X}_n \leq \theta + a/\sqrt{n}).$$

Define

$$S_n = \#\{i \leq n : X_i \leq \theta + a/\sqrt{n}\}.$$

The key to this derivation is the observation that $\tilde{X}_n \leq \theta + a/\sqrt{n}$ if and only if $S_n \geq m$. Also, by viewing observation $i$ as a success if $X_i \leq \theta + a/\sqrt{n}$, it is evident that

$$S_n \sim \text{Binomial}\big(n, F(\theta + a/\sqrt{n})\big).$$

The next step involves normal approximation for the distribution of $S_n$. First note that if $Y_n \sim \text{Binomial}(n,p)$, then $Y_n/n$ can be viewed as the average of $n$ i.i.d. Bernoulli variables. Therefore by the central limit theorem,

$$\sqrt{n}\left(\frac{Y_n}{n} - p\right) = \frac{Y_n - np}{\sqrt{n}} \Rightarrow N\left(0, p(1-p)\right),$$

and hence

$$\begin{aligned} P\left(\frac{Y_n - np}{\sqrt{n}} > y\right) &= 1 - P\left(\frac{Y_n - np}{\sqrt{n}} \le y\right) \\ &\to 1 - \Phi\left(\frac{y}{\sqrt{p(1-p)}}\right) = \Phi\left(\frac{-y}{\sqrt{p(1-p)}}\right), \end{aligned}$$

as $n \to \infty$. In fact, this approximation for the binomial distribution holds uniformly in $y$ and uniformly for $p$ in any compact subset of $(0,1)$.[5]

The normal approximation for the binomial distribution just discussed gives

$$\begin{aligned} &P\big(\sqrt{n}(\tilde{X}_n - \theta) \le a\big) \\ &\quad = P(S_n > m-1) \\ &\quad = P\left(\frac{S_n - nF(\theta + a/\sqrt{n})}{\sqrt{n}} > \frac{m - 1 - nF(\theta + a/\sqrt{n})}{\sqrt{n}}\right) \\ &\quad = \Phi\left(\frac{\big[nF(\theta + a/\sqrt{n}) - m + 1\big]/\sqrt{n}}{\sqrt{F(\theta + a/\sqrt{n})\big(1 - F(\theta + a/\sqrt{n})\big)}}\right) + o(1). \end{aligned} \tag{8.4}$$

Here "$o(1)$" is used to denote a sequence that tends to zero as $n \to \infty$. See Section 8.6 for a discussion of notation and various notions of scales of magnitude. Since $F$ is continuous at $\theta$,

$$\sqrt{F(\theta + a/\sqrt{n})\big(1 - F(\theta + a/\sqrt{n})\big)} \to 1/2,$$

as $n \to \infty$. And because $F$ is differentiable at $\theta$,

$$\begin{aligned} \frac{nF(\theta + a/\sqrt{n}) - m + 1)}{\sqrt{n}} &= a\frac{F(\theta + a/\sqrt{n}) - F(\theta)}{a/\sqrt{n}} + \frac{nF(\theta) - m + 1}{\sqrt{n}} \\ &= a\frac{F(\theta + a/\sqrt{n}) - F(\theta)}{a/\sqrt{n}} + \frac{1}{2\sqrt{n}} \to aF'(\theta). \end{aligned}$$

---

[5] "Uniformity" here means that the difference between the two sides will tend to zero as $n \to \infty$, even if $y$ and $p$ both vary with $n$, provided $p$ stays away from zero and one ($\limsup p < 1$ and $\liminf p > 0$). This can be easily proved using the Berry–Esséen bound (8.2).

Since the numerator and denominator of the argument of $\Phi$ in (8.4) both converge,

$$P\big(\sqrt{n}(\tilde{X}_n - \theta) \le a\big) \to \Phi\big(2aF'(\theta)\big).$$

The limit here is the cumulative distribution function for the normal distribution with mean zero and variance $1/\big(4[F'(\theta)]^2\big)$ evaluated at $a$, and so

$$\sqrt{n}(\tilde{X}_n - \theta) \Rightarrow N\left(0, \frac{1}{4[F'(\theta)]^2}\right). \tag{8.5}$$

A similar derivation leads to the following central limit theorem for other quantiles.

**Theorem 8.18.** *Let $X_1, X_2, \dots$ be i.i.d. with common cumulative distribution function $F$, let $\gamma \in (0,1)$, and let $\tilde{\theta}_n$ be the $\lfloor \gamma n \rfloor$th order statistic for $X_1, \dots, X_n$ (or a weighted average of the $\lfloor \gamma n \rfloor$th and $\lceil \gamma n \rceil$th order statistics).*[6] *If $F(\theta) = \gamma$, and if $F'(\theta)$ exists and is finite and positive, then*

$$\sqrt{n}(\tilde{\theta}_n - \theta) \Rightarrow N\left(0, \frac{\gamma(1-\gamma)}{\big[F'(\theta)\big]^2}\right),$$

*as $n \to \infty$.*

## 8.5 Asymptotic Relative Efficiency

A comparison of the mean and median will only be natural if they both estimate the same parameter. In a location family this will happen naturally if the error distribution is symmetric. So let us assume that our data are i.i.d. and have common density $f(x-\theta)$ with $f$ symmetric about zero, $f(u) = f(-u)$, $u \in \mathbb{R}$. Then $P_\theta(X_i < \theta) = P_\theta(X_i > \theta) = 1/2$, and $E_\theta X_i = \theta$ (provided the mean exists). By the central limit theorem,

$$\sqrt{n}(\overline{X}_n - \theta) \Rightarrow N(0, \sigma^2),$$

where

$$\sigma^2 = \int x^2 f(x)\, dx,$$

and by (8.5),

$$\sqrt{n}(\tilde{X}_n - \theta) \Rightarrow N\left(0, \frac{1}{4f^2(0)}\right).$$

(Here we naturally take $f(0) = F'(0)$.) Suppose $f$ is the standard normal density, $f(x) = e^{-x^2/2}/\sqrt{2\pi}$. Then $\sigma^2 = 1$ and $1/\big(4f^2(0)\big) = \pi/2$. Since the

---

[6] Here $\lfloor x \rfloor$, called the *floor* of $x$, is the largest integer $y$ with $y \le x$. Also, $\lceil x \rceil$ is the smallest integer $y \ge x$, called the *ceiling* of $x$.

variance of the limiting distribution is larger for the median than the mean, the median is less efficient than the mean. To understand the import of this difference in efficiency, define $m = m_n = \lfloor \pi n/2 \rfloor$, and note that $\sqrt{n/m} \to \sqrt{2/\pi}$ as $n \to \infty$. Using Theorem 8.13,

$$\sqrt{n}(\tilde{X}_m - \theta) = \sqrt{\frac{n}{m}}\sqrt{m}(\tilde{X}_m - \theta) \Rightarrow N(0,1).$$

This shows that the error distribution for the median of $m$ observations is approximately the same as the error distribution for the mean of $n$ observations. As $n \to \infty$, $m/n \to \pi/2$, and this limiting ratio $\pi/2$ is called the *asymptotic relative efficiency* (ARE) of the mean $\overline{X}_n$ with respect to the median $\tilde{X}_n$. In general, if $\hat{\theta}_n$ and $\tilde{\theta}_n$ are sequences of estimators, and if

$$\sqrt{n}(\hat{\theta}_n - \theta) \Rightarrow N(0, \sigma^2_{\hat{\theta}})$$

and

$$\sqrt{n}(\tilde{\theta}_n - \theta) \Rightarrow N(0, \sigma^2_{\tilde{\theta}}),$$

then the asymptotic relative efficiency of $\hat{\theta}_n$ with respect to $\tilde{\theta}_n$ is $\sigma^2_{\tilde{\theta}}/\sigma^2_{\hat{\theta}}$. This relative efficiency can be interpreted as the ratio of sample sizes necessary for comparable error distributions.

In our first comparison of the mean and median the data were a random sample from $N(\theta, 1)$. In this case, the mean is UMVU, so it should be of no surprise that it is more efficient than the median. If instead

$$f(x) = \frac{1}{2}e^{-|x|},$$

then

$$\sigma^2 = \int x^2 \frac{1}{2} e^{-|x|}\,dx = \int_0^\infty x^2 e^{-x}\,dx = \Gamma(3) = 2! = 2.$$

So here $\sqrt{n}(\overline{X}_n - \theta) \Rightarrow N(0,2)$, $\sqrt{n}(\tilde{X}_n - \theta) \Rightarrow N(0,1)$, and the asymptotic relative efficiency of $\overline{X}_n$ with respect to $\tilde{X}_n$ is $1/2$. Now the median is more efficient than the mean, and roughly twice as many observations will be needed for a comparable error distribution if the mean is used instead of the median. In this case, the median is the maximum likelihood estimator of $\theta$. Later results in Sections 9.3 and 16.6 show that maximum likelihood estimators are generally fully efficient.

*Example 8.19.* Suppose $X_1, \dots, X_n$ is a random sample from $N(\theta, 1)$, and we are interested in estimating

$$p = P_\theta(X_i \le a) = \Phi(a - \theta).$$

One natural estimator is

$$\hat{p} = \Phi(a - \overline{X}),$$

where $\overline{X} = (X_1 + \cdots + X_n)/n$. (This is the maximum likelihood estimator.) Another natural estimator is the proportion of the observations that are at most $a$,

$$\tilde{p} = \frac{1}{n}\#\{i \le n : X_i \le a\} = \frac{1}{n}\sum_{i=1}^{n} I\{X_i \le a\}.$$

By the central limit theorem,

$$\sqrt{n}(\tilde{p} - p) \Rightarrow N(0, \tilde{\sigma}^2),$$

as $n \to \infty$, where

$$\tilde{\sigma}^2 = \mathrm{Var}_\theta\big(I\{X_i \le a\}\big) = \Phi(a-\theta)\big(1 - \Phi(a-\theta)\big).$$

Because the first estimator is a function of the average $\overline{X}$, by the delta method, Proposition 8.14,

$$\sqrt{n}(\hat{p} - p) \Rightarrow N(0, \hat{\sigma}^2),$$

as $n \to \infty$, where

$$\hat{\sigma}^2 = \left[\frac{d}{dx}\Phi(a-x)\bigg|_{x=\theta}\right]^2 = \phi^2(a-\theta).$$

The asymptotic relative efficiency of $\hat{p}$ with respect to $\tilde{p}$ is

$$\mathrm{ARE} = \frac{\Phi(a-\theta)\big(1 - \Phi(a-\theta)\big)}{\phi^2(a-\theta)}.$$

In this example, the asymptotic relative efficiency depends on the unknown parameter $\theta$. When $\theta = a$, ARE $= \pi/2$, and the ARE increases without bound as $|\theta - a|$ increases. Note, however, that $\tilde{p}$ is a sensible estimator even if the stated model is wrong, provided the data are indeed i.i.d. In contrast, $\hat{p}$ is only reasonable if the model is correct. Gains in efficiency using $\hat{p}$ should be balanced against the robustness of $\tilde{p}$ to departures from the model.

## 8.6 Scales of Magnitude

In many asymptotic calculations it is convenient to have a standard notation indicating orders of magnitudes of variables in limiting situations. We begin with a definition for sequences of constants.

**Definition 8.20.** *Let $a_n$ and $b_n$, $n \ge 1$, be constants. Then*

1. *$a_n = o(b_n)$ as $n \to \infty$ means that $a_n/b_n \to 0$ as $n \to \infty$;*
2. *$a_n = O(b_n)$ as $n \to \infty$ means that $|a_n/b_n|$ remains bounded, i.e., that $\limsup_{n\to\infty} |a_n/b_n| < \infty$; and*

*3.* $a_n \sim b_n$ *means that* $a_n/b_n \to 1$ *as* $n \to \infty$.

Thus $a_n = o(b_n)$ when $a_n$ is of smaller order of magnitude than $b_n$, $a_n = O(b_n)$ when the magnitude of $a_n$ is at most comparable to the magnitude of $b_n$, and $a_n \sim b_n$ when $a_n$ is asymptotic to $b_n$. Note that $a_n = o(1)$ means that $a_n \to 0$.

Large oh and small oh notation may also be used in equations or inequalities. For instance, $a_n = b_n + O(c_n)$ means that $a_n - b_n = O(c_n)$, and $a_n \le b_n + o(c_n)$ means that $a_n \le b_n + d_n$ for some sequence $d_n$ with $d_n = o(c_n)$. Exploiting this idea, $a_n \sim b_n$ can be written as $a_n = b_n\big(1 + o(1)\big)$.

Although Definition 8.20 is stated for sequences indexed by a discrete variable $n$, analogous notation can be used for functions indexed by a continuous variable $x$. For instance, $a(x) = o\big(b(x)\big)$ as $x \to x_0$ would mean that $a(x)/b(x) \to 0$ as $x \to x_0$. The limit $x_0$ here could be finite or infinite. As an example, if $f$ has two derivatives at $x$, then the two-term Taylor expansion for $f$ can be expressed as

$$f(x+\epsilon) = f(x) + \epsilon f'(x) + \frac{1}{2}\epsilon^2 f''(x) + o(\epsilon^2)$$

as $\epsilon \to 0$. If $f'''$ is exists and is finite at $x$, this can be strengthened to

$$f(x+\epsilon) = f(x) + \epsilon f'(x) + \frac{1}{2}\epsilon^2 f''(x) + O(\epsilon^3)$$

as $\epsilon \to 0$.

In the following stochastic extension, the basic idea is that the original notion can fail, but only on a set with arbitrarily small probability.

**Definition 8.21.** *Let* $X_n$ *and* $Y_n$, $n \ge 1$, *be random variables, and let* $b_n$, $n \ge 1$, *be constants. Then*

*1.* $X_n = o_p(b_n)$ *as* $n \to \infty$ *means that* $X_n/b_n \xrightarrow{p} 0$ *as* $n \to \infty$;
*2.* $X_n = O_p(1)$ *as* $n \to \infty$ *means that*

$$\sup_n P\big(|X_n| > K\big) \to 0$$

*as* $K \to \infty$; *and*
*3.* $X_n = O_p(b_n)$ *means that* $X_n/b_n = O_p(1)$ *as* $n \to \infty$.

The definition for $O_p(1)$ is equivalent to a notion called *tightness* for the distributions of the $X_n$. Tightness is necessary for convergence in distribution, and so, if $X_n \Rightarrow X$, then $X_n = O_p(1)$.

Here are a few useful propositions about stochastic scales of magnitude.

**Proposition 8.22.** *If* $X_n = O_p(a_n)$ *and* $Y_n = O_p(b_n)$, *then*

$$X_n Y_n = O_p(a_n b_n).$$

*Also, if $\alpha > 0$ and $X_n = O_p(a_n)$, then $X_n^\alpha = O_p(a_n^\alpha)$. Similarly, if $X_n = O_p(a_n)$, $\alpha > 0$, and $Y_n = o_p(b_n)$, then*

$$X_nY_n = o_p(a_nb_n) \text{ and } Y_n^\alpha = o_p(b_n^\alpha).$$

**Proposition 8.23.** *Let $\alpha$ and $\beta$ be constants with $\alpha > 0$. If $E|X_n|^\alpha = O(n^\beta)$ as $n \to \infty$, then $X_n = O_p(n^{\beta/\alpha})$ as $n \to \infty$.*

**Proposition 8.24.** *If $X_n = O_p(a_n)$ with $a_n \to 0$, and if $f(\epsilon) = o(\epsilon^\alpha)$ as $\epsilon \to 0$ with $\alpha > 0$, then*

$$f(X_n) = o_p(a_n^\alpha).$$

This result is convenient for delta method derivations such as Proposition 8.14. By the central limit theorem, $\overline{X}_n = \mu + O_p(1/\sqrt{n})$, and by Taylor expansion

$$f(\mu + \epsilon) = f(\mu) + \epsilon f'(\mu) + o(\epsilon)$$

as $\epsilon \to 0$, whenever $f$ is differentiable at $\mu$. So by Proposition 8.24,

$$f(\overline{X}_n) - f(\mu) = (\overline{X}_n - \mu)f'(\mu) + o_p(1/\sqrt{n}),$$

and rearranging terms,

$$\sqrt{n}\big(f(\overline{X}_n) - f(\mu)\big) = \sqrt{n}(\overline{X}_n - \mu)f'(\mu) + o_p(1) \Rightarrow N\big(0, [f'(\mu)]^2\sigma^2\big).$$

## 8.7 Almost Sure Convergence[7]

In this section, we consider a notion of convergence for random variables called *almost sure convergence* or *convergence with probability one.*

**Definition 8.25.** *Random variables $Y_1, Y_2, \dots$ defined on a common probability space* converge almost surely *to a random variable $Y$ on the same space if*

$$P(Y_n \to Y) = 1.$$

The statistical implications of this mode of convergence are generally similar to the implications of convergence in probability, and in the rest of this book we refer to almost sure convergence only when the distinction seems statistically relevant. To understand the difference between these modes of convergence, introduce

$$M_n = \sup_{k \ge n} |Y_k - Y|,$$

and note that $Y_n \to Y$ if and only if $M_n \to 0$. Now $M_n \to 0$ if for every $\epsilon > 0$, $M_n < \epsilon$ for all $n$ sufficiently large. Define $B_\epsilon$ as the event that $M_n < \epsilon$ for all

[7] Results in this section are used only in Chapter 20.

$n$ sufficiently large. An outcome is in $B_\epsilon$ if and only if it is in one of the sets $\{M_n < \epsilon\}$, and thus

$$B_\epsilon = \bigcup \{M_n < \epsilon\}.$$

If an outcome gives a convergent sequence, it must be in $B_\epsilon$ for every $\epsilon$, and so

$$\{Y_n \to Y\} = \bigcap_{\epsilon>0} B_\epsilon,$$

and we have almost sure convergence if and only if

$$P\left[\bigcap_{\epsilon>0} B_\epsilon\right] = 1.$$

Since the $B_\epsilon$ decrease as $\epsilon \to 0$, using the continuity property of probability measures (1.1), this will happen if and only if $P(B_\epsilon) = 1$ for all $\epsilon > 0$. But because the events $\{M_n < \epsilon\}$ increase with $n$, $P(B_\epsilon) = \lim_{n\to\infty} P(M_n < \epsilon)$. Putting this all together, $Y_n \to Y$ almost surely if and only if for every $\epsilon > 0$, $P(M_n \geq \epsilon) \to 0$, that is, if and only if $M_n \xrightarrow{p} 0$. In words, almost sure convergence means the largest difference after stage $n$ tends to zero in probability as $n \to \infty$.

*Example 8.26.* If $Y_n \sim \text{Bernoulli}(p_n)$, then $Y_n \xrightarrow{p} 0$ if and only if $p_n \to 0$. Almost sure convergence will also depend on the joint distribution of these variables. If they are independent, then $M_n = \sup_{k\geq n} |Y_n - 0| \sim \text{Bernoulli}(\pi_n)$ with

$$1 - \pi_n = P(M_n = 0) = P(Y_k = 0, k \geq n) = \prod_{k=n}^{\infty}(1 - p_k).$$

This product tends to 1 as $n \to \infty$ if and only if $\sum p_n < \infty$. So in this independent case, $Y_n \to 0$ almost surely if and only if $\sum p_n < \infty$. If instead $U$ is uniformly distributed on $(0,1)$ and $Y_n = I\{U \leq p_n\}$, then $Y_n \to 0$ almost surely if and only if $p_n \to 0$, that is, if and only if $Y_n \xrightarrow{p} 0$.

The following result is the most famous result on almost sure convergence. For a proof, see Billingsley (1995) or any standard text on probability.

**Theorem 8.27 (Strong Law of Large Numbers).** *If $X_1, X_2, \ldots$ are i.i.d. with finite mean $\mu = EX_i$, and if $\overline{X}_n = (X_1 + \cdots + X_n)/n$, then $\overline{X}_n \to \mu$ almost surely as $n \to \infty$.*

## 8.8 Problems[8]

*1. Random variables $X_1$, $X_2, \ldots$ are called "$m$-dependent" if $X_i$ and $X_j$ are independent whenever $|i - j| \geq m$. Suppose $X_1$, $X_2$, $\ldots$ are $m$-dependent

[8] Solutions to the starred problems are given at the back of the book.

with $EX_1 = EX_2 = \cdots = \xi$ and $\text{Var}(X_1) = \text{Var}(X_2) = \cdots = \sigma^2 < \infty$. Let $\overline{X}_n = (X_1 + \cdots + X_n)/n$. Show that $\overline{X}_n \xrightarrow{p} \xi$ as $n \to \infty$. Hint: You should be able to bound $\text{Cov}(X_i, X_j)$ and $\text{Var}(\overline{X}_n)$.

*2. Let $X_1, \dots, X_n$ be i.i.d. from an exponential distribution with failure rate $\lambda$, and let $M_n = \max\{X_1, \dots, X_n\}$. Is $\log(n)/M_n$ a consistent estimator of $\lambda$?

*3. If $X_1, \dots, X_n$ are i.i.d. from the uniform distribution on $(0, \theta)$ with maximum $M_n = \max\{X_1, \dots, X_n\}$, then the UMVU estimator of $\theta$ is $\hat{\theta}_n = (n+1)M_n/n$. Determine the limiting distribution of $n(\hat{\theta}_n - \theta)$ as $n \to \infty$.

*4. Let $X_1, \dots, X_n$ be i.i.d. Bernoulli variables with success probability $p$. Let $\hat{p}_n = (X_1 + \cdots + X_n)/n$.
   a) Show that $\sqrt{n}(\hat{p}_n^2 - p^2) \Rightarrow N\bigl(0, 4p^3(1-p)\bigr)$.
   b) Find the UMVU estimator $\delta_n$ of $\sigma^2 = 4p^3(1-p)$, the asymptotic variance in (a).
   c) Determine the limiting distribution of $n(\delta_n - \sigma^2)$ when $p = 3/4$. Hint: The maximum likelihood estimator of $\sigma^2$ is $\hat{\sigma}^2 = 4\hat{p}_n^3(1 - \hat{p}_n)$. Show that $n(\delta_n - \hat{\sigma}_n^2)$ converges in probability to a constant, and use a two-term Taylor expansion to find the limiting distribution of $n(\hat{\sigma}^2 - \sigma^2)$.

*5. Let $X_1, \dots, X_n$ be i.i.d. with common density $f_\theta(x) = (x - \theta)^+ e^{\theta - x}$. Show that $M_n = \min\{X_1, \dots, X_n\}$ is a consistent estimator of $\theta$, and determine the limiting distribution for $\sqrt{n}(M_n - \theta)$.

*6. Prove that if $A_n \xrightarrow{p} 1$ and $Y_n \Rightarrow Y$, then $A_n Y_n \Rightarrow Y$. (This is a special case of Theorem 8.13.)

7. Suppose $X_1, X_2, \dots$ are i.i.d. with common density

$$f(x) = \begin{cases} \dfrac{1}{(1+x)^2}, & x > 0; \\ 0, & \text{otherwise}, \end{cases}$$

and let $M_n = \max\{X_1, \dots, X_n\}$. Show that $M_n/n$ converges in distribution, and give a formula for the limiting distribution function.

8. If $\epsilon > 0$ and $\sup E|X_n|^{1+\epsilon} < \infty$, show that $X_n$, $n \ge 1$, are uniformly integrable.

9. Suppose $X_1, X_2, \dots$ are integrable and

$$\lim_{t \to \infty} \limsup_{n \to \infty} E\bigl[|X_n| I\{|X_n| \ge t\}\bigr] = 0.$$

Show that $X_n$, $n \ge 1$, are uniformly integrable.

10. Suppose $X_n \Rightarrow X$, $x_n \to x$, and the cumulative distribution function for $X$ is continuous at $x$. Show that $P(X_n \le x_n) \to P(X \le x)$.

11. Let $X_1, X_2, \dots$ be i.i.d. variables uniformly distributed on $(0, 1)$, and let $\tilde{X}_n$ denote the geometric average of the first $n$ of these variables; that is,

$$\tilde{X}_n = (X_1 \times \cdots \times X_n)^{1/n}.$$

a) Show that $\tilde{X}_n \xrightarrow{p} 1/e$ as $n \to \infty$.
b) Show that $\sqrt{n}(\tilde{X}_n - 1/e)$ converges in distribution, and identify the limit.

12. Let $X_1, X_2, \dots$ be i.i.d. from the uniform distribution on $(1, 2)$, and let $H_n$ denote the harmonic average of the first $n$ variables:

$$H_n = \frac{n}{X_1^{-1} + \cdots + X_n^{-1}}.$$

a) Show that $H_n \xrightarrow{p} c$ as $n \to \infty$, identifying the constant $c$.
b) Show that $\sqrt{n}(H_n - c)$ converges in distribution, and identify the limit.

13. Show that if $Y_n \xrightarrow{p} c$ as $n \to \infty$, then $Y_n \Rightarrow Y$ as $n \to \infty$. Give the distribution or cumulative distribution function for $Y$.
14. Let $X_1, X_2, \dots$ be i.i.d. from a uniform distribution on $(0, e)$, and define

$$Y_n = \sqrt[n]{\textstyle\prod_{i=1}^{n^2} X_i}.$$

Show that $Y_n \Rightarrow Y$ as $n \to \infty$, giving the cumulative distribution function for $Y$.

15. Let $X_1, X_2, \dots$ be i.i.d. from $N(\mu, \sigma^2)$, let $w_1, w_2, \dots$ be positive weights, and define weighted averages

$$Y_n = \frac{\sum_{i=1}^n w_i X_i}{\sum_{i=1}^n w_i}, \qquad n = 1, 2, \dots.$$

a) Suppose $w_k = 1/k$, $k = 1, 2, \dots$. Show that $Y_n \xrightarrow{p} c$, identifying the limiting value $c$.
b) Suppose $w_k = 1/(2k-1)^2$. Show that $Y_n \Rightarrow Y$, giving the distribution for $Y$. Hint:

$$\sum_{k=1}^{\infty} \frac{1}{(2k-1)^2} = \frac{\pi^2}{8} \quad\text{and}\quad \sum_{k=1}^{\infty} \frac{1}{(2k-1)^4} = \frac{\pi^4}{96}.$$

*16. Let $Y_1, \dots, Y_n$ be independent with $Y_i \sim N(\alpha + \beta x_i, \sigma^2)$, $i = 1, \dots, n$, where $x_1, \dots, x_n$ are known constants, and $\alpha$, $\beta$, and $\sigma^2$ are unknown parameters. Find the maximum likelihood estimators of these parameters, $\alpha$, $\beta$, and $\sigma^2$.

*17. Let $X_1, \dots, X_n$ be jointly distributed. The first variable $X_1 \sim N(0, 1)$, and, for $j = 1, \dots, n-1$, the conditional distribution of $X_{j+1}$ given $X_1 = x_1, \dots, X_j = x_j$ is $N(\rho x_j, 1)$. Find the maximum likelihood estimator of $\rho$.

18. Distribution theory for order statistics in the tail of the distribution can behave differently than order statistics such as the median, that are near the middle of the distribution. Let $X_1, \dots, X_n$ be i.i.d. from an exponential distribution with unit failure rate.

a) Suppose we are interested in the limiting distribution for $X_{(2)}$, the second order statistic. Naturally, $X_{(2)} \xrightarrow{p} 0$ as $n \to \infty$. For an interesting limit theory we should scale $X_{(2)}$ by an appropriate power of $n$, but the correct power is not $1/2$. Suppose $x > 0$. Find a value $p$ so that $P(n^p X_{(2)} \le x)$ converges to a value between 0 and 1. (If $p$ is too small, the probability will tend to 1, and if $p$ is too large the probability will tend to 0.)

b) Determine the limiting distribution for $X_{(n)} - \log n$.

*19. Let $X_1, \dots, X_n$ be i.i.d. from an exponential distribution with failure rate $\theta$. Let $\hat{p}_n = \#\{i \le n : X_i \ge 1\}/n$ and $\overline{X}_n = (X_1 + \cdots + X_n)/n$. Determine the asymptotic relative efficiency of $-\log \hat{p}_n$ with respect to $1/\overline{X}_n$.

*20. Let $X_1, \dots, X_n$ be i.i.d. from $N(\theta, \theta)$, with $\theta > 0$ an unknown parameter, and consider estimating $\theta(\theta + 1)$. Determine the asymptotic relative efficiency of $\overline{X}_n(\overline{X}_n + 1)$ with respect to $\delta_n = (X_1^2 + \cdots + X_n^2)/n$, where, as usual, $\overline{X}_n = (X_1 + \cdots + X_n)/n$.

*21. Let $Q_n$ denote the upper quartile (or 75th percentile) for a random sample $X_1, \dots, X_n$ from $N(0, \sigma^2)$. If $\Phi(c) = 3/4$, then $Q_n \xrightarrow{p} c\sigma$, and so $\tilde{\sigma}_n = Q_n/c$ is a consistent estimator of $\sigma$. Let $\hat{\sigma}$ be the maximum likelihood estimator of $\sigma$. Determine the asymptotic relative efficiency of $\tilde{\sigma}$ with respect to $\hat{\sigma}$.

22. If $X_1, \dots, X_n$ are i.i.d. from $N(\theta, \theta)$, then two natural estimators of $\theta$ are the sample mean $\overline{X}$ and the sample variance $S^2$. Determine the asymptotic relative efficiency of $S^2$ with respect to $\overline{X}$.

23. Suppose $X_1, \dots, X_n$ are i.i.d. Poisson variables with mean $\lambda$ and we are interested in estimating $p = P_\lambda(X_i = 0) = e^{-\lambda}$.

a) One estimator for $p$ is the proportion of zeros in the sample, $\tilde{p} = \#\{i \le n : X_i = 0\}/n$. Find the limiting distribution for $\sqrt{n}(\tilde{p} - p)$.

b) Another estimator would be the maximum likelihood estimator $\hat{p}$. Give a formula for $\hat{p}$ and determine the limiting distribution for $\sqrt{n}(\hat{p} - p)$.

c) Find the asymptotic relative efficiency of $\tilde{p}$ with respect to $\hat{p}$.

*24. Suppose $X_1, \dots, X_n$ are i.i.d. $N(0, \sigma^2)$, and let $M$ be the median of $|X_1|, \dots, |X_n|$.

a) Find $c \in \mathbb{R}$ so that $\tilde{\sigma} = cM$ is a consistent estimator of $\sigma$.

b) Determine the limiting distribution for $\sqrt{n}(\tilde{\sigma} - \sigma)$.

c) Find the maximum likelihood estimator $\hat{\sigma}$ of $\sigma$ and determine the limiting distribution for $\sqrt{n}(\hat{\sigma} - \sigma)$.

d) Determine the asymptotic relative efficiency of $\tilde{\sigma}$ with respect to $\hat{\sigma}$.

25. Suppose $X_1, X_2, \dots$ are i.i.d. from the beta distribution with parameters $\alpha > 0$ and $\beta > 0$. The mean of this distribution is $\mu = \alpha/(\alpha + \beta)$. Solving, $\alpha = \beta\mu/(1 - \mu)$. If $\beta$ is known, this suggests

$$\tilde{\alpha} = \frac{\beta\overline{X}}{1 - \overline{X}}$$

as a natural estimator for $\alpha$. Determine the asymptotic relative efficiency of this estimator $\tilde{\alpha}$ with respect to the maximum likelihood estimator $\hat{\alpha}$.

26. Let $X_1, \dots, X_n$ be i.i.d. Poisson with mean $\lambda$, and consider estimating

$$g(\lambda) = P_\lambda(X_i = 1) = \lambda e^{-\lambda}.$$

One natural estimator might be the proportion of ones in the sample:

$$\hat{p}_n = \frac{1}{n}\#\{i \le n : X_i = 1\}.$$

Another choice would be the maximum likelihood estimator, $g(\overline{X}_n)$, with $\overline{X}_n$ the sample average.
   a) Find the asymptotic relative efficiency of $\hat{p}_n$ with respect to $g(\overline{X}_n)$.
   b) Determine the limiting distribution of

$$n\bigl[g(\overline{X}_n) - 1/e\bigr]$$

   when $\lambda = 1$.
27. Let $X_1, \dots, X_n$ be i.i.d. from $N(\theta, 1)$, and let $U_1, \dots, U_n$ be i.i.d. from a uniform distribution on $(0, 1)$, with all $2n$ variables independent. Define $Y_i = X_i U_i$, $i = 1, \dots, n$. If the $X_i$ and $U_i$ are both observed, then $\overline{X}$ would be a natural estimator for $\theta$. If only the products $Y_1, \dots, Y_n$ are observed, then $2\overline{Y}$ may be a reasonable estimator. Determine the asymptotic relative efficiency of $2\overline{Y}$ with respect to $\overline{X}$.
28. Definition 8.21 for $O_p(1)$ does not refer explicitly to limiting values as $n \to \infty$. But in fact the conclusion only depends on the behavior of the sequence for large $n$. Show that if

$$\limsup_{n\to\infty} P\bigl(|X_n| > K\bigr) \to 0$$

as $K \to \infty$, then $X_n = O_p(1)$, so that "sup" in the definition could be changed to "lim sup."
29. Prove Proposition 8.22.
30. *Markov's inequality.* Show that for any constant $c > 0$ and any random variable $X$,

$$P(|X| \ge c) \le E|X|/c.$$

31. Use Markov's inequality from the previous problem to prove Proposition 8.23.
32. If $X_n \Rightarrow X$ as $n \to \infty$, show that $X_n = O_p(1)$ as $n \to \infty$. Also, show that the converse fails, finding a sequence of random variables $X_n$ that are $O_p(1)$ but do not converge in distribution.
33. Show that if $X_n = O_p(1)$ as $n \to \infty$ and $f$ is a continuous function on $\mathbb{R}$, then $f(X_n) = O_p(1)$ as $n \to \infty$. Also, give an example showing that this result can fail if $f$ is discontinuous at some point $x$.
34. Let $M_n$, $n \ge 1$, be positive, integer-valued random variables.
   a) Show that if $M_n \to \infty$ almost surely as $n \to \infty$, and $X_n \to 0$ almost surely as $n \to \infty$, then $X_{M_n} \to 0$ almost surely as $n \to \infty$.

b) Show that if $M_n \xrightarrow{p} \infty$, and $X_n \to 0$ almost surely, then $X_{M_n} \xrightarrow{p} 0$.

35. Let $X_1, X_2, \ldots$ be independent Bernoulli variables with $P(X_n = 1) = 1/n$. Then $X_n \xrightarrow{p} 0$, but almost sure convergence fails. Find positive, integer-valued random variables $M_n$, $n \geq 1$, such that $M_n \to \infty$ almost surely with $X_{m_n} = 1$. This shows that the almost sure convergence for $X_n$ in the previous problem is essential.

# 9

# Estimating Equations and Maximum Likelihood

Many estimators in statistics are specified implicitly as solutions to equations or as values maximizing some function. In this chapter we study why these methods work and learn ways to approximate distributions. Although we focus on methods for i.i.d. observations, many of the ideas can be extended. Results for stationary time series are sketched in Section 9.9.

A first example, introduced in Section 8.3, concerns maximum likelihood estimation. The maximum likelihood estimator $\hat{\theta}$ maximizes the likelihood function $L(\cdot)$ or log-likelihood $l(\cdot) = \log L(\cdot)$. And if $l$ is differentiable and the maximum occurs in the interior of the parameter space, then $\hat{\theta}$ solves $\nabla l(\theta) = 0$. Method of moments estimators, considered in Problem 9.2, provide a second example. If $X_1, \dots, X_n$ are i.i.d. observations with average $\overline{X}$, and if $\mu(\theta) = E_\theta X_i$, then the method of moments estimator of $\theta$ solves $\mu(\theta) = \overline{X}$. A final example would be $M$-estimators, considered in Section 9.8.

## 9.1 Weak Law for Random Functions[1]

In this section we develop a weak law of large numbers for averages of random functions. This is used in the rest of the chapter to establish consistency and asymptotic normality of maximum likelihood and other estimators.

Let $X_1, X_2, \dots$ be i.i.d., let $K$ be a compact set in $\mathbb{R}^p$, and define

$$W_i(t) = h(t, X_i), \qquad t \in K,$$

where $h(t, x)$ is a continuous function of $t$ for all $x$. Then $W_1, W_2, \dots$ are i.i.d. random functions taking values in $C(K)$, the space of continuous functions on $K$.

Functions in $C(K)$ behave in many ways like vectors. They can be added, subtracted, and multiplied by constants, with these operations satisfying the

[1] The theory developed in this section is fairly technical, but uniform convergence is important for applications developed in later sections.

R.W. Keener, *Theoretical Statistics: Topics for a Core Course*, Springer Texts in Statistics, DOI 10.1007/978-0-387-93839-4_9, © Springer Science+Business Media, LLC 2010

usual properties. Sets with these properties are called *linear spaces.* In addition, notions of convergence can be introduced for functions in $C(K)$. There are various possibilities. The one we use in this section is based on a notion of length. For $w \in C(K)$ define

$$\|w\|_\infty = \sup_{t \in K} |w(t)|,$$

called the *supremum norm* of $w$. Functions $w_n$ converge to $w$ in this norm if $\|w_n - w\|_\infty \to 0$. With this norm, $C(K)$ is complete (all Cauchy sequences converge), and a complete linear space with a norm is called a *Banach space.*

A final nice property of $C(K)$ is separability. A subset of some set is called *dense* if every element in the set is arbitrarily close to some point in the subset. For instance, the rational numbers are a dense subset of $\mathbb{R}$ because there are rational numbers arbitrarily close to any real number $x \in \mathbb{R}$. A space is *separable* if it has a countable dense subset. We state the law of large numbers in this section for i.i.d. random functions in $C(K)$, but the result also holds for i.i.d. random elements in an arbitrary separable Banach space.[2]

**Lemma 9.1.** *Let $W$ be a random function in $C(K)$ and define*

$$\mu(t) = EW(t), \qquad t \in K.$$

*(This function $\mu$ is called the* mean *of $W$.) If $E\|W\|_\infty < \infty$, then $\mu$ is continuous. Also,*

$$\sup_{t \in K} E \sup_{s:\|s-t\|<\epsilon} |W(s) - W(t)| \to 0$$

*as $\epsilon \downarrow 0$.*

*Proof.* Let $t_n$, $n \geq 1$, be a sequence of constants in $K$ converging to $t$. Because $W$ is continuous, the random variables $W(t_n)$ converge to $W(t)$ as $n \to \infty$. They are also dominated by $\|W\|_\infty$, which has a finite expectation. Thus

$$\mu(t_n) = EW(t_n) \to EW(t) = \mu(t)$$

as $n \to \infty$ by dominated convergence, and $\mu$ is continuous.

For the second part, define

$$M_\epsilon(t) = \sup_{s:\|s-t\|<\epsilon} |W(s) - W(t)|,$$

[2] As usual, we are not giving much attention to issues of measurability, and the notion of what we mean by a "random function" is a bit vague. To be more specific, define $B_a(w) = \{f \in C(K) : \|f - w\|_\infty < a\}$, called the open ball with radius $a$ centered at $w$. The Borel $\sigma$-field $\mathcal{B}$ can then be defined as the smallest $\sigma$-field that contains all open balls. If probability is defined on a measurable space $(\mathcal{X}, \mathcal{A})$, then $W : \mathcal{X} \to C(K)$ is *measurable* and would be called a random function if $W^{-1}(B) \in \mathcal{A}$ for any Borel set $B \in \mathcal{B}$. Aside from defining Borel sets using open balls instead of intervals, this definition is essentially the same as the definition of measurability for random variables given in Definition 1.7.

and let $\lambda_\epsilon$ be the mean of $M_\epsilon$,

$$\lambda_\epsilon(t) = EM_\epsilon(t).$$

Because $W$ is continuous, $M_\epsilon$ is continuous. Also, since $|M_\epsilon(t)| \le 2\|W\|_\infty$, $E\|M_\epsilon\|_\infty < \infty$, and by the first part of this theorem, $\lambda_\epsilon$ is continuous. By continuity, $M_\epsilon(t) \to 0$ as $\epsilon \to 0$, and by dominated convergence, $\lambda_\epsilon(t) \to 0$ as $\epsilon \to 0$. Since the functions $\lambda_\epsilon$ are decreasing as $\epsilon \downarrow 0$, by Dini's theorem (Theorem A.5) the convergence is uniform, that is, $\sup_{t\in K} \lambda_\epsilon(t) \to 0$ as $\epsilon \downarrow 0$. ☐

**Theorem 9.2.** *Let $W, W_1, W_2, \dots$ be i.i.d. random functions in $C(K)$, $K$ compact, with mean $\mu$ and $E\|W\|_\infty < \infty$, and let $\overline{W}_n = (W_1 + \cdots + W_n)/n$. Then*

$$\|\overline{W}_n - \mu\|_\infty \xrightarrow{p} 0$$

*as $n \to \infty$.*

By the weak law of large numbers, for any $t \in K$, $\overline{W}_n(t) \xrightarrow{p} \mu(t)$. But the theorem is stronger, asserting that this convergence holds with uniformity in $t$.

*Proof.* Fix $\epsilon > 0$. For notation, let

$$M_{\delta,j}(t) = \sup_{s:\|s-t\|<\delta} \bigl|W_j(s) - W_j(t)\bigr|$$

with mean $\lambda_\delta(t)$. Choose $\delta$ using the second assertion of Lemma 9.1 so that

$$\lambda_\delta(t) = E \sup_{s:\|s-t\|<\delta} |W(s) - W(t)| < \epsilon, \qquad \forall t \in K,$$

and note that with this choice of $\delta$, if $\|t - s\| < \delta$, then

$$|\mu(t) - \mu(s)| = \bigl|E\bigl[W(t) - W(s)\bigr]\bigr| \le E\bigl|W(t) - W(s)\bigr| \le \epsilon.$$

Let $B_\delta(t) = \bigl\{s : \|s - t\| < \delta\bigr\}$, the open ball with radius $\delta$ about $t$. Since $K$ is compact, the open sets $B_\delta(t)$, $t \in K$, covering $K$ have a finite subcover $O_i = B_\delta(t_i)$, $i = 1, \dots, m$. Then

$$\begin{aligned}
&\|\overline{W}_n - \mu\|_\infty \\
&\quad = \max_{i=1,\dots,m} \sup_{t\in O_i} \bigl|\overline{W}_n(t) - \mu(t)\bigr| \\
&\quad \le \max_i \sup_{t\in O_i} \Bigl[\bigl|\overline{W}_n(t) - \overline{W}_n(t_i)\bigr| + \bigl|\overline{W}_n(t_i) - \mu(t_i)\bigr| + \bigl|\mu(t_i) - \mu(t)\bigr|\Bigr] \\
&\quad \le \max_i \sup_{t\in O_i} \bigl|\overline{W}_n(t) - \overline{W}_n(t_i)\bigr| + \max_i \bigl|\overline{W}_n(t_i) - \mu(t_i)\bigr| + \epsilon.
\end{aligned}$$

Now

$$\begin{aligned}\sup_{t\in O_i}\left|\overline{W}_n(t)-\overline{W}_n(t_i)\right| &= \frac{1}{n}\sup_{t\in O_i}\left|\sum_{j=1}^n [W_j(t)-W_j(t_i)]\right| \\ &\le \frac{1}{n}\sum_{j=1}^n \sup_{t\in O_i}\left|W_j(t)-W_j(t_i)\right| \\ &= \frac{1}{n}\sum_{j=1}^n M_{\delta,j}(t_i) \stackrel{\text{def}}{=} \overline{M}_{\delta,n}(t_i).\end{aligned}$$

By the law of large numbers,

$$\overline{M}_{\delta,n}(t_i) \xrightarrow{p} \lambda_\delta(t_i) < \epsilon.$$

Using these bounds,

$$\|\overline{W}_n-\mu\|_\infty < 2\epsilon + \max_i\bigl(\overline{M}_{\delta,n}(t_i)-\lambda_\delta(t_i)\bigr) + \max_i\bigl|\overline{W}_n(t_i)-\mu(t_i)\bigr|.$$

The two maximums in this equation both converge to zero in probability and using this it is easy to argue that $P\bigl(\|\overline{W}_n-\mu\|_\infty > 3\epsilon\bigr) \to 0$ as $n\to\infty$. □

*Remark 9.3.* The same proof coupled with the strong law of large numbers, stated in Section 8.7, shows that $\|\overline{W}_n-\mu\|_\infty \to 0$ almost surely.

The following result shows the usefulness of uniform convergence. None of the conclusions follow from pointwise convergence in probability.

**Theorem 9.4.** *Let $G_n$, $n\ge 1$, be random functions in $C(K)$, $K$ compact, and suppose $\|G_n-g\|_\infty \xrightarrow{p} 0$ with $g$ a nonrandom function in $C(K)$.*

1. *If $t_n$, $n\ge 1$, are random variables converging in probability to a constant $t^*\in K$, $t_n \xrightarrow{p} t^*$, then $G_n(t_n)\xrightarrow{p} g(t^*)$.*
2. *If $g$ achieves its maximum at a unique value $t^*$, and if $t_n$ are random variables maximizing $G_n$, so that*

$$G_n(t_n) = \sup_{t\in K} G_n(t),$$

   *then $t_n \xrightarrow{p} t^*$.*
3. *If $K\subset\mathbb{R}$ and $g(t)=0$ has a unique solution $t^*$, and if $t_n$ are random variables solving $G_n(t_n)=0$, then $t_n\xrightarrow{p} t^*$.*

*Proof.* For the first assertion, since

$$\begin{aligned}\bigl|G_n(t_n)-g(t^*)\bigr| &\le \bigl|G_n(t_n)-g(t_n)\bigr| + \bigl|g(t_n)-g(t^*)\bigr| \\ &\le \|G_n-g\|_\infty + \bigl|g(t_n)-g(t^*)\bigr|,\end{aligned}$$

and since $g(t_n)\xrightarrow{p} g(t^*)$,

$$\begin{aligned}
P\big(\big|G_n(t_n) - g(t^*)\big| > \epsilon\big) \\
&\le P\left(\|G_n - g\|_\infty + \big|g(t_n) - g(t^*)\big| > \epsilon\right) \\
&\le P\left(\|G_n - g\|_\infty > \epsilon/2\right) + P\left(\big|g(t_n) - g(t^*)\big| > \epsilon/2\right) \\
&\to 0.
\end{aligned}$$

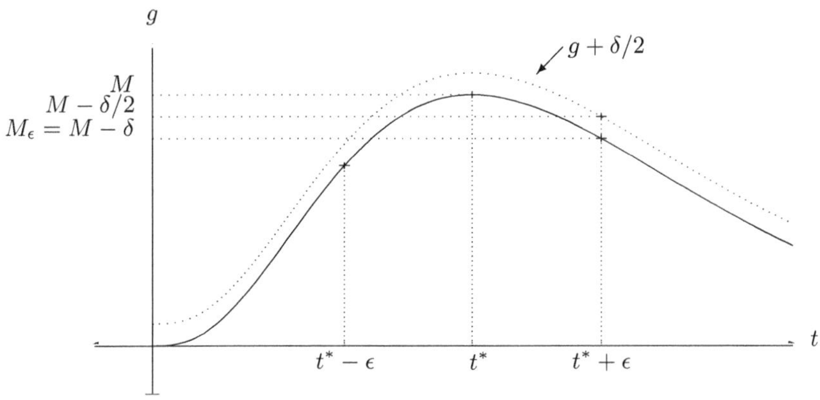


**Fig. 9.1.** $g$ and $g + \delta/2$.

For the second assertion, fix $\epsilon$ and let $K_\epsilon = K - B_\epsilon(t^*)$. This set is compact; it is bounded because $K$ is bounded, and it is closed because it is the intersection of two closed sets, $K$ and the complement of $B_\epsilon(t^*)$. Let $M = g(t^*) = \sup_K g$ and let $M_\epsilon = \sup_{K_\epsilon} g$. Since $K_\epsilon$ is compact, $M_\epsilon = g(t^*_\epsilon)$ for some $t^*_\epsilon \in K_\epsilon$, and since $g$ has a unique maximum over $K$, $M_\epsilon < M$. Define $\delta = M - M_\epsilon > 0$. See Figure 9.1. Suppose $\|G_n - g\|_\infty < \delta/2$. Then

$$\sup_{K_\epsilon} G_n < \sup_{K_\epsilon} g + \frac{\delta}{2} = M - \frac{\delta}{2}$$

and

$$\sup_K G_n \ge G_n(t^*) > g(t^*) - \frac{\delta}{2} = M - \frac{\delta}{2},$$

and $t_n$ must lie in $B_\epsilon(t^*)$. Thus

$$P\big(\|G_n - g\|_\infty < \delta/2\big) \le P\big(\|t_n - t^*\| < \epsilon\big).$$

Taking complements,

$$P\big(\|t_n - t^*\| \ge \epsilon\big) \le P\big(\|G_n - g\|_\infty \ge \delta/2\big) \to 0,$$

and so $t_n \xrightarrow{p} t^*$. The third assertion in the theorem can be established in a similar fashion. □

*Remark 9.5.* The law of large numbers and the first and third assertions in Theorem 9.4 can be easily extended to multivariate situations where the random functions are vector-valued, mapping a compact set $K$ into $\mathbb{R}^p$.

*Remark 9.6.* In the approach to the weak law here, continuity plays a key role in proving uniform convergence. Uniform convergence without continuity is also possible. One important result concerns empirical distribution functions. If $X_1, \dots, X_n$ are i.i.d., then a natural estimator for the common cumulative distribution function $F$ would be the *empirical cumulative distribution function* $\hat{F}_n$, defined as

$$\hat{F}_n(x) = \frac{1}{n}\#\{i \le n : X_i \le x\}, \qquad x \in \mathbb{R}.$$

The Glivenko–Cantelli theorem asserts that $\|\hat{F}_n - F\|_\infty \xrightarrow{p} 0$ as $n \to \infty$. In the proof of this result, monotonicity replaces continuity as the key regularity used to establish uniform convergence.

## 9.2 Consistency of the Maximum Likelihood Estimator

For this section let $X, X_1, X_2, \dots$ be i.i.d. with common density $f_\theta$, $\theta \in \Omega$, and let $l_n$ be the log-likelihood function for the first $n$ observations:

$$l_n(\omega) = \log \prod_{i=1}^n f_\omega(X_i) = \sum_{i=1}^n \log f_\omega(X_i).$$

(We use $\omega$ as the dummy argument here, reserving $\theta$ to represent the true value of the unknown parameter in the sequel.) Then the maximum likelihood estimator $\hat{\theta}_n = \hat{\theta}_n(X_1, \dots, X_n)$ from the first $n$ observations will maximize $l_n$. For regularity, assume $f_\theta(x)$ is continuous in $\theta$.

**Definition 9.7.** *The* Kullback–Leibler information *is defined as*

$$I(\theta, \omega) = E_\theta \log\bigl[f_\theta(X)/f_\omega(X)\bigr].$$

*It can be viewed as a measure of the information discriminating between $\theta$ and $\omega$ when $\theta$ is the true value of the unknown parameter.*

**Lemma 9.8.** *If $P_\theta \ne P_\omega$, then $I(\theta, \omega) > 0$.*

*Proof.* By Jensen's inequality,

$$\begin{aligned} -I(\theta, \omega) &= E_\theta \log\bigl[f_\omega(X)/f_\theta(X)\bigr] \\ &\le \log E_\theta\bigl[f_\omega(X)/f_\theta(X)\bigr] \\ &= \log \int_{f_\theta > 0} \frac{f_\omega(x)}{f_\theta(x)} f_\theta(x)\, d\mu(x) \\ &\le \log 1 \\ &= 0. \end{aligned}$$

Strict equality will occur only if $f_\omega(X)/f_\theta(X)$ is constant a.e. But then the densities will be proportional and hence equal a.e., and $P_\theta$ and $P_\omega$ will be the same. □

The next result gives consistency for the maximum likelihood estimator when $\Omega$ is compact. The result following is an extension when $\Omega = \mathbb{R}^p$. Define

$$W(\omega) = \log\left[\frac{f_\omega(X)}{f_\theta(X)}\right].$$

**Theorem 9.9.** *If $\Omega$ is compact, $E_\theta\|W\|_\infty < \infty$, $f_\omega(x)$ is a continuous function of $\omega$ for a.e. $x$, and $P_\omega \neq P_\theta$ for all $\omega \neq \theta$, then under $P_\theta$, $\hat{\theta}_n \xrightarrow{p} \theta$.*

*Proof.* If $W_i(\omega) = \log\bigl(f_\omega(X_i)/f_\theta(X_i)\bigr)$, then under $P_\theta$, $W_1, W_2, \dots$ are i.i.d. random functions in $C(\Omega)$ with mean $\mu(\omega) = -I(\theta,\omega)$. Note that $\mu(\theta) = 0$ and $\mu(\omega) < 0$ for $\omega \neq \theta$ by Lemma 9.8, and so $\mu$ has a unique maximum at $\theta$. Since

$$\overline{W}_n(\omega) = \frac{1}{n}\sum_{j=1}^n W_i(\omega) = \frac{l_n(\omega) - l_n(\theta)}{n},$$

$\hat{\theta}_n$ maximizes $\overline{W}_n$. By Theorem 9.2, $\|\overline{W}_n - \mu\|_\infty \to 0$, and the result follows from the second assertion of Theorem 9.4. □

*Remark 9.10.* The argument used to prove consistency here is based on the proof in Wald (1949). In this paper, the one-sided condition that $E_\theta \sup_\Omega W < \infty$ replaces $E_\theta\|W\|_\infty < \infty$. Inspecting the proof here, it is not hard to see that Wald's weaker condition is sufficient.

**Theorem 9.11.** *Suppose $\Omega = \mathbb{R}^p$, $f_\omega(x)$ is a continuous function of $\omega$ for a.e. $x$, $P_\omega \neq P_\theta$ for all $\omega \neq \theta$, and $f_\omega(x) \to 0$ as $\omega \to \infty$. If $E_\theta\|1_K W\|_\infty < \infty$ for any compact set $K \subset \mathbb{R}^p$, and if $E_\theta \sup_{\|\omega\|>a} W(\omega) < \infty$ for some $a > 0$, then under $P_\theta$, $\hat{\theta}_n \xrightarrow{p} \theta$.*

*Proof.* Since $f_\omega(x) \to 0$ as $\omega \to \infty$, if $f_\theta(X) > 0$,

$$\sup_{\|\omega\|>b} W(\omega) \to -\infty$$

as $b \to \infty$. By a dominated convergence argument the expectation of this variable will tend to $-\infty$ as $b \to \infty$, and we can choose $b$ so that

$$E_\theta \sup_{\|\omega\|>b} W(\omega) < 0.$$

Note that $b$ must exceed $\|\theta\|$, because $W(\theta) = 0$. Since

$$\sup_{\|\omega\|>b} \overline{W}_n(\omega) \le \frac{1}{n}\sum_{j=1}^n \sup_{\|\omega\|>b} W_i(\omega) \xrightarrow{p} E_\theta \sup_{\|\omega\|>b} W(\omega),$$

$$P_\theta\Bigl(\sup_{\|\omega\|>b} \overline{W}_n(\omega) \geq 0\Bigr) \to 0,$$

as $n \to \infty$. Let $K$ be the closed ball of radius $b$, and let $\tilde{\theta}_n$ be variables maximizing $\overline{W}_n$ over $K$.[3] By Theorem 9.9, $\tilde{\theta}_n \xrightarrow{p} \theta$. Since $\hat{\theta}_n$ must lie in $K$ whenever

$$\sup_{\|\omega\|>b} \overline{W}_n(\omega) < \overline{W}_n(\theta) = 0,$$

$P_\theta(\hat{\theta}_n = \tilde{\theta}_n) \to 1$. It then follows that $\hat{\theta}_n \xrightarrow{p} \theta$. □

*Remark 9.12.* A similar result can be obtained when $\Omega$ is an arbitrary open set. The corresponding conditions would be that $f_\omega(x) \to 0$ as $\omega$ approaches the boundary of $\Omega$, and that $E_\theta \sup_{\omega\in K^c} W(\omega) < \infty$ for some compact set $K$. Although conditions for consistency are fairly mild, counterexamples are possible when they fail. Problem 9.4 provides one example.

*Example 9.13.* Suppose we have a location family with densities $f_\theta(x) = g(x-\theta)$, $\theta \in \mathbb{R}$, and that

1. $g$ is continuous and bounded, so $\sup_{x\in\mathbb{R}} g(x) = K < \infty$,
2. $g(x) \to 0$ as $x \to \pm\infty$, and
3. $\int |\log g(x)| g(x)\, dx < \infty$.

Then

$$\begin{aligned} E_\theta \sup_{\omega\in\mathbb{R}} W(\omega) &= E_\theta \sup_{\omega\in\mathbb{R}} \log \frac{g(X-\omega)}{g(X-\theta)} \\ &= \log K - E_\theta \log g(X-\theta) \\ &= \log K - \int [\log g(x)] g(x)\, dx \\ &< \infty. \end{aligned}$$

Hence $\hat{\theta}_n$ is consistent by the one-sided adaptation of our consistency theorems mentioned in Remark 9.10. The third condition here is not very stringent; it holds for most densities, including the Cauchy and other $t$-densities, that decay algebraically near infinity.

## 9.3 Limiting Distribution for the MLE

**Theorem 9.14.** *Assume:*

1. *Variables $X, X_1, X_2, \ldots$ are i.i.d. with common density $f_\theta$, $\theta \in \Omega \subset \mathbb{R}$.*

---

[3] To be careful, as we define $\tilde{\theta}_n$, we should also insist that $\tilde{\theta}_n = \hat{\theta}_n$ whenever $\hat{\theta}_n \in K$, to cover cases with multiple maxima.

2. *The set* $A = \{x : f_\theta(x) > 0\}$ *is independent of* $\theta$.
3. *For every* $x \in A$, $\partial^2 f_\theta(x)/\partial\theta^2$ *exists and is continuous in* $\theta$.
4. *Let* $W(\theta) = \log f_\theta(X)$. *The Fisher information* $I(\theta)$ *from a single observation exists, is finite, and can be found using either*

$$I(\theta) = E_\theta W'(\theta)^2 \quad \text{or} \quad I(\theta) = -E_\theta W''(\theta).$$

*Also,*

$$E_\theta W'(\theta) = 0.$$

5. *For every* $\theta$ *in the interior of* $\Omega$ *there exists* $\epsilon > 0$ *such that*

$$E_\theta \|1_{[\theta-\epsilon,\theta+\epsilon]} W''\|_\infty < \infty.$$

6. *The maximum likelihood estimator* $\hat{\theta}_n$ *is consistent.*

*Then for any* $\theta$ *in the interior of* $\Omega$,

$$\sqrt{n}(\hat{\theta}_n - \theta) \Rightarrow N\big(0, 1/I(\theta)\big)$$

*under* $P_\theta$ *as* $n \to \infty$.

The assumptions in this theorem are fairly mild, although similar results, such as those in Chapter 16, are possible under weaker conditions. Assumption 2 usually precludes families of uniform distributions or truncated families. Assumptions 3 and 4 are the same as assumptions discussed for the Cramér–Rao bound, and Assumption 5 strengthens 4. Concerning the final assumption, for the proof $\hat{\theta}_n$ needs to be consistent, but it is not essential that it maximizes the likelihood. What matters is that $\sqrt{n}\overline{W}'_n(\hat{\theta}_n) \xrightarrow{P_\theta} 0$. In regular cases this will hold for Bayes estimators. There may also be models satisfying the other assumptions for this theorem in which the maximum likelihood estimator does not exist or is not consistent. In these examples there is often a consistent $\hat{\theta}_n$ solving $\overline{W}'_n(\hat{\theta}_n) = 0$, with this consistent root of the likelihood equation asymptotically normal.

The following technical lemma shows that, when proving convergence in distribution, we only need consider what happens on a sequence of events with probabilities converging to one.

**Lemma 9.15.** *Suppose* $Y_n \Rightarrow Y$, *and* $P(B_n) \to 1$ *as* $n \to \infty$. *Then for arbitrary random variables* $Z_n$, $n \ge 1$,

$$Y_n 1_{B_n} + Z_n 1_{B_n^c} \Rightarrow Y$$

*as* $n \to \infty$.

*Proof.* For any $\epsilon > 0$,

$$P\big(|Z_n 1_{B_n^c}| > \epsilon\big) \le P(B_n^c) = 1 - P(B_n) \to 0$$

as $n \to \infty$. So $Z_n 1_{B_n^c} \xrightarrow{p} 0$ as $n \to \infty$. Also,

$$P\big(|1_{B_n} - 1| > \epsilon\big) \le P(B_n^c) = 1 - P(B_n) \to 0$$

as $n \to \infty$, and so $1_{B_n} \xrightarrow{p} 1$ as $n \to \infty$. With these observations, the lemma now follows from Theorem 8.13. □

*Proof of Theorem 9.14.* Choose $\epsilon > 0$ using Assumption 5 small enough that $[\theta - \epsilon, \theta + \epsilon] \subset \Omega^0$ and $E_\theta \| 1_{[\theta-\epsilon,\theta+\epsilon]} W'' \|_\infty < \infty$, and let $B_n$ be the event $\hat{\theta}_n \in [\theta - \epsilon, \theta + \epsilon]$. Because $\hat{\theta}_n$ is consistent, $P_\theta(B_n) \to 1$, and since $\hat{\theta}_n$ maximizes $n\overline{W}_n(\cdot) = l_n(\cdot)$, on $B_n$ we have $\overline{W}'_n(\hat{\theta}_n) = 0$. Taylor expansion of $\overline{W}'_n$ about $\theta$ gives

$$\overline{W}'_n(\hat{\theta}_n) = \overline{W}'_n(\theta) + \overline{W}''_n(\tilde{\theta}_n)(\hat{\theta}_n - \theta),$$

where $\tilde{\theta}_n$ is an intermediate value between $\hat{\theta}_n$ and $\theta$. Setting the left-hand side of this equation to zero and solving, on $B_n$,

$$\sqrt{n}(\hat{\theta}_n - \theta) = \frac{\sqrt{n}\overline{W}'_n(\theta)}{-\overline{W}''_n(\tilde{\theta}_n)}. \tag{9.1}$$

By Assumption 4, the variables averaged in $\overline{W}'_n(\theta)$ are i.i.d., mean zero, with variance $I(\theta)$. By the central limit theorem,

$$\sqrt{n}\,\overline{W}'_n(\theta) \Rightarrow Z \sim N\big(0, I(\theta)\big).$$

Turning to the denominator, since $|\tilde{\theta}_n - \theta| \le |\hat{\theta}_n - \theta|$, at least on $B_n$, and $\hat{\theta}_n$ is consistent, $\tilde{\theta}_n \xrightarrow{p} \theta$. By Theorem 9.2,

$$\|1_{[\theta-\epsilon,\theta+\epsilon]}(\overline{W}''_n - \mu)\|_\infty \xrightarrow{p} 0,$$

where $\mu(\omega) = E_\theta W''(\omega)$, and so, by second assertion of Theorem 9.4,

$$\overline{W}''_n(\tilde{\theta}_n) \xrightarrow{p} \mu(\theta) = -I(\theta).$$

Since the behavior of $\hat{\theta}_n$ on $B_n^c$ cannot affect convergence in distribution (by Lemma 9.15),

$$\sqrt{n}(\hat{\theta}_n - \theta) \Rightarrow \frac{Z}{I(\theta)} \sim N\big(0, 1/I(\theta)\big),$$

as $n \to \infty$ by Theorem 8.13. □

*Remark 9.16.* The argument that

$$\overline{W}''_n(\tilde{\theta}_n) = \frac{1}{n} l''_n(\tilde{\theta}_n) \xrightarrow{p} -I(\theta)$$

holds for any variables $\tilde{\theta}_n$ converging to $\theta$ in probability. This is exploited later as we study asymptotic confidence intervals.

## 9.4 Confidence Intervals

A point estimator $\delta$ for an unknown parameter $g(\theta)$ provides no information about accuracy. Confidence intervals address this deficiency by seeking two statistics, $\delta_0$ and $\delta_1$, that bracket $g(\theta)$ with high probability.

**Definition 9.17.** *If $\delta_0$ and $\delta_1$ are statistics, then the random interval $(\delta_0, \delta_1)$ is called a $1-\alpha$* confidence interval *for $g(\theta)$ if*

$$P_\theta\big(g(\theta) \in (\delta_0, \delta_1)\big) \geq 1 - \alpha,$$

*for all $\theta \in \Omega$. Also, a random set $S = S(X)$ constructed from data $X$ is called a $1-\alpha$* confidence region *for $g(\theta)$ if*

$$P_\theta\big(g(\theta) \in S\big) \geq 1 - \alpha,$$

*for all $\theta \in \Omega$.*

*Remark 9.18.* In many examples, coverage probabilities equal $1-\alpha$ for all $\theta \in \Omega$, in which case the interval or region might be called an *exact* confidence interval or an exact confidence region.

*Example 9.19.* Let $X_1, \dots, X_n$ be i.i.d. from $N(\mu, \sigma^2)$. Then from the results in Section 4.3, $\overline{X} = (X_1 + \cdots + X_n)/n$ and $S^2 = \sum_{i=1}^n (X_i - \overline{X})^2/(n-1)$ are independent, with $\overline{X} \sim N(\mu, \sigma^2/n)$ and $(n-1)S^2/\sigma^2 \sim \chi^2_{n-1}$. Define

$$Z = \frac{\overline{X} - \mu}{\sigma/\sqrt{n}} \sim N(0,1)$$

and

$$V = \frac{(n-1)S^2}{\sigma^2} \sim \chi^2_{n-1}.$$

These variables $Z$ and $V$ are called *pivots*, since their distribution does not depend on the unknown parameters $\mu$ and $\sigma^2$. This idea is similar to ancillarity, but $Z$ and $V$ are not statistics since both variables depend explicitly on unknown parameters. Since $Z$ and $V$ are independent, the variable

$$T = \frac{Z}{\sqrt{V/(n-1)}} \tag{9.2}$$

is also a pivot. Its distribution is called the $t$-distribution on $n-1$ degrees of freedom, denoted $T \sim t_{n-1}$. The density for $T$ is

$$f_T(x) = \frac{\Gamma\big((\nu+1)/2\big)}{\sqrt{\nu\pi}\,\Gamma(\nu/2)(1+x^2/\nu)^{(\nu+1)/2}}, \qquad x \in \mathbb{R},$$

where $\nu = n-1$, the number of degrees of freedom.

Pivots can be used to set confidence intervals. For $p \in (0,1)$, let $t_{p,\nu}$ denote the upper $p$th quantile for the $t$-distribution on $\nu$ degrees of freedom, so that

$$P(T \geq t_{p,\nu}) = \int_{t_{p,\nu}}^{\infty} f_T(x)\,dx = p.$$

By symmetry,

$$P(T \geq t_{\alpha/2,n-1}) = P(T \leq -t_{\alpha/2,n-1}) = \alpha/2,$$

and so

$$P(-t_{\alpha/2,n-1} < T < t_{\alpha/2,n-1}) = 1 - \alpha.$$

Now

$$T = \frac{Z}{\sqrt{V/(n-1)}} = \frac{\overline{X} - \mu}{S/\sqrt{n}},$$

and so

$$-t_{\alpha/2,n-1} < T < t_{\alpha/2,n-1}$$

if and only if

$$\frac{|\overline{X} - \mu|}{S/\sqrt{n}} < t_{\alpha/2,n-1}$$

if and only if

$$|\overline{X} - \mu| < t_{\alpha/2,n-1}\frac{S}{\sqrt{n}}$$

if and only if

$$\mu \in \left(\overline{X} - t_{\alpha/2,n-1}\frac{S}{\sqrt{n}}, \overline{X} + t_{\alpha/2,n-1}\frac{S}{\sqrt{n}}\right) \stackrel{\text{def}}{=} (\delta_0, \delta_1).$$

Thus for any $\theta = (\mu, \sigma^2)$,

$$P_\theta\big(\mu \in (\delta_0, \delta_1)\big) = 1 - \alpha$$

and $(\delta_0, \delta_1)$ is a $1 - \alpha$ confidence interval for $\mu$.

The pivot $V$ can be used in a similar fashion to set confidence intervals for $\sigma^2$. Let $\chi^2_{p,\nu}$ denote the upper $p$th quantile for the chi-square distribution on $\nu$ degrees of freedom. Then

$$P(V \geq \chi^2_{\alpha/2,n-1}) = P(V \leq \chi^2_{1-\alpha/2,n-1}) = \alpha/2,$$

and

$$\begin{aligned}
1 - \alpha &= P_\theta\left(\chi^2_{1-\alpha/2,n-1} < V = \frac{(n-1)S^2}{\sigma^2} < \chi^2_{\alpha/2,n-1}\right) \\
&= P_\theta\left[\sigma^2 \in \left(\frac{(n-1)S^2}{\chi^2_{\alpha/2,n-1}}, \frac{(n-1)S^2}{\chi^2_{1-\alpha/2,n-1}}\right)\right].
\end{aligned}$$

Thus

$$\left(\frac{(n-1)S^2}{\chi^2_{\alpha/2,n-1}}, \frac{(n-1)S^2}{\chi^2_{1-\alpha/2,n-1}}\right)$$

is a $1-\alpha$ confidence interval for $\sigma^2$.

## 9.5 Asymptotic Confidence Intervals

Suppose the conditions of Theorem 9.14 hold, so that under $P_\theta$,

$$\sqrt{n}(\hat{\theta}_n - \theta) \Rightarrow N\bigl(0, 1/I(\theta)\bigr)$$

as $n \to \infty$, where $\hat{\theta}_n$ is the maximum likelihood estimator of $\theta$ based on $n$ observations. Multiplying by $\sqrt{I(\theta)}$, this implies

$$\sqrt{nI(\theta)}(\hat{\theta}_n - \theta) \Rightarrow N(0,1). \tag{9.3}$$

Since the limiting distribution here is independent of $\theta$, $\sqrt{nI(\theta)}(\hat{\theta}_n - \theta)$ is called an *approximate pivot.* If we define $z_p = \Phi^{\leftarrow}(1-p)$, the upper $p$th quantile of $N(0,1)$, then

$$P_\theta\bigl(\sqrt{nI(\theta)}|\hat{\theta}_n - \theta| < z_{\alpha/2}\bigr) \to 1-\alpha$$

as $n \to \infty$. If we define the (random) set

$$S = \bigl\{\theta \in \Omega : \sqrt{nI(\theta)}|\hat{\theta}_n - \theta| < z_{\alpha/2}\bigr\}, \tag{9.4}$$

then $\theta \in S$ if and only if $\sqrt{nI(\theta)}|\hat{\theta}_n - \theta| < z_{\alpha/2}$, and so

$$P_\theta(\theta \in S) \to 1-\alpha$$

as $n \to \infty$. This set $S$ is called a $1-\alpha$ *asymptotic confidence region* for $\theta$.

Practical considerations may make the confidence region $S$ in (9.4) undesirable. It need not be an interval, which may make the region hard to describe and difficult to interpret. Also, if the Fisher information $I(\cdot)$ is a complicated function, the inequalities defining the region may be difficult to solve. To avoid these troubles, note that if $I(\cdot)$ is continuous, then $\sqrt{I(\hat{\theta}_n)/I(\theta)} \xrightarrow{P_\theta} 1$, and so by Theorem 8.13 and (9.3),

$$\sqrt{nI(\hat{\theta}_n)}(\hat{\theta}_n - \theta) = \sqrt{\frac{I(\hat{\theta}_n)}{I(\theta)}}\sqrt{nI(\theta)}(\hat{\theta}_n - \theta) \Rightarrow N(0,1)$$

as $n \to \infty$. From this,

$$\begin{aligned}
P_\theta\Big(\sqrt{nI(\hat\theta_n)}|\hat\theta_n-\theta|<z_{\alpha/2}\Big)
&= P_\theta\left[\theta\in\left(\hat\theta_n-\frac{z_{\alpha/2}}{\sqrt{nI(\hat\theta_n)}},\hat\theta_n+\frac{z_{\alpha/2}}{\sqrt{nI(\hat\theta_n)}}\right)\right]\\
&\to 1-\alpha
\end{aligned}$$

as $n\to\infty$. So

$$\left(\hat\theta_n-\frac{z_{\alpha/2}}{\sqrt{nI(\hat\theta_n)}},\hat\theta_n+\frac{z_{\alpha/2}}{\sqrt{nI(\hat\theta_n)}}\right) \tag{9.5}$$

is a $1-\alpha$ asymptotic confidence interval for $\theta$.

The interval (9.5) requires explicit calculation of the Fisher information. In addition, it might be argued that confidence intervals should be based solely on the shape of the likelihood function, and not on quantities that involve an expectation, such as $I(\hat\theta_n)$. Using Remark 9.16, $-l_n''(\hat\theta_n)/n \xrightarrow{P_\theta} I(\theta)$. So $\sqrt{-l_n''(\hat\theta_n)}/\sqrt{nI(\theta)} \xrightarrow{P_\theta} 1$, and multiplying (9.3) by this ratio,

$$\sqrt{-l_n''(\hat\theta_n)}(\hat\theta_n-\theta)\Rightarrow N(0,1) \tag{9.6}$$

under $P_\theta$ as $n\to\infty$. From this,

$$\left(\hat\theta_n-\frac{z_{\alpha/2}}{\sqrt{-l_n''(\hat\theta_n)}},\hat\theta_n+\frac{z_{\alpha/2}}{\sqrt{-l_n''(\hat\theta_n)}}\right) \tag{9.7}$$

is a $1-\alpha$ asymptotic confidence interval for $\theta$. The statistic $-l_n''(\hat\theta_n)$ used to set the width of this interval is called the *observed* or *sample Fisher information*.

The interval (9.7) relies on the log-likelihood only through $\hat\theta_n$ and the curvature at $\hat\theta_n$. Our final confidence regions are called *profile regions* as they take more account of the actual shape of the likelihood function. By Taylor expansion about $\hat\theta_n$,

$$2l_n(\hat\theta_n)-2l_n(\theta)=\Big[\sqrt{-l_n''(\theta_n^*)}(\theta-\hat\theta_n)\Big]^2,$$

where $\theta_n^*$ is an intermediate value between $\theta$ and $\hat\theta_n$ (provided $l_n'(\hat\theta_n)=0$, which happens with probability approaching one if $\theta\in\Omega^o$). By the argument leading to (9.6),

$$\sqrt{-l_n''(\theta_n^*)}(\hat\theta_n-\theta)\Rightarrow Z\sim N(0,1),$$

and so, using Corollary 8.11,

$$2l_n(\hat\theta_n)-2l_n(\theta)\Rightarrow Z^2\sim\chi_1^2.$$

Noting that $P(Z^2<z_{\alpha/2}^2)=P(z_{\alpha/2}<Z<z_{\alpha/2})=1-\alpha$,

$$P_\theta\big(2l_n(\hat\theta_n) - 2l_n(\theta) < z^2_{\alpha/2}\big) \to 1-\alpha.$$

If we define

$$S = \big\{\theta \in \Omega : 2l_n(\hat\theta_n) - 2l_n(\theta) < z^2_{\alpha/2}\big\}, \tag{9.8}$$

then $P_\theta(\theta \in S) \to 1-\alpha$ and $S$ is a $1-\alpha$ asymptotic confidence region for $\theta$. Figure 9.2 illustrates how this set $S = (\delta_0, \delta_1)$ would be found from the log-likelihood function $l_n(\cdot)$.

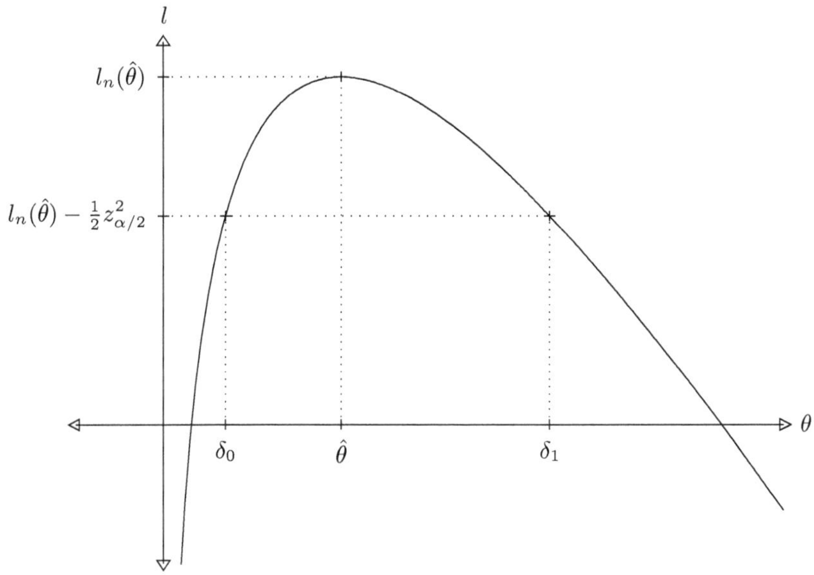


**Fig. 9.2.** Profile confidence interval $(\delta_0, \delta_1)$.

*Example 9.20.* Suppose $X_1, \dots, X_n$ are i.i.d. from a Poisson distribution with mean $\theta$. Then

$$l_n(\theta) = n\overline{X}\log\theta - n\theta - \log\left(\prod_{i=1}^n X_i!\right),$$

where $\overline{X} = (X_1 + \cdots + X_n)/n$. Since

$$l_n'(\theta) = \frac{n\overline{X}}{\theta} - n,$$

the maximum likelihood estimator of $\theta$ is $\hat\theta = \overline{X}$. Also, $I(\theta) = 1/\theta$. The first confidence region considered, (9.4), is

$$\begin{aligned} S &= \big\{\theta > 0 : \sqrt{n/\theta}|\hat\theta - \theta| < z_{\alpha/2}\big\} \\ &= \big\{\theta > 0 : \hat\theta^2 - 2\hat\theta\theta + \hat\theta^2 < z^2_{\alpha/2}\theta/n\big\} = (\hat\theta^-, \hat\theta^+), \end{aligned}$$

where

$$\hat{\theta}^{\pm} = \hat{\theta} + \frac{z_{\alpha/2}^2}{2n} \pm \sqrt{\left(\hat{\theta} + \frac{z_{\alpha/2}^2}{2n}\right)^2 - \hat{\theta}^2}.$$

The next confidence interval, (9.5), based on $I(\hat{\theta}) = 1/\overline{X}$ is

$$\left(\overline{X} - z_{\alpha/2}\sqrt{\overline{X}/n}, \overline{X} + z_{\alpha/2}\sqrt{\overline{X}/n}\right).$$

For this example, the third confidence interval is the same because the observed Fisher information, $-l_n''(\hat{\theta}) = n\overline{X}/\hat{\theta} = n/\overline{X}$, agrees with $nI(\hat{\theta})$. Note that the lower endpoint for this confidence interval will be negative if $\overline{X}$ is close enough to zero. Finally, the profile confidence interval (9.8) is

$$\left\{\theta > 0 : \theta - \overline{X}\log(\theta/\overline{X}) - \overline{X} < \frac{z_{\alpha/2}^2}{2n}\right\}.$$

This set will be an interval, because the left-hand side of the inequality is a convex function of $\theta$, but the endpoints cannot be given explicitly and must be computed numerically.

*Example 9.21.* Imagine an experiment in which $X$ is either 1 or 2, according to the toss of a fair coin, and that

$$Y|X = x \sim N(\theta, x).$$

Multiplying the marginal density (mass function) of $X$ by the conditional density of $Y$ given $X$, the joint density of $X$ and $Y$ is

$$f_\theta(x, y) = \frac{1}{2\sqrt{2\pi x}} \exp\left[-\frac{(y-\theta)^2}{2x}\right].$$

The Fisher information is

$$I(\theta) = -E_\theta \frac{\partial^2}{\partial\theta^2} \log f_\theta(X, Y) = E_\theta\left[\frac{1}{X}\right] = \frac{3}{4}.$$

If $(X_1, Y_1), \dots, (X_n, Y_n)$ is a random sample from this distribution, then

$$l_n(\theta) = \sum_{i=1}^n \log f_\theta(X_i, Y_i) = \sum_{i=1}^n \left[-\frac{(Y_i - \theta)^2}{2X_i} - \frac{1}{2}\log(8\pi X_i)\right]$$

and

$$l_n'(\theta) = \sum_{i=1}^n \frac{Y_i - \theta}{X_i}.$$

Equating this to zero, the maximum likelihood estimator is

$$\hat{\theta}_n = \frac{\sum_{i=1}^n (Y_i/X_i)}{\sum_{i=1}^n (1/X_i)}.$$

Also, $l_n''(\theta) = -\sum_{i=1}^n (1/X_i)$. Here the first two confidence intervals, (9.4) and (9.5), are the same (since the Fisher information is constant), namely

$$\left(\hat{\theta}_n - z_{\alpha/2}\sqrt{\frac{4}{3n}}, \hat{\theta}_n + z_{\alpha/2}\sqrt{\frac{4}{3n}}\right).$$

The last two intervals are also the same (because the log-likelihood is exactly quadratic), namely

$$\left(\hat{\theta}_n - \frac{z_{\alpha/2}}{\sqrt{\sum_{i=1}^n (1/X_i)}}, \hat{\theta}_n + \frac{z_{\alpha/2}}{\sqrt{\sum_{i=1}^n (1/X_i)}}\right). \tag{9.9}$$

In this example, the latter, likelihood-based intervals are clearly superior. Given $X_1 = x_1, \dots, X_n = x_n$, $\hat{\theta}_n$ is exactly $N\bigl(\theta, 1/\sum_{i=1}(1/x_i)\bigr)$, and by smoothing, the coverage probability for (9.9) is exactly $1-\alpha$. Also, the width of (9.9) varies in an appropriate fashion: it is shorter when many of the $X_i$ are 1s, increasing in length when more of the $X_i$ are 2s.

## 9.6 EM Algorithm: Estimation from Incomplete Data

The EM algorithm (Dempster et al. (1977)) is a recursive method to calculate maximum likelihood estimators from incomplete data. The "full data" $X$ has density from an exponential family, but is not observed. Instead, the observed data $Y$ is a known function of $X$, $Y = g(X)$, with $g$ many-to-one (so that $X$ cannot be recovered from $Y$). Here we assume for convenience the density for $X$ is in canonical form, given by

$$h(x)e^{\eta T(x) - A(\eta)}.$$

We also assume that $\eta \in \Omega \subset \mathbb{R}$, although the algorithm works in higher dimensions, and that $Y$ is discrete. (The full data $X$ can be discrete or continuous.)

The EM algorithm may be useful when data are partially observed in some sense. For instance, $X_1, \dots, X_n$ could be a random sample from some exponential family, and $Y_i$ could be $X_i$ rounded to the nearest integer. Similar possibilities could include censored or truncated data.

The EM algorithm can also be used in situations with missing data. For instance, we may be studying answers for two multiple choice questions on some survey. The full data $X$ gives information on answers for both questions for every subject. The incomplete data $Y$ may provide counts for all answer combinations for respondents who answered both questions, along with counts for the first question for respondents who skipped the second question, and counts for the second question for respondents who skipped the first question.

Let $\mathcal{X}$ denote the sample space for $X$, $\mathcal{Y}$ the sample space for $Y$, and $\mathcal{X}(y)$ the cross-section

$$\mathcal{X}(y) = \{x \in \mathcal{X} : g(x) = y\}.$$

Then $Y = y$ if and only if $X \in \mathcal{X}(y)$.

**Proposition 9.22.** *The joint density of $X$ and $Y$ (with respect to $\mu \times \nu$ with $\mu$ the dominating measure for $X$ and $\nu$ counting measure on $\mathcal{Y}$) is*

$$1_{\mathcal{X}(y)}(x)h(x)e^{\eta T(x)-A(\eta)}.$$

*Proof.* Let $f$ be an arbitrary nonnegative function on $\mathcal{X} \times \mathcal{Y}$. Then $f(X,Y) = \sum_{y\in\mathcal{Y}} f(X,y)I\{Y = y\}$. Since expectation is linear (or by Fubini's theorem) and $Y = g(X)$,

$$\begin{aligned} Ef(X,Y) &= \sum_{y\in\mathcal{Y}} Ef(X,y)I\{g(X) = y\} \\ &= \sum_{y\in\mathcal{Y}} \int_{\mathcal{X}} f(x,y)I\{g(x) = y\}h(x)e^{\eta T(x)-A(\eta)}\,d\mu(x), \end{aligned}$$

and the proposition follows because $I\{g(x) = y\} = 1_{\mathcal{X}(y)}(x)$. ☐

To define the algorithm, recall that the maximum likelihood estimate of $\eta$ from the full data $X$ is $\psi(T)$, where $\psi$ is the inverse of $A'$. Also, define

$$e(y,\eta) = E_\eta\big[T(X) \mid Y = y\big].$$

This can be computed as an integral against the conditional density of $X$ given $Y = y$. Dividing the joint density of $X$ and $Y$ by the marginal density of $Y$, this conditional density is

$$\frac{1_{\mathcal{X}(y)}(x)h(x)e^{\eta T(x)-A(\eta)}}{f_\eta(y)},$$

where

$$f_\eta(y) = P_\eta(Y = y) = P_\eta\big(X \in \mathcal{X}(y)\big) = \int_{\mathcal{X}(y)} h(x)e^{\eta T(x)-A(\eta)}\,d\mu(x).$$

The algorithm begins with an initial guess $\hat\eta_0$ for the true maximum likelihood estimate $\hat\eta$. Using this initial guess and data $Y$, the value of $T(X)$ is imputed to be $T_1 = e(Y,\hat\eta_0)$ (this is called an E-step). The refined estimate for $\hat\eta$ is $\hat\eta_1 = \psi(T_1)$ (an M-step). These E- and M-steps are repeated as necessary, starting with the current estimate for $\hat{\eta}$ instead of the initial guess, until the values converge.

If the exponential family is not specified in canonical form, so the density is $h(x)e^{\eta(\theta)T(x)-B(\theta)}$, the E-step of the EM algorithm stays the same,

$$T_{k+1} = E_{\hat{\theta}_k}[T(X)|Y],$$

and for the M-step, $\hat{\theta}_k$ maximizes $\eta(\theta)T_k - B(\theta)$ over $\theta \in \Omega$.

If the EM algorithm converges to $\tilde{\eta}$, then $\tilde{\eta}$ will satisfy

$$\tilde{\eta} = \psi\bigl(e(Y, \tilde{\eta})\bigr),$$

or, equivalently,

$$A'(\tilde{\eta}) = e(Y, \tilde{\eta}).$$

Since

$$\begin{aligned}
\frac{\partial}{\partial\eta} \log f_\eta(Y) &= \frac{\frac{\partial}{\partial\eta} \int_{\mathcal{X}(Y)} h(x)e^{\eta T(x)-A(\eta)}\, d\mu(x)}{f_\eta(Y)} \\
&= \frac{\int_{\mathcal{X}(Y)} \bigl[T(x) - A'(\eta)\bigr] h(x)e^{\eta T(x)-A(\eta)}\, d\mu(x)}{f_\eta(Y)} \\
&= e(Y, \eta) - A'(\eta),
\end{aligned}$$

the log-likelihood has zero slope when $\eta = \tilde{\eta}$.

*Example 9.23 (Rounding).* Suppose $X_1, \dots, X_n$ are i.i.d. exponential variables with common density $f_\eta(x) = \eta e^{-\eta x}$, $x > 0$; $f_\eta(x) = 0$, $x \le 0$, and let $Y_i = \lfloor X_i \rfloor$, the greatest integer less than or equal to $X_i$, so we only observe the variables rounded down to the nearest integer. The joint distributions of $X_1, \dots, X_n$ form an exponential family with canonical parameter $\eta$ and complete sufficient statistic $T = -(X_1 + \cdots + X_n)$. The maximum likelihood estimator of $\eta$ based on $X$ is $\psi(T) = -n/T$. Arguing as in Proposition 9.22,

$$\begin{aligned}
E_\eta[X_i|Y_i = y_i] &= E_\eta[X_i | y_i \le X_i < y_i + 1] \\
&= \frac{\int_{y_i}^{y_i+1} x\eta e^{-\eta x}\, dx}{\int_{y_i}^{y_i+1} \eta e^{-\eta x}\, dx} = y_i + \frac{e^\eta - 1 - \eta}{\eta(e^\eta - 1)},
\end{aligned}$$

and by the independence,

$$E_\eta[X_i|Y_1 = y_1, \dots, Y_n = y_n] = E_\eta[X_i|Y_i = y_i].$$

Thus

$$\begin{aligned}
e(y, \eta) &= E_\eta[T|Y_1 = y_1, \dots, Y_n = y_n] \\
&= -\sum_{i=1}^n E_\eta[X_i|Y_i = y_i] = -n\left[\overline{y} + \frac{e^\eta - 1 - \eta}{\eta(e^\eta - 1)}\right].
\end{aligned}$$

The EM algorithm is given by

$$\hat{\eta}_j = -\frac{n}{T_j} \text{ and } T_{j+1} = -n\left[\overline{Y} + \frac{e^{\hat{\eta}_j} - 1 - \hat{\eta}_j}{\hat{\eta}_j(e^{\hat{\eta}_j} - 1)}\right].$$

In this example, the mass function for $Y_i$ can be computed explicitly:

$$P_\eta(Y_i = y) = P_\eta(y \le X_i < y+1) = (1 - e^{-\eta})(e^{-\eta})^y, \qquad y = 0, 1, \dots,$$

and we see that $Y_1, \dots, Y_n$ are i.i.d. from a geometric distribution with $p = 1 - e^{-\eta}$. The maximum likelihood estimator for $p$ is

$$\hat{p} = \frac{1}{1 + \overline{Y}},$$

and since $\eta = -\log(1-p)$, the maximum likelihood estimator for $\eta$ is

$$\hat{\eta} = -\log(1 - \hat{p}) = \log(1 + 1/\overline{Y}).$$

To study the convergence of the EM iterates $\hat{\eta}_j$, $j \ge 1$, to the maximum likelihood estimator $\hat{\eta}$, suppose $\hat{\eta}_j = \hat{\eta} + \epsilon$. By Taylor expansion,

$$\begin{aligned} T_{j+1} &= -n\left[\overline{Y} + \frac{1}{\hat{\eta} + \epsilon} - \frac{1}{e^{\hat{\eta}+\epsilon} - 1}\right] \\ &= -\frac{n}{\hat{\eta}}\left[1 - \frac{\epsilon}{\hat{\eta}} + \frac{\epsilon\hat{\eta}e^{\hat{\eta}}}{(e^{\hat{\eta}} - 1)^2} + O(\epsilon^2)\right], \end{aligned}$$

and from this,

$$\hat{\eta}_{j+1} = \hat{\eta} + \epsilon\left[1 - \frac{\hat{\eta}^2 e^{\hat{\eta}}}{(e^{\hat{\eta}} - 1)^2}\right] + O(\epsilon^2). \tag{9.10}$$

In particular, if $\hat{\eta}_j = \hat{\eta}$, so $\epsilon = 0$, $\hat{\eta}_{j+1}$ also equals $\hat{\eta}$. This shows that $\hat{\eta}$ is a fixed point of the recursion.

As a numerical routine for optimization, the EM algorithm is generally stable and reliable. One appealing property is that the likelihood increases with each successive iteration. This follows because it is in the class of MM algorithms, discussed in Lange (2004). But convergence is not guaranteed: if the likelihood has multiple modes, the algorithm may converge to a local maximum. Sufficient conditions for convergence are given by Wu (1983). Although the EM algorithm is stable, convergence can be slow. By (9.10), there is linear convergence in our example, with the convergence error $\hat{\eta}_j - \hat{\eta}$ decreasing by a constant factor (approximately) with each iteration. Linear convergence is typical for the EM algorithm. If the likelihood for $Y$ is available, quadratic convergence, with $\hat{\eta}_{j+1} - \hat{\eta} = O\big((\hat{\eta}_j - \hat{\eta})^2\big)$, may be possible by Newton–Raphson or another search algorithm, but faster routines are generally less stable and often require information about derivatives of the objective function.

The EM algorithm can be developed without the exponential family structure assumed here. It can also be supplemented to provide numerical approximations for observed Fisher information. For these and other extensions, see McLachlan and Krishnan (2008).

## 9.7 Limiting Distributions in Higher Dimensions

Most of the results presented earlier in this chapter have natural extensions in higher dimensions. If $x = (x_1, \dots, x_p)$ and $y = (y_1, \dots, y_p)$ are vectors in $\mathbb{R}^p$, then $x \le y$ will mean that $x_i \le y_i$, $i = 1, \dots, p$. The cumulative distribution function $H$ of a random vector $Y$ in $\mathbb{R}^p$ is defined by $H(y) = P(Y \le y)$.

**Definition 9.24.** *Let $Y, Y_1, Y_2, \dots$ be random vectors taking values in $\mathbb{R}^p$, with $H$ the cumulative distribution of $Y$ and $H_n$ the cumulative distribution function of $Y_n$, $n = 1, 2, \dots$. Then $Y_n$* converges in distribution *to $Y$ as $n \to \infty$, written $Y_n \Rightarrow Y$, if $H_n(y) \to H(y)$ as $n \to \infty$ at any continuity point $y$ of $H$.*

For a set $S \subset \mathbb{R}^p$, let $\partial S = \overline{S} - S^o$ denote the boundary of $S$. The following result lists conditions equivalent to convergence in distribution.

**Theorem 9.25.** *If $Y, Y_1, Y_2, \dots$ are random vectors in $\mathbb{R}^p$, then the following conditions are equivalent.*

1. *$Y_n \Rightarrow Y$ as $n \to \infty$.*
2. *$Eu(Y_n) \to Eu(Y)$ for every bounded continuous function $u : \mathbb{R}^p \to \mathbb{R}$.*
3. *$\liminf_{n\to\infty} P(Y_n \in G) \ge P(Y \in G)$ for every open set $G$.*
4. *$\limsup_{n\to\infty} P(Y_n \in F) \le P(Y \in F)$ for every closed set $F$.*
5. *$P(Y_n \in S) \to P(Y \in S)$ for any Borel set $S$ such that $P(Y \in \partial S) = 0$.*

This result is called the *portmanteau theorem*. The second condition in this result is often taken as the definition of convergence in distribution. As is the case for one dimension, the following result is an easy corollary.

**Corollary 9.26.** *If $f : \mathbb{R}^p \to \mathbb{R}^m$ is a continuous function, and if $Y_n \Rightarrow Y$ (a random vector in $\mathbb{R}^p$), then*

$$f(Y_n) \Rightarrow f(Y).$$

In the multivariate extension of the central limit theorem, averages of i.i.d. random vectors, after suitable centering and scaling, will converge to a limit, called the multivariate normal distribution. One way to describe this distribution uses moment generating functions. The moment generating function $M_Y$ for a random vector $Y$ in $\mathbb{R}^p$ is given by

$$M_Y(u) = Ee^{u'Y}, \qquad u \in \mathbb{R}^p.$$

As in the univariate case, if the moment generating functions of two random vectors $X$ and $Y$ agree on any nonempty open set, then $X$ and $Y$ have the same distribution. Suppose $Z = (Z_1, \dots, Z_p)'$ with $Z_1, \dots, Z_p$ a random sample from $N(0, 1)$. By independence,

$$Ee^{u'Z} = (Ee^{u_1 Z_1}) \times \dots \times (Ee^{u_p Z_p}) = e^{u_1^2/2} \times \dots \times e^{u_p^2/2} = e^{u'u/2}.$$

Suppose we define $X = \mu + AZ$. Then

$$EX = \mu \text{ and } \mathrm{Cov}(X) = \Sigma = AA'.$$

Taking $u = A't$ in the formula above for $Ee^{u'Z}$, $X$ has moment generating function

$$\begin{aligned} M_X(t) &= Ee^{t'X} = Ee^{t'\mu + t'AZ} = e^{t'\mu} Ee^{u'Z} \\ &= e^{t'\mu + u'u/2} = e^{t'\mu + t'AA't/2} = e^{t'\mu + t'\Sigma t/2}. \end{aligned}$$

Note that this function depends on $A$ only through the covariance $\Sigma = AA'$. The distribution for $X$ is called the *multivariate normal distribution* with mean $\mu$ and covariance matrix $\Sigma$, written

$$X \sim N(\mu, \Sigma).$$

Linear transformations preserve normality. If $X \sim N(\mu, \Sigma)$ and $Y = AX + b$, then

$$\begin{aligned} M_Y(u) &= Ee^{u'(AX+b)} = e^{u'b} Ee^{u'AX} \\ &= e^{u'b} M_X(A'u) = \exp[u'b + u'A\mu + u'A\Sigma A'u/2], \end{aligned}$$

and so

$$Y \sim N(b + A\mu, A\Sigma A').$$

Naturally, the parameters for this distribution are the mean and covariance of $Y$.

In the construction for $N(\mu, \Sigma)$, any nonnegative definite matrix $\Sigma$ is possible. One suitable matrix $A$ would be a symmetric square root of $\Sigma$. This can be found writing $\Sigma = ODO'$ with $O$ an orthogonal matrix (so $O'O = I$) and $D$ diagonal, and defining $\Sigma^{1/2} = OD^{1/2}O'$, where $D^{1/2}$ is diagonal with entries the square roots of the diagonal entries of $D$. Then $\Sigma^{1/2}$ is symmetric and

$$\Sigma^{1/2}\Sigma^{1/2} = OD^{1/2}O'OD^{1/2}O' = OD^{1/2}D^{1/2}O' = ODO' = \Sigma. \tag{9.11}$$

As a side note, the construction here can be used to define other powers, including negative powers, of a symmetric positive definite matrix $\Sigma$. In this case, the diagonal entries $D_{ii}$ of $D$ are all positive, $D^\alpha$ can be taken as the diagonal matrix with diagonal entries $D_{ii}^\alpha$, and $\Sigma^\alpha \overset{\text{def}}{=} OD^\alpha O'$. This construction gives $\Sigma^0 = I$, and the powers of $\Sigma$ satisfy

$$\Sigma^\alpha \Sigma^\beta = \Sigma^{\alpha+\beta}.$$

When $\Sigma$ is positive definite ($\Sigma > 0$), $N(\mu, \Sigma)$ is absolutely continuous. To derive the density, note that the density of $Z$ is

$$\prod_{i=1}^{p} \frac{e^{-z_i^2/2}}{\sqrt{2\pi}} = \frac{e^{-z'z/2}}{(2\pi)^{p/2}}.$$

Also, the linear transformation $z \rightsquigarrow \mu + \Sigma^{1/2}z$ is one-to-one with inverse $x \rightsquigarrow \Sigma^{-1/2}(x-\mu)$. (Here $\Sigma^{-1/2}$ is the inverse of $\Sigma^{1/2}$.) The Jacobian of the inverse transformation is $\det(\Sigma^{-1/2}) = 1/\sqrt{\det \Sigma}$. So if $X = \mu + \Sigma^{1/2}Z \sim N(\mu, \Sigma)$, a multivariate change of variables gives

$$\begin{aligned} P(X \in B) = P(\mu + \Sigma^{1/2}Z \in B) &= \int \cdots \int 1_B(\mu + \Sigma^{1/2}z) \frac{e^{-z'z/2}}{(2\pi)^{p/2}}\, dz \\ &= \int \cdots \int 1_B(x) \frac{\exp\Bigl(-\frac{1}{2}\bigl(\Sigma^{-1/2}(x-\mu)\bigr)'\bigl(\Sigma^{-1/2}(x-\mu)\bigr)\Bigr)}{(2\pi)^{p/2}\sqrt{\det \Sigma}}\, dx. \end{aligned}$$

From this, $X \sim N(\mu, \Sigma)$ has density

$$\frac{\exp\Bigl(-\frac{1}{2}(x-\mu)'\Sigma^{-1}(x-\mu)\Bigr)}{(2\pi)^{p/2}\sqrt{\det \Sigma}}.$$

The following result generalizes the central limit theorem (Theorem 8.12) to higher dimensions. For a proof, see Billingsley (1995) or any standard text on probability.

**Theorem 9.27 (Multivariate Central Limit Theorem).** *If $X_1, X_2, \dots$ are i.i.d. random vectors with common mean $\mu$ and common covariance matrix $\Sigma$, and if $\overline{X}_n = (X_1 + \cdots + X_n)/n$, $n \ge 1$, then*

$$\sqrt{n}(\overline{X} - \mu) \Rightarrow Y \sim N(0, \Sigma).$$

Asymptotic normality of the maximum likelihood estimator will involve random matrices. The most convenient way to deal with convergence in probability of random matrices is to treat them as vectors, introducing the Euclidean (or Frobenius) norm

$$\|M\| = \Bigl\{\sum_{i,j} M_{ij}^2\Bigr\}^{1/2}.$$

**Definition 9.28.** *A sequence of random matrices $M_n$, $n \ge 1$ converges in probability to a random matrix $M$, written $M_n \xrightarrow{p} M$, if for every $\epsilon > 0$,*

$$P\bigl(\|M_n - M\| > \epsilon\bigr) \to 0$$

*as $n \to \infty$. Equivalently, $M_n \xrightarrow{p} M$ as $n \to \infty$ if $[M_n]_{ij} \xrightarrow{p} M_{ij}$ as $n \to \infty$, for all $i$ and $j$.*

The following results are natural extensions of the corresponding results in one dimension.

**Theorem 9.29.** *If* $M_n \xrightarrow{p} M$ *as* $n \to \infty$, *with* $M$ *a constant matrix, and if* $f$ *is continuous at* $M$, *then* $f(M_n) \xrightarrow{p} f(M)$.

**Theorem 9.30.** *If* $Y, Y_1, Y_2, \dots$ *are random vectors in* $\mathbb{R}^p$ *with* $Y_n \Rightarrow Y$ *as* $n \to \infty$, *and if* $M_n$ *are random matrices with* $M_n \xrightarrow{p} M$ *as* $n \to \infty$, *with* $M$ *a constant matrix, then*

$$M_n Y_n \Rightarrow MY$$

*as* $n \to \infty$

Technical details establishing asymptotic normality of the maximum likelihood estimator in higher dimensions are essentially the same as the details in one dimension, so the presentation here just highlights the main ideas in an informal fashion. Let $X, X_1, X_2, \dots$ be i.i.d. with common density $f_\theta$, $\theta \in \Omega \subset \mathbb{R}^p$. As in one dimension, the log-likelihood can be written as a sum,

$$l_n(\theta) = \sum_{i=1}^n \log f_\theta(X_i).$$

As in Section 4.6, the Fisher information is a matrix,

$$I(\theta) = \text{Cov}_\theta\bigl(\nabla_\theta \log f_\theta(X)\bigr) = -E_\theta \nabla_\theta^2 \log f_\theta(X),$$

and

$$E_\theta \nabla_\theta \log f_\theta(X) = 0.$$

The maximum likelihood estimator based on $X_1, \dots, X_n$ maximizes $l_n$. If $\hat\theta_n$ is consistent and $\theta$ lies in the interior of $\Omega$, then with probability tending to one,

$$\nabla_\theta l_n(\hat\theta_n) = 0.$$

Taylor expansion of $\nabla_\theta l_n(\cdot)$ about $\theta$ gives the following approximation:

$$\nabla_\theta l_n(\hat\theta_n) \approx \nabla_\theta l_n(\theta) + \nabla_\theta^2 l_n(\theta)(\hat\theta_n - \theta).$$

Setting this expression to zero, solving, and introducing powers of $n$,

$$\sqrt{n}(\hat\theta_n - \theta) \approx \left[-\frac{1}{n}\nabla_\theta^2 l_n(\theta)\right]^{-1} \frac{1}{\sqrt{n}}\nabla_\theta l_n(\theta). \tag{9.12}$$

By the multivariate central limit theorem,

$$\frac{1}{\sqrt{n}}\nabla_\theta l_n(\theta) = \sqrt{n}\left[\frac{1}{n}\sum_{i=1}^n \nabla_\theta \log f_\theta(X_i) - 0\right] \Rightarrow Y \sim N\bigl(0, I(\theta)\bigr)$$

as $n \to \infty$. Also, by the law of large numbers,

$$-\frac{1}{n}\nabla_\theta^2 l_n(\theta) = \frac{1}{n}\sum_{i=1}^n \bigl[-\nabla_\theta^2 \log f_\theta(X_i)\bigr] \xrightarrow{P_\theta} I(\theta)$$

as $n \to \infty$. Since the function $A \rightsquigarrow A^{-1}$ is continuous for nonsingular matrices $A$, if $I(\theta) > 0$,

$$\left[-\frac{1}{n}\nabla_\theta^2 l_n(\theta)\right]^{-1} \xrightarrow{P_\theta} I(\theta)^{-1}.$$

The error in (9.12) tends to zero in probability, and then using Theorem 9.30,

$$\sqrt{n}(\hat\theta_n - \theta) \Rightarrow I(\theta)^{-1}Y \sim N\bigl(0, I(\theta)^{-1}\bigr).$$

To verify the stated distribution for $I(\theta)^{-1}Y$, note that $Y$ has the same distribution as $I(\theta)^{1/2}Z$ with $Z$ a vector of i.i.d. standard normal variates, and so

$$I(\theta)^{-1}Y \sim I(\theta)^{-1}I(\theta)^{1/2}Z = I(\theta)^{-1/2}Z \sim N\bigl(0, I(\theta)^{-1}\bigr).$$

The following proposition is a multivariate extension of the delta method.

**Proposition 9.31.** *If $g : \Omega \to \mathbb{R}$ is differentiable at $\theta$, $I(\theta)$ is positive definite, and $\sqrt{n}(\hat\theta_n - \theta) \Rightarrow N\bigl(0, I(\theta)^{-1}\bigr)$, then*

$$\sqrt{n}\bigl(g(\hat\theta_n) - g(\theta)\bigr) \Rightarrow N\bigl(0, \nu^2(\theta)\bigr)$$

*with*

$$\nu^2(\theta) = \bigl(\nabla g(\theta)\bigr)' I(\theta)^{-1}\nabla g(\theta).$$

As an application of this result, if $\hat\nu_n$ is a consistent estimator of $\nu(\theta)$ and $\nu(\theta) > 0$, then

$$\left(g(\hat\theta_n) - \frac{z_{\alpha/2}\hat\nu_n}{\sqrt{n}},\ g(\hat\theta_n) + \frac{z_{\alpha/2}\hat\nu_n}{\sqrt{n}}\right) \tag{9.13}$$

is a $1-\alpha$ asymptotic confidence interval for $g(\theta)$.

Finally, the delta method can be extended to vector-valued functions. In this result, $Dg(\theta)$ denotes a matrix of partial derivatives of $g$, with entries $[Dg(\theta)]_{ij} = \partial g_i(\theta)/\partial\theta_j$.

**Proposition 9.32.** *If $g : \Omega \to \mathbb{R}^m$ is differentiable at $\theta$, $I(\theta)$ is positive definite, and $\sqrt{n}(\hat\theta_n - \theta) \Rightarrow N\bigl(0, I(\theta)^{-1}\bigr)$, then*

$$\sqrt{n}\bigl(g(\hat\theta_n) - g(\theta)\bigr) \Rightarrow N\bigl(0, \Sigma(\theta)\bigr)$$

*with*

$$\Sigma(\theta) = Dg(\theta)I(\theta)^{-1}[Dg(\theta)]'.$$

## 9.8 $M$-Estimators for a Location Parameter

Let $X, X_1, X_2, \dots$ be i.i.d. from some distribution $Q$, and let $\rho$ be a convex function[4] on $\mathbb{R}$ with $\rho(x) \to \infty$ as $x \to \pm\infty$. The *$M$-estimator* $T_n$ associated

[4] These conditions on $\rho$ are convenient, because with them $H$ must have a minimum. But they could be relaxed.

with $\rho$ minimizes

$$H(t) = \sum_{i=1}^{n} \rho(X_i - t)$$

over $t \in \mathbb{R}$. If $\rho$ is continuously differentiable and $\psi = \rho'$, then $T_n$ is also a root of the estimating equation

$$\overline{W}_n(t) \stackrel{\text{def}}{=} \frac{1}{n} \sum_{i=1}^{n} \psi(X_i - t) = 0.$$

Several common estimates of location are $M$-estimators. If $\rho(x) = x^2$, then $T_n = \overline{X}_n$, the sample average, and if $\rho(x) = |x|$, then $T_n$ is the median. Finally, if $Q$ lies in a location family of absolutely continuous distributions with log-concave densities $f(x - \theta)$, then taking $\rho = -\log f$, $T_n$ is the maximum likelihood estimator of $\theta$.

To study convergence, let us assume $\rho$ is continuously differentiable, and define

$$\lambda(t) = E\psi(X - t) = \int \psi(x - t)\, dQ(x).$$

Since $\rho$ is convex, $\psi$ is nondecreasing and $\lambda$ is nonincreasing. Also, $\lambda(t)$ will be negative for $t$ sufficiently large and positive for $t$ sufficiently small.

**Lemma 9.33.** *If $\lambda(t)$ is finite for all $t \in \mathbb{R}$ and $\lambda(t) = 0$ has a unique root $c$, then $T_n \xrightarrow{p} c$.*

Using part 3 of Theorem 9.4, this lemma follows fairly easily from our law of large numbers for random functions, Theorem 9.2. The monotonicity of $\psi$ can be used both to restrict attention to a compact set $K$ and to argue that the envelope of the summands over $K$ is integrable.

If $\rho$ is symmetric, $\rho(x) = \rho(-x)$ for all $x \in \mathbb{R}$, and if the distribution of $X$ is symmetric about some value $\theta$, so that $X - \theta \sim \theta - X$, then in this lemma the limiting value $c$ is $\theta$.

Asymptotic normality for $T_n$ can be established with an argument similar to that used to show asymptotic normality for the maximum likelihood estimator. If $\psi$ is continuously differentiable, then Taylor expansion of $\overline{W}_n$ about $c$ gives

$$\overline{W}_n(T_n) = \overline{W}_n(c) + (T_n - c)\overline{W}_n'(t_n^*),$$

with $t_n^*$ an intermediate value between $c$ and $T_n$. Since $\overline{W}_n(T_n)$ is zero,

$$\sqrt{n}(T_n - c) = -\frac{\sqrt{n}\,\overline{W}_n(c)}{\overline{W}_n'(t_n^*)}.$$

By the central limit theorem,

$$\sqrt{n}\,\overline{W}_n(c) \Rightarrow N\bigl(0, \operatorname{Var}[\psi(X - c)]\bigr),$$

and since $t_n^* \xrightarrow{p} c$, under suitable regularity[5]

$$-\overline{W}'_n(t_n^*) \xrightarrow{p} -E\psi'(X-c) = \lambda'(c).$$

Thus

$$\sqrt{n}(T_n - c) \Rightarrow N\bigl(0, V(\psi, Q)\bigr),$$

where

$$V(\psi, Q) = \frac{E\psi^2(X-c)}{[\lambda'(c)]^2}.$$

$M$-estimation was introduced by Huber (1964) to consider *robust* estimation of a location parameter. As noted above, if $\rho$ is symmetric, $\rho(x) = \rho(-x)$ for all $x \in \mathbb{R}$, and if the distribution of $X$ is symmetric about $\theta$, $X - \theta \sim \theta - X$, then $T_n$ is a consistent estimator of $\theta$. For instance, we might have $Q = N(\theta, 1)$, so that $X - \theta \sim N(0,1)$. Taking $\rho$ the square function, $\rho(x) = x^2$, $T_n$ is the sample average $\overline{X}_n$, which is consistent and fully efficient. In a situation like this it may seem foolish to base $M$-estimation on any other function $\rho$, an impression that seems entirely reasonable if we have complete confidence in a normal model for the data. Unfortunately, doubts arise if we entertain the possibility that our normal distribution for $X$ is even slightly "contaminated" by some other distribution. Perhaps

$$X \sim (1-\epsilon)N(\theta, 1) + \epsilon Q^*, \tag{9.14}$$

with $Q^*$ some other distribution symmetric about $\theta$. Then

$$\text{Var}(X) = 1 - \epsilon + \epsilon \int (x-\theta)^2 \, dQ^*(x).$$

By the central limit theorem, the asymptotic performance of $\overline{X}_n$ is driven by the variance of the summands, and even a small amount of contamination $\epsilon$ can significantly degrade the performance of $\overline{X}_n$ if the variance of $Q^*$ is large. If $Q^*$ has infinite variance, $\sqrt{n}(\overline{X}_n - \theta)$ will not even converge in distribution.

Let $\mathcal{C} = \mathcal{C}_\epsilon$ be the class of all distributions for $X$ with the form in (9.14). If one is confident that $Q \in \mathcal{C}_\epsilon$ it may be natural to use an $M$-estimator with

$$\sup_{Q \in \mathcal{C}_\epsilon} V(\psi, Q)$$

as small as possible. The following result shows that this is possible and describes the optimal function $\psi_0$. The optimal function $\psi_0$ and $\rho_0 = \psi_0'$ are plotted in Figure 9.3.

**Theorem 9.34 (Huber).** *The asymptotic variance $V(\psi, Q)$ has a saddle point: There exists $Q_0 = (1-\epsilon)N(\theta, 1) + \epsilon Q_0^* \in \mathcal{C}_\epsilon$ and $\psi_0$ such that*

---

[5] The condition $E \sup_{t \in [c-\epsilon, c+\epsilon]} |\psi'(X-t)| < \infty$ for some $\epsilon > 0$ is sufficient.

$$\sup_{Q \in \mathcal{C}_\epsilon} V(\psi_0, Q) = V(\psi_0, Q_0) = \inf_{\psi} V(\psi, Q_0).$$

*If $k$ solves*

$$\frac{1}{1-\epsilon} = P\big(|Z| < k\big) + \frac{2\phi(k)}{k},$$

*where $Z \sim N(0,1)$, and if*

$$\rho_0(t) = \begin{cases} \frac{1}{2}t^2, & |t| \le k; \\ k|t| - \frac{1}{2}k^2, & |t| \ge k, \end{cases}$$

*then $\psi_0 = \rho_0'$ and $Q_0^*$ is any distribution symmetric about $\theta$ with*

$$Q_0^*\big([\theta - k, \theta + k]\big) = 0.$$

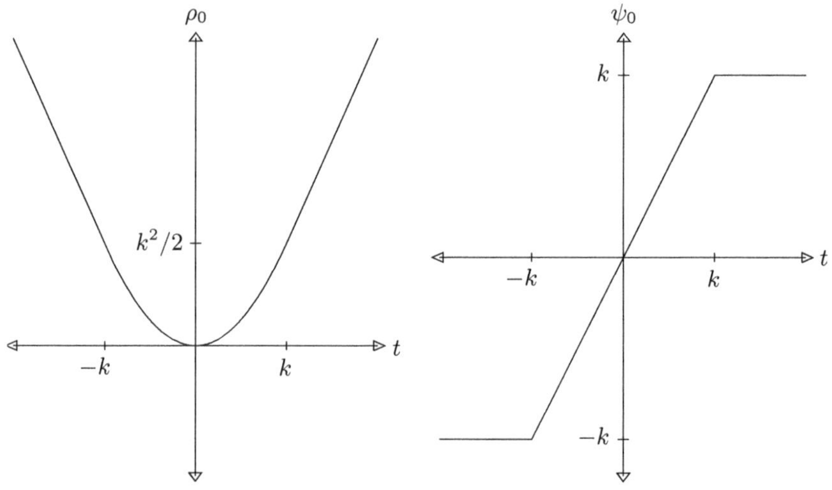


**Fig. 9.3.** Functions $\rho_0$ and $\psi_0$.

## 9.9 Models with Dependent Observations[6]

The asymptotic theory developed earlier in this chapter is based on models with i.i.d. observations. Extensions in which the observations need not have the same distribution and may exhibit dependence are crucial in various applications, and there is a huge literature extending basic results in various directions. In our discussion of the i.i.d. case, the law of large numbers and

[6] Results in this section are somewhat technical and are not used in later chapters.

the central limit theorem were our main tools from probability. Extensions typically rely on more general versions of these results, but the overall nature of the argument is similar to that for the i.i.d. case in most other ways. As a single example, this section sketches how large-sample theory can be developed for models for stationary time series. For extensions in a variety of other directions, see DasGupta (2008).

Time series analysis concerns inference for observations observed over time. Dependence is common and is allowed in the models considered here, but we restrict attention to observations that are stationary according to the following definition. A sequence of random variables, $X_n$, $n \in \mathbb{Z}$, will be called a (stochastic) *process*, and can be viewed as an infinite-dimensional random vector taking values in $\mathbb{R}^{\mathbb{Z}}$.

**Definition 9.35.** *The process $X$ is (strictly)* stationary *if*

$$(X_1, \dots, X_k) \sim (X_{n+1}, \dots, X_{n+k}),$$

*for all $k \geq 1$ and $n \in \mathbb{Z}$.*

Taking $k = 1$ in this definition, observations $X_i$ from a stationary process are identically distributed, and it feels natural to expect information to accumulate fairly linearly over time, as it would with i.i.d. data.

Viewing a sequence $x_n$, $n \in \mathbb{Z}$, as a single point $x \in \mathbb{R}^{\mathbb{Z}}$, we can define a shift operator $T$ that acts on $x$ by incrementing time. Specifically, $y = T(x)$ if $y_n = x_{n+1}$ for all $n \in \mathbb{Z}$. Using $T$, a process $X$ is stationary if $X \sim T(X)$, where $X \sim Y$ means that the finite-dimensional distributions for $X$ and $Y$ agree:

$$(X_i, X_{i+1}, \dots X_j) \sim (Y_i, Y_{i+1}, \dots Y_j)$$

for all $i \leq j$ in $\mathbb{Z}$.

*Example 9.36.* If $X_n$, $n \in \mathbb{Z}$, are i.i.d. from some distribution $Q$, then $X$ is stationary. More generally, a mixture model in which, given $Y = y$, the variables $X_n$, $n \in \mathbb{Z}$, are i.i.d. from $Q_y$ also gives a stationary process $X$.

If $\epsilon_n$, $n \in \mathbb{Z}$, are i.i.d. from some distribution $Q$ with $E\epsilon_n = 0$ and $E\epsilon_n^2 < \infty$, and if $c_n$, $n \geq 1$, are square summable constants, then

$$X_n = \sum_{j=0}^{\infty} c_j \epsilon_{n-j}, \qquad n \in \mathbb{Z},$$

defines a stationary process $X$, called a *linear process*. If $c_n = \rho^n$ with $|\rho| < 1$ then

$$X_{n+1} = \rho X_n + \epsilon_{n+1},$$

and if $Q = N(0, \sigma^2)$ we have the autoregressive model introduced in Example 6.4.

The ergodic theorem is a generalization of the law of large numbers useful in this setting. To describe this result, let $\mathcal{B}_{\mathbb{Z}}$ denote the Borel sets of $\mathbb{R}^{\mathbb{Z}}$.[7] A set $B \in \mathcal{B}_{\mathbb{Z}}$ is called *shift invariant* if $x \in B$ if and only if $T(x) \in B$. Changing the value for $x_n$ at any fixed $n$ will not change whether $x$ lies in a shift invariant set; inclusion can only depend on how the sequence behaves as $|n| \to \infty$. For instance, sets

$$\Bigl\{x : \limsup_{n\to-\infty} x_n \le c\Bigr\} \text{ and } \Biggl\{x : \frac{1}{n}\sum_{i=1}^{n} x_i \to c \text{ as } n \to \infty\Biggr\}$$

are shift invariant, but $\{x : x_3 + x_7 \le x_4\}$ is not.

**Definition 9.37.** *A stationary process $X_n$, $n \in \mathbb{Z}$, is* ergodic *if*

$$P(X \in B) = 0 \text{ or } 1$$

*whenever $B$ is a shift invariant set in $\mathcal{B}_{\mathbb{Z}}$.*

In Example 9.36, i.i.d. variables and linear processes can be shown to be ergodic. But i.i.d. mixtures generally are not; see Problem 9.38.

With this definition we can now state the ergodic theorem. Let $T_j$ denote $T$ composed with itself $j$ times.

**Theorem 9.38 (Ergodic Theorem).** *If $X$ is a stationary ergodic process and $E|g(X)| < \infty$, then*

$$\frac{1}{n}\sum_{j=1}^{n} g\bigl(T_j(X)\bigr) \to \mu_g \stackrel{\text{def}}{=} Eg(X)$$

*almost surely as $n \to \infty$. The convergence here also holds in mean,*

$$E\left|\frac{1}{n}\sum_{j=1}^{n} g\bigl(T_j(X)\bigr) - \mu_g\right| \to 0.$$

As noted, if the $X_n$ are i.i.d. from some distribution $Q$, then $X$ is ergodic. If $g$ is defined by $g(x) = x_0$, then $\mu_g = EX_n$, $g\bigl(T_j(X)\bigr) = X_j$, and the ergodic theorem gives the strong law of large numbers.

For convergence in distribution we use an extension of the ordinary central limit theorem to martingales.

**Definition 9.39.** *For $n \ge 1$, let $M_n$ be a function of $X_1, \dots, X_n$. The process $M_n$, $n \ge 1$, is a (zero mean) martingale with respect to $X_n$, $n \ge 1$, if $EM_1 = 0$ and*

$$E[M_{n+1}|X_1, \dots, X_n] = M_n, \qquad n \ge 1.$$

[7] Formally, $\mathcal{B}_{\mathbb{Z}}$ is the smallest $\sigma$-field that contains all (finite) rectangles of form $\{x \in \mathbb{R}^{\mathbb{Z}} : x_k \in (a_k, b_k), i \le k \le j\}$.

If $M_n$, $n \ge 1$, is a martingale, then by smoothing

$$\begin{aligned} E[M_{n+2}|X_1,\dots,X_n] &= E\bigl[E[M_{n+2}|X_1,\dots,X_{n+1}] \bigm| X_1,\dots,X_n\bigr] \\ &= E[M_{n+1}|X_1,\dots,X_n] = M_n. \end{aligned}$$

With further iteration it is easy to see that

$$E[M_{n+k}|X_1,\dots,X_n] = M_n, \qquad n \ge 1, \quad k \ge 1. \tag{9.15}$$

Defining differences $Y_{n+1} = M_{n+1} - M_n$, $n \ge 1$, with $Y_1 = M_1$,

$$M_n = \sum_{i=1}^n Y_i, \qquad n \ge 1.$$

Using (9.15),

$$\begin{aligned} E[Y_{n+k}|X_1,\dots,X_n] &= E[M_{n+k+1}|X_1,\dots,X_n] - E[M_{n+k}|X_1,\dots,X_n] \\ &= M_n - M_n = 0, \qquad n \ge 1, \ k \ge 1. \end{aligned}$$

If the $X_i$ are i.i.d. with mean $\mu$ and $Y_i = X_i - \mu$, it is easy to check that $M_n = Y_1 + \cdots + Y_n$, $n \ge 1$, is a martingale. By the ordinary central limit theorem, $M_n/\sqrt{n}$ is approximately normal. In the more general case, the summands $Y_i$ may be dependent. But by smoothing,

$$EY_{n+k}Y_n = EE[Y_{n+k}Y_n|X_1,\dots,X_n] = E\bigl[Y_n E[Y_{n+k}|X_1,\dots,X_n]\bigr] = 0,$$

and so they remain uncorrelated, as in the i.i.d. case. Let $\sigma_n^2 \stackrel{\text{def}}{=} \text{Var}(Y_n) = EY_n^2$, and note that since the summands are uncorrelated,

$$\text{Var}(M_n) = \sigma_1^2 + \cdots + \sigma_n^2.$$

For convenience, we assume that $\sigma_n^2 \to \sigma^2$ as $n \to \infty$. (For more general results, see Hall and Heyde (1980).) Then

$$\text{Var}\bigl(M_n/\sqrt{n}\bigr) = \frac{1}{n}\sum_{i=1}^n \sigma_i^2 \to \sigma^2.$$

In contrast with the ordinary central limit theorem, the result for martingales requires some control of the conditional variances

$$s_n^2 \stackrel{\text{def}}{=} \text{Var}(Y_n|X_1,\dots,X_{n-1}) = E[Y_n^2|X_1,\dots,X_{n-1}].$$

Specifically, the following result from Brown (1971a) assumes that

$$\frac{1}{n}\sum_{1=1}^n s_i^2 \xrightarrow{p} \sigma^2, \tag{9.16}$$

and that

$$\frac{1}{n}\sum_{i=1}^{n} E\big[Y_i^2 I\{|Y_i| \geq \epsilon\sqrt{n}\}\big] \to 0 \tag{9.17}$$

as $n \to \infty$ for all $\epsilon > 0$. Requirement (9.17) is called the *Lindeberg condition.* Since $Es_i^2 = \sigma_i^2$, (9.16) might be considered a law of large numbers.

**Theorem 9.40.** *If* $M_n$*,* $n \geq 1$*, is a mean zero martingale satisfying* (9.16) *and* (9.17)*, then*

$$\frac{M_n}{\sqrt{n}} \Rightarrow N(0, \sigma^2).$$

Turning now to inference, let $\theta \in \Omega \subset \mathbb{R}$ be an unknown parameter, and let $P_\theta$ be the distribution for a process $X$ that is stationary and ergodic for all $\theta \in \Omega$. Also, assume that finite-dimensional joint distributions for $X$ are dominated, and let $f_\theta(x_1, \dots, x_n)$ denote the density of $X_1, \dots, X_n$ under $P_\theta$. As usual, this density can be factored using conditional densities as

$$f_\theta(x_1, \dots, x_n) = \prod_{i=1}^{n} f_\theta(x_i | x_1, \dots, x_{i-1}).$$

The log-likelihood function is then

$$l_n(\omega) = \sum_{i=1}^{n} \log f_\omega(X_i | X_1, \dots, X_{i-1}),$$

where, as before, we let $\omega$ denote a generic value for the unknown parameter, reserving $\theta$ for the true value.

With dependent observations, the conditional distributions for $X_n$ change as we condition on past observations. But for most models of interest the amount of change decrease as we condition further into the past, with these distributions converging to the conditional distribution given the entire history of the process. Specifically, we assume that the conditional densities

$$f_\omega(X_n | X_{n-1}, \dots, X_{n-m}) \to f_\omega(X_n | X_{n-1}, \dots) \tag{9.18}$$

in an appropriate sense as $m \to \infty$. The autoregressive model, for instance, has Markov structure with the conditional distributions for $X_n$ depending only on the previous observation, $f_\omega(X_n | X_{n-1}, \dots, X_{n-m}) = f_\omega(X_n | X_{n-1})$. So in this case (9.18) is immediate.

In Section 9.2, our first step towards understanding consistency of the maximum likelihood estimator $\hat\theta_n$ was to argue that if $\omega \neq \theta$, $l_n(\theta)$ will exceed $l_n(\omega)$ with probability tending to one as $n \to \infty$. To understand why that will happen in this case, define

$$g(X) = \log \frac{f_\theta(X_0 | X_{-1}, X_{-2}, \dots)}{f_\omega(X_0 | X_{-1}, X_{-2}, \dots)}.$$

Note that

$$E_\theta\big[g(X)\bigm| X_{-1}, X_{-2}, \dots\big]$$

is the Kullback–Leibler information between $\theta$ and $\omega$ for the conditional distributions of $X_0$ given the past, and is positive unless these conditional distributions coincide. Assuming this is not almost surely the case,

$$\mu_g = E_\theta g(X) = E_\theta E_\theta[g(X)|X_{-1}, X_{-2}, \dots]$$

is positive. Using (9.18), if $j$ is large,

$$\log\frac{f_\theta(X_j|X_1,\dots,X_{j-1})}{f_\omega(X_j|X_1,\dots,X_{j-1})} \approx \log\frac{f_\theta(X_j|X_{j-1},X_{j-2},\dots)}{f_\omega(X_j|X_{j-1},X_{j-2},\dots)} = g\big(T_j(X)\big).$$

Using this approximation,

$$\frac{l_n(\theta)-l_n(\omega)}{n} = \frac{1}{n}\sum_{j=1}^n \log\frac{f_\theta(X_j|X_1,\dots,X_{j-1})}{f_\omega(X_j|X_1,\dots,X_{j-1})} \approx \frac{1}{n}\sum_{i=1}^n g\big(T_j(X)\big), \tag{9.19}$$

converging to $\mu_g > 0$ as $n \to \infty$. If the approximation error here tends to zero in probability (see Problem 9.40 for a sufficient condition), then the likelihood at $\theta$ will be greater than the likelihood at $\omega$ with probability tending to one. Building on this basic idea, consistency of $\hat\theta_n$ can be established in regular cases using the same arguments as those for the i.i.d. case, changing likelihood at a point $\omega$ to the supremum of the likelihood in a neighborhood of $\omega$ (or a neighborhood of infinity).

In the univariate i.i.d. case, asymptotic normality followed using Taylor approximation to show that

$$\sqrt{n}(\hat\theta_n - \theta) = \frac{\dfrac{1}{\sqrt{n}}l_n'(\theta)}{-\dfrac{1}{n}l_n''(\theta)} + o_p(1) \tag{9.20}$$

with

$$\frac{1}{\sqrt{n}}l_n'(\theta) \Rightarrow N\big(0, I(\theta)\big) \text{ and } -\frac{1}{n}l_n''(\theta) \xrightarrow{p} I(\theta). \tag{9.21}$$

The same Taylor expansion argument can be used in this setting, so we mainly need to understand why the limits in (9.21) hold. Convergence for $-l_n''(\theta)/n$ is similar to the argument for consistency above. If we define $h$ as

$$h(X) = -\frac{\partial^2}{\partial\theta^2}\log f_\theta(X_0|X_{-1},X_{-2},\dots),$$

and assume for large $j$,

$$\begin{aligned}-\frac{\partial^2}{\partial\theta^2}\log f_\theta(X_j|X_1,\dots,X_{j-1}) &\approx -\frac{\partial^2}{\partial\theta^2}\log f_\theta(X_j|X_{j-1},X_{j-2},\dots)\\ &= h\big(T_j(X)\big),\end{aligned}$$

which is essentially that (9.18) holds in a differentiable sense, then

$$-\frac{1}{n}l_n''(\theta) = -\frac{1}{n}\sum_{j=1}^{n}\frac{\partial^2}{\partial\theta^2}\log f_\theta(X_j|X_1,\dots,X_{j-1}) \approx \frac{1}{n}\sum_{j=1}^{n}h(T_j(X)),$$

converging to $E_\theta h(X)$ by the ergodic theorem. Since Fisher information here for all $n$ observations is $I_n(\theta) = -E_\theta l_n''(\theta)$, if the approximation error tends to zero in probability, then

$$\frac{1}{n}I_n(\theta) \xrightarrow{p} E_\theta h(X). \tag{9.22}$$

So it is natural to define $I(\theta) = E_\theta h(X)$, interpreted with large samples as average Fisher information per observation.

Asymptotic normality for the score function $l_n'(\theta)$ is based on the martingale central limit theorem. Define

$$Y_j = \frac{\partial}{\partial\theta}\log f_\theta(X_j|X_1,\dots,X_{j-1})$$

so that

$$l_n'(\theta) = \sum_{j=1}^{n}Y_j.$$

The martingale structure needed will hold if we can pass derivatives inside integrals, as in the Cramér–Rao bound, but now with conditional densities. Specifically, we want

$$\begin{aligned}0 &= \int\frac{\partial}{\partial\theta}f_\theta(x_j|x_1,\dots,x_{j-1})\,d\mu(x_j)\\ &= \int\left[\frac{\partial}{\partial\theta}\log f_\theta(x_j|x_1,\dots,x_{j-1})\right]f_\theta(x_j|x_1,\dots,x_{j-1})\,d\mu(x_j).\end{aligned}$$

Viewing this integral as an expectation, we see that

$$E_\theta\bigl[Y_j \bigm| X_1,\dots,X_{j-1}\bigr] = 0,$$

which shows that $l_n'(\theta)$, $n \ge 1$, is a martingale. We also assume

$$\begin{aligned}s_j^2 &\stackrel{\text{def}}{=} \operatorname{Var}_\theta\bigl(Y_j \bigm| X_1,\dots,X_{j-1}\bigr)\\ &= -E_\theta\left[\frac{\partial^2}{\partial\theta^2}\log f_\theta(X_j|X_1,\dots,X_{j-1})\,\middle|\,X_1,\dots,X_{j-1}\right],\end{aligned}$$

which holds if a second derivative can be passed inside the integral above. By smoothing, $\sigma_j^2 = E_\theta s_j^2 = \operatorname{Var}_\theta(Y_j)$, converging to $I(\theta)$ as $j \to \infty$ by an argument like that for (9.22). Therefore if the Lindeberg condition holds, then

$$\frac{1}{\sqrt{n}}l_n'(\theta) \Rightarrow N\bigl(0, I(\theta)\bigr)$$

as $n \to \infty$ by the martingale central limit theorem. Thus with suitable regularity (9.21) should hold in this setting as it did with i.i.d. observations. Then using (9.20)

$$\sqrt{n}(\hat{\theta}_n - \theta) \Rightarrow N\big(0, 1/I(\theta)\big)$$

as $n \to \infty$.

The derivation above is sketchy, but can be made rigorous with suitable regularity. Some possibilities are explored in the problems, but good conditions may also depend on the context. Martingale limit theory is developed in Hall and Heyde (1980). The martingale structure of the score function does not depend on stationarity or ergodicity, and Hall and Heyde's book has a chapter on large-sample theory for the maximum likelihood estimator without these restrictions. Results for stationary ergodic Markov chains are given in Billingsley (1961) and Roussas (1972).

## 9.10 Problems[8]

1. Let $Z_1, Z_2, \dots$ be i.i.d. standard normal, and define random functions $G_n$, $n \ge 1$, taking values in $\mathcal{C}(K)$ with $K = [0, 1]$ by

$$G_n(t) = nZ_n(1-t)t^n - t, \qquad t \in [0, 1].$$

   Finally, take $g(t) = EG_n(t) = -t$.
   a) Show that for any $t \in [0, 1]$, $G_n(t) \xrightarrow{p} g(t)$.
   b) Compute $\sup_{t\in[0,1]} n(1-t)t^n$ and find its limit as $n \to \infty$.
   c) Show that $\|G_n - g\|_\infty$ does not converge in probability to zero as $n \to \infty$.
   d) Let $T_n$ maximize $G_n$ over $[0, 1]$. Show that $T_n$ does not converge to zero in probability.
2. *Method of moments estimation.* Let $X_1, X_2, \dots$ be i.i.d. observations from some family of distributions indexed by $\theta \in \Omega \subset \mathbb{R}$. Let $\overline{X}_n$ denote the average of the first $n$ observations, and let $\mu(\theta) = E_\theta X_i$ and $\sigma^2(\theta) = \mathrm{Var}_\theta(X_i)$. Assume that $\mu$ is strictly monotonic and continuously differentiable. The method of moments estimator $\hat{\theta}_n$ solves $\mu(\theta) = \overline{X}_n$. If $\mu'(\theta) \ne 0$, find the limiting distribution for $\sqrt{n}(\hat{\theta}_n - \theta)$.
3. Take $K = [0, 1]$, let $W_n$, $n \ge 1$, be random functions taking values in $\mathcal{C}(K)$, and let $f$ be a constant function in $\mathcal{C}(K)$. Consider the following conjecture. If $\|W_n - f\|_\infty \xrightarrow{p} 0$ as $n \to \infty$, then $\int_0^1 W_n(t)\,dt \xrightarrow{p} \int_0^1 f(t)\,dt$. Is this conjecture true or false? If true, give a proof; if false, find a counterexample.
4.

[8] Solutions to the starred problems are given at the back of the book.

5. maximum likelihood estimation!inconsistent example If $Z \sim N(\mu, \sigma^2)$, then $X = e^Z$ has the lognormal distribution with parameters $\mu$ and $\sigma^2$. In some situations a threshold $\gamma$, included by taking

$$X = \gamma + e^Z,$$

may be desirable, and in this case $X$ is said to have the three-parameter lognormal distribution with parameters $\gamma$, $\mu$, and $\sigma^2$. Let data $X_1, \dots, X_n$ be i.i.d. from this three-parameter lognormal distribution.
   a) Find the common marginal density for the $X_i$.
   b) Suppose the threshold $\gamma$ is known. Find the maximum likelihood estimators $\hat{\mu} = \hat{\mu}(\gamma)$ and $\hat{\sigma}^2 = \hat{\sigma}^2(\gamma)$ of $\mu$ and $\sigma^2$. (Assume $\gamma < X_{(1)}$.)
   c) Let $l(\gamma, \mu, \sigma^2)$ denote the log-likelihood function. The maximum likelihood estimator for $\gamma$, if it exists, will maximize $l\big(\gamma, \hat{\mu}(\gamma), \hat{\sigma}^2(\gamma)\big)$ over $\gamma$. Determine

$$\lim_{\gamma \uparrow X_{(1)}} l\big(\gamma, \hat{\mu}(\gamma), \hat{\sigma}^2(\gamma)\big).$$

   Hint: Show first that as $\gamma \uparrow X_{(1)}$,

$$\hat{\mu}(\gamma) \sim \frac{1}{n} \log(X_{(1)} - \gamma) \text{ and } \hat{\sigma}^2(\gamma) \sim \frac{n-1}{n^2} \log^2(X_{(1)} - \gamma).$$

   Remark: This thought-provoking example is considered in Hill (1963).
6. Let $X_1, X_2, \dots$ be i.i.d. from a uniform distribution on $(0,1)$, and let $T_n \in [0,1]$ be the unique solution of the equation

$$\sum_{i=1}^n t^{X_i} = \sum_{i=1}^n X_i^2.$$

   a) Show that $T_n \xrightarrow{p} c$ as $n \to \infty$, identifying the constant $c$.
   b) Find the limiting distribution for $\sqrt{n}(T_n - c)$ as $n \to \infty$.
7. Let $X_1, X_2, \dots$ be i.i.d. from a uniform distribution on $(0,1)$ and let $T_n$ maximize

$$\sum_{i=1}^n \frac{\log(1 + t^2 X_i)}{t}$$

over $t > 0$.
   a) Show that $T_n \xrightarrow{p} c$ as $n \to \infty$, identifying the constant $c$.
   b) Find the limiting distribution for $\sqrt{n}(T_n - c)$ as $n \to \infty$.

*8. If $V$ and $W$ are independent variables with $V \sim \chi^2_j$ and $W \sim \chi^2_k$, then the ratio $(V/j)/(W/k)$ has an $F$ distribution with $j$ and $k$ degrees of freedom. Suppose $X_1, \dots, X_m$ is a random sample from $N(\mu_x, \sigma_x^2)$ and $Y_1, \dots, Y_n$ is an independent random sample from $N(\mu_y, \sigma_y^2)$. Find a pivotal quantity with an $F$ distribution. Use this quantity to set a $1-\alpha$ confidence interval for the ratio $\sigma_x/\sigma_y$.

*9. Let $X_1, \dots, X_n$ be i.i.d. from a uniform distribution on $(0, \theta)$.

a) Find the maximum likelihood estimator $\hat{\theta}$ of $\theta$.
b) Show that $\hat{\theta}/\theta$ is a pivotal quantity and use it to set a $1-\alpha$ confidence interval for $\theta$.

*10. Let $X_1, \dots, X_n$ be i.i.d. exponential variables with failure rate $\theta$. Then $T = X_1 + \cdots + X_n$ is complete sufficient. Determine the density of $\theta T$, showing that it is a pivot. Use this pivot to derive a $1 - \alpha$ confidence interval for $\theta$.

*11. Consider a location/scale family of distributions with densities $f_{\theta,\sigma}$ given by

$$f_{\theta,\sigma}(x) = \frac{g\left(\frac{x-\theta}{\sigma}\right)}{\sigma}, \qquad x \in \mathbb{R},$$

where $g$ is a known probability density.
a) Find the density of $(X - \theta)/\sigma$ if $X$ has density $f_{\theta,\sigma}$.
b) If $X_1$ and $X_2$ are independent variables with the same distribution from this family, show that

$$W = \frac{X_1 + X_2 - 2\theta}{|X_1 - X_2|}$$

is a pivot.
c) Derive a confidence interval for $\theta$ using the pivot from part (b).
d) Give a confidence interval for $\sigma$ based on an appropriate pivot.

*12. Suppose $S_1(X)$ and $S_2(X)$ are both $1-\alpha$ confidence regions for the same parameter $g(\theta)$. Show that the intersection $S_1(X) \cap S_2(X)$ is a confidence region with coverage probability at least $1 - 2\alpha$.

*13. Let $(X_1, Y_1)$, …, $(X_n, Y_n)$ be i.i.d. with $X_i \sim N(0, 1)$ and $Y_i | X_i = x \sim N(x\theta, 1)$.
a) Find the maximum likelihood estimate $\hat{\theta}$ of $\theta$.
b) Find the Fisher information $I(\theta)$ for a single observation $(X_i, Y_i)$.
c) Determine the limiting distribution of $\sqrt{n}(\hat{\theta} - \theta)$.
d) Give a $1 - \alpha$ asymptotic confidence interval for $\theta$ based on $I(\hat{\theta})$.
e) Compare the interval in part (d) with a $1 - \alpha$ asymptotic confidence interval based on observed Fisher information.
f) Determine the (exact) distribution of $\sqrt{\sum X_i^2}(\hat{\theta} - \theta)$ and use it to find the true coverage probability for the interval in part (e). Hint: Condition on $X_1, \dots, X_n$ and use smoothing.

14. Let $X_1, \dots, X_n$ be a random sample from $N(\theta, \theta^2)$. Give or describe four asymptotic confidence intervals for $\theta$.

*15. Suppose $X$ has a binomial distribution with $n$ trials and success probability $p$. Give or describe four asymptotic confidence intervals or regions for $p$. Find these regions numerically if $1 - \alpha = 95\%$, $n = 100$, and $X = 30$.

16. Let $X_1, \dots, X_n$ be i.i.d. from a geometric distribution with success probability $p$. Describe four asymptotic confidence regions for $p$.

17. *A variance stabilizing approach.* Let $X_1, X_2, \dots$ be i.i.d. from a Poisson distribution with mean $\theta$, and let $\hat{\theta}_n = \overline{X}_n$ be the maximum likelihood estimator of $\theta$.
    a) Find a function $g : [0, \infty) \to \mathbb{R}$ such that
    $$Z_n = \sqrt{n}\big[g(\hat{\theta}_n) - g(\theta)\big] \Rightarrow N(0, 1).$$
    b) Find a $1 - \alpha$ asymptotic confidence interval for $\theta$ based on the approximate pivot $Z_n$.
18. Let $X_1, X_2, \dots$ be i.i.d. from $N(\mu, \sigma^2)$. Suppose we know that $\sigma$ is a known function of $\mu$, $\sigma = g(\mu)$. Let $\hat{\mu}_n$ denote the maximum likelihood estimator for $\mu$ under this assumption, based on $X_1, \dots, X_n$.
    a) Give a $1-\alpha$ asymptotic confidence interval for $\mu$ centered at $\hat{\mu}_n$. Hint: If $Z \sim N(0, 1)$, then $\mathrm{Var}(Z^2) = 2$ and $\mathrm{Cov}(Z, Z^2) = 0$.
    b) Compare the width of the asymptotic confidence interval in part (a) with the width of the $t$-confidence interval that would be appropriate if $\mu$ and $\sigma$ were not functionally related. Specifically, show that the ratio of the two widths converges in probability as $n \to \infty$, identifying the limiting value. (The limit should be a function of $\mu$.)
19. Suppose that the density for our data $X$ comes from an exponential family with density
    $$h(x)e^{\eta(\theta)T(x) - B(\theta)}, \qquad \theta \in \Omega \subset \mathbb{R}.$$
    If $\hat{\theta}$ is the maximum likelihood estimator of $\theta$, show that $-l''(\hat{\theta})$ and $I(\hat{\theta})$ agree. (Assume that $\eta$ is differentiable and monotonic.) So in this case, the asymptotic confidence intervals (9.5) and (9.7) are the same.

*20. Suppose electronic components are independent and work properly with probability $p$, and that components are tested successively until one fails. Let $X_1$ denote the number that work properly. In addition, suppose devices are constructed using two components connected in series. For proper performance, both components need to work properly, and these devices will work properly with probability $p^2$. Assume these devices are made with different components and are also tested successively until one fails, and let $X_2$ denote the number of devices that work properly.
    a) Determine the maximum likelihood estimator of $p$ based on $X_1$ and $X_2$.
    b) Give the EM algorithm to estimate $p$ from $Y = X_1 + X_2$.
    c) If $Y = 5$ and the initial guess for $p$ is $\hat{p}_0 = 1/2$, give the next two estimates, $\hat{p}_1$ and $\hat{p}_2$, from the EM algorithm.

*21. Suppose $X_1, \dots, X_n$ are i.i.d. with common (Lebesgue) density
    $$f_\theta(x) = \frac{\theta e^{\theta x}}{2 \sinh \theta}, \qquad x \in (-1, 1),$$
    and let $Y_i = I\{X_i > 0\}$. If $\theta = 0$ the $X_i$ are uniformly distributed on $(-1, 1)$.

a) Give an equation for the maximum likelihood estimator $\hat{\theta}_x$ based on $X_1, \dots, X_n$.
b) Find the maximum likelihood estimator $\hat{\theta}_y$ based on $Y_1, \dots, Y_n$.
c) Determine the EM algorithm to compute $\hat{\theta}_y$.
d) Show directly that $\hat{\theta}_y$ is a fixed point for the EM algorithm.
e) Give the first two iterates, $\hat{\theta}_1$ and $\hat{\theta}_2$, of the EM algorithm if the initial guess is $\hat{\theta}_0 = 0$ and there are 5 observations with $Y_1 + \cdots + Y_5 = 3$.

22. Consider a multinomial model for a two-way contingency table with independence, so that $N = (N_{11}, N_{12}, N_{21}, N_{22})$ is multinomial with $n$ trials and success probabilities $\bigl(pq, p(1-q), q(1-p), (1-p)(1-q)\bigr)$. Here $p \in (0,1)$ and $q \in (0,1)$ are unknown parameters.
a) Find the maximum likelihood estimators of $p$ and $q$ based on $N$.
b) Suppose we misplace the off-diagonal entries of the table, so our observed data are $X_1 = N_{11}$ and $X_2 = N_{22}$. Describe in detail the EM algorithm used to compute the maximum likelihood estimators of $p$ and $q$ based on $X_1$ and $X_2$.
c) If the initial guess for $p$ is $2/3$, the initial guess for $q$ is $1/3$, the number of trials is $n = 12$, $X_1 = 4$, and $X_2 = 2$, what are the revised estimates for $p$ after one and two complete iterations of the EM algorithm?

23. Let $X_1, \dots, X_n$ be i.i.d. exponential variables with failure rate $\lambda$. Also, for $i = 1, \dots, n$, let $Y_i = I\{X_i > c_i\}$, where the thresholds $c_1, \dots, c_n$ are known constants in $(0, \infty)$.
a) Derive the EM recursion to compute the maximum likelihood estimator of $\lambda$ based on $Y_1, \dots, Y_n$.
b) Give the first two iterates, $\hat{\lambda}_1$ and $\hat{\lambda}_2$, if the initial guess is $\hat{\lambda}_0 = 1$ and there are three observations, $Y_1 = 1$, $Y_2 = 1$, and $Y_3 = 0$, with $c_1 = 1$, $c_2 = 2$, and $c_3 = 3$.

24. *Contingency tables with missing data.* Counts indicating responses to two binary questions, $A$ and $B$, in a survey are commonly presented in a two-by-two contingency table. In practice, some respondents may only answer one of the questions. If $m$ respondents answer both questions, then cross-classified counts $N = (N_{11}, N_{12}, N_{21}, N_{22})$ for these respondents would be observed, and would commonly be modeled as having a multinomial distribution with $m$ trials and success probability $p = (p_{11}, p_{12}, p_{21}, p_{22})$. Count information for the $n_A$ respondents that only answer question $A$ could be summarized by a variable $R$ representing the number of these respondents who gave the first answer to question $A$. Under the "missing at random" assumption that population proportions for these individuals are the same as proportions for individuals who answer both questions, $R$ would have a binomial distribution with success probability $p_{1+} = p_{11} + p_{12}$. Similarly, for the $n_B$ respondents who only answer question $B$, the variable $C$ counting the number who give the first answer to question $B$ would have a binomial distribution with success probability $p_{+1} = p_{11} + p_{21}$.

a) Develop an EM algorithm to find the maximum likelihood estimator of $p$ from these data, $N, R, C$. The complete data $X$ should be three independent tables $N$, $N^A$, and $N^B$, with sample sizes $m$, $n_A$, and $n_B$, respectively, and common success probability $p$, related to the observed data by $R = N^A_{1+}$ and $C = N^B_{+1}$.

b) Suppose the observed data are

$$N = \begin{pmatrix} 5 & 10 \\ 10 & 5 \end{pmatrix}, \qquad R = 5, \qquad C = 10,$$

with $m = 30$ and $n_A = n_B = 15$. If the initial guess for $p$ is $\hat{p}_0 = N/30$, find the first two iterates for the EM algorithm, $\hat{p}_1$ and $\hat{p}_2$.

25. *A simple hidden Markov model.* Let $X_1, X_2, \dots$ be Bernoulli variables with $EX_1 = 1/2$ and the joint mass function determined recursively by

$$P(X_{k+1} \neq x_k | X_1 = x_1, \dots, X_k = x_k) = \theta, \qquad n = 1, 2, \dots.$$

Viewed as a process in time, $X_n$, $n \geq 1$, is a Markov chain on $\{0, 1\}$ that changes state at each stage with probability $\theta$. Suppose these variables are measured with error. Specifically, let $Y_1, Y_2, \dots$ be Bernoulli variables that are conditionally independent given the $X_i$, satisfying

$$P(Y_i \neq X_i | X_1, X_2, \dots) = \gamma.$$

Assume that the error probability $\gamma$ is a known constant, and $\theta \in (0, 1)$ is an unknown parameter.

a) Show that the joint mass functions for $X_1, \dots, X_n$ form an exponential family.

b) Find the maximum likelihood estimator for $\theta$ based on $X_1, \dots, X_n$.

c) Give formulas for the EM algorithm to compute the maximum likelihood estimator of $\theta$ based on $Y_1, \dots, Y_n$.

d) Give the first two iterates for the EM algorithm, $\hat{\theta}_1$ and $\hat{\theta}_2$, if the initial guess is $\hat{\theta}_0 = 1/2$, the error probability $\gamma$ is 10%, and there are four observations: $Y_1 = 1$, $Y_2 = 1$, $Y_3 = 0$ and $Y_4 = 1$.

26. *Probit analysis.* Let $Y_1, \dots, Y_n$ be independent Bernoulli variables with

$$P(Y_i = 1) = \Phi(\alpha + \beta t_i),$$

where $t_1, \dots, t_n$ are known constants and $\alpha$ and $\beta$ in $\mathbb{R}$ are unknown parameters. Also, let $X_1, \dots, X_n$ be independent with $X_i \sim N(\alpha + \beta t_i, 1)$, $i = 1, \dots, n$.

a) Describe a function $g : R^n \to \{0, 1\}^n$ such that $Y \sim g(X)$ for any $\alpha$ and $\beta$.

b) Find the maximum likelihood estimator for $\theta = (\alpha, \beta)$ based on $X$.

c) Give formulas for the EM algorithm to compute the maximum likelihood estimator of $\theta$ based on $Y$.

d) Suppose we have five observations, $Y = (0, 0, 1, 0, 1)$ and $t_i = i$ for $i = 1, \dots, 5$. Give the first two iterates for the EM algorithm, $\hat{\theta}_1$ and $\hat{\theta}_2$, if the initial guess is $\hat{\theta}_0 = (-2, 1)$.

*27. Suppose $X_1, X_2, \dots$ are i.i.d. with common density $f_\theta$, where $\theta = (\eta, \lambda) \in \Omega \subset \mathbb{R}^2$. Let $I = I(\theta)$ denote the Fisher information matrix for the family, and let $\hat{\theta}_n = (\hat{\eta}_n, \hat{\lambda}_n)$ denote the maximum likelihood estimator from the first $n$ observations.

a) Show that $\sqrt{n}(\hat{\eta}_n - \eta) \Rightarrow N(0, \tau^2)$ under $P_\theta$ as $n \to \infty$, giving an explicit formula for $\tau^2$ in terms of the Fisher information matrix $I$.

b) Let $\tilde{\eta}_n$ denote the maximum likelihood estimator of $\eta$ from the first $n$ observations when $\lambda$ has a known value. Then $\sqrt{n}(\tilde{\eta}_n - \eta) \Rightarrow N(0, \nu^2)$ under $P_\theta$ as $n \to \infty$. Give an explicit formula for $\nu^2$ in terms of the Fisher information matrix $I$, and show that $\nu^2 \le \tau^2$. When is $\nu^2 = \tau^2$?

c) Assume $I(\cdot)$ is a continuous function, and derive a $1 - \alpha$ asymptotic confidence interval for $\eta$ based on the plug-in estimator $I(\hat{\theta}_n)$ of $I(\theta)$.

d) The observed Fisher information matrix for a model with several parameters can be defined as $-\nabla^2 l(\hat{\theta}_n)$, where $\nabla^2$ is the Hessian matrix of partial derivatives (with respect to $\eta$ and $\lambda$). Derive a $1-\alpha$ asymptotic confidence interval for $\eta$ based on observed Fisher information instead of $I(\hat{\theta}_n)$.

28. Let $(X_1, Y_1), (X_2, Y_2), \dots$ be i.i.d. with common Lebesgue density

$$\frac{\exp\left\{-\dfrac{(x-\mu_x)^2 - 2\rho(x-\mu_x)(y-\mu_y) + (y-\mu_y)^2}{2(1-\rho^2)}\right\}}{2\pi\sqrt{1-\rho^2}},$$

where $\theta = (\mu_x, \mu_y, \rho) \in \mathbb{R}^2 \times (-1, 1)$ is an unknown parameter. (This is a bivariate normal density with both variances equal to one.)

a) Give formulas for the maximum likelihood estimators of $\mu_x$, $\mu_y$, and $\rho$.

b) Find the $(3\times3)$ Fisher information matrix $I(\theta)$ for a single observation.

c) Derive asymptotic confidence intervals for $\mu_x$ and $\rho$ based on $I(\hat{\theta})$ and based on observed Fisher information (so you should give four intervals, two for $\mu_x$ and two for $\rho$).

*29. Suppose $W$ and $X$ have a known joint density $q$, and that

$$Y|W = w,\ X = x \sim N(\alpha w + \beta x, 1).$$

Let $(W_1, X_1, Y_1), \dots, (W_n, X_n, Y_n)$ be i.i.d., each with the same joint distribution as $(W, X, Y)$.

a) Find the maximum likelihood estimators $\hat{\alpha}$ and $\hat{\beta}$ of $\alpha$ and $\beta$. Determine the limiting distribution of $\sqrt{n}(\hat{\alpha} - \alpha)$. (The answer will depend on moments of $W$ and $X$.)

b) Suppose $\beta$ is known. What is the maximum likelihood estimator $\tilde{\alpha}$ of $\alpha$? Find the limiting distribution of $\sqrt{n}(\tilde{\alpha} - \alpha)$. When will the limiting distribution for $\sqrt{n}(\tilde{\alpha} - \alpha)$ be the same as the limiting distribution in part (a)?

*30. Prove Proposition 9.31 and show that (9.13) is a $1 - \alpha$ asymptotic confidence interval for $g(\theta)$. Suggest two estimators for $\nu(\theta)$.

*31. Let $N_{11}, \dots, N_{22}$ be cell counts for a two-way table with independence. Specifically, $N$ has a multinomial distribution on $n$ trials, and the cell probabilities satisfy $p_{ij} = p_{i+}p_{+j}$, $i = 1, 2$, $j = 1, 2$. The distribution of $N$ is determined by $\theta = (p_{+1}, p_{1+})$. Find the maximum likelihood estimator $\hat{\theta}$, and give an asymptotic confidence intervals for $p_{11} = \theta_1\theta_2$.

32. Suppose $(X_1, Y_1), \dots, (X_n, Y_n)$ are i.i.d. random vectors in $\mathbb{R}^2$ with common density

$$\frac{\exp\{-(x-\theta_x)^2 + \sqrt{2}(x-\theta_x)(y-\theta_y) - (y-\theta_y)^2\}}{\pi\sqrt{2}}.$$

In polar coordinates we can write $\theta_x = \|\theta\| \cos\omega$ and $\theta_y = \|\theta\| \sin\omega$, with $\omega \in (-\pi, \pi]$. Derive asymptotic confidence intervals for $\|\theta\|$ and $\omega$.

33. Suppose $X_1, X_2, \dots$ are i.i.d. from some distribution $Q_\theta$, with $\theta \in \Omega \subset \mathbb{R}^p$. Assume that the Fisher information matrix $I(\theta)$ exists and is positive definite and continuous as a function of $\theta$. Also, assume that the family $\{Q_\theta : \theta \in \Omega\}$ is regular enough that the maximum likelihood estimators $\hat{\theta}_n$ are consistent, and that

$$\sqrt{n}(\hat{\theta}_n - \theta) \Rightarrow N\bigl(0, I(\theta)^{-1}\bigr).$$

a) Find the limiting distribution for

$$\sqrt{n} I(\theta)^{1/2}(\hat{\theta}_n - \theta).$$

b) Find the limiting distribution for

$$n(\hat{\theta}_n - \theta)' I(\hat{\theta})(\hat{\theta}_n - \theta).$$

Hint: This variable should almost be a function of the random vector in part (a).

c) The variable in part (b) should be an asymptotic pivot. Use this pivot to find an asymptotic $1 - \alpha$ confidence region for $\theta$. (Use the upper quantile for the limiting distribution only.)

d) If $p = 2$ and $I(\theta)$ is diagonal, describe the shape of your asymptotic confidence region. What is the shape of the region if $I(\theta)$ is not diagonal?

34. *Simultaneous confidence intervals.* Suppose $X_1, X_2, \dots$ are i.i.d. from some distribution $Q_\theta$ with $\theta$ two-dimensional:

$$\theta = \begin{pmatrix} \beta \\ \lambda \end{pmatrix} \in \Omega \subset \mathbb{R}^2.$$

Assume that the Fisher information matrix $I(\theta)$ exists, is positive definite, is a continuous function of $\theta$, and is diagonal,

$$I(\theta) = \begin{pmatrix} I_\beta(\theta) & 0 \\ 0 & I_\lambda(\theta) \end{pmatrix}.$$

Finally, assume the family $\{Q_\theta : \theta \in \Omega\}$ is regular enough that the maximum likelihood estimators are asymptotically normal:

$$\sqrt{n}(\hat{\theta}_n - \theta) = \sqrt{n}\left(\begin{pmatrix} \hat{\beta}_n \\ \hat{\lambda}_n \end{pmatrix} - \begin{pmatrix} \beta \\ \lambda \end{pmatrix}\right) \Rightarrow N\bigl(0, I(\theta)^{-1}\bigr).$$

a) Let

$$M_n = \max\left\{\sqrt{nI_\beta(\hat{\theta}_n)}|\hat{\beta}_n - \beta|, \sqrt{nI_\lambda(\hat{\theta}_n)}|\hat{\lambda}_n - \lambda|\right\}.$$

Show that $M_n \Rightarrow M$ as $n \to \infty$. Does the distribution of $M$ depend on $\theta$?

b) Let $q$ denote the upper $\alpha$th quantile for $M$. Derive a formula relating $q$ to quantiles for the standard normal distribution.

c) Use $M_n$ to find a $1-\alpha$ asymptotic confidence region $S$ for $\theta$. (You should only use the upper quantile for the limiting distribution.) Describe the shape of the confidence region $S$.

d) Find intervals $CI_\beta$ and $CI_\lambda$ based on the data, such that

$$P(\beta \in CI_\beta \text{ and } \lambda \in CI_\lambda) \to 1 - \alpha.$$

From this, it is natural to call intervals $CI_\beta$ and $CI_\lambda$ asymptotic simultaneous confidence intervals for $\lambda$ and $\beta$, because the chance they simultaneously cover $\beta$ and $\lambda$ is approximately $1-\alpha$.

35. *Multivariate confidence regions.* Let $X_1, \dots, X_m$ be a random sample from $N(\mu_x, 1)$ and $Y_1, \dots, Y_n$ be a random sample from $N(\mu_y, 1)$, with all $m+n$ variables independent, and let $\overline{X} = (X_1 + \cdots + X_m)/m$ and $\overline{Y} = (Y_1 + \cdots + Y_n)/n$.

a) Find the cumulative distribution function for

$$V = \max\bigl\{|\overline{X} - \mu_x|, |\overline{Y} - \mu_y|\bigr\}.$$

b) Assume $n = m$ and use the pivot from part (a) to find a $1-\alpha$ confidence region for $\theta = (\mu_x, \mu_y)$. What is the shape of this region?

36. Let $X_1, X_2, \dots$ be i.i.d. from a distribution $Q$ that is symmetric about $\theta$. Let $\overline{X}_n$ and $\tilde{X}_n$ denote the mean and median of the first $n$ observations, and let $T_n$ be the $M$-estimator from the first $n$ observations using the function $\rho$ given in Theorem 9.34 with $k = 1$.

a) Determine the asymptotic relative efficiency of $T_n$ with respect to $\overline{X}_n$ if $Q = N(\theta, 1)$.

b) Determine the asymptotic relative efficiency of $T_n$ with respect to $\tilde{X}_n$ if $Q$ is absolutely continuous with density

$$\frac{1}{\pi[(x-\theta)^2+1]},$$

a Cauchy density with location $\theta$.

37. Let $X_1, X_2, \ldots$ be i.i.d. from a distribution $Q$ that is symmetric about $\theta$ and absolutely continuous with density $q$. Fix $\epsilon$ (or $k$), and let $T_n$ be the $M$-estimator from the first $n$ observations using the function $\rho$ given in Theorem 9.34. Take $\psi = \rho'$.
    a) Suggest a consistent estimator for $\lambda'(\theta)$. You can assume that $\lambda'(\theta) = -E\psi'(X-\theta)$.
    b) Suggest a consistent estimator for $E\psi^2(X-\theta)$.
    c) Using the estimators in parts (a) and (b), find an asymptotic $1-\alpha$ confidence interval for $\theta$.
38. Suppose $Y \sim N(0,1)$ and that, given $Y = y$, $X_n$, $n \in \mathbb{Z}$, are i.i.d. from $N(y,1)$. Find a shift invariant set $B$ with $P(X \in B) = 1/2$.
39. Show that if $E|Y_i|^{2+\epsilon}$ is bounded for some $\epsilon > 0$, then the Lindeberg condition (9.17) holds.
40. Show that if

$$E_\theta\left(g(x) - \log\frac{f_\theta(X_0|X_{-1},\ldots,X_{-k})}{f_\omega(X_0|X_{-1},\ldots,X_{-k})}\right)^2, \qquad k = 1, 2, \ldots,$$

are finite[9] and tend to zero as $k \to \infty$, then the approximation error in (9.19) tends to zero in probability as $n \to \infty$.

41. Let $X_n$, $n \in \mathbb{Z}$, be a stationary process with $X_0$ uniformly distributed on $(0,1)$, satisfying

$$X_n = \langle 2X_{n+1}\rangle, \qquad n \in \mathbb{Z}.$$

Here $\langle x\rangle \stackrel{\text{def}}{=} x - \lfloor x\rfloor$ denotes the fractional part of $x$. Show that $X - 1/2$ is a linear process. Identify a distribution $Q$ for the innovations $\epsilon_i$ and coefficients $c_n$, $n \geq 1$.

[9] Actually, it is not hard to argue that the conclusion will still hold if some of these moments are infinite, provided they still converge to zero.

# 10

# Equivariant Estimation

In our study of UMVU estimation, we discovered that, for some models, if we restrict attention to the class of unbiased estimators there may be a best choice. Equivariant estimation is similar, but now we restrict attention to estimators that satisfy symmetry restrictions. At an abstract level, these restrictions are imposed using group theory. The basic ideas are developed here only for estimation of a location parameter, but we try to proceed in a fashion that illustrates the role of group theory.

## 10.1 Group Structure

For estimation of a location parameter, the group of interest is the real line, $\mathcal{G} = \mathbb{R}$, with group multiplication, denoted by $*$, taken to be addition. So $g_1 * g_2 = g_1 + g_2$. This group *acts*, denoted by "$\star$", on points $\theta \in \mathbb{R}$ (the parameter space) by $g \star \theta = g + \theta$, and acts on points $x \in \mathbb{R}^n$ (the data space) by

$$g \star x = \begin{pmatrix} g + x_1 \\ \vdots \\ g + x_n \end{pmatrix} = g\mathbf{1} + x,$$

where $\mathbf{1} \in \mathbb{R}^n$ denotes a column vector of 1s.

Location models arise when each datum $X_i$ can be thought of as the true quantity of interest, $\theta \in \mathbb{R}$, plus measurement error $\epsilon_i$. So

$$X_i = \theta + \epsilon_i, \qquad i = 1, \ldots, n.$$

Writing

$$X = \begin{pmatrix} X_1 \\ \vdots \\ X_n \end{pmatrix} \text{ and } \epsilon = \begin{pmatrix} \epsilon_1 \\ \vdots \\ \epsilon_n \end{pmatrix},$$

these equations can be written in vector form as

R.W. Keener, *Theoretical Statistics: Topics for a Core Course*, Springer Texts in Statistics,
DOI 10.1007/978-0-387-93839-4_10, © Springer Science+Business Media, LLC 2010

$$X = \theta\mathbf{1} + \epsilon.$$

In a location model, the distribution of the error vector $\epsilon$ is fixed, $\epsilon \sim P_0$. This assumption allows dependence between the $\epsilon_i$, but they are often taken to be i.i.d., in which case $P_0 = Q^n$ with $Q$ the common marginal distribution, $\epsilon_i \sim Q$. Letting $P_\theta$ denote the distribution of $X$, so

$$X = \theta\mathbf{1} + \epsilon \sim P_\theta,$$

the family $\mathcal{P} = \{P_\theta, \theta \in \mathbb{R}\}$ is called a *location family*, and $\theta$ is called a *location parameter.*

The symmetry restriction imposed on estimators, called *equivariance*, is defined as follows.

**Definition 10.1.** *An estimator $\delta$ for the location $\theta$ in a location family is called* equivariant *if*

$$\delta(x_1 + g, \dots, x_n + g) = \delta(x_1, \dots, x_n) + g,$$

*or, using vector notation,*

$$\delta(x + g\mathbf{1}) = \delta(x) + g,$$

*for all $g \in \mathbb{R}$, $x \in \mathbb{R}^n$. Using the actions of $g$ on points in $\mathbb{R}$ and $\mathbb{R}^n$, this equation can be written succinctly as*

$$\delta(g \star x) = g \star \delta(x).$$

Examples of equivariant estimators include the sample mean and the median. An optimality theory for equivariant estimation requires considerable structure. The family of distributions must behave naturally under group actions, and the loss function must be invariant, defined below. For location families, since $\theta\mathbf{1} + g\mathbf{1} + \epsilon = g \star (\theta\mathbf{1} + \epsilon)$ has distribution $P_{g\star\theta}$,

$$P_{g\star\theta}(X \in B) = P(\theta\mathbf{1} + g\mathbf{1} + \epsilon \in B) = P\big(g \star (\theta\mathbf{1} + \epsilon\big) \in B) = P_\theta(g \star X \in B).$$

**Definition 10.2.** *A loss function $L$ for the location $\theta$ in a location family is called* invariant *if*

$$L(g \star \theta, g \star d) = L(\theta, d),$$

*for all $g \in \mathbb{R}$, $\theta \in \mathbb{R}$, $d \in \mathbb{R}$. Defining $\rho(x) = L(0, x)$ and taking $g = -\theta$, $L$ is invariant if*

$$L(\theta, d) = \rho(d - \theta)$$

*for all $\theta \in \mathbb{R}$, $d \in \mathbb{R}$.*

Suppose $\delta$ is equivariant and $L$ is invariant. Then the risk of $\delta$ is

$$R(\theta, \delta) = E_\theta \rho\big(\delta(X) - \theta\big) = E\rho\big(\delta(\theta\mathbf{1} + \epsilon) - \theta\big) = E\rho\big(\delta(\epsilon)\big).$$

With the structure imposed, the risk function is a constant, independent of $\theta$. This means that graphs of risk functions for equivariant estimators cannot cross, and we anticipate that there will be a best equivariant estimator $\delta^*$, called the *minimum risk equivariant* estimator. The technical issue here is simply whether the infimum of the risks as $\delta$ varies over the class of equivariant estimators is achieved.

As we proceed, it is convenient to add the assumption that $P_0$ is absolutely continuous with density $f$.

**Proposition 10.3.** *If $P_0$ is absolutely continuous with density $f$, then $P_\theta$ is absolutely continuous with density $f(x_1 - \theta, \dots, x_n - \theta) = f(x - \theta\mathbf{1})$. Conversely, if distributions $P_\theta$ are absolutely continuous with densities $f(x - \theta\mathbf{1})$, then if $\epsilon \sim P_0$, $\theta\mathbf{1} + \epsilon \sim P_\theta$ and $\mathcal{P} = \{P_\theta : \theta \in \mathbb{R}\}$ is a location family.*

*Proof.* Since $\theta\mathbf{1} + \epsilon \sim P_\theta$, the change of variables $e_i = x_i - \theta$, $i = 1, \dots, n$, gives

$$\begin{aligned} P_\theta(B) &= P(\theta\mathbf{1} + \epsilon \in B) \\ &= E 1_B(\theta\mathbf{1} + \epsilon) \\ &= \int \cdots \int 1_B(\theta + e_1, \dots, \theta + e_n) f(e_1, \dots, e_n)\, de_1 \cdots de_n \\ &= \int \cdots \int 1_B(x_1, \dots, x_n) f(x_1 - \theta, \dots, x_n - \theta)\, dx_1 \cdots dx_n \\ &= \underset{B}{\int \cdots \int} f(x_1 - \theta, \dots, x_n - \theta)\, dx_1 \cdots dx_n. \end{aligned}$$

The converse follows similarly. ☐

A function $h$ on $\mathbb{R}^n$ is called *invariant* if $h(g \star x) = h(x)$ for all $x \in \mathbb{R}^n$, $g \in \mathbb{R}$. One invariant function of particular interest is

$$Y = Y(X) = \begin{pmatrix} X_1 - X_n \\ \vdots \\ X_{n-1} - X_n \end{pmatrix}.$$

If $h$ is an arbitrary invariant function, then taking $g = -X_n$,

$$\begin{aligned} h(X) &= h(X - X_n\mathbf{1}) \\ &= h(X_1 - X_n, \dots, X_{n-1} - X_n, 0) \\ &= h(Y_1, \dots, Y_{n-1}, 0). \end{aligned}$$

This shows that any invariant function must be a function of $Y$. For this reason $Y$ is called a *maximal invariant.* This functional relationship means that $Y$ contains at least as much information about $X$ as any other invariant function $h(X)$. Suppose $\delta_0$ and $\delta$ are equivariant estimators. Then their difference $\delta_0 - \delta$ is an invariant function because

$$\delta_0(g \star x) - \delta(g \star x) = [\delta_0(x) + g] - [\delta(x) + g] = \delta_0(x) - \delta(x).$$

So the difference must be a function of $Y$,

$$\delta_0(X) - \delta(X) = v(Y).$$

Conversely, if $\delta_0$ is equivariant and $v$ is an arbitrary function, then

$$\delta(X) = \delta_0(X) - v(Y)$$

is an equivariant estimator, because

$$\delta(g \star x) = \delta_0(g \star x) - v\big(Y(g \star x)\big) = \delta_0(x) + g - v\big(Y(x)\big) = \delta(x) + g.$$

## 10.2 Estimation

The next result shows that optimal estimators are constructed by conditioning on the maximal invariant $Y$ introduced in the previous section.

**Theorem 10.4.** *Consider equivariant estimation of a location parameter with an invariant loss function. Suppose there exists an equivariant estimator $\delta_0$ with finite risk, and that for a.e. $y \in \mathbb{R}^{n-1}$ there is a value $v^* = v^*(y)$ that minimizes*

$$E_0\big[\rho\big(\delta_0(X) - v\big) \bigm| Y = y\big]$$

*over $v \in \mathbb{R}$. Then there is a minimum risk equivariant estimator given by*

$$\delta^*(X) = \delta_0(X) - v^*(Y).$$

*Proof.* From the discussion above, $\delta^*$ is equivariant. Let $\delta(X) = \delta_0(X) - v(Y)$ be an arbitrary equivariant estimator. Then by smoothing, using the fact that risk functions for equivariant estimators are constant,

$$\begin{aligned} R(\theta, \delta) &= E_0 \rho\big(\delta_0(X) - v(Y)\big) \\ &= E_0 E_0\big[\rho\big(\delta_0(X) - v(Y)\big) \bigm| Y\big] \\ &\geq E_0 E_0\big[\rho\big(\delta_0(X) - v^*(Y)\big) \bigm| Y\big] \\ &= E_0 \rho\big(\delta_0(X) - v^*(Y)\big) \\ &= E_0 \rho\big(\delta^*(X)\big) \\ &= R(\theta, \delta^*). \end{aligned}$$

□

To calculate the minimum risk equivariant estimator in this theorem explicitly, let us assume that the equivariant estimator

$$\delta_0(X) = X_n$$

has finite risk. To evaluate the conditional expectation in the theorem we need the conditional distribution of $X_n$ given $Y$ (under $P_0$), which we can obtain from the joint density. Using a change of variables $y_i = x_i - x_n$, $i = 1, \dots, n-1$, in the integrals against $dx_i$,

$$\begin{aligned} P_0\left[\begin{pmatrix} Y \\ X_n \end{pmatrix} \in B\right] &= E_0 1_B(Y_1, \dots, Y_{n-1}, X_n) \\ &= E_0 1_B(X_1 - X_n, \dots, X_{n-1} - X_n, X_n) \\ &= \int \cdots \int 1_B(x_1 - x_n, \dots, x_n) f(x_1, \dots, x_n)\, dx_1 \cdots dx_n \\ &= \underset{B}{\int \cdots \int} f(y_1 + x_n, \dots, y_{n-1} + x_n, x_n)\, dy_1 \cdots dy_{n-1}\, dx_n. \end{aligned}$$

So the joint density of $Y$ and $X_n$ under $P_0$ is

$$f(y_1 + x_n, \dots, y_{n-1} + x_n, x_n).$$

Integration against $x_n$ gives the marginal density of $Y$,

$$f_Y(y) = \int f(y_1 + t, \dots, y_{n-1} + t, t)\, dt.$$

So the conditional density of $X_n = \delta_0$ given $Y = y$ is

$$f_{X_n|Y}(t|y) = \frac{f(y_1 + t, \dots, y_{n-1} + t, t)}{f_Y(y)}.$$

From the theorem, $v^* = v^*(y)$ should be chosen to minimize

$$\frac{\int \rho(t - v) f(y_1 + t, \dots, y_{n-1} + t, t)\, dt}{\int f(y_1 + t, \dots, y_{n-1} + t, t)\, dt}.$$

This is simplified by a change of variables in both integrals taking $t = x_n - u$. Here $x_n$ is viewed as a constant, and we define $x_i$ by $x_i - x_n = y_i$, so that $y_i + t = x_i - u$. Then this expression equals

$$\frac{\int \rho(x_n - v - u) f(x_1 - u, \dots, x_n - u)\, du}{\int f(x_1 - u, \dots, x_n - u)\, du}.$$

Since $\delta^*(x) = x_n - v^*(y)$, it must be the value that minimizes

$$\frac{\int \rho(d - u) f(x_1 - u, \dots, x_n - u)\, du}{\int f(x_1 - u, \dots, x_n - u)\, du}. \tag{10.1}$$

Formally, this looks very similar to the calculations to compute a posterior risk in a Bayesian model. The likelihood at $\theta = u$ is $f(x_1 - u, \dots, x_n - u)$, and $\rho$ is the loss function. If the prior density were taken as one, so the prior distribution is Lebesgue measure $\lambda$, then formally we would be choosing $\delta$ to minimize our posterior risk. Of course, as precise mathematics this is suspect because Lebesgue measure (or any multiple of Lebesgue measure) is not a probability and cannot serve as a proper prior distribution for $\theta$ in a Bayesian model. But the posterior distribution obtained from formal calculations with prior distribution Lebesgue measure will be a probability measure, and the minimum risk equivariant estimator can be viewed informally as Bayes with Lebesgue measure as a prior.

If we define the action of group elements $g$ on Borel sets $B$ by

$$g \star B \stackrel{\text{def}}{=} \{g \star x : x \in B\},$$

then Lebesgue measure is invariant, $\lambda(B) = \lambda(g \star B)$. Measures invariant under the action of some group are called *Haar measures*, and in this setting multiples of Lebesgue measure are the only invariant measures. The structure we have discovered here persists in more general settings. With suitable structure, best equivariant estimators are formally Bayes with Haar measure as the prior distribution for the unknown parameter. For further discussion, see Eaton (1983, 1989).

With squared error loss, $\rho(d - u) = (d - u)^2$, minimization to find the minimum risk equivariant estimator can be done explicitly. If $W$ is an arbitrary random variable, then $E(W - d)^2 = EW^2 - 2dEW + d^2$, and this quadratic function of $d$ is minimized when $d = EW$. If $W$ has density $h$, then $E(W - d)^2 = \int (u - d)^2 h(u)\,du$ and the minimizing value for $d$ is $\int uh(u)\,du$. The minimization of (10.1) has this form, with $h$ (the formal posterior density) given by

$$h(u) = \frac{f(x_1 - u, \dots, x_n - u)}{\int f(x_1 - t, \dots, x_n - t)\,dt}.$$

So with squared error loss, the minimum risk equivariant estimator is

$$\delta^*(X) = \frac{\int u f(X_1 - u, \dots, X_n - u)\,du}{\int f(X_1 - u, \dots, X_n - u)\,du}. \tag{10.2}$$

This estimator $\delta^*$ is called the *Pitman estimator.*

*Example 10.5.* Suppose the measurement errors $\epsilon_1, \dots, \epsilon_n$ are i.i.d. standard exponential variables. Then the density $f(e)$ of $\epsilon$ will be positive when $e_i > 0$, $i = 1, \dots, n$, i.e., when $\min\{e_1, \dots, e_n\} > 0$, and so this density is

$$f(e) = \begin{cases} e^{-(e_1 + \cdots + e_n)}, & \min\{e_1, \dots, e_n\} > 0; \\ 0, & \text{otherwise.} \end{cases}$$

Letting $M = \min\{X_1, \dots, X_n\}$ and noting that $\min\{X_1 - u, \dots, X_n - u\} > 0$ if and only if $u < M$,

$$f(X_1-u,\dots,X_n-u)=\begin{cases}e^{nu-(X_1+\cdots+X_n)}, & u<M;\\ 0, & \text{otherwise.}\end{cases}$$

Thus the Pitman estimator (10.2) in this example is

$$\delta^* = \frac{\int_{-\infty}^{M} u e^{nu-(X_1+\cdots+X_n)}\,du}{\int_{-\infty}^{M} e^{nu-(X_1+\cdots+X_n)}\,du} = \frac{\int_{-\infty}^{0}(t+M)e^{nt}\,dt}{\int_{-\infty}^{0} e^{nt}\,dt} = M-\frac{1}{n}.$$

## 10.3 Problems[1]

*1. Let $X_1 \sim N(\theta,1)$, and suppose that for $j=1,\dots,n-1$ the conditional distribution of $X_{j+1}$ given $X_1=x_1,\dots,X_j=x_j$ is $N\big((x_j+\theta)/2,1\big)$. Show that the joint distributions for $X_1,\dots,X_n$ form a location family and determine the minimum risk equivariant estimator for $\theta$ under squared error loss.

*2. Let $X$ have cumulative distribution function $F$, and assume that $F$ is continuous.
   a) Show that $g(c)=E|X-c|$ is minimized when $c$ is a median of $F$, so $F(c)=1/2$.
   b) Generalizing part (a), define

   $$g(c)=E\big[a(X-c)^+ + b(c-X)^+\big],$$

   where $a$ and $b$ are positive constants. Find the quantile of $F$ that minimizes $g$.

3. Let $\epsilon_1,\dots,\epsilon_n$ be i.i.d. standard exponential variables, and let $X_i=\theta+\epsilon_i$, $i=1,\dots,n$. Using the result in Problem 10.2, determine the minimum risk equivariant estimator of $\theta$ based on $X_1,\dots,X_n$ if the loss function is $L(\theta,d)=|\theta-d|$.

*4. Suppose $X$ has density

$$\frac{1}{2}e^{-|x-\theta|}.$$

Using the result in Problem 10.2, determine the minimum risk equivariant estimator of $\theta$ when the loss for estimating $\theta$ by $d$ is

$$L(\theta,d)=a(d-\theta)^+ + b(\theta-d)^+,$$

with $a$ and $b$ positive constants.

5. Suppose $X$ and $Y$ are independent, with $X\sim N(\theta,1)$ and $Y$ absolutely continuous with density $e^{\theta-y}$ for $y>\theta$, 0 for $y\le\theta$. Determine the minimum risk equivariant estimator of $\theta$ based on $X$ and $Y$ under squared error loss.

[1] Solutions to the starred problems are given at the back of the book.

6. Suppose $\hat{\theta}$ is minimum risk equivariant under squared error loss and that the risk of $\hat{\theta}$ is finite. Is $\hat{\theta}$ then unbiased? Prove or give a counterexample.
7. Suppose $X$ and $Y$ are independent random variables, $X$ with density $\frac{1}{2}e^{-|x-\theta|}$, $x \in \mathbb{R}$, and $Y$ with density $e^{-2|y-\theta|}$, $y \in \mathbb{R}$. Find the minimum risk equivariant estimator of $\theta$ under squared error loss based on $X$ and $Y$.
8. *Equivariant estimation for scale parameters.* Let $\epsilon_1, \dots, \epsilon_n$ be positive random variables with joint distribution $P_1$. If $\sigma > 0$ is an unknown parameter, and $X \sim \sigma\epsilon \sim P_\sigma$, then $\{P_\sigma : \sigma > 0\}$ is called a *scale family*, and $\sigma$ is a scale parameter. (Similar developments are possible without the restriction to positive variables.) The transformation group for equivariant scale estimation is $\mathcal{G} = (0, \infty)$ with $g_1 * g_2 = g_1 g_2$, and group elements act on data values $x \in \mathcal{X} = (0, \infty)^n$ and parameters $\sigma$ by multiplication, $g \star x = gx$ and $g \star \sigma = g\sigma$.
   a) A loss function $L(\sigma, d)$ is *invariant* if $L(g \star \sigma, g \star d) = L(\sigma, d)$ for all $g, \sigma, d$ in $(0, \infty)$. For instance, $L(\sigma, d) = \rho(d/\sigma)$ is invariant. Show that any invariant loss function $L$ must have this form.
   b) A function $h$ on $\mathcal{X}$ is *invariant* if $h(g \star x) = h(x)$ for all $g \in \mathcal{G}$, $x \in \mathcal{X}$. The function
   $$Y(x) = \begin{pmatrix} x_1/x_n \\ \vdots \\ x_{n-1}/x_n \end{pmatrix}$$
   is invariant. If $h$ is invariant, show that $h(x) = \nu\bigl(Y(x)\bigr)$ for some function $\nu$.
   c) An estimator $\delta : \mathcal{X} \to (0, \infty)$ is *equivariant* if $\delta(g \star x) = g \star \delta(x) = g\delta(x)$ for all $g > 0$, $x \in (0, \infty)^n$. Show that the risk function $R(\sigma, \delta)$ for an equivariant estimator of scale, with an invariant loss function, is constant in $\sigma$.
   d) If $\delta$ is an arbitrary equivariant estimator and $\delta_0$ is a fixed equivariant estimator, then $\delta_0/\delta$ is invariant. So $\delta(X) = \delta_0(X)/\nu(Y)$ for some $\nu$. Use this representation to prove a result similar to Theorem 10.4 identifying the minimum risk equivariant estimator in regular cases.
   e) If the distribution $P_1$ for $\epsilon$ is absolutely continuous with density $f$, find the density for $X \sim P_\sigma$.
9. Let $U_1$, $U_2$, and $V$ be independent variables with $U_1$ and $U_2$ uniformly distributed on $(-1, 1)$ and $P(V = 2) = P(V = -2) = 1/2$. Suppose our data $X_1$ and $X_2$ are given by
   $$X_1 = \theta + U_1 + V \text{ and } X_2 = \theta + U_2,$$
   with $\theta \in \mathbb{R}$ and unknown location parameter.
   a) Find the minimum risk equivariant estimator $\delta$ for $\theta$ under squared error loss based on $X_1$ and $X_2$.

b) The best equivariant estimator if we observe only a single observation $X_i$ is that observation $X_i$. Will the estimator $\delta$ from both observations lie between $X_1$ and $X_2$? Explain your answer.

# 11

# Empirical Bayes and Shrinkage Estimators

Many of the classical ideas in statistics become less reliable when there are many parameters. Results in this chapter suggest a natural approach in some situations and illustrate one way in which classical ideas may fail.

## 11.1 Empirical Bayes Estimation

Suppose several objects are measured using some device and that the measurement errors are i.i.d. from $N(0,1)$. (Results here can easily be extended to the case where the errors are from $N(0,\sigma^2)$ with $\sigma^2$ a known constant.) If we measure $p$ objects then our data $X_1,\dots,X_p$ are independent with

$$X_i \sim N(\theta_i, 1), \qquad i = 1,\dots,p,$$

where $\theta_1,\dots,\theta_p$ are the unknown true values. Let $X$ denote the vector $(X_1,\dots,X_p)'$ and $\theta$ the vector $(\theta_1,\dots,\theta_p)'$. If we estimate $\theta_i$ by $\delta_i(X)$ and incur squared error loss, then our total loss, called *compound squared error loss*, is

$$L(\theta,\delta) = \sum_{i=1}^{p} \bigl(\theta_i - \delta_i(X)\bigr)^2.$$

Note that the framework here allows the estimator $\delta_i(X)$ of $\theta_i$ to depend on $X_j$ for $j \neq i$. This is deliberate, and although it may seem unnecessary or unnatural, some estimators for $\theta_i$ in this section depend on $X_i$ and to some extent the other observations. This may be an interesting enigma to ponder as you read this section. Letting $\delta(X)$ denote the vector $\bigl(\delta_1(X),\dots,\delta_p(X)\bigr)'$, the compound loss $L(\theta,\delta)$ equals $\bigl\|\delta(X)-\theta\bigr\|^2$, and the risk function for $\delta$ is given by

$$R(\theta,\delta) = E_\theta\bigl\|\delta(X)-\theta\bigr\|^2.$$

At this stage, let us consider a Bayesian formulation in which the unknown parameter is taken to be a random variable $\Theta$. For a prior distribution, let

R.W. Keener, *Theoretical Statistics: Topics for a Core Course*, Springer Texts in Statistics, DOI 10.1007/978-0-387-93839-4_11, © Springer Science+Business Media, LLC 2010

$\Theta_1, \dots, \Theta_p$ be i.i.d. from $N(0, \tau^2)$. Given $\Theta = \theta$, $X_1, \dots, X_p$ are independent with $X_i \sim N(\theta_i, 1)$, $i = 1, \dots, p$. Then the conditional density of $X$ given $\Theta = \theta$ is

$$\frac{1}{(2\pi)^{p/2}} \exp\left[-\frac{1}{2}\sum_{i=1}^{p}(x_i - \theta_i)^2\right],$$

the marginal density of $\Theta$ is

$$\frac{1}{(2\pi\tau^2)^{p/2}} \exp\left[-\frac{1}{2\tau^2}\sum_{i=1}^{p}\theta_i^2\right],$$

and, multiplying these together, the joint density of $X$ and $\Theta$ is

$$\frac{1}{(2\pi\tau)^p} \exp\left[-\frac{1}{2}\sum_{i=1}^{p}(x_i - \theta_i)^2 - \frac{1}{2\tau^2}\sum_{i=1}^{p}\theta_i^2\right].$$

Completing the square,

$$(x_i - \theta_i)^2 + \frac{\theta_i^2}{\tau^2} = \left(1 + \frac{1}{\tau^2}\right)\left(\theta_i - \frac{x_i}{1 + 1/\tau^2}\right)^2 + \frac{x_i^2}{1 + \tau^2}.$$

Integrating against $\theta$, the marginal density of $X$ is

$$\int \cdots \int \frac{\exp\left[-\frac{1}{2}\left(1 + \frac{1}{\tau^2}\right)\sum_{i=1}^{p}\left(\theta_i - \frac{x_i}{1 + 1/\tau^2}\right)^2 - \frac{1}{2}\sum_{i=1}^{p}\frac{x_i^2}{1+\tau^2}\right]}{(2\pi\tau)^p}\, d\theta$$
$$= \frac{1}{\bigl(2\pi(1+\tau^2)\bigr)^{p/2}} \exp\left[-\frac{1}{2}\sum_{i=1}^{p}\frac{x_i^2}{1+\tau^2}\right].$$

This is a product of densities for $N(0, 1+\tau^2)$, and so $X_1, \dots, X_p$ are i.i.d. from $N(0, 1+\tau^2)$. Dividing the joint density of $X$ and $\Theta$ in the Bayesian model by the marginal density of $X$, the conditional density of $\Theta$ given $X = x$ is

$$\frac{1}{\left(2\pi\frac{\tau^2}{1+\tau^2}\right)^{p/2}} \exp\left[-\frac{1}{2}\sum_{i=1}^{p}\frac{\left(\theta_i - \frac{x_i}{1+1/\tau^2}\right)^2}{\tau^2/(1+\tau^2)}\right].$$

Noting that this factors into a product of normal densities, we see that given $X = x$, $\Theta_1, \dots, \Theta_p$ are independent with

$$\Theta_i | X = x \sim N\left(\frac{x_i}{1 + 1/\tau^2}, \frac{\tau^2}{1+\tau^2}\right).$$

From this

$$E[\Theta_i|X] = \frac{X_i}{1+1/\tau^2} \text{ and } \operatorname{Var}(\Theta_i|X) = \frac{\tau^2}{1+\tau^2}.$$

The expected loss under the Bayesian model is

$$\begin{aligned}
EL\bigl(\Theta, \delta(X)\bigr) &= E\sum_{i=1}^{p}\bigl(\delta_i(X) - \Theta_i\bigr)^2 \\
&= EE\left[\sum_{i=1}^{p}\bigl(\delta_i(X) - \Theta_i\bigr)^2 \;\middle|\; X\right] \\
&= E\sum_{i=1}^{p} E\left[\bigl(\delta_i(X) - \Theta_i\bigr)^2 \;\middle|\; X\right] \\
&= E\sum_{i=1}^{p}\left[\operatorname{Var}(\Theta_i|X) + \left(\frac{X_i}{1+1/\tau^2} - \delta_i(X)\right)^2\right].
\end{aligned}$$

This risk is minimized taking

$$\delta_i(X) = \frac{X_i}{1+1/\tau^2} = \left(1 - \frac{1}{1+\tau^2}\right)X_i. \tag{11.1}$$

In the Bayesian approach to this problem here, the choice of $\tau$ is crucial. In an *empirical Bayes* approach to estimation, the data are used to estimate parameters of the prior distribution. To do this in the current setting, recall that under the Bayesian model, $X_1, \dots, X_p$ are i.i.d. from $N(0, 1+\tau^2)$. The UMVU estimate of $1+\tau^2$ is $\sum_{i=1}^p X_i^2/p$, and slightly different multiples of $\|X\|^2$ may be sensible. The James–Stein estimator of $\theta$ is based on estimating $1/(1+\tau^2)$ by $(p-2)/\|X\|^2$ in (11.1). The resulting estimator is

$$\delta_{JS}(X) = \left(1 - \frac{p-2}{\|X\|^2}\right)X. \tag{11.2}$$

The next section considers the risk of this estimator.

Although the derivation above has a Bayesian feel, the standard deviation $\tau$ that specifies the marginal prior distributions for the $\Theta_i$ is not modeled as a random variable. This deviation $\tau$ might be called a *hyperparameter*, and a fully Bayesian approach to this problem would treat $\tau$ as a random variable with its own prior distribution. Then given $\tau$, $\Theta_1, \dots, \Theta_p$ would be conditionally i.i.d. from $N(0, \tau^2)$. This approach, specifying the prior by coupling a marginal prior distribution for hyperparameters with conditional distributions for the regular parameters, leads to hierarchical Bayes models. With modern computing, estimators based on these models can be practical and have gained popularity in recent years. Hierarchical models are considered in greater detail in Section 15.1.

## 11.2 Risk of the James–Stein Estimator[1]

The following integration by parts identity is an important tool in our study of the risk of the James–Stein estimator. Fubini's theorem provides a convenient way to establish an appropriate regularity condition for this identity.

**Lemma 11.1 (Stein).** *Suppose* $x \sim N(\mu, \sigma^2)$, $h : \mathbb{R} \to \mathbb{R}$ *is differentiable (absolutely continuous is also sufficient), and*

$$E|h'(X)| < \infty. \tag{11.3}$$

*Then*

$$E(X-\mu)h(X) = \sigma^2 Eh'(X).$$

*Proof.* Assume for now that $\mu = 0$ and $\sigma^2 = 1$. If the result holds for a function $h$ it also holds for $h$ plus a constant, and so we can assume without loss of generality that $h(0) = 0$. By Fubini's theorem,

$$\begin{aligned}
\int_0^\infty xh(x)e^{-x^2/2}\,dx &= \int_0^\infty x\left[\int_0^x h'(y)\,dy\right] e^{-x^2/2}\,dx \\
&= \int_0^\infty \int_0^\infty I\{y<x\}xh'(y)e^{-x^2/2}\,dy\,dx \\
&= \int_0^\infty h'(y)\left[\int_y^\infty xe^{-x^2/2}\,dx\right] dy \\
&= \int_0^\infty h'(y)e^{-y^2/2}\,dy.
\end{aligned}$$

The regularity necessary in Fubini's theorem to justify the interchange of the order of integration follows from (11.3). A similar calculation shows that

$$\int_{-\infty}^0 xh(x)e^{-x^2/2}\,dx = \int_{-\infty}^0 h'(y)e^{-y^2/2}\,dy.$$

Adding these together and dividing by $\sqrt{2\pi}$, $EXh(X) = Eh'(X)$ when $X \sim N(0,1)$.

For the general case, let $Z = (X-\mu)/\sigma \sim N(0,1)$. Then $X = \mu + \sigma Z$ and

$$\begin{aligned}
E(X-\mu)h(X) &= \sigma EZh(\mu+\sigma Z) \\
&= \sigma^2 Eh'(\mu+\sigma Z) \\
&= \sigma^2 Eh'(X).
\end{aligned}$$

□

The next lemma generalizes the previous result to higher dimensions. If $h : \mathbb{R}^p \to \mathbb{R}^p$, let $Dh$ denote the $p \times p$ matrix of partial derivatives,

[1] This section covers optional material not used in later chapters.

$$\left[Dh(x)\right]_{ij} = \frac{\partial h_i(x)}{\partial x_j}.$$

Also, let $\|Dh\|$ denote the Euclidean norm of this matrix,

$$\|Dh(x)\| = \left\{\sum_{i,j}\left[Dh(x)\right]_{ij}^2\right\}^{1/2}.$$

**Lemma 11.2.** *Let $X_1, \ldots, X_p$ be independent with $X_i \sim N(\theta_i, 1)$, $i = 1, \ldots, p$. If*

$$E\|Dh(X)\| < \infty,$$

*then*

$$E(X-\theta)'h(X) = E\operatorname{tr}\{Dh(X)\}.$$

*Proof.* Using Stein's lemma (Lemma 11.1) and smoothing,

$$\begin{aligned} E(X_i-\theta_i)h_i(X) &= EE\left[(X_i-\theta_i)h_i(X) \mid X_1,\ldots,X_{i-1},X_{i+1},\ldots,X_p\right] \\ &= EE\left[\frac{\partial h_i(X)}{\partial X_i} \,\middle|\, X_1,\ldots,X_{i-1},X_{i+1},\ldots,X_p\right] \\ &= E\frac{\partial h_i(X)}{\partial X_i} \\ &= E\left[Dh(X)\right]_{ii}. \end{aligned}$$

Summation over $i$ gives the stated result. ☐

The final result provides an unbiased estimator of the risk. Let $X_1, \ldots, X_p$ be independent with $X_i \sim N(\theta_i, 1)$. Given an estimator $\delta(X)$ of $\theta$, define $h(X)$ as $X - \delta(X)$ so that

$$\delta(X) = X - h(X). \tag{11.4}$$

For the James–Stein estimator (11.2),

$$h(X) = \frac{p-2}{\|X\|^2}X.$$

**Theorem 11.3.** *Suppose $X_1, \ldots, X_p$ are independent with $X_i \sim N(\theta_i, 1)$ and that $h$ and $\delta$ are related as in (11.4). Assume that $h$ is differentiable and define*

$$\hat{R} = p + \left\|h(X)\right\|^2 - 2\operatorname{tr}\{Dh(X)\}.$$

*Then*

$$R(\theta,\delta) = E_\theta\left\|\delta(X)-\theta\right\|^2 = E_\theta\hat{R},$$

*provided $E_\theta\left\|Dh(X)\right\| < \infty$.*

*Proof.* Using Lemma 11.2,

$$\begin{aligned} R(\theta,\delta) &= E_\theta \sum_{i=1}^p \bigl(X_i-\theta_i-h_i(X)\bigr)^2 \\ &= E_\theta\left[\sum_{i=1}^p (X_i-\theta_i)^2 + \sum_{i=1}^p h_i^2(X) - 2\sum_{i=1}^p (X_i-\theta_i)h_i(X)\right] \\ &= p + E_\theta\bigl\|h(X)\bigr\|^2 - 2E_\theta(X-\theta)\cdot h(X) \\ &= p + E_\theta\bigl\|h(X)\bigr\|^2 - 2E_\theta \mathrm{tr}\bigl\{Dh(X)\bigr\}. \end{aligned}$$

□

For the James–Stein estimator (11.2),

$$h(X) = \frac{p-2}{\|X\|^2}X,$$

and so

$$h_i(x) = \frac{(p-2)x_i}{x_1^2+\cdots+x_p^2}.$$

Since

$$\begin{aligned} \frac{\partial h_i(x)}{\partial x_i} &= \frac{p-2}{x_1^2+\cdots+x_p^2} - \frac{(p-2)x_i(2x_i)}{(x_1^2+\cdots+x_p^2)^2} \\ &= \frac{p-2}{\|x\|^2} - \frac{2(p-2)x_i^2}{\|x\|^4}, \\ \mathrm{tr}\bigl\{Dh(x)\bigr\} &= \frac{p(p-2)}{\|x\|^2} - \frac{2(p-2)\sum_{i=1}^p x_i^2}{\|x\|^4} = \frac{(p-2)^2}{\|x\|^2}. \end{aligned}$$

Also,

$$\bigl\|h(x)\bigr\|^2 = \sum_{i=1}^p \frac{\bigl[(p-2)x_i\bigr]^2}{\|x\|^4} = \frac{(p-2)^2}{\|x\|^2}.$$

Thus, for the James–Stein estimator,

$$\hat R = p + \frac{(p-2)^2}{\|X\|^2} - 2\frac{(p-2)^2}{\|X\|^2} = p - \frac{(p-2)^2}{\|X\|^2}. \tag{11.5}$$

By Theorem 11.3,

$$R(\theta,\delta_{JS}) = E_\theta \hat R = E_\theta\left[p - \frac{(p-2)^2}{\|X\|^2}\right] < p = R(\theta, X).$$

Hence when $p>2$, the James–Stein estimator always has smaller compound risk than the estimator $X$. Because the risk function for $\delta_{JS}$ is better than the risk function for $X$, in the language of decision theory, developed in the next section, $X$ is called inadmissible.

When $\|\theta\|$ is large, $\|X\|$ will be large with high probability. Then the James–Stein estimator and $X$ will be very similar and will have similar risk. But when $\|\theta\|$ is small there can be a substantial decrease in risk using the James–Stein estimator instead of $X$. If $\theta = 0$, then

$$\|X\|^2 = \sum_{i=1}^{p} X_i^2 \sim \chi_p^2.$$

Integrating against the chi-square density, as in (4.10),

$$E_0 \frac{1}{\|X\|^2} = \frac{1}{p-2}.$$

Using this and (11.5),

$$R(0,\delta) = E_0 \left[ p - \frac{(p-2)}{\|X\|^2} \right] = p - \frac{(p-2)^2}{p-2} = 2.$$

Regardless of the dimension of $\theta$ and $X$, at the origin $\theta = 0$, the James–Stein estimator has risk equal to two.

The results in this section can be extended in various ways. James and Stein (1961) derived the estimator (11.2) and also consider estimation when $\sigma^2$ is unknown. Extensions to ridge regression are reviewed in Draper and van Nostrand (1979). Stein's identity in Lemma 11.1 can be developed for other families of distributions, and these identities have been used in various interesting ways. Chen (1975) and Stein (1986) use them to obtain Poisson limit theorems, and Woodroofe (1989) uses them for interval estimation and to approximate posterior distributions.

## 11.3 Decision Theory[2]

The calculations in the previous section show that $X$ is inadmissible when the dimension $p$ is three or higher, leaving open the natural question of what happens in one or two dimensions. In this section, several results from decision theory are presented and used to characterize admissible procedures and show that for compound estimation $X$ is admissible when $p = 1$. A similar argument shows that $X$ is also admissible when $p = 2$, although the necessary calculations in that case are quite delicate.

Formal decision theory begins with a parameter space $\Omega$, an action space $\mathcal{A}$, a data space $\mathcal{X}$, a model $\mathcal{P} = \{P_\theta : \theta \in \Omega\}$, and a loss function $L : \Omega \times \mathcal{A} \to [-\infty, \infty]$. For simplicity and convenience, we assume that $\mathcal{X} = \mathbb{R}^n$, that $\Omega$ and $\mathcal{A}$ are Borel subsets of Euclidean spaces, and that the loss function $L$ is nonnegative and measurable, with $L(\theta, a)$ lower semicontinuous in $a$.

[2] This section covers optional material not used in later chapters.

A measurable function $\delta : \mathcal{X} \to \mathcal{A}$ is called a *nonrandomized decision rule*, and its *risk function* is defined as

$$R(\theta, \delta) = \int L\big(\theta, \delta(x)\big)\, dP_\theta(x) = E_\theta L\big(\theta, \delta(X)\big), \qquad \theta \in \Omega.$$

The set of all nonrandomized decision rules is denoted $\mathcal{D}_n$.

A nonrandomized decision rule associates with each $x$ an action $\delta(x)$. In contrast, a randomized decision rule associates with each $x$ a probability distribution $\delta_x$, the idea being that if $X = x$ is observed, a random action $A$ will be drawn from $\delta_x$. So $A|X = x \sim \delta_x$. Formally, $\delta$ is should be a stochastic transition kernel, satisfying the regularity condition that $\delta_x(A)$ is a measurable function of $x$ for any Borel set $A$. By smoothing, the risk function for $\delta$ can be defined as

$$\begin{aligned} R(\theta, \delta) = E_\theta L(\theta, A) &= E_\theta E_\theta\big[L(\theta, A)|X\big] \\ &= \iint L(\theta, a)\, d\delta_x(a)\, dP_\theta(x), \qquad \theta \in \Omega. \end{aligned}$$

The set of all randomized decision rules is denoted $\mathcal{D}$.

*Example 11.4 (Estimation).* For estimating a univariate parameter $g(\theta)$ it is natural to take $\mathcal{A} = \mathbb{R}$ as the action space, and a decision rule $\delta$ would be called an estimator. Representative loss functions include squared error loss with $L(\theta, a) = [a - g(\theta)]^2$ and weighted squared error loss with $L(\theta, a) = w(\theta)[a - g(\theta)]^2$.

*Example 11.5 (Testing).* In testing problems, the action space is $\mathcal{A} = \{0, 1\}$, with action "0" associated with accepting $H_0 : \theta \in \Omega_0$ and action "1" associated with accepting $H_1 : \theta \in \Omega_1$. For each $x$, $\delta_x$ is a Bernoulli distribution, which can be specified by its "success" probability $\varphi(x) = \delta_x\big(\{1\}\big)$. This provides a one-to-one correspondence between test functions $\varphi$ and randomized decision rules. A representative loss function now might be zero-one loss in which there is unit loss for accepting the wrong hypothesis:

$$L(\theta, a) = I\{a = 1, \theta \in \Omega_0\} + I\{a = 0, \theta \in \Omega_1\}.$$

If the power function $\beta$ is defined as

$$\beta(\theta) = P_\theta(A = 1) = E_\theta P_\theta(A = 1|X) = E_\theta \delta_X\big(\{1\}\big) = E_\theta \varphi(X),$$

then the risk function with this loss is

$$\begin{aligned} R(\theta, \delta) &= \begin{cases} P_\theta(A = 1), & \theta \in \Omega_0; \\ P_\theta(A = 0), & \theta \in \Omega_0, \end{cases} \\ &= \begin{cases} \beta(\theta), & \theta \in \Omega_0; \\ 1 - \beta(\theta), & \theta \in \Omega_1. \end{cases} \end{aligned}$$

A decision rule $\delta$ is called *inadmissible* if a competing rule $\delta^*$ has a better risk function, specifically if $R(\theta,\delta^*) \le R(\theta,\delta)$ for all $\theta \in \Omega$ with $R(\theta,\delta^*) < R(\theta,\delta)$ for some $\theta \in \Omega$. If this happens we say that $\delta^*$ *dominates* $\delta$. All other rules are called *admissible*. If minimizing risk is the sole concern, no one would ever want to use an inadmissible rule, and there has been considerable interest in characterizing admissible rules. Our first results below show that Bayes rules are typically admissible. More surprising perhaps are extensions, such as Theorem 11.8 below, showing that the remaining admissible rules are almost Bayes in a suitable sense. For notation, for a prior distribution $\Lambda$ let

$$R(\Lambda,\delta) = \int R(\theta,\delta)\,d\Lambda(\theta), \tag{11.6}$$

the integrated risk of $\delta$ under $\Lambda$, and let

$$R(\Lambda) = \inf_{\delta\in\mathcal{D}} R(\Lambda,\delta), \tag{11.7}$$

the minimal integrated risk. Finally, the decision rule $\delta$ is called *Bayes* for a prior $\Lambda$ if it minimizes the integrated risk, that is, if

$$R(\Lambda,\delta) = R(\Lambda). \tag{11.8}$$

At this stage it is worth noting that in definitions (11.6) and (11.7) the prior $\Lambda$ does not really need to be a probability measure; the equations make sense as long as $\Lambda$ is finite, or even if it is infinite but $\sigma$-finite. The definition of Bayes for $\Lambda$ also makes sense for these $\Lambda$. But if the prior $\Lambda$ is not specified, $\delta$ is called proper Bayes only if (11.8) holds for some probability distribution $\Lambda$. Of course, if $\Lambda$ is finite and $\delta$ is Bayes for $\Lambda$ it is also Bayes for the probability distribution $\Lambda(\cdot)/\Lambda(\Omega)$. Thus we are only disallowing rules that are Bayes with respect to an "improper" prior with $\Lambda(\Omega) = \infty$ in this designation.

The next two results address the admissibility of Bayes rules.

**Theorem 11.6.** *If a Bayes rule $\delta$ for $\Lambda$ is essentially unique, then $\delta$ is admissible.*

*Proof.* Suppose $R(\theta,\delta^*) \le R(\theta,\delta)$ for all $\theta \in \Omega$. Then, by (11.6), $R(\Lambda,\delta^*) \le R(\Lambda,\delta)$, and $\delta^*$ must also be Bayes for $\Lambda$. But then, by the essential uniqueness, $\delta = \delta^*$, a.e. $\mathcal{P}$, and so $R(\theta,\delta) = R(\theta,\delta^*)$ for all $\theta \in \Omega$. □

The next result refers to the *support* of the prior distribution $\Lambda$, defined as the smallest closed set $B$ with $\Lambda(B) = 1$. Note that if the support of $\Lambda$ is $\Omega$ and $B$ is an open set with $\Lambda(B) = 0$, then $B$ must be empty, since otherwise $B^c$ would be a closed set smaller than $\Omega$ with $\Lambda(B^c) = 1$.

**Theorem 11.7.** *If risk functions for all decision rules are continuous in $\theta$, if $\delta$ is Bayes for $\Lambda$ and has finite integrated risk $R(\Lambda,\delta) < \infty$, and if the support of $\Lambda$ is the whole parameter space $\Omega$, then $\delta$ is admissible.*

*Proof.* Suppose again that $R(\theta,\delta^*) \le R(\theta,\delta)$ for all $\theta \in \Omega$. Then, as before, $\delta^*$ is Bayes for $\Lambda$ and $\delta$ and $\delta^*$ must have the same integrated risk, $R(\Lambda,\delta) = R(\Lambda,\delta^*$. Hence

$$\int \big(R(\theta,\delta) - R(\theta,\delta^*)\big)\, d\Lambda(\theta) = 0.$$

Since the integrand here is nonnegative, by integration fact 2 in Section 1.4 the set

$$\{\theta : R(\theta,\delta) - R(\theta,\delta^*) > 0\}$$

has $\Lambda$ measure zero. But since risk functions are continuous, this set is open and must then be empty since $\Lambda$ has support $\Omega$. So the risk functions for $\delta$ and $\delta^*$ must be the same, $R(\theta,\delta) = R(\theta,\delta^*)$ for all $\theta \in \Omega$. □

A collection of decision rules is called a *complete class* if all rules outside the class are inadmissible. A complete class will then contain all of the admissible rules. In various situations suitable limits of Bayes procedures form a complete class. Because randomized decision rules are formally stochastic transition functions, a proper statement of most of these results involves notions of convergence for these objects, akin to our notion of convergence in distribution for probability distributions, but complicated by the functional dependence on $X$. An exception arises if the loss function $L(\theta,a)$ is strictly convex in $a$. In this case, admissible rules must be nonrandomized by the Rao–Blackwell theorem (Theorem 3.28), and we have the following result, which can be stated without reference to complicated notions of convergence. This result appears with a careful proof as Theorem 4A.12 of Brown (1986). Let $\mathcal{B}_0$ denote the class of Bayes rules for priors $\Lambda$ concentrated on *finite* subsets of $\Omega$.

**Theorem 11.8.** *Let $\mathcal{P}$ be a dominated family of distributions with $p_\theta$ as the density for $P_\theta$, and assume that $p_\theta(x) > 0$ for all $x \in \mathcal{X}$ and all $\theta \in \Omega$. If $L(\theta,\cdot)$ is nonnegative and strictly convex for all $\theta \in \Omega$, and if $L(\theta,a) \to \infty$ as $\|a\| \to \infty$, again for all $\theta \in \Omega$, then the set of pointwise limits of rules in $\mathcal{B}_0$ forms a complete class.*

This and similar results show that in regular cases any admissible rule will be a limit of Bayes rules. Unfortunately, some limits may give inadmissible rules, and these results cannot be used to show that a given rule is admissible. The final theoretical result of this section gives a sufficient condition for admissibility. For regularity, it assumes that all risk functions are continuous, but similar results are available in different situations. Let

$$\overline{B}_r(x) = \big\{y : \|y - x\| \le r\big\},$$

the closed ball of radius $r$ about $x$.

**Theorem 11.9.** *Assume that risk functions for all decision rules are continuous in $\theta$. Suppose that for any closed ball $\overline{B}_r(x)$ there exist finite measures $\Lambda_m$ such that $R(\Lambda_m,\delta) < \infty$, $m \ge 1$,*

$$\liminf \Lambda_m\bigl(\overline{B}_r(x)\bigr) > 0,$$

*and*

$$R(\delta, \Lambda_m) - R(\Lambda_m) \to 0.$$

*Then $\delta$ is admissible.*

*Proof.* Suppose $\delta^*$ dominates $\delta$. Then $R(\theta_0, \delta^*) < R(\theta_0, \delta)$ for some $\theta_0 \in \Omega$. By continuity

$$\inf_{\theta \in \overline{B}_r(\theta_0)} \bigl[R(\theta, \delta) - R(\theta, \delta^*)\bigr] \to R(\theta_0, \delta) - R(\theta_0, \delta^*) > 0$$

as $r \downarrow 0$, and so there exist values $\epsilon > 0$ and $r_0 > 0$ such that

$$R(\theta, \delta) \geq R(\theta, \delta^*) + \epsilon, \qquad \forall \theta \in \overline{B}_{r_0}(\theta_0).$$

Since $\delta^*$ dominates $\delta$, this implies

$$R(\theta, \delta) \geq R(\theta, \delta^*) + \epsilon I\bigl\{\theta \in \overline{B}_{r_0}\bigr\}.$$

Integrating this against $\Lambda_m$,

$$R(\Lambda_m, \delta) \geq R(\Lambda_m, \delta^*) + \epsilon \Lambda_m\bigl(\overline{B}_{r_0}\bigr) \geq R(\Lambda_m) + \epsilon \Lambda_m\bigl(\overline{B}_{r_0}\bigr),$$

contradicting the assumptions of the theorem. □

Stein (1955) gives a necessary and sufficient condition for admissibility, and using this result the condition in this theorem is also necessary. Related results are given in Blyth (1951), Le Cam (1955), Farrell (1964, 1968a,b), Brown (1971b), and Chapter 8 of Berger (1985).

*Example 11.10.* Consider a Bayesian formulation of the one-sample problem in which $\Theta \sim N(0, \tau^2)$ and given $\Theta = \theta$, $X_1, \dots, X_n$ are i.i.d. from $N(\theta, \sigma^2)$ with $\sigma^2$ a known constant. By the calculation for Problem 6.21, the posterior distribution for $\Theta$ is

$$\Theta | X = x \sim N\left(\frac{\overline{x}}{1 + \sigma^2/(n\tau^2)}, \frac{\sigma^2 \tau^2}{\sigma^2 + n\tau^2}\right),$$

where $\overline{x} = (x_1 + \cdots + x_n)/n$. So the Bayes estimator under squared error loss is

$$\frac{\overline{X}}{1 + \sigma^2/(n\tau^2)}$$

with integrated risk

$$\frac{\sigma^2 \tau^2}{\sigma^2 + n\tau^2}.$$

Since the Bayes estimator converges to $\overline{X}$ as $\tau \to \infty$, if we are hoping to use Theorem 11.9 to show that the sample average $\delta = \overline{X} = (X_1 + \cdots + X_n)/n$

is admissible, it may seem natural to take $\Lambda_m = N(0, m)$. But this does not quite work; since densities for these distributions tend to zero, with this choice $\Lambda_m(\overline{B}_r(x))$ tends to zero as $m \to \infty$. The problem can be simply fixed by rescaling, taking $\Lambda_m = \sqrt{m}N(0, m)$. The density for this measure is $\phi(\theta/\sqrt{m})$, converging pointwise to $\phi(0) = 1/\sqrt{2\pi}$. So by dominated convergence,

$$\Lambda_m(\overline{B}_r(x)) = \int_{x-r}^{x+r} \phi(\theta/\sqrt{m})\, d\theta \to \frac{2r}{\sqrt{2\pi}}.$$

Scaling the prior by $\sqrt{m}$ scales risks and expectations by the same factor $\sqrt{m}$, and so

$$R(\delta, \Lambda_m) = \frac{\sqrt{m}\sigma^2}{n} \quad \text{and} \quad R(\Lambda_m) = \sqrt{m}\sigma^2 \frac{m}{\sigma^2 + nm}.$$

Then

$$R(\delta, \Lambda_m) - R(\Lambda_m) = \frac{\sqrt{m}\sigma^4}{n(\sigma^2 + nm)} \to 0$$

as $m \to \infty$, and by Theorem 11.9 $\overline{X}$ is admissible.

Stein (1956) shows admissibility of the sample average $\overline{X}$ in $p = 2$ dimensions. The basic approach is similar to that pursued in this example, but the priors $\Lambda_n$ must be chosen with great care; it is not hard to see that scaled conjugate normal distributions will not work.

For a more complete introduction to decision theory, see Chernoff and Moses (1986) or Bickel and Doksum (2007), and for a more substantial treatment, see Berger (1985), Ferguson (1967), or Miescke and Liese (2008).

## 11.4 Problems[3]

*1. Consider estimating the failure rates $\lambda_1, \dots, \lambda_p$ for independent exponential variables $X_1, \dots, X_p$. So $X_i$ has density $\lambda_i e^{-\lambda_i x}$, $x > 0$.
   a) Following a Bayesian approach, suppose the unknown parameters are modeled as random variables $\Lambda_1, \dots, \Lambda_p$. For a prior distribution, assume these variables are i.i.d. from a gamma distribution with shape parameter $\alpha$ and unit scale parameter, so $\Lambda_i$ has density $\lambda^{\alpha-1}e^{-\lambda}/\Gamma(\alpha)$, $\lambda > 0$. Determine the marginal density of $X_i$ in this model.
   b) Find the Bayes estimate of $\Lambda_i$ in the Bayesian model with squared error loss.
   c) The Bayesian model gives a family of joint distributions for $X_1, \dots, X_p$ indexed solely by the parameter $\alpha$ (the joint distribution does not depend on $\lambda_1, \dots, \lambda_p$). Determine the maximum likelihood estimate of $\alpha$ for this family.

[3] Solutions to the starred problems are given at the back of the book.

d) Give an empirical Bayes estimator for $\lambda_i$ combining the "empirical" estimate for $\alpha$ in part (c) with the Bayes estimate for $\lambda_i$ when $\alpha$ is known in part (b).

*2. Consider estimation of regression slopes $\theta_1, \dots, \theta_p$ for $p$ pairs of observations, $(X_1, Y_1)$, ..., $(X_p, Y_p)$, modeled as independent with $X_i \sim N(0, 1)$ and $Y_i | X_i = x \sim N(\theta_i x, 1)$.

a) Following a Bayesian approach, let the unknown parameters $\Theta_1$, ..., $\Theta_p$ be i.i.d. random variables from $N(0, \tau^2)$. Find the Bayes estimate of $\Theta_i$ in this Bayesian model with squared error loss.

b) Determine $EY_i^2$ in the Bayesian model. Using this, suggest a simple method of moments estimator for $\tau^2$.

c) Give an empirical Bayes estimator for $\theta_i$ combining the simple "empirical" estimate for $\tau$ in (b) with the Bayes estimate for $\theta_i$ when $\tau$ is known in (a).

*3. Consider estimation of the means $\theta_1, \dots, \theta_p$ of $p$ independent Poisson random variables $X_1, \dots, X_p$ under compound squared error loss, $L(\theta, d) = \sum_{i=1}^p (\theta_i - d_i)^2$.

a) Following a Bayesian approach, let the unknown parameters be modeled as random variables $\Theta_1, \dots, \Theta_p$ that are i.i.d. with common density $\lambda e^{-\lambda x}$ for $x > 0$, $0$ for $x \le 0$. Determine the Bayes estimators of $\Theta_1, \dots, \Theta_p$.

b) Determine the marginal density (mass function) of $X_i$ in the Bayesian model.

c) In the Bayesian model, $X_1, \dots, X_p$ are i.i.d. with the common density in part (b). Viewing this joint distribution as a family of distributions parameterized by $\lambda$, what is the maximum likelihood estimator of $\lambda$.

d) Suggest empirical Bayes estimators for $\theta_1, \dots, \theta_p$ based on the Bayesian estimators in part (a) with an empirical estimator of $\lambda$ from part (c).

4. Consider estimating success probabilities $\theta_1, \dots, \theta_p$ for $p$ independent binomial variables $X_1, \dots, X_p$, each based on $m$ trials, under compound squared error loss, $L(\theta, d) = \sum_{i=1}^p (\theta_i - d_i)^2$.

a) Following a Bayesian approach, model the unknown parameters as random variables $\Theta_1, \dots, \Theta_p$ that are i.i.d. from a beta distribution with parameters $\alpha$ and $\beta$. Determine the Bayes estimators of $\Theta_1, \dots, \Theta_p$.

b) In the Bayesian model, $X_1, \dots, X_p$ are i.i.d. Determine the first two moments for their common marginal distribution, $EX_i$ and $EX_i^2$. Using these, suggest simple method of moments estimators for $\alpha$ and $\beta$.

c) Give empirical Bayes estimators for $\theta_i$ combining the simple "empirical" estimates for $\alpha$ and $\beta$ in (b) with the Bayes estimate for $\theta_i$ when $\alpha$ and $\beta$ are known in (a).

5. Consider estimation of unknown parameters $\theta_1, \dots, \theta_p$ based on data $X_1, \dots, X_p$ that are independent with $X_i \sim N(\theta_i, 1)$ under compound squared error loss.

a) Following a Bayesian approach, model the unknown parameters as random variables $\Theta_1, \dots, \Theta_p$ that are i.i.d. from $N(\nu, \tau^2)$. Find Bayes estimators for the random parameters $\Theta_1, \dots, \Theta_p$.
b) Suggest "empirical" estimates for $\nu$ and $\tau^2$ based on $\overline{X}$ and $S^2$, the mean and sample variance of the $X_i$.
c) Give empirical Bayes estimators for $\theta_1, \dots, \theta_p$ based on the Bayesian estimators in (a) and the estimates for $\nu$ and $\tau^2$ in (b).

6. Consider estimation of unknown parameters $\theta_1, \dots, \theta_p$ based on data $X_1, \dots, X_p$ that are independent with $X_i \sim \text{Unif}(0, \theta_i)$, $i = 1, \dots, p$, under compound squared error loss.
   a) Following a Bayesian approach, model the unknown parameters as random variables $\Theta_1, \dots, \Theta_p$ which are i.i.d. and absolutely continuous with common density
   $$x\lambda^2 1_{(0,\infty)}(x) e^{-\lambda x}.$$
   Find Bayes estimators for $\Theta_1, \dots, \Theta_p$.
   b) Suggest an empirical estimate for $\lambda$ based on the sample average $\overline{X}$.
   c) Give empirical Bayes estimators for $\theta_1, \dots, \theta_p$ based on the Bayes estimators in (a) and the estimator for $\lambda$ in (b).

# 12

# Hypothesis Testing

In hypothesis testing data are used to infer which of two competing hypotheses, $H_0$ or $H_1$, is correct. As before, $X \sim P_\theta$ for some $\theta \in \Omega$, and the two competing hypotheses are that the unknown parameter $\theta$ lies in set $\Omega_0$ or in set $\Omega_1$, written

$$H_0 : \theta \in \Omega_0 \text{ versus } H_1 : \theta \in \Omega_1.$$

We assume that $\Omega_0$ and $\Omega_1$ partition $\Omega$, so $\Omega = \Omega_0 \bigcup \Omega_1$ and $\Omega_0 \bigcap \Omega_1 = \emptyset$. This chapter derives optimal tests when the parameter $\theta$ is univariate. Extensions to higher dimensions are given in Chapter 13.

## 12.1 Test Functions, Power, and Significance

A *nonrandomized test* of $H_0$ versus $H_1$ can be specified by a *critical region* $S$ with the convention that we accept $H_1$ when $X \in S$ and accept $H_0$ when $X \notin S$. The performance of this test is described by its *power function* $\beta(\cdot)$, which gives the chance of rejecting $H_0$ as a function of $\theta \in \Omega$:

$$\beta(\theta) = P_\theta(X \in S).$$

Ideally, we would want $\beta(\theta) = 0$ for $\theta \in \Omega_0$ and $\beta(\theta) = 1$ for $\theta \in \Omega_1$, but in practice this is generally impossible.

In the mathematical formulation for hypothesis testing just presented, the hypotheses $H_0$ and $H_1$ have a symmetric role. But in applications $H_0$ generally represents the status quo, or what someone would believe about $\theta$ without compelling evidence to the contrary. In view of this, attention is often focused on tests that have a small chance of error when $H_0$ is correct. This can be quantified by the *significance level* $\alpha$ defined as

$$\alpha = \sup_{\theta \in \Omega_0} P_\theta(X \in S).$$

In words, the level $\alpha$ is the worst chance of falsely rejecting $H_0$.

R.W. Keener, *Theoretical Statistics: Topics for a Core Course*, Springer Texts in Statistics, DOI 10.1007/978-0-387-93839-4_12, © Springer Science+Business Media, LLC 2010

For technical reasons it is convenient to allow external randomization to "help" the researcher decide between $H_0$ and $H_1$. Randomized tests are characterized by a *test or critical function* $\varphi$ with range a subset of $[0,1]$. Given $X = x$, $\varphi(x)$ is the chance of rejecting $H_0$. The power function $\beta$ still gives the chance of rejecting $H_0$, and by smoothing,

$$\beta(\theta) = P_\theta(\text{reject } H_0) = E_\theta P_\theta(\text{reject } H_0 | X) = E_\theta \varphi(X).$$

Note that a nonrandomized test with critical region $S$ can be viewed as a randomized test with $\varphi = 1_S$. Conversely, if $\varphi(x)$ is always 0 or 1, then the randomized test with critical function $\varphi$ can be considered a nonrandomized test with critical region $S = \{x : \varphi(x) = 1\}$.

The set of all critical functions is convex, for if $\varphi_1$ and $\varphi_2$ are critical functions and $\gamma \in (0,1)$, then $\gamma\varphi_1 + (1-\gamma)\varphi_2$ is also a critical function. Convex combinations of nonrandomized tests are not possible, and this is the main advantage of allowing randomization. For randomized tests the level $\alpha$ is defined as

$$\alpha = \sup_{\theta \in \Omega_0} \beta(\theta) = \sup_{\theta \in \Omega_0} E_\theta \varphi(X).$$

## 12.2 Simple Versus Simple Testing

A hypothesis is called *simple* if it completely specifies the distribution of the data, so $H_i : \theta \in \Omega_i$ is simple when $\Omega_i$ contains a single parameter value $\theta_i$. When both hypotheses, $H_0$ and $H_1$ are simple, the Neyman–Pearson lemma (Proposition 12.2 below) provides a complete characterization of all reasonable tests. This result makes use of *Lagrange multipliers*, an important idea in optimization theory of independent interest.

Suppose $H_0$ and $H_1$ are both simple, and let $p_0$ and $p_1$ denote densities for $X$ under $H_0$ and $H_1$, respectively.[1] Since there are only two distributions for the data $X$, the power function for a test $\varphi$ has two values,

$$\alpha = E_0\varphi = \int \varphi(x)p_0(x)\,d\mu(x)$$

and

$$E_1\varphi = \int \varphi(x)p_1(x)\,d\mu(x).$$

Ideally, the first of these values $\alpha$ is near zero, and the other value $\beta$ is near one. These objectives are in conflict. To do as well as possible we consider the constrained maximization problem of maximizing $E_1\varphi$ among all test $\varphi$ with $E_0\varphi = \alpha$. The following proposition shows that solutions of unconstrained optimization problems with Lagrange multipliers ($k$) also solve optimization problems with constraints.

[1] As a technical note, there is no loss of generality in assuming densities $p_0$ and $p_1$, since the two distributions $P_0$ and $P_1$ are both absolutely continuous with respect to their sum $P_0 + P_1$.

**Proposition 12.1.** *Suppose $k \geq 0$, $\varphi^*$ maximizes*

$$E_1\varphi - kE_0\varphi$$

*among all critical functions, and $E_0\varphi^* = \alpha$. Then $\varphi^*$ maximizes $E_1\varphi$ over all $\varphi$ with level at most $\alpha$.*

*Proof.* Suppose $\varphi$ has level at most $\alpha$, $E_0\varphi \leq \alpha$. Then

$$\begin{aligned} E_1\varphi &\leq E_1\varphi - kE_0\varphi + k\alpha \\ &\leq E_1\varphi^* - kE_0\varphi^* + k\alpha \\ &= E_1\varphi^*. \end{aligned}$$ □

Maximizing $E_1\varphi - kE_0\varphi$ is fairly easy because

$$\begin{aligned} E_1\varphi - kE_0\varphi &= \int \bigl[p_1(x) - kp_0(x)\bigr]\varphi(x)\,d\mu(x) \\ &= \int_{p_1(x)>kp_0(x)} \bigl|p_1(x) - kp_0(x)\bigr|\varphi(x)\,d\mu(x) \\ &\quad - \int_{p_1(x)<kp_0(x)} \bigl|p_1(x) - kp_0(x)\bigr|\varphi(x)\,d\mu(x). \end{aligned} \tag{12.1}$$

Clearly, any test $\varphi^*$ maximizing this expression must have

$$\varphi^*(x) = 1, \text{ when } p_1(x) > kp_0(x),$$

and

$$\varphi^*(x) = 0, \text{ when } p_1(x) < kp_0(x).$$

When division by zero is not an issue, these tests are based on the *likelihood ratio* $L(x) = p_1(x)/p_0(x)$, with $\varphi^*(x) = 1$ if $L(x) > k$ and $\varphi^*(x) = 0$ if $L(x) < k$. When $L(x) = k$, $\varphi(x)$ can take any value in $[0,1]$. Any test of this form is called a *likelihood ratio test.* In addition, the test $\varphi = I\{p_0 = 0\}$ is also considered a likelihood ratio test. (This can be viewed as the test that arises when $k = \infty$.)

**Proposition 12.2 (Neyman–Pearson Lemma).** *Given any level $\alpha \in [0,1]$, there exists a likelihood ratio test $\varphi_\alpha$ with level $\alpha$, and any likelihood ratio test with level $\alpha$ maximizes $E_1\varphi$ among all tests with level at most $\alpha$.*

The fact that likelihood ratio tests maximize $E_1\varphi$ among tests with the same or smaller level follows from the discussion above. A formal proof that any desired level $\alpha \in [0,1]$ can be achieved with a likelihood ratio test is omitted, but similar issues are addressed in the proof of the first part of Theorem 12.9. Also, Example 12.6 below illustrates the type of adjustments that are needed to achieve level $\alpha$ in a typical situation. The next result shows that if a test is optimal, it must be a likelihood ratio test.

**Proposition 12.3.** *Fix $\alpha \in [0,1]$, let $k$ be the critical value for a likelihood ratio test $\varphi_\alpha$ described in Proposition 12.2, and define $B = \{x : p_1(x) \neq kp_0(x)\}$. If $\varphi^*$ maximizes $E_1\varphi$ among all tests with level at most $\alpha$, then $\varphi^*$ and $\varphi_\alpha$ must agree on $B$, $1_B\varphi^* = 1_B\varphi_\alpha$, a.e. $\mu$.*

*Proof.* Assume $k \in (0,\infty)$ and let $B_1 = \{p_1 > kp_0\}$ and $B_2 = \{p_1 < kp_0\}$, so that $B = B_1 \bigcup B_2$. Since $\varphi^*$ and $\varphi_\alpha$ both maximize $E_1\varphi$, we have $E_1\varphi^* = E_1\varphi_\alpha$. And since $\varphi_\alpha$ maximizes $E_1\varphi - kE_0\varphi$, $kE_0\varphi_\alpha = k\alpha \leq kE_0\varphi^*$. So $E_0\varphi^*$ must equal $\alpha$, and $\varphi_\alpha$ and $\varphi^*$ both have level $\alpha$. Thus they both give the same value in (12.1). Since $\varphi_\alpha$ is 1 on $B_1$ and 0 on $B_2$, using (12.1),

$$\int 1_{B_1}|p_1 - kp_0|(1-\varphi^*)\,d\mu + \int 1_{B_2}|p_1 - kp_0|\varphi^*\,d\mu = 0.$$

Since the arguments of both integrands are nonnegative, both integrands must be zero a.e. $\mu$, and since $|p_1 - kp_0|$ is positive on $B_1$ and $B_2$, we must have

$$1_{B_1}(1-\varphi^*) + 1_{B_2}\varphi^* = 1_{B_1}|\varphi^* - \varphi_\alpha| + 1_{B_2}|\varphi^* - \varphi_\alpha| = 0$$

a.e. $\mu$.

When $k = 0$, $\varphi_\alpha = 1$ on $p_1 > 0$, and $\varphi_\alpha$ has power $E_1\varphi_\alpha = 1$. If $\varphi^*$ has power 1, then $0 = E_1(\varphi_\alpha - \varphi^*) = \int_B |\varphi^* - \varphi_\alpha| p_1\,d\mu$, so again $\varphi^*$ and $\varphi_\alpha$ agree a.e. $\mu$ on $B$.

For the final degenerate case, "$k = \infty$," $B$ should be defined as $\{p_0 > 0\}$. In this case $\varphi_\alpha = 0$ on $p_0 > 0$, and so $\alpha = 0$. If $\varphi^*$ has level $\alpha = 0$, $0 = E_0(\varphi^* - \varphi_\alpha) = \int_B |\varphi^* - \varphi_\alpha| p_0\,d\mu$, and once again $\varphi^*$ and $\varphi_\alpha$ agree a.e. $\mu$ on $B$. ☐

**Corollary 12.4.** *If $P_0 \neq P_1$ and $\varphi_\alpha$ is a likelihood ratio test with level $\alpha \in (0,1)$, then $E_1\varphi_\alpha > \alpha$.*

*Proof.* Consider the test $\varphi^*$ which is identically $\alpha$, regardless of the value of $x$. Since $\varphi_\alpha$ maximizes $E_1\varphi$ among tests with level $\alpha$, $E_1\varphi_\alpha \geq E_1\varphi^* = \alpha$. Suppose $E_1\varphi_\alpha = \alpha$. Then $\varphi^*$ also maximizes $E_1\varphi$ among tests with level $\alpha$, and by Proposition 12.3, $\varphi_\alpha$ and $\varphi^*$ must agree a.e. on $B$. But since $\alpha \in (0,1)$ and $\varphi_\alpha$ is 0 or 1 on $B$, they cannot agree on $B$. Thus $B$ must be a null set and $p_1 = kp_0$ a.e. $\mu$. Integrating this against $\mu$, $k$ must equal 1, so the densities agree a.e. $\mu$ and $P_0 = P_1$. ☐

*Example 12.5.* Suppose $X$ is absolutely continuous with density

$$p_\theta(x) = \begin{cases} \theta e^{-\theta x}, & x > 0; \\ 0, & \text{otherwise}, \end{cases}$$

and that we would like to test $H_0 : \theta = 1$ versus $H_1 : \theta = \theta_1$, where $\theta_1$ is a specified constant greater than one. A likelihood ratio test $\varphi$ is one if

$$\frac{p_1(X)}{p_0(X)} = \frac{\theta_1 e^{-\theta_1 X}}{e^{-X}} > k,$$

or equivalently if

$$X < \frac{\log(\theta_1/k)}{\theta_1 - 1} = k'.$$

The test is zero if $X > k'$. When $X = k'$ the test can take any value in $[0,1]$, but the choice will not affect any power calculations since $P_\theta(X = k') = 0$. The level of this likelihood ratio test is

$$\alpha = P_0(X < k') = \int_0^{k'} e^{-x}\, dx = 1 - e^{-k'}.$$

Solving,

$$k' = -\log(1-\alpha)$$

gives a test with level $\alpha$. If $\varphi_\alpha$ is a test with

$$\varphi_\alpha(X) = \begin{cases} 1, & X < -\log(1-\alpha); \\ 0, & X > -\log(1-\alpha), \end{cases}$$

then by Proposition 12.1, $\varphi_\alpha$ maximizes $E_{\theta_1}\varphi$ among all tests with level $\alpha$. Something surprising and remarkable has happened here. This test $\varphi_\alpha$, which is optimal for testing $H_0 : \theta = 1$ versus $H_1 : \theta = \theta_1$, does not depend on the value $\theta_1$. If $\varphi$ is any competing test with level $\alpha$, then

$$E_{\theta_1}\varphi \le E_{\theta_1}\varphi_\alpha, \qquad \text{for all } \theta_1 > 1.$$

Features of this example that give the same optimal test regardless of the value of $\theta_1$ are detailed and exploited in the next section.

*Example 12.6.* Suppose $X$ has a binomial distribution with success probability $\theta$ and $n = 2$ trials. If we are interested in testing $H_0 : \theta = 1/2$ versus $H_1 : \theta = 3/4$, then

$$L(X) = \frac{p_1(X)}{p_0(X)} = \frac{\binom{2}{X}(3/4)^X(1/4)^{2-X}}{\binom{2}{X}(1/2)^X(1/2)^{2-X}} = \frac{3^X}{4}.$$

Under $H_0$,

$$L(X) = \begin{cases} 1/4, & \text{with probability } 1/4; \\ 3/4, & \text{with probability } 1/2; \\ 9/4, & \text{with probability } 1/4. \end{cases}$$

Suppose the desired significance level is $\alpha = 5\%$. If $k$ is less than $9/4$, then $L(2) = 9/4 > k$ and $\varphi(2) = 1$. But then $E_0\varphi(X) \ge \varphi(2)P_0(X = 2) = 1/4$. If instead $k$ is greater than $9/4$, $\varphi$ is identically zero. So $k$ must equal $9/4$, and $\varphi(0) = \varphi(1) = 0$. Then to achieve the desired level we must have

$$5\% = E_0\varphi(X) = \tfrac{1}{4}\varphi(0) + \tfrac{1}{2}\varphi(1) + \tfrac{1}{4}\varphi(2) = \tfrac{1}{4}\varphi(2).$$

Solving, $\varphi(2) = 1/5$ gives a test with level $\alpha = 5\%$.

The assertion in Proposition 12.2 that there exists a likelihood ratio test with any desired level $\alpha \in [0,1]$ is established in a similar fashion. First $k$ is adjusted so that $P_0\bigl(L(X) > k\bigr)$ and $P_0\bigl(L(X) \geq k\bigr)$ bracket $\alpha$, and then a value $\gamma \in [0,1]$ is chosen for $\varphi(X)$ when $L(X) = k$ to achieve level $\alpha$.

## 12.3 Uniformly Most Powerful Tests

A test $\varphi^*$ with level $\alpha$ is called *uniformly most powerful* if

$$E_\theta\varphi^* \geq E_\theta\varphi, \qquad \forall\theta \in \Omega_1,$$

for all $\varphi$ with level at most $\alpha$. Uniformly most powerful tests for composite hypotheses generally only arise when the parameter of interest is univariate, $\theta \in \Omega \subset \mathbb{R}$ and the hypotheses are of the form $H_0 : \theta \leq \theta_0$ versus $H_1 : \theta > \theta_0$, where $\theta_0$ is a fixed constant.[2] In addition, the family of densities needs to have an appropriate structure.

**Definition 12.7.** *A family of densities $p_\theta(x)$, $\theta \in \Omega \subset \mathbb{R}$ has* monotone likelihood ratios *if there exists a statistic $T = T(x)$ such that whenever $\theta_1 < \theta_2$, the likelihood ratio $p_{\theta_2}(x)/p_{\theta_1}(x)$ is a nondecreasing function of $T$. Also, the distributions should be* identifiable, *$P_{\theta_1} \neq P_{\theta_2}$ whenever $\theta_1 \neq \theta_2$. Natural conventions concerning division by zero are used here, with the likelihood ratio interpreted as $+\infty$ when $p_{\theta_2} > 0$ and $p_{\theta_1} = 0$. On the null set where both densities are zero the likelihood ratio is not defined and monotonic dependence on $T$ is not required.*

*Example 12.8.* If the densities $p_\theta$ form an exponential family,

$$p_\theta(x) = \exp\bigl\{\eta(\theta)T(x) - B(\theta)\bigr\}h(x),$$

with $\eta(\cdot)$ strictly increasing, then if $\theta_2 > \theta_1$,

$$\frac{p_{\theta_2}(x)}{p_{\theta_1}(x)} = \exp\bigl\{[\eta(\theta_2) - \eta(\theta_1)]T(x) + B(\theta_1) - B(\theta_2)\bigr\},$$

which is increasing in $T(x)$.

[2] Minor variants are possible here: $H_0$ could be $\theta = \theta_0$, $\theta < \theta_0$, $\theta \geq \theta_0$, etc. Uniformly most powerful tests are also possible when the null hypothesis $H_0$ is two-sided, but this case sees little application.

**Theorem 12.9.** *Suppose the family of densities has monotone likelihood ratios. Then*

1. *The test $\varphi^*$ given by*

$$\varphi^*(x) = \begin{cases} 1, & T(x) > c; \\ \gamma, & T(x) = c; \\ 0, & T(x) < c, \end{cases}$$

*is uniformly most powerful testing $H_0 : \theta \le \theta_0$ versus $H_1 : \theta > \theta_0$ and has level $\alpha = E_{\theta_0}\varphi^*$. Also, the constants $c \in \mathbb{R}$ and $\gamma \in [0,1]$ can be adjusted to achieve any desired significance level $\alpha \in (0,1)$.*

2. *The power function $\beta(\theta) = E_\theta\varphi^*$ for this test is nondecreasing and strictly increasing whenever $\beta(\theta) \in (0,1)$.*
3. *If $\theta_1 < \theta_0$, then this test $\varphi^*$ minimizes $E_{\theta_1}\varphi$ among all tests with $E_{\theta_0}\varphi = \alpha = E_{\theta_0}\varphi^*$.*

*Proof.* Suppose $\theta_1 < \theta_2$ and let

$$L(x) = \frac{p_{\theta_2}(x)}{p_{\theta_1}(x)}.$$

Since the family has monotone likelihood ratios, $L$ is a nondecreasing function of $T$. If $k$ is the value of $L$ when $T = c$, then (see Figure 12.1)

$$\varphi^*(x) = \begin{cases} 1, & \text{when } L > k; \\ 0, & \text{when } L < k. \end{cases}$$

Thus $\varphi^*$ is a likelihood ratio test of $\theta = \theta_1$ versus $\theta = \theta_2$. By Corollary 12.4, $E_{\theta_2}\varphi^* \ge E_{\theta_1}\varphi^*$, with strict inequality unless both expectations are zero or one. So the second assertion of the theorem holds, and $\varphi^*$ has level $\alpha = E_{\theta_0}\varphi^*$.

To show that $\varphi^*$ is uniformly most powerful, suppose $\tilde{\varphi}$ has level at most $\alpha$ and $\theta_1 > \theta_0$. Then $E_{\theta_0}\tilde{\varphi} \le \alpha$, and since $\varphi^*$ is a likelihood ratio test of $\theta = \theta_0$ versus $\theta = \theta_1$ maximizing $E_{\theta_1}\varphi$ among all tests with $E_{\theta_0}\varphi \le E_{\theta_0}\varphi^* = \alpha$, $E_{\theta_1}\varphi^* \ge E_{\theta_1}\tilde{\varphi}$. Similarly, if $\theta_1 < \theta_0$, since $\varphi^*$ is a likelihood ratio test of $\theta = \theta_1$ versus $\theta = \theta_0$ with some critical value $k$, it must maximize $E_{\theta_0}\varphi - kE_{\theta_1}\varphi$. Thus if $\tilde{\varphi}$ is a competing test with $E_{\theta_0}\tilde{\varphi} = \alpha = E_{\theta_0}\varphi^*$, then $E_{\theta_1}\tilde{\varphi} \ge E_{\theta_1}\varphi^*$, proving the third assertion in the theorem.

To finish, we must show that $c$ and $\gamma$ can be adjusted so that $E_{\theta_0}\varphi^* = \alpha$. Let $F$ denote the cumulative distribution function for $T$ when $\theta = \theta_0$. Define

$$c = \sup\{x : F(x) \le 1 - \alpha\}.$$

If $x > c$, then $F(x) > 1 - \alpha$. Because $F$ is right continuous,

$$F(c) = \lim_{x \downarrow c} F(x) \ge 1 - \alpha.$$

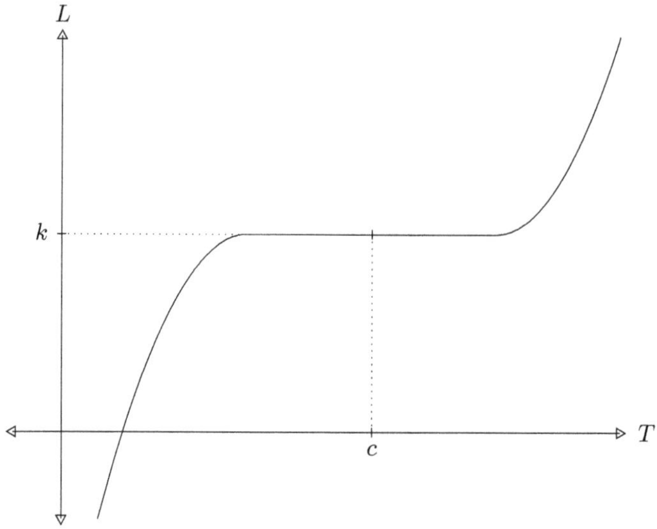


**Fig. 12.1.** The likelihood ratio $L$ as a function of $T$.

But for $x < c$, $F(x) \le 1 - \alpha$, and so

$$F(c-) \stackrel{\text{def}}{=} \lim_{x \uparrow c} F(x) \le 1 - \alpha.$$

Now let $q = F(c) - F(c-) = P_{\theta_0}(T = c)$ (see Problem 1. 16), and define

$$\gamma = \frac{F(c) - (1 - \alpha)}{q}.$$

(If $q = 0$, $\gamma$ can be any value in $[0, 1]$.) By the bounds for $F(c)$ and $F(c-)$, $\gamma$ must lie in $[0, 1]$, and then

$$\begin{aligned} E_{\theta_0} \varphi^* &= \gamma P_{\theta_0}(T = c) + P_{\theta_0}(T > c) \\ &= F(c) - (1 - \alpha) + \bigl(1 - F(c)\bigr) \\ &= \alpha. \end{aligned}$$

□

*Example 12.10.* Suppose our data $X_1, \dots, X_n$ are i.i.d. from the uniform distribution on $(0, \theta)$. The joint density $p_\theta(x)$ is positive if and only if $x_i \in (0, \theta)$, $i = 1, \dots, n$, and this happens if and only if $M(x) = \min\{x_1, \dots, x_n\} > 0$ and $T(x) = \max\{x_1, \dots, x_n\} < \theta$. Thus

$$p_\theta(x) = \begin{cases} 1/\theta^n, & M(x) > 0 \text{ and } T(x) < \theta; \\ 0, & \text{otherwise.} \end{cases}$$

Suppose $\theta_2 > \theta_1$, $M(x) > 0$, and $T(x) < \theta_2$. Then

$$\frac{p_{\theta_2}(x)}{p_{\theta_1}(x)} = \begin{cases} \theta_1^n/\theta_2^n, & T(x) < \theta_1; \\ +\infty, & T(x) \geq \theta_1. \end{cases}$$

This shows that the family of joint densities has monotone likelihood ratios. (The behavior of the likelihood ratio when both densities are zero does not matter; this is why there is no harm assuming $M(x) > 0$ and $T(x) < \theta_2$.) If we are interested in testing $H_0 : \theta \leq 1$ versus $H_1 : \theta > 1$, the test function $\varphi$ given by

$$\varphi = \begin{cases} 1, & T \geq c; \\ 0, & \text{otherwise.} \end{cases}$$

is uniformly most powerful. This test has level

$$P_1(T \geq c) = 1 - c^n,$$

and a specified level $\alpha$ can be achieved taking

$$c = (1 - \alpha)^{1/n}.$$

The power of this test is

$$\beta_\varphi(\theta) = P_\theta(T \geq c) = \begin{cases} 0, & \theta < c; \\ 1 - \dfrac{1-\alpha}{\theta^n}, & \theta \geq c. \end{cases}$$

In this example, one competing test $\tilde{\varphi}$ is given by

$$\tilde{\varphi} = \begin{cases} \alpha, & T < 1; \\ 1, & T \geq 1. \end{cases}$$

For $\theta < 1$, $E_\theta \tilde{\varphi} = \alpha$, so this test also has level $\alpha$. For $\theta > 1$, this test has power

$$\begin{aligned} \beta_{\tilde{\varphi}}(\theta) = E_\theta \tilde{\varphi} &= \alpha P_\theta(T < 1) + P_\theta(T \geq 1) \\ &= \frac{\alpha}{\theta^n} + 1 - \frac{1}{\theta^n} \\ &= \beta_\varphi(\theta). \end{aligned}$$

The power functions $\beta_\varphi$ and $\beta_{\tilde{\varphi}}$ are plotted in Figure 12.2. Because the power functions for $\tilde{\varphi}$ and $\varphi$ are the same under $H_1$, these two tests are both uniformly most powerful. Under $H_0$, the power function for $\varphi$ is smaller than the power function for $\tilde{\varphi}$, so $\varphi$ is certainly the better test. The test $\tilde{\varphi}$ here is an example of an *inadmissible*[3] uniformly most powerful test.

[3] A test $\tilde{\varphi}$ is called inadmissible if a competing test $\varphi$ has a better power function: $\beta_{\tilde{\varphi}}(\theta) \geq \beta_\varphi(\theta)$ for all $\theta \in \Omega_0$, and $\beta_{\tilde{\varphi}}(\theta) \leq \beta_\varphi(\theta)$ for all $\theta \in \Omega_1$, with strict inequality in one of these bounds for some $\theta \in \Omega$.

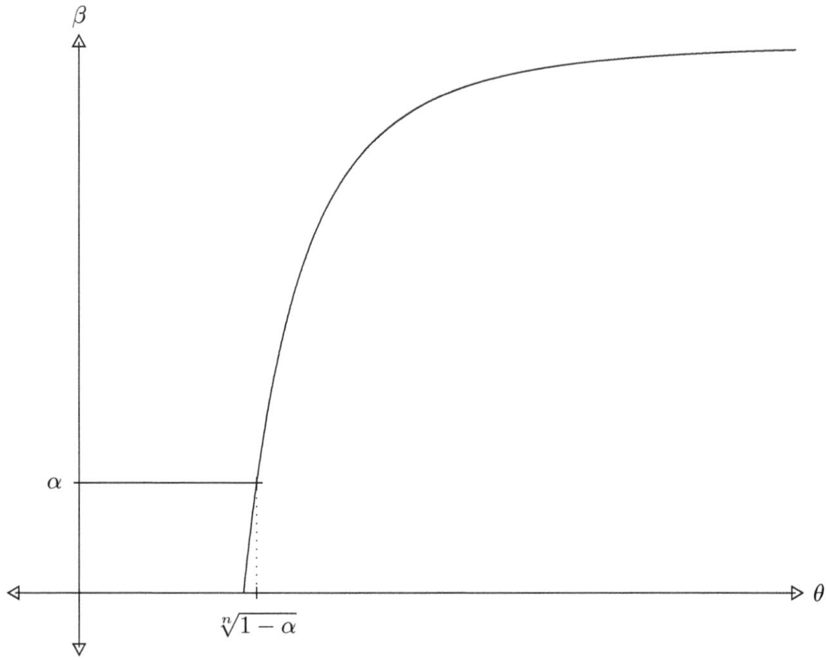


**Fig. 12.2.** Power functions $\beta_\varphi$ and $\beta_{\tilde{\varphi}}$.

## 12.4 Duality Between Testing and Interval Estimation

Recall that a random set $S(X)$ is a $1-\alpha$ confidence region for a parameter $\xi = \xi(\theta)$ if

$$P_\theta\bigl(\xi \in S(X)\bigr) \geq 1-\alpha, \qquad \forall \theta \in \Omega.$$

For every $\xi_0$, let $A(\xi_0)$ be the acceptance region for a nonrandomized level $\alpha$ test of $H_0 : \xi(\theta) = \xi_0$ versus $H_1 : \xi(\theta) \neq \xi_0$, so that

$$P_\theta\bigl[X \in A\bigl(\xi(\theta)\bigr)\bigr] \geq 1-\alpha, \qquad \forall \theta \in \Omega.$$

Define

$$S(x) = \bigl\{\xi : x \in A(\xi)\bigr\}.$$

Then $\xi(\theta) \in S(X)$ if and only if $X \in A\bigl(\xi(\theta)\bigr)$, and so

$$P_\theta\bigl(\xi(\theta) \in S(X)\bigr) = P_\theta\bigl(X \in A\bigl(\xi(\theta)\bigr)\bigr) \geq 1-\alpha.$$

This shows that $S(X)$ is a $1-\alpha$ confidence region for $\xi$.

The construction above can be used to construct confidence regions from a family of nonrandomized tests. Conversely, a $1-\alpha$ confidence region $S(X)$ can be used to construct a family of tests. If we define

$$\varphi = \begin{cases} 1, & \xi_0 \notin S(X); \\ 0, & \text{otherwise,} \end{cases}$$

then if $\xi(\theta) = \xi_0$,

$$E_\theta \varphi = P_\theta\big(\xi_0 \notin S(X)\big) = P_\theta\big(\xi(\theta) \notin S(X)\big) \le \alpha.$$

This shows that this test has level at most $\alpha$ testing $H_0 : \xi(\theta) = \xi_0$ versus $H_1 : \xi(\theta) \neq \xi_0$. If the coverage probability for $S(X)$ is exactly $1-\alpha$, $P_\theta\big(\xi(\theta) \in S(X)\big) = 1 - \alpha$, for all $\theta \in \Omega$, then $\varphi$ will have level exactly $\alpha$.

*Example 12.11.* Suppose the densities for a model have monotone likelihood ratios. Also, for convenience assume $F_\theta(t) = P_\theta(T \le t)$ is continuous and strictly increasing in $t$, for all $\theta \in \Omega$. For each $\theta \in \Omega$, define $u(\theta)$ so that

$$P_\theta\big(T < u(\theta)\big) = F_\theta\big(u(\theta)\big) = 1 - \alpha.$$

Then

$$\varphi = \begin{cases} 1, & T \ge u(\theta_0); \\ 0, & \text{otherwise,} \end{cases}$$

is uniformly most powerful testing $H_0 : \theta = \theta_0$ versus $H_1 : \theta > \theta_0$ and has level

$$E_{\theta_0}\varphi = P_{\theta_0}\big(T \ge u(\theta_0)\big) = \alpha.$$

This test has acceptance region

$$A(\theta_0) = \big\{x : T(x) < u(\theta_0)\big\}.$$

**Proposition 12.12.** *The function $u(\cdot)$ is strictly increasing.*

*Proof.* Suppose $\theta > \theta_0$. By the second part of Theorem 12.9, the power function for $\varphi$ is strictly increasing at $\theta_0$, and so

$$E_\theta\varphi = P_\theta\big(T \ge u(\theta_0)\big) > E_{\theta_0}\varphi = \alpha.$$

Thus $P_\theta\big(T < u(\theta_0)\big) < 1-\alpha$. But from the definition of $u(\cdot)$, $P_\theta\big(T < u(\theta)\big) = 1 - \alpha$, and so $u(\theta) > u(\theta_0)$. Since $\theta > \theta_0$ are arbitrary parameter values, $u$ is strictly increasing. □

The confidence set dual to the family of tests with acceptance regions $A(\theta)$, $\theta \in \Omega$, is

$$S(X) = \big\{\theta : X \in A(\theta)\big\} = \big\{\theta : T(X) < u(\theta)\big\}.$$

Because $u$ is strictly increasing, this region is the interval (see Figure 12.3)

$$S(X) = \big(u^{\leftarrow}(T), \infty\big) \bigcap \Omega.$$

Here $u^{\leftarrow}$ is the inverse function of $u$.

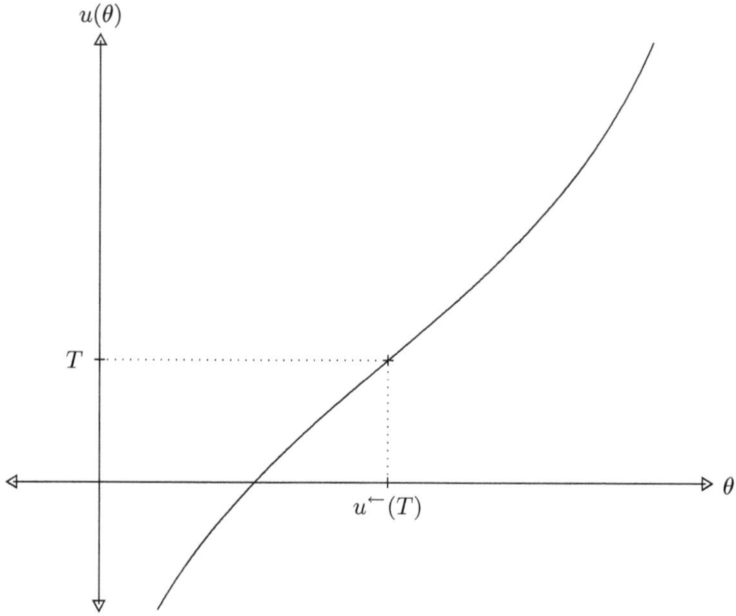


**Fig. 12.3.** The increasing function $u$.

For a concrete example, suppose $X$ is exponential with mean $\theta$, so

$$p_\theta(x) = \frac{1}{\theta}e^{-x/\theta}, \qquad x > 0.$$

The densities for $X$ form an exponential family with $\eta = -1/\theta$, an increasing function of $\theta$. So we have monotone likelihood ratios with $T = X$. The function $u$ is defined so that

$$P_\theta\bigl(X < u(\theta)\bigr) = 1 - \alpha$$

or

$$P_\theta\bigl(X \ge u(\theta)\bigr) = \alpha.$$

Because $P_\theta(X \ge x) = e^{-x/\theta}$,

$$e^{-u(\theta)/\theta} = \alpha.$$

Solving,

$$u(\theta) = -\theta \log \alpha.$$

Since $u^{\leftarrow}(X)$ is the value $\theta$ solving

$$X = -\theta \log \alpha,$$

we have

$$u^{\leftarrow}(X) = \frac{X}{-\log \alpha},$$

and the $1 - \alpha$ confidence set for $\theta$ is

$$S(X) = \left( \frac{X}{-\log \alpha}, \infty \right).$$

As a check,

$$\begin{aligned} P_\theta\bigl(\theta \in S(X)\bigr) &= P_\theta \left( \frac{X}{-\log \alpha} < \theta \right) \\ &= \int_0^{-\theta \log \alpha} \frac{1}{\theta} e^{-x/\theta}\, dx \\ &= 1 - \alpha. \end{aligned}$$

Our construction of confidence sets by duality works with any family of level $\alpha$ tests. But intuition suggests that better tests should give better confidence intervals. In the example just considered, tests in the family are uniformly most powerful, and a natural conjecture would be that the dual confidence interval $S(X)$ should be optimal in some related sense. This is indeed the case. To deduce the proper notion of optimality, let $S^*(X)$ be a competing confidence set, and take

$$\varphi^* = \begin{cases} 1, & \theta_0 \notin S^*(X); \\ 0, & \text{otherwise.} \end{cases}$$

Then $\varphi^*$ is a test of $H_0 : \theta = \theta_0$ with level at most $\alpha$. The corresponding test $\varphi$ dual to $S(X)$ is

$$\varphi = \begin{cases} 1, & \theta_0 \notin S(X); \\ 0, & \text{otherwise.} \end{cases}$$

If the tests dual to $S$ are uniformly most powerful testing $H_0 : \theta = \theta_0$ versus $H_1 : \theta > \theta_0$, then for any $\theta > \theta_0$,

$$E_\theta \varphi \ge E_\theta \varphi^*.$$

The left- and right-hand sides of this equation are $P_\theta\bigl(\theta_0 \notin S(X)\bigr)$ and $P_\theta\bigl(\theta_0 \notin S^*(X)\bigr)$, respectively, and so

$$P_\theta\bigl(\theta_0 \in S(X)\bigr) \le P_\theta\bigl(\theta_0 \in S^*(X)\bigr). \tag{12.2}$$

This shows that if $\theta$ is the true value of the parameter, then $S(X)$ has a smaller chance of containing any incorrect value $\theta_0 < \theta$.

In practice, a researcher may be most concerned with the length of a confidence interval, and the optimality for $S$ in (12.2) may seem less relevant.

However, using Fubini's theorem, there is a relation between the expected length and the probabilities $P_\theta\bigl(\theta_0 \in S(X)\bigr)$. Let $\lambda$ denote Lebesgue measure on $\mathbb{R}$, so that $\lambda(A) = \int_A dx$ is the length of $A$. Also, assume for convenience that $\Omega$ is the interval $(\underline{\omega}, \overline{\omega})$ (we allow $\underline{\omega} = -\infty$ and/or $\overline{\omega} = \infty$). Then, by Fubini's theorem,

$$\begin{aligned} E_\theta \lambda\bigl(S(X) \cap (\underline{\omega}, \theta)\bigr) &= E_\theta \int_{\underline{\omega}}^{\theta} I\bigl(\theta_0 \in S(X)\bigr)\, d\theta_0 \\ &= \int \int_{\underline{\omega}}^{\theta} I\bigl(\theta_0 \in S(x)\bigr)\, d\theta_0\, dP_\theta(x) \\ &= \int_{\underline{\omega}}^{\theta} P_\theta\bigl(\theta_0 \in S(X)\bigr)\, d\theta_0. \end{aligned}$$

Similarly,

$$E_\theta \lambda\bigl(S^*(X) \cap (\underline{\omega}, \theta)\bigr) = \int_{\underline{\omega}}^{\theta} P_\theta\bigl(\theta_0 \in S(X)\bigr)\, d\theta_0,$$

and so, by (12.2),

$$E_\theta \lambda\bigl(S(X) \cap (\underline{\omega}, \theta)\bigr) \le E_\theta \lambda\bigl(S^*(X) \cap (\underline{\omega}, \theta)\bigr).$$

So the expected length of $S(X)$ below $\theta$ is minimal among all $1-\alpha$ confidence intervals.

## 12.5 Generalized Neyman–Pearson Lemma[4]

Treatment of two-sided hypotheses in the next section relies on an extension of the Neyman–Pearson lemma in which the test function must satisfy several constraints. Let $g(x)$ take values in $\mathbb{R}^m$, and consider maximizing

$$\int \varphi f \, d\mu$$

over all test functions $\varphi$ satisfying

$$\int \varphi g \, d\mu = c, \tag{12.3}$$

where $c$ is a specified vector in $\mathbb{R}^m$. Introducing a Lagrange multiplier $k \in \mathbb{R}^m$, consider maximizing

$$\int (f - k \cdot g)\varphi \, d\mu \tag{12.4}$$

[4] Results in the rest of this chapter and Chapter 13 are more technical and are not used in subsequent chapters.

without constraint. A test function maximizing (12.4) will have form

$$\varphi(x) = \begin{cases} 1, & f(x) > k \cdot g(x); \\ 0, & f(x) < k \cdot g(x), \end{cases} \tag{12.5}$$

for a.e. $x$ ($\mu$). As in our discussion of the Neyman–Pearson lemma, if a function of this form satisfies (12.3), it clearly solves the constrained maximization problem.

Let $\mathcal{K}$ denote the the set of all test functions (measurable functions with range a subset of $[0,1]$).

**Theorem 12.13.** *Assume $f$ and $g$ are both integrable with respect to $\mu$ and that the class $\mathcal{C}$ of all test functions $\varphi \in \mathcal{K}$ satisfying* (12.3) *is not empty. Then*

1. *There exists a test function $\varphi^*$ maximizing $\int \varphi f\, d\mu$ over $\mathcal{C}$.*
2. *If $\varphi^* \in \mathcal{C}$ satisfies* (12.5) *for some $k \in \mathbb{R}^m$, then $\varphi^*$ maximizes $\int \varphi f\, d\mu$ over $\varphi \in \mathcal{C}$.*
3. *If $\varphi^* \in \mathcal{C}$ has form* (12.5) *with $k \geq 0$, then $\varphi^*$ maximizes $\int \varphi f\, d\mu$ over all $\varphi$ satisfying $\int \varphi g\, d\mu \leq c$.*
4. *Let $\mathcal{L}_g$ be the linear mapping from test functions $\varphi \in \mathcal{K}$ to vectors in $\mathbb{R}^m$ given by $\mathcal{L}_g(\varphi) = \int \varphi g\, d\mu$, and let $M$ denote the range of $\mathcal{L}_g$. Then $M$ is closed and convex. If $c$ lies in the interior of $M$, there exists a Lagrange multiplier $k \in \mathbb{R}^m$ and a test function $\varphi^* \in \mathcal{C}$ maximizing $\int (f - k \cdot g)\varphi\, d\mu$ over $\varphi \in \mathcal{K}$. Also, if any $\varphi \in \mathcal{C}$ maximizes $\int \varphi f\, d\mu$ over $\mathcal{C}$, then* (12.5) *must hold a.e. $\mu$.*

The proof of this result relies on an important and useful result from functional analysis, the weak compactness theorem. In functional analysis, functions are viewed as points in a vector space, much as vectors are viewed as points in $\mathbb{R}^n$. But notions of convergence for functions are much richer. For instance, functions $f_n$, $n \geq 1$, converge pointwise to $f$ if $\lim_{n\to\infty} f_n(x) = f(x)$ for all $x$. In contrast, uniform convergence would hold if $\lim_{n\to\infty} \sup_x |f_n(x) - f(x)| = 0$. Uniform convergence implies pointwise convergence, but not vice versa. (For instance, the functions $1_{(n,n+1)}$ converge pointwise to the zero function, but the convergence is not uniform.) The notion of convergence of interest here is called *weak convergence.*

**Definition 12.14.** *A sequence of uniformly bounded measurable functions $\varphi_n$,* $n \geq 1$, converge weakly *to $\varphi$, written $\varphi_n \xrightarrow{w} \varphi$, if*

$$\int \varphi_n f\, d\mu \to \int \varphi f\, d\mu$$

*whenever $\int |f|\, d\mu < \infty$.*

If the functions $\varphi_n$ converge pointwise to $\varphi$, then weak convergence follows from dominated convergence, but pointwise convergence is not necessary for weak convergence. With this notion of convergence, the objective function in Theorem 12.13,

$$\mathcal{L}_f(\varphi) = \int \varphi f \, d\mu,$$

is a continuous function of $\varphi$; that is, $\mathcal{L}_f(\varphi_n) \to \mathcal{L}_f(\varphi)$ whenever $\varphi_n \xrightarrow{w} \varphi$. The linear constraint function $\mathcal{L}_g$ introduced in Theorem 12.13, is also continuous.

**Theorem 12.15 (Weak Compactness Theorem).** *The set $\mathcal{K}$ is weakly compact:*[5] *any sequence of functions $\varphi_n$, $n \geq 1$, in $\mathcal{K}$ has a convergent subsequence, $\varphi_{n(j)} \xrightarrow{w} \varphi \in \mathcal{K}$ as $j \to \infty$.*

See Appendix A.5 of Lehmann and Romano (2005) for a proof. In the proof of Theorem 12.13, we also need the following result, called the *supporting hyperplane theorem.* For this and other results in convex analysis, see Rockafellar (1970).

**Theorem 12.16 (Supporting Hyperplane Theorem).** *If $x$ is a point on the boundary of a convex set $K \subset \mathbb{R}^m$, then there exists a nonzero vector $v \in \mathbb{R}^m$ such that*

$$v \cdot y \leq v \cdot x, \qquad \forall y \in K.$$

*Proof of Theorem 12.13.* The first assertion follows by weak compactness of $\mathcal{K}$. Take

$$K_{\mathcal{C}} = \sup_{\varphi \in \mathcal{C}} \mathcal{L}_f(\varphi),$$

and let $\varphi_n$, $n \geq 1$, be a sequence of test functions in $\mathcal{C}$ such that

$$\mathcal{L}_f(\varphi_n) \to K_{\mathcal{C}}.$$

By the weak compactness theorem (Theorem 12.15), there must be a subsequence $\varphi_{n(m)}$, $m \geq 1$, with

$$\varphi_{n(m)} \xrightarrow{w} \varphi^* \in \mathcal{K},$$

and since $\mathcal{L}_f$ is continuous, $\mathcal{L}_f(\varphi^*) = K_{\mathcal{C}}$. If $\varphi^* \in \mathcal{C}$ we are done. But this follows by continuity of $\mathcal{L}_g$ because $\mathcal{L}_g(\varphi^*) = \lim_{m\to\infty} \mathcal{L}_g(\varphi_{n(m)})$ and $\mathcal{L}_g(\varphi_{n(m)}) = c$ for all $m \geq 1$.

For the second assertion, if $\varphi^* \in \mathcal{C}$ has form (12.5), then $\varphi^*$ maximizes $\int \varphi(f - k \cdot g)\, d\mu = \mathcal{L}_f(\varphi) - k \cdot \mathcal{L}_g(\varphi)$ over all $\mathcal{K}$, and hence $\varphi^*$ maximizes $\mathcal{L}_f(\varphi) - k \cdot \mathcal{L}_g(\varphi)$ over $\varphi \in \mathcal{C}$. But when $\varphi \in \mathcal{C}$, $\mathcal{L}_f(\varphi) - k \cdot \mathcal{L}_g(\varphi) = \mathcal{L}_f(\varphi) - k \cdot c$, and so $\varphi^*$ maximizes $\mathcal{L}_f(\varphi)$ over $\varphi \in \mathcal{C}$.

[5] The topology of weak convergence has a countable base, and so compactness and sequential compactness (stated in this theorem) are equivalent.

Suppose $\varphi^* \in \mathcal{C}$ has form (12.5), so it maximizes $\mathcal{L}_f(\varphi) - k \cdot \mathcal{L}_g(\varphi)$ over all $\varphi \in \mathcal{K}$. Then if $k \geq 0$ and $\mathcal{L}_g(\varphi) \leq c$,

$$\begin{aligned}\mathcal{L}_f(\varphi) &\leq \mathcal{L}_f(\varphi) - k \cdot \mathcal{L}_g(\varphi) + k \cdot c \\ &\leq \mathcal{L}_f(\varphi^*) - k \cdot \mathcal{L}_g(\varphi^*) + k \cdot c = \mathcal{L}_f(\varphi^*).\end{aligned}$$

This proves the third assertion.

The final assertion is a bit more involved. First, $M$ is convex, for if $x_0 = \mathcal{L}_g(\varphi_0)$ and $x_1 = \mathcal{L}_g(\varphi_1)$ are arbitrary points in $M$, and if $\gamma \in [0, 1]$, using the linearity of $\mathcal{L}_g$,

$$\begin{aligned}\gamma x_0 + (1-\gamma)x_1 &= \gamma \mathcal{L}_g(\varphi_0) + (1-\gamma)\mathcal{L}_g(\varphi_1) \\ &= \mathcal{L}_g\big(\gamma\varphi_0 + (1-\gamma)\varphi_1\big) \in M.\end{aligned}$$

Closure of $M$ follows by weak compactness and continuity of $\mathcal{L}_g$. Suppose $x$ is a limit point of $M$, so that $x = \lim_{n\to\infty} \mathcal{L}_g(\varphi_n)$ for some sequence of test functions $\varphi_n$, $n \geq 1$. Letting $\varphi_{n(m)}$, $m \geq 1$, be a subsequence converging weakly to $\varphi$,

$$\mathcal{L}_g(\varphi) = \lim_{m\to\infty} \mathcal{L}_g(\varphi_{n(m)}) = \lim_{m\to\infty} x_{n(m)} = x,$$

which shows that $x \in M$.

For the final part of the theorem, assume that $c$ lies in the interior of $M$. Let $\varphi^* \in \mathcal{C}$ maximize $\mathcal{L}_f(\varphi)$ over $\varphi \in \mathcal{C}$, and take $K_{\mathcal{C}} = \mathcal{L}_f(\varphi^*)$. Define $\mathcal{L} : \mathcal{K} \to \mathbb{R}^{m+1}$ by

$$\mathcal{L}(\varphi) = \begin{pmatrix} \mathcal{L}_f(\varphi) \\ \mathcal{L}_g(\varphi) \end{pmatrix}.$$

The arguments showing that $M$ is closed and convex also show that the range $\tilde{M}$ of $\mathcal{L}$ is closed and convex. The point

$$x = \mathcal{L}(\varphi^*) = \begin{pmatrix} \mathcal{L}_f(\varphi^*) \\ \mathcal{L}_g(\varphi^*) \end{pmatrix} = \begin{pmatrix} \mathcal{L}_f(\varphi^*) \\ c \end{pmatrix}$$

lies in $\tilde{M}$. Because $\varphi^*$ maximizes $\mathcal{L}_f(\varphi^*)$ over $\varphi \in \mathcal{C}$, if $\epsilon > 0$, the point

$$\begin{pmatrix} \mathcal{L}_f(\varphi^*) + \epsilon \\ c \end{pmatrix}$$

cannot lie in $\tilde{M}$, and thus $x$ lies on the boundary of $\tilde{M}$. By the supporting hyperplane theorem (Theorem 12.16), there is a nonzero vector $v = \binom{a}{b}$ such that

$$v \cdot y \leq v \cdot x, \qquad \forall y \in \tilde{M},$$

or, equivalently, such that

$$a\mathcal{L}_f(\varphi) + b \cdot \mathcal{L}_g(\varphi) \leq a\mathcal{L}_f(\varphi^*) + b \cdot \mathcal{L}_g(\varphi^*), \qquad \forall \varphi \in \mathcal{K}.$$

Here $a$ cannot be zero, for then this bound would assert that $b \cdot \mathcal{L}_g(\varphi) \le b \cdot c$ for all $\varphi \in \mathcal{K}$, contradicting the assumption that $c$ lies in the interior of $M$. For $\varphi \in \mathcal{C}$ this bound becomes $a\mathcal{L}_f(\varphi) \le a\mathcal{L}_f(\varphi^*)$. Because $\varphi^*$ maximizes $\mathcal{L}_f(\varphi)$ over $\varphi \in \mathcal{C}$, $a$ must be positive, unless we are in a degenerate situation in which $\mathcal{L}_f(\varphi) = \mathcal{L}_f(\varphi^*)$ for all $\varphi \in \mathcal{C}$. And if $a$ is positive, we are done, for then the bound is

$$\mathcal{L}_f(\varphi) + (b/a) \cdot \mathcal{L}_g(\varphi) \le \mathcal{L}_f(\varphi^*) + (b/a) \cdot \mathcal{L}_g(\varphi^*), \qquad \forall \varphi \in \mathcal{K},$$

and we can take $k = -b/a$.

To handle the degenerate case, suppose $\mathcal{L}_g(\varphi_1) = \mathcal{L}_g(\varphi_2) = \tilde{c} \neq c$. Because $c$ is an interior point of $M$, it can be expressed as a nontrivial convex combination of $\tilde{c}$ and some other point $\mathcal{L}_g(\varphi_3) \neq c$ in $M$; that is,

$$c = \gamma\tilde{c} + (1-\gamma)\mathcal{L}_g(\varphi_3),$$

for some $\gamma \in (0,1)$. Since $\mathcal{L}_g$ is linear,

$$\gamma\varphi_1 + (1-\gamma)\varphi_3 \text{ and } \gamma\varphi_2 + (1-\gamma)\varphi_3$$

both lie in $\mathcal{C}$, and so

$$\begin{aligned}\mathcal{L}_f\bigl(\gamma\varphi_1 + (1-\gamma)\varphi_3\bigr) &= \gamma\mathcal{L}_f(\varphi_1) + (1-\gamma)\mathcal{L}_f(\varphi_3)\\ &= \mathcal{L}_f\bigl(\gamma\varphi_2 + (1-\gamma)\varphi_3\bigr)\\ &= \gamma\mathcal{L}_f(\varphi_2) + (1-\gamma)\mathcal{L}_f(\varphi_3).\end{aligned}$$

So we must have $\mathcal{L}_f(\varphi_1) = \mathcal{L}_f(\varphi_2)$. Thus, if $\binom{\ell_0}{\tilde{c}}$ and $\binom{\ell_1}{\tilde{c}}$ both lie in $\tilde{M}$, then $\ell_0 = \ell_1$. Since $\tilde{M}$ is convex and contains the origin, the only way this can happen is if $\mathcal{L}_f(\varphi)$ is a linear function of $\mathcal{L}_g(\varphi)$,

$$\mathcal{L}_f(\varphi) = k \cdot \mathcal{L}_g(\varphi), \qquad \varphi \in \mathcal{K}.$$

In this case, $\varphi^*$ trivially maximizes $\mathcal{L}_f(\varphi) - k \cdot \mathcal{L}_g(\varphi)$.

To finish, if $\varphi$ maximizes $\mathcal{L}_f$ over $\mathcal{C}$, then $\mathcal{L}(\varphi) = \mathcal{L}(\varphi^*)$ and $\varphi$ also maximizes $\mathcal{L}_f - k \cdot \mathcal{L}_g$ over $\mathcal{K}$. It is then clear that (12.5) must hold a.e. $\mu$; if not, a function satisfying (12.5) would give a larger value for $\mathcal{L}_f - k \cdot \mathcal{L}_g$. ☐

## 12.6 Two-Sided Hypotheses

This section focuses on testing $H_0 : \theta = \theta_0$ versus $H_1 : \theta \neq \theta_0$ with data from a one-parameter exponential family. Generalization to families satisfying a condition analogous to the monotone likelihood ratio condition is possible. Tests of $H_0 : \theta \in [\theta_1, \theta_2]$ versus $H_1 : \theta < \theta_1$ or $\theta > \theta_2$ can be developed in a similar fashion, but results about the point null hypothesis seem more useful in practice. Also, uniformly most powerful tests when $H_0$ is two-sided can be obtained (see Problem 12.39), but these tests are not used very often.

With data from an exponential family there will be a sufficient statistic, and the next result shows that we can then restrict attention to tests based on the sufficient statistic.

**Theorem 12.17.** *Suppose that $T$ is sufficient for the model $\mathcal{P} = \{P_\theta : \theta \in \Omega\}$. Then for any test $\varphi = \varphi(X)$, the test*

$$\psi = \psi(T) = E_\theta\big[\varphi(X) \mid T\big]$$

*has the same power function as $\varphi$,*

$$E_\theta\psi(T) = E_\theta\varphi(X), \qquad \forall\theta \in \Omega.$$

*Proof.* This follows immediately from smoothing,

$$E_\theta\varphi(X) = E_\theta E_\theta\big[\varphi(X) \mid T\big] = E_\theta\psi(T).$$ □

The next theorem shows that if the densities for $X$ come from an exponential family, then the densities for $T$ will also be from an exponential family. This is established using the following fundamental lemma, which shows how likelihood ratios can be introduced to write an expectation under one distribution as an expectation under a different distribution. This lemma is quite useful in a variety of situations.

**Lemma 12.18.** *Let $P_0$ and $P_1$ be possible distributions for a random vector $X$ with densities $p_0$ and $p_1$ with respect to $\mu$. If $p_1(x) = 0$ whenever $p_0(x) = 0$, then $P_1 \ll P_0$ and $P_1$ has density*

$$\frac{dP_1}{dP_0}(x) = L(x) = \frac{p_1(x)}{p_0(x)}$$

*with respect to $P_0$. (The value for $L(x)$ when $p_0(x) = 0$ does not matter; for definiteness, take $L(x) = 1$ when $p_0(x) = 0$.) Introducing this likelihood ratio, we can write expectations under $P_1$ as expectations under $P_0$ using the formula*

$$E_1 h(X) = E_0 h(X)L(X),$$

*valid whenever the expectations exist. When $h$ is an indicator function, we have*

$$P_1(B) = E_0 1_B(X)L(X).$$

*Proof.* First note that $N = \{x : p_0(x) = 0\}$ is a null set for $P_0$ because

$$P_0(N) = \int_N p_0\,d\mu = \int 1_N p_0\,d\mu = \int 0\,d\mu = 0.$$

Similarly, $\{x : p_1(x) = 0\}$ is a null set for $P_1$, and since $N \subset \{x : p_1(x) = 0\}$, $N$ is also a null set for $P_1$. So $1_{N^c} = 1$ a.e. $P_0$ and $P_1$, and multiplication by this function cannot change the value of an integral against either distribution. Suppose $M$ is a null set for $P_0$. Then

$$\int 1_M p_0\,d\mu = \int 1_M 1_{N^c} p_0\,d\mu = \int 1_{M\cap N^c} p_0\,d\mu,$$

which implies that $1_{M\cap N^c}p_0 = 0$ a.e. $\mu$ (by the second fact about integration in Section 1.4). Because $p_0 > 0$ whenever the indicator is 1, $M\cap N^c$ must be a null set for $\mu$. But $P_1$ is dominated by $\mu$, and so $M\cap N^c$ is a null set for $P_1$. But $M \subset N \cup (M\cap N^c)$, which is a union of two null sets for $P_1$, showing that $M$ must be a null set for $P_1$. To write expectations under $P_1$ as expectations under $P_0$,

$$\begin{aligned} E_1 h(X) &= \int h p_1\,d\mu = \int 1_{N^c} h p_1\,d\mu = \int 1_{N^c} h \frac{p_1}{p_0} p_0\,d\mu \\ &= \int 1_{N^c} h L p_0\,d\mu = \int h L p_0\,d\mu = E_0 h(X) L(X). \qquad \square \end{aligned}$$

**Theorem 12.19.** *If the distribution for $X$ comes from an exponential family with densities*

$$p_\theta(x) = h(x)e^{\eta(\theta)\cdot T(x) - B(\theta)}, \qquad \theta\in\Omega,$$

*then the induced distribution for $T = T(X)$ has density*

$$q_\theta(t) = e^{\eta(\theta)\cdot t - B(\theta)}, \qquad \theta\in\Omega,$$

*with respect to some measure $\nu$.*

*Proof.* Two ideas are used. First, using Lemma 12.18, we can introduce a likelihood ratio to write probabilities under $P_\theta$ as expectations under $P_{\theta_0}$, where $\theta_0$ is a fixed point in $\Omega$. This likelihood ratio is

$$L = \frac{p_\theta(X)}{p_{\theta_0}(X)} = e^{[\eta(\theta)-\eta(\theta_0)]\cdot T + B(\theta_0) - B(\theta)}.$$

The second is that expectations of functions of $T$ can be written as integrals against the density of $X$, or as integrals against the marginal distribution of $T$. Let $\nu^*$ denote the marginal distribution of $T$ when $\theta = \theta_0$. Then

$$\begin{aligned} P_\theta(T\in B) &= E_{\theta_0} I\{T\in B\} e^{[\eta(\theta)-\eta(\theta_0)]\cdot T + B(\theta_0) - B(\theta)} \\ &= \int I\{t\in B\} e^{[\eta(\theta)-\eta(\theta_0)]\cdot t + B(\theta_0) - B(\theta)}\,d\nu^*(t) \\ &= \int_B q_\theta(t) e^{-\eta(\theta_0)\cdot t + B(\theta_0)}\,d\nu^*(t). \end{aligned}$$

If we define $\nu$ by

$$\nu(A) = \int_A e^{-\eta(\theta_0)\cdot t + B(\theta_0)}\,d\nu^*(t),$$

then $\nu$ has density $e^{-\eta(\theta_0)\cdot t+B(\theta_0)}$ with respect to $\nu^*$, and

$$P_\theta(T \in B) = \int_B q_\theta(t)\,d\nu(t),$$

completing the proof. □

Consider testing $H_0 : \theta = \theta_0$ versus $H_1 : \theta \neq \theta_0$ based on data $X$ with density

$$h(x)e^{\eta(\theta)T(x)-B(\theta)}, \qquad \theta \in \Omega, \tag{12.6}$$

where $\eta$ is strictly increasing and differentiable. From results in Section 12.3, there are level $\alpha$ tests $\varphi_\pm$ with form

$$\varphi_+ = \begin{cases} 1, & T > c_+; \\ \gamma_+, & T = c_+; \\ 0, & T < c_+, \end{cases} \text{ and } \varphi_- = \begin{cases} 1, & T < c_-; \\ \gamma_-, & T = c_-; \\ 0, & T > c_-. \end{cases}$$

These tests are most powerful for one-sided alternatives. If $\theta_- < \theta_0 < \theta_+$, then $\varphi_+$ will have maximal power at $\theta_+$, and $\varphi_-$ will have maximal power at $\theta_-$. Since these tests are different, this shows that there cannot be a uniformly most powerful level $\alpha$ test. To achieve uniformity, we must restrict the class of tests under consideration. We do this by constraining the derivative of the power function at $\theta_0$. The formula in the following theorem is useful.

**Theorem 12.20.** *If $\eta$ is differentiable at $\theta$ and $\theta$ lies in the interior of $\Omega$, then the derivative of the power function $\beta$ for a test $\varphi$ is given by*

$$\beta'(\theta) = \eta'(\theta)E_\theta T\varphi - B'(\theta)\beta(\theta).$$

*Proof.* If differentiation under the integral sign works, then

$$\begin{aligned}
\beta'(\theta) &= \frac{\partial}{\partial\theta}\int \varphi(x)e^{\eta(\theta)T(x)-B(\theta)}h(x)\,d\mu(x) \\
&= \int \varphi(x)\frac{\partial}{\partial\theta}e^{\eta(\theta)T(x)-B(\theta)}h(x)\,d\mu(x) \\
&= \int \varphi(x)\bigl[\eta'(\theta)T(x) - B'(\theta)\bigr]e^{\eta(\theta)T(x)-B(\theta)}h(x)\,d\mu(x) \\
&= \eta'(\theta)E_\theta\varphi T - B'(\theta)E_\theta\varphi,
\end{aligned}$$

and the result follows. Differentiation under the integral sign can be justified using Theorem 2.4 and the chain rule, or by dominated convergence directly. □

Note that because $\varphi_+$ has maximal power for $\theta > \theta_0$,

$$\begin{aligned}\beta'_\varphi(\theta_0) &= \lim_{\epsilon\downarrow 0} \frac{\beta_\varphi(\theta_0+\epsilon)-\beta_\varphi(\theta_0)}{\epsilon} \\ &\le \lim_{\epsilon\downarrow 0} \frac{\beta_{\varphi_+}(\theta_0+\epsilon)-\beta_{\varphi_+}(\theta_0)}{\epsilon} = \beta'_{\varphi_+}(\theta_0) \stackrel{\text{def}}{=} m_+. \end{aligned} \tag{12.7}$$

Similarly, $\beta'_\varphi(\theta_0) \ge \beta'_{\varphi_-}(\theta_0) \stackrel{\text{def}}{=} m_-$. For $m \in [m_-, m_+]$, let $\mathcal{C}_m$ denote the class of all level $\alpha$ tests $\varphi$ with $\beta'_\varphi(\theta_0) = m$. The next theorem shows that when $m \in (m_-, m_+)$ there is a uniformly most powerful test in $\mathcal{C}_m$. This test is two-sided, according to the following definition.

**Definition 12.21.** *A test $\varphi$ is called* two-sided *if there are finite constants $t_1 \le t_2$ such that*

$$\varphi = \begin{cases} 1, & T < t_1 \text{ or } T > t_2; \\ 0, & T \in (t_1, t_2). \end{cases}$$

*In addition, the test should not be one-sided. Specifically, $E\varphi I\{T \ge t_2\}$ and $E\varphi I\{T \le t_1\}$ should both be positive.*

**Theorem 12.22.** *If $\theta_0$ lies in the interior of $\Omega$, $\alpha \in (0,1)$, $X$ has density (12.6), and $\eta$ is differentiable and strictly increasing with $0 < \eta'(\theta_0) < \infty$, then for any value $m \in (m_-, m_+)$ there is a two-sided test $\varphi^* \in \mathcal{C}_m$. Any such test is uniformly most powerful in class $\mathcal{C}_m$: for any competing test $\varphi \in \mathcal{C}_m$,*

$$E_\theta \varphi \le E_\theta \varphi^*, \qquad \forall \theta \in \Omega.$$

*Proof.* Using Theorem 12.20, if $\varphi \in \mathcal{C}_m$, then

$$\beta'_\varphi(\theta_0) = \eta'(\theta_0) E_{\theta_0} T\varphi - \alpha B'(\theta_0) = m,$$

which happens if and only if

$$E_{\theta_0} T\varphi = \int \varphi T p_{\theta_0}\, d\mu = \frac{m + \alpha B'(\theta_0)}{\eta'(\theta_0)}.$$

If we define

$$g(x) = \begin{pmatrix} p_{\theta_0}(x) \\ T(x) p_{\theta_0}(x) \end{pmatrix} \quad \text{and} \quad c = \begin{pmatrix} \alpha \\ \dfrac{m + \alpha B'(\theta_0)}{\eta'(\theta_0)} \end{pmatrix},$$

then a test function $\varphi$ lies in $\mathcal{C}_m$ if and only if $\mathcal{L}_g(\varphi) = \int \varphi g \, d\mu = c$. Because $m_+ > m$ and $m_- < m$, the point $c$ lies in the interior of the convex hull of the four points $\mathcal{L}_g(\varphi_+)$, $\mathcal{L}_g(\varphi_-)$, $\mathcal{L}_g(\alpha - \epsilon)$, and $\mathcal{L}_g(\alpha + \epsilon)$. (Here "$\alpha \pm \epsilon$" denotes a test function that equals $\alpha \pm \epsilon$ regardless of the value of $X$.) Thus $c$ lies in the interior of the range $M$ of $\mathcal{L}_g$.

With this background on the nature of the constraints for test functions $\varphi$ in $\mathcal{C}_m$, we can now use the last assertion in Theorem 12.13 to show that there exists a two-sided test $\varphi^*$ in $\mathcal{C}_m$. Let $\tilde{\theta}$ be some fixed point in $\Omega$ different from $\theta_0$ and consider maximizing

$$E_{\tilde{\theta}}\varphi = \int \varphi p_{\tilde{\theta}}\, d\mu$$

over $\varphi$ in $\mathcal{C}_m$. Using the fourth assertion in Theorem 12.13 there is a Lagrange multiplier $k \in \mathbb{R}^2$ and a test $\varphi^* \in \mathcal{C}_m$ maximizing $\int \varphi(p_{\tilde{\theta}} - k \cdot g)\, d\mu$ with form

$$\varphi^* = \begin{cases} 1, & p_{\tilde{\theta}} > (k_1 + k_2 T)p_{\theta_0}; \\ 0, & p_{\tilde{\theta}} < (k_1 + k_2 T)p_{\theta_0}. \end{cases}$$

Dividing through by $p_{\theta_0}$,

$$\varphi^* = \begin{cases} 1, & \exp\bigl\{\bigl(\eta(\tilde{\theta}) - \eta(\theta_0)\bigr)T - B(\tilde{\theta}) + B(\theta_0)\bigr\} > k_1 + k_2 T; \\ 0, & \exp\bigl\{\bigl(\eta(\tilde{\theta}) - \eta(\theta_0)\bigr)T - B(\tilde{\theta}) + B(\theta_0)\bigr\} < k_1 + k_2 T. \end{cases}$$

The line $k_1 + k_2 t$ must intersect the exponential function $\exp\bigl\{\bigl(\eta(\tilde{\theta}) - \eta(\theta_0)\bigr)t - B(\tilde{\theta}) + B(\theta_0)\bigr\}$, for otherwise $\varphi^*$ would be identically one. Because the exponential function is strictly convex, the line and exponential function intersect either once, if the line is tangent to the curve, or twice. Let $t_1^* < t_2^*$ denote the two points of intersection when the line is not tangent, and let $t_1^* = t_2^*$ be the single point of intersection when the line is tangent to the exponential function. Since the exponential function is convex, $\varphi^*$ has form

$$\varphi^* = \begin{cases} 1, & T < t_1^* \text{ or } T > t_2^*; \\ 0, & T \in (t_1^*, t_2^*). \end{cases}$$

To finish showing that $\varphi^*$ is a two-sided test, we need to verify $\varphi^*$ is not one-sided. Suppose $E_\theta \varphi I\{T \le t_1\} = 0$. By Theorem 12.9 $\psi = E_\theta(\varphi^*|T)$ has the same power function as $\varphi^*$, and this test is uniformly most powerful testing $\theta \le \theta_0$ against $\theta > \theta_0$ by Theorem 12.9. Because $\varphi_+$ is also uniformly most powerful, the power functions for $\varphi^*$ and $\varphi_+$ must agree for $\theta \ge \theta_0$, and the slope of the power function for $\varphi^*$ at $\theta_0$ must be $m_+$. This is a contradiction for if $\varphi^*$ lies in $\mathcal{C}_m$ this slope must be $m < m_+$. Similarly, $E_\theta \varphi I\{T \ge t_2\} = 0$, and so $\varphi^*$ is a two-sided test.

To conclude, we show that any two-sided test $\tilde{\varphi} \in \mathcal{C}_m$ is uniformly most powerful. So assume

$$\tilde{\varphi} = \begin{cases} 1, & T < \tilde{t}_1 \text{ or } T > \tilde{t}_2; \\ 0, & T \in (\tilde{t}_1, \tilde{t}_2), \end{cases}$$

and let $\theta$ be an arbitrary point in $\Omega$ not equal to $\theta_0$. Define $\kappa \in \mathbb{R}^2$ so that the line $\kappa_1 + \kappa_2 t$ passes through the points

$$\left(\tilde{t}_1, e^{[\eta(\theta)-\eta(\theta_0)]\tilde{t}_1 - B(\theta)+B(\theta_0)}\right)$$

and

$$\left(\tilde{t}_2, e^{[\eta(\theta)-\eta(\theta_0)]\tilde{t}_2 - B(\theta)+B(\theta_0)}\right).$$

If $\tilde{t}_1 = \tilde{t}_2$, so these points are the same, then the line should also have slope $(\theta-\theta_0)e^{(\theta-\theta_0)\tilde{t}_1 - A(\theta)+A(\theta_0)}$ so that it lies tangent to the exponential curve. By convexity of the exponential function and algebra similar to that used above, $\tilde{\varphi}$ has form

$$\tilde{\varphi} = \begin{cases} 1, & p_\theta > (\kappa_1 + \kappa_2 T)p_{\theta_0}; \\ 0, & p_\theta < (\kappa_1 + \kappa_2 T)p_{\theta_0}. \end{cases}$$

From this, $\tilde{\varphi}$ clearly maximizes $\int \varphi(p_\theta - \kappa \cdot g)\,d\mu$ over all $\varphi \in \mathcal{K}$. But for test function $\varphi \in \mathcal{C}_m$,

$$\int \varphi(p_\theta - \kappa \cdot g)\,d\mu = E_\theta\varphi - \kappa \cdot c.$$

Thus $E_\theta\tilde{\varphi} \geq E_\theta\varphi$ for any $\varphi \in \mathcal{C}_m$, and, since $\theta$ is arbitrary, $\tilde{\varphi}$ is uniformly most powerful in $\mathcal{C}_m$. □

*Remark 12.23.* A similar result can be obtained testing $H_0 : \theta \in [\theta_1, \theta_2]$ versus $H_1 : \theta \notin [\theta_1, \theta_2]$. Suppose $\varphi^*$ is a two-sided test with $E_{\theta_1}\varphi^* = \alpha_1$ and $E_{\theta_2}\varphi^* = \alpha_2$. Then $\varphi^*$ has level $\alpha = \max\{\alpha_1, \alpha_2\}$ and is uniformly most powerful among all tests $\varphi$ with $E_{\theta_1}\varphi = \alpha_1$ and $E_{\theta_2}\varphi = \alpha_2$.

*Remark 12.24.* If the slope $m$ for the power function for a test $\varphi$ at $\theta_0$ differs from zero, then there will be points $\theta \neq \theta_0$ where the power for the test is less than $\alpha$. If this happens, the test $\varphi$ is called *biased*. If an unbiased test is desired, the slope $m$ should be constrained to equal zero. This idea is developed and extended in the next section of this chapter.

## 12.7 Unbiased Tests

In the previous section we encountered a situation in which uniformly most powerful tests cannot exist unless we constrain the class of test functions under consideration. One appealing constraint restricts attention to tests that are unbiased according to the following definition. Theorem 12.26 below finds uniformly most powerful unbiased tests for one-parameter exponential families, and Chapter 13 has extensions to higher dimensions.

**Definition 12.25.** *A test $\varphi$ for $H_0 : \theta \in \Omega_0$ versus $H_1 : \theta \in \Omega_1$ with level $\alpha$ is* unbiased *if its power $\beta_\varphi(\theta) = E_\theta\varphi$ satisfies*

$$\beta_\varphi(\theta) \leq \alpha, \quad \forall\theta \in \Omega_0 \ \text{and}\ \beta_\varphi(\theta) \geq \alpha, \quad \forall\theta \in \Omega_1.$$

If there is a uniformly most powerful test $\varphi^*$, then it is automatically unbiased because $\beta_{\varphi^*}(\theta) \ge \beta_\varphi(\theta)$, for all $\theta \in \Omega_1$, and the right-hand side of this inequality is identically $\alpha$ for the degenerate test, which equals $\alpha$ regardless of the observed data.

**Theorem 12.26.** *If $\alpha \in (0,1)$, $\theta_0$ lies in the interior of $\Omega$, $X$ has density (12.6), and $\eta$ is differentiable and strictly increasing with $0 < \eta'(\theta_0) < \infty$, then there is a two-sided, level $\alpha$ test $\varphi^*$ with $\beta'_{\varphi^*}(\theta_0) = 0$. Any such test is uniformly most powerful testing $H_0 : \theta = \theta_0$ versus $H_1 : \theta \ne \theta_0$ among all unbiased tests with level $\alpha$.*

Changing the sign of $T$ and $\eta$, this result is also true if $\eta$ is differentiable and strictly decreasing with $-\infty < \eta'(\theta_0) < 0$.

*Proof.* Since $\theta_0$ lies in the interior of $\Omega$, the power function for any unbiased test $\varphi$ must have zero slope at $\theta_0$, and so $\varphi \in \mathcal{C}_0$. The theorem is essentially a corollary of Theorem 12.22, provided $0 \in (m_-, m_+)$. This is established in the following lemma. ☐

**Lemma 12.27.** *Under the assumptions of Theorem 12.26,*

$$m_+ = \beta'_{\varphi_+}(\theta_0) > 0 \text{ and } m_- = \beta'_{\varphi_-}(\theta_0) < 0.$$

*Proof.* Let us begin showing that $E_{\theta_0} T\varphi_+ > \alpha E_{\theta_0} T$. The argument is similar to the proof of Proposition 12.3. From the form of $\varphi_+$,

$$\begin{aligned} E_{\theta_0} T\varphi_+ - \alpha E_{\theta_0} T &= E_{\theta_0}(T - c_+)\varphi_+ - E_{\theta_0}(T - c_+)\alpha \\ &= E_{\theta_0}(T - c_+)(\varphi_+ - \alpha) \\ &= E_{\theta_0}|T - c_+|\,|\varphi_+ - \alpha|. \end{aligned}$$

Since $\alpha \in (0,1)$, this expression is strictly positive unless $P_{\theta_0}(T = c_+) = 1$. But if $P_{\theta_0}(T = c_+) = 1$, then $P_\theta(T = c_+) = 1$ for all $\theta \in \Omega$ and all distributions in the family are the same. Thus $E_{\theta_0} T\varphi_+ > \alpha E_{\theta_0} T$. Using this in the formula in Theorem 12.20,

$$\beta'_{\varphi_+}(\theta_0) = \eta'(\theta_0) E_{\theta_0} T\varphi_+ - \alpha B'(\theta_0) > \alpha\big(\eta'(\theta_0) E_{\theta_0} T - B'(\theta_0)\big).$$

The lower bound here is zero because $B'(\theta) = \eta'(\theta) E_\theta T$, which follows from Theorem 12.20, with $\varphi$ identically one. ☐

*Remark 12.28.* Because $B'(\theta) = \eta'(\theta) E_\theta T$, using Theorem 12.20 the constraint $\beta'_{\varphi^*}(\theta_0) = 0$ in Theorem 12.26 becomes

$$0 = \eta'(\theta_0)\big[E_{\theta_0} T\varphi^* - \alpha E_{\theta_0} T\big] = \eta'(\theta_0)\operatorname{Cov}_{\theta_0}(\varphi^*, T).$$

So any two-sided test $\varphi^*$ with level $\alpha$ that is uncorrelated with $T$ is uniformly most powerful unbiased.

*Example 12.29.* Suppose $X$ has an exponential distribution with failure rate $\theta$, so

$$p_\theta(x) = \begin{cases} \theta e^{-\theta x}, & x > 0; \\ 0, & \text{otherwise,} \end{cases}$$

and consider testing $H_0 : \theta = 1$ versus $H_1 : \theta \neq 1$. Let

$$\varphi = \begin{cases} 0, & X \in (c_1, c_2); \\ 1, & X \le c_1 \text{ or } X \ge c_2. \end{cases}$$

By Theorem 12.26, $\varphi$ is uniformly most powerful unbiased provided

$$E_1\varphi = 1 - \int_{c_1}^{c_2} e^{-x}\,dx = 1 - e^{-c_1} + e^{-c_2} = \alpha \tag{12.8}$$

and

$$\begin{aligned} E_1 X\varphi = E_1 X - E_1 X(1-\varphi) &= 1 - \int_{c_1}^{c_2} x e^{-x}\,dx \\ &= 1 - (1+c_1)e^{-c_1} + (1+c_2)e^{-c_2} = \alpha E_1 X = \alpha. \end{aligned}$$

Using (12.8), this equation simplifies to

$$c_1 e^{-c_1} = c_2 e^{-c_2}. \tag{12.9}$$

Isolating $c_2$ in (12.8),

$$c_2 = -\log\bigl(e^{-c_1} - 1 + \alpha\bigr).$$

Using this in (12.9),

$$c_1 e^{-c_1} = -\bigl(e^{-c_1} - 1 + \alpha\bigr)\log\bigl(e^{-c_1} - 1 + \alpha\bigr).$$

The solution to this equation must be found numerically. Note that as $c_1$ varies from 0 to $-\log(1-\alpha) > 0$, the left-hand side increases from 0 to $-(1-\alpha)\log(1-\alpha) > 0$, as the right-hand side decreases from $-\alpha\log\alpha > 0$ to 0, and so, by continuity, a solution must exist. For $\alpha = 5\%$, $c_1 = 0.042363$ and $c_2 = 4.7652$.

In practice, numerical issues can be eliminated by choosing $c_1$ and $c_2$ so that $P_1(X \le c_1) = P_1(X \ge c_2) = \alpha/2$, for then $c_1 = -\log(1-\alpha/2)$ and $c_2 = -\log(\alpha/2)$. But the resulting test is biased. For instance, if $\alpha = 5\%$, $c_1 = 0.025318$ and $c_2 = 3.6889$, quite different from the critical values above for the best unbiased test.

## 12.8 Problems[6]

*1. Suppose $X \sim P_\theta$ for some $\theta \in \Omega$, and that $U$ is uniformly distributed on $(0,1)$ and is independent of $X$. Let $\varphi(X)$ be a randomized test based on $X$. Find a nonrandomized test based on $X$ and $U$, so $\psi(X,U) = 1_S(X,U)$ for some critical region $S$, with the same power function as $\varphi$, $E_\theta\varphi(X) = E_\theta\psi(X,U)$, for all $\theta \in \Omega$.

*2. Suppose $\sup|h(x)| = M$ and $Eh(Z) = 0$, where $Z \sim N(0,1)$. Give a sharp upper bound for $Eh(2Z)$.

*3. Determine the density of $Z_1/Z_2$ when $Z_1$ and $Z_2$ are independent standard normal random variables. (This should be useful in the next problem.)

*4. Let $X_1$ and $X_2$ be independent, and let $\sigma_1^2 > 0$ and $\sigma_2^2 > 0$ be known variances. Find the error rate for the best symmetric test of $H_0 : X_1 \sim N(0,\sigma_1^2), X_2 \sim N(0,\sigma_2^2)$ versus $H_1 : X_1 \sim N(0,\sigma_2^2), X_2 \sim N(0,\sigma_1^2)$. A symmetric test here is a test that takes the opposite action if the two data values are switched, so $\varphi(x_1,x_2) = 1 - \varphi(x_2,x_1)$. For a symmetric test the error probabilities under $H_0$ and $H_1$ will be equal.

5. Suppose $\sup_{x\geq 0}|h(x)| = M$ and $Eh(X) = 0$, where $X$ has a standard exponential distribution. Give a sharp upper bound for $Eh(2X)$.

*6. Suppose $X$ is uniformly distributed on $(0,2)$.
   a) Suppose $\sup_{(0,2)}|h(x)| \leq M$ and $Eh(X) = 0$. Give an upper bound for $Eh(X^2/2)$, and a function $h$ that achieves the bound.
   b) Suppose instead that $|h(x)| \leq Mx$, $0 < x < 2$, but we still have $Eh(X) = 0$. Now what is the best upper bound for $Eh(X^2/2)$? What function achieves the bound?

*7. Consider a model in which $X$ has density

$$p_\theta(x) = \frac{\theta}{(1+\theta x)^2}, \qquad x > 0.$$

   a) Show that the derivative of the power function $\beta$ of a test $\varphi$ is given by

$$\beta'(\theta) = E_\theta\left[\frac{1-\theta X}{\theta(1+\theta X)}\varphi(X)\right].$$

   b) Among all tests with $\beta(1) = \alpha$, which one maximizes $\beta'(1)$?

8. Let $X$ have a Poisson distribution with mean one. Suppose $|h(x)| \leq 1$, $x = 0,1,2,\ldots$, and $Eh(X) = 0$. Find the largest possible value for $Eh(2X)$, and the function $h$ that achieves the maximum.

*9. Suppose data $X$ has density $p_\theta$, $\theta \in \Omega \subset \mathbb{R}$, and that these densities are regular enough that the derivative of the power function of any test function $\varphi$ can be evaluated differentiating under the integral sign,

$$\beta'_\varphi(\theta) = \int \varphi(x)\frac{\partial p_\theta(x)}{\partial\theta}\,d\mu(x).$$

[6] Solutions to the starred problems are given at the back of the book.

A test $\varphi^*$ is called *locally most powerful* testing $H_0 : \theta = \theta_0$ versus $H_1 : \theta > \theta_0$ if it maximizes $\beta'_\varphi(\theta_0)$ among all tests $\varphi$ with level $\alpha$. Determine the form of the locally most powerful test.

*10. Suppose $X = (X_1, \dots, X_n)$ with the $X_i$ i.i.d. with common density $f_\theta$. The locally most powerful test of $H_0 : \theta = \theta_0$ versus $H_1 : \theta > \theta_0$ from Problem 12.9 should reject $H_0$ if an appropriate statistic $T$ exceeds a critical value $c$. Use the central limit theorem to describe how the critical level $c$ can be chosen when $n$ is large to achieve a level approximately $\alpha$. The answer should involve Fisher information at $\theta = \theta_0$.

11. Laplace's law of succession gives a distribution for Bernoulli variables $X_1, X_2, \dots$ in which $P(X_1 = 1) = 1/2$, and

$$P(X_{j+1} = 1 | X_1 = x_1, \dots, X_j = x_j) = \frac{1 + x_1 + \dots + x_j}{j+2}, \qquad j \geq 1.$$

Consider testing the hypothesis $H_1$ that $X_1, \dots, X_n$ have this distribution against the null hypothesis $H_0$ that the variables are i.i.d. with $P(X_i = 1) = 1/2$. If $n = 10$, find the best test with size $\alpha = 5\%$. What is the power of this test?

12. An entrepreneur would like to sell a fixed amount $M$ of some product through online auctions. Let $R(t) \geq 0$ denote his selling rate at time $t$. Assuming all of the merchandise is sold eventually,

$$\int_0^\infty R(t)\, dt = M.$$

The sales rate and price should be related, with the sales rate increasing as the price decreases. Assume that price is inversely proportional to $\sqrt{R}$, so that the rate of return (price times selling rate) at time $t$ is $c\sqrt{R(t)}$. Discounting future profits, the entrepreneur would like to maximize

$$\int_0^\infty c\sqrt{R(t)}e^{-\delta t}\, dt,$$

where $\delta > 0$ denotes the discount rate. Use a Lagrange multiplier approach to find the best rate function $R$ for the entrepreneur.

13. Consider simple versus simple testing from a Bayesian perspective. Let $\Theta$ have a Bernoulli distribution with $P(\Theta = 1) = p$ and $P(\Theta = 0) = 1 - p$. Given $\Theta = 0$, $X$ will have density $p_0$, and given $\Theta = 1$, $X$ will have density $p_1$.
   a) Show that the chance of accepting the wrong hypothesis in the Bayesian model using a test function $\varphi$ is
   $$R(\varphi) = E\big[I\{\Theta = 0\}\varphi(X) + I\{\Theta = 1\}(1 - \varphi(X))\big].$$
   b) Use smoothing to relate $R(\varphi)$ to $E_0\varphi = E\big[\varphi(X) \bigm| \Theta = 0\big]$ and $E_1\varphi = E\big[\varphi(X) \bigm| \Theta = 1\big]$.

c) Find the test function $\varphi^*$ minimizing $R(\varphi)$. Show that $\varphi^*$ is a likelihood ratio test, identifying the critical value $k$.

*14. Let $X$ denote the number of tails before the first heads if a coin is tossed repeatedly. If successive tosses are independent and $p$ is the chance of heads, determine the uniformly most powerful test of $H_0 : p = 1/2$ versus $H_1 : p < 1/2$ with level $\alpha = 5\%$. What is the power of this test if $p$ is 40%?

*15. Suppose $X$ and $Y$ are jointly distributed from a bivariate normal distribution with correlation $\rho$, means $EX = EY = 0$, and $\text{Var}(X) = \text{Var}(Y) = 1/(1-\rho^2)$. Determine the uniformly most powerful test of $H_0 : \rho \le 0$ versus $H_1 : \rho > 0$ based on $(X, Y)$.

*16. Consider a location family with densities $p_\theta(x) = g(x-\theta)$. Show that if $g$ is twice differentiable and $d^2 \log g(x)/dx^2 \le 0$ for all $x$, then the densities have monotone likelihood ratios in $x$. Give an analogous differential condition sufficient to ensure that densities for a scale family $p_\theta(x) = g(x/\theta)/\theta$, $x > 0$, have monotone likelihood ratios in $x$.

*17. *p-values.* Suppose we have a family of tests $\varphi_\alpha$, $\alpha \in (0,1)$ indexed by level (so $\varphi_\alpha$ has level $\alpha$), and that these tests are "nested" in the sense that $\varphi_\alpha(x)$ is nondecreasing as a function of $\alpha$. We can then define the "p-value" or "attained significance" for observed data $x$ as $\inf\{\alpha : \varphi_\alpha(x) = 1\}$, thought of as the smallest value for $\alpha$ where test $\varphi_\alpha$ rejects $H_0$. Suppose we are testing $H_0 : \theta \le \theta_0$ versus $H_1 : \theta > \theta_0$ and that the densities for data $X$ have monotone likelihood ratios in $T$. Further suppose $T$ has a continuous distribution.

a) Show that the family of uniformly most powerful tests are nested in the sense described.

b) Show that if $X = x$ is observed, the p-value $P(x)$ is

$$P_{\theta_0}\big[T(X) > t\big],$$

where $t = T(x)$ is the observed value of $T$.

c) Determine the distribution of the p-value $P(X)$ when $\theta = \theta_0$.

18. Let $F$ be a cumulative distribution function that is continuous and strictly increasing on $[0, \infty)$ with $F(0) = 0$, and let $q_\alpha$ denote the upper $\alpha$th quantile for $F$, so $F(q_\alpha) = 1 - \alpha$. Suppose we have a single observation $X$ with

$$P_\theta(X \le x) = F(x/\theta), \qquad x \in \mathbb{R}, \quad \theta > 0.$$

a) Consider testing $H_0 : \theta \le \theta_0$ versus $H_1 : \theta > \theta_0$. Find the significance level for the test $\varphi = 1_{(c,\infty)}$. What choice for $c$ will give a specified level $\alpha$?

b) Let $\varphi_\alpha$ denote the test with level $\alpha$ in part (a). Show that the tests $\varphi_\alpha$, $\alpha \in (0,1)$, are nested in the sense described in Problem 12.17, and give a formula to compute the p-value $P(X)$.

19. Suppose $X$ has a Poisson distribution with parameter $\lambda$. Determine the uniformly most powerful test of $H_0 : \lambda \le 1$ versus $H_1 : \lambda > 1$ with level $\alpha = 5\%$.

*20. Do the densities $p_\theta(x) = (1 + \theta x)/2$, $x \in (-1, 1)$, $\theta \in [-1, 1]$, have monotone likelihood ratios in $T(x) = x$?

*21. Let $f$ be a specified probability density on $(0, 1)$ and let

$$p_\theta(x) = \theta + (1 - \theta)f(x), \qquad x \in (0, 1),$$

where $\theta \in [0, 1]$ is an unknown parameter. Show that these densities have monotone likelihood ratios, identifying the statistic $T(x)$.

22. Suppose we observe a single observation $X$ from $N(\theta, \theta^2)$.
   a) Do the densities for $X$ have monotone likelihood ratios?
   b) Let $\phi^*$ be the best level $\alpha$ test of $H_0 : \theta = 1$ versus $H_1 : \theta = 2$. Is $\phi^*$ also the best level $\alpha$ test of $H_0 : \theta = 1$ versus $H_1 : \theta = 4$?

23. Consider tests for $H_0 : \theta = 0$ versus $H_1 : \theta \neq 0$ based on a single observation $X$ from $N(\theta, 1)$. Using the apparent symmetry of this testing problem, it seems natural to base a test on $Y = |X|$.
   a) Find densities $q_\theta$ for $Y$ and show that the distribution for $Y$ depends only on $|\theta|$.
   b) Show that the densities $q_\theta$, $\theta \geq 0$, have monotone likelihood ratios.
   c) Find the uniformly most powerful level $\alpha$ test of $H_0$ versus $H_1$ based on $Y$.
   d) The uniformly most powerful test $\varphi^*(Y)$ in part (c) is not most powerful compared with tests based on $X$. Find a level $\alpha$ test $\varphi(X)$ with better power at $\theta = -1$,

$$E_{-1}\varphi(X) > E_{-1}\varphi^*(Y).$$

   What is the difference in power at $\theta = -1$ if $\alpha = 5\%$?

24. Let $P_0$ and $P_1$ be two probability distributions, and for $\epsilon \in (0, 1)$, let $P_\epsilon$ denote the mixture distribution $(1-\epsilon)P_0+\epsilon P_1$. Let $E_0$, $E_1$, and $E_\epsilon$ denote expectation when $X \sim P_0$, $X \sim P_1$, and $X \sim P_\epsilon$, respectively.
   a) Let $\varphi$ be a test function with $\alpha = E_0\varphi(X)$ and $\beta = E_1\varphi(X)$. Express $E_\epsilon\varphi(X)$ as a function of $\epsilon$, $\alpha$, and $\beta$.
   b) Using the result in part (a), argue directly that if $\varphi$ is the most powerful level $\alpha$ test of $H_0 : X \sim P_0$ versus $H_1 : X \sim P_1$, then it is also the most powerful level $\alpha$ test of $H_0 : X \sim P_0$ versus $H_1 : X \sim P_\epsilon$.
   c) Suppose $P_0$ and $P_1$ have densities $p_0$ and $p_1$, respectively, with respect to a measure $\mu$. Find the density for $P_\epsilon$.
   d) Using part (c), show that if $\varphi$ is a likelihood ratio test of $H_0 : X \sim P_0$ versus $H_1 : X \sim P_1$, then it is also a likelihood ratio test of $H_0 : X \sim P_0$ versus $H_1 : X \sim P_\epsilon$.

*25. Suppose $X$ has a Poisson distribution with mean $\lambda$, and that $U$ is uniformly distributed on $(0, 1)$ and is independent of $X$.
   a) Show that the joint densities of $X$ and $U$ have monotone likelihood ratios in $T = X + U$.
   b) Describe how to construct level $\alpha$ uniformly most powerful tests of $H_0 : \lambda = \lambda_0$ versus $H_1 : \lambda > \lambda_0$ based on $X$ and $U$. Specify the resulting test explicitly if $\alpha = 5\%$ and $\lambda_0 = 2$.

c) Describe confidence intervals dual to the family of tests in part (b). Give the confidence interval if the data are $X = 2$ and $U = 0.7$.

26. Suppose $X$ has a geometric distribution with success probability $\theta$, so $P_\theta(X = x) = \theta(1 - \theta)^x$, $x = 0, 1, \ldots$; and that $U$ is uniformly distributed on $(0, 1)$ and is independent of $X$.
    a) Show that the joint densities of $X$ and $U$ have monotone likelihood ratios in $T = -(X + U)$.
    b) Describe how to construct level $\alpha$ uniformly most powerful tests of $H_0 : \theta = \theta_0$ versus $H_1 : \theta > \theta_0$ based on $X$ and $U$. Specify the resulting test explicitly if $\alpha = 5\%$ and $\theta_0 = 1/20$.
    c) Describe confidence intervals dual to the family of tests in part (b). Give the confidence interval if the data are $X = 2$ and $U = 0.7$.
27. Suppose $X_1, \ldots, X_n$ are i.i.d. from $N(0, \sigma^2)$.
    a) Determine the uniformly most powerful test of $H_0 : \sigma = \sigma_0$ versus $H_1 : \sigma > \sigma_0$.
    b) Find a confidence interval for $\sigma$ using duality from the tests in part (a).

*28. Let $X_1, \ldots, X_n$ be i.i.d. observations from a uniform distribution on the interval $(0, \theta)$. Find confidence intervals $S_1$ dual to the family of uniformly most powerful tests of $\theta = \theta_0$ versus $\theta > \theta_0$ and $S_2$ dual to the family of uniformly most powerful tests of $\theta = \theta_0$ versus $\theta < \theta_0$. Then use the result from the Problem 9.12 to find a 95% confidence interval for $\theta$. This interval should have finite length and exclude zero.

29. Suppose $Y_1$ and $Y_2$ are independent variables, both uniformly distributed on $(0, \theta)$, but our observation is $X = Y_1 + Y_2$.
    a) Show that the densities for $X$ have monotone likelihood ratios.
    b) Find the UMP level $\alpha$ test of $H_0 : \theta = \theta_0$ versus $H_1 : \theta > \theta_0$ based on $X$.
    c) Find a confidence set for $\theta$ dual to the tests in part (b).
30. Let $X$ and $Y$ be independent with $X \sim N(\mu_x, 1)$ and $Y \sim N(\mu_y, 1)$. Take $\|\mu\|^2 = \mu_x^2 + \mu_y^2$, and consider testing $H_0 : \mu_x = \mu_y = 0$ versus $H_1 : \|\mu\| > 0$. For rotational symmetry, a test based on $T = X^2 + Y^2$ may seem natural. The density of $T$ is

$$f_{\|\mu\|}(t) = \begin{cases} \frac{1}{2} I_0\bigl(\sqrt{t}\|\mu\|\bigr) \exp\bigl\{-\frac{1}{2}\bigl(t + \|\mu\|^2\bigr)\bigr\}, & t > 0; \\ 0, & \text{otherwise}, \end{cases}$$

where $I_0$ is a modified Bessel function given by

$$I_0(x) = \frac{1}{\pi} \int_0^\pi e^{x \cos \omega}\, d\omega.$$

a) Show that $I_0(x) > I_0'(x)$ and that $x I_0''(x) + I_0'(x) = x I_0(x)$.

b) Show that $x I_0'(x)/I_0(x)$ is increasing in $x$. Use this to show that for $c > 1$, $I_0(cx)/I_0(x)$ is an increasing function of $x$. Hint:

$$\log I_0(cx) - \log I_0(x) = \int_1^c \frac{\partial \log I_0(ux)}{\partial u}\, du.$$

c) Show that the densities $f_{\|\mu\|}$ have monotone likelihood ratios.
d) Find the uniformly most powerful level $\alpha$ test of $H_0$ versus $H_1$ based on $T$.
e) Find a level $\alpha$ test of $H_0$ versus $H_1$ based on $X$ and $Y$ that has power as high as possible if $\mu_x = \mu_y = 1$. Is this the same test as the test in part (d)?
f) Suggest a level $\alpha$ test of $\tilde{H}_0 : \mu_x = c_x,\ \mu_y = c_y$, versus $\tilde{H}_1 : \mu_x \neq c_x$ or $\mu_y \neq c_y$, based on $\tilde{T} = (X - c_x)^2 + (Y - c_y)^2$.
g) Find a $1 - \alpha$ confidence region for $(\mu_x, \mu_y)$ dual to the family of tests in part (f). What is the shape of your confidence region?

*31. Suppose $X \sim N(\theta, 1)$, and let $\varphi$ be a test function with power $\beta(\theta) = E_\theta \varphi(X)$.
a) Show that $\beta'(0) = E_0 X\varphi(X)$.
b) What test function $\varphi$ maximizes $\beta(1)$ subject to constraints $\beta(0) = \alpha$ and $\beta'(0) = 0$?

*32. Suppose $X_1$ and $X_2$ are independent positive random variables with common Lebesgue density $p_\theta(x) = \theta/(1 + \theta x)^2$, $x > 0$.
a) Use dominated convergence to write the derivative $\beta'(\theta)$ of the power function for a test $\varphi$ as an expectation.
b) Determine the locally most powerful test $\varphi$ of $H_0 : \theta \le \theta_0$ versus $H_1 : \theta > \theta_0$ with $\beta_\varphi(\theta_0) = 5\%$. As in Problem 12.9, a locally most powerful test here would maximize $\beta'(\theta_0)$ among all tests with level $\alpha$. Hint: The relevant test statistic can be written as the sum of two independent variables. First find the $P_{\theta_0}$ marginal distribution of these variables.
c) Determine a 95% confidence region for $\theta$ by duality, inverting the family of tests in part (b).

*33. Suppose we have a single observation from an exponential distribution with failure rate $\lambda$, and consider testing $H_0 : \lambda = 2$ versus $H_1 : \lambda \neq 2$. Find a test $\varphi^*$ with minimal level $\alpha$ among all tests $\varphi$ with 50% power at $\lambda = 1$ and $\lambda = 3$, $E_1\varphi = E_3\varphi = 1/2$.

34. Suppose $X$ has a uniform distribution on $(0, 1)$. Find the test function $\varphi$ that maximizes $E\varphi(X)$ subject to constraints

$$E\varphi(X^2) = E\varphi(1 - X^2) = 1/2.$$

35. Define $\varphi_1 = 1_{[0,1]}$, $\varphi_2 = 1_{[0,1/2]}$, $\varphi_3 = 1_{[1/2,1]}$, $\varphi_4 = 1_{[0,1/3]}$, $\varphi_5 = 1_{[1/3,2/3]}$, $\varphi_6 = 1_{[2/3,1]}$, etc.
a) If $x \in [0, 1]$, what is $\limsup_{n\to\infty} \varphi_n(x)$?
b) What is $\lim_{n\to\infty} \int_0^1 \varphi(x)\,dx$?
c) Suppose $f$ is bounded and nonnegative. Find

$$\lim_{n\to\infty} \int_0^1 \varphi(x) f(x)\,dx.$$

d) Suppose $f \ge 0$ and $\int_0^1 f(x)\,dx < \infty$. Use dominated convergence to show that

$$\lim_{k\to\infty} \int_0^1 \bigl(f(x) - k\bigr)^+ \,dx = 0.$$

e) Suppose $f \ge 0$ and $\int_0^1 f(x)\,dx < \infty$. Show that

$$\lim_{n\to\infty} \int_0^1 f(x)\varphi_n(x)\,dx = 0.$$

Hint: Note that for any $k$,

$$f(x) = \min\{f(x), k\} + \bigl(f(x) - k\bigr)^+.$$

Use this to find an upper bound for

$$\limsup_{n\to\infty} \int_0^1 f(x)\varphi_n(x)\,dx.$$

f) Let $\varphi$ be the "zero" test function, $\varphi(x) = 0$, for all $x$. Do the functions $\varphi_n$ converge pointwise to $\varphi$?

36. For $n \ge 1$ and $x \in (0,1)$, define

$$\phi_n(x) = I\bigl(\lfloor 2^n x^2 \rfloor \in \{0, 2, 4, \ldots\}\bigr),$$

and let $\mu$ be Lebesgue measure on $(0,1)$. Find the weak limit of these functions, that is, a function $\phi$ on $(0,1)$ such that $\phi_n \xrightarrow{w} \phi$.

*37. Let $\varphi_n$ be a sequence of test functions converging pointwise to $\varphi$, $\varphi_n(x) \to \varphi(x)$ for all $x$.
  a) Does it follow that $\varphi_n^2 \xrightarrow{w} \varphi^2$? Prove or give a counterexample.
  b) Does it follow that $1/\varphi_n \xrightarrow{w} 1/\varphi$? Prove or give a counterexample.

38. Let $X$ have a Cauchy distribution with location $\theta$, so

$$p_\theta(x) = \frac{1}{\pi\bigl[1 + (x-\theta)^2\bigr]}, \qquad x \in \mathbb{R},$$

and consider testing $H_0 : \theta = 0$ versus $H_1 : \theta \ne 0$. Find a test $\varphi$ with level $\alpha = 5\%$ that maximizes $E_1\varphi$ subject to the constraint $E_1\varphi = E_{-1}\varphi$. Is this test uniformly most powerful unbiased?

39. Suppose $X$ has an exponential distribution with failure rate $\lambda$, so $p_\lambda(x) = \lambda e^{-\lambda x}$ for $x > 0$. Determine the most powerful test of $H_0 : \lambda = 1$ or $\lambda = 4$ versus $H_1 : \lambda = 2$ with level $\alpha = 5\%$. The test you derive is in fact the uniformly most powerful test of $H_0 : \lambda \le 1$ or $\lambda \ge 4$ versus $H_1 : \lambda \in (1,4)$ with level $\alpha = 5\%$.

40. *Locally most powerful tests in two-sided situations.* Suppose we have a single observation $X$ with density

$$p_\theta(x) = \begin{cases} \dfrac{\theta}{(1+\theta x)^2}, & x > 0; \\ 0, & \text{otherwise,} \end{cases}$$

where $\theta > 0$. Find a test $\varphi^*$ of $H_0 : \theta = 1$ versus $H_1 : \theta \neq 1$ with level $\alpha = 5\%$ that maximizes $\beta''_\varphi(1)$, subject to the constraint $\beta'_\varphi(1) = 0$.

*41. Suppose $X$ has a binomial distribution with two trials and success probability $p$. Determine the uniformly most powerful unbiased test of $H_0 : p = 2/3$ versus $H_1 : p \neq 2/3$. Assume $\alpha < 4/9$.

*42. Let $X_1 \dots X_4$ be i.i.d. from $N(0, \sigma^2)$. Determine the uniformly most powerful unbiased test of $H_0 : \sigma = 1$ versus $H_1 : \sigma \neq 1$ with size $\alpha = 5\%$.

43. Suppose we observe a single observation $X$ with density

$$f_\theta(x) = c(\theta)|x|e^{-(x-\theta)^2/2}, \qquad x \in \mathbb{R}.$$

a) Give a formula for $c(\theta)$ in terms of the cumulative distribution function $\Phi$ for the standard normal distribution.
b) Find the uniformly most powerful unbiased test of $H_0 : \theta = 0$ versus $H_1 : \theta \neq 0$.

*44. Consider testing $H_0 : \theta = \theta_0$ versus $H_1 : \theta \neq \theta_0$ based on a single observation $X$ with density

$$p_\theta(x) = \begin{cases} \dfrac{\theta e^{\theta x}}{2\sinh\theta}, & |x| < 1; \\ 0, & |x| \geq 1. \end{cases}$$

When $\theta = 0$, $p_\theta$ should be $1/2$ if $|x| < 1$, and zero otherwise.
a) Specify the form of the uniformly most powerful unbiased test with level $\alpha$, and give equations to determine constants needed to specify the test.
b) Specify the uniformly most powerful unbiased test explicitly when $\theta_0 = 0$.

45. Let $X_1, \dots, X_n$ be independent with

$$X_i \sim N(t_i\theta, 1), \qquad i = 1, \dots, n,$$

where $t_1, \dots, t_n$ are known constants and $\theta$ is an unknown parameter.
a) Determine the uniformly most powerful unbiased test of $H_0 : \theta = \theta_0$ versus $H_1 : \theta \neq \theta_0$.
b) Find a confidence region for $\theta$ inverting the family of tests in part (a).

46. Suppose our data consist of two independent observations, $X$ and $Y$, from a Poisson distribution with mean $\lambda$. Find the uniformly most powerful unbiased test of $H_0 : \lambda = 1$ versus $H_1 : \lambda \neq 1$ with level $\alpha = 10\%$.

47. A random angle $X$ has density

$$p_\theta(x) = \begin{cases} \dfrac{\exp\bigl[\theta\cos x\bigr]}{2\pi I_0(\theta)}, & x \in [0, 2\pi); \\ 0, & \text{otherwise,} \end{cases}$$

where $\theta \in \mathbb{R}$ and $I_0$ is a modified Bessel function ($I_0(0) = 1$). Find the uniformly most powerful unbiased test of $H_0 : \theta = 0$ versus $H_1 : \theta \neq 0$ with level $\alpha$.

48. Suppose $X$ has density

$$p_\theta(x) = \frac{x^2 \exp\bigl[-\frac{1}{2}(x-\theta)^2\bigr]}{\sqrt{2\pi}(1+\theta^2)}, \qquad x \in \mathbb{R}.$$

Find the uniformly most powerful unbiased test of $H_0 : \theta = 0$ versus $H_1 : \theta \neq 0$ with level $\alpha = 5\%$.

49. Because a good test of $H_0 : \theta \in \Omega_0$ versus $H_1 : \theta \in \Omega_1$ should have high power on $\Omega_1$ and small power on $\Omega_0$, a test function $\phi$ might be chosen to minimize

$$\int_{\Omega_0} \beta_\phi(\theta) w(\theta)\, d\Lambda(\theta) + \int_{\Omega_1} \bigl(1 - \beta_\phi(\theta)\bigr) w(\theta)\, d\Lambda(\theta),$$

where $\Lambda$ is a measure on $\Omega = \Omega_0 \cup \Omega_1$ and $w \geq 0$ is a weight function. (With a natural loss structure, Bayes risks would have this form.)

a) Describe a test function $\phi^*$ that minimizes this criterion. Assume that $\mathcal{P}$ is a dominated family with densities $p_\theta$, $\theta \in \Omega$.
b) Find the optimal test function $\phi^*$ explicitly if $w$ is identically one, $\Lambda$ is Lebesgue measure on $(0, \infty)$, $P_\theta$ is the exponential distribution with failure rate $\theta$, $\Omega_0 = (0, 1]$, and $\Omega_1 = (1, \infty)$.

# 13

# Optimal Tests in Higher Dimensions

In Section 13.2, uniformly most powerful unbiased tests are considered for multiparameter exponential families. The discussion involves marginal and conditional distributions described in Section 13.1. The $t$-test and Fisher's exact test are considered as examples in Section 13.3.

## 13.1 Marginal and Conditional Distributions

The main result of this section uses the following technical lemma about conditional and marginal distributions when a joint density factors, but the dominating measure is not a product measure.

**Lemma 13.1.** *Let $Y$ be a random vector in $\mathbb{R}^n$ and $T$ a random vector in $\mathbb{R}^m$, and let $P_0$ and $P_1$ be two possible joint distributions for $Y$ and $T$. Introduce marginal distributions*

$$Q_0(B) = P_0(T \in B) \text{ and } Q_1(B) = P_1(T \in B),$$

*and conditional distributions*

$$R_{0t}(B) = P_0(Y \in B|T = t) \text{ and } R_{1t}(B) = P_1(Y \in B|T = t).$$

*Assume $P_1 \ll P_0$ and that the density for $P_1$ has form*

$$\frac{dP_1}{dP_0}(t, y) = a(y)b(t)$$

*with $a(y) > 0$ for all $y \in \mathbb{R}^n$. Then $Q_1 \ll Q_0$ and $R_{1t} \ll R_{0t}$, for a.e. $t$ ($Q_1$) with densities given by*

$$\frac{dQ_1}{dQ_0}(t) = b(t)E_0\big[a(Y)|T = t\big] = b(t)\int a\,dR_{0t}$$

*and*

$$\frac{dR_{1t}}{dR_{0t}}(y) = \frac{a(y)}{\int a\,dR_{0t}}.$$

R.W. Keener, *Theoretical Statistics: Topics for a Core Course*, Springer Texts in Statistics,
DOI 10.1007/978-0-387-93839-4_13, © Springer Science+Business Media, LLC 2010

*Proof.* The formula for the marginal density is established by first using Lemma 12.18 to write $P_1(T \in B)$ as an expectation under $P_0$ involving the likelihood ratio $a(Y)b(T)$, followed by a smoothing argument to write this expectation as an integral against $Q_0$. Thus

$$\begin{aligned} P_1(T \in B) &= E_0\bigl[I\{T \in B\}a(Y)b(T)\bigr] \\ &= E_0 E_0\bigl[I\{T \in B\}a(Y)b(T) \mid T\bigr] \\ &= E_0\Bigl[I\{T \in B\}b(T)E_0\bigl[a(Y) \mid T\bigr]\Bigr] \\ &= \int_B b(t)E_0\bigl[a(Y) \mid T = t\bigr]\,dQ_0(t). \end{aligned}$$

Next, if the stated density for $R_{1t}$ is correct,

$$R_{1t}(C) = \int_C \frac{a(y)}{\int a\,dR_{0t}}\,dR_{0t}(y),$$

and so, according to Definition 6.2 of conditional distributions, we must show that

$$P_1(T \in B, Y \in C) = \int_B \int_C \frac{a(y)}{\int a\,dR_{0t}}\,dR_{0t}(y)\,dQ_1(t).$$

Using the formula for the density of $Q_1$ with respect to $Q_0$, the right-hand side of this equation equals

$$\begin{aligned} &\int_B \int I\{y \in C\}a(y)\,dR_{0t}(y)b(t)\,dQ_0(t) \\ &\qquad = E_0\Bigl[I\{T \in B\}b(T)E_0\bigl[I\{Y \in C\}a(Y) \mid T\bigr]\Bigr] \\ &\qquad = E_0\bigl[a(Y)b(T)I\{T \in B, Y \in C\}\bigr] \\ &\qquad = P_1(T \in B, Y \in C), \end{aligned}$$

where the last equality follows from Lemma 12.18. □

Suppose the distribution for data $X$ comes from an $(r+s)$-parameter canonical exponential family with densities

$$p_{\theta,\eta}(x) = h(x)\exp\bigl[\theta \cdot U(x) + \eta \cdot T(x) - A(\theta,\eta)\bigr], \tag{13.1}$$

where $\theta$ is $r$-dimensional and $\eta$ is $s$-dimensional. The following theorem gives the form of marginal and conditional distributions for the sufficient statistics $U$ and $T$.

**Theorem 13.2.** *If $X$ has density $p_{\theta,\eta}$ in (13.1), then there exist measures $\lambda_\theta$ and $\nu_t$ such that:*

1. *With $\theta$ fixed, the marginal distributions of $T$ will form an $s$-parameter exponential family with densities*

   $$\exp\bigl[\eta \cdot t - A(\theta,\eta)\bigr]$$

   *with respect to $\lambda_\theta$.*

2. *The conditional distributions of $U$ given $T = t$ form an exponential family with densities*

$$\exp\bigl[\theta \cdot u - A_t(\theta)\bigr]$$

*with respect to $\nu_t$. These densities are independent of $\eta$.*

*Proof.* Fix some point $(\theta_0, \eta_0) \in \Omega$, and let $\nu$ be the joint distribution of $T$ and $U$ under $P_{\theta_0,\eta_0}$. Arguing as in the proof of Theorem 12.19, under $P_{\theta,\eta}$, $T$ and $U$ have joint density

$$\exp\bigl[(\theta - \theta_0) \cdot u + (\eta - \eta_0) \cdot t + A(\theta_0, \eta_0) - A(\theta, \eta)\bigr]$$

with respect to $\nu$. If $R_0$ denotes the conditional distribution of $U$ given $T$ under $P_{\theta_0,\eta_0}$, then by Lemma 13.1 the marginal density of $T$ under $P_{\theta,\eta}$ is

$$e^{\eta \cdot t - A(\theta,\eta)} \int \exp\bigl((\theta - \theta_0) \cdot u - \eta_0 \cdot t + A(\theta_0, \eta_0)\bigr)\, dR_{0t}(u)$$

with respect to $Q_0$, the marginal density of $T$ under $P_{\theta_0,\eta_0}$. This is of the correct form provided we choose $\lambda_\theta$ so that

$$\frac{d\lambda_\theta}{dQ_0}(t) = \int \exp\bigl((\theta - \theta_0) \cdot u - \eta_0 \cdot t + A(\theta_0, \eta_0)\bigr)\, dR_{0t}(u).$$

By the second formula in Lemma 13.1, under $P_{\theta,\eta}$ the conditional density of $U$ given $T = t$ with respect to $R_{0t}$ is

$$\frac{e^{(\theta - \theta_0) \cdot u}}{\int e^{(\theta - \theta_0) \cdot v}\, dR_{0t}(v)}.$$

This density has the desired form. □

## 13.2 UMP Unbiased Tests in Higher Dimensions

If the power function $\beta_\varphi$ for an unbiased test $\varphi$ is continuous, then $\beta_\varphi(\theta) \le \alpha$ for $\theta$ in $\overline{\Omega}_0$, the closure of $\Omega_0$, and $\beta_\varphi(\theta) \ge \alpha$ for $\theta \in \overline{\Omega}_1$. If we take $\omega = \overline{\Omega}_0 \cap \overline{\Omega}_1$, the common boundary of $\Omega_0$ and $\Omega_1$, then

$$\beta_\varphi(\theta) = \alpha, \qquad \forall \theta \in \omega.$$

Tests satisfying this equation are called $\alpha$-*similar*. Here $\alpha$ need not denote the level of the tests, because $\beta_\varphi$ may exceed $\alpha$ at points $\theta$ in $\Omega_0 - \omega$.

**Lemma 13.3.** *Suppose $\varphi^*$ is $\alpha$-similar and has level $\alpha$, and that power functions $\beta_\varphi$ for all test functions $\varphi$ are continuous. If $\varphi^*$ is uniformly most powerful among all $\alpha$-similar tests, then it is unbiased and uniformly most powerful among all unbiased tests.*

*Proof.* The degenerate test that equals $\alpha$ regardless of the observed data is $\alpha$-similar. Since $\varphi^*$ has better power, $\beta_{\varphi^*}(\theta) \geq \alpha$, $\theta \in \Omega_1$. Because $\varphi^*$ has level $\alpha$, $\beta_{\varphi^*}(\theta) \leq \alpha$, $\theta \in \Omega_0$. Thus $\varphi^*$ is unbiased. If a competing test $\varphi$ is unbiased, then since its power function is continuous it is $\alpha$-similar. Then $\beta_\varphi \leq \beta_{\varphi^*}$ on $\Omega_1$ because $\varphi^*$ is uniformly most powerful among all $\alpha$-similar tests. □

The tests we develop use conditioning to reduce to the univariate case. Part of why this works is that the tests have the structure in the following definition.

**Definition 13.4.** *Suppose $T$ is sufficient for the subfamily $\{P_\theta : \theta \in \omega\}$. An $\alpha$-similar test $\varphi$ has* Neyman structure *if*

$$E_\theta[\varphi|T = t] = \alpha, \qquad \textit{for a.e. } t, \quad \forall \theta \in \omega.$$

**Theorem 13.5.** *If $T$ is complete and sufficient for $\{P_\theta : \theta \in \omega\}$, then every similar test has Neyman structure.*

*Proof.* For $\theta \in \omega$, let $h(T) = E_\theta(\varphi|T)$. (Because $T$ is sufficient, $h$ is independent of $\theta \in \omega$.) By smoothing,

$$E_\theta \varphi = E_\theta h(T) = \alpha, \qquad \forall \theta \in \omega.$$

By completeness, $h(T) = E_\theta(\varphi|T) = \alpha$, a.e., for all $\theta \in \omega$. □

Suppose data $X$ has density

$$p_{\theta,\eta}(x) = h(x) \exp\bigl\{\theta U(x) + \eta \cdot T(x) - A(\theta, \eta)\bigr\}. \tag{13.2}$$

Here $\theta$ is univariate, but $\eta$ can be $s$-dimensional. The tests of interest are derived by conditioning on $T$. By Theorem 13.2, the conditional distributions of $U$ given $T = t$ form a one-parameter exponential family with canonical parameter $\theta$. Theorem 12.9 gives a uniformly most powerful conditional test of $H_0 : \theta \leq \theta_0$ versus $H_1 : \theta > \theta_0$, given by

$$\varphi_1 = \begin{cases} 1, & U > c(T); \\ \gamma(T), & U = c(T); \\ 0, & U < c(T), \end{cases}$$

with $c(\cdot)$ and $\gamma(\cdot)$ adjusted so that

$$P_{\theta_0,\eta}\bigl(U > c(t) \bigm| T = t\bigr) + \gamma(t) P_{\theta_0,\eta}\bigl(U = c(t) \bigm| T = t\bigr) = \alpha.$$

Similarly, Theorem 12.26 gives a uniformly most powerful unbiased conditional test of $H_0 : \theta = \theta_0$ versus $H_1 : \theta \neq \theta_0$, given by

$$\varphi_2 = \begin{cases} 1, & U < c_-(T); \\ 1, & U > c_+(t); \\ \gamma_-(T), & U = c_-(T); \\ \gamma_+(T), & U = c_+(T); \\ 0, & U \in \bigl(c_-(T), c_+(T)\bigr), \end{cases}$$

with $c_\pm(\cdot)$ and $\gamma_\pm(\cdot)$ adjusted so that

$$E_{\theta_0,\eta}[\varphi_2|T=t]=\alpha$$

and

$$E_{\theta_0,\eta}[\varphi_2 U|T=t]=\alpha E_{\theta_0,\eta}[U|T=t].$$

**Theorem 13.6.** *If the exponential family* (13.2) *is of full rank and $\Omega$ is open, then $\varphi_1$ is a uniformly most powerful unbiased test of $H_0:\theta\le\theta_0$ versus $H_1:\theta>\theta_0$, and $\varphi_2$ is a uniformly most powerful unbiased test of $H_0:\theta=\theta_0$ versus $H_1:\theta\neq\theta_0$.*

*Proof.* Let us begin proving the assertion about $\varphi_1$. First note that the conditions on the exponential family ensure that the densities with $\theta=\theta_0$ form a full rank exponential family with $T$ as a complete sufficient statistic. Also, from the construction, $E_{\theta_0,\eta}[\varphi_1|T]=\alpha$, so by smoothing $E_{\theta_0,\eta}\varphi_1=\alpha$ and $\varphi_1$ is $\alpha$-similar. Suppose $\varphi$ is a competing $\alpha$-similar test. Then $\varphi$ has Neyman structure by Theorem 13.5 and $E_{\theta_0,\eta}[\varphi|T=t]=\alpha$. Because $\varphi_1$ is the most powerful conditional test of $\theta=\theta_0$ versus $\theta>\theta_0$, if $\theta>\theta_0$, then

$$E_{\theta,\eta}(\varphi_1|T=t)\ge E_{\theta,\eta}(\varphi|T=t),$$

and by smoothing,[1]

$$E_{\theta,\eta}\varphi_1=E_{\theta,\eta}E_{\theta,\eta}(\varphi_1|T)\ge E_{\theta,\eta}E_{\theta,\eta}(\varphi|T)=E_{\theta,\eta}\varphi.$$

This shows that $\varphi_1$ is uniformly most powerful $\alpha$-similar. By Theorem 12.9, the conditional power function for $\varphi_1$ is increasing in $\theta$, and so if $\theta<\theta_0$,

$$E_{\theta,\eta}\varphi_1=E_{\theta,\eta}E_{\theta,\eta}(\varphi_1|T)\le E_{\theta_0,\eta}E_{\theta,\eta}(\varphi_1|T)=\alpha.$$

Thus $\varphi_1$ has level $\alpha$. By Theorem 2.4, power functions for all test functions are continuous, so by Lemma 13.3 $\varphi_1$ is uniformly most powerful unbiased.

The argument for the assertion about $\varphi_2$ is a bit more involved. Let

$$m(\theta,\eta)=E_{\theta,\eta}U=\frac{\partial A(\theta,\eta)}{\partial\eta},$$

by (2.4). By dominated convergence (as in Theorem 12.20),

---

[1] There is a presumption here that $(\theta_0,\eta)$ lies in $\Omega$ regardless the choice of $(\theta,\eta)\in\Omega$. Unfortunately, this does not have to be the case. This issue can be resolved by reparameterization. If $(\theta_0,\eta_0)\in\Omega$ and we are concerned with power at $(\theta_1,\eta_1)\in\Omega$, define new parameters $\tilde\theta=\theta$ and $\tilde\eta=\eta+(\eta_0-\eta_1)(\theta-\theta_0)/(\theta_1-\theta_0)$. Then the original parameters of interest, $(\theta_0,\eta_0)$ and $(\theta_1,\eta_1)$ become $(\tilde\theta_0,\tilde\eta_0)=(\theta_0,\eta_0)$ and $(\tilde\theta_1,\tilde\eta_1)=(\theta_1,\eta_0)$. The canonical statistics for the reparameterized family are $\tilde T=T$ and $\tilde U=U+(\eta_0-\eta_1)\cdot T/(\theta_1-\theta_0)$. Since we condition on $T$, it is easy to see that the test $\varphi_1$ will be the same, regardless of the parameterization.

$$\begin{aligned}\frac{\partial \beta_\varphi(\theta,\eta)}{\partial \theta} &= \frac{\partial}{\partial \theta}\int \varphi(x) p_{\theta,\eta}(x)\,d\mu(x)\\ &= \int \varphi(x)\frac{\partial}{\partial \theta} p_{\theta,\eta}(x)\,d\mu(x)\\ &= \int \varphi(x)\bigl(U(x)-m(\theta,\eta)\bigr)p_{\theta,\eta}(x)\,d\mu(x)\\ &= E_{\theta,\eta}\varphi\bigl(U-m(\theta,\eta)\bigr).\end{aligned}$$

Suppose $\varphi$ is unbiased. Then this derivative must be zero when $\theta = \theta_0$, and thus

$$E_{\theta_0,\eta}\varphi U - \alpha m(\theta_0,\eta) = E_{\theta_0,\eta}[\varphi U - \alpha U] = 0.$$

Conditioning on $T$,

$$0 = E_{\theta_0,\eta} E_{\theta_0,\eta}\bigl[\varphi U - \alpha U \bigm| T\bigr],$$

and since $T$ is complete for the family of distributions with $\theta = \theta_0$, this implies that

$$E_{\theta_0,\eta}\bigl[\varphi U - \alpha U \bigm| T\bigr] = 0.$$

But $\varphi$ is $\alpha$-similar and has Neyman structure, implying $E_{\theta_0,\eta}[\varphi|T] = \alpha$, and so

$$E_{\theta_0,\eta}[\varphi U|T] = \alpha E_{\theta_0,\eta}[U|T].$$

By Theorem 12.20, this constraint ensures that the conditional power $E_{\theta,\eta}[\varphi|T]$ has zero slope at $\theta = \theta_0$. By Theorem 12.22, $\varphi_2$ is the uniformly most powerful conditional test satisfying this condition, and so

$$E_{\theta,\eta}[\varphi_2|T] \ge E_{\theta,\eta}[\varphi|T].$$

Taking expectations, by smoothing

$$E_{\theta,\eta}\varphi_2 \ge E_{\theta,\eta}\varphi.$$

Thus $\varphi$ is uniformly most powerful unbiased. (Again, reparameterization can be used to treat cases where $(\theta,\eta) \in \Omega$ but $(\theta_0,\eta) \notin \Omega$.) □

## 13.3 Examples

*Example 13.7 (The t-test).* The theory developed in the previous section can be used to test the mean of a normal distribution. Suppose $X_1,\dots,X_n$ is a random sample from $N(\mu,\sigma^2)$, and consider testing $H_0 : \mu \le 0$ versus $H_1 : \mu > 0$. The joint density from Example 2.3 is

$$\frac{1}{(2\pi)^{n/2}}\exp\left[\frac{\mu}{\sigma^2}U(x) - \frac{1}{2\sigma^2}T(x) - \frac{n\mu^2}{2\sigma^2} - n\log\sigma\right],$$

with $U(x) = x_1 + \cdots + x_n$ and $T(x) = x_1^2 + \cdots + x_n^2$. This has form (13.2) with $\theta = \mu/\sigma^2$ and $\eta = -1/(2\sigma^2)$. Note that the hypotheses expressed using the canonical parameters are $H_0 : \theta \le 0$ versus $H_1 : \theta > 0$. By Theorem 13.6, the uniformly most powerful unbiased test has form

$$\varphi = \begin{cases} 1, & U > c(T); \\ 0, & \text{otherwise,} \end{cases}$$

with $c(\cdot)$ chosen so that

$$P_{\mu=0}\bigl[U > c(t) \bigm| T = t\bigr] = \alpha.$$

To proceed we need the conditional distribution of $U$ given $T = t$ when $\mu = 0$. Note that the family of distributions with $\mu = 0$ is an exponential family with complete sufficient statistic $T$. Also, if we define $Z = X/\sigma$, so that $Z_1, \dots, Z_n$ are i.i.d. standard normal, or $Z \sim N(0, I)$, then $W = X/\|X\| = Z/\|Z\|$ is ancillary. By Basu's theorem (Theorem 3.21), $T$ and $W$ are independent. Because $\|X\| = \sqrt{T}$, $X = W\sqrt{T}$, and using independence between $T$ and $W$,

$$E\bigl[h(X)|T = t\bigr] = E\bigl[h(W\sqrt{t})|T = t\bigr] = Eh(W\sqrt{t}).$$

This shows that

$$X|T = t \sim W\sqrt{t}. \tag{13.3}$$

The vector $W$ is said to be uniformly distributed on the unit sphere. Note that if $O$ is an arbitrary orthogonal matrix ($OO' = I$), then $OZ \sim N(0, OO') = N(0, I)$. Also $\|OZ\|^2 = (OZ)'(OZ) = Z'O'OZ = Z'Z = \|Z\|^2$. Thus $Z$ and $OZ$ have the same length and distribution. Then

$$OW = \frac{OZ}{\|Z\|} = \frac{OZ}{\|OZ\|} \sim \frac{Z}{\|Z\|} = W.$$

So $W$ and $OW$ have the same distribution, which shows that the uniform distribution on the unit sphere is invariant under orthogonal transformations. In fact, this is the only probability distribution on the unit sphere that is invariant under orthogonal transformations. Using (13.3), since $U = \mathbf{1}'X$ where $\mathbf{1}$ denotes a column of 1s,

$$P_{\mu=0}\bigl[U > c(t) \bigm| T = t\bigr] = P(\mathbf{1}'W > c(t)/\sqrt{t}).$$

This equals $\alpha$ if we take $c(t)/\sqrt{t} = q$, the upper $\alpha$th quantile for $\mathbf{1}'W$. Thus the uniformly most powerful unbiased test rejects $H_0$ if

$$\frac{U}{\sqrt{T}} \ge q.$$

Although it may not be apparent, this is equivalent to the usual test based on the $t$-statistic

$$t = \frac{\overline{X}}{S/\sqrt{n}}.$$

To see this, note that

$$\overline{X} = U/n$$

and

$$\begin{aligned} S^2 &= \frac{1}{n-1}\sum_{i=1}^{n}(X_i - \overline{X})^2 \\ &= \frac{1}{n-1}\sum_{i=1}^{n}X_i^2 - \frac{n}{n-1}\overline{X}^2 = \frac{T}{n-1} - \frac{U^2}{n(n-1)}, \end{aligned}$$

and so

$$t = \frac{U/\sqrt{n}}{\sqrt{(T-U^2/n)/(n-1)}} = \frac{\sqrt{n-1}\,\mathrm{Sign}(U)}{\sqrt{nT/U^2-1}} = g(U/\sqrt{T}).$$

The function $g(\cdot)$ here is strictly increasing, and so $U/\sqrt{T} > q$ if and only if $t > g(q)$. When $\mu = 0$, $t$ has the $t$-distribution on $n-1$ degrees of freedom, and so level $\alpha$ is achieved taking $g(q) = t_{\alpha,n-1}$, the upper $\alpha$th quantile of this distribution. So our test then rejects $H_0$ when

$$t > t_{\alpha,n-1}.$$

Details for the two-sided case, testing $H_0 : \mu = 0$ versus $H_1 : \mu \neq 0$, are similar. The uniformly most powerful level $\alpha$ test rejects $H_0$ when $|t| > t_{\alpha/2,n-1}$.

*Example 13.8 (Fisher's Exact Test).* A second example of unbiased testing concerns dependence in a two-way contingency table. Consider two questions on a survey, $A$ and $B$, and suppose each of these questions has two answers. Responses to these questions might be coded with variables $X_1, \ldots, X_n$ and $Y_1, \ldots, Y_n$ taking $X_k = 1$ if respondent $k$ gives the first answer to question $A$, $X_k = 2$ if respondent $k$ gives the second answer to question $A$; and $Y_k = 1$ if respondent $k$ gives the first answer to question $B$, $Y_k = 2$ if respondent $k$ gives the second answer to question $B$. If the pairs $(X_k, Y_k)$, $k = 1, \ldots, n$, are i.i.d., and if

$$p_{ij} = P(X_k = i, Y_k = j), \qquad i = 1, 2, \quad j = 1, 2,$$

then the joint density is

$$P(X_1 = x_1, \ldots, X_n = x_n, Y_1 = y_1, \ldots, Y_n = y_n) = \prod_{i=1}^{2}\prod_{j=1}^{2} p_{ij}^{n_{ij}}, \tag{13.4}$$

where $n_{ij} = \#\{k : x_k = i, y_k = j\}$. So if we take

$$N_{ij} = \#\{k : X_k = i, Y_k = j\}, \qquad i = 1, 2, \quad j = 1, 2,$$

then $N = (N_{11}, N_{12}, N_{21}, N_{22})$ is a sufficient statistic. Based on these data, we may want to test whether there is positive dependence between the two questions. But first we need to resolve what we mean by "positive dependence." There seem to be various possibilities.

Let $(X, Y)$ be a generic variable distributed as $(X_k, Y_k)$. Perhaps we should define positive dependence between the questions to mean that the correlation between $X$ and $Y$ is positive. Because

$$E(X - 1) = P(X = 2) = p_{21} + p_{22} \stackrel{\text{def}}{=} p_{2+}$$

and

$$E(Y - 1) = P(Y = 2) = p_{12} + p_{22} \stackrel{\text{def}}{=} p_{+2},$$

$$\begin{aligned} \text{Cov}(X, Y) &= \text{Cov}(X - 1, Y - 1) = E(X - 1)(Y - 1) - p_{2+}p_{+2} \\ &= P(X = 2, Y = 2) - p_{2+}p_{+2} = p_{22} - p_{2+}p_{+2} = p_{22}p_{11} - p_{12}p_{21}. \end{aligned}$$

So the covariance between $X$ and $Y$ is positive if and only if

$$p_{22}p_{11} > p_{12}p_{21}.$$

Another notion of positive dependence might be that the chance $X$ equals the larger value 2 increases if we learn that $Y$ equals its larger value, that is, if

$$P(X = 2 | Y = 2) > P(X = 2).$$

Equivalently,

$$\frac{p_{22}}{p_{12} + p_{22}} > p_{21} + p_{22}.$$

Cross-multiplication and a bit of algebra show that this happens if and only if $p_{22}p_{11} > p_{12}p_{21}$, so this notion of positive dependence is the same as the notion based on correlation.

The distribution of $N$ is multinomial,

$$P(N_{11} = n_{11}, \dots, N_{22} = n_{22}) = \binom{n}{n_{11}, \dots, n_{22}} p_{11}^{n_{11}} p_{12}^{n_{12}} p_{21}^{n_{21}} p_{22}^{n_{22}}.$$

It is convenient to introduce new variables $U = N_{11}$, $T_1 = N_{11} + N_{12}$, and $T_2 = N_{11} + N_{21}$. If $N$ is given in a two-way table as in Example 5.5, then $T_1$ and $T_2$ determine the marginal sums. Given $U$ and $T$, we can solve for $N$, specifically,

$$N_{11} = U, \qquad N_{12} = T_1 - U, \qquad N_{21} = T_2 - U,$$

and

$$N_{22} = n + U - T_1 - T_2.$$

Thus there is a one-to-one relation between $N$ and variables $T$ and $U$, and

$$\begin{aligned}
&P(T=t,U=u)\\
&\quad=P\bigl(N_{11}=u,N_{12}=t_1-u,N_{21}=t_2-u,N_{22}=n+u-t_1-t_2\bigr)\\
&\quad=h(u,t)p_{11}^u p_{12}^{t_1-u}p_{21}^{t_2-u}p_{22}^{n+u-t_1-t_2}\\
&\quad=h(u,t)\left(\frac{p_{11}p_{22}}{p_{12}p_{21}}\right)^u\left(\frac{p_{12}}{p_{22}}\right)^{t_1}\left(\frac{p_{21}}{p_{22}}\right)^{t_2}p_{22}^n\\
&\quad=h(u,t)\exp\bigl\{\theta u+\eta\cdot t-A(\theta,\eta)\bigr\}, \qquad (13.5)
\end{aligned}$$

where

$$\theta=\log\left(\frac{p_{11}p_{22}}{p_{12}p_{21}}\right),\qquad \eta=\begin{pmatrix}\log(p_{12}/p_{22})\\ \log(p_{21}/p_{22})\end{pmatrix},$$

$$h(u,t)=\binom{n}{u,t_1-u,t_2-u,n+u-t_1-t_2},$$

and

$$A(\theta,\eta)=-n\log p_{22}.$$

Using Theorem 13.6, a uniformly most powerful unbiased test is given by

$$\varphi=\begin{cases}1, & U>c(T);\\ \gamma(T), & U=c(T);\\ 0, & U<c(T),\end{cases}$$

with $c(\cdot)$ and $\gamma(\cdot)$ adjusted so that

$$\alpha=P_{\theta=0}\bigl(U>c(t)|T=t\bigr)+\gamma(t)P_{\theta=0}\bigl(U=c(t)|T=t\bigr).$$

To describe the test in a more explicit fashion, we need the conditional distribution of $U$ given $T=t$ when $\theta=0$. This distribution does not depend on $\eta$. It is convenient to assume that $p_{11}=p_{12}=p_{21}=p_{22}=1/4$ and to denote probability in this case by $P_0$. Then $P_0(X_k=1)=P_0(X_k=2)=P_0(Y_k=1)=P_0(Y_k=2)=1/2$, so the joint density in (13.4) equals the product of the marginal densities and under $P_0$ the variables $X_k$, $k=1,\dots,n$, and $Y_k$, $k=1,\dots,n$, are all independent. Since $T_1$ depends on $X_k$, $k=1,\dots,n$, and $T_2$ depends on $Y_k$, $k=1,\dots,n$, $T_1$ and $T_2$ are independent under $P_0$, each from a binomial distribution with $n$ trials and success probability 1/2. Thus

$$P_0(T=t)=P_0(T_1=t_1)P_0(T_2=t_2)=\binom{n}{t_1}\binom{n}{t_2}\frac{1}{4^n}.$$

Using (13.5),

$$P_0(U=u,T=t)=\binom{n}{u,t_1-u,t_2-u,n+u-t_1-t_2}\frac{1}{4^n}.$$

Dividing these expressions, after a bit of algebra,

$$P_0(U = u|T = t) = \frac{P_0(U = u, T = t)}{P_0(T = t)} = \frac{\binom{t_1}{u}\binom{n-t_1}{t_2-u}}{\binom{n}{t_2}}.$$

This is the hypergeometric distribution, which arises in sampling theory. Consider drawing $t_2$ times without replacement from an urn containing $t_1$ red balls and $n - t_1$ white balls, and let $U$ denote the number of red balls in the sample. Then there are $\binom{n}{t_2}$ samples, and the number of samples for which $U = u$ is $\binom{t_1}{u}\binom{n-t_1}{t_2-u}$. If the chances for all possible samples are the same, then the chance $U = u$ is given by the formula above.

The two-sided case can be handled in a similar fashion. Direct calculation shows that $\theta = 0$ if and only if $X$ and $Y$ are independent, and so testing $H_0 : \theta = 0$ versus $H_1 : \theta \neq 0$ amounts to testing whether answers for the two questions are independent. Again the best test conditions on the margins $T$, and probability calculations are based on the hypergeometric distribution. Calculations to set the constants $c_\pm$ and $\gamma_\pm$ are messy and need to be done numerically. These tests for two-way contingency tables were introduced by Fisher and are called *Fisher's exact tests.*

## 13.4 Problems[2]

*1. Consider a two-parameter exponential family with Lebesgue density

$$p_{\theta,\phi}(x, y) = (x + y)e^{\theta x+\phi y-A(\theta,\phi)}, \qquad x \in (0, 1), \quad y \in (0, 1).$$

   a) Find $A(\theta, \phi)$.
   b) Find the marginal density of $X$. Check that the form of this distribution agrees with Theorem 13.2.
   c) Find the conditional density of $X$ given $Y = y$. Again, check that this agrees with Theorem 13.2.
   d) Determine the uniformly most powerful unbiased test of $H_0 : \theta \le 0$ versus $H_1 : \theta > 0$.
   e) Determine the uniformly most powerful unbiased test of $H_0 : \theta = 0$ versus $H_1 : \theta \neq 0$.

2. Suppose $X$ and $Y$ are absolutely continuous with joint density

$$p_{\theta,\eta}(x, y) = \begin{cases} \eta(\theta + \eta)e^{\theta x+\eta y}, & 0 < x < y; \\ 0, & \text{otherwise,} \end{cases}$$

   where $\eta < 0$ and $\eta + \theta < 0$.

[2] Solutions to the starred problems are given at the back of the book.

a) Determine the marginal density of $Y$. Show that for fixed $\theta$ these densities form an exponential family with parameter $\eta$.
b) Determine the conditional density of X given $Y = y$. Show that for fixed $y$, these densities form an exponential family with parameter $\theta$.

*3. Let $X$ and $Y$ be independent variables, both with gamma distributions. The parameters for the distribution of $X$ are $\alpha_x$ and $\lambda_x$; the parameters for $Y$ are $\alpha_y$ and $\lambda_y$; and both shape parameters, $\alpha_x$ and $\alpha_y$, are known constants.
   a) Determine the uniformly most powerful unbiased test of $H_0 : \lambda_x \le \lambda_y$ versus $H_1 : \lambda_x > \lambda_y$. Hint: You should be able to relate the critical value for the conditional test to a quantile for the beta distribution.
   b) If $X_1, \dots, X_n$ is a random sample from $N(0, \sigma_x^2)$ and $Y_1, \dots, Y_m$ is a random sample from $N(0, \sigma_y^2)$, then one common test of $H_0 : \sigma_x^2 \le \sigma_y^2$ versus $H_1 : \sigma_x^2 > \sigma_y^2$ rejects $H_0$ if and only if $F = s_x^2/s_y^2$ exceeds the upper $\alpha$th quantile of the $F$-distribution on $n$ and $m$ degrees of freedom, where $s_x^2 = \sum_{i=1}^n X_i^2/n$ and $s_y^2 = \sum_{j=1}^m Y_j^2/m$. Show that this test is the same as the test in part (a). Give a formula relating quantiles for the $F$-distribution to quantiles for the beta distribution.

4. Consider a regression model in which data $Y_1, \dots, Y_n$ are independent with $Y_i \sim N(\alpha + \beta x_i, 1)$, $i = 1, \dots, n$. Here $\alpha$ and $\beta$ are unknown parameters, and $x_1, \dots, x_n$ are known constants. Determine the uniformly most powerful unbiased test of $H_0 : \beta = 0$ versus $H_1 : \beta \ne 0$.

*5. Consider a regression model in which data $Y_1, \dots, Y_n$ are independent with $Y_i \sim N(\beta x_i + \gamma w_i, 1)$, $i = 1, \dots, n$. Here $\beta$ and $\gamma$ are unknown parameters, and $x_1, \dots, x_n$ and $w_1, \dots, w_n$ are known constants. Determine the uniformly most powerful unbiased test of $H_0 : \beta \le \gamma$ versus $H_1 : \beta > \gamma$.

*6. Let $X_1, \dots, X_m$ be a random sample from the Poisson distribution with mean $\lambda_x$, and let $Y_1, \dots, Y_n$ be an independent random sample from the Poisson distribution with mean $\lambda_y$.
   a) Describe the uniformly most powerful unbiased test of $H_0 : \lambda_x \le \lambda_y$ versus $H_1 : \lambda_x > \lambda_y$.
   b) Suppose $\alpha = 5\%$ and the observed data are $X_1 = 3$, $X_2 = 5$, and $Y_1 = 1$. What is the chance the uniformly most powerful test will reject $H_0$?
   c) Give an approximate version of the test, valid if $m$ and $n$ are large.

7. Consider a two-parameter exponential family with density

$$p_{\theta,\eta}(x, y) = \begin{cases} \dfrac{\theta^2\eta^2(x+y)}{\theta+\eta} e^{-\theta x - \eta y}, & x > 0 \text{ and } y > 0; \\ 0, & \text{otherwise.} \end{cases}$$

   Determine the level $\alpha$ uniformly most powerful unbiased test of $H_0 : \theta \le \eta$ versus $H_1 : \theta > \eta$.

8. Suppose $X$ and $Y$ are absolutely continuous with joint density

$$p_\theta(x,y) = \begin{cases} (x+y)e^{-\theta_1 x^2 - \theta_2 y^2 - A(\theta)}, & x > 0, y > 0; \\ 0, & \text{otherwise}. \end{cases}$$

a) Find $A(\theta)$.
b) Find the uniformly most powerful unbiased test of $H_0 : \theta_1 \le \theta_2$ versus $H_1 : \theta_1 > \theta_2$.

9. Let $X$ have a normal distribution with mean $\mu$ and variance $\sigma^2$.
   a) For $x > 0$, find

$$p(x) = \lim_{\epsilon \downarrow 0} P\bigl(x < X < x + \epsilon \bigm| x^2 < X^2 < (x+\epsilon)^2\bigr).$$

   In part (b) you can assume that the conditional distribution of $X$ given $X^2$ is given by $P(X = x | X^2 = x^2) = p(x)$ and $P(X = -x | X^2 = x^2) = 1 - p(x)$, $x > 0$.
   b) Find the uniformly most powerful unbiased level $\alpha$ test $\varphi$ of $H_0 : \mu \le 0$ versus $H_1 : \mu > 0$ based on the single observation $X$.
10. Let $X_1, \dots, X_n$ be i.i.d. from $N(\mu, \sigma^2)$. Determine the uniformly most powerful unbiased test of $H_0 : \sigma^2 \le 1$ versus $H_1 : \sigma^2 > 1$.

# 14

# General Linear Model

The general linear model incorporates many of the most popular and useful models that arise in applied statistics, including models for multiple regression and the analysis of variance. The basic model can be written succinctly in matrix form as

$$Y = X\beta + \epsilon, \tag{14.1}$$

where $Y$, our observed data, is a random vector in $\mathbb{R}^n$, $X$ is an $n \times p$ matrix of known constants, $\beta \in \mathbb{R}^p$ is an unknown parameter, and $\epsilon$ is a random vector in $\mathbb{R}^n$ of unobserved errors. We usually assume that $\epsilon_1, \dots, \epsilon_n$ are a random sample from $N(0, \sigma^2)$, with $\sigma > 0$ an unknown parameter, so that

$$\epsilon \sim N(0, \sigma^2 I). \tag{14.2}$$

But some of our results hold under the less restrictive conditions that $E\epsilon_i = 0$ for all $i$, $\mathrm{Var}(\epsilon_i) = \sigma^2$ for all $i$, and $\mathrm{Cov}(\epsilon_i, \epsilon_j) = 0$ for all $i \neq j$. In matrix notation, $E\epsilon = 0$ and $\mathrm{Cov}(\epsilon) = \sigma^2 I$. Since $Y$ is $\epsilon$ plus a constant vector and $E\epsilon = 0$, we have $EY = X\beta$ and $\mathrm{Cov}(Y) = \mathrm{Cov}(\epsilon) = \sigma^2 I$. With the normal distribution for $\epsilon$ in (14.2),

$$Y \sim N(X\beta, \sigma^2 I). \tag{14.3}$$

*Example 14.1 (Quadratic Regression).* In quadratic regression, a response variable $Y$ is modeled as a quadratic function of some explanatory variable $x$ plus a random error. Specifically,

$$Y_i = \beta_1 + \beta_2 x_i + \beta_3 x_i^2 + \epsilon_i, \qquad i = 1, \dots, n.$$

Here the explanatory variables $x_1, \dots, x_n$ are taken to be known constants, $\beta_1$, $\beta_2$, and $\beta_3$ are unknown parameters, and $\epsilon_1, \dots, \epsilon_n$ are i.i.d. from $N(0, \sigma^2)$. If we define the design matrix $X$ as

R.W. Keener, *Theoretical Statistics: Topics for a Core Course*, Springer Texts in Statistics, DOI 10.1007/978-0-387-93839-4_14, © Springer Science+Business Media, LLC 2010

$$X = \begin{pmatrix} 1 & x_1 & x_1^2 \\ 1 & x_2 & x_2^2 \\ \vdots & \vdots & \vdots \\ 1 & x_n & x_n^2 \end{pmatrix},$$

then $Y = X\beta + \epsilon$, as in (14.1).

*Example 14.2 (One-Way ANOVA).* Suppose we have independent random samples from three normal populations with common variance $\sigma^2$, so

$$Y_i \sim \begin{cases} N(\beta_1, \sigma^2), & i = 1, \dots, n_1; \\ N(\beta_2, \sigma^2), & i = n_1 + 1, \dots, n_1 + n_2; \\ N(\beta_3, \sigma^2), & i = n_1 + n_2 + 1, \dots, n_1 + n_2 + n_3 \stackrel{\text{def}}{=} n. \end{cases}$$

If we define

$$X = \begin{pmatrix} 1 & 0 & 0 \\ \vdots & \vdots & \vdots \\ 1 & 0 & 0 \\ 0 & 1 & 0 \\ \vdots & \vdots & \vdots \\ 0 & 1 & 0 \\ 0 & 0 & 1 \\ \vdots & \vdots & \vdots \\ 0 & 0 & 1 \end{pmatrix},$$

then $EY = X\beta$ and the model has form (14.3).

In applications the parameters $\beta_1, \dots, \beta_p$ usually arise naturally when formulating the model. As a consequence they are generally easy to interpret. But for technical reasons it is often more convenient to view the unknown mean of $Y$, namely,

$$\xi \stackrel{\text{def}}{=} EY = X\beta$$

in $\mathbb{R}^n$ as the unknown parameter. If $c_1, \dots, c_p$ are the columns of $X$, then

$$\xi = X\beta = \beta_1 c_1 + \cdots + \beta_p c_p,$$

which shows that $\xi$ must be a linear combination of the columns of $X$. So $\xi$ must lie in the vector space

$$\omega \stackrel{\text{def}}{=} \text{span}\{c_1, \dots, c_p\} = \{X\beta : \beta \in \mathbb{R}^p\}.$$

Using $\xi$ instead of $\beta$, the vector of unknown parameters is $\theta = (\xi, \sigma)$ taking values in $\Omega = \omega \times (0, \infty)$.

Since $Y$ has mean $\xi$, it is fairly intuitive that our data must provide some information distinguishing between any two values for $\xi$, since the distributions for $Y$ under two different values for $\xi$ must be different. Whether this also holds for $\beta$ depends on the rank $r$ of $X$. Since $X$ has $p$ columns, this rank $r$ is at most $p$. If the rank of $X$ equals $p$ then the mapping $\beta \rightsquigarrow X\beta$ is one-to-one, and each value $\xi \in \omega$ is the image of a unique value $\beta \in \mathbb{R}^p$. But if the columns of $X$ are linearly dependent, then a nontrivial linear combination of the columns of $X$ will equal zero, so $Xv = 0$ for some $v \neq 0$. But then $X(\beta + v) = X\beta + Xv = X\beta$, and parameters $\beta$ and $\beta^* = \beta + v$ both give the same mean $\xi$. Here our data $Y$ provides no information to distinguish between parameter values $\beta$ and $\beta^*$.

*Example 14.3.* Suppose

$$X = \begin{pmatrix} 1 & 0 & 1 \\ 1 & 0 & 1 \\ 1 & 1 & 0 \\ 1 & 1 & 0 \end{pmatrix}.$$

Here the three columns of $X$ are linearly dependent because the first column is the sum of the second and third columns, and the rank of $X$ is 2, $r = 2 < p = 3$. Note that parameter values

$$\beta = \begin{pmatrix} 1 \\ 0 \\ 0 \end{pmatrix} \quad \text{and} \quad \beta^* = \begin{pmatrix} 0 \\ 1 \\ 1 \end{pmatrix}$$

both give

$$\xi = \begin{pmatrix} 1 \\ 1 \\ 1 \\ 1 \end{pmatrix}.$$

## 14.1 Canonical Form

Many results about testing and estimation in the general linear model follow easily once the data are expressed in a canonical form. Let $v_1, \dots, v_n$ be an orthonormal basis for $\mathbb{R}^n$, chosen so that $v_1, \dots, v_r$ span $\omega$. Entries in the canonical data vector $Z$ are coefficients expressing $Y$ as a linear combination of these basis vectors,

$$Y = Z_1 v_1 + \cdots + Z_n v_n. \tag{14.4}$$

Algebraically, $Z$ can be found introducing an $n \times n$ matrix $O$ with columns $v_1, \dots, v_n$. Then $O$ is an orthogonal matrix, $O'O = OO' = I$, and $Y$ and $Z$ are related by

$$Z = O'Y \quad \text{or} \quad Y = OZ.$$

Since $Y = \xi + \epsilon$, $Z = O'(\xi + \epsilon) = O'\xi + O'\epsilon$. If we define $\eta = O'\xi$ and $\epsilon^* = O'\epsilon$, then

$$Z = \eta + \epsilon^*.$$

Because $E\epsilon^* = EO'\epsilon = O'E\epsilon = 0$ and

$$\mathrm{Cov}(\epsilon^*) = \mathrm{Cov}(O'\epsilon) = O'\mathrm{Cov}(\epsilon)O = O'(\sigma^2 I)O = \sigma^2 O'O = \sigma^2 I,$$

$\epsilon^* \sim N(0, \sigma^2 I)$ and $\epsilon_1^*, \dots, \epsilon_n^*$ are i.i.d. from $N(0, \sigma^2)$. Since $Z = \eta + \epsilon^*$,

$$Z \sim N(\eta, \sigma^2 I). \tag{14.5}$$

Next, let $c_1, \dots, c_p$ denote the columns of the design matrix $X$. Then $\xi = X\beta = \sum_{i=1}^p \beta_i c_i$ and

$$\eta = O'\xi = \begin{pmatrix} v_1' \\ \vdots \\ v_n' \end{pmatrix} \sum_{i=1}^p \beta_i c_i = \begin{pmatrix} \sum_{i=1}^p \beta_i v_1' c_i \\ \vdots \\ \sum_{i=1}^p \beta_i v_n' c_i \end{pmatrix}.$$

Since $c_1, \dots, c_p$ all lie in $\omega$, and $v_{r+1}, \dots, v_n$ all lie in $\omega^\perp$, we have $v_k' c_i = 0$ for $k > r$, and thus

$$\eta_{r+1} = \cdots = \eta_n = 0. \tag{14.6}$$

Because $\eta = O'\xi$,

$$\xi = O\eta = (v_1 \dots v_n) \begin{pmatrix} \eta_1 \\ \vdots \\ \eta_r \\ 0 \\ \vdots \\ 0 \end{pmatrix} = \sum_{i=1}^r \eta_i v_i.$$

These equations establish a one-to-one relation between points $\xi \in \omega$ and $(\eta_1, \dots, \eta_r) \in \mathbb{R}^r$.

Since $Z \sim N(\eta, \sigma^2 I)$, the variables $Z_1, \dots, Z_n$ are independent with $Z_i \sim N(\eta_i, \sigma^2)$. The density of $Z$, taking advantage of the fact that $\eta_{r+1} = \cdots = \eta_n = 0$, is

$$\begin{aligned} &\frac{1}{\sqrt{2\pi\sigma^2}^n} \exp\left[-\frac{1}{2\sigma^2}\sum_{i=1}^r (z_i - \eta_i)^2 - \frac{1}{2\sigma^2}\sum_{i=r+1}^n z_i^2\right] \\ &\qquad = \exp\left[-\frac{1}{2\sigma^2}\sum_{i=1}^n z_i^2 + \frac{1}{\sigma^2}\sum_{i=1}^r \eta_i z_i - \sum_{i=1}^r \frac{\eta_i^2}{2\sigma^2} - \frac{n}{2}\log(2\pi\sigma^2)\right]. \end{aligned}$$

These densities form a full rank $(r+1)$-parameter exponential family with complete sufficient statistic

$$\left(Z_1, \dots, Z_r, \sum_{i=1}^n Z_i^2\right). \tag{14.7}$$

## 14.2 Estimation

Exploiting the canonical form, many parameters are easy to estimate. Because $EZ_i = \eta_i$, $i = 1, \dots, r$, $Z_i$ is the UMVU estimator of $\eta_i$, $i = 1, \dots, r$. Also, since $\xi = \sum_{i=1}^r \eta_i v_i$,

$$\hat{\xi} = \sum_{i=1}^{r} Z_i v_i \tag{14.8}$$

is a natural estimator of $\xi$. Noting that

$$E\hat{\xi} = \sum_{i=1}^{r} EZ_i v_i = \sum_{i=1}^{r} \eta_i v_i = \xi,$$

$\hat{\xi}$ is unbiased. Since it is a function of the complete sufficient statistic, it should be optimal in some sense. One measure of optimality might be the expected squared distance from the true value $\xi$. If $\tilde{\xi}$ is a competing unbiased estimator, then

$$E\|\tilde{\xi} - \xi\|^2 = \sum_{j=1}^{n} E(\tilde{\xi}_j - \xi_j)^2 = \sum_{j=1}^{n} \mathrm{Var}(\tilde{\xi}_j). \tag{14.9}$$

Because $\hat{\xi}_j$ is unbiased for $\xi_j$ and is a function of the complete sufficient statistic, $\mathrm{Var}(\hat{\xi}_j) \le \mathrm{Var}(\tilde{\xi}_j)$, $j = 1, \dots, n$. So $\hat{\xi}$ minimizes each term in the sum in (14.9), and hence

$$E\|\hat{\xi} - \xi\|^2 \le E\|\tilde{\xi} - \xi\|^2.$$

A more involved argument shows that $\hat{\xi}$ also minimizes the expectation of any other nonnegative quadratic form in the estimation error, $E(\tilde{\xi} - \xi)'A(\tilde{\xi} - \xi)$, among all unbiased estimators.

From (14.4), we can write $Y$ as

$$Y = \sum_{i=1}^{r} Z_i v_i + \sum_{i=r+1}^{n} Z_i v_i = \hat{\xi} + \sum_{i=r+1}^{n} Z_i v_i.$$

In this expression the first summand, $\hat{\xi}$, lies in $\omega$, and the second, $Y - \hat{\xi} = \sum_{i=r+1}^{n} Z_i v_i$, lies in $\omega^\perp$. This difference $Y - \hat{\xi}$ is called the *vector of residuals*, denoted by $e$:

$$e \stackrel{\text{def}}{=} Y - \hat{\xi} = \sum_{i=r+1}^{n} Z_i v_i. \tag{14.10}$$

Since $Y = \hat{\xi} + e$, by the Pythagorean theorem, if $\tilde{\xi}$ is any point in $\omega$, then

$$\|Y - \tilde{\xi}\|^2 = \|\hat{\xi} - \tilde{\xi} + e\|^2 = \|\hat{\xi} - \tilde{\xi}\|^2 + \|e\|^2,$$

because $\hat{\xi} - \tilde{\xi} \in \omega$ is orthogonal to $e \in \omega^\perp$. From this equation, it is apparent that $\hat{\xi}$ is the unique point in $\omega$ closest to the data vector $Y$. This closest point is called the *projection* of $Y$ onto $\omega$. The mapping $Y \rightsquigarrow \hat{\xi}$ is linear and can be represented by an $n \times n$ matrix $P$,

$$\hat{\xi} = PY,$$

with $P$ called the (orthogonal) *projection matrix* onto $\omega$. Since $\hat{\xi} \in \omega$, $P\hat{\xi} = \hat{\xi}$, and so $P^2Y = P(PY) = P\hat{\xi} = \hat{\xi} = PY$. Because $Y$ can take arbitrary values in $\mathbb{R}^n$, this shows that $P^2 = P$. (Matrices that satisfy this equation are called *idempotent*.) Using the orthonormal basis, $P$ can be written as $P = v_1v_1' + \cdots + v_rv_r'$. (To check that this works, just multiply (14.4) by this sum.) But for explicit calculation, formulas that do not rely on construction of the basis vectors $v_1, \dots, v_n$ are more convenient, and are developed below.

Since arbitrary points in $\omega$ can be written as $X\beta$ for some $\beta \in \mathbb{R}^p$, if $\hat{\xi} = X\hat{\beta}$, then $\hat{\beta}$ must minimize

$$\|Y - X\beta\|^2 = \sum_{i=1}^n \bigl[Y_i - (X\beta)_i\bigr]^2 \tag{14.11}$$

over $\beta \in \mathbb{R}^p$. For this reason, $\hat{\beta}$ is called the *least squares estimator* of $\beta$. Of course, when the rank $r$ of $X$ is less than $p$, $\hat{\beta}$ will not be unique. But unique or not, all partial derivatives of the least squares criterion (14.11) must vanish at $\beta = \hat{\beta}$. This often provides a convenient way to calculate $\hat{\beta}$ and then $\hat{\xi}$.

Another approach to explicit calculation proceeds directly from geometric considerations. Since the columns $c_i$, $i = 1, \dots, p$, of $X$ lie in $\omega$, and $e = Y - \hat{\xi}$ lies in $\omega^\perp$, we must have $c_i'e = 0$, which implies

$$X'e = 0.$$

Since $Y = \hat{\xi} + e$,

$$X'Y = X'(\hat{\xi} + e) = X'\hat{\xi} + X'e = X'\hat{\xi} = X'X\hat{\beta}. \tag{14.12}$$

If $X'X$ is invertible, then this equation gives

$$\hat{\beta} = (X'X)^{-1}X'Y. \tag{14.13}$$

The matrix $X'X$ is invertible if $X$ has full rank, $r = p$. In fact, $X'X$ is positive definite in this case. To see this, let $v$ be an eigenvector of $X'X$ with $\|v\| = 1$ and eigenvalue $\lambda$. Then

$$\|Xv\|^2 = v'X'Xv = \lambda v'v = \lambda,$$

which must be strictly positive since $Xv = c_1v_1 + \cdots + c_pv_p$ cannot be zero if $X$ has full rank. When $X$ has full rank

$$PY = \hat{\xi} = X\hat{\beta} = X(X'X)^{-1}X'Y,$$

and so the projection matrix $P$ onto $\omega$ can be written as

$$P = X(X'X)^{-1}X'. \tag{14.14}$$

Since $\hat{\xi}$ is unbiased, $a'\hat{\xi}$ is an unbiased estimator of $a'\xi$. This estimator is UMVU because $\hat{\xi}$ is a function of the complete sufficient statistic. By (14.12), $X'Y = X'\hat{\xi}$, and so by (14.13), when $X$ is full rank

$$\hat{\beta} = (X'X)^{-1}X'\hat{\xi}.$$

This equation shows that $\hat{\beta}_i$ is a linear function of $\hat{\xi}$, and so $\hat{\beta}_i$ is UMVU for $\beta_i$.

## 14.3 Gauss–Markov Theorem

For this section we relax the assumptions for the general linear model. The model still has $Y = X\beta+\epsilon$, but now the $\epsilon_i$, $i = 1, \dots, n$, need not be a random sample from $N(0, \sigma^2)$. Instead, we assume the $\epsilon_i$, $i = 1, \dots, n$, have zero mean, $E\epsilon_i = 0$, $i = 1, \dots, n$; a common variance, $\text{Var}(\epsilon_i) = \sigma^2$, $i = 1, \dots, n$; and are uncorrelated, $\text{Cov}(\epsilon_i, \epsilon_j) = 0$, $i \neq j$. In matrix form, these assumptions can be written as

$$E\epsilon = 0 \text{ and } \text{Cov}(\epsilon) = \sigma^2 I.$$

Then

$$EY = X\beta = \xi \text{ and } \text{Cov}(Y) = \sigma^2 I.$$

Any estimator of the form $a'Y = a_1Y_1+\cdots+a_nY_n$, with $a \in \mathbb{R}^n$ a constant vector, is called a *linear estimate*. Using (1.15),

$$\begin{aligned} \text{Var}(a'Y) &= \text{Cov}(a'Y) = a'\text{Cov}(Y)a \\ &= a'(\sigma^2 I)a = \sigma^2 a'a = \sigma^2\|a\|^2. \end{aligned} \tag{14.15}$$

Because $EY = \xi$, the estimator $a'\hat{\xi}$ is unbiased for $a'\xi$. Since $\hat{\xi} = PY$, $a'\hat{\xi} = a'PY = (Pa)'Y$, and so by (14.15)

$$\text{Var}(a'\hat{\xi}) = \sigma^2\|Pa\|^2. \tag{14.16}$$

Also, by (1.15) and since $P$ is symmetric with $P^2 = P$,

$$\text{Cov}(\hat{\xi}) = \text{Cov}(PY) = P\text{Cov}(Y)P = P(\sigma^2 I)P = \sigma^2 P.$$

When $X$ has full rank, we can compute the covariance of the least squares estimator $\hat{\beta}$ using (1.15) as

$$\begin{aligned} \text{Cov}(\hat{\beta}) &= \text{Cov}\big((X'X)^{-1}X'Y\big) \\ &= (X'X)^{-1}X'\text{Cov}(Y)X(X'X)^{-1} = \sigma^2(X'X)^{-1}. \end{aligned} \tag{14.17}$$

**Theorem 14.4 (Gauss–Markov).** *Suppose*

$$EY = X\beta \text{ and } \operatorname{Cov}(Y) = \sigma^2 I.$$

*Then the (least squares) estimator* $a'\hat{\xi}$ *of* $a'\xi$ *is unbiased and has uniformly minimum variance among all linear unbiased estimators.*

*Proof.* Let $\delta = b'Y$ be a competing unbiased estimator. By (14.15) and (14.16), the variances of $\delta$ and $a'\hat{\xi}$ are

$$\operatorname{Var}(\delta) = \sigma^2\|b\|^2 \text{ and } \operatorname{Var}(a'\hat{\xi}) = \sigma^2\|Pa\|^2.$$

If $\epsilon$ happens to come from a normal distribution, since both of these estimators are unbiased and $a'\hat{\xi}$ is UMVU, $\operatorname{Var}(a'\hat{\xi}) \le \operatorname{Var}(\delta)$, or

$$\sigma^2\|Pa\|^2 \le \sigma^2\|b\|^2.$$

But formulas for the variances of the estimators do not depend on normality, and thus $\operatorname{Var}(a'\hat{\xi}) \le \operatorname{Var}(\delta)$ in general. □

Although $a'\hat{\xi}$ is the "best" linear estimate, in some examples nonlinear estimates can be more precise.

*Example 14.5.* Suppose

$$Y_i = \beta + \epsilon_i, \qquad i = 1, \dots, n,$$

where $\epsilon_1, \dots, \epsilon_n$ are i.i.d. with common density

$$f(x) = \frac{e^{-\sqrt{2}|x|/\sigma}}{\sigma\sqrt{2}}, \qquad x \in \mathbb{R}.$$

By the symmetry, $E\epsilon_i = 0$, $i = 1, \dots, n$, and

$$\begin{aligned}\operatorname{Var}(\epsilon_i) = E\epsilon_i^2 &= 2\int_0^\infty \frac{x^2 e^{-\sqrt{2}x/\sigma}}{\sigma\sqrt{2}}\,dx \\ &= \frac{\sigma^2}{2}\int_0^\infty u^2 e^{-u}\,du = \frac{\sigma^2}{2}\Gamma(3) = \sigma^2, \qquad i = 1, \dots, n.\end{aligned}$$

So $\operatorname{Cov}(Y) = \operatorname{Cov}(\epsilon) = \sigma^2 I$, and if we take $X = (1, \dots, 1)'$, then $EY = X\beta$. This shows that the conditions of the Gauss–Markov theorem are satisfied. If $a = n^{-1}X$, then $a'\xi = n^{-1}X'X\beta = \beta$. By the Gauss–Markov theorem, the best linear estimator of $\beta$ is

$$\hat{\beta} = \frac{1}{n}X'\hat{\xi} = \frac{1}{n}X'X(X'X)^{-1}X'Y = \frac{1}{n}X'Y = \overline{Y},$$

the sample average. This estimator has variance $\sigma^2/n$. A competing estimator might be the sample median,

$$\tilde{Y} = \operatorname{med}\{Y_1, \dots, Y_n\} = \beta + \operatorname{med}\{\epsilon_1, \dots, \epsilon_n\}.$$

By (8.5), $\sqrt{n}(\tilde{Y} - \beta) \Rightarrow N(0, \sigma^2/2)$. This result suggests that

$$\operatorname{Var}\bigl(\sqrt{n}(\tilde{Y} - \beta)\bigr) \to \sigma^2/2.$$

This can be established formally using Theorem 8.16 and showing that the variables $n(\tilde{Y} - \beta)^2$ are uniformly integrable. Since $\operatorname{Var}\bigl(\sqrt{n}(\overline{Y} - \beta)\bigr) = \sigma^2$, for large $n$ the variance of $\tilde{Y}$ is roughly half the variance of $\overline{Y}$.

## 14.4 Estimating $\sigma^2$

From the discussion in Section 14.1, $Z_{r+1}, \dots, Z_n$ are i.i.d. from $N(0, \sigma^2)$. Thus $EZ_i^2 = \sigma^2$, $i = r+1. \dots, n$, and the average of these variables,

$$S^2 = \frac{1}{n-r} \sum_{i=r+1}^{n} Z_i^2, \tag{14.18}$$

is an unbiased estimator of $\sigma^2$. But $S^2$ is a function of the complete sufficient statistic $(Z_1, \dots, Z_r, \sum_{i=1}^n Z_i^2)$ in (14.7), and so $S^2$ is the UMVU estimator of $\sigma^2$. The estimator $S^2$ can be computed from the length of the residual vector $e$ in (14.10). To see this, first write

$$\|e\|^2 = e'e = \left(\sum_{i=r+1}^{n} Z_i v_i'\right)\left(\sum_{j=r+1}^{n} Z_j v_j\right) = \sum_{i=r+1}^{n}\sum_{j=r+1}^{n} Z_i Z_j v_i' v_j.$$

Because $v_1, \dots, v_n$ is an orthonormal basis, $v_i'v_j$ equals zero when $i \neq j$ and equals one when $i = j$. So the double summation in this equation simplifies giving

$$\|e\|^2 = \sum_{i=r+1}^{n} Z_i^2, \tag{14.19}$$

and so

$$S^2 = \frac{\|e\|^2}{n-r} = \frac{\|Y - \hat{\xi}\|^2}{n-r}. \tag{14.20}$$

Because $\hat{\xi}$ in (14.8) is a function of $Z_1, \dots, Z_r$, and $e$ in (14.10) is a function of $Z_{r+1}, \dots, Z_n$, $S^2$ and $\hat{\xi}$ are independent. Also, using (14.19), (14.20), and the definition of the chi-square distribution,

$$\frac{(n-r)S^2}{\sigma^2} = \sum_{i=r+1}^{n} (Z_i/\sigma)^2 \sim \chi^2_{n-r}, \tag{14.21}$$

since $Z_i/\sigma \sim N(0,1)$.

The distribution theory just presented can be used to set confidence intervals for linear estimators. If $a$ is a constant vector in $\mathbb{R}^n$, then from (14.16) the standard deviation of least squares estimator $a'\hat{\xi}$ of $a'\xi$ is $\sigma\|Pa\|$. This standard deviation is naturally estimated as

$$\hat{\sigma}_{a'\hat{\xi}} \stackrel{\text{def}}{=} S\|Pa\|.$$

**Theorem 14.6.** *In the general linear model with $Y \sim N(\xi, \sigma^2 I)$, $\xi \in \omega$, and $\sigma^2 > 0$,*

$$\bigl(a'\hat{\xi} - \hat{\sigma}_{a'\hat{\xi}} t_{\alpha/2,n-r}, a'\hat{\xi} + \hat{\sigma}_{a'\hat{\xi}} t_{\alpha/2,n-r}\bigr)$$

*is a $1-\alpha$ confidence interval for $a'\xi$.*

*Proof.* Because $a'\hat{\xi} \sim N\bigl(a'\xi, \sigma^2\|Pa\|^2\bigr)$,

$$\frac{a'\hat{\xi} - a'\xi}{\sigma\|Pa\|} \sim N(0,1).$$

This variable is independent of $(n-r)S^2/\sigma^2$ because $S^2$ and $\hat{\xi}$ are independent. Using (9.2), the definition of the $t$-distribution,

$$\frac{\dfrac{a'\hat{\xi} - a'\xi}{\sigma\|Pa\|}}{\sqrt{\dfrac{1}{n-r}\dfrac{(n-r)S^2}{\sigma^2}}} = \frac{a'\hat{\xi} - a'\xi}{S\|Pa\|} \sim t_{n-r}.$$

The coverage probability of the stated interval is

$$\begin{aligned} P\bigl(a'\hat{\xi} - S\|Pa\|t_{\alpha/2,n-r} < a'\xi < a'\hat{\xi} + S\|Pa\|t_{\alpha/2,n-r}\bigr) \\ = P\left(-t_{\alpha/2,n-r} < \frac{a'\hat{\xi} - a'\xi}{S\|Pa\|} < t_{\alpha/2,n-r}\right) \\ = 1-\alpha. \end{aligned}$$

□

When $X$ has full rank, $\beta_i$ is a linear function of $\xi$, estimated by $\hat{\beta}_i$ with variance $\sigma\bigl[(X'X)^{-1}\bigr]_{ii}$. So the estimated standard deviation of $\hat{\beta}_i$ is

$$\hat{\sigma}_{\hat{\beta}_i} = S\sqrt{\bigl[(X'X)^{-1}\bigr]_{ii}},$$

and

$$\bigl(\hat{\beta}_i - \hat{\sigma}_{\hat{\beta}_i} t_{\alpha/2,n-p}, \hat{\beta}_i + \hat{\sigma}_{\hat{\beta}_i} t_{\alpha/2,n-p}\bigr) \tag{14.22}$$

is a $1-\alpha$ confidence interval for $\beta_i$.

## 14.5 Simple Linear Regression

To illustrate the ideas developed, we consider simple linear regression in which a response variable $Y$ is a linear function of an independent variable $x$ plus random error. Specifically,

$$Y_i = \beta_1 + \beta_2(x_i - \overline{x}) + \epsilon_i, \qquad i = 1, \dots, n.$$

The independent variables $x_1, \dots, x_n$ with average $\overline{x}$ are taken to be known constants, $\beta_1$ and $\beta_2$ are unknown parameters, and $\epsilon_1, \dots, \epsilon_n$ are i.i.d. from $N(0, \sigma^2)$. This gives a general linear model with design matrix

$$X = \begin{pmatrix} 1 & x_1 - \overline{x} \\ \vdots & \vdots \\ 1 & x_n - \overline{x} \end{pmatrix}.$$

In parameterizing the mean of $Y$ (called the *regression function*) as $\beta_1 + \beta_2(x - \overline{x})$, $\beta_1$ would be interpreted not as an intercept, but as the value of the regression when $x = \overline{x}$. Note that $\sum_{i=1}^n (x_i - \overline{x}) = \sum_{i=1}^n x_i - n\overline{x} = 0$, which means that the two columns of $X$ are orthogonal. This will simplify many later results. For instance, $X$ will have rank 2 unless all entries in the second column are zero, which can only occur if $x_1 = \cdots = x_n$. Also, since the entries in $X'X$ are inner products of the columns of $X$, this matrix and $(X'X)^{-1}$ are both diagonal:

$$X'X = \begin{pmatrix} n & 0 \\ 0 & \sum_{i=1}^n (x_i - \overline{x})^2 \end{pmatrix}$$

and

$$(X'X)^{-1} = \begin{pmatrix} 1/n & 0 \\ 0 & 1/\sum_{i=1}^n (x_i - \overline{x})^2 \end{pmatrix}.$$

Since

$$X'Y = \begin{pmatrix} \sum_{i=1}^n Y_i \\ \sum_{i=1}^n Y_i(x_i - \overline{x}) \end{pmatrix},$$

$$\hat{\beta} = (X'X)^{-1}X'Y = \begin{pmatrix} \frac{1}{n}\sum_{i=1}^n Y_i \\ \sum_{i=1}^n Y_i(x_i - \overline{x}) / \sum_{i=1}^n (x_i - \overline{x})^2 \end{pmatrix}.$$

Also,

$$\operatorname{Cov}(\hat{\beta}) = \sigma^2 (X'X)^{-1} = \begin{pmatrix} \sigma^2/n & 0 \\ 0 & \sigma^2/\sum_{i=1}^n (x_i - \overline{x})^2 \end{pmatrix}. \tag{14.23}$$

To estimate $\sigma^2$, since $\hat{\xi}_i = \hat{\beta}_1 + \hat{\beta}_2(x_i - \overline{x})$,

$$e_i = Y_i - \hat{\beta}_1 - \hat{\beta}_2(x_i - \overline{x}),$$

and then

$$S^2 = \frac{1}{n-2}\sum_{i=1}^{n} e_i^2.$$

This equation can be rewritten in various ways. For instance,

$$S^2 = \frac{1}{n-2}\sum_{i=1}^{n}(Y_i - \overline{Y})^2(1-\hat{\rho}^2),$$

where $\hat{\rho}$ is the *sample correlation* defined as

$$\hat{\rho} = \frac{\sum_{i=1}^{n}(Y_i - \overline{Y})(x_i - \overline{x})}{\left[\sum_{i=1}^{n}(Y_i - \overline{Y})^2 \sum_{i=1}^{n}(x_i - \overline{x})^2\right]^{1/2}}.$$

This equation shows that $\hat{\rho}^2$ may be viewed as the proportion of the variation of $Y$ that is "explained" by the linear relation between $Y$ and $x$.

Using (14.22),

$$\left(\hat{\beta}_1 - \frac{St_{\alpha/2,n-2}}{\sqrt{n}}, \hat{\beta}_1 + \frac{St_{\alpha/2,n-2}}{\sqrt{n}}\right)$$

is a $1-\alpha$ confidence interval for $\beta_1$, and

$$\left(\hat{\beta}_2 - \frac{St_{\alpha/2,n-2}}{\sqrt{\sum_{i=1}^{n}(x_i - \overline{x})^2}}, \hat{\beta}_2 + \frac{St_{\alpha/2,n-2}}{\sqrt{\sum_{i=1}^{n}(x_i - \overline{x})^2}}\right)$$

is a $1-\alpha$ confidence interval for $\beta_2$.

## 14.6 Noncentral *F* and Chi-Square Distributions

Distribution theory for testing in the general linear model relies on noncentral $F$ and chi-square distributions.

**Definition 14.7.** *If $Z_1, \dots, Z_p$ are independent and $\delta \geq 0$ with*

$$Z_1 \sim N(\delta, 1) \quad \text{and} \quad Z_j \sim N(0,1), \quad j = 2, \dots, p,$$

*then $W = \sum_{i=1}^{p} Z_i^2$ has the* noncentral chi-square distribution *with noncentrality parameter $\delta^2$ and $p$ degrees of freedom, denoted*

$$W \sim \chi_p^2(\delta^2).$$

**Lemma 14.8.** *If $Z \sim N_p(\mu, I)$, then $Z'Z \sim \chi_p^2(\|\mu\|^2)$.*

*Proof.* Let $O$ be an orthogonal matrix where the first row is $\mu'/\|\mu\|$, so that

$$O\mu = \tilde{\mu} = \begin{pmatrix} \|\mu\| \\ 0 \\ \vdots \\ 0 \end{pmatrix}.$$

Then

$$\tilde{Z} = OZ \sim N_p(\tilde{\mu}, I_p).$$

From the definition, $\tilde{Z}'\tilde{Z} = \sum_{i=1}^p \tilde{Z}^2 \sim \chi_p^2(\|\mu\|^2)$, and the lemma follows because

$$\tilde{Z}'\tilde{Z} = Z'O'OZ = Z'Z. \qquad \square$$

The next lemma shows that certain quadratic forms for multivariate normal vectors have noncentral chi-square distributions.

**Lemma 14.9.** *If $\Sigma$ is a $p \times p$ positive definite matrix and if $Z \sim N_p(\mu, \Sigma)$, then*

$$Z'\Sigma^{-1}Z \sim \chi_p^2(\mu'\Sigma^{-1}\mu).$$

*Proof.* Let $A = \Sigma^{-1/2}$, the symmetric square root of $\Sigma^{-1}$, defined in (9.11). Then $AZ \sim N_p(A\mu, I_p)$, and so

$$Z'\Sigma^{-1}Z = (AZ)'(AZ) \sim \chi_p^2(\|A\mu\|^2).$$

The lemma follows because $\|A\mu\|^2 = (A\mu)'(A\mu) = \mu'AA\mu = \mu'\Sigma^{-1}\mu$. $\square$

**Definition 14.10.** *If $V$ and $W$ are independent variables with $V \sim \chi_k^2(\delta^2)$ and $W \sim \chi_m^2$, then*

$$\frac{V/k}{W/m} \sim F_{k,m}(\delta^2),$$

*the* noncentral $F$-distribution *with degrees of freedom $k$ and $m$ and noncentrality parameter $\delta^2$. When $\delta^2 = 0$ this distribution is simply called the $F$ distribution, $F_{k,m}$.*

## 14.7 Testing in the General Linear Model

In the general linear model, $Y \sim N(\xi, \sigma^2 I)$ with the mean $\xi$ in a linear subspace $\omega$ with dimension $r$. In this section we consider testing $H_0 : \xi \in \omega_0$ versus $H_1 : \xi \in \omega - \omega_0$ with $\omega_0$ a $q$-dimensional linear subspace of $\omega$, $0 \le q < r$. Null hypotheses of this form arise when $\beta$ satisfies linear constraints. For instance we might have $H_0 : \beta_1 = \beta_2$ or $H_0 : \beta_1 = 0$. (Similar ideas can be used to test $\beta_1 = c$ or other affine constraints; see Problem 14.13.)

Let $\hat{\xi}$ and $\hat{\xi}_0$ denote least squares estimates for $\xi$ under the full model and under $H_0$. Specifically, $\hat{\xi} = PY$ and $\hat{\xi}_0 = P_0 Y$, where $P$ and $P_0$ are the projection matrices onto $\omega$ and $\omega_0$. The test statistic of interest is based on $\|Y - \hat{\xi}\|$, the distance between $Y$ and $\omega$, and $\|Y - \hat{\xi}_0\|$, the distance between $Y$ and $\omega_0$. Because $\omega_0 \subset \omega$, the former distance must be smaller, but if the distances are comparable, then at least qualitatively $H_0$ may seem adequate. The test statistic is

$$T = \frac{n-r}{r-q} \frac{\|Y - \hat{\xi}_0\|^2 - \|Y - \hat{\xi}\|^2}{\|Y - \hat{\xi}\|^2},$$

and $H_0$ will be rejected if $T$ exceeds a suitable constant. Noting that $Y - \hat{\xi} \in \omega^\perp$ and $\hat{\xi} - \hat{\xi}_0 \in \omega$, the vectors $Y - \hat{\xi}$ and $\hat{\xi} - \hat{\xi}_0$ are orthogonal, and the squared length of their sum, by the Pythagorean theorem, is

$$\|Y - \hat{\xi}_0\|^2 = \|Y - \hat{\xi}\|^2 + \|\hat{\xi} - \hat{\xi}_0\|^2.$$

Using this, the formula for $T$ can be rewritten as

$$T = \frac{n-r}{r-q} \frac{\|\hat{\xi} - \hat{\xi}_0\|^2}{\|Y - \hat{\xi}\|^2} = \frac{\|\hat{\xi} - \hat{\xi}_0\|^2}{(r-q)S^2}. \tag{14.24}$$

This test statistic is equivalent to the generalized likelihood ratio test statistic that will be introduced and studied in Chapter 17. When $r - q = 1$ the test is uniformly most powerful unbiased, and when $r - q > 1$ the test is most powerful among tests satisfying symmetry restrictions.

For level and power calculations we need the distribution of $T$ given in the next theorem.

**Theorem 14.11.** *Under the general linear model,*

$$T \sim F_{r-q,n-r}(\delta^2),$$

*where*

$$\delta^2 = \frac{\|\xi - P_0 \xi\|^2}{\sigma^2}. \tag{14.25}$$

*Proof.* Write

$$Y = \sum_{i=1}^{n} Z_i v_i,$$

where $v_1, \dots, v_n$ is an orthonormal basis chosen so that $v_1, \dots, v_q$ span $\omega_0$ and $v_1, \dots, v_r$ span $\omega$. Then, as in (14.8),

$$\hat{\xi}_0 = \sum_{i=1}^{q} Z_i v_i \text{ and } \hat{\xi} = \sum_{i=1}^{r} Z_i v_i.$$

Also, as in (14.5) and (14.6), $Z \sim N(\eta, \sigma^2 I)$ with $\eta_{r+1} = \cdots = \eta_n = 0$. Since $v_i'v_j$ is zero when $i \neq j$ and one when $i = j$,

$$\|Y - \hat{\xi}\|^2 = \left\| \sum_{i=r+1}^{n} Z_i v_i \right\|^2 = \left( \sum_{i=r+1}^{n} Z_i v_i' \right) \left( \sum_{j=r+1}^{n} Z_j v_j \right)$$
$$= \sum_{i=r+1}^{n} \sum_{j=r+1}^{n} Z_i Z_j v_i' v_j = \sum_{i=r+1}^{n} Z_i^2.$$

Similarly

$$\|Y - \hat{\xi}_0\|^2 = \sum_{i=q+1}^{n} Z_i^2,$$

and so

$$T = \frac{\frac{1}{r-q} \sum_{i=q+1}^{r} (Z_i/\sigma)^2}{\frac{1}{n-r} \sum_{i=r+1}^{n} (Z_i/\sigma)^2}.$$

The variables $Z_i$ are independent, and so the numerator and denominator in this formula for $T$ are independent. Because $Z_i/\sigma \sim N(\eta_i/\sigma, 1)$, by Lemma 14.8,

$$\sum_{i=q+1}^{r} \left( \frac{Z_i}{\sigma} \right)^2 \sim \chi^2_{r-q}(\delta^2),$$

where

$$\delta^2 = \sum_{i=q+1}^{r} \frac{\eta_i^2}{\sigma^2}. \tag{14.26}$$

Also, since $\eta_i = 0$ for $i = r+1, \ldots, n$, $Z_i/\sigma \sim N(0, 1)$, $i = r+1, \ldots, n$, and so $\sum_{i=r+1}^{n} (Z_i/\sigma)^2 \sim \chi^2_{n-r}$. So by Definition 14.10 for the noncentral $F$-distribution, $T \sim F_{r-q,n-r}(\delta^2)$ with $\delta^2$ given in (14.26). To finish we must show that (14.25) and (14.26) agree, or that

$$\sum_{i=q+1}^{r} \eta_i^2 = \|\xi - P_0 \xi\|^2.$$

Since

$$\xi = E\hat{\xi} = \sum_{i=1}^{r} \eta_i v_i$$

and

$$P_0 \xi = E P_0 Y = E\hat{\xi}_0 = \sum_{i=1}^{q} \eta_i v_i,$$

$$\xi - P_0\xi = \sum_{i=q+1}^{r} \eta_i v_i.$$

Then, by the Pythagorean theorem,

$$\|\xi - P_0\xi\|^2 = \sum_{i=q+1}^{r} \eta_i^2,$$

as desired. ☐

*Example 14.12.* Consider a model in which

$$Y_i = \begin{cases} x_i\beta_1 + \epsilon_i, & i = 1, \dots, n_1; \\ x_i\beta_2 + \epsilon_i, & i = n_1 + 1, \dots, n_1 + n_2 = n, \end{cases}$$

with $\epsilon_1, \dots, \epsilon_n$ i.i.d. from $N(0, \sigma^2)$ and independent variables $x_1, \dots, x_n$ taken as known constants. This model might be appropriate if you have bivariate data from two populations, each satisfying simple linear regression through the origin. In such a situation, the most interesting hypothesis to consider may be that the two populations are the same, and so let us test $H_0 : \beta_1 = \beta_2$ versus $H_1 : \beta_1 \neq \beta_2$. The design matrix under the full model is

$$X = \begin{pmatrix} x_1 & 0 \\ \vdots & \vdots \\ x_{n_1} & 0 \\ 0 & x_{n_1+1} \\ \vdots & \vdots \\ 0 & x_n \end{pmatrix},$$

and straightforward algebra gives

$$X'X = \begin{pmatrix} \sum_{i=1}^{n_1} x_i^2 & 0 \\ 0 & \sum_{i=n_1+1}^{n} x_i^2 \end{pmatrix} \text{ and } X'Y = \begin{pmatrix} \sum_{i=1}^{n_1} x_iY_i \\ \sum_{i=n_1+1}^{n} x_iY_i \end{pmatrix}.$$

So

$$\hat\beta = (X'X)^{-1}X'Y = \begin{pmatrix} \sum_{i=1}^{n_1} x_iY_i / \sum_{i=1}^{n_1} x_i^2 \\ \sum_{i=n_1+1}^{n} x_iY_i / \sum_{i=n_1+1}^{n} x_i^2 \end{pmatrix}$$

and

$$\|Y - \hat\xi\|^2 = \sum_{i=1}^{n_1} (Y_i - x_i\hat\beta_1)^2 + \sum_{i=n_1+1}^{n} (Y_i - x_i\hat\beta_2)^2.$$

Here $X$ is of full rank (unless $x_1 = \cdots = x_{n_1} = 0$ or $x_{n_1} = \cdots = x_n = 0$), and so $r = p = 2$ and

$$S^2 = \frac{\|Y - \hat{\xi}\|^2}{n-2}.$$

Under $H_0$ we also have a general linear model. If $\beta_0$ denotes the common value of $\beta_1$ and $\beta_2$, then $Y_i = x_i\beta_0 + \epsilon_i$, $i = 1, \dots, n$, and the design matrix is

$$X_0 = \begin{pmatrix} x_1 \\ \vdots \\ x_n \end{pmatrix}.$$

Then $X_0'X_0 = \sum_{i=1}^n x_i^2$ and $X_0'Y = \sum_{i=1}^n x_iY_i$, and so

$$\hat{\beta}_0 = (X_0'X_0)^{-1}X_0'Y = \frac{\sum_{i=1}^n x_iY_i}{\sum_{i=1}^n x_i^2}.$$

Thus

$$\begin{aligned}\|\hat{\xi} - \hat{\xi}_0\|^2 &= \sum_{i=1}^{n_1}(x_i\hat{\beta}_1 - x_i\hat{\beta}_0)^2 + \sum_{i=n_1+1}^{n}(x_i\hat{\beta}_2 - x_i\hat{\beta}_0)^2 \\ &= (\hat{\beta}_1 - \hat{\beta}_0)^2\sum_{i=1}^{n_1}x_i^2 + (\hat{\beta}_2 - \hat{\beta}_0)^2\sum_{i=n_1+1}^{n}x_i^2.\end{aligned}$$

Noting that $\hat{\beta}_0$ is a weighted average of $\hat{\beta}_1$ and $\hat{\beta}_2$,

$$\hat{\beta}_0 = \frac{\sum_{i=1}^{n_1}x_i^2}{\sum_{i=1}^n x_i^2}\hat{\beta}_1 + \frac{\sum_{i=n_1+1}^{n}x_i^2}{\sum_{i=1}^n x_i^2}\hat{\beta}_2,$$

this expression simplifies to

$$\|\hat{\xi} - \hat{\xi}_0\|^2 = \frac{\sum_{i=1}^{n_1}x_i^2\sum_{i=n_1+1}^{n}x_i^2}{\sum_{i=1}^n x_i^2}(\hat{\beta}_1 - \hat{\beta}_2)^2. \tag{14.27}$$

So the test statistic $T$ is given by

$$T = \frac{\sum_{i=1}^{n_1}x_i^2\sum_{i=n_1+1}^{n}x_i^2}{S^2\sum_{i=1}^n x_i^2}(\hat{\beta}_1 - \hat{\beta}_2)^2.$$

Under $H_0$, $T \sim F_{1,n-2}$, and the level-$\alpha$ test will reject $H_0$ if $T$ exceeds $F_{\alpha,1,n-2}$, the upper $\alpha$th quantile of this $F$-distribution.

For power calculations we need the noncentrality parameter $\delta^2$ in (14.25). Given the calculations we have done, the easiest way to find $\delta^2$ is to note that if our data were observed without error, i.e., if $\epsilon$ were zero, then $\hat{\xi}$ would be $\xi$, $\hat{\beta}_1$ and $\hat{\beta}_2$ would be $\beta_1$ and $\beta_2$, and $\hat{\xi}_0$ would be $P_0\xi$. Using this observation and (14.27),

$$\delta^2 = \frac{\sum_{i=1}^{n_1}x_i^2\sum_{i=n_1+1}^{n}x_i^2}{\sigma^2\sum_{i=1}^n x_i^2}(\beta_1 - \beta_2)^2.$$

The power is then the chance that a variable from $F_{1,n-2}(\delta^2)$ exceeds $F_{\alpha,1,n-2}$.

## 14.8 Simultaneous Confidence Intervals

A researcher studying a complex data set may construct confidence intervals for many parameters. Even with a high coverage probability for each interval there may be a substantial chance some of them will fail, raising a concern that the ones that fail may be reported as meaningful when all that is really happening is natural chance variation. Simultaneous confidence intervals have been suggested to protect against this possibility. A few basic ideas are developed here, first in the context of one-way ANOVA models, introduced before in Example 14.2.

The model under consideration has

$$Y_{kl} = \beta_k + \epsilon_{kl}, \qquad 1 \le l \le c, \qquad 1 \le k \le p.$$

This can be viewed as a model for independent random samples from $p$ normal populations all with a common variance. The design here has the same number of observations $c$ from each population. Listing the variables $Y_{kl}$ in a single vector, as in Example 14.2, this is a general linear model. The least squares estimator of $\beta$ should minimize

$$\sum_{l=1}^{c}\sum_{k=1}^{p}(Y_{kl} - \beta_k)^2.$$

The partial derivative of this criterion with respect to $\beta_m$ is

$$-2\sum_{l=1}^{c}(Y_{ml} - \beta_m)$$

which vanishes when $\beta_m = \hat{\beta}_m$ given by

$$\hat{\beta}_m = \overline{Y}_{m\cdot} \stackrel{\text{def}}{=} \frac{1}{c}\sum_{l=1}^{c} Y_{ml}, \qquad m = 1, \dots, p.$$

These are the least squares estimators. Here $r = p$ and $n = pc$, so

$$S^2 = \frac{\|Y - \hat{\xi}\|^2}{pc - p} = \frac{1}{p(c-1)}\sum_{l=1}^{c}\sum_{k=1}^{p}(Y_{kl} - \hat{\beta}_k)^2.$$

The least squares estimators are averages of different collections of the $Y_{kl}$. Thus $\hat{\beta}_1, \dots, \hat{\beta}_p$ are independent with

$$\hat{\beta}_k \sim N(\beta_k, \sigma^2/c), \qquad k = 1, \dots, p.$$

Also

$$\frac{p(c-1)S^2}{\sigma^2} \sim \chi^2_{p(c-1)},$$

and $S^2$ is independent of $\hat{\beta}$.

To start, let us try to find intervals $I_1, \dots, I_p$ that simultaneously cover parameters $\beta_1, \dots, \beta_p$ with specified probability $1-\alpha$. Specifically, we want

$$P(\beta_k \in I_k, k = 1, \dots, p) = 1 - \alpha.$$

The confidence intervals in Theorem 14.6 or (14.22) are

$$\left(\hat{\beta}_k - \frac{S}{\sqrt{c}} t_{\alpha/2, p(c-1)}, \hat{\beta}_k + \frac{S}{\sqrt{c}} t_{\alpha/2, p(c-1)}\right),$$

and intuition suggests that we may be able to achieve our objective taking

$$I_k = \left(\hat{\beta}_k - \frac{S}{\sqrt{c}} q, \hat{\beta}_k + \frac{S}{\sqrt{c}} q\right), \qquad k = 1, \dots, p,$$

if $q$ is chosen suitably. Now

$$\begin{aligned} P(\beta_k \in I_k, k = 1, \dots, p) &= P\left(|\hat{\beta}_k - \beta_k| < \frac{S}{\sqrt{c}} q, k = 1, \dots, p\right) \\ &= P\left(\max_{1 \le k \le p} \frac{|\hat{\beta}_k - \beta_k|}{S/\sqrt{c}} < q\right) \\ &= P\left(\max_{1 \le k \le p} \frac{|Z_k|}{\sqrt{W}} < q\right), \end{aligned}$$

where

$$Z_k = \frac{\hat{\beta}_k - \beta_k}{\sigma/\sqrt{c}} \sim N(0,1), \qquad k = 1, \dots, p$$

and $W = S^2/\sigma^2$. Because $Z_1, \dots, Z_p$ and $W$ are independent and $mW \sim \chi^2_m$ with $m = p(c-1)$, the simultaneous coverage probability here does not depend on $\beta$ or $\sigma$.

**Definition 14.13.** *If $Z_1, \dots, Z_p$ and $W$ are independent variables with $Z_k \sim N(0,1)$, $k = 1, \dots, p$, and $mW \sim \chi^2_m$, then*

$$\frac{\max_{1 \le k \le p} |Z_k|}{\sqrt{W}}$$

*has the* studentized maximum modulus distribution *with parameters $p$ and $m$.*

If $q$ is the upper $\alpha$th quantile of this studentized maximum modulus distribution, then the intervals $I_1, \dots, I_p$ have simultaneous coverage probability $1 - \alpha$.

In practice, researchers will often be more interested in comparing populations than estimating individual means, and so confidence intervals for differences $\beta_j - \beta_i$ may be of interest. So let us now seek intervals $I_{ij}$ such that

$$P(\beta_j - \beta_i \in I_{ij}, \forall i \neq j) = 1 - \alpha.$$

Now we may naturally suspect we can achieve this objective with intervals of the form

$$I_{ij} = \left( \hat{\beta}_j - \hat{\beta}_i - \frac{S}{\sqrt{c}} q, \hat{\beta}_j - \hat{\beta}_i + \frac{S}{\sqrt{c}} q \right)$$

with $q$ adjusted suitably. Then

$$\begin{aligned}
P(\beta_j - \beta_i \in I_{ij}, \forall i \neq j) \\
&= P\left( \left| (\hat{\beta}_j - \beta_j) - (\hat{\beta}_i - \beta_i) \right| < \frac{S}{\sqrt{c}} q, \forall i \neq j \right) \\
&= P\left( \frac{\left| \sqrt{c}(\hat{\beta}_j - \beta_j) - \sqrt{c}(\hat{\beta}_i - \beta_i) \right|}{S} < q, \forall i \neq j \right) \\
&= P\left( \frac{|Z_j - Z_i|}{\sqrt{W}} < q, \forall i \neq j \right) \\
&= P\left( \frac{\max_{1 \le k \le p} Z_k - \min_{1 \le k \le p} Z_k}{\sqrt{W}} < q \right).
\end{aligned}$$

This approach works because this probability does not depend on $\beta$ or $\sigma$.

**Definition 14.14.** *If $Z_1, \dots, Z_p$ and $W$ are independent variables with $Z_k \sim N(0,1)$, $k = 1, \dots, p$ and $mW \sim \chi^2_m$, then*

$$\frac{\max_{1 \le k \le p} Z_k - \min_{1 \le k \le p} Z_k}{\sqrt{W}}$$

*has the* studentized range distribution *with parameters $p$ and $m$.*

If $q$ is the upper $\alpha$th quantile of the studentized range distribution, then the intervals $I_{ij}$ will have simultaneous coverage probability $1 - \alpha$.

The derivations of simultaneous confidence intervals just presented relies heavily on the structure of the ANOVA model under consideration. More general results are possible using a method due to Scheffé. This method is based on confidence sets for a parameter $\psi \in \mathbb{R}^q$, with $q \le r$, which is a linear function of the mean $\xi$; that is,

$$\psi = A\xi = AX\beta$$

for some $q \times n$ matrix $A$. When $X$ is full rank $\beta = (X'X)^{-1}X'\xi$, and so $A = (X'X)^{-1}X'$ will give $\psi = \beta$. Other linear functions of $\beta$ are also possible. Because $P\xi = \xi$, we have $AP\xi = A\xi = \psi$, and replacing $A$ by $A^* = AP$ will not change $\psi$. Then $A^*P = APP = AP = A^*$. Changing $A$ to $A^*$, if necessary, we can assume without loss of generality that $A = AP$. This is convenient because then the least squares estimator of $\psi$ is

$$\hat{\psi} = A\hat{\xi} = APY = AY.$$

Finally, we assume that the rows of $AX$ are linearly independent. Since $AX$ is then of full rank and $\psi = AX\beta$, this ensures that $\psi$ can assume arbitrary values in $\mathbb{R}^q$. Also, note that then the rank of $AX$ will be less than the rank of $A$, for if the rows of $A$ satisfy a nontrivial linear constraint, $v'A = 0$, then $v'AX = 0$, and the rows of $AX$ will satisfy the same linear constraint. If we define $B = AA'$, then $B$ is positive definite because

$$q \le \text{rank}(AX) \le \text{rank}(A) \le q,$$

showing that $A$ and $AX$ are both of full rank, and $v'Bv = v'AA'v = \|A'v\|^2$, which is positive unless $v = 0$ as $A$ is full rank. Using (1.15),

$$\hat{\psi} \sim N(\psi, \sigma^2 B),$$

and so by Lemma 14.9,

$$\frac{(\hat{\psi} - \psi)'B^{-1}(\hat{\psi} - \psi)}{\sigma^2} \sim \chi_q^2.$$

Because $\hat{\psi}$ is a function of $\hat{\xi}$, and $\hat{\xi}$ is independent of $S^2$, this quadratic form is independent of

$$\frac{(n-r)S^2}{\sigma^2} \sim \chi_{n-r}^2.$$

Then by Definition 14.10,

$$\frac{(\hat{\psi} - \psi)'B^{-1}(\hat{\psi} - \psi)/(q\sigma^2)}{S^2/\sigma^2} = \frac{(\hat{\psi} - \psi)'B^{-1}(\hat{\psi} - \psi)}{qS^2} \sim F_{q,n-r}.$$

From this

$$P\bigl((\hat{\psi} - \psi)'B^{-1}(\hat{\psi} - \psi) \le qS^2 F_{\alpha,q,n-r}\bigr) = 1 - \alpha.$$

The set of values for $\psi$ where this event obtains is a multivariate ellipse centered at $\hat{\psi}$. This random ellipse is a $1 - \alpha$ confidence set for $\psi$.

To form simultaneous confidence intervals from the confidence ellipse described, note that

$$(\hat{\psi} - \psi)'B^{-1}(\hat{\psi} - \psi) = \|B^{-1/2}(\hat{\psi} - \psi)\|^2.$$

Also, for any $h \in \mathbb{R}^q$,

$$h'Bh = h'B^{1/2}B^{1/2}h = \|B^{1/2}h\|^2.$$

By the Schwarz inequality,

$$\|B^{1/2}h\|^2\|B^{-1/2}(\hat{\psi} - \psi)\|^2 \ge \bigl[h'B^{1/2}B^{-1/2}(\hat{\psi} - \psi)\bigr]^2 = \bigl[h'(\hat{\psi} - \psi)\bigr]^2.$$

So

$$\begin{aligned}
P\Big\{\big[h'(\hat{\psi}-\psi)\big]^2 &\le qS^2h'BhF_{\alpha,q,n-r}, \forall h \in \mathbb{R}^q\Big\}\\
&\ge P\big(\|B^{1/2}h\|^2\|B^{-1/2}(\hat{\psi}-\psi)\|^2 \le qS^2h'BhF_{\alpha,q,n-r}, \forall h \in \mathbb{R}^q\big)\\
&= P\big(\|B^{-1/2}(\hat{\psi}-\psi)\|^2 \le qS^2F_{\alpha,q,n-r}\big)\\
&= 1-\alpha.
\end{aligned}$$

But taking $h = B^{-1}(\hat{\psi}-\psi)$, this probability can be at most $1-\alpha$, and so we must have equality. Since

$$\operatorname{Var}\big(h'(\hat{\psi}-\psi)\big) = \sigma^2 h'Bh,$$

naturally estimated by

$$\hat{\sigma}^2_{h'\hat{\psi}} = S^2h'Bh,$$

we can write this identity as

$$P\Big\{\big[h'(\hat{\psi}-\psi)\big]^2 \le \hat{\sigma}^2_{h'\hat{\psi}}qF_{\alpha,q,n-r}, \forall h \in \mathbb{R}^q\Big\} = 1-\alpha.$$

So the intervals

$$\Big(h'\hat{\psi} - \hat{\sigma}_{h'\hat{\psi}}\sqrt{qF_{\alpha,q,n-r}},\ h'\hat{\psi} + \hat{\sigma}_{h'\hat{\psi}}\sqrt{qF_{\alpha,q,n-r}}\Big)$$

contain $h'\psi$ simultaneously for all $h \in \mathbb{R}^q$, with probability $1-\alpha$.

*Example 14.15.* In the model for simple linear regression considered in Section 14.5, the value for the regression function at a specified value $x$ for the independent variable is $\beta_1 + \beta_2(x-\overline{x})$, estimated by $\hat{\beta}_1 + \hat{\beta}_2(x-\overline{x})$. Using (14.23), the variance of this estimator is

$$\sigma^2\left[\frac{1}{n} + \frac{(x-\overline{x})^2}{s_{xx}}\right],$$

where $s_{xx} = \sum_{i=1}^n (x_i - \overline{x})^2$. This variance is estimated by

$$S^2\left[\frac{1}{n} + \frac{(x-\overline{x})^2}{s_{xx}}\right].$$

Taking $\psi = \beta$, the regression at $x$, $\beta_1 + \beta_2(x-\overline{x})$, will lie in

$$\begin{aligned}
\Bigg(\hat{\beta}_1 + \hat{\beta}_2(x-\overline{x}) - S\sqrt{\frac{1}{n} + \frac{(x-\overline{x})^2}{s_{xx}}}\sqrt{2F_{\alpha,2,n-2}},&\\
\hat{\beta}_1 + \hat{\beta}_2(x-\overline{x}) + S\sqrt{\frac{1}{n} + \frac{(x-\overline{x})^2}{s_{xx}}}\sqrt{2F_{\alpha,2,n-2}}\Bigg)&,
\end{aligned}$$

simultaneously for all $x \in \mathbb{R}$, with probability at least $1-\alpha$. These confidence bands are plotted along with the regression line in Figure 14.1.

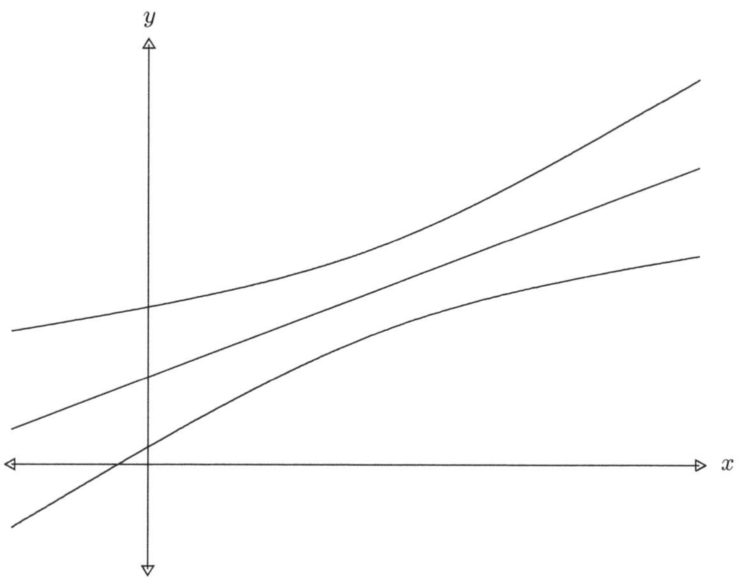


**Fig. 14.1.** Confidence bands.

*Example 14.16.* In one-way ANOVA comparisons between weighted averages of the means of certain populations and weighted averages of the means of different populations may be of interest. Scaled differences between weighted averages are called *contrasts.*

**Definition 14.17.** *A* contrast *in one-way ANOVA is any parameter of the form*

$$a'\beta = a_1\beta_1 + \cdots + a_p\beta_p$$

*with* $a_1 + \cdots + a_p = 0$.

Examples of contrasts include $\beta_3 - \beta_1$ or $\frac{1}{2}(\beta_1 + \beta_3) - \left(\frac{1}{6}\beta_2 + \frac{1}{3}\beta_4 + \frac{1}{2}\beta_5\right)$. Taking

$$\psi = \begin{pmatrix} \beta_1 - \beta_p \\ \vdots \\ \beta_{p-1} - \beta_p \end{pmatrix},$$

a contrast $a'\beta$ equals $h'\psi$ with $h' = (a_1, \ldots, a_{p-1})$. The least squares estimate of this contrast is $\sum_{i=1}^p a_i\hat{\beta}_i$ which has variance $\sum_{i=1}^p a_i^2\sigma^2/c = \|a\|^2\sigma^2/c$, estimated replacing $\sigma^2$ by $S^2$. Thus with probability $1 - \alpha$,

$$a'\beta \in \left(a'\hat{\beta} - \frac{S\|a\|}{\sqrt{c}}\sqrt{(p-1)F_{\alpha,p-1,p(c-1)}},\right.$$
$$\left. a'\hat{\beta} + \frac{S\|a\|}{\sqrt{c}}\sqrt{(p-1)F_{\alpha,p-1,p(c-1)}}\right),$$

simultaneously for all $a \in \mathbb{R}^p$ with $a_1 + \cdots + a_p = 0$. If instead we were interested in all linear combinations of $\beta$, we would need to take $\psi = \beta$. For this case, $q$ equals $p$, and we have simultaneous confidence intervals

$$a'\beta \in \left(a'\hat{\beta} - \frac{S\|a\|}{\sqrt{c}}\sqrt{pF_{\alpha,p,p(c-1)}}, a'\hat{\beta} + \frac{S\|a\|}{\sqrt{c}}\sqrt{pF_{\alpha,p,p(c-1)}}\right),$$

for all $a \in \mathbb{R}^p$. These intervals will be a bit wider than the simultaneous confidence intervals for contrasts.

## 14.9 Problems[1]

*1. Consider a model in which our data are

$$Y_i = \beta_1 + w_i\beta_2 + x_i\beta_3 + \epsilon_i, \qquad i = 1, \dots, n,$$

where $w_1, \dots, w_n$ and $x_1, \dots, x_n$ are observed constants; $\beta_1$, $\beta_2$, and $\beta_3$ are unobserved parameters; and $\epsilon_1, \dots, \epsilon_n$ are unobserved random variables which are i.i.d. from $N(0, \sigma^2)$. Assume that through accident or design, $w_1 + \cdots + w_n = 0$ and $x_1 + \cdots + x_n = 0$. For notation, let $S_{ww} = \sum_{i=1}^n w_i^2$, $S_{xx} = \sum_{i=1}^n x_i^2$, $S_{wx} = \sum_{i=1}^n w_i x_i$, and so on.
a) Write the model in matrix form as $Y = X\beta + \epsilon$ describing entries in the matrix $X$.
b) If $n > 3$, when will $X$ be of full rank?
c) Assuming $X$ is of full rank, give an explicit formula for $\hat{\beta}$. (It will involve terms such as $S_{xx}$, $S_{xY}$, etc.)
d) Find the covariance matrix of $\hat{\beta}$.
e) Give a formula for the UMVU estimator of $\sigma^2$.
f) Find confidence intervals for $\beta_1$ and for $\beta_3 - \beta_2$.

2. Consider a general linear model with $n = 2m$ in which $Y_i = \beta_1 + \beta_3 x_i + \epsilon_i$, $i = 1, \dots, m$, and $Y_i = \beta_2 + \beta_3 x_i + \epsilon_i$, $i = m+1, \dots, n$. Here $\epsilon_1, \dots, \epsilon_n$ are i.i.d. from $N(0, \sigma^2)$; $\beta = (\beta_1, \beta_2, \beta_3)'$ and $\sigma^2$ are unknown parameters, and $x_1, \dots, x_n$ are known constants with $x_1 + \cdots + x_m = x_{m+1} + \cdots + x_n = 0$.
a) Write the model in vector form as $Y = X\beta + \epsilon$ describing entries in the design matrix $X$.
b) Determine the UMVU estimator $\hat{\beta}$ of $\beta$.
c) Write $\beta_1$ as a linear function of $\xi = EY$; that is, find a vector $w$ such that $\beta_1 = w'\xi$.
d) Find the UMVU estimator $\hat{\xi}_i$ of $\xi_i$.
e) Show that $\hat{\beta}_1$ from part (b) equals $w'\hat{\xi}$ from parts (c) and (d).
f) Determine the variance of $\hat{\beta}_2 - \hat{\beta}_1$.
g) Determine the UMVU estimator $S^2$ of $\sigma^2$.

[1] Solutions to the starred problems are given at the back of the book.

h) Derive a $1-\alpha$ confidence interval for $\beta_2 - \beta_1$.
i) If the parameter $\beta$ were known, the best estimate for a future observation $Y$ from the first population (the population for the first half of the data observed) at some specified level $x$ for the independent variable would be $\beta_1 + \beta_3 x$. Give a 95% confidence interval for this quantity. Sketch the upper and lower limits of this interval as a function of $x$ if $\hat\beta_1 = 0$ and $\hat\beta_3 = 1/2$.
j) Give an explicit formula for the test statistic $T$ used to test $H_0 : \beta_1 = \beta_2$ versus $H_1 : \beta_1 \neq \beta_2$, and explain how this statistic would be used to test $H_0$ versus $H_1$ at level $\alpha$.
k) Determine the distribution for $T$ with an explicit formula for the noncentrality parameter.
l) Show that the test of $\beta_1 = \beta_2$ rejects $H_0$ if and only if the confidence interval for $\beta_2 - \beta_1$ does not contain zero.

3. Consider a general linear model with $n = 2m$ in which

$$Y_i = \beta_1 + \beta_2 x_i + \epsilon_i, \qquad i = 1, \dots, m,$$

and

$$Y_i = \beta_1 + \beta_3 x_i + \epsilon_i, \qquad i = m+1, \dots, n.$$

Here $\epsilon_1, \dots, \epsilon_n$ are i.i.d. from $N(0, \sigma^2)$; $\beta = (\beta_1, \beta_2, \beta_3)'$ and $\sigma^2$ are unknown parameters; and $x_1, \dots, x_n$ are known constants with $x_1 + \cdots + x_m = x_{m+1} + \cdots + x_n = 0$.

a) Write the model in vector form as $Y = X\beta + \epsilon$ describing entries in the design matrix $X$.
b) Determine the UMVU estimator $\hat\beta$ of $\beta$.
c) Write $\beta_2$ as a linear function of $\xi = EY$; that is, find a vector $w$ such that $\beta_2 = w'\xi$.
d) Find the UMVU estimator $\hat\xi_i$ of $\xi_i$.
e) Show that $\hat\beta_2$ from part (b) equals $w'\hat\xi$ from parts (c) and (d).
f) Determine the UMVU estimator $S^2$ of $\sigma^2$.
g) Determine the variance of $\hat\beta_3 - \hat\beta_2$.
h) Derive a 95% confidence interval for $\beta_3 - \beta_2$.
i) If the parameter $\beta$ were known, the best estimate for a future observation $Y$ from the first population (the population for the first half of the data observed) at some specified level $x$ for the independent variable would be $\beta_1 + \beta_2 x$. Give a 95% confidence interval for this quantity. Sketch the upper and lower limits of this interval as a function of $x$ if $\hat\beta_1 = 0$ and $\hat\beta_2 = 1/2$.
j) Use the fact that a suitable multiple of $(\hat\beta_3 - \hat\beta_2)/S$ has a $t$-distribution to derive a test of $H_0 : \beta_2 = \beta_3$ versus $H_1 : \beta_2 \neq \beta_3$ with level $\alpha = 5\%$.
k) Show that this test of $H_0 : \beta_2 = \beta_3$ rejects $H_0$ if and only if the confidence interval for $\beta_3 - \beta_2$ does not contain zero.

l) Derive the $F$-test of $H_0 : \beta_2 = \beta_3$ versus $H_1 : \beta_2 \neq \beta_3$ with level $\alpha = 5\%$. Show that this test is the same as the test above based on the $t$-distribution.

*4. *Two-way analysis of variance without replication.* Suppose a researcher wishes to study the effects of two factors $A$ and $B$ on some response variable. If $A$ and $B$ both occur at $m$ levels, then there are $m^2$ possible combinations of factors. If observations are expensive, a design in which there is a single observation for each treatment combination may be desirable. Let $Y_{ij}$ denote the response when factor $A$ is at level $i$ and factor $B$ is at level $j$. A common (additive) model for these data has

$$Y_{ij} = \alpha_i + \gamma_j + \epsilon_{ij},$$

where the $\epsilon_{ij}$ are i.i.d. from $N(0, \sigma^2)$. This can be considered a general linear model with $\beta = (\alpha_1, \dots, \alpha_m, \gamma_1, \dots, \gamma_m)'$.

a) Increasing every parameter $\alpha_i$ by an amount $\Delta$ and simultaneously decreasing every parameter $\gamma_j$ by the same amount $\Delta$ leaves the mean $\xi$ of $Y$ unchanged. So it is clear that $r = \dim(\omega)$ is less than $p = 2m$. Determine the dimension $r$ of $\omega$ in this model.
b) Find the least squares estimate for $\xi_{ij} = EY_{ij}$ in this model.
c) Determine $S^2$, the usual unbiased estimator of $\sigma^2$.
d) Show that $\alpha_i - \alpha_j$ is a linear function of $\xi$ and determine the least squares estimate of this difference.
e) Let $\overline{Y}_{i\cdot} = (Y_{i1} + \cdots + Y_{im})/m$, the average response with factor $A$ at level $i$. Use the studentized range distribution and the fact that $\overline{Y}_{1\cdot}, \dots, \overline{Y}_{m\cdot}$ are independent to derive simultaneous $1 - \alpha$ confidence intervals for all differences $\alpha_i - \alpha_j$, $1 \le i < j \le m$.
f) Derive a level $1 - \alpha$ test of $H_0 : \alpha_1 = \cdots = \alpha_m$ versus $H_1 : \alpha_i \neq \alpha_j$ for some $i \neq j$.
g) Give the power for this test of $\alpha_1 = \cdots = \alpha_m$. Your answer should involve the cumulative distribution function for a noncentral $F$ distribution. Give the degrees of freedom and provide a formula for the noncentrality parameter.
h) Use the Scheffé method to derive simultaneous confidence intervals for all contrasts of the $\alpha_i$, that is, all linear combinations $a_1\alpha_1 + \cdots + a_m\alpha_m$ with $a_1 + \cdots + a_m = 0$. Note that these contrasts are estimable because they can be written as linear combinations of differences $\alpha_1 - \alpha_m, \dots, \alpha_{m-1} - \alpha_m$.

5. Time series regression models often incorporate regular oscillation over time, and in some cases this structure can be incorporated into a general linear model. Let $t_j$, $j = 1, \dots, n$, be known time points, and assume

$$Y_j = r \sin(t_j + \theta) + \epsilon_j, \qquad j = 1, \dots, n.$$

Here the errors $\epsilon_i$ are i.i.d. from $N(0, \sigma^2)$, and $r > 0$, $\theta \in [-\pi, \pi)$, and $\sigma^2 > 0$ are unknown parameters. Assume for convenience that the time points are evenly spaced with $4k$ observations per cycle and $m$ cycles, so that $n = 4km$ and $t_j = j\pi/(2k)$. With these assumptions,

$$\sum_{j=1}^{n} \sin^2(t_j) = \sum_{j=1}^{n} \cos^2(t_j) = n/2$$

and

$$\sum_{j=1}^{n} \sin(t_j) = \sum_{j=1}^{n} \cos(t_j) = \sum_{j=1}^{n} \sin(t_j)\cos(t_j) = 0.$$

a) Introduce new parameters $\beta_1 = r\sin\theta$ and $\beta_2 = r\cos\theta$. Show that after replacing $r$ and $\theta$ with these parameters, we have a general linear model.
b) Find UMVU estimators $\hat{\beta}_1$ and $\hat{\beta}_2$ for $\beta_1$ and $\beta_2$.
c) Find the UMVU estimator of $\sigma^2$.
d) Derive 95% confidence intervals for $\beta_1$ and $\beta_2$.
e) Show that a suitable multiple of $\hat{r}^2 = \hat{\beta}_1^2 + \hat{\beta}_2^2$ has a noncentral chi-square distribution. Identify the degrees of freedom and the noncentrality parameter.
f) Derive a test of $H_0 : \theta = \theta_0$ versus $H_1 : \theta \neq \theta_0$ with level $\alpha = 5\%$.

6. Let $\beta_1, \dots, \beta_3$ be the angles for a triangle in degrees, so $\beta_1 + \beta_2 + \beta_3 = 180$; and let $Y_1, \dots, Y_3$ be measurements of these angles. Assume that the measurement errors, $\epsilon_i = Y_i - \beta_i$, $i = 1, \dots, 3$, are i.i.d. $N(0, \sigma^2)$.
   a) Find UMVU estimates $\hat{\beta}_1$ and $\hat{\beta}_2$ for $\beta_1$ and $\beta_2$.
   b) Find the covariance matrix for $(\hat{\beta}_1, \hat{\beta}_2)$ and compare the variance of $\hat{\beta}_1$ with $Y_1$.
   c) Find an unbiased estimator for $\sigma^2$.
   d) Derive confidence intervals for $\beta_1$ and $\beta_2 - \beta_1$.
7. *Side conditions when $r < p$.* When $r < p$, different values for $\beta$ will give the same mean $\xi = X\beta$, and various values for $\beta$ will minimize $\|Y - X\beta\|^2$. One approach to force a unique answer is to impose side conditions on $\beta$. Because the row span and column span of a matrix are the same, the space $V \subset \mathbb{R}^p$ spanned by the *rows* of $X$ will have dimension $r < p$, and $V^\perp$ will have dimension $p - r$.
   a) Show that $\beta \in V^\perp$ if and only if $X\beta = 0$.
   b) Let $\omega = \{X\beta : \beta \in \mathbb{R}^p\}$. Show that the map $\beta \rightsquigarrow X\beta$ from $V$ to $\omega$ is one-to-one and onto.
   c) Let $h_1, \dots, h_{p-r}$ be linearly independent vectors spanning $V^\perp$. Show that $\beta \in V$ if and only if $h_i \cdot \beta = 0$, $i = 1, \dots, p - r$. Equivalently, $\beta \in V$ if and only if $H\beta = 0$, where $H' = (h_1, \dots, h_{p-r})$.
   d) From part (b), there should be a unique $\hat{\beta}$ in $V$ with $X\hat{\beta} = \hat{\xi}$, and $\hat{\beta}$ will then minimize $\|Y - X\beta\|^2$ over $\beta \in V$. Using part (c), $\hat{\beta}$ can be characterized as the unique value minimizing $\|Y - X\beta\|^2$ over $\beta \in \mathbb{R}^p$ satisfying the side condition $H\hat{\beta} = 0$. Show that $\hat{\beta}$ minimizes

$$\left\| \begin{pmatrix} Y \\ 0 \end{pmatrix} - \begin{pmatrix} X \\ H \end{pmatrix} \beta \right\|^2$$

over $\beta \in \mathbb{R}^p$. Use this to derive an explicit equation for $\hat{\beta}$.

8. A variable $Y$ has a log-normal distribution with parameters $\mu$ and $\sigma^2$ if $\log Y \sim N(\mu, \sigma^2)$.
   a) Find the mean and density for the log-normal distribution.
   b) If $Y_1, \dots, Y_n$ are i.i.d. from the log-normal distribution with unknown parameters $\mu$ and $\sigma^2$, find the UMVU for $\mu$.
   c) If $Y_1, \dots, Y_n$ are i.i.d. from the log-normal distribution with parameters $\mu$ and $\sigma^2$, with $\sigma^2$ a known constant, find the UMVU for the common mean $\nu = EY_i$.
   d) In simple linear regression, $Y_1, \dots, Y_n$ are independent with $Y_i \sim N(\beta_1 + \beta_2 x_i, \sigma^2)$. In some applications this model may be inappropriate because the $Y_i$ are positive; perhaps $Y_i$ is the weight or volume of the $i$th unit. Suggest a similar model without this defect based on the log-normal distribution. Explain how you would estimate $\beta_1$ and $\beta_2$ in your model.
9. Consider the general linear model with normality:

$$Y \sim N(X\beta, \sigma^2 I), \qquad \beta \in \mathbb{R}^p, \quad \sigma^2 > 0.$$

If the rank $r$ of $X$ equals $p$, show that $(\hat{\beta}, S^2)$ is a complete sufficient statistic.

10. Consider a regression version of the two-sample problem in which

$$Y_i = \begin{cases} \beta_1 + \beta_2 x_i + \epsilon_i, & i = 1, \dots, n_1; \\ \beta_3 + \beta_4 x_i + \epsilon_i, & i = n_1 + 1, \dots, n_1 + n_2 = n, \end{cases}$$

with $\epsilon_1, \dots, \epsilon_n$ i.i.d. from $N(0, \sigma^2)$. Derive a $1 - \alpha$ confidence interval for $\beta_4 - \beta_2$, the difference between the two regression slopes.

11. *Inverse linear regression.* Consider the model for simple linear regression,

$$Y_i = \beta_1 + \beta_2(x_i - \overline{x}) + \epsilon_i, \qquad i = 1, \dots, n,$$

studied in Section 14.5.
   a) Derive a level $\alpha$-test of $H_0 : \beta_2 = 0$ versus $H_1 : \beta_2 \neq 0$.
   b) Let $y_0$ denote a "target" value for the mean of $Y$. The regression line $\beta_1 + \beta_2(x - \overline{x})$ achieves this value when the independent variable $x$ equals

$$\theta = \overline{x} + \frac{y_0 - \beta_1}{\beta_2}.$$

   Derive a level-$\alpha$ test of $H_0 : \theta = \theta_0$ versus $H_1 : \theta \neq \theta_0$. Hint: You may want to find a test first assuming $y_0 = 0$. After a suitable transformation, the general case should be similar.

c) Use duality to find a confidence region, first discovered by Fieller (1954), for $\theta$. Show that this region is an interval if the test in part (a) rejects $\beta_2 = 0$.

12. Find the mean and variance of the noncentral chi-square distribution on $p$ degrees of freedom with noncentrality parameter $\delta^2$.

*13. Consider a general linear model $Y \sim N(\xi, \sigma^2 I_n)$, $\xi \in \omega$, $\sigma^2 > 0$ with $\dim(\omega) = r$. Define $\psi = A\xi \in \mathbb{R}^q$ where $q < r$, and assume $A = AP$ where $P$ is the projection onto $\omega$, so that $\hat{\psi} = A\hat{\xi} = AY$, and that $A$ has full rank $q$.
   a) The $F$ test derived in Section 14.7 allows us to test $\psi = 0$ versus $\psi \neq 0$. Modify that theory and give a level-$\alpha$ test of $H_0 : \psi = \psi_0$ versus $H_1 : \psi \neq \psi_0$ with $\psi_0$ some constant vector in $\mathbb{R}^q$. Hint: Let $Y^* = Y - \xi_0$ with $\xi_0 \in \omega$ and $A\xi_0 = \psi_0$. Then the null hypothesis will be $H_0^* : A\xi^* = 0$.
   b) In the discussion of the Sheffé method for simultaneous confidence intervals,

$$\{\psi : (\psi - \hat{\psi})'(AA')^{-1}(\psi - \hat{\psi}) \leq qS^2 F_{\alpha,q,n-r}\}$$

   was shown to be a level $1 - \alpha$ confidence ellipse for $\psi$. Show that this confidence region can be obtained using duality from the family of tests in part (a).

*14. *Analysis of covariance.* Suppose

$$Y_{kl} = \beta_k + \beta_0 x_{kl} + \epsilon_{kl}, \qquad 1 \leq l \leq c, \quad 1 \leq k \leq p,$$

with the $\epsilon_{kl}$ i.i.d. from $N(0, \sigma^2)$ and the $x_{kl}$ known constants.
   a) If $\sum_{l=1}^c x_{kl} = 0$, $k = 1, \dots, c$, use the studentized maximum modulus distribution to derive simultaneous confidence intervals for $\beta_1, \dots, \beta_p$.
   b) If $\sum_{l=1}^c x_{1l} = \sum_{l=1}^c x_{2l} = \cdots = \sum_{l=1}^c x_{pl}$, use the studentized range distribution to derive simultaneous confidence intervals for all differences $\beta_i - \beta_j$, $1 \leq i < j \leq p$. Hint: The algebra will be simpler if you first reparameterize adding an appropriate multiple of $\beta_0$ to $\beta_1, \dots, \beta_p$.

*15. *Unbalanced one-way layout.* Suppose we have samples from $p$ normal populations with common variance, but that the sample sizes from the different populations are not the same, so that

$$Y_{ij} = \beta_i + \epsilon_{ij}, \qquad 1 \leq i \leq p, \quad j = 1, \dots, n_i,$$

with the $\epsilon_{ij}$ i.i.d. from $N(0, \sigma^2)$.
   a) Derive a level-$\alpha$ test of $H_0 : \beta_1 = \cdots = \beta_p$ versus $H_1 : \beta_i \neq \beta_j$ for some $i \neq j$.
   b) Use the Scheffé method to derive simultaneous confidence intervals for all contrasts $a_1\beta_1 + \cdots + a_p\beta_p$ with $a_1 + \cdots + a_p = 0$.

16. *Factorial experiments.* A "$2^4$" experiment is a factorial experiment to study the effects of four factors, each at two levels. The experiment has

$n = 16$ as the sample size (called the number of runs), with each run one of the 16 possible combinations of the four factors. Letting "+" and "−" be shorthand for $+1$ and $-1$, define vectors

$$x'_A = (+,+,+,+,+,+,+,+,-,-,-,-,-,-,-,-),$$

$$x'_B = (+,+,+,+,-,-,-,-,+,+,+,+,-,-,-,-),$$

$$x'_C = (+,+,-,-,+,+,-,-,+,+,-,-,+,+,-,-),$$

$$x'_D = (+,-,+,-,+,-,+,-,+,-,+,-,+,-,+,-),$$

and let $\mathbf{1}$ denote a column of ones. A "+1" for the $j$th entry of one of these vectors means that factor is set to the high level on the $j$th run, and a "−1" means the factor is set to the low level. So, for instance, on run 5, factors $A$, $C$, and $D$ are at the high level, and factor $B$ is at the low level. The vector $Y$ gives the responses for the 16 runs. In an additive model for the experiment,

$$Y = \mu\mathbf{1} + \theta_A x_A + \theta_B x_B + \theta_C x_C + \theta_D x_D + \epsilon,$$

with the unobserved error $\epsilon \sim N(0, \sigma^2 I)$. Parameters $\theta_A$, $\theta_B$, $\theta_C$, and $\theta_D$ are called the main effects for the factors.

a) Find the least squares estimates for the main effects, and give the covariance matrix for these estimators.
b) Find the UMVU estimator for $\sigma^2$.
c) Derive simultaneous confidence intervals for the main effects using the studentized maximum modulus distribution.

17. Consider the $2^4$ factorial experiment described in Problem 14.16. Let $x_{AB}$ be the elementwise product of $x_A$ with $x_B$,

$$x'_{AB} = (+,+,+,+,-,-,-,-,-,-,-,-,+,+,+,+),$$

and define $x_{AC}$, $x_{AD}$, $x_{BC}$, $x_{BD}$, and $x_{CD}$ similarly. A model with two-way interactions has

$$\begin{aligned} Y = \mu\mathbf{1} &+ \theta_A x_A + \theta_B x_B + \theta_C x_C + \theta_D x_D + \theta_{AB} x_{AB} \\ &+ \theta_{AC} x_{AC} + \theta_{AD} x_{AD} + \theta_{BC} x_{BC} + \theta_{BD} x_{BD} + \theta_{CD} x_{CD} + \epsilon, \end{aligned}$$

still with $\epsilon \sim N(0, \sigma^2 I)$. The additional parameters in this model are called two-way interaction effects. For instance, $\theta_{BD}$ is the interaction effect of factors $B$ and $D$.

a) Find least squares estimators for the $\binom{4}{2}$ interaction effects, and give the covariance matrix for these estimators.
b) Derive a test for the null hypothesis that all of the interaction effects are null, that is,

$$H_0 : x_{AB} = x_{AC} = x_{AD} = x_{BC} = x_{BD} = x_{CD} = 0,$$

versus the alternative that at least one of these effects is nonzero.

c) Use the Scheffé method to derive simultaneous confidence intervals for all contrasts of the two-way interaction effects.

18. *Bonferroni approach to simultaneous confidence intervals.*
   a) Suppose $I_1, \dots, I_k$ are $1 - \alpha$ confidence intervals for parameters $\eta_1 = g_1(\theta), \dots, \eta_k = g_k(\theta)$, and let $\gamma$ be the simultaneous coverage probability,
   $$\gamma = \inf_\theta P\big[\eta_i \in I_i, \forall i = 1, \dots, k\big].$$
   Use Boole's inequality (see Problem 1.7) to derive a lower bound for $\gamma$. For a fixed value $\alpha^*$, what choice for $\alpha$ will ensure $\gamma \ge 1 - \alpha^*$?
   b) Suppose the confidence intervals in part (a) are independent. In this case, what choice for $\alpha$ will ensure $\gamma \ge 1 - \alpha^*$?
   c) Consider one-way ANOVA with $c = 6$ observations from each of $p = 4$ populations. Compare the Bonferroni approach to simultaneous estimation of the differences $\beta_i - \beta_j$, $1 \le i < j \le 4$, with the approach based on the studentized range distribution. Because the Bonferroni approach is conservative, the intervals should be wider. What is the ratio of the lengths when $1 - \alpha^* = 95\%$? The 95th percentile for the studentized range distribution with parameters 4 and 20 is 3.96.

# 15

# Bayesian Inference: Modeling and Computation

This chapter explores several practical issues for a Bayesian approach to inference. The first section explores an approach used to specify prior distributions called *hierarchical modeling*, based on hyperparameters and conditioning. Section 15.2 discusses the robustness to the choice of prior distribution. Sections 15.4 and 15.5 deal with the Metropolis–Hastings algorithm and the Gibbs sampler, simulation methods that can be used to approximate posterior expectations numerically. As background, Section 15.3 provides a brief introduction to Markov chains. Finally, Section 15.6 illustrates how Gibbs sampling can be used in a Bayesian approach to image processing.

## 15.1 Hierarchical Models

Hierarchical modeling is a mixture approach to setting a prior distribution in stages. It arises when there is a natural family of prior distributions $\{\Lambda_\tau : \tau \in \mathbb{T}\}$ for our unknown parameter $\Theta$. If the value $\tau$ characterizing the prior distribution is viewed as an unknown parameter, then for a proper Bayesian analysis $\tau$ should be viewed as a realization of an unknown random variable $\mathcal{T}$. With this approach, there are two random parameters, $\mathcal{T}$ and $\Theta$. Because the distribution for our data $X$ depends only on $\Theta$, $\mathcal{T}$ is called a *hyperparameter*. If $G$ is the prior distribution for $\mathcal{T}$, then the Bayesian model is completely specified by

$$\mathcal{T} \sim G, \qquad \Theta | \mathcal{T} = \tau \sim \Lambda_\tau,$$

and

$$X | \mathcal{T} = \tau, \Theta = \theta \sim P_\theta.$$

Note that in this model,

$$P(\Theta \in B) = EP(\Theta \in B | \mathcal{T}) = E\Lambda_{\mathcal{T}}(B) = \int \Lambda_\tau(B)\, dG(\tau),$$

R.W. Keener, *Theoretical Statistics: Topics for a Core Course*, Springer Texts in Statistics, DOI 10.1007/978-0-387-93839-4_15, © Springer Science+Business Media, LLC 2010 

which shows that the prior distribution for $\Theta$ is now a mixture of distributions in the family $\{\Lambda_\tau : \tau \in \mathbb{T}\}$. Using this mixture, the hyperparameter $\mathcal{T}$ could be eliminated from the model, although in some situations this may be counterproductive.

*Example 15.1 (Compound Estimation).* As a first example, let us consider the compound estimation problem considered from an empirical Bayes perspective in Section 11.1. In that section, the parameters $\Theta_1, \ldots, \Theta_n$ were i.i.d. from $N(0, \tau^2)$, with the hyperparameter $\tau$ viewed as a constant. For a hierarchical Bayesian analysis, a prior distribution $G$ would be specified for $\mathcal{T}$. Then

$$\Theta|\mathcal{T} = \tau = N(0, \tau^2 I) \text{ and } X|\mathcal{T} = \tau, \Theta = \theta \sim N(\theta, I).$$

In this example, if we eliminated the hyperparameter $\mathcal{T}$ then the prior distribution for $\Theta$ would not be conjugate and we would not be able to take advantage of exact formulas based on that structure. If the dimension $n$ is large, numerical calculations may be a challenge. Keeping $\mathcal{T}$, smoothing leads to some simplifications. Using (11.1) the Bayes estimator for $\Theta$ is

$$\delta = E[\Theta|X] = E\big[E[\Theta|X, \mathcal{T}] \mid X\big] = XE\left[\frac{\mathcal{T}^2}{1+\mathcal{T}^2} \,\middle|\, X\right].$$

As noted in Section 11.1, given $\mathcal{T} = \tau$, $X_1, \ldots, X_n$ are i.i.d. from $N(0, 1+\tau^2)$, so the likelihood for $\tau$ has a simple form and compound inference can be accomplished using standard conditioning formulas to compute $E\big[\mathcal{T}^2/(1+\mathcal{T}^2) \mid X\big]$. Note that all of the integrals involved are one-dimensional; therefore a numerical approach is quite feasible.

*Example 15.2 (General Linear Model).* The general linear model was introduced in Chapter 14. Here we consider Bayesian inference with the error variance $\sigma^2$ assumed known. This leaves $\beta$ as the sole unknown parameter, viewed as a random vector $\mathcal{B}$, and

$$Y|\mathcal{B} = \beta \sim N(X\beta, \sigma^2 I),$$

with $X$ a known $n \times p$ matrix. For a prior distribution, proceeding as in the last example we might take

$$\mathcal{B} \sim N(0, \tau^2 I).$$

If $\kappa$ is the variance ratio $\sigma^2/\tau^2$, the posterior distribution is

$$\mathcal{B}|Y = y \sim N\left((X'X + \kappa I)^{-1}X'Y, \sigma^2(X'X + \kappa I)^{-1}\right).$$

As in the previous example, the posterior mean here still shrinks the UMVU $(X'X)^{-1}X'Y$ towards the origin, although the shrinkage "factor" is now the matrix $(X'X + \kappa I)^{-1}X'X$ instead of a constant.

The Bayes estimator $(X'X + \kappa I)^{-1}X'Y$ is also called a *ridge regression estimator*, originally suggested for numerical reasons to help regularize $X'X$.

As in the first example, the prior distributions $N(0, \tau^2 I)$ for $\mathcal{B}$ here are indexed by a hyperparameter $\tau$. A hierarchical approach would model $\tau$ as a random variable $\mathcal{T}$ from some distribution $G$.

*Example 15.3 (Testing).* Another class of examples arises if natural prior distributions for $\theta$ seem appropriate under competing scientific hypotheses. In these examples the hyperparameter $\mathcal{T}$ can be a discrete variable indexing the competing theories. As a concrete example, the standard model for one-way ANOVA has $Y_{ij} \sim N(\theta_i, \sigma^2)$, $i = 1, \dots, I$, $j = 1, \dots, n_i$, with all $n = n_1 + \cdots + n_I$ observations independent. If there is reason to believe that all the $\theta_i$ may be equal, then a prior distribution in which $\Theta_1 \sim N(\mu_\theta, \sigma_\theta^2)$ with all other $\Theta_i$ equal to $\Theta_1$ may be reasonable, so

$$\Theta \sim N(\mu_\theta \mathbf{1}, \sigma_\theta^2 \mathbf{1}\mathbf{1}').$$

If instead the means differ, the prior

$$\Theta \sim N(\mu_\theta \mathbf{1}, \sigma_\theta^2 I)$$

may be more natural. If $\mathcal{T} = 1$ or $2$ indexes these possibilities, then the mixture prior for $\Theta$ in a hierarchical model would be the convex combination

$$\Theta \sim P(\mathcal{T} = 1) N(\mu_\theta \mathbf{1}, \sigma_\theta^2 \mathbf{1}\mathbf{1}') + P(\mathcal{T} = 2) N(\mu_\theta \mathbf{1}, \sigma_\theta^2 I).$$

## 15.2 Bayesian Robustness

Ideally, in a Bayesian analysis the prior distribution is chosen to reflect a researcher's knowledge and beliefs about the unknown parameter $\Theta$. But in practice the choice is often dictated to some degree by convenience. Conjugate priors are particularly appealing here due to simple formulas for the posterior mean. Unfortunately, the convenience of such priors entails some risk.

To explore robustness issues related to the choice of the prior distribution in a very simple setting, consider a measurement error model with

$$X | \Theta = \theta \sim N(\theta, 1).$$

Suppose the true prior distribution $\Lambda$ is a $t$-distribution on three degrees of freedom with density

$$\lambda(\theta) = \frac{2}{\sqrt{3}\pi(1 + \theta^2/3)^2}.$$

Calculations with this prior distribution are a challenge, so it is tempting to use a conjugate normal distribution instead. The normal distribution $\Lambda_N =$

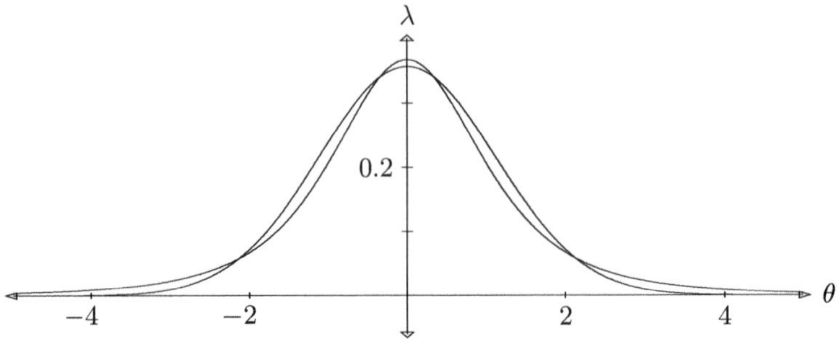


**Fig. 15.1.** Prior densities $\lambda$ and $\lambda_N$.

$N(0, 5/4)$ seems close to $\Lambda$. Densities $\lambda$ and $\lambda_N$ for $\Lambda$ and $\Lambda_N$, graphed in Figure 15.1, are quite similar; the largest difference between them is

$$\sup_{\theta} |\lambda(\theta) - \lambda_N(\theta)| = 0.0333,$$

and $|\Lambda(B) - \Lambda_N(B)| \leq 7.1\%$ for any Borel set $B$.

With squared error loss, the Bayes estimator with $\Lambda_N$ as the prior is

$$\delta_N(X) = \frac{5X}{9},$$

and its risk function is

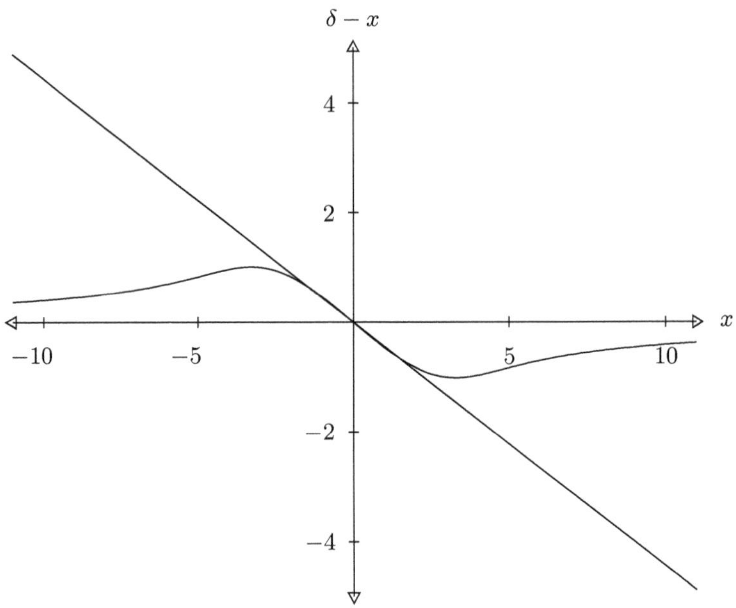


**Fig. 15.2.** Differences $\delta(x) - x$ and $\delta_N(x) - x$.

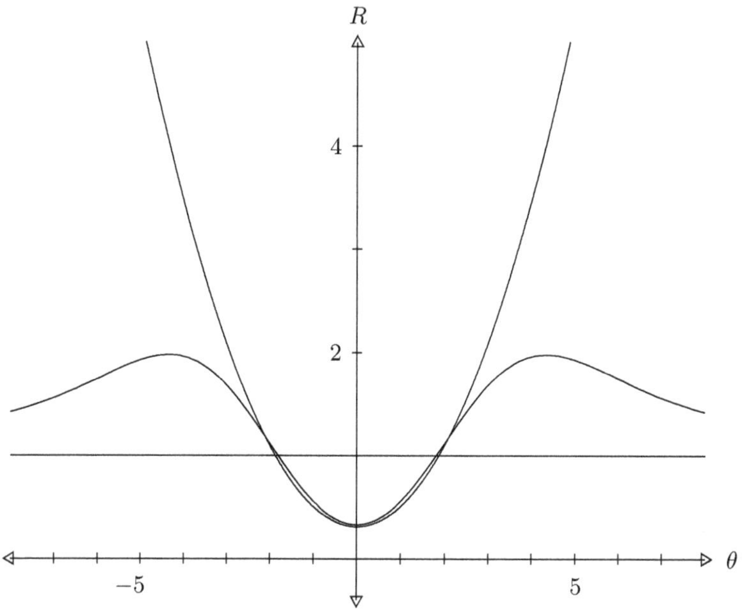


**Fig. 15.3.** Risk functions $R(\theta, \delta_U)$, $R(\theta, \delta_N)$, and $R(\theta, \delta)$.

$$R(\theta, \delta_N) = E\big[\big(\delta_N(X) - \theta\big)^2 \mid \Theta = \theta\big] = \frac{25 + 16\theta^2}{81}.$$

The integrated risk of $\delta_N$ under $\Lambda_N$ is

$$\int R(\theta, \delta_N)\, d\Lambda_N(\theta) = \frac{45}{81},$$

much better than the integrated risk of 1 for the UMVU $\delta_U(X) = X$. This improvement is achieved by the shrinkage towards zero, which improves the variance of the estimator and introduces little bias when $\theta$ is near zero.

The true prior $\Lambda$ has heavier tails, placing more weight on the region where $\delta_N$ is more heavily biased and its risk $R(\theta, \delta_N)$ is large. As one might guess, the Bayes estimator works to minimize risk for large $\theta$ with less shrinkage when $|X|$ is large. This can be seen in Figure 15.2, which graphs the differences $\delta - \delta_U$ and $\delta_N - \delta_U$. The estimators $\delta(X)$ and $\delta_N(X)$ are very similar if $|X| < 2$, but as $|X|$ increases, $\delta(X)$ moves closer to $X$. In fact,

$$\delta(x) = x - 4/x + o(1/x) \tag{15.1}$$

as $x \to \pm\infty$. Thus as $\theta \to \pm\infty$, the bias of $\delta$ tends to 0 and its risk function approaches 1, the risk of $\delta_U$, instead of increasing without bound as the quadratic risk of $\delta_N$. Figure 15.3 shows risk functions for $\delta_U$, $\delta_N$, and $\delta$.

With the true prior $\Lambda$, the integrated risk for $\delta_N$ is

$$\int R(\theta, \delta_N)\, d\Lambda(\theta) = \frac{73}{81} = 0.901,$$

almost as high as the risk of $X$. The best possible integrated risk with the true prior, achieved using the Bayes estimator $\delta$, is

$$\int R(\theta, \delta)\, d\Lambda(\theta) = 0.482,$$

which is 46.5% smaller than the integrated risk for $\delta_N$.

## 15.3 Markov Chains

**Definition 15.4.** *A sequence of random vectors $X_0$, $X_1$, $X_2$, ... taking values in a state space $(\mathcal{X}, \mathcal{A})$ form a (time homogeneous)* Markov chain *with* transition kernel[1] $Q$, *if*

$$P(X_{n+1} \in B | X_0 = x_0, \dots, X_n = x_n) = P(X_{n+1} \in B | X_n = x_n) = Q_{x_n}(B),$$

*for all $n \geq 1$, all $B \in \mathcal{A}$, and almost all $x_1, \dots, x_n$.*

Using smoothing, the joint distribution of the vectors in a Markov chain can be found from the initial distribution for $X_0$ and the transition kernel $Q$. The algebra involved can be easily described introducing some convenient notation. For a probability measure $\pi$ on $(\mathcal{X}, \mathcal{A})$, define a probability measure $\pi Q$ by

$$\pi Q(B) = \int Q_x(B)\, d\pi(x). \tag{15.2}$$

Note that if $\pi_n$ denotes the distribution for $X_n$, then by smoothing,

$$\begin{aligned}\pi_{n+1}(B) = P(X_{n+1} \in B) &= EP(X_{n+1} \in B | X_n) \\ &= EQ_{X_n}(B) = \int Q_x(B)\, d\pi_n(x) = \pi_n Q(B),\end{aligned}$$

and so

$$\pi_{n+1} = \pi_n Q. \tag{15.3}$$

A distribution $\pi$ is called *stationary* if $\pi = \pi Q$. Using (15.3), if the initial distribution $\pi_0$ for $X_0$ is stationary, then $\pi_2 = \pi_1 Q = \pi_1$. Further iteration shows that $\pi_1 = \pi_2 = \pi_3 = \cdots$, so in this case the random vectors in the chain are identically distributed.

If $Q$ and $\tilde{Q}$ are transition kernels on $\mathcal{X}$, define the product kernel $Q\tilde{Q}$ by

[1] The kernel $Q$ should satisfy the usual regularity conditions for stochastic transition kernels: $Q_x$ should be a probability measure on $(\mathcal{X}, \mathcal{A})$ for all $x \in \mathcal{X}$, and $Q_x(B)$ should be a measurable function of $x$ for all $B \in \mathcal{A}$.

$$(Q\tilde{Q})_x = Q_x\tilde{Q}. \tag{15.4}$$

Taking $Q^2 \stackrel{\text{def}}{=} QQ$, by smoothing,

$$\begin{aligned}
&P(X_{n+2} \in B|X_0 = x_0, \dots, X_n = x_n) \\
&\qquad = \int P(X_{n+2} \in B|X_0 = x_0, \dots, X_{n+1} = x_{n+1})\, dQ_{x_n}(x_{n+1}) \\
&\qquad = Q^2_{x_n}(B),
\end{aligned}$$

and so $Q^2$ can be viewed as the two-step transition kernel for the Markov chain. Similarly, the $k$-fold product $Q^k$ gives chances for $k$-step transitions:

$$Q^k_{x_n}(B) = P(X_{n+k} \in B|X_0 = x_0, \dots, X_n = x_n).$$

*Example 15.5.* If the state space $\mathcal{X}$ is finite, $\mathcal{X} = \{1, \dots, m\}$ say, then a distribution $\pi$ on $\mathcal{X}$ can be specified through its mass function given as a row vector

$$\boldsymbol{\pi} = \big[\pi\big(\{1\}\big), \dots, \pi\big(\{m\}\big)\big],$$

and the transition kernel $Q$ can be specified by a matrix $\mathbf{Q}$ with

$$\mathbf{Q}_{ij} = Q_i\big(\{j\}\big) = P(X_{n+1} = j|X_n = i).$$

If we let $\boldsymbol{\pi}_n$ be the row vector for the distribution of $X_n$, then

$$\begin{aligned}
[\boldsymbol{\pi}_{n+1}]_j = P(X_{n+1} = j) &= EP(X_{n+1} = j|X_n) = EQ_{X_n}(\{j\}) \\
&= \sum_i P(X_n = i)Q_i(\{j\}) = \sum_i [\boldsymbol{\pi}_n]_i \mathbf{Q}_{ij} = [\boldsymbol{\pi}_n\mathbf{Q}]_j.
\end{aligned}$$

So

$$\boldsymbol{\pi}_{n+1} = \boldsymbol{\pi}_n\mathbf{Q}.$$

Thus, if distributions are represented as row vectors and transition kernels are represented as matrices, the "multiplication" in (15.3) becomes ordinary matrix multiplication. Similarly, if $\mathbf{Q}$ and $\tilde{\mathbf{Q}}$ are matrices corresponding to transition kernels $Q$ and $\tilde{Q}$, then

$$(Q\tilde{Q})_i(\{j\}) = \int \tilde{Q}_x(\{j\} dQ_i(x) = \sum_x \mathbf{Q}_{ix}\tilde{\mathbf{Q}}_{xj} = [\mathbf{Q}\tilde{\mathbf{Q}}]_{ij}$$

and the matrix representing $Q\tilde{Q}$ is simply the matrix product $\mathbf{Q}\tilde{\mathbf{Q}}$.

For a finite Markov chain, because the mass function for $Q_i$ sums to 1, we have

$$\sum_j \mathbf{Q}_{ij} = [\mathbf{Q1}]_i = 1,$$

and hence

$$\mathbf{Q1} = \mathbf{1}.$$

This shows that $\mathbf{1}$ is a right-eigenvector for $\mathbf{Q}$ with unit eigenvalue. Since $\pi Q$ is given by matrix multiplication, a probability distribution $\pi$ will be stationary if

$$\pi = \pi \mathbf{Q},$$

that is, if $\pi$ is a left-eigenvector of $\mathbf{Q}$ with unit eigenvalue. In general, if $\lambda$ is an eigenvalue for $\mathbf{Q}$, then $|\lambda| \le 1$,

Convergence properties for finite Markov chains are commonly related to the Frobenius theory for positive matrices, covered in Appendix 2 of Karlin and Taylor (1975). If $A$ is an $n \times n$ matrix with eigenvalues $\lambda_1, \ldots, \lambda_n$, then its *spectral radius* is defined as $r = \max\{|\lambda_1|, \ldots, |\lambda_n|\}$.

**Theorem 15.6 (Perron–Frobenius).** *Let $A$ be an $n \times n$ matrix with non-negative entries and spectral radius $r$, and assume that $A^m > 0$ for some $m > 0$. Then*

1. *The spectral radius $r$ is a simple[2] eigenvalue for $A$, $r > 0$, and if $\lambda$ is any other eigenvalue, $|\lambda| < r$.*
2. *There are left- and right-eigenvectors associated with $r$ with positive entries. Specifically, there is a row vector $v$ and a column vector $w$ with $v > 0$, $w > 0$, $vA = rv$, and $Aw = rw$.*
3. *If $v$ is normalized so that its entries sum to 1 and $w$ is normalized so that $vw = 1$, then*
$$r^{-n} A^n \to wv$$
*as $n \to \infty$.*
4. *The spectral radius $r$ satisfies*
$$\min_i \sum_j A_{ij} \le r \le \max_i \sum_j A_{ij}.$$

To characterize eigenvalues and convergence properties for finite chains in regular cases, we need a few definitions. Let

$$L_x(A) = P(X_n \in A, \exists n \ge 1 | X_0 = x),$$

the chance the chain ever visits $A$ if it starts at $x$. States $i$ and $j$ for a finite chain are said to *communicate* if the chain can move from either of the states to the other; that is, if $L_i(\{j\}) > 0$ and $L_j(\{i\}) > 0$. If all of the states communicate, the chain is called *irreducible.* The chain is called *periodic* if $\mathcal{X}$ can be partitioned into sets $\mathcal{X}_1, \ldots, \mathcal{X}_k$, $k \ge 2$, and the process cycles between these sets: if $i \in \mathcal{X}_j$, $1 \le j \le k-1$,

$$Q_i(\mathcal{X}_{j+1}) = P(X_1 \in \mathcal{X}_{j+1} | X_0 = i) = 1,$$

and if $i \in \mathcal{X}_k$, $Q_i(\mathcal{X}_1) = 1$.

---

[2] An eigenvalue is simple if it is a simple root of the characteristic equation. In this case, eigenspaces (left or right) will be one-dimensional.

For finite chains, properties needed for simulation arise when the chain is irreducible and aperiodic. In this case, it is not hard to argue that $\mathbf{Q}^m > 0$, so $\mathbf{Q}$ satisfies the conditions of the Perron–Frobenius theorem. Because $\mathbf{Q1} = \mathbf{1}$, the spectral radius $r$ for $\mathbf{Q}$ must be $r = 1$ by the fourth assertion of the theorem. Because $r$ is a simple eigenvalue, there will exist a unique corresponding left-eigenvector $\boldsymbol{\pi}$ with entries summing to 1, corresponding to a unique stationary distribution $\pi$ for the chain. The mass function for this distribution can be found by solving the linear equations

$$\boldsymbol{\pi} = \boldsymbol{\pi}\mathbf{Q} \text{ and } \boldsymbol{\pi}\mathbf{1} = 1.$$

By the third assertion in the theorem,

$$\mathbf{Q}^n \to \mathbf{1}\boldsymbol{\pi}$$

as $n \to \infty$. In probabilistic terms, this means that

$$X_n \Rightarrow \pi$$

as $n \to \infty$, regardless of the initial state (or distribution) for the chain. The stationary distribution also gives the long run proportion of time the process spends in the various states, and the following law of large numbers holds:

$$\frac{1}{n}\sum_{i=1}^{n} f(X_i) \to \int f\,d\pi,$$

with probability one, regardless of the initial distribution $\pi_0$. Using this, the value for $\int f\,d\pi$ can be approximated by simulation, having a computer generate the chain numerically and averaging the values for $f$. For an extended discussion of finite chains, see Kemeny and Snell (1976).

The theory for Markov chains when $\mathcal{X}$ is infinite but denumerable is similar, although irreducible and aperiodic chains without a stationary distribution are possible; see Karlin and Taylor (1975) or another introduction to stochastic processes. When $\mathcal{X}$ is not denumerable, the relevant theory, presented in Nummelin (1984) or Meyn and Tweedie (1993), is much more complicated. For simulation, the most appealing notion of regularity might be Harris recurrence. Tierney (1994) gives convergence results for the Metropolis–Hastings algorithm and the Gibbs sampler, discussed in the next two sections.

## 15.4 Metropolis–Hastings Algorithm

In a Bayesian model, if $\Theta$ has density $\lambda$ and the conditional density of $X$ given $\Theta = \theta$ is $p_\theta$, then the posterior density of $\Theta$ given $X = x$ is proportional (in $\theta$) to

$$\lambda(\theta)p_\theta(x).$$

To compute the posterior density $\lambda(\theta|x)$ we should divide this function by its integral,

$$\int \lambda(\theta)p_\theta(x)\,d\theta.$$

In practice, this integral may be difficult to evaluate, explicitly or numerically, especially if $\Theta$ is multidimensional. The Metropolis–Hastings algorithm is a simulation method that allows approximate sampling from this posterior distribution without computing the normalizing constant. Specifically, the algorithm gives a Markov chain that has the target law as its stationary distribution.

To describe the transition kernel $Q$ for the Markov chain, let $\pi$ denote a target distribution on some state space $\mathcal{X}$ with density $f$ with respect to a dominating measure $\mu$. The chain runs by accepting or rejecting potential states generated using a transition kernel $J$ with densities $j_x = dJ_x/d\mu$. The chances for accepting or rejecting a new value are based on a function $r$ given by

$$r(x_0, x^*) = \frac{f(x^*)/j_{x_0}(x^*)}{f(x_0)/j_{x^*}(x_0)}.$$

Note that $r$ can be computed if $f$ is only known up to a proportionality constant. Let $X_0$ denote the initial state of the chain. Given $X_0 = x_0$, a variable $X^*$ is drawn from $J_{x_0}$, so

$$X^*|X_0 = x_0 \sim J_{x_0}.$$

The next state for the Markov chain, $X_1$, will be either $X_0$ or $X^*$, with

$$P(X_1 = X^*|X_0 = x_0, X^* = x^*) = \min\{r(x_0, x^*), 1\}.$$

Thus

$$\begin{aligned} P(X_1 \in A|X_0 &= x_0, X^* = x^*) \\ &= 1_A(x^*)\min\{r(x_0, x^*), 1\} + 1_A(x_0)\bigl(1 - \min\{r(x_0, x^*), 1\}\bigr). \end{aligned}$$

Integrating against the conditional distribution for $X^*$ given $X_0 = x_0$, by smoothing

$$\begin{aligned} Q_{x_0}(A) &\stackrel{\text{def}}{=} P(X_1 \in A|X_0 = x_0) \\ &= 1_A(x_0) + \int_A \min\{r(x_0, x^*), 1\} j_{x_0}(x^*)\,d\mu(x^*) \\ &\quad - 1_A(x_0)\int \min\{r(x_0, x^*), 1\} j_{x_0}(x^*)\,d\mu(x^*). \end{aligned}$$

To check that $\pi$ is a stationary distribution for the chain with transition kernel $Q$ we need to show that if $X_0 \sim \pi$, then $X_1 \sim \pi$. If $X_0 \sim \pi$, then by smoothing,

$$P(X_1 \in A) = EP(X_1 \in A|X_0) = EQ_{X_0}(A) = \int Q_{x_0}(A)\,d\pi(x_0).$$

Because $\int 1_A(x_0)\,d\pi(x_0) = \pi(A)$, this will hold if

$$\iint \min\{r(x_0,x^*),1\}1_A(x^*)f(x_0)j_{x_0}(x^*)\,d\mu(x_0)\,d\mu(x^*)$$
$$= \iint \min\{r(x_0,x^*),1\}1_A(x_0)f(x_0)j_{x_0}(x^*)\,d\mu(x_0)\,d\mu(x^*).$$

Using the formula for $r$, this equation becomes

$$\iint \min\{f(x^*)j_{x^*}(x_0), f(x_0)j_{x_0}(x^*)\}1_A(x^*)\,d\mu(x_0)\,d\mu(x^*)$$
$$= \iint \min\{f(x^*)j_{x^*}(x_0), f(x_0)j_{x_0}(x^*)\}1_A(x_0)\,d\mu(x_0)\,d\mu(x^*),$$

which holds by Fubini's theorem.

Convergence of the Metropolis–Hastings algorithm is discussed in Tierney (1994). Turning to practical considerations, several things should be considered in choosing the jump kernel $J$. First, it should be easy to sample values from $J_x$, and the formula to compute $r$ should be simple. In addition, $J$ should move easily to all relevant areas of the state space and jumps should not be rejected too often.

## 15.5 Gibbs Sampler

The Gibbs sampler is based on alternate sampling from the conditional distributions for the target distribution $\pi$. If $(X,Y) \sim \pi$, let $R$ denote the conditional distribution of $X$ given $Y$, and let $\tilde{R}$ denote the conditional distribution of $Y$ given $X$. If $(X_0,Y_0)$ is the initial state for the Markov chain, then we find $(X_1,Y_1)$ by first sampling $X_1$ from $R$ and then drawing $Y_1$ from $\tilde{R}$. Specifically,

$$X_1|X_0 = x_0, Y_0 = y_0 \sim R_{y_0}$$

and

$$Y_1|X_0 = x_0, Y_0 = y_0, X_1 = x_1 \sim \tilde{R}_{x_1}.$$

Continuing in this fashion, $(X_i,Y_i)$, $i \ge 0$ is a Markov chain.

The Gibbs sampler can be easily extended to joint distributions for more than two variables (or vectors). If we are interested in simulating the joint distribution of $X$, $Y$, and $Z$, say, we could generate a new $X$ sampling from the conditional distribution for $X$ given $Y$ and $Z$, then generate a new $Y$ from the conditional distribution of $Y$ given $X$ and $Z$, then generate a new $Z$ from the conditional distribution of $Z$ given $X$ and $Y$, and so on.

The Gibbs sampler is useful in simulation mainly through dimension reduction. Sampling from a univariate distribution is typically much easier than multivariate sampling. If a better approach is not available, univariate simulation is possible using the probability integral transformation whenever the cumulative distribution function is available. Also, note that if the target distribution $\pi$ is absolutely continuous with density $f$ proportional to a known function $g$, then, in order to compute $f$ from $g$, we would normalize $g$ dividing by its integral,

$$\iint g(x,y)\,dx\,dy.$$

In contrast, to find conditional densities we would normalize $g$ dividing by

$$\int g(x,y)\,dx \quad \text{or} \quad \int g(x,y)\,dy.$$

The normalization for the conditional distributions needed for Gibbs sampling involves univariate integration instead of the multiple integration needed to find the joint density.

To check that the Gibbs sampler has $\pi$ as a stationary distribution, let $\pi_X$ and $\pi_Y$ denote the marginal distributions of $X$ and $Y$ when $(X,Y) \sim \pi$. By smoothing,

$$\begin{aligned} P\big[(X,Y) \in A\big] &= EE\big[1_A(X,Y) \bigm| Y\big] \\ &= \iint 1_A(x,y)\,dR_y(x)\,d\pi_Y(y), \end{aligned} \tag{15.5}$$

and reversing $X$ and $Y$,

$$P\big[(X,Y) \in A\big] = \iint 1_A(x,y)\,d\tilde{R}_x(y)\,d\pi_X(x). \tag{15.6}$$

Suppose we start the chain with distribution $\pi$, so $(X_0,Y_0) \sim (X,Y)$. Then by smoothing, since $Y_0 \sim \pi_Y$ and the conditional distribution of $X_1$ given $Y_0$ is $R$,

$$P\big[(X_1,Y_0) \in A\big] = EE\big[1_A(X_1,Y_0) \bigm| Y_0\big] = \iint 1_A(x,y)\,dR_y(x)\,d\pi_Y(y).$$

Comparing this with (15.5), $(X_1,Y_0) \sim \pi$. In particular, $X_1 \sim \pi_X$. Smoothing again, since $\tilde{R}$ is the conditional distribution of $Y_1$ given $X_1$,

$$\begin{aligned} P\big[(X_1,Y_1) \in A\big] &= EE\big[1_A(X_1,Y_1) \bigm| X_1\big] \\ &= \iint 1_A(x,y)\,d\tilde{R}_x(y)\,d\pi_X(x). \end{aligned}$$

Comparing this with (15.6), $(X_1,Y_1) \sim (X,Y) \sim \pi$, which shows that $\pi$ is stationary.

## 15.6 Image Restoration

Gibbs sampling was introduced in a landmark paper by Geman and Geman (1984) on Bayesian image restoration. The example here is based on this work with a particularly simple form for the prior distribution. The unknown image is represented by unknown greyscale values $\Theta_z$ at $nm$ pixels $z = (i,j)$ in a rectangular grid $\mathcal{T}$:

$$z = (i,j) \in \mathcal{T} \stackrel{\text{def}}{=} \{1,\dots,m\} \times \{1,\dots,n\}.$$

Given $\Theta = \theta$, observed data $X_z$, $z \in \mathcal{T}$, are independent with

$$X_z \sim N(\theta_z, \sigma^2), \qquad z \in \mathcal{T}, \tag{15.7}$$

and $\sigma^2$ considered known.

In real images, greyscale values at nearby pixels are generally highly correlated, whereas well-separated pixels are nearly uncorrelated. For good performance, correlations for the prior distribution for $\Theta$ should have similar form. For simplicity we restrict attention here to normal distributions. In one dimension, the autoregressive model in Example 6.4 has these features; it is not hard to show that $\mathrm{Cor}(X_i, X_j) = \rho^{|i-j|}$. The joint density in that example is proportional to

$$\exp\left[-\frac{a}{2}\sum_{i=1}^{n} x_i^2 + b\sum_{i=1}^{n-1} x_i x_{i+1}\right],$$

where the constants $a$ and $b$ satisfy $|b| < a/2$, which ensures that this expression is integrable. To construct a prior density for $\Theta$ with a similar form, call a set of two pixels $\{z_1, z_2\} \in \mathcal{T}^2$ an *edge* if $\|z_1 - z_2\| = 1$ and let $\mathcal{E}$ denote the set of all edges. The priors of interest here have form

$$\lambda(\theta) \propto_\theta \exp\left[-\frac{a}{2}\sum_z \theta_z^2 + b \sum_{\{z_1,z_2\}\in\mathcal{E}} \theta_{z_1}\theta_{z_2}\right]. \tag{15.8}$$

For integrability, assume $|b| < a/4$. Neglecting effects that arise near the edge of the image,

$$\sigma_\Theta^2 \stackrel{\text{def}}{=} \mathrm{Var}(\Theta_z) = \frac{2K\big(\sqrt{\eta}\big)}{a\pi},$$

where $\eta = 4b/a$ and $K$ is the complete elliptic integral of the first kind, given[3] by

$$K(x) = \int_0^{\pi/2} \frac{d\phi}{\sqrt{1 - x^2\sin^2(\phi)}} = \frac{\pi}{2}\sum_{n=0}^{\infty}\left(\frac{x^2}{16}\right)^n \binom{2n}{n}^2.$$

Also, if $z_1$ and $z_2$ are adjacent pixels, then

[3] Different sources use slightly different definitions.

$$\rho \stackrel{\text{def}}{=} \operatorname{Cor}(\Theta_{z_1}, \Theta_{z_2}) = \frac{2K(\sqrt{\eta}) - \pi}{2\eta K(\sqrt{\eta})}.$$

Solving,

$$2K(\sqrt{\eta}) = \frac{\pi}{1 - \rho\eta}.$$

As $\eta$ increases to 1, $K(\sqrt{\eta})$ increases without bound, and so $\rho \uparrow 1$ as $\eta \uparrow 1$. By results in Cody (1965), $K(\sqrt{\eta}) \sim \log[4/(1-\eta)]$ as $\eta \uparrow 1$. From this, when $\rho$ is near 1,

$$\eta \approx 1 - 4\exp\left[-\frac{\pi}{2(1-\rho)}\right],$$

and so

$$a \approx \frac{1}{\sigma_\Theta^2(1-\rho)} \quad \text{and} \quad b \approx \frac{a}{4} - a\exp\left[-\frac{\pi}{2(1-\rho)}\right], \tag{15.9}$$

relating $a$ and $b$ in the prior density to $\sigma_\Theta^2$ and $\rho$.

The prior density in (15.8) has an interesting and useful structure. Suppose we let $z_1, \dots, z_4$ denote the pixels adjacent to some pixel $z$. If we fix the values for $\theta$ at these pixels, then $\lambda$ factors into a function of $\theta_z$ and a function of $\theta$s at the remaining $nm - 5$ pixels. From this, given $\Theta_{z_1}, \dots, \Theta_{z_4}$, $\Theta_z$ is conditionally independent of the image values at the remaining pixels. This conditional independence might be considered a Markov property, and the distribution for $\Theta$ here is called a *Markov random field.* Building on this idea, let us divide $\mathcal{T}$ into "even" and "odd" pixels:

$$\mathcal{T}_1 = \{(i,j) \in \mathcal{T} : i + j \text{ odd}\} \quad \text{and} \quad \mathcal{T}_2 = \{(i,j) \in \mathcal{T} : i + j \text{ even}\}.$$

If we fix the values for $\theta_z$, $z \in \mathcal{T}_1$, then $\lambda(\theta)$ has form $\prod_{z \in \mathcal{T}_2} f_z(\theta_z)$. Thus given $\Theta_z$, $z \in \mathcal{T}_1$, the $\Theta_z$, $z \in \mathcal{T}_2$, are conditionally independent. Below we show that posterior distributions have this same structure.

Taking $\tau = 1/\sigma^2$, the "precision" of the $X_z$, by (15.7) the density for $X$ given $\Theta = \theta$ is

$$p_\theta(x) \propto_\theta \exp\left[-\frac{\tau}{2}\sum_{z \in \mathcal{T}} \theta_z^2 + \tau \sum_{z \in \mathcal{T}} \theta_z x_z\right].$$

Therefore the conditional density for $\Theta$ given $X = x$ is

$$\lambda(\theta|x) = c(x)\exp\left[-\frac{a+\tau}{2}\sum_{z \in \mathcal{T}} \theta_z^2 + \tau \sum_{z \in \mathcal{T}} \theta_z x_z + b \sum_{\{z_1,z_2\} \in \mathcal{E}} \theta_{z_1}\theta_{z_2}\right], \tag{15.10}$$

with $c(x)$ chosen as usual so that $\int \cdots \int \lambda(\theta|x)\, d\theta = 1$. With the quadratic structure, this conditional density must be normal. The mean can be found solving linear equations to minimize the quadratic function of $\theta$ in the exponent, and the covariance is minus one half the inverse of the matrix defining the quadratic form in $\theta$. With modern computing these calculations may be possible, but the $nm \times nm$ matrices involved, although sparse, are very large. Thus Gibbs sampling may be an attractive alternative.

To implement Gibbs sampling, we need to determine the relevant conditional distributions. Let $\mathcal{N}_z$ denote the pixels neighboring $z \in \mathcal{T}$,

$$\mathcal{N}_z = \{\tilde{z} \in \mathcal{T} : \|\tilde{z} - z\| = 1\},$$

and define

$$S_z = \sum_{\tilde{z} \in \mathcal{N}_z} \theta_{\tilde{z}},$$

the sum of the $\theta$ values at pixels neighboring $z$. Isolating the terms in (15.10) that depend on $\theta_z$, $\lambda(\theta|x)$ is

$$\exp\left[-\frac{a+\tau}{2}\theta_z^2 + \theta_z\bigl(\tau x_z + bS_z\bigr)\right] \tag{15.11}$$

times a term that is functionally independent of $\theta_z$. As a function of $\theta_z$, the expression in (15.11) is proportional to a normal density with variance $1/(a+\tau)$ (or precision $a+\tau$) and mean

$$\frac{\tau x_z + bS_z}{a+\tau}.$$

With the product structure, given $\Theta_{\tilde{z}} = \theta_{\tilde{z}}$, $\tilde{z} \in \mathcal{T}_1$, the $\Theta_z$ for $z \in \mathcal{T}_2$ are conditionally independent with

$$\Theta_z | \Theta_{\tilde{z}} = \theta_{\tilde{z}}, \tilde{z} \in \mathcal{T}_1, X = x \sim N\left(\frac{\tau x_z + bS_z}{a+\tau}, \frac{1}{a+\tau}\right), \qquad z \in \mathcal{T}_2, \tag{15.12}$$

with a similar result for the conditional distribution of the image given values at pixels in $\mathcal{T}_2$.

The conditional distributions just described are exactly what we need to implement Gibbs sampling from the posterior distribution of the image. Starting with image values at pixels in $\mathcal{T}_1$, independent values at pixels in $\mathcal{T}_2$ would be drawn using the conditional marginal distributions in (15.12). Reversing the sets $\mathcal{T}_1$ and $\mathcal{T}_2$, values at pixels in $\mathcal{T}_1$ would next be drawn independently from the appropriate normal distributions. Iterating, the posterior mean should be close to the average values in the simulation.

To illustrate how this approach works in practice, let us consider a numerical example. The true $\theta$ is a $99 \times 64$ image of the letter A, displayed as the first image in Figure 15.4. The value for $\theta$ at "dark" pixels is 0 and the value at "light" pixels is 5. The second image in Figure 15.4 shows the raw data $X$, drawn from a normal distribution with mean $\theta$ and covariance $9I$, so $\sigma = 3$.

By symmetry, the mean for the prior distribution $\lambda$ in (15.8) is zero. But the average greyscale value in the true image $\theta$ is 4.1477, which is significantly different. It seems natural, although a bit ad hoc, to center the raw data by

**Fig. 15.4.** Left to right: True image $\theta$, raw data $X$, undersmoothed, matched covariance, oversmoothed.

subtracting the overall average $\overline{X} = \sum_{z\in\mathcal{T}} X_z/(nm)$ from the greyscale value at each pixel, before proceeding with our Bayesian analysis. After processing, we can add $\overline{X}$ back to the posterior mean. Results doing this should be similar to those obtained using a normal prior with mean $\overline{X}$ for $\Theta$, or (more properly) following a hierarchical approach in which the mean for $\Theta$ is an additional hyperparameter and this parameter has a reasonably diffuse distribution.

Empirical estimates for $\sigma_\Theta^2$ and $\rho$, based on the true $\theta$, are 3.5350 and 0.8565, respectively. Using (15.9) these values correspond to $a = 1.971$ and $b = 0.493$. With these values, the prior variance and covariance between values at adjacent pixels will match the empirical values for the true image, and it seems reasonable to hope for excellent restoration using this prior. Of course in practice the true image and associated moments are unknown. For comparison, we have also done an analysis with two other priors. In both, we take $\sigma_\Theta^2 = 3.5350$, matching the empirical variance for the true $\theta$. But in one of the priors we take $\rho = 0.70$ and in the other we take $\rho = 0.95$. Since higher values for $\rho$ give smoother images $\Theta$, we anticipate that the posterior mean will undersmooth $X$ when $\rho = 0.70$ and oversmooth $X$ when $\rho = 0.95$. The final three images in Figure 15.4 show posterior means for these three prior distributions, which can be found by Gibbs sampling.[4]

Evaluating the performance of an image restoration method is perhaps a bit subtle. In Figure 15.4, the three posterior means look more like the true image than the raw data, but the raw data $X$ seems visually almost as "clear." One measure of performance might be the average mean square error, MSE $= \sum_z(\hat{\theta} - \theta)^2/(nm)$. The raw data $X$ is the UMVU estimator here and has MSE $= 9.0016$, very close to the expected MSE which is exactly 9, the common variance of the $X_z$. Mean squared errors for the posterior means are 0.9786, 1.0182, and 1.5127 for the undersmoothed, matched, and oversmoothed priors, respectively. Surprisingly, by this measure the image using the undersmoothed prior is a bit better than the image that matches the covariance between values at adjacent pixels.

[4] Actually, taking advantage of the normal structure in this particular example, means for the Gibbs sampling Markov chain can be found recursively and converge to the posterior mean. This approach was used to produce the images in Figure 15.4, iterating the recursion numerically. See Problem 15.9.

## 15.7 Problems

1. Consider Example 15.3 in the balanced case where $n_1 = \cdots = n_I$. Derive a formula for
$$P[\mathcal{T} = 1 | Y_{ij}, i = 1, \dots, I, j = 1, \dots, n_j].$$
2. Verify the relation (15.1).
3. Find the stationary distribution for a Markov chain on $\mathcal{X} = \mathbb{R}$ with kernel $Q$ given by
$$Q_x = N(cx, 1), \qquad x \in \mathbb{R},$$
where $c$ is a fixed constant with $|c| < 1$.
4. Let $Q$ be the transition kernel for a Markov chain on $\mathcal{X} = \{0, 1, 2, \dots\}$ given by
$$\mathbf{Q}_{ij} = \begin{cases} 1, & i = 0, j = 1; \\ 1/2, & i > 0, j = i + 1; \\ 1/2, & i > 0, j = 0; \\ 0, & \text{otherwise}, \end{cases}$$
where $\mathbf{Q}_{ij} \stackrel{\text{def}}{=} Q_i(\{j\})$. (Here $\mathbf{Q}$ might naturally be viewed as an infinite-dimensional transition matrix.) So at each stage, this chain has an equal chance of increasing by one or falling back to zero. The Markov chain with transition kernel $Q$ has a unique stationary distribution $\pi$. Find the mass function for $\pi$.
5. Consider using the Metropolis–Hastings algorithm to sample from the standard normal distribution. Assume that the jump kernel $J$ is given by
$$J_x = N(x/2, 1), \qquad x \in \mathbb{R}.$$
Give a formula for $r$ and find the chance the chain does not move when it is at position $x$; that is, $P(X_1 = x | X_0 = x)$.
6. Consider using the Metropolis–Hastings algorithm to sample from a discrete distribution on $\mathcal{X} = \{1, \dots, 5\}$ with mass function
$$f(x) = cx, \qquad x = 1, \dots, 5,$$
for some constant $c$. Suppose the transition matrix for the jump kernel $J$ is
$$\mathbf{J} = \begin{pmatrix} 1/2 & 1/2 & 0 & 0 & 0 \\ 1/2 & 0 & 1/2 & 0 & 0 \\ 0 & 1/2 & 0 & 1/2 & 0 \\ 0 & 0 & 1/2 & 0 & 1/2 \\ 0 & 0 & 0 & 1/2 & 1/2 \end{pmatrix}.$$
Find the transition matrix $\mathbf{Q}$ for the Metropolis–Hastings chain. Check that the vector $\boldsymbol{\pi}$ corresponding to the mass function $f$ is a left-eigenvector for $\mathbf{Q}$ with unit eigenvalue.

7. Consider Gibbs sampling with target distribution

$$\pi = N\left(\begin{pmatrix}0\\0\end{pmatrix}, \begin{pmatrix}1 & \rho\\ \rho & 1\end{pmatrix}\right).$$

Find $\pi_1$ and $\pi_2$ if $X_0 = x$ and $Y_0 = y$, so that $\pi_0$ is a point mass at $(x, y)$.

8. Consider Gibb's sampling for an absolutely continuous distribution with density

$$f(x,y) = \begin{cases} ce^{-x-y-xy}, & x > 0, y > 0; \\ 0, & \text{otherwise,} \end{cases}$$

for some constant $c$. Find the joint density of $X_1$ and $Y_1$ if $X_0 = Y_0 = 1$.

9. Consider using Gibbs sampling for the posterior distribution of an image in the model considered in Section 15.6. Let $\Theta(n)$, $n \geq 0$, be images generated by Gibbs simulation, and let $\mu(n)$ denote the mean of $\Theta(n)$. Use smoothing and (15.12) to derive an equation expressing $\mu(n+1)$ as a function of $\mu(n)$. These means $\mu(n)$ converge to the true mean of the posterior distribution as $n \to \infty$, so the equation you derive can be used to find the posterior mean by numerical recursion.

10. Consider Bayesian image restoration for the model considered in Section 15.6 when the prior density has form

$$\lambda(\theta) \propto_\theta \exp\left[-\frac{a}{2}\sum_z \theta_z^2 + b \sum_{\{z_1,z_2\}\in\mathcal{E}_1} \theta_{z_1}\theta_{z_2} + c \sum_{\{z_1,z_2\}\in\mathcal{E}_2} \theta_{z_1}\theta_{z_2}\right],$$

where

$$\mathcal{E}_1 = \{(z_1,z_2) \in \mathcal{T}^2 : \|z_1 - z_2\| = 1\}$$

and

$$\mathcal{E}_2 = \{(z_1,z_2) \in \mathcal{T}^2 : \|z_1 - z_2\| = \sqrt{2}\}.$$

With this prior, it is natural to partition the pixels $\mathcal{T}$ into four sets, $\mathcal{T}_{00}$, $\mathcal{T}_{01}$, $\mathcal{T}_{10}$, and $\mathcal{T}_{11}$, given by

$$\mathcal{T}_{ab} = \{(i,j) \in \mathcal{T} : i \equiv a \pmod 2,\ j \equiv b \pmod 2\},$$

for $a = 0, 1$ and $b = 0, 1$. Describe how to implement Gibb's sampling from the posterior distribution in this case. As in (16.11), for $z \in \mathcal{T}_{00}$ find the conditional distribution of $\Theta_z$ given $\Theta_{\tilde z} = \theta_{\tilde z}$, $\tilde z \in \mathcal{T}_{01} \cup \mathcal{T}_{10} \cup \mathcal{T}_{11}$, and $X = x$.

# 16

# Asymptotic Optimality[1]

In a rough sense, Theorem 9.14 shows that the maximum likelihood estimator achieves the Cramér–Rao lower bound asymptotically, which suggests that it is asymptotically fully efficient. In this chapter we explore results on asymptotic optimality formalizing notions of asymptotic efficiency and showing that maximum likelihood or similar estimators are efficient in regular cases. Notions of asymptotic efficiency are quite technical and involved, and the treatment here is limited. Our main goal is to derive a result from Hájek (1972), Theorem 16.25 below, which shows that the maximum likelihood estimator is locally asymptotically minimax.

To motivate later results, the first section begins with a curious example that shows why simple approaches in this area fail.

## 16.1 Superefficiency

Suppose $X_1, X_2, \dots$ are i.i.d. with common density $f_\theta$, $\theta \in \Omega$. By the Cramér–Rao lower bound, if $\delta_n = \delta_n(X_1, \dots, X_n)$ is an unbiased estimator of $g(\theta)$, then

$$\mathrm{Var}_\theta(\delta_n) \geq \frac{\left[g'(\theta)\right]^2}{nI(\theta)},$$

where

$$I(\theta) = E_\theta\left(\frac{\partial}{\partial\theta}\log f_\theta(X_i)\right)^2$$

is the Fisher information for a single observation. Suppose we drop the assumption that $\delta_n$ is unbiased, but assume it is asymptotically normal:

[1] From Section 16.3 on, the results in this chapter are very technical. The material on contiguity in Section 16.2 is needed for the discussion of generalized likelihood ratio tests in Chapter 17, but results from the remaining sections are not used in later chapters.

R.W. Keener, *Theoretical Statistics: Topics for a Core Course*, Springer Texts in Statistics, DOI 10.1007/978-0-387-93839-4_16, © Springer Science+Business Media, LLC 2010

$$\sqrt{n}\big(\delta_n - g(\theta)\big) \Rightarrow N\big(0, \sigma^2(\theta)\big).$$

This seems to suggest that $\delta_n$ is almost unbiased and that

$$\operatorname{Var}_\theta(\sqrt{n}\delta_n) \to \sigma^2(\theta).$$

(This supposition is not automatic, but does hold if the sequence $n\big(\delta_n - g(\theta)\big)^2$ is uniformly integrable.) It seems natural to conjecture that

$$\sigma^2(\theta) \geq \frac{\big[g'(\theta)\big]^2}{I(\theta)}.$$

But the following example, discovered by Hodges in 1951, shows that this conjecture can fail. The import of this counterexample is that a proper formulation of asymptotic optimality will need to consider features of an estimator's distribution beyond the asymptotic variance.

*Example 16.1.* Let $X_1, X_2, \dots$ be i.i.d. from $N(\theta, 1)$ and take $\overline{X}_n = (X_1 + \cdots + X_n)/n$. Define $\delta_n$, graphed in Figure 16.1, by

$$\delta_n = \begin{cases} \overline{X}_n, & |\overline{X}_n| \geq 1/n^{1/4}; \\ a\overline{X}_n, & |\overline{X}_n| < 1/n^{1/4}, \end{cases}$$

where $a$ is some constant in $(0, 1)$. Let us compute the limiting distribution of $\sqrt{n}(\delta_n - \theta)$.

Suppose $\theta < 0$. Fix $x$ and consider

$$P_\theta\big(\sqrt{n}(\delta_n - \theta) \leq x\big) = P(\delta_n \leq \theta + x/\sqrt{n}).$$

Since $\theta + x/\sqrt{n} \to \theta < 0$ and $-1/n^{1/4} \to 0$, for $n$ sufficiently large, $\theta + x/\sqrt{n} < -1/n^{1/4}$, and then

$$P_\theta\big(\sqrt{n}(\delta_n - \theta) \leq x\big) = P_\theta(\overline{X}_n \leq \theta + x/\sqrt{n}) = \Phi(x).$$

So in this case, $\sqrt{n}(\delta_n - \theta) \Rightarrow N(0, 1)$. A similar calculation shows that $\sqrt{n}(\delta_n - \theta) \Rightarrow N(0, 1)$ when $\theta > 0$.

Suppose now that $\theta = 0$. Fix $x$ and consider

$$P_\theta(\sqrt{n}\delta_n \leq x) = P_\theta(\delta_n \leq x/\sqrt{n}).$$

For $n$ sufficiently large, $a|x|$ will be less than $n^{1/4}$, or, equivalently, $a|x\sqrt{n}| < 1/n^{1/4}$, and then

$$P_0(\sqrt{n}\delta_n \leq x) = P_0(a\overline{X}_n \leq x/\sqrt{n}) = \Phi(x/a).$$

This is the cumulative distribution function for $N(0, a^2)$. So when $\theta = 0$, $\sqrt{n}(\delta_n - \theta) \Rightarrow N(0, a^2)$.

These calculations show that in general,

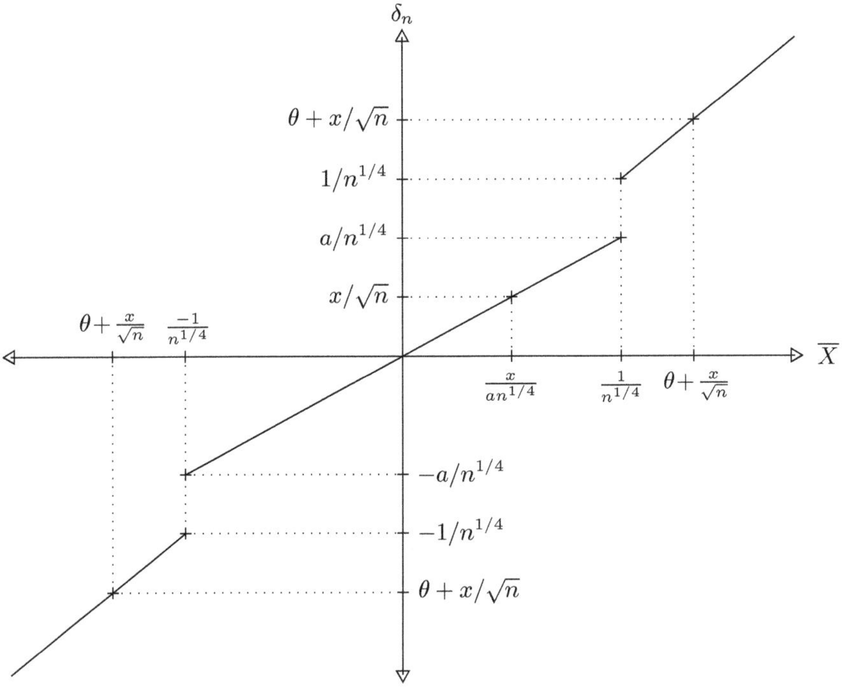


**Fig. 16.1.** The Hodges' estimator $\delta_n$.

$$\sqrt{n}(\delta_n - \theta) \Rightarrow N\big(0, \sigma^2(\theta)\big), \tag{16.1}$$

where

$$\sigma^2(\theta) = \begin{cases} 1, & \theta \neq 0; \\ a^2, & \theta = 0. \end{cases}$$

This estimator is called "superefficient" since the variance of the limiting distribution when $\theta = 0$ is smaller than $1/I(\theta) = 1$.

Because $\sqrt{n}(\overline{X}_n - \theta) \sim N(0, 1)$, (16.1) seems to suggest that $\delta_n$ may be a better estimator than $\overline{X}_n$ when $n$ is large. To explore what is going on, let us consider the risk functions for these estimators under squared error loss. Since $R(\theta, \overline{X}_n) = E_\theta(\overline{X}_n - \theta)^2 = 1/n$, $nR(\theta, \overline{X}_n) = 1$. It can be shown that

$$nR(\theta, \delta_n) \to \begin{cases} 1, & \theta \neq 0; \\ a^2, & \theta = 0, \end{cases}$$

as one might expect from (16.1). But comparison of $\delta_n$ and $\overline{X}_n$ by pointwise convergence of their risk functions scaled up by $n$ does not give a complete picture, because the convergence is not uniform in $\theta$. One simple way to see this is to note (see Figure 16.1) that $\delta_n$ never takes values in the interval

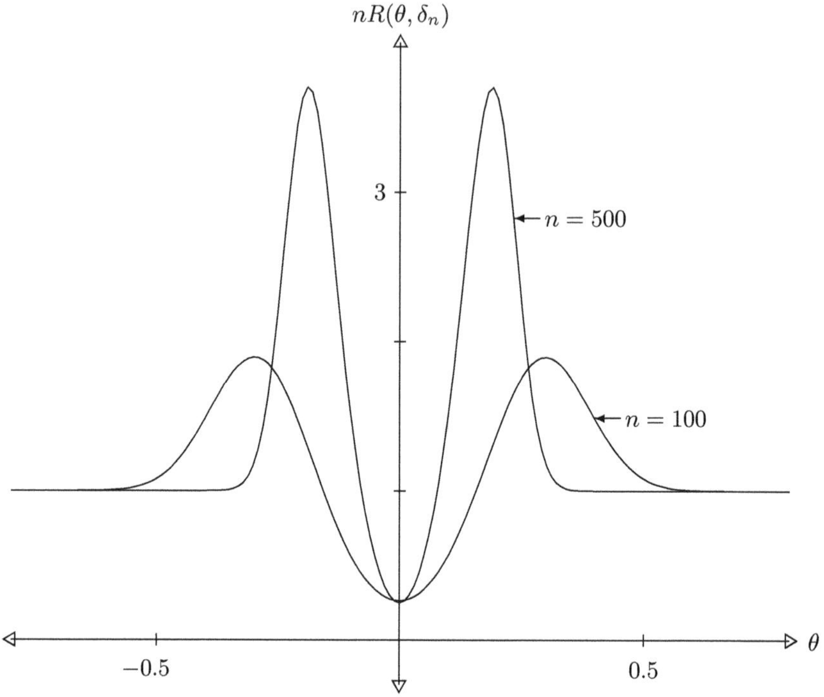


**Fig. 16.2.** Scaled risk of $\delta_n$ with $n = 100$ and $n = 500$ ($a = 1/2$).

$$\left(\frac{a}{n^{1/4}}, \frac{1}{n^{1/4}}\right).$$

If we define

$$\theta_n = \frac{1+a}{2n^{1/4}}$$

to be the midpoint of this interval, then $\delta_n$ will always miss $\theta_n$ by at least half the width of the interval, and so

$$(\delta_n - \theta_n)^2 \geq \left(\frac{1-a}{2n^{1/4}}\right)^2 = \frac{(1-a)^2}{4\sqrt{n}}.$$

From this,

$$nR(\theta_n, \delta_n) \geq n\frac{(1-a)^2}{4\sqrt{n}} = \frac{(1-a)^2}{4}\sqrt{n} \to \infty,$$

as $n \to \infty$. This shows that for large $n$ the risk of $\delta_n$ at $\theta_n$ will be much worse than the risk of $\overline{X}_n$ at $\theta_n$. Figure 16.2 plots $n$ times the risk function for $\delta_n$ when $n = 100$ and $n = 500$ with $a$ fixed at $1/2$. As $n$ increases, the improved risk near zero does not seem sufficient compensation for the worsening risk nearby.

## 16.2 Contiguity

Recall that a measure $\tilde{Q}$ is absolutely continuous with respect to another measure $Q$ if $\tilde{Q}(N) = 0$ whenever $Q(N) = 0$. To impart a statistical consequence to this notion, suppose we are interested in testing $H_0 : X \sim Q$ versus $H_1 : X \sim \tilde{Q}$. If the level $\alpha = E_0\varphi(X)$ is zero, then $N = \{x : \varphi(x) > 0\}$ is a null set under $Q$, and must then also be a null set under $\tilde{Q}$. So the power $\beta = E_1\varphi(X)$ must also be zero. Conversely, $\beta > 0$ implies $\alpha > 0$. If the measures $Q$ and $\tilde{Q}$ are mutually absolutely continuous, $\alpha < 1$ implies $\beta < 1$. In this sense the competing distributions $Q$ and $\tilde{Q}$ are hard to distinguish, at least perfectly.

Contiguity might be viewed as an asymptotic notion of absolute continuity. It concerns two sequences of distributions, $\tilde{Q}_n$ and $Q_n$, $n = 1, 2, \dots$. These might be viewed as competing joint distributions for data, with $n$ representing the sample size. So $\tilde{Q}_n$ and $Q_n$ are defined on a common measurable space $(\mathcal{X}_n, \mathcal{A}_n)$, but these spaces generally vary with $n$. For instance, $\mathcal{X}_n$ would be $\mathbb{R}^n$ if $n$ is the sample size and the individual observations are univariate.

**Definition 16.2.** *The measures $\tilde{Q}_n$ are contiguous to the measures $Q_n$ if $\tilde{Q}_n(A_n) \to 0$ whenever $Q_n(A_n) \to 0$.*

Contiguity can also be framed in the statistical context of simple versus simple testing. Suppose $\varphi_n$, $n \ge 1$, are tests of $H_0 : X \sim Q_n$ versus $H_1 : X \sim \tilde{Q}_n$ with levels $\alpha_n = E_0\varphi_n(X)$ and powers $\beta_n = E_1\varphi_n(X)$. If $\tilde{Q}_n$ are contiguous to $Q_n$ and $\alpha_n \to 0$ then $\beta_n \to 0$. If the sequences are mutually contiguous (i.e., $Q_n$ is also contiguous to $\tilde{Q}_n$), then $\beta_n \to 1$ implies $\alpha_n \to 1$. In this sense the competing hypotheses remain hard to distinguish as $n$ increases without bound.

*Example 16.3.* Suppose

$$Q_n = N_n(\mu_n, \sigma^2 I) \text{ and } \tilde{Q}_n = N_n(\nu_n, \sigma^2 I)$$

where $\mu_n$ and $\nu_n$ in $\mathbb{R}^n$, $n \ge 1$. By the Neyman–Pearson lemma, the best level $\alpha$ test of $H_0 : X \sim Q_n$ versus $H_1 : X \sim \tilde{Q}_n$ rejects $H_0$ if

$$(\nu_n - \mu_n) \cdot (X - \mu_n) > \sigma \|\mu_n - \nu_n\| z_\alpha$$

and has power

$$\Phi\left(\frac{\|\mu_n - \nu_n\|}{\sigma} - z_\alpha\right).$$

Suppose $M = \limsup \|\mu_n - \nu_n\| < \infty$, and let $A_n$, $n \ge 1$, be Borel sets with $A_n \subset \mathbb{R}^n$. The function $1_{A_n}$, viewed as a test of $H_0$ versus $H_1$ has level $\alpha_n = E_0 1_{A_n} = Q_n(A_n)$ and power $E_1 1_{A_n} = \tilde{Q}_n(A_n)$. But the power of this test is at most that of the optimal test, and so

$$\tilde{Q}_n(A_n) \le \Phi\left(\frac{\|\mu_n - \nu_n\|}{\sigma} - z_{\alpha_n}\right).$$

But if $Q_n(A_n) = \alpha_n \to 0$, then $z_{\alpha_n} \to \infty$, and from this bound, $\tilde{Q}_n(A_n)$ must also tend to zero. This shows that $\tilde{Q}_n$, $n \geq 1$, are contiguous to $Q_n$, $n \geq 1$. And because we could reverse the roles of $\mu_n$ and $\nu_n$ without changing $\|\mu_n - \nu_n\|$, when $\limsup \|\mu_n - \nu_n\| < \infty$ the sequences will be mutually contiguous.

Suppose instead that $\|\mu_n - \nu_n\| \to \infty$. If we take

$$A_n = \Bigl\{x \in \mathbb{R}^n : (\nu_n - \mu_n)\cdot\bigl(x - \tfrac{1}{2}(\mu_n + \nu_n)\bigr) > 0\Bigr\},$$

corresponding to the critical region for a symmetric likelihood ratio test, then

$$Q_n(A_n) = \Phi\left(-\frac{1}{2\sigma}\|\mu_n - \nu_n\|\right) \to 0$$

and

$$\tilde{Q}_n(A_n) = \Phi\left(\frac{1}{2\sigma}\|\mu_n - \nu_n\|\right) \to 1.$$

So in this case the measures are not contiguous. Taking subsequences, they are also not contiguous if $\limsup \|\mu_n - \nu_n\| = \infty$.

If $\mu_n = \theta\mathbf{1}$ and $\nu_n = (\theta + \delta_n)\mathbf{1}$, then under $Q_n$ the entries of $X$ are i.i.d. from $N(\theta, \sigma^2)$, and under $\tilde{Q}_n$ the entries are i.i.d. from $N(\theta + \delta_n, \sigma^2)$. In this case the sequences will be contiguous if $\limsup n\delta_n^2 < \infty$. This may be interpreted as meaning that shifts in the common mean of order $1/\sqrt{n}$ cannot be detected with probability approaching one. This sort of behavior is typical in regular models; see Theorem 16.10 below.

Considering the role of the Neyman–Pearson lemma in this example, it seems natural that contiguity should be related to likelihood ratios. The notion of uniform integrability also plays a role. If $X$ is integrable, then $E|X|I\{|X| \geq t\} \to 0$, as $t \to \infty$, by dominated convergence, and our Definition 8.15 of uniform integrability asserts that this holds uniformly over a collection of random variables.

**Lemma 16.4.** *Suppose $\tilde{Q}_n \ll Q_n$ with $L_n$ the density (or likelihood ratio) $d\tilde{Q}_n/dQ_n$. Let $X_n \sim Q_n$. If the likelihood ratios $L_n(X_n)$, $n \geq 1$, are uniformly integrable, then the measures $\tilde{Q}_n$ are contiguous to the measures $Q_n$.*

As in Lemma 12.18, the likelihood ratios $L_n$ would usually be computed as $\tilde{p}_n/p_n$, with $\tilde{p}_n$ and $p_n$ the densities of $\tilde{Q}_n$ and $Q_n$ with respect to some measure $\mu_n$.

*Proof.* Using $L_n$, and letting $E$ denote expectation when $X_n \sim Q_n$, we have the following bound, valid for any $t > 0$:

$$\begin{aligned}\tilde{Q}_n(A_n) = \int_{A_n} L_n\,dQ_n &= EL_n(X_n)I\{X_n \in A_n\}\\ &\leq tEI\{X_n \in A_n\} + EL_n(X_n)I\{L_n(X_n) \geq t\}.\end{aligned}$$

The first term in this bound is $tQ_n(A_n)$, which tends to zero if $Q_n(A_n) \to 0$. So in this case

$$\limsup \tilde{Q}_n(A_n) \le \limsup EL_n(X_n)I\{L_n(X_n) \ge t\}.$$

Using uniform integrability, the right-hand side here tends to zero as $t \to \infty$, and so $\limsup \tilde{Q}_n(A_n)$ must be zero, proving the lemma. ☐

The next result shows that convergence in probability remains in effect following a shift to a contiguous sequence of distributions.

**Proposition 16.5.** *Suppose $X_n \sim Q_n$ and $\tilde{X}_n \sim \tilde{Q}_n$, let $T_n$ be an arbitrary sequence of estimators, and assume $\tilde{Q}_n$, $n \ge 1$, are contiguous to $Q_n$, $n \ge 1$. If $T_n(X_n) \xrightarrow{p} c$, then $T_n(\tilde{X}_n) \xrightarrow{p} c$. Similarly, if $T_n(X_n) = O_p(1)$, then $T_n(\tilde{X}_n) = O_p(1)$.*

*Proof.* From the definition of convergence in probability, for any $\epsilon > 0$,

$$P\bigl(|T_n(X_n) - c| \ge \epsilon\bigr) \to 0.$$

Viewing this probability as the $Q_n$-measure of a set, by contiguity

$$P\bigl(|T_n(\tilde{X}_n) - c| \ge \epsilon\bigr) \to 0,$$

and since $\epsilon$ is an arbitrary positive number, $T_n(\tilde{X}_n) \xrightarrow{p} c$. The second assertion can be established with a similar argument. ☐

To state the final result about contiguity in its proper generality, we again want to view functions as points in a vector space, as we did in Section 12.5, now with a different notion of convergence. Given a measure $\mu$ on $(\mathcal{X}, \mathcal{B})$, let

$$\mathcal{L}_2(\mu) = \left\{f : \int f^2\,d\mu < \infty\right\},$$

and define the $\mathcal{L}_2$-*length* of a function $f \in \mathcal{L}_2(\mu)$ as

$$\|f\|_2 = \left(\int f^2\,d\mu\right)^{1/2}.$$

Then $\|f-g\|_2$ represents the distance between two functions $f$ and $g$ in $\mathcal{L}_2(\mu)$, and with this distance $\mathcal{L}_2(\mu)$ is a metric space,[2] similar in many respects to $\mathbb{R}^n$. Using this distance we have the following natural notions of convergence and differentiation.

[2] To be more precise, $\mathcal{L}_2(\mu)$ is a pseudometric space, because $\|f - g\|$ can be zero for functions $f \ne g$ if they differ only on a null set. It would be a proper metric space if we were to introduce equivalence classes of functions and consider two functions the same if they agreed almost everywhere.

**Definition 16.6.** *A sequence of functions* $f_n \in \mathcal{L}_2(\mu)$ converges in $\mathcal{L}_2$ *to* $f \in \mathcal{L}_2(\mu)$, *denoted* $f_n \overset{\mathcal{L}_2}{\to} f$, *if* $\|f_n - f\|_2 \to 0$.

**Definition 16.7.** *A mapping* $\theta \rightsquigarrow f_\theta$ *from* $\mathbb{R}$ *to* $\mathcal{L}_2(\mu)$ *is* differentiable in quadratic mean *at* $\theta_0$ *with derivative* $V$ *if* $V \in \mathcal{L}_2(\mu)$ *and*

$$\frac{f_{\theta_0+\epsilon} - f_{\theta_0}}{\epsilon} \overset{\mathcal{L}_2}{\to} V,$$

*as* $\epsilon \to 0$.

When the domain of the map is $\mathbb{R}^p$, the derivative, analogous to the gradient in multivariate calculus, will be a vector-valued function. First-order Taylor approximation should give $f_{\theta_0+\epsilon} \approx f_{\theta_0} + \epsilon \cdot V$, which motivates the following definition.

**Definition 16.8.** *A mapping* $\theta \rightsquigarrow f_\theta$ *from* $\mathbb{R}^p$ *to* $\mathcal{L}_2(\mu)$ *is* differentiable in quadratic mean *at* $\theta_0$ *with derivative* $V$ *if* $\int \|V\|^2\,d\mu < \infty$ *and*

$$\frac{f_{\theta_0+\epsilon} - f_{\theta_0} - \epsilon \cdot V}{\|\epsilon\|} \overset{\mathcal{L}_2}{\to} 0,$$

*as* $\epsilon \to 0$.

This notion of differentiation is generally weaker than pointwise differentiability. In most cases the following lemma allows us to compute this derivative as the gradient, provided it exists.

**Lemma 16.9.** *Let* $\theta \rightsquigarrow f_\theta$ *be a mapping from* $\mathbb{R}^p$ *to* $\mathcal{L}_2(\mu)$. *If* $\nabla_\theta f_\theta(x)$ *exists for almost all* $x$, *for* $\theta$ *in some neighborhood of* $\theta_0$, *and if*

$$\int \|\nabla_\theta f_\theta\|^2\,d\mu$$

*is continuous at* $\theta_0$, *then the mapping is differentiable in quadratic mean at* $\theta_0$ *with derivative the gradient* $\nabla_\theta f_\theta$ *evaluated at* $\theta = \theta_0$.

Returning to statistics, let $\mathcal{P} = \{P_\theta : \theta \in \Omega\}$ be a dominated family with densities $p_\theta$, $\theta \in \Omega$, and assume for now that $\Omega \subset \mathbb{R}$. Then the functions $\sqrt{p_\theta}$ can be viewed as points in $\mathcal{L}_2(\mu)$. When ordinary derivatives exist, (4.14) and the chain rule give

$$I(\theta) = \int \left(\frac{\partial \log p_\theta}{\partial \theta}\right)^2 p_\theta\,d\mu = 4\int \left(\frac{\partial \sqrt{p_\theta}}{\partial \theta}\right)^2 d\mu.$$

This suggests the following definition for Fisher information:

$$I(\theta) = 4\|V_\theta\|_2^2, \tag{16.2}$$

with $V_\theta$ the quadratic mean derivative of $\theta \rightsquigarrow \sqrt{p_\theta}$. This definition is more general and proper than the formulas given earlier that require extra regularity. And the next result shows that the regularity necessary to define Fisher information in this way gives contiguity in i.i.d. models with parameter shifts of order $1/\sqrt{n}$.

**Theorem 16.10.** *If $\mathcal{P} = \{P_\theta : \theta \in \Omega\}$ is a dominated family with densities $p_\theta$, and if $\theta \rightsquigarrow \sqrt{p_\theta}$ is differentiable in quadratic mean at $\theta_0$, then the measures $P^n_{\theta_0+\Delta/\sqrt{n}}$, $n \geq 1$, are contiguous to measures $P^n_{\theta_0}$, $n \geq 1$.*

## 16.3 Local Asymptotic Normality

Previous chapters provide a fair amount of information about optimal estimation sampling from a normal distribution with unknown mean. For instance, if $X \sim N(\theta, 1)$, $\theta \in \mathbb{R}$, then $X$ is complete sufficient, and it is the UMVU and best equivariant estimator of $\theta$ under squared error loss. And under squared error loss, it is also minimax, minimizing $\sup_\theta E_\theta(\delta - \theta)^2$. This is established and generalized in Section 16.6.

In large samples, the maximum likelihood estimator $\hat{\theta}$ is approximately normal (after suitable rescaling). If $\hat{\theta}$ provides most of the information from the data, it would be natural to hope that inference from large samples may be similar to inference sampling from a normal distribution. Naturally, this will involve some rescaling, because with large samples small changes in parameter values will be noticeable from our data. This notion is made precise by considering likelihood ratios; a sequence of distribution families is called *locally asymptotically normal* if the likelihood ratios for the families are close to those for normal distributions, in an appropriate sense.

Suppose $X \sim P_\theta = N_p(\theta, \Sigma)$, $\theta \in \mathbb{R}^p$, with $\Sigma$ a fixed positive definite covariance matrix. Then the log-likelihood ratio between parameter values $t$ and 0 is

$$\begin{aligned} \ell(t, 0) &= \log \frac{dP_t}{dP_0}(X) \\ &= -\frac{1}{2}(X - t)'\Sigma^{-1}(X - t) + \frac{1}{2}X'\Sigma^{-1}X \\ &= t'\Sigma^{-1}X - \frac{1}{2}t'\Sigma^{-1}t, \end{aligned} \tag{16.3}$$

which is a quadratic function of $t$ with the linear coefficients random and the quadratic coefficients constant.

To see why likelihood ratios are approximately of this form in large samples, suppose $X_i$, $i \geq 1$, are i.i.d. with common density $f_\theta$, $\theta \in \Omega \subset \mathbb{R}$. Fix $\theta$ and define $W_i(\omega) = \log\bigl(f_\omega(X_i)/f_\theta(X_i)\bigr)$. Under sufficient regularity (see Section 4.5),

$$E_\theta W_i'(\theta) = E_\theta \frac{\partial \log f_\theta(X_i)}{\partial \theta} = 0,$$

$$\operatorname{Var}_\theta\bigl(W_i'(\theta)\bigr) = E_\theta \left(\frac{\partial \log f_\theta(X_i)}{\partial \theta}\right)^2 = I(\theta),$$

and

$$E_\theta W_i''(\theta) = E_\theta \frac{\partial^2 \log f_\theta(X_i)}{\partial \theta^2} = -I(\theta).$$

By the central limit theorem,

$$S_n = \frac{1}{\sqrt{n}} \sum_{i=1}^n W_i'(\theta) \Rightarrow N\bigl(0, I(\theta)\bigr),$$

and by the law of large numbers,

$$\frac{1}{n} \sum_{i=1}^n W_i''(\theta) \xrightarrow{p} -I(\theta).$$

A two-term Taylor expansion then suggests the following approximation for log-likelihood ratios between $\theta$ and "contiguous" alternative $\theta + t/\sqrt{n}$:

$$\begin{aligned}
\ell_n(\theta + t/\sqrt{n}, \theta) &= \log\left[\frac{\prod_{i=1}^n f_{\theta+t/\sqrt{n}}(X_i)}{\prod_{i=1}^n f_\theta(X_i)}\right] \\
&= \sum_{i=1}^n W_i(\theta + t/\sqrt{n}) \\
&\approx \frac{t}{\sqrt{n}} \sum_{i=1}^n W_i'(\theta) + \frac{t^2}{2n} \sum_{i=1}^n W_i''(\theta) \\
&\approx tS_n - \frac{1}{2} t^2 I(\theta). \qquad (16.4)
\end{aligned}$$

This is quite similar in form to (16.3)

The following definition formalizes conditions on likelihood ratios sufficient for the notions of asymptotic optimality developed in Section 16.6, yet weak enough to hold in a wide class of applications. These applications include cases where the data are not identically distributed and cases where the data are dependent.

**Definition 16.11.** *Consider a sequence of models,*

$$\mathcal{P}_n = \{P_{\theta,n} : \theta \in \Omega \subset \mathbb{R}^p\}, \qquad n \geq 1,$$

*and let $\ell_n$ denote log-likelihood ratios for $\mathcal{P}_n$,*

$$\ell_n(\omega,\theta) = \log\left[\frac{dP_{\omega,n}}{dP_{\theta,n}}\right].$$

*These models are* locally asymptotically normal *(LAN) at a parameter value $\theta$ in the interior of $\Omega$ if there exist random vectors $S_n = S_n(\theta)$ and a positive definite matrix $J = J(\theta)$ such that:*

1. *If $t_n$ is any convergent sequence, $t_n \to t \in \mathbb{R}^p$,*
$$\ell_n(\theta + t_n/\sqrt{n}, \theta) - \left[t'S_n - \tfrac{1}{2}t'Jt\right] \overset{P_{\theta,n}}{\to} 0$$
*as $n \to \infty$.*
2. *Under $P_{\theta,n}$, $S_n \Rightarrow Z \sim N(0, J)$ as $n \to \infty$.*

*Remark 16.12.* The second condition in this definition can be replaced by the condition that the measures $P_{\theta,n}$ and $P_{\theta_n,n}$ are contiguous whenever $\sqrt{n}(\theta_n - \theta)$ remains bounded. Mixtures of these measures are also contiguous if the mixing distributions concentrate appropriately near $\theta$. Specifically, if $B(r)$ denotes the ball of radius $r$ about $\theta$ and if $\pi_n$ are probability distributions on $\Omega$ such that $\liminf \pi_n\big(B(c/\sqrt{n})\big) \uparrow 1$ as $c \to \infty$, then $P_{\theta,n}$ and $\int P_{\omega,n} d\pi_n(\omega)$ are contiguous.

*Remark 16.13.* If the models are LAN and $t_n \to t$, the distributions of $S_n$ under $P_{\theta+t_n/\sqrt{n},n}$ are also approximately normal. Specifically, under $P_{\theta+t_n/\sqrt{n},n}$, $S_n \Rightarrow N(Jt, J)$. To understand the nature of the argument, assume

$$P_{\theta+t_n/\sqrt{n},n} \ll P_{\theta,n},$$

and let $f$ be a bounded continuous function. With suitable uniform integrability, one would then expect

$$\begin{aligned} E_{\theta+t_n/\sqrt{n},n} f(S_n) &= E_{\theta,n} f(S_n) e^{\ell_n(\theta+t_n/\sqrt{n},\theta)} \\ &\approx E f(Z) e^{t'Z - t'Jt/2} = E f(Jt + Z). \end{aligned}$$

**Theorem 16.14.** *Suppose $X_1, X_2, \dots$ are i.i.d. with common density $p_\theta$, and let $P_{\theta,n}$ be the joint distribution of $X_1, \dots, X_n$. If the mapping $\omega \rightsquigarrow \sqrt{p_\omega}$ is differentiable in quadratic mean at a parameter value $\theta$ in the interior of the parameter space $\Omega \subset \mathbb{R}^p$, then the families $\mathcal{P}_n = \{P_{\theta,n} : \theta \in \Omega\}$ are locally asymptotically normal at $\theta$ with $J$ the Fisher information given in* (16.2), *$J = I(\theta)$.*

The quadratic approximation $t'S_n - \frac{1}{2}t'Jt$ in the LAN definition is maximized at $\hat{t} = J^{-1}S_n$. Because the maximum likelihood estimator $\hat{\theta}_n$ maximizes

$$l_n(\omega) - l_n(\theta) = \ell_n\left(\theta + \frac{\sqrt{n}(\omega - \theta)}{\sqrt{n}}, \theta\right)$$

over $\omega \in \Omega$, the LAN approximation suggests that

$$\sqrt{n}(\hat{\theta}_n - \theta) \approx J^{-1} S_n.$$

Regularity conditions akin to those of Theorem 9.14 but extended to the multivariate case ensure that

$$\sqrt{n}(\hat{\theta}_n - \theta) - J^{-1} S_n \xrightarrow{P_\theta} 0, \tag{16.5}$$

as $n \to \infty$. We assume as we proceed that (16.5) holds for suitable estimators $\hat{\theta}_n$, but these estimators need not be maximum likelihood.

*Example 16.15.* Suppose $X_1, X_2, \dots$ are i.i.d. absolutely continuous random vectors in $\mathbb{R}^4$ with common density

$$p_\theta(x) = c\frac{e^{-\|x-\theta\|^2}}{\|x-\theta\|}.$$

The families of joint distributions here are LAN. With the pole in the density, the likelihood function is infinite at each data point, so a maximum likelihood estimator will be one of the observed data. Section 6.3 of Le Cam and Yang (2000) details a general method to find estimators $\hat{\theta}_n$ satisfying (16.5). This method is based on using the LAN approximation to improve a reasonable preliminary estimator, such as $\overline{X}_n$ in this example.

## 16.4 Minimax Estimation of a Normal Mean

An estimator $\delta$ is called *minimax* if it minimizes $\sup_{\theta \in \Omega} R(\theta, \delta)$. In this section we find minimax estimates for the mean of a normal distribution. These results are used in the next section when a locally asymptotically minimax notion of asymptotic optimality is developed.

As an initial problem, suppose $X \sim N_p(\theta, I)$, $\theta \in \mathbb{R}^p$, and consider a Bayesian model with prior distribution

$$\Theta \sim N(0, \sigma^2 I).$$

Then

$$\Theta | X = x \sim N\left(\frac{\sigma^2 x}{1+\sigma^2}, \frac{\sigma^2}{1+\sigma^2} I\right),$$

and the Bayes estimator under compound squared error loss is

$$\tilde{\theta} = \frac{\sigma^2 X}{1+\sigma^2},$$

with Bayes risk

$$E\|\tilde{\theta} - \Theta\|^2 = EE\big[\|\tilde{\theta} - \Theta\|^2 \mid X\big] = \frac{p\sigma^2}{1+\sigma^2}.$$

Because $\tilde{\theta}$ is Bayes, conditioning on $\Theta$ gives

$$\begin{aligned} E\|\tilde{\theta}-\Theta\|^2 &= EE\big[\|\tilde{\theta}-\Theta\|^2 \mid \Theta\big] \\ &= ER(\Theta,\tilde{\theta}) \le ER(\Theta,\delta) \le \sup_\theta R(\theta,\delta), \end{aligned}$$

for any competing estimator $\delta$. So for any $\delta$,

$$\sup_\theta R(\theta,\delta) \ge p\frac{\sigma^2}{1+\sigma^2}. \tag{16.6}$$

But this holds for any $\sigma$, so letting $\sigma \to \infty$ we have

$$\sup_\theta R(\theta,\delta) \ge p$$

for any $\delta$. But the risk of $\delta(X) = X$ equals $p$, and so $X$ is minimax.

As an extension, we show that $X$ is also minimax when the loss function has form $L(\theta,d) = W(d-\theta)$ with $W$ "bowl-shaped" according to the following definition.

**Definition 16.16.** *A function $W : \mathbb{R}^p \to [0,\infty]$ is* bowl-shaped *if $\{x : W(x) \le \alpha\}$ is convex and symmetric about zero for every $\alpha \ge 0$.*

The following result due to Anderson (1955) is used to find Bayes estimators with bowl-shaped loss functions.

**Theorem 16.17 (Anderson's lemma).** *If $f$ is a Lebesgue density on $\mathbb{R}^p$ with $\{x : f(x) \ge \alpha\}$ convex and symmetric about zero for every $\alpha \ge 0$, and if $W$ is bowl-shaped, then*

$$\int W(x-c)f(x)\,dx \ge \int W(x)f(x)\,dx,$$

*for every $c \in \mathbb{R}^p$.*

The proof of this result relies on the following inequality.

**Theorem 16.18 (Brunn–Minkowski).** *If $A$ and $B$ are nonempty Borel sets in $\mathbb{R}^p$ with sum $A+B = \{x+y : x \in A, y \in B\}$ (the Minkowski sum of $A$ and $B$), and $\lambda$ denotes Lebesgue measure, then*

$$\lambda(A+B)^{1/p} \ge \lambda(A)^{1/p} + \lambda(B)^{1/p}.$$

*Proof.* Let a *box* denote a bounded Cartesian product of intervals and suppose $A$ and $B$ are both boxes with $a_1,\dots,a_p$ the lengths of the sides of $A$ and $b_1,\dots,b_p$ the lengths of the sides of $B$. Then

$$\lambda(A) = \prod_{i=1}^p a_i \text{ and } \lambda(B) = \prod_{i=1}^p b_i.$$

The sum $A+B$ is also a box, and the lengths of the sides of this box are $a_1+b_1,\dots,a_p+b_p$. Thus

$$\lambda(A+B)=\prod_{i=1}^{p}(a_i+b_i).$$

Since arithmetic averages bound geometric averages (see Problem 3.32),

$$\left(\prod_{i=1}^{p}\frac{a_i}{a_i+b_i}\right)^{1/p}+\left(\prod_{i=1}^{p}\frac{b_i}{a_i+b_i}\right)^{1/p}\le\frac{1}{p}\sum_{i=1}^{p}\frac{a_i}{a_i+b_i}+\frac{1}{p}\sum_{i=1}^{p}\frac{b_i}{a_i+b_i}=1,$$

which gives the desired inequality for boxes.

We next show that the inequality holds when $A$ and $B$ are both finite unions of disjoint boxes. The proof is based on induction on the total number of boxes in $A$ and $B$, and there is no harm assuming that $A$ has at least two boxes (if not, just switch $A$ and $B$). Translating $A$ (if necessary) we can assume that some coordinate hyperplane, $\{x : x_k=0\}$ separates two of the boxes in $A$. Define half-spaces $H_+=\{x : x_k\ge 0\}$, $H_-=\{x : x_k<0\}$ and let $A_\pm$ be intersections of $A$ with these half spaces, $A_+=A\cap H_+$ and $A_-=A\cap H_-$. Note that $A_\pm$ are both finite intersections of boxes with the total number of boxes in each of them less than the number of boxes in $A$. The proportion of the volume of $A$ in $H_+$ is $\lambda(A_+)/\lambda(A)$, and by translating $B$ we make $\lambda(B\cap H_+)/\lambda(B)$ match this proportion. Defining $B_\pm=B\cap H_\pm$ we then have

$$\frac{\lambda(A_\pm)}{\lambda(A)}=\frac{\lambda(B_\pm)}{\lambda(B)}. \tag{16.7}$$

Because intersection with a half-plane cannot increase the number of boxes in a set, the number of boxes in $A_+$ and $B_+$ and the number of boxes in $A_-$ and $B_-$ are both less than the number of boxes in $A$ and $B$, and by the inductive hypothesis we can assume that the inequality holds for both of these pairs. Also note that since $A_++B_+\subset H_+$ and $A_-+B_-\subset H_-$,

$$\lambda\big((A_++B_+)\cup(A_-+B_-)\big)=\lambda(A_++B_+)+\lambda(A_-+B_-),$$

and that $A+B\supset(A_++B_+)\cup(A_-+B_-)$. Using these, (16.7), and the inductive hypothesis,

$$\begin{aligned}\lambda(A+B)&\ge\lambda(A_++B_+)+\lambda(A_-+B_-)\\&\ge\big(\lambda(A_+)^{1/p}+\lambda(B_+)^{1/p}\big)^p+\big(\lambda(A_-)^{1/p}+\lambda(B_-)^{1/p}\big)^p\\&=\lambda(A_+)\left(1+\frac{\lambda(B)^{1/p}}{\lambda(A)^{1/p}}\right)^p+\lambda(A_-)\left(1+\frac{\lambda(B)^{1/p}}{\lambda(A)^{1/p}}\right)^p\\&=\lambda(A)\left(1+\frac{\lambda(B)^{1/p}}{\lambda(A)^{1/p}}\right)^p\\&=\big(\lambda(A)^{1/p}+\lambda(B)^{1/p}\big)^p.\end{aligned}$$

This proves the theorem when $A$ and $B$ are finite unions of boxes. The general case follows by an approximation argument. As a starting point, we use the fact that Lebesgue measure is *regular*, which means that $\lambda(K) < \infty$ for all compact $K$, and for any $B$,

$$\lambda(B) = \inf\{\lambda(V) : V \supset B, V \text{ open}\}$$

and

$$\lambda(B) = \sup\{\lambda(K) : K \subset B, K \text{ compact}\}.$$

Suppose $A$ is open. Fix $\epsilon > 0$ and let $K \subset A$ be a compact set with $\lambda(K) \geq \lambda(A) - \epsilon$. Because the distance between $K$ and $A^c$ is positive, we can cover $K$ with open boxes centered at all points of $K$ so that each box in the cover lies in $A$. The union $\tilde{A}$ of a finite subcover will then contain $K$ and lie in $A$, so $\lambda(\tilde{A}) \geq \lambda(A) - \epsilon$, and $\tilde{A}$ will be a finite union of disjoint boxes. Similarly, if $B$ is open there is a set $\tilde{B} \subset B$ that is a finite union of disjoint boxes with $\lambda(\tilde{B}) \geq \lambda(B) - \epsilon$. Because $A + B \supset \tilde{A} + \tilde{B}$,

$$\begin{aligned}\lambda(A+B)^{1/p} \geq \lambda(\tilde{A}+\tilde{B})^{1/p} &\geq \lambda(\tilde{A})^{1/p} + \lambda(\tilde{B})^{1/p} \\ &\geq \bigl(\lambda(A) - \epsilon\bigr)^{1/p} + \bigl(\lambda(B) - \epsilon\bigr)^{1/p}. \qquad (16.8)\end{aligned}$$

Letting $\epsilon \to 0$, the inequality holds for nonempty open sets. Next, suppose $A$ and $B$ are both compact. Define open sets

$$A_n = \{x : \|x - y\| < 1/n, \exists y \in A\}$$

and

$$B_n = \{x : \|x - y\| < 1/n, \exists y \in B\}.$$

Then

$$A + B = \bigcap_{n \geq 1} (A_n + B_n),$$

for if $s$ lies in the intersection, then $s = a_n + b_n$ with $a_n \in A_n$ and $b_n \in B_n$, and along a subsequence $(a_n, b_n) \to (a, b) \in A \times B$. Then $s = a + b \in A + B$. Using this,

$$\begin{aligned}\lambda(A+B)^{1/p} &= \lim_{n\to\infty} \lambda(A_n + B_n)^{1/p} \\ &\geq \lim_{n\to\infty} \bigl(\lambda(A_n)^{1/p} + \lambda(B_n)^{1/p}\bigr) \\ &= \lambda(A)^{1/p} + \lambda(B)^{1/p}.\end{aligned}$$

Finally, if $A$ and $B$ are arbitrary Borel sets with positive and finite measure, and if $\epsilon > 0$, there are compact subsets $\tilde{A} \subset A$ and $\tilde{B} \subset B$ such that $\lambda(\tilde{A}) \geq \lambda(A) - \epsilon$ and $\lambda(\tilde{B}) \geq \lambda(B) - \epsilon$. The inequality then follows by the argument used in (16.8) ☐

**Corollary 16.19.** *If $A$ and $B$ are symmetric convex subsets of $\mathbb{R}^p$ and $c$ is any vector in $\mathbb{R}^p$, then*

$$\lambda\bigl((c+A)\cap B\bigr) \le \lambda(A\cap B).$$

*Proof.* Let $K_+ = (c+A)\cap B$ and $K_- = (-c+A)\cap B$. By symmetry $K_- = -K_+$ and so $\lambda(K_+) = \lambda(K_-)$. Define $K = \frac{1}{2}(K_+ + K_-)$, and note that $K \subset A\cap B$. By the Brunn–Minkowski inequality,

$$\begin{aligned}\lambda(K)^{1/p} &\ge \lambda\left(\frac{1}{2}K_+\right)^{1/p} + \lambda\left(\frac{1}{2}K_-\right)^{1/p}\\ &= \frac{1}{2}\lambda(K_+)^{1/p} + \frac{1}{2}\lambda(K_-)^{1/p} = \lambda(K_+)^{1/p}.\end{aligned}$$

So

$$\lambda(A\cap B) \ge \lambda(K) \ge \lambda(K_+) = \lambda\bigl((c+A)\cap B\bigr). \qquad \square$$

*Proof of Theorem 16.17.* For $u \ge 0$ define convex symmetric sets $A_u = \{x : W(x) \le u\}$ and $B_u = \{x : f(x) \ge u\}$. Using Fubini's theorem and Corollary 16.19,

$$\begin{aligned}\int W(x-c)f(x)\,dx &= \iint_0^\infty \int_0^\infty I\bigl[W(x-c) > u, f(x) \ge v\bigr]\,du\,dv\,dx\\ &= \int_0^\infty \int_0^\infty \int I(x \notin c + A_u, x \in B_v)\,dx\,du\,dv\\ &= \int_0^\infty \int_0^\infty \bigl[\lambda(B_v) - \lambda\bigl((c+A_u)\cap B_v\bigr)\bigr]\,du\,dv\\ &\ge \int_0^\infty \int_0^\infty \bigl[\lambda(B_v) - \lambda(A_u\cap B_v)\bigr]\,du\,dv\\ &= \int W(x)f(x)\,dx.\end{aligned} \qquad \square$$

**Theorem 16.20.** *Suppose $X \sim N_p(\theta, \Sigma)$ with $\Sigma$ a known positive definite matrix, and consider estimating the mean $\theta$ with loss function $L(\theta, d) = W(\theta - d)$ and $W$ bowl-shaped. Then $X$ is minimax.*

*Proof.* Consider a Bayesian formulation in which the prior distribution for $\Theta$ is $N(0, \sigma^2\Sigma)$. Let $\tilde{\Theta} = \Sigma^{-1/2}\Theta$ and $\tilde{X} = \Sigma^{-1/2}X$ and note that

$$\tilde{\Theta} \sim N(0, \sigma^2 I)$$

and

$$\tilde{X} | \tilde{\Theta} = \tilde{\theta} \sim N(\tilde{\theta}, I).$$

As before,

$$\tilde{\Theta}|\tilde{X} = \tilde{x} \sim N\left(\frac{\sigma^2\tilde{x}}{1+\sigma^2}, \frac{\sigma^2}{1+\sigma^2}I\right).$$

Since conditioning on $\tilde{X}$ is the same as conditioning on $X$, multiplication by $\Sigma^{1/2}$ gives

$$\Theta|X = x \sim N\left(\frac{\sigma^2 x}{1+\sigma^2}, \frac{\sigma^2}{1+\sigma^2}\Sigma\right).$$

If $Z \sim N\bigl((0, \sigma^2\Sigma/(1+\sigma^2)\bigr)$, with density $f$, then the posterior risk of an estimator $\delta$ is

$$\begin{aligned} E\bigl[W(\Theta - \delta(X)) \mid X = x\bigr] &= EW\left(Z + \frac{\sigma^2 x}{1+\sigma^2} - \delta(x)\right) \\ &= \int W\left(z + \frac{\sigma^2 x}{1+\sigma^2} - \delta(x)\right) f(z)\,dz. \end{aligned}$$

By Theorem 16.17, this is minimized if $\delta(x) = \sigma^2 x/(1+\sigma^2)$, and so again the Bayes estimator is

$$\tilde{\theta} = \frac{\sigma^2 X}{1+\sigma^2}.$$

If $\epsilon = X - \Theta$, then

$$\epsilon|\Theta = \theta \sim N(0, \Sigma),$$

and so $\epsilon$ and $\Theta$ are independent and the marginal distribution of $\epsilon$ is $N(0, \Sigma^2)$. Using this, the Bayes risk is

$$\begin{aligned} EW(\Theta - \tilde{\theta}) &= EW\left(\Theta - \frac{\sigma^2(\Theta + \epsilon)}{1+\sigma^2}\right) \\ &= EW\left(\frac{\Theta}{1+\sigma^2} - \frac{\sigma^2\epsilon}{1+\sigma^2}\right) = EW\left(\frac{\sigma\epsilon}{\sqrt{1+\sigma^2}}\right), \end{aligned}$$

which converges to $EW(\epsilon)$ as $\sigma \to \infty$ by monotone convergence. Arguing as we did for (16.6), for any $\delta$

$$\sup_\theta R(\theta, \delta) \ge EW(\epsilon).$$

Since this is the risk of $X$ (for any $\theta$), $X$ is minimax. ☐

## 16.5 Posterior Distributions

In this section we derive normal approximations for posterior distributions for LAN families. Local asymptotic optimality is derived using these approximations with arguments similar to those in the preceding section.

The approximations for posterior distributions are developed using a notion of convergence stronger than convergence in distribution, based on the following norm.

**Definition 16.21.** *The* total variation *norm of the difference between two probability measures $P$ and $Q$ is defined as*

$$\|P - Q\| = \sup\{|\int f\,dP - \int f\,dQ| : |f| \le 1\}.$$

If $P$ and $Q$ have densities $p$ and $q$ with respect to a measure $\mu$, and $|f| \le 1$, then

$$\left|\int f\,dP - \int f\,dQ\right| = \left|\int f(p-q)\,d\mu\right| \le \int |p-q|\,d\mu.$$

This bound is achieved when $f = \mathrm{Sign}(p-q)$, and so

$$\|P - Q\| = \int |p-q|\,d\mu.$$

Taking advantage of the fact that $p$ and $q$ both integrate to one,

$$\begin{aligned} \|P - Q\| &= \int_{p>q} (p-q)\,d\mu + \int_{q>p} (q-p)\,d\mu = 2\int_{p>q} (p-q)\,d\mu \\ &= 2\int_{p>q} (1-L)\,dP = 2\int \bigl(1 - \min\{1, L\}\bigr)\,dP, \end{aligned} \tag{16.9}$$

where $L = q/p$ is the likelihood ratio $dQ/dP$.

If $f$ is a bounded function, $\sup|f| = M$, and we take $f^* = f/M$, then $|f^*| \le 1$ and so

$$\left|\int f\,dP - \int f\,dQ\right| = M\left|\int f^*\,dP - \int f^*\,dQ\right| \le M\|P - Q\|. \tag{16.10}$$

Strong convergence is defined using the total variation norm. Distributions $P_n$ *converge strongly* to $P$ if $\|P_n - P\| \to 0$. If this happens, then by (16.10) $\int f\,dP_n \to \int f\,dP$ for any bounded measurable $f$. This can be compared with convergence in distribution where, by Theorem 8.9, $\int f\,dP_n \to \int f\,dP$ for bounded *continuous* functions, but convergence can fail if $f$ is discontinuous. So strong convergence implies convergence in distribution.

**Lemma 16.22.** *Let $\tilde{P}$ and $P$ be two possible joint distributions for random vectors $X$ and $Y$. Suppose $\tilde{P} \ll P$, and let $f$ denote the density $d\tilde{P}/dP$. Let $\tilde{Q}$ and $Q$ denote the marginal distributions for $X$ when $(X,Y) \sim \tilde{P}$ and $(X,Y) \sim P$, and let $\tilde{R}_x$ and $R_x$ denote the conditional distributions for $Y$ given $X = x$ when $(X,Y) \sim \tilde{P}$ and $(X,Y) \sim P$. Then $\tilde{Q} \ll Q$ with density*

$$h(x) = \frac{d\tilde{Q}}{dQ}(x) = \int f(x,y)\,dR_x(y),$$

*and $\tilde{R}_x \ll R_x$ (a.e.) with density*

$$\frac{d\tilde{R}_x}{dR_x}(y) = \frac{f(x,y)}{h(x)}.$$

*Proof.* Using Lemma 12.18 and smoothing,

$$\tilde{E}g(X) = Eg(X)f(X,Y) = Eg(X)E[f(X,Y)|X] = \int g(x)h(x)\,dQ(x).$$

To show that the stated densities for the conditional distributions are correct we need to show that iterated integration gives the integral against $\tilde{P}$. This is the case because

$$\begin{aligned}\iint g(x,y)\frac{f(x,y)}{h(x)}\,dR_x(y)\,d\tilde{Q}(x) &= \iint g(x,y)f(x,y)\,dR_x(y)dQ(x)\\ &= EE[g(X,Y)f(X,Y)|X]\\ &= Eg(X,Y)f(X,Y) = \tilde{E}g(X,Y).\end{aligned}$$ □

To motivate the main result, approximating posterior distributions, suppose our family is LAN at $\theta_0$, and that the prior distribution for $\Theta$ is $N(\theta_0, \Gamma^{-1}/n)$.[3] If the LAN approximation and the approximation (16.5) for $\hat{\theta}$ were exact, then the likelihood function would be proportional to

$$\exp\Bigl[n(\theta-\theta_0)'J(\hat{\theta}_n-\theta_0) - \frac{n}{2}(\theta-\theta_0)'J(\theta-\theta_0)\Bigr],$$

and the posterior distribution for $\Theta$ would be

$$G_{x,n} = N\bigl(\theta_0 + (\Gamma+J)^{-1}J(\hat{\theta}_n-\theta_0), (\Gamma+J)^{-1}/n\bigr). \tag{16.11}$$

For convenience, as we proceed dependence on $n$ is suppressed from the notation.

**Theorem 16.23.** *Suppose our families are LAN at $\theta_0$ in the interior of $\Omega$ and that $\hat{\theta}$ satisfies* (16.5). *Consider a sequence of Bayesian models in which $\Theta \sim N(\theta_0, \Gamma^{-1}/n)$ (truncated to $\Omega$) with $\Gamma$ a fixed positive definite matrix. Let $F_x$ denote the conditional distribution of $\Theta$ given $X = x$, and let $G_x$ denote the normal approximation for this distribution given in* (16.11). *Then $\|F_X - G_X\|$ converges to zero in probability as $n \to \infty$.*

*Proof.* (Sketch) Let $P$ denote the joint distribution of $X$ and $\Theta$ (in the Bayesian model), and let $Q$ denote the marginal distribution for $X$. Introduce another model in which $X$ has the same marginal distribution and $\tilde{\Theta}|X = x \sim G_x$, the normal approximation for the posterior. Let $\tilde{P}$ denote the joint distribution for $X$ and $\tilde{\Theta}$. Finally, let $\overline{P} = \frac{1}{2}(P+\tilde{P})$, so that $P \ll \overline{P}$ and $\tilde{P} \ll \overline{P}$, and introduce densities

$$f(x,\theta) = \frac{dP}{d\overline{P}}(x,\theta) \text{ and } \tilde{f}(x,\theta) = \frac{d\tilde{P}}{d\overline{P}}(x,\theta).$$

[3] If $\Omega$ is not all $\mathbb{R}^p$, then we should truncate the prior to $\Omega$. If $\theta_0$ lies in the interior of $\Omega$, only minor changes result and the theorem is correct as stated. But to keep the presentation accessible we do not worry about this issue in the proof.

The marginal distributions for $X$ are the same under $P$, $\tilde{P}$, and (thus) $\overline{P}$. So both marginal densities for $X$ must be one, and by Lemma 16.22, $f(x,\cdot)$ and $\tilde{f}(x,\cdot)$ are densities for $F_x$ and $G_x$. Using (16.9),

$$\|F_x - G_x\| = 2\int\bigl[1 - \min\{1, L(x,\theta)\}\bigr]dF_x(\theta),$$

where $L$ is the likelihood ratio

$$L(x,\theta) = \frac{\tilde{f}(x,\theta)}{f(x,\theta)}.$$

Integrating against the marginal distribution of $X$,

$$E\|F_X - G_X\| = 2E\bigl[1 - \min\{1, L(X,\Theta)\}\bigr],$$

and the theorem will follow if $L(X,\Theta) \xrightarrow{p} 1$.

We next want to rewrite $L$ to take advantage of the things we know about the likelihood and $G_x$. Suppose $P \ll \tilde{P}$. Then $P$ has density $f(x,\theta)/\tilde{f}(x,\theta)$ with respect to $\tilde{P}$. Because the marginal distributions of $X$ are the same under $P$ and $\tilde{P}$, the marginal density must be one, and the formula in Lemma 16.22 then gives

$$\int \frac{f(x,\tilde{\theta})}{\tilde{f}(x,\tilde{\theta})}\,dG_x(\tilde{\theta}) = 1. \tag{16.12}$$

When $P \ll \tilde{P}$ fails, this need not hold exactly, but remains approximately true.[4] Assuming, for convenience, that (16.12) holds exactly, we have

$$L(x,\theta) = \int \frac{\tilde{f}(x,\theta)}{\tilde{f}(x,\tilde{\theta})}\frac{f(x,\tilde{\theta})}{f(x,\theta)}\,dG_x(\tilde{\theta}),$$

and from this the theorem will follow if

$$\frac{\tilde{f}(X,\Theta)}{\tilde{f}(X,\tilde{\Theta})}\frac{f(X,\tilde{\Theta})}{f(X,\Theta)} \xrightarrow{P} 1. \tag{16.13}$$

The two fractions here can both be viewed as likelihood ratios, since $f$ and $\tilde{f}$ are both joint densities. Specifically, viewing $\tilde{f}$ as proportional to a density for $X$ times the normal conditional density $G_x$,

$$\begin{aligned}\frac{\tilde{f}(X,\Theta)}{\tilde{f}(X,\tilde{\Theta})} = \exp\Bigl[&-\frac{n}{2}(\Theta-\theta_0)'(\Gamma+J)(\Theta-\theta_0)\\ &+\frac{n}{2}(\tilde{\Theta}-\theta_0)'(\Gamma+J)(\tilde{\Theta}-\theta_0)\\ &+n(\hat{\theta}-\theta_0)'J(\Theta-\tilde{\Theta})\Bigr],\end{aligned}$$

---

[4] If $H$ is the conditional distribution given $X$ under $\overline{P}$, then $\int f(x,\tilde{\theta})\,dH_x(\tilde{\theta}) = 1$. Since $dG_x(\tilde{\theta}) = \tilde{f}(x,\tilde{\theta})dH_x(\tilde{\theta})$, the true value for the integral is $1 - P\bigl(\tilde{f}(X,\Theta) = 0 \bigm| X = x\bigr)$. This approaches one by a contiguity argument.

and viewing $f$ as proportional to the normal density for $\Theta$ times a conditional density for $X$ given $\theta$,

$$\frac{f(X,\tilde{\Theta})}{f(X,\Theta)} = \exp\Big[-\frac{n}{2}(\tilde{\Theta}-\theta_0)'\Gamma(\tilde{\Theta}-\theta_0) + \frac{n}{2}(\Theta-\theta_0)'\Gamma(\Theta-\theta_0) + \ell_n(\tilde{\Theta},\theta_0) - \ell_n(\Theta,\theta_0)\Big].$$

If the LAN approximation for $\ell_n$ and approximation (16.5) held exactly, then the left-hand side of (16.13) would be one. The proof is completed by arguing that the approximations imply convergence in probability. □

## 16.6 Locally Asymptotically Minimax Estimation

Our first lemma uses the approximations of the previous section and Anderson's lemma (Theorem 16.17) to approximate Bayes risks with the loss function a bounded bowl-shaped function.

**Lemma 16.24.** *Suppose our families are LAN at $\theta$ in the interior of $\Omega$ and that $\hat{\theta}$ satisfies* (16.5). *Consider Bayesian models in which the prior distribution for $\Theta$ is $N(\theta, \sigma^2 J^{-1}/n)$. If $W$ is a bounded bowl-shaped function, then*

$$\lim_{n\to\infty} \inf_{\delta} E_n W\big(\sqrt{n}(\delta-\Theta)\big) = EW\left(\frac{\sigma Z}{\sqrt{1+\sigma^2}}\right),$$

*where $Z \sim N(0, J^{-1})$.*

*Proof.* Let

$$G_{x,n} = N\left(\theta + \frac{\sigma^2(\hat{\theta}_n - \theta)}{1+\sigma^2}, \frac{\sigma^2 J^{-1}}{n(1+\sigma^2)}\right),$$

the approximation for the posterior distribution from Theorem 16.23, and let $F_{n,x}$ denote the true posterior distribution of $\theta$. Define

$$\begin{aligned}\rho_n(x) &= \inf_d E_n\big[W\big(\sqrt{n}(d-\Theta)\big) \mid X = x\big] \\ &= \inf_d \int W\big(\sqrt{n}(d-\theta)\big)\, dF_{x,n}(\theta).\end{aligned}$$

Then, as in Theorem 7.1,

$$\inf_{\delta} E_n W\big(\sqrt{n}(\delta-\Theta)\big) = E_n \rho_n(X).$$

Using (16.10), for any $d$,

$$\left| \int W\big(\sqrt{n}(d-\theta)\big)\,dF_{x,n}(\theta) - \int W\big(\sqrt{n}(d-\theta)\big)\,dG_{x,n}(\theta) \right| \\ \le M\|F_{x,n} - G_{x,n}\|,$$

and it then follows that

$$\left| \rho_n(x) - \inf_d \int W\big(\sqrt{n}(d-\theta)\big)\,dG_{x,n}(\theta) \right| \le M\|F_{x,n} - G_{x,n}\|,$$

where $M = \sup |W|$. But by Anderson's lemma (Theorem 16.17),

$$\inf_d \int W\big(\sqrt{n}(d-\theta)\big)\,dG_{x,n}(\theta) = EW\left(\frac{\sigma Z}{\sqrt{1+\sigma^2}}\right).$$

So

$$\left| E_n\rho_n(X) - EW\left(\frac{\sigma Z}{\sqrt{1+\sigma^2}}\right)\right| \le E_n\left|\rho_n(X) - EW\left(\frac{\sigma Z}{\sqrt{1+\sigma^2}}\right)\right| \\ \le ME_n\|F_{X,n} - G_{X,n}\|.$$

The lemma follows because this expectation tends to zero by Theorem 16.23. ☐

**Theorem 16.25.** *Suppose our families are LAN at $\theta_0$ in the interior of $\Omega$ and that $\hat{\theta}$ satisfies* (16.5), *and let $W$ be a bowl-shaped function. Then for any sequence of estimators $\delta_n$,*

$$\lim_{b\to\infty}\lim_{c\to\infty}\liminf_{n\to\infty}\sup_{\|\theta-\theta_0\|\le c/\sqrt{n}} E_\theta \min\big\{b, W\big(\sqrt{n}(\delta_n-\theta)\big)\big\} \ge EW(Z),$$

*where $Z \sim N(0, J^{-1})$. The asymptotic lower bound here is achieved if $\delta_n = \hat{\theta}_n$.*

*Proof.* Let $\pi_n = N(\theta_0, \sigma^2 J^{-1}/n)$, the prior distribution for $\Theta$ in Lemma 16.24, and note that $\theta_0 + \sigma Z/\sqrt{n} \sim \pi_n$. Also, let $W_b = \min\{b, W\}$, a bounded bowl-shaped function. Then

$$\begin{aligned}
&\inf_\delta EW_b\big(\sqrt{n}(\delta-\Theta)\big) \\
&\qquad \le EW_b\big(\sqrt{n}(\delta_n-\Theta)\big) \\
&\qquad = \int E_\theta W_b\big(\sqrt{n}(\delta_n-\theta)\big)\,d\pi_n(\theta) \\
&\qquad \le \sup_{\|\theta-\theta_0\|\le c/\sqrt{n}} E_\theta W_b\big(\sqrt{n}(\delta_n-\theta)\big)\pi_n\big(\{\theta : \|\theta-\theta_0\| \le c/\sqrt{n}\}\big) \\
&\qquad\qquad + b\pi_n\big(\{\theta : \|\theta-\theta_0\| > c/\sqrt{n}\}\big).
\end{aligned}$$

But

$$\pi_n\big(\{\theta : \|\theta-\theta_0\| \le c/\sqrt{n}\}\big) = P(\|Z\| \le c/\sigma).$$

Solving the inequality to bound the supremum and taking the lim inf as $n \to \infty$, using Lemma 16.24,

$$\liminf_{n\to\infty} \sup_{\|\theta-\theta_0\|\le c/\sqrt{n}} E_\theta W_b\bigl(\sqrt{n}(\delta_n-\theta)\bigr) \ge \frac{EW_b\bigl(\sigma Z/\sqrt{1+\sigma^2}\bigr) - bP\bigl(\|Z\| > c/\sigma\bigr)}{P\bigl(\|Z\| \le c/\sigma\bigr)}.$$

Letting $c \to \infty$ the denominator on the right-hand side tends to one, leaving a lower bound of $EW_b\bigl(\sigma Z/\sqrt{1+\sigma^2}\bigr)$. Because this lower bound must hold for $\sigma$ arbitrarily large,

$$\lim_{c\to\infty} \liminf_{n\to\infty} \sup_{\|\theta-\theta_0\|\le c/\sqrt{n}} E_\theta W_b\bigl(\sqrt{n}(\delta_n-\theta)\bigr) \ge EW_b(Z).$$

The first part of the theorem now follows because $EW_b(Z) \to EW(Z)$ as $b \to \infty$ by monotone convergence. The second part that the asymptotic bound is achieved if $\delta_n = \hat\theta_n$ holds because $\sqrt{n}(\hat\theta_n - \theta) \Rightarrow N(0, J^{-1})$ uniformly over $\|\theta - \theta_0\| \le c/\sqrt{n}$. Using contiguity, this follows from (16.5) and normal approximation for the distributions of $S_n$ mentioned in Remark 16.13. ☐

In addition to the local risk optimality of $\hat\theta$ one can also argue that $\hat\theta$ is asymptotically sufficient, as described in the next result. For a proof see Le Cam and Yang (2000).

**Theorem 16.26.** *Suppose the families $\mathcal{P}_n$ are locally asymptotically normal at every $\theta$ and that estimators $\hat\theta_n$ satisfy (16.5). Then $\hat\theta_n$ is asymptotically sufficient. Specifically, there are other families $\mathcal{Q}_n = \{Q_{\theta,n} : \theta \in \Omega\}$ such that:*

1. *Statistic $\hat\theta_n$ is (exactly) sufficient for $\mathcal{Q}_n$.*
2. *For every $b > 0$ and all $\theta \in \Omega$,*

$$\sup_{|\omega-\theta|\le b/\sqrt{n}} \|Q_{\omega,n} - P_{\omega,n}\| \to 0$$

*as $n \to \infty$.*

For a more complete discussion of asymptotic methods in statistics, see van der Vaart (1998), Le Cam and Yang (2000), or Le Cam (1986).

## 16.7 Problems

1. Consider a regression model in which $Y_i = x_i\beta + \epsilon_i$, $i = 1, 2, \dots$, with the $\epsilon_i$ i.i.d. from $N(0, \sigma^2)$, and assume that $\sum_{i=1}^\infty x_i^2 < \infty$. Let $Q_n$ denote the joint distribution of $Y_1, \dots, Y_n$ if $\beta = \beta_0$, and let $\tilde Q_n$ denote the joint distribution if $\beta = \beta_1$.

a) Show that the distributions $Q_n$ and $\tilde{Q}_n$ are mutually contiguous.
b) Let $L_n$ denote the likelihood ratio $d\tilde{Q}_n/dQ_n$. Find limiting distributions for $L_n$ when $\beta = \beta_0$ and when $\beta = \beta_1$. Are the limiting distributions the same?

2. Let $X_1, X_2, \dots$ be i.i.d. from a uniform distribution on $(0, \theta)$. Let $Q_n$ denote the joint distribution for $X_1, \dots, X_n$ when $\theta = 1$, and let $\tilde{Q}_n$ denote the joint distribution when $\theta = 1 + 1/n^p$ with $p$ a fixed positive constant. For which values of $p$ are $Q_n$ and $\tilde{Q}_n$ mutually contiguous?
3. Prove the second assertion of Proposition 16.5: If the distributions for $\tilde{X}_n$ are contiguous to those for $X_n$, and if $T_n(X_n) = O_p(1)$, then $T_n(\tilde{X}_n) = O_p(1)$.
4. Let $X$ and $Y$ be random vectors with distributions $P_X$ and $P_Y$. If $h$ is a one-to-one function, show that

$$\|P_X - P_Y\| = \|P_{h(X)} - P_{h(Y)}\|.$$

In particular, if $X$ and $Y$ are random variables and $a \neq 0$,

$$\|P_X - P_Y\| = \|P_{aX+b} - P_{aY+b}\|.$$

5. Show that $X_n \xrightarrow{p} 0$ if and only if $E\min\{1, |X_n|\} \to 0$.
6. Define $g(x) = \min\{1, |x|\}$ and let $Z = E\big[|Y| \bigm| X\big]$. Show that $Eg(Z) \geq Eg(Y)$. Use this and the result from Problem 16.5 to show that $L(\Theta, X) \xrightarrow{p} 1$ when (16.13) holds.
7. Let $Y_n$ be integrable random variables. Show that if $E\big[|Y_n| \bigm| Z_n\big] \xrightarrow{p} 0$, then $Y_n \xrightarrow{p} 0$.

# 17

# Large-Sample Theory for Likelihood Ratio Tests

The tests in Chapters 12 and 13 have strong optimality properties but require conditions on the densities for the data and the form of the hypotheses that are rather special and can fail for many natural models. By contrast, the generalized likelihood ratio test introduced in this chapter requires little structure, but it does not have exact optimality properties. Use of this test is justified by large-sample theory. In Section 17.2 we derive approximations for its level and power. Wald tests and score tests are popular alternatives to generalized likelihood ratio tests with similar asymptotic performance. They are discussed briefly in Section 17.4.

## 17.1 Generalized Likelihood Ratio Tests

Let the data $X_1, \dots, X_n$ be i.i.d. with common density $f_\theta$ for $\theta \in \Omega$. The likelihood function is

$$L(\theta) = L(\theta|X_1, \dots, X_n) = \prod_{i=1}^{n} f_\theta(X_i).$$

The (generalized) likelihood ratio statistic for testing $H_0 : \theta \in \Omega_0$ versus $H_1 : \theta \in \Omega_1$ is defined as

$$\lambda = \lambda(X_1, \dots, X_n) = \frac{\sup_{\Omega_1} L(\theta)}{\sup_{\Omega_0} L(\theta)}.$$

The likelihood ratio test rejects $H_0$ if $\lambda > k$. When $H_0$ and $H_1$ are both simple hypotheses, this test is the optimal test described in the Neyman–Pearson lemma.

Typical situations where likelihood ratio tests are used have $\Omega_0$ a smooth manifold of smaller dimension than $\Omega = \Omega_0 \cup \Omega_1$. In this case, if $L(\theta)$ is continuous, $\lambda$ can be computed as

R.W. Keener, *Theoretical Statistics: Topics for a Core Course*, Springer Texts in Statistics, DOI 10.1007/978-0-387-93839-4_17, © Springer Science+Business Media, LLC 2010

$$\lambda = \lambda(X_1, \ldots, X_n) = \frac{\sup_{\Omega} L(\theta)}{\sup_{\Omega_0} L(\theta)}.$$

Furthermore, if these supremum are attained, then

$$\lambda = \frac{L(\hat{\theta})}{L(\tilde{\theta})}, \tag{17.1}$$

where $\hat{\theta}$ is the maximum likelihood estimate of $\theta$ under the full model, and $\tilde{\theta}$ is the maximum likelihood under $H_0$, with $\theta$ varying over $\Omega_0$.

*Example 17.1.* Suppose $X_1, \ldots, X_n$ are a random sample from $N(\mu, \sigma^2)$ and $\theta = (\mu, \sigma)$. The log-likelihood function is

$$l(\theta) = \log L(\theta) = -\frac{n}{2} \log(2\pi\sigma^2) - \sum_{i=1}^{n} \frac{(X_i - \mu)^2}{2\sigma^2}.$$

The partial derivative with respect to $\mu$ is

$$\frac{1}{\sigma^2} \sum_{i=1}^{n} (X_i - \mu).$$

Setting this equal to zero gives

$$\hat{\mu} = \overline{X} = \frac{1}{n} \sum_{i=1}^{n} X_i$$

as the value for $\mu$ that maximizes $l$, regardless of the value of $\sigma$, so $\hat{\mu}$ is the maximum likelihood estimate for $\mu$. We can find the maximum likelihood estimate $\hat{\sigma}$ of $\sigma$ by maximizing $l(\hat{\mu}, \sigma)$ over $\sigma > 0$. Setting

$$\frac{\partial}{\partial \sigma} l(\hat{\mu}, \sigma) = -\frac{n}{\sigma} + \sum_{i=1}^{n} \frac{(X_i - \overline{X})^2}{\sigma^3}$$

equal to zero gives

$$\hat{\sigma}^2 = \frac{1}{n} \sum_{i=1}^{n} (X_i - \overline{X})^2.$$

Using these values in the formula for $l$, after some simplification

$$L(\hat{\mu}, \hat{\sigma}) = \frac{e^{-n/2}}{(2\pi\hat{\sigma}^2)^{n/2}}. \tag{17.2}$$

Suppose we wish to test $H_0 : \mu = 0$ against $H_1 : \mu \neq 0$. The maximum likelihood estimate for $\mu$ under the null hypothesis is $\tilde{\mu} = 0$ (of course). Setting

$$\frac{\partial}{\partial\sigma} l(0,\sigma) = -\frac{n}{\sigma} + \sum_{i=1}^{n} \frac{X_i^2}{\sigma^3}$$

equal to zero gives

$$\tilde{\sigma}^2 = \frac{1}{n}\sum_{i=1}^{n} X_i^2$$

as the maximum likelihood estimate for $\sigma^2$ under $H_0$. After some algebra,

$$L(\tilde{\mu},\tilde{\sigma}) = \frac{e^{-n/2}}{(2\pi\tilde{\sigma}^2)^{n/2}}. \tag{17.3}$$

Using (17.2) and (17.3) in (17.1), the likelihood ratio statistic is

$$\lambda = \frac{\tilde{\sigma}^n}{\hat{\sigma}^n}.$$

Using the identity

$$\sum_{i=1}^{n}(X_i - \overline{X})^2 = \sum_{i=1}^{n} X_i^2 - n\overline{X}^2,$$

we have

$$\begin{aligned}
\lambda &= \left[\frac{\sum_{i=1}^n X_i^2}{\sum_{i=1}^n (X_i-\overline{X})^2}\right]^{n/2} \\
&= \left[\frac{\sum_{i=1}^n (X_i-\overline{X})^2 + n\overline{X}^2}{\sum_{i=1}^n (X_i-\overline{X})^2}\right]^{n/2} \\
&= \left[1 + \frac{n\overline{X}^2}{\sum_{i=1}^n (X_i-\overline{X})^2}\right]^{n/2} \\
&= \left[1 + \frac{T^2}{n-1}\right]^{n/2},
\end{aligned}$$

where $T = \sqrt{n}\,\overline{X}/S$ is the $t$-statistic usually used to test $H_0$ against $H_1$. Since the function relating $\lambda$ to $|T|$ is strictly increasing, the likelihood ratio test is equivalent to the usual $t$-test, which rejects if $|T|$ exceeds a constant.

*Example 17.2.* Let $(X_1, Y_1), \dots, (X_n, Y_n)$ be a sample from a bivariate normal distribution. The log-likelihood is

$$\begin{aligned}
l(\mu_x,\mu_y,\sigma_x,\sigma_y,\rho) = -\frac{1}{2(1-\rho^2)}\Bigg[&\sum_{i=1}^n\left(\frac{X_i-\mu_x}{\sigma_x}\right)^2 + \sum_{i=1}^n\left(\frac{Y_i-\mu_y}{\sigma_y}\right)^2 \\
&- 2\rho\sum_{i=1}^n\left(\frac{X_i-\mu_x}{\sigma_x}\right)\left(\frac{Y_i-\mu_y}{\sigma_y}\right)\Bigg] \\
&- n\log\bigl(2\pi\sigma_x\sigma_y\sqrt{1-\rho^2}\bigr).
\end{aligned}$$

We derive the likelihood ratio test of $H_0 : \rho = 0$ versus $H_1 : \rho \neq 0$. When $\rho = 0$, we have independent samples from two normal distributions, and using results from the previous example,

$$\tilde{\mu}_x = \overline{X}, \qquad \tilde{\mu}_y = \overline{Y}$$

and

$$\tilde{\sigma}_x^2 = \frac{1}{n}\sum_{i=1}^n (X_i - \overline{X})^2, \qquad \tilde{\sigma}_y^2 = \frac{1}{n}\sum_{i=1}^n (Y_i - \overline{Y})^2.$$

The easiest way to find the maximum likelihood estimates for the full model is to note that the family of distributions is a five-parameter exponential family, so the canonical sufficient statistic is the maximum likelihood estimate for its mean. This gives

$$\sum_{i=1}^n X_i = n\hat{\mu}_x, \qquad \sum_{i=1}^n Y_i = n\hat{\mu}_y,$$
$$\sum_{i=1}^n X_i^2 = n(\hat{\mu}_x^2 + \hat{\sigma}_x^2), \qquad \sum_{i=1}^n Y_i^2 = n(\hat{\mu}_y^2 + \hat{\sigma}_y^2),$$

and

$$\sum_{i=1}^n X_i Y_i = n(\hat{\mu}_x\hat{\mu}_y + \hat{\rho}\hat{\sigma}_x\hat{\sigma}_y).$$

Solving these equations gives $\hat{\mu}_x = \tilde{\mu}_x$, $\hat{\mu}_y = \tilde{\mu}_y$, $\hat{\sigma}_x = \tilde{\sigma}_x$, $\hat{\sigma}_y = \tilde{\sigma}_y$, and

$$\hat{\rho} = \frac{\sum_{i=1}^n (X_i - \overline{X})(Y_i - \overline{Y})}{\sqrt{\sum_{i=1}^n (X_i - \overline{X})^2}\sqrt{\sum_{i=1}^n (Y_i - \overline{Y})^2}}.$$

Using (17.1),

$$\begin{aligned}\log\lambda &= \log L(\overline{X}, \overline{Y}, \hat{\sigma}_x, \hat{\sigma}_y, \hat{\rho}) - \log L(\overline{X}, \overline{Y}, \hat{\sigma}_x, \hat{\sigma}_y, 0)\\ &= -\frac{n}{2}\log(1 - \hat{\rho}^2).\end{aligned}$$

Equivalent test statistics are $|\hat{\rho}|$ or $|T|$, where

$$T = \frac{\hat{\rho}\sqrt{n-2}}{\sqrt{1-\hat{\rho}^2}}.$$

Under $H_0$, $T$ has a $t$-distribution on $n-2$ degrees of freedom. In fact, the conditional distribution of $T$ given the $Y_i$ is $t$ on $n-2$ degrees of freedom. To see this, let $Z_i = (X_i - \mu_x)/\sigma_x$, and let $V_i = Y_i - \overline{Y}$. Since $X_i - \overline{X} = \sigma_x(Z_i - \overline{Z})$, we can write

$$\hat{\rho} = \frac{\sum_{i=1}^n (Z_i - \overline{Z})V_i}{\sqrt{\sum_{i=1}^n (Z_i - \overline{Z})^2}\sqrt{\sum_{i=1}^n V_i^2}}.$$

Let $a = (1,\dots,1)'/\sqrt{n}$ and $b = V/\|V\|$. Then $\|a\| = 1$, $\|b\| = 1$, and

$$a\cdot b = \frac{1}{\sqrt{n}\|V\|}\sum_{i=1}^{n}(Y_i - \overline{Y}) = 0.$$

Hence we can find an orthogonal matrix $O$ where the first two columns are $a$ and $b$. Because $O$ is constructed from $Y$, $Z$ and $O$ are independent under $H_0$. By this independence, if we define transformed variables $\check{Z} = O'Z$, then $\check{Z}|O \sim N(0,I)$, which implies that $\check{Z}_1, \dots, \check{Z}_n$ are i.i.d. standard normal. Note that $\check{Z}_1 = a\cdot Z = \sqrt{n}\,\overline{Z}$ and $\check{Z}_2 = b\cdot Z$. Since $\|Z\| = \|\check{Z}\|$,

$$\begin{aligned}\sum_{i=1}^{n}(Z_i - \overline{Z})^2 &= \sum_{i=1}^{n} Z_i^2 - n\overline{Z}^2\\ &= \|Z\|^2 - \check{Z}_1^2\\ &= \sum_{i=2}^{n}\check{Z}_i^2,\end{aligned}$$

and hence

$$\hat{\rho} = \frac{Z\cdot b}{\sqrt{\sum_{i=2}^{n}\check{Z}_i^2}} = \frac{\check{Z}_2}{\sqrt{\sum_{i=2}^{n}\check{Z}_i^2}}.$$

From this, $1-\hat{\rho}^2 = \sum_{i=3}^{n}\check{Z}_i^2/\sum_{i=2}^{n}\check{Z}_i^2$, so

$$T = \frac{\check{Z}_2}{\sqrt{\frac{1}{n-2}\sum_{i=3}^{n}\check{Z}_i^2}}.$$

The sum in the denominator has the chi-square distribution on $n-2$ degrees of freedom, and the numerator and denominator are independent. Therefore this agrees with the usual definition for the $t$-distribution.

## 17.2 Asymptotic Distribution of $2\log\lambda$

In this section we derive the asymptotic distribution of $2\log\lambda$ when $\theta\in\Omega_0$ or $\theta$ is near $\Omega_0$. A rigorous treatment requires considerable attention to detail and deep mathematics, at least if one is concerned with getting the best regularity conditions. To keep the presentation here as accessible as possible, we keep the treatment somewhat informal and base it on assumptions stronger than necessary. Specifically, we assume that conditions necessary for a multivariate version of Theorem 9.14 are in force: the maximum likelihood estimators $\hat{\theta}_n$ are consistent, and the densities $f_\theta(x)$ are regular enough to allow us to define the Fisher information matrix $I(\theta)$ (positive definite for all $\theta\in\Omega$ and continuous as a function of $\theta$) and use Taylor expansion to show that

$$\sqrt{n}(\hat{\theta}_n - \theta) = I^{-1}(\theta)\frac{1}{\sqrt{n}}\nabla l_n(\theta) + o_p(1) \Rightarrow N\bigl(0, I^{-1}(\theta)\bigr). \tag{17.4}$$

The parameter space $\Omega$ is an open subset of $\mathbb{R}^r$, and $\Omega_0$ is a smooth submanifold of $\Omega$ with dimension $q < r$. Finally, we assume that $\tilde{\theta}_n$ is consistent if $\theta \in \Omega_0$.

To use likelihood ratio tests in applications, we need to know the size, so it is natural to want an approximation for the power $\beta_n(\theta)$ of the test when $\theta \in \Omega_0$. Also, to design experiments it is useful to approximate the power at other points $\theta \notin \Omega_0$. Now for fixed $\theta \in \Omega_1$, if $n$ is large enough any reasonable test will likely reject $H_0$ and the power $\beta_n(\theta)$ should tend to one as $n \to \infty$. But a theorem detailing this would not be very useful in practice. For a more interesting theory we study the power at points near $\Omega_0$. Specifically, we study the distribution of $2\log\lambda_n$ along a sequence of parameter values $\theta_n = \theta_0 + \Delta/\sqrt{n}$, where $\theta_0 \in \Omega_0$ and $\Delta$ is a fixed constant, and show that

$$P_{\theta_n}[2\log\lambda_n < t] \to F(t)$$

as $n \to \infty$, where $F$ is the cumulative distribution function for a noncentral chi-square distribution with $r - q$ degrees of freedom. When $\Delta = 0$, $\theta_n = \theta_0 \in \Omega_0$ and this result approximates the cumulative distribution function of $2\log\lambda_n$ under $H_0$. In this case, the noncentrality parameter is zero, so the likelihood ratio test, which rejects if $2\log\lambda$ exceeds the upper $\alpha$th quantile of the chi-square distribution on $r-q$ degrees of freedom, has size approximately $\alpha$. Other choices for $\Delta$ allow one to approximate the power of this test.

The assumptions for Theorems 16.10 or 16.14 are weaker than those above, so the joint distributions for $X_1, \dots, X_n$ under $\theta_n$ are contiguous to the joint distributions under $\theta_0$, and, by Remark 16.13, under $P_{\theta_n}$

$$\frac{1}{\sqrt{n}}\nabla l_n(\theta_0) \Rightarrow N\bigl(I(\theta_0)\Delta, I(\theta_0)\bigr).$$

By Proposition 16.5, a sequence that is $o_p(1)$ under $P_{\theta_0}$ will also be $o_p(1)$ under $P_{\theta_n}$. If we define

$$Z_n = \sqrt{n}(\hat{\theta}_n - \theta_0) = \sqrt{n}(\hat{\theta}_n - \theta_n) + \Delta,$$

then by (17.4), under $P_{\theta_n}$,

$$Z_n \Rightarrow N\bigl(\Delta, I^{-1}(\theta_0)\bigr). \tag{17.5}$$

Equivalently, $\sqrt{n}(\hat{\theta}_n - \theta_n) \Rightarrow N\bigl(0, I^{-1}(\theta_0)\bigr)$, showing in some sense that the usual normal approximation for the distribution of the maximum likelihood estimator holds uniformly over contiguous parameter values, which seems natural.

The rest of the argument is based on using Taylor expansion and the normal equations to relate $2\log\lambda_n$ to $Z_n$. The $o_p$ and $O_p$ notations for scales of magnitude, introduced in Section 8.6, provide a convenient way to keep track of the size of errors in these expansions. Note that $Z_n = O_p(1)$ by (17.5), and so $o_p(Z_n) = o_p(1)$. The normal equations for $\hat\theta_n$ are simple, the gradient of $l_n$ must vanish at $\hat\theta_n$; that is,

$$\nabla l_n(\hat\theta_n) = 0. \tag{17.6}$$

The normal equations for $\tilde\theta_n$ involve the local geometry of $\Omega_0$ and are more delicate. For $\theta \in \Omega_0$, let $V_\theta$ denote the tangent space[1] at $\theta$, and let $P(\theta)$ denote the projection matrix onto $V_\theta$. As $x \in \Omega_0$ approaches $\theta$, $x - \theta$ should almost lie in $V_\theta$. Specifically,

$$x - \theta = P(\theta)(x-\theta) + o\bigl(\|x-\theta\|\bigr). \tag{17.7}$$

Also, the matrices $P(\theta)$ should vary continuously with $\theta$. When $\tilde\theta_n$ lies in the interior of $\Omega$, the directional derivatives of $l_n$ for vectors in the tangent space at $\tilde\theta_n$ must vanish; otherwise we could move a little in $\Omega_0$ and increase the likelihood. So if $\tilde P_n \overset{\text{def}}{=} P(\tilde\theta_n)$, then

$$\tilde P_n \nabla l_n(\tilde\theta_n) = 0. \tag{17.8}$$

Also, by continuity, because $\tilde\theta_n \to \theta_0$, $\tilde P_n \xrightarrow{p} P_0 \overset{\text{def}}{=} P(\theta_0)$, the projection matrix onto the tangent plane of $\Omega_0$ at $\theta_0$. Using this,

$$P_0 \nabla l_n(\tilde\theta_n) = o_p\bigl(\nabla l_n(\tilde\theta_n)\bigr). \tag{17.9}$$

Since $\tilde\theta_n$ and $\theta_0$ are close to each other and both lie in $\Omega_0$, by (17.7) their normalized difference $Y_n = \sqrt{n}(\tilde\theta_n - \theta_0)$ satisfies

$$Y_n = P_0 Y_n + o_p(Y_n). \tag{17.10}$$

Let $\nabla^2 l_n$ denote the Hessian matrix of second partial derivatives of $l_n$. By the weak law of large numbers,

$$\frac{1}{n}\nabla^2 l_n(\theta) = \frac{1}{n}\sum_{i=1}^n \nabla_\theta^2 \log f_\theta(X_i) \xrightarrow{p} -I(\theta)$$

in $P_\theta$-probability as $n \to \infty$, since

$$E_\theta \nabla_\theta^2 \log f_\theta(X_1) = -I(\theta).$$

By contiguity, $\nabla^2 l_n(\theta_0)/n \to -I(\theta_0)$ in $P_{\theta_n}$-probability as $n \to \infty$. Also, using Theorem 9.2, our weak law for random functions, convergence in probability also holds if the Hessian is evaluated at intermediate values approaching $\theta_0$. Using this observation, one-term Taylor expansions of $\nabla l_n/\sqrt{n}$ about $\theta_0$ give

[1] See Appendix A.4 for an introduction to manifolds and tangent spaces.

$$\frac{1}{\sqrt{n}}\nabla l_n(\hat{\theta}_n) - \frac{1}{\sqrt{n}}\nabla l_n(\theta_0) = \bigl(-I(\theta_0) + o_p(1)\bigr)\sqrt{n}(\hat{\theta}_n - \theta_0) \tag{17.11}$$

and

$$\frac{1}{\sqrt{n}}\nabla l_n(\tilde{\theta}_n) - \frac{1}{\sqrt{n}}\nabla l_n(\theta_0) = \bigl(-I(\theta_0) + o_p(1)\bigr)\sqrt{n}(\tilde{\theta}_n - \theta_0). \tag{17.12}$$

With the definition of $Z_n$ and (17.6), the first Taylor approximation above becomes

$$\frac{1}{\sqrt{n}}\nabla l_n(\theta_0) = I(\theta_0)Z_n. + o_p(1). \tag{17.13}$$

Multiplying the second Taylor approximation by $P_0$ and using (17.9),

$$P_0\frac{1}{\sqrt{n}}\nabla l_n(\theta_0) = P_0 I(\theta_0)Y_n + o_p(Y_n) + o_p(1),$$

and these last two equations give

$$P_0 I(\theta_0)Z_n = P_0 I(\theta_0)Y_n + o_p(Y_n) + o_p(1). \tag{17.14}$$

We have now obtained three key equations: (17.5), (17.10), and (17.14). We also need an equation relating $2\log\lambda_n$ to $Y_n$ and $Z_n$. This follows from a two-term Taylor expansion, again equating $\nabla^2 l_n$ at intermediate values with $-nI(\theta_0) + o_p(n)$, which gives

$$\begin{aligned}
2\log\lambda_n &= 2l_n(\hat{\theta}_n) - 2l_n(\tilde{\theta}_n)\\
&= 2(\hat{\theta}_n - \theta_0)'\nabla l_n(\theta_0) - (\hat{\theta}_n - \theta_0)'\bigl(nI(\theta_0) + o_p(n)\bigr)(\hat{\theta}_n - \theta_0)\\
&\quad - 2(\tilde{\theta}_n - \theta_0)'\nabla l_n(\theta_0) + (\tilde{\theta}_n - \theta_0)'\bigl(nI(\theta_0) + o_p(n)\bigr)(\tilde{\theta}_n - \theta_0)\\
&= 2Z_n'\frac{1}{\sqrt{n}}\nabla l_n(\theta_0) - Z_n'\bigl(I(\theta_0) + o_p(1)\bigr)Z_n\\
&\quad - 2Y_n'\frac{1}{\sqrt{n}}\nabla l_n(\theta_0) + Y_n'\bigl(I(\theta_0) + o_p(1)\bigr)Y_n.
\end{aligned}$$

Using (17.13)

$$\begin{aligned}
2\log\lambda_n &= 2Z_n'I(\theta_0)Z_n - Z_n'I(\theta_0)Z_n + o_p(1)\\
&\quad - 2Y_n'I(\theta_0)Z_n + Y_n'I(\theta_0)Y_n + o_p\bigl(\|Y_n\|^2\bigr)\\
&= (Z_n - Y_n)'I(\theta_0)(Z_n - Y_n) + o_p\bigl(\|Y_n\|^2\bigr) + o_p(1).
\end{aligned} \tag{17.15}$$

The approximation for the $P_{\theta_n}$ distributions of $2\log\lambda_n$, mentioned above, follows eventually from (17.5), (17.10), (17.14), and (17.15). The algebra for this derivation is easier if we write the quantities involved in a convenient basis. Let $V = V_{\theta_0}$ denote the tangent space of $\Omega_0$ at $\theta_0$, and let $V^\perp$ denote its orthogonal complement. Then for $v \in V$, $P_0 v = v$, and for $v \in V^\perp$, $P_0 v = 0$. Let $e_1, \ldots, e_q$ be an orthonormal basis for $V$, and let $e_{q+1}, \ldots, e_r$ be an orthonormal basis for $V^\perp$. Because $e_1, \dots, e_r$ is an orthonormal basis for $\mathbb{R}^r$, $O = (e_1, \dots, e_r)$ is an orthogonal matrix. Also, $P_0 O = (e_1, \dots, e_q, 0, \dots, 0)$, so

$$O' P_0 O \stackrel{\text{def}}{=} \check{P} = \begin{pmatrix} I_q & 0 \\ 0 & 0 \end{pmatrix},$$

where $I_q$ denotes the $q \times q$ identity matrix and the zeros are zero matrices with suitable dimensions. In the new basis the key variables are $\check{Y} = O'Y_n$, $\check{Z} = O'Z_n$, $\check{I} = O'I(\theta_0)O$, and $\check{\Delta} = O'\Delta$. By (17.5), $\check{Z} \Rightarrow N_r\bigl(O'\Delta, O'I^{-1}(\theta_0)O\bigr)$, and since $O'I^{-1}(\theta_0)O = \check{I}^{-1}$, this becomes

$$\check{Z} \Rightarrow N_r(\check{\Delta}, \check{I}^{-1}). \tag{17.16}$$

Premultiplying (17.10) by $O'$ and inserting $O'O$ at useful places gives

$$O'P_0OO'Y_n = O'Y_n + o_p(Y_n),$$

or

$$\check{P}\check{Y} = \check{Y} + o_p(\check{Y}). \tag{17.17}$$

Similarly, premultiplying (17.14) by $O'$ gives

$$O'P_0OO'I(\theta_0)OO'Y_n = O'P_0OO'I(\theta_0)OO'Z_n + o_p(Y_n) + o_p(1),$$

or

$$\check{P}\check{I}\check{Y} = \check{P}\check{I}\check{Z} + o_p(\check{Y}) + o_p(1). \tag{17.18}$$

Finally, (17.15) gives

$$2\log\lambda_n = (Z_n - Y_n)OO'I(\theta_0)OO'(Z_n - Y_n) + o_p\bigl(\|Y_n\|\bigr) + o_p(1),$$

which becomes

$$2\log\lambda_n = (\check{Z} - \check{Y})'\check{I}(\check{Z} - \check{Y}) + o_p\bigl(\|\check{Y}\|\bigr) + o_p(1). \tag{17.19}$$

To continue we need to partition $\check{Z}$, $\check{Y}$, and $\check{I}$ as

$$\check{Z} = \begin{pmatrix} \check{Z}_1 \\ \check{Z}_2 \end{pmatrix}, \qquad \check{Y} = \begin{pmatrix} \check{Y}_1 \\ \check{Y}_2 \end{pmatrix}, \qquad \check{I} = \begin{pmatrix} \check{I}_{11} & \check{I}_{12} \\ \check{I}_{21} & \check{I}_{22} \end{pmatrix},$$

where $\check{Z}_1 \in \mathbb{R}^q$, $\check{Y}_1 \in \mathbb{R}^q$, and $\check{I}_{11}$ is $q \times q$. Formula (17.17) gives

$$\check{Y}_2 = o_p(\check{Y}) = o_p(\check{Y}_1) + o_p(\check{Y}_2) \ \text{ or } \ \bigl(1 + o_p(1)\bigr)\check{Y}_2 = o_p(\check{Y}_1),$$

which implies

$$\check{Y}_2 = o_p(\check{Y}_1).$$

Thus $o_p(\check{Y}) = o_p(\check{Y}_1)$, and (17.18) gives

$$\begin{aligned}\check{P}\check{I}\check{Z} &= \begin{pmatrix}\check{I}_{11}\check{Z}_1 + \check{I}_{12}\check{Z}_2 \\ 0\end{pmatrix} \\ &= \check{P}\check{I}\check{Y} + o_p(\|\check{Y}_1\|) + o_p(1) \\ &= \begin{pmatrix}\check{I}_{11}\check{Y}_1 \\ 0\end{pmatrix} + o_p(\|\check{Y}_1\|) + o_p(1).\end{aligned}$$

This can be written as

$$\big(\check{I}_{11} + o_p(1)\big)\check{Y}_1 = \check{I}_{11}\check{Z}_1 + \check{I}_{12}\check{Z}_2 + o_p(1),$$

which implies (since $\check{I}_{11}$ is positive definite)

$$\check{Y}_1 = \check{Z}_1 + \check{I}_{11}^{-1}\check{I}_{12}\check{Z}_2 + o_p(1).$$

Note that since $\check{Z} = O_p(1)$, this equation shows that $\check{Y} = O_p(1)$, which allows us to express errors more simply in what follows. Because

$$\check{Z} - \check{Y} = \begin{pmatrix}-\check{I}_{11}^{-1}\check{I}_{12}\check{Z}_2 \\ \check{Z}_2\end{pmatrix} + o_p(1),$$

(17.19) gives

$$\begin{aligned}2\log\lambda_n &= (-\check{Z}_2'\check{I}_{21}I_{11}^{-1}, \check{Z}_2')\begin{pmatrix}\check{I}_{11} & \check{I}_{12} \\ \check{I}_{21} & \check{I}_{22}\end{pmatrix}\begin{pmatrix}-\check{I}_{11}^{-1}\check{I}_{12}\check{Z}_2 \\ \check{Z}_2\end{pmatrix} + o_p(1) \\ &= \check{Z}_2'(\check{I}_{22} - \check{I}_{21}\check{I}_{11}^{-1}\check{I}_{12})\check{Z}_2 + o_p(1).\end{aligned}$$

Letting $\Sigma = \check{I}^{-1}$, from the formula for inverting partitioned matrices,[2]

$$\Sigma_{22} = (\check{I}_{22} - \check{I}_{21}\check{I}_{11}^{-1}\check{I}_{12})^{-1},$$

and so

$$2\log\lambda_n = \check{Z}_2\Sigma_{22}^{-1}\check{Z}_2 + o_p(1).$$

From (17.16), $\check{Z}_2 \Rightarrow N_{r-q}(\check{\Delta}_2, \Sigma_{22})$. Using Lemma 14.9,

$$2\log\lambda_n \Rightarrow \chi^2_{r-q}(\check{\Delta}_2'\Sigma_{22}^{-1}\check{\Delta}_2).$$

This is the desired result, but for explicit computation it is convenient to express the noncentrality parameter in the original basis. Let $\check{Q} = I - \check{P}$ and $Q_0 = I - P_0$. Then

$$(\check{P} + \check{Q}\Sigma\check{Q})^{-1} = \begin{pmatrix}I_q & 0 \\ 0 & \Sigma_{22}\end{pmatrix}^{-1} = \begin{pmatrix}I_q & 0 \\ 0 & \Sigma_{22}^{-1}\end{pmatrix},$$

and we can express the noncentrality parameter as

[2] See Appendix A.6.

$$\begin{aligned}
&\Delta_2'\Sigma_{22}^{-1}\Delta_2\\
&= \check{\Delta}'\check{Q}\begin{pmatrix} I_q & 0 \\ 0 & \Sigma_{22}^{-1}\end{pmatrix}\check{Q}\check{\Delta}\\
&= \check{\Delta}'\check{Q}(\check{P}+\check{Q}\check{I}^{-1}\check{Q})^{-1}\check{Q}\check{\Delta}\\
&= \Delta'OO'Q_0O\big(O'P_0O+O'Q_0O(O'I(\theta_0)^{-1}O)^{-1}O'Q_0O\big)^{-1}O'Q_0OO'\Delta\\
&= \Delta'Q_0\big(P_0+Q_0I(\theta_0)^{-1}Q_0\big)^{-1}Q_0\Delta.
\end{aligned}$$

When this formula is used in practice, it may be more convenient to substitute $I^{-1}(\theta_n)$ for $I^{-1}(\theta_0)$. Since $\theta_n$ converges to $\theta_0$, for large $n$ this has negligible impact on power calculations.

## 17.3 Examples

*Example 17.3.* As a first example, suppose that $X_1, \dots, X_n, Y_1, \dots, Y_n, Z_1, \dots, Z_n$ are independent with $X_i \sim \text{Poisson}(\theta_1)$, $Y_i \sim \text{Poisson}(\theta_2)$, and $Z_i \sim \text{Poisson}(\theta_3)$, for $i = 1, \dots, n$. We can view this as a random sample of random vectors in $\mathbb{R}^3$ with density

$$f_\theta(x,y,z) = \frac{\theta_1^x\theta_2^y\theta_3^z}{x!y!z!}e^{-\theta_1-\theta_2-\theta_3}.$$

Then

$$\log f_\theta(x,y,z) = x\log\theta_1+y\log\theta_2+z\log\theta_3-\theta_1-\theta_2-\theta_3-\log(x!y!z!),$$

and

$$\nabla_\theta^2 \log f_\theta(x,y,z) = \begin{pmatrix} -x/\theta_1^2 & 0 & 0 \\ 0 & -y/\theta_2^2 & 0 \\ 0 & 0 & -z/\theta_3^2\end{pmatrix}.$$

Hence

$$I(\theta) = -E_\theta\nabla_\theta^2\log f_\theta(X_1,Y_1,Z_1) = \begin{pmatrix} 1/\theta_1 & 0 & 0 \\ 0 & 1/\theta_2 & 0 \\ 0 & 0 & 1/\theta_3\end{pmatrix}.$$

Suppose we want to test $H_0 : \theta_1+\theta_2 = \theta_3$ versus $H_1 : \theta_1+\theta_2 \neq \theta_3$. The log-likelihood is

$$\begin{aligned} l(\theta) = \log(\theta_1)\sum_{i=1}^n X_i + \log(\theta_2)\sum_{i=1}^n Y_i + \log(\theta_3)\sum_{i=1}^n Z_i\\ - n\theta_1 - n\theta_2 - n\theta_3 - \sum_{i=1}^n \log(X_i!Y_i!Z_i!). \end{aligned}$$

Maximizing $l$ gives

$$\hat{\theta}_1 = \overline{X}, \qquad \hat{\theta}_2 = \overline{Y}, \qquad \hat{\theta}_3 = \overline{Z}.$$

Also, $\tilde{\theta}_1$ and $\tilde{\theta}_2$ must maximize

$$\begin{aligned} l(\theta_1, \theta_2, \theta_1 + \theta_2) &= \log(\theta_1) \sum_{i=1}^n X_i + \log(\theta_2) \sum_{i=1}^n Y_i + \log(\theta_1 + \theta_2) \sum_{i=1}^n Z_i \\ &\quad - n\theta_1 - n\theta_2 - n(\theta_1 + \theta_2) - \sum_{i=1}^n \log(X_i! Y_i! Z_i!) \end{aligned}$$

or (dividing by $n$ and dropping the term that is independent of $\theta$)

$$\overline{X} \log \theta_1 + \overline{Y} \log \theta_2 + \overline{Z} \log(\theta_1 + \theta_2) - 2(\theta_1 + \theta_2).$$

Setting partial derivatives with respect to $\theta_1$ and $\theta_2$ equal to zero gives

$$\frac{\overline{X}}{\tilde{\theta}_1} + \frac{\overline{Z}}{\tilde{\theta}_1 + \tilde{\theta}_2} = 2 \text{ and } \frac{\overline{Y}}{\tilde{\theta}_2} + \frac{\overline{Z}}{\tilde{\theta}_1 + \tilde{\theta}_2} = 2.$$

From these equations, $\overline{X}/\tilde{\theta}_1 = \overline{Y}/\tilde{\theta}_2$. So

$$\frac{\overline{Z}}{\tilde{\theta}_1 + \tilde{\theta}_2} = \frac{\overline{Z}}{\tilde{\theta}_2(1 + \overline{X}/\overline{Y})}.$$

Using this in the second normal equation,

$$2\tilde{\theta}_2 = \overline{Y} + \frac{\overline{Z}}{1 + \overline{X}/\overline{Y}} = \overline{Y}\left(\frac{\overline{Z} + \overline{X} + \overline{Y}}{\overline{X} + \overline{Y}}\right).$$

Hence

$$\tilde{\theta}_1 = \frac{\overline{X}}{2}\left(\frac{\overline{Z} + \overline{X} + \overline{Y}}{\overline{X} + \overline{Y}}\right), \qquad \tilde{\theta}_2 = \frac{\overline{Y}}{2}\left(\frac{\overline{Z} + \overline{X} + \overline{Y}}{\overline{X} + \overline{Y}}\right),$$

and

$$\tilde{\theta}_3 = \frac{\overline{Z} + \overline{X} + \overline{Y}}{2}.$$

Now

$$\begin{aligned} 2 \log \lambda &= 2\big(l(\hat{\theta}) - l(\tilde{\theta})\big) \\ &= 2n\Big[\overline{X} \log(\hat{\theta}_1/\tilde{\theta}_1) + \overline{Y} \log(\hat{\theta}_2/\tilde{\theta}_2) + \overline{Z} \log(\hat{\theta}_3/\tilde{\theta}_3) \\ &\qquad + \tilde{\theta}_1 + \tilde{\theta}_2 + \tilde{\theta}_3 - \hat{\theta}_1 - \hat{\theta}_2 - \hat{\theta}_3\Big]. \end{aligned}$$

Since $\tilde{\theta}_1 + \tilde{\theta}_2 + \tilde{\theta}_3 = \hat{\theta}_1 + \hat{\theta}_2 + \hat{\theta}_3$, this simplifies to

$$2 \log \lambda = -2n\left[(\overline{X} + \overline{Y}) \log\left(\frac{\overline{Z} + \overline{X} + \overline{Y}}{2\overline{X} + 2\overline{Y}}\right) + \overline{Z} \log\left(\frac{\overline{Z} + \overline{X} + \overline{Y}}{2\overline{Z}}\right)\right].$$

In this example, $r = 3$ and $q = 2$, so under $H_0$, $2\log\lambda$ is approximately $\chi_1^2$. If $c$ is the $1-\alpha$ quantile of $\chi_1^2$, then the test that rejects if $2\log\lambda > c$ has size approximately $\alpha$. To approximate the power of this test using the results from the last section we need to identify the projection matrices that arise. Because $\Omega_0$ is linear, the tangent space $V = V_{\theta_0}$ is the same for all $\theta_0 \in \Omega_0$. The vectors

$$v_1 = \frac{1}{\sqrt{2}}\begin{pmatrix}-1\\1\\0\end{pmatrix},\quad v_2 = \frac{1}{\sqrt{6}}\begin{pmatrix}1\\1\\2\end{pmatrix},\quad \text{and } v_3 = \frac{1}{\sqrt{3}}\begin{pmatrix}1\\1\\-1\end{pmatrix}$$

form an orthonormal basis for $\mathbb{R}^3$. Both $v_1$ and $v_2$ lie in the tangent space $V$, and $v_3$ lies in $V^\perp$. So

$$P_0 = v_1v_1' + v_2v_2' = \frac{1}{3}\begin{pmatrix}2 & -1 & 1\\ -1 & 2 & 1\\ 1 & 1 & 2\end{pmatrix}$$

and

$$Q_0 = v_3v_3' = \frac{1}{3}\begin{pmatrix}1 & 1 & -1\\ 1 & 1 & -1\\ -1 & -1 & 1\end{pmatrix}.$$

Suppose $\theta = \theta_0 + \Delta/\sqrt{n}$, where $\theta_0 \in \Omega_0$. Then

$$Q_0\theta = Q_0\theta_0 + Q_0\Delta/\sqrt{n},$$

which implies

$$Q_0\Delta = \sqrt{n}Q_0\theta = \sqrt{n}v_3(v_3'\theta).$$

Since

$$I^{-1}(\theta) = \begin{pmatrix}\theta_1 & 0 & 0\\ 0 & \theta_2 & 0\\ 0 & 0 & \theta_3\end{pmatrix},$$

we have

$$\begin{aligned}Q_0I^{-1}(\theta)Q_0 &= v_3\big(v_3'I^{-1}(\theta)v_3\big)v_3'\\ &= \frac{\theta_1+\theta_2+\theta_3}{3}v_3v_3'\\ &= \frac{\theta_1+\theta_2+\theta_3}{3}Q_0.\end{aligned}$$

Hence

$$\begin{aligned}\big(P_0 + Q_0I^{-1}(\theta)Q_0\big)^{-1} &= \left(P_0 + \frac{\theta_1+\theta_2+\theta_3}{3}Q_0\right)^{-1}\\ &= P_0 + \frac{3}{\theta_1+\theta_2+\theta_3}Q_0.\end{aligned}$$

The formula for the noncentrality parameter (substituting $I^{-1}(\theta)$ for $I^{-1}(\theta_0)$) is

$$\begin{aligned}\Delta' Q_0 \left(P_0 + \frac{3}{\theta_1+\theta_2+\theta_3} Q_0\right) Q_0 \Delta &= \frac{3\Delta' Q_0 Q_0 \Delta}{\theta_1+\theta_2+\theta_3} \\ &= \frac{3\|Q_0\Delta\|^2}{\theta_1+\theta_2+\theta_3} \\ &= \frac{3n(v_3'\theta)^2}{\theta_1+\theta_2+\theta_3} \\ &= \frac{n(\theta_1+\theta_2-\theta_3)^2}{\theta_1+\theta_2+\theta_3}.\end{aligned}$$

To be concrete, suppose a test with size 5% is desired. Then one would take $c = 1.96^2 = 3.84$. If $\theta_1 = \theta_2 = 1$, $\theta_3 = 2.3$, and $n = 100$, then the noncentrality parameter comes out $\delta^2 = 9/4.3 = 2.09 = 1.45^2$. If we let $Z \sim N(0,1)$, then $(Z+1.45)^2 \sim \chi_1^2(2.09)$ and the power of the test is approximately

$$\begin{aligned}P(2\log\lambda > 3.84) &\approx P\{(Z+1.45)^2 > 1.96^2\} \\ &= P(Z > .51) + P(Z < -3.41) = 0.3053.\end{aligned}$$

*Example 17.4.* Our final example concerns the classic problem of testing independence in two-way contingency tables. The data are

$$\begin{pmatrix} N_{11} \\ N_{12} \\ N_{21} \\ N_{22} \end{pmatrix} \sim \text{Multinomial}(n; p_{11}, p_{12}, p_{21}, p_{22}).$$

Because the $p_{ij}$ must sum to one, they lie on the unit simplex in $\mathbb{R}^4$. This set is not open, so to apply our results directly we take

$$\theta = \begin{pmatrix} p_{11} \\ p_{12} \\ p_{21} \end{pmatrix}.$$

The maximum likelihood estimates for the $p_{ij}$ are

$$\hat{p}_{ij} = \frac{N_{ij}}{n}.$$

Here we follow the common convention where a "+" as a subscript indicates that terms should be summed; so $p_{i+} = p_{i1} + p_{i2}$ and $N_{+j} = N_{1j} + N_{2j}$, for example. The null hypothesis of independence in the table is

$$H_0 : p_{ij} = p_{i+}p_{+j}, \quad \text{for } i = 1,2 \text{ and } j = 1,2.$$

Equivalently,

$$H_0 : p_{11} = p_{1+}p_{+1}. \tag{17.20}$$

(For instance, if (17.20) holds, then $p_{12} = p_{1+} - p_{11} = p_{1+} - p_{1+}p_{+1} = p_{1+}(1 - p_{+1}) = p_{1+}p_{+2}$.) The log-likelihood function is

$$l = \sum_{i,j} N_{ij} \log p_{ij} + \log \binom{n}{N_{11}, \ldots, N_{22}}.$$

Under $H_0$,

$$\begin{aligned}
l &= \sum_{ij} N_{ij} \log(p_{i+}p_{+j}) + \log \binom{n}{N_{11}, \ldots, N_{22}} \\
&= \sum_{ij} N_{ij} \log(p_{i+}) + \sum_{ij} N_{ij} \log(p_{+j}) + \log \binom{n}{N_{11}, \ldots, N_{22}} \\
&= \sum_{i} N_{i+} \log(p_{i+}) + \sum_{j} N_{+j} \log(p_{+j}) + \log \binom{n}{N_{11}, \ldots, N_{22}} \\
&= N_{1+} \log(p_{1+}) + N_{2+} \log(1 - p_{1+}) + N_{+1} \log(p_{+1}) \\
&\quad + N_{+2} \log(1 - p_{+1}) + \log \binom{n}{N_{11}, \ldots, N_{22}}.
\end{aligned}$$

Setting partial derivatives with respect to $p_{+1}$ and $p_{1+}$ to zero gives the following normal equations for $\tilde{p}_{1+}$ and $\tilde{p}_{+1}$:

$$\frac{N_{1+}}{\tilde{p}_{1+}} - \frac{N_{2+}}{1 - \tilde{p}_{1+}} = 0$$

and

$$\frac{N_{1+}}{\tilde{p}_{+1}} - \frac{N_{+2}}{1 - \tilde{p}_{+1}} = 0.$$

Solving these equations,

$$\tilde{p}_{1+} = \frac{N_{1+}}{n} = \hat{p}_{1+}$$

and

$$\tilde{p}_{+1} = \frac{N_{+1}}{n} = \hat{p}_{+1}.$$

It follows that $\tilde{p}_{+j} = \hat{p}_{+j}$ for $j = 1, 2$ and $\tilde{p}_{i+} = \hat{p}_{i+}$ for $i = 1, 2$. Therefore

$$\tilde{p}_{ij} = \hat{p}_{i+}\hat{p}_{+j}$$

for $i = 1, 2$ and $j = 1, 2$. Plugging in the maximum likelihood estimates derived gives

$$\begin{aligned} 2\log\lambda &= 2l(\hat\theta) - 2l(\tilde\theta) \\ &= 2\sum_{i,j} N_{ij}\log(\hat p_{ij}) - 2\sum_{i,j} N_{ij}\log(\hat p_{i+}\hat p_{+j}) \\ &= 2\sum_{i,j} N_{ij}\log\left(\frac{\hat p_{ij}}{\hat p_{i+}\hat p_{+j}}\right). \end{aligned}$$

Let us now turn our attention to the approximate distribution of the likelihood ratio test statistic. Because $\Omega_0$ has dimension $q = 2$ and $\Omega$ is an open set in $\mathbb{R}^3$, under $H_0$,

$$2\log\lambda \mathrel{\dot\sim} \chi_1^2.$$

To approximate the power at contiguous alternatives we need the Fisher information. We know that

$$\sqrt{n}(\hat\theta - \theta) \Rightarrow N_3\bigl(0, I^{-1}(\theta)\bigr)$$

as $n \to \infty$. Since $\hat\theta$ can be viewed as an average of $n$ i.i.d. vectors, by the central limit theorem,

$$\sqrt{n}(\hat\theta - \theta) \Rightarrow N_3(0, \Sigma)$$

as $n \to \infty$, where $\Sigma$ is the covariance of $\hat\theta$ when $n = 1$. With $n = 1$,

$$\begin{aligned} \operatorname{Cov}(\hat p_{ij}, \hat p_{kl}) &= \operatorname{Cov}(N_{ij}, N_{kl}) \\ &= EN_{ij}N_{kl} - p_{ij}p_{kl} \\ &= \begin{cases} -p_{ij}p_{kl}, & (i,j) \neq (k,l); \\ p_{ij}(1-p_{ij}), & i=k, j=l. \end{cases} \end{aligned}$$

Letting $q_{ij} = 1 - p_{ij}$, we have

$$\Sigma = I^{-1}(\theta) = \begin{pmatrix} p_{11}q_{11} & -p_{11}p_{12} & -p_{11}p_{21} \\ -p_{11}p_{12} & p_{12}q_{12} & -p_{12}p_{21} \\ -p_{11}p_{21} & -p_{12}p_{21} & p_{21}q_{21} \end{pmatrix}.$$

Fix $\theta_0 \in \Omega_0$ and let $V_0$ be the tangent space for $\Omega_0$ at $\theta_0$. To identify the projection matrices $P_0$ and $Q_0$ onto $V_0$ and $V_0^\perp$, note that parameters $\theta \in \Omega_0$ must satisfy the constraint

$$\theta_1 - (\theta_1 + \theta_2)(\theta_1 + \theta_3) \stackrel{\text{def}}{=} g(\theta) = 0.$$

Using results from Appendix A.4, $V_0^\perp$ is the space spanned by the rows of $Dg(\theta_0)$, or the columns of $Dg(\theta_0)' = \nabla g(\theta_0)$. Direct calculation gives

$$\nabla g(\theta) = \begin{pmatrix} 1 - \theta_2 - \theta_3 - 2\theta_1 \\ -(\theta_1 + \theta_3) \\ -(\theta_1 + \theta_2) \end{pmatrix}.$$

Let

$$v_3 = \frac{\nabla g(\theta_0)}{\|\nabla g(\theta_0)\|},$$

and choose $v_1$ and $v_2$ so that $\{v_1, v_2, v_3\}$ is an orthonormal basis for $\mathbb{R}^3$. As in Example 17.3, $v_1$ and $v_2$ span $V_0$, $v_3$ spans $V_0^\perp$,

$$P_0 = v_1 v_1' + v_2 v_2'$$

and

$$Q_0 = v_3 v_3'.$$

The noncentrality parameter is

$$\begin{aligned}
\delta^2 &= \Delta' Q_0 (P_0 + Q_0 I^{-1}(\theta_0) Q_0)^{-1} Q_0 \Delta \\
&= \Delta' v_3 v_3' (P_0 + v_3 v_3' I^{-1}(\theta_0) v_3 v_3')^{-1} v_3 v_3' \Delta \\
&= (v_3'\Delta)^2 v_3' \bigl(P_0 + [v_3' I^{-1}(\theta_0) v_3] Q_0\bigr)^{-1} v_3 \\
&= (v_3'\Delta)^2 v_3' \left(P_0 + \frac{Q_0}{v_3' I^{-1}(\theta_0) v_3}\right) v_3 \\
&= \frac{(v_3'\Delta)^2}{v_3' I^{-1}(\theta_0) v_3} \\
&= \frac{n\bigl(\nabla g(\theta_0)\cdot(\theta_n - \theta_0)\bigr)^2}{\nabla g(\theta_0)' I^{-1}(\theta_0) \nabla g(\theta_0)}.
\end{aligned}$$

The derivation leading to this formula works whenever $r - q = 1$ with $\Omega_0$ the parameters $\theta \in \Omega$ satisfying a single differentiable constraint $g(\theta) = 0$.

To illustrate use of the distributional results in a more concrete setting, let us consider the following design question. How large should the sample size be to achieve a test with (approximate) level $\alpha = 5\%$ and power 90% when $p_{11} = p_{22} = 0.3$ and $p_{12} = p_{21} = 0.2$? The parameter value associated with these cell probabilities is

$$\theta_n = \begin{pmatrix} 0.3 \\ 0.2 \\ 0.2 \end{pmatrix}.$$

Under $H_0$, $2\log\lambda \sim Z^2$, where $Z \sim N(0,1)$. Since $P\bigl(Z^2 > (1.96)^2\bigr) = 1 - P(-1.96 < Z < 1.96) = 5\% = \alpha$, the test should reject if $2\log\lambda > (1.96)^2$. Since $(Z+\delta)^2 \sim \chi_1^2(\delta^2)$, under the alternative $\theta_n$,

$$2\log\lambda \,\dot\sim\, (Z+\delta)^2.$$

To meet the design objective, we need

$$\begin{aligned}
90\% &\approx P\bigl((Z+\delta)^2 > (1.96)^2\bigr) \\
&= P(Z > 1.96 - \delta) + P(Z < -1.96 - \delta).
\end{aligned}$$

The second term here is negligible, so we require

$$90\% \approx P(Z > 1.96 - \delta).$$

This holds if $1.96-\delta$ is the 10th percentile of the standard normal distribution. This percentile is $-1.282$ which gives $\delta = 1.96+1.282 = 3.242$ and $\delta^2 = 10.51$. The marginal cell probabilities under $\theta_n$ are $p_{i+} = p_{+j} = 1/2$, so the natural choice for $\theta_0$ is

$$\theta_0 = \begin{pmatrix} 1/4 \\ 1/4 \\ 1/4 \end{pmatrix}.$$

Then

$$\nabla g(\theta_0) = \begin{pmatrix} 0 \\ -1/2 \\ -1/2 \end{pmatrix}$$

and

$$\nabla g(\theta_0) \cdot (\theta_n - \theta_0) = \begin{pmatrix} 0 \\ -1/2 \\ -1/2 \end{pmatrix} \cdot \begin{pmatrix} 1/20 \\ -1/20 \\ -1/20 \end{pmatrix} = \frac{1}{20}.$$

Also,

$$I^{-1}(\theta_0) = \frac{1}{4}\begin{pmatrix} 3/4 & -1/4 & -1/4 \\ -1/4 & 3/4 & -1/4 \\ -1/4 & -1/4 & 3/4 \end{pmatrix},$$

and so

$$\nabla g(\theta_0)' I^{-1}(\theta_0) \nabla g(\theta_0) = \frac{1}{4}\left(\frac{3}{4} - \frac{1}{4} - \frac{1}{4} + \frac{3}{4}\right) = \frac{1}{16}.$$

Hence

$$\delta^2 = \frac{n(1/20)^2}{(1/4)^2} = \frac{n}{25}.$$

Setting this equal to 10.51 gives $n = 263$ as the sample size required for the level and power specified.

In practice many statisticians test independence in $2 \times 2$ tables using Pearson's chi-square test statistic,

$$T = \sum_{i,j} \frac{(N_{ij} - n\hat{p}_{i+}\hat{p}_{+j})^2}{n\hat{p}_{i+}\hat{p}_{+j}}.$$

For large $n$, $T$ and $2\log\lambda$ are asymptotically equivalent. To demonstrate this equivalence (without any serious attempt at mathematical rigor), we write

$$2\log\lambda = 2n\sum_{i,j} \hat{p}_{ij} \log\left(\frac{\hat{p}_{ij}}{\hat{p}_{i+}\hat{p}_{+j}}\right).$$

We view this as a function of the $\hat{p}_{ij}$ with the marginal probabilities $\hat{p}_{i+}$ and $\hat{p}_{+j}$ considered *fixed constants* and Taylor expand about $\hat{p}_{ij} = \hat{p}_{i+}\hat{p}_{+j}$ (equality here is approximately correct under both $H_0$ and contiguous alternatives). To compute the gradient of the function,

$$\frac{\partial}{\partial \hat{p}_{kl}}(2\log\lambda) = 2n\left[\log\left(\frac{\hat{p}_{kl}}{\hat{p}_{k+}\hat{p}_{+l}}\right)+1\right]. \tag{17.21}$$

Then

$$\frac{\partial}{\partial \hat{p}_{ij}}(2\log\lambda)\bigg|_{\hat{p}_{ij}=\hat{p}_{i+}\hat{p}_{+j}} = 2n,$$

and the gradient at the point of expansion is

$$\begin{pmatrix} 2n \\ 2n \\ 2n \\ 2n \end{pmatrix}.$$

Taylor expansion through the gradient term gives

$$2\log\lambda \approx 2n\sum_{i,j}(\hat{p}_{ij}-\hat{p}_{i+}\hat{p}_{+j}) = 0.$$

To get an interesting answer we need to keep an extra term in our Taylor expansion. Because (17.21) only depends on $\hat{p}_{ij}$, the Hessian matrix is diagonal. Now

$$\frac{\partial^2}{\partial \hat{p}_{ij}^2}(2\log\lambda) = \frac{2n}{\hat{p}_{ij}},$$

so

$$\frac{\partial^2}{\partial \hat{p}_{ij}^2}(2\log\lambda)\bigg|_{\hat{p}_{ij}=\hat{p}_{i+}\hat{p}_{+j}} = \frac{2n}{\hat{p}_{i+}\hat{p}_{+j}}.$$

Taylor expansion through the Hessian term gives

$$2\log\lambda \approx \frac{1}{2}\sum_{i,j}\frac{2n}{\hat{p}_{i+}\hat{p}_{+j}}(\hat{p}_{ij}-\hat{p}_{i+}\hat{p}_{+j})^2 = T.$$

## 17.4 Wald and Score Tests

The Wald and score (or Lagrange multiplier) tests are alternatives to the generalized likelihood ratio tests with similar properties. Assume that the null hypothesis can be written as $H_0 : g(\theta) = 0$, with the constraint function $g : \Omega \to \mathbb{R}^{r-q}$ continuously differentiable and $Dg(\theta)$ of full rank for $\theta \in \Omega_0$.

The basic idea behind the Wald test (Wald (1943)) is simply that if $H_0$ is correct, then $g(\hat{\theta}_n)$ should be close to zero. By Proposition 9.32, if $\theta \in \Omega_0$, then

$$\sqrt{n}g(\hat{\theta}_n) \Rightarrow N\left[0, Dg(\theta)I(\theta)^{-1}\big(Dg(\theta)\big)'\right].$$

By Lemma 14.9 and results on weak convergence,

$$T_W \stackrel{\text{def}}{=} n\big(g(\hat{\theta}_n)\big)'\left[Dg(\hat{\theta}_n)I(\hat{\theta}_n)^{-1}\big(Dg(\hat{\theta}_n)\big)'\right]^{-1} g(\hat{\theta}_n) \Rightarrow \chi^2_{r-q}$$

when $\theta \in \Omega_0$.

Rao's score test (Rao (1948)) is based on the notion that if $\theta \in \Omega_0$, then $\tilde{\theta}_n$ should be a good estimate of $\theta$ and the gradient of the log-likelihood should not be too large at $\tilde{\theta}_n$. Differencing (17.12) and (17.11), for $\theta \in \Omega_0$

$$\frac{1}{\sqrt{n}}\nabla l_n(\tilde{\theta}_n) = I(\theta)(Z_n - Y_n) + o_p(1)$$

as $n \to \infty$. Using (17.13) it is then not hard to show that

$$T_S \stackrel{\text{def}}{=} \frac{1}{n}\big(\nabla l_n(\tilde{\theta}_n)\big)'I(\tilde{\theta}_n)^{-1}\nabla l_n(\tilde{\theta}_n) = 2\log\lambda_n + o_p(1) \Rightarrow \chi^2_{r-q}.$$

The three test statistics, $2\log\lambda$, $T_W$, and $T_S$, have different strengths and weaknesses. Although the derivation here only considers the asymptotic null distributions of $T_W$ and $T_S$, with the methods and regularity assumed in Section 17.2 it is not hard to argue that all three tests are asymptotically equivalent under distributions contiguous to a null distribution; specifically, differences between any two of the statistics will tend to zero in probability. Furthermore, variants of $T_W$ and $T_S$ in which the Fisher information is estimated consistently in a different fashion, perhaps using observed Fisher information, are also equivalent under distributions contiguous to a null distribution.

The score test only relies on $\tilde{\theta}_n$. The maximum likelihood estimator $\hat{\theta}_n$ under the full model is not needed. This may be advantageous if the full model is difficult to fit. Unfortunately, it also means that although the test will have good power at alternatives near a null distribution, the power may not be high at more distant alternatives. In fact, there are examples where the power of the score test does not tend to one at fixed alternatives as $n \to \infty$. See Freedman (2007).

In contrast to the score test, the Wald test statistic relies only on the maximum likelihood estimator under the full model $\hat{\theta}_n$, and there is no need to compute $\tilde{\theta}_n$. This may make $T_W$ easier to compute than $2\log\lambda$.

With a fixed nominal level $\alpha$, at a fixed alternative $\theta \in \Omega_1$ the powers of the generalized likelihood, Wald, and the score test generally tend to one quickly. With sufficient regularity, the convergence occurs exponentially quickly, with the generalized likelihood ratio test having the best possible rate of convergence. This rate of convergence is called the *Bahadur slope* for the test, and the generalized likelihood ratio is thus *Bahadur efficient*. But from a practical standpoint, the ability of a test to detect smaller differences may be more important, and in this regard it is harder to say which of these tests is best.

## 17.5 Problems[3]

1. Consider three samples: $W_1, \ldots, W_k$ from $N(\mu_1, \sigma_1^2)$; $X_1, \ldots, X_m$ from $N(\mu_1, \sigma_2^2)$; and $Y_1, \ldots, Y_n$ from $N(\mu_2, \sigma_2^2)$, all independent, where $\mu_1$, $\mu_2$, $\sigma_1$, and $\sigma_2$ are unknown parameters. Derive the generalized likelihood test statistic $\lambda$ to test $H_0 : \sigma_1 = \sigma_2$ versus $H_1 : \sigma_1 \neq \sigma_2$. You should be able to reduce the normal equations under the full model to a single cubic equation. Explicit solution of this cubic equation is not necessary.
2. Consider data for a two-way contingency table $N_{11}$, $N_{12}$, $N_{21}$, $N_{22}$ from a multinomial distribution with $n$ trials and success probabilities $p_{11}$, $p_{12}$, $p_{21}$, $p_{22}$. Derive the generalized likelihood test statistic $\lambda$ to test "symmetry," $H_0 : p_{12} = p_{21}$ versus $H_1 : p_{12} \neq p_{21}$.
*3. *Random effects models.* One model that is used to analyze a blocked experiment comparing $p$ treatments has $Y_{ij} = \alpha_i + \beta_j + \epsilon_{ij}$, $i = 1, \ldots, p$, $j = 1, \ldots, n$, with the $\alpha_i$ and $\beta_j$ viewed as unknown constant parameters and the $\epsilon_{ij}$ unobserved and i.i.d. from $N(0, \sigma^2)$. In some circumstances, it may be more natural to view the blocking variables $\beta_j$ as random, perhaps as i.i.d. from $N(0, \tau^2)$ and independent of the $\epsilon_{ij}$. This gives a model in which the vectors $Y_j = (Y_{1j}, \ldots, Y_{pj})'$, $j = 1, \ldots, n$, are i.i.d. from $N(\alpha, \sigma^2 I + \tau^2 \mathbf{1}\mathbf{1}')$. Here "$\mathbf{1}$" denotes a column of 1s in $\mathbb{R}^p$, and the unknown parameters are $\alpha \in \mathbb{R}^p$, $\sigma^2 > 0$, and $\tau^2 \geq 0$.
   a) Derive the likelihood ratio test statistic to test $H_0 : \tau^2 = 0$ versus $H_1 : \tau^2 > 0$.
   b) Derive the likelihood ratio test statistic to test $H_0 : \alpha_1 = \cdots = \alpha_p$.
*4. Let $X$ and $Y$ be independent exponential variables with failure rates $\theta_x$ and $\theta_y$, respectively.
   a) Find the generalized likelihood ratio test statistic $\lambda$, based on $X$ and $Y$, to test $H_0 : \theta_x = 2\theta_y$ versus $H_1 : \theta_x \neq 2\theta_y$.
   b) Suppose the test rejects if $\lambda \geq c$. How should the critical level $c$ be adjusted to give level $\alpha$?
5. Let $X$ and $Y$ be independent normal variables, both with variance one and means $\theta_X = EX$ and $\theta_Y = EY$.
   a) Derive the generalized likelihood test statistic $\lambda$ (or $\log \lambda$) to test $H_0 : \theta_X = 0$ or $\theta_Y = 0$ versus $H_1 : \theta_X \neq 0$ and $\theta_y \neq 0$.
   b) The likelihood test using $\lambda$ from part (a) rejects $H_0$ if $\log \lambda > k$. Derive a formula for the power of this test when $\theta_X = 0$.
   c) Find the significance level $\alpha$ as a function of $k$. How should $k$ be chosen to achieve a desired level $\alpha$?
6. Suppose $X \sim N_p(\theta, I)$ and consider testing $H_0 : \theta \in \Omega_0$ versus $H_1 : \theta \notin \Omega_0$.
   a) Show that the likelihood ratio test statistic $\lambda$ is equivalent to the distance $D$ between $X$ and $\Omega_0$, defined as

$$D = \inf\{\|X - \theta\| : \theta \in \Omega_0\}.$$

[3] Solutions to the starred problems are given at the back of the book.

(Equivalent here means there is a one-to-one increasing relationship between the two statistics.)

b) Using part (a), a generalized likelihood ratio test will reject $H_0$ if $D > c$. What is the significance level $\alpha$ for this test if $p = 2$ and $\Omega_0 = \{\theta : \theta_1 \le 0, \theta_2 \le 0\}$?

7. Show that in the general linear model there is an increasing one-to-one relationship between the generalized likelihood ratio statistic $\lambda$ and the test statistic $T$ in (14.24), so that tests based on $\lambda$ and $T$ are equivalent.

*8. Let $X_1, \dots, X_n$ be a random sample from an exponential distribution with mean $\theta_1$, and let $Y_1, \dots, Y_n$ be an independent sample from an exponential distribution with mean $\theta_2$.

a) Find the likelihood ratio test statistic for testing $H_0 : \theta_1/\theta_2 = c_0$ versus $H_1 : \theta_1/\theta_2 \ne c_0$, where $c_0$ is a constant.

b) Use the large-sample approximation for the null distribution of $2\log\lambda$ and the duality between testing and interval estimation to describe a confidence set for $\theta_1/\theta_2$ with coverage probability approximately 95% (the set is an interval, but you do not have to demonstrate this fact). If $n = 100$, $\overline{X} = 2$, and $\overline{Y} = 1$, determine whether the parameter ratio 2.4 lies in the confidence set.

c) How large should the sample size $n$ be if we want the likelihood ratio test for testing $H_0 : \theta_1 = \theta_2$ versus $H_1 : \theta_1 \ne \theta_2$ at level 5% to have power 90% when $\theta_1 = 0.9$ and $\theta_2 = 1.1$?

*9. Let $W_1, \dots, W_n$, $X_1, \dots, X_n$, and $Y_1, \dots, Y_n$ be independent random samples from $N(\mu_w, \sigma_w^2)$, $N(\mu_x, \sigma_x^2)$, and $N(\mu_y, \sigma_y^2)$, respectively.

a) Find the likelihood ratio test statistic for testing $H_0 : \sigma_w = \sigma_x = \sigma_y$ versus the alternative that at least two of the standard deviations differ.

b) What is the approximate power of the likelihood ratio test with level $\alpha = 5\%$ if $n = 200$, $\sigma_w = 1.8$, $\sigma_x = 2.2$ and $\sigma_y = 2.0$? You can express the answer in terms of a noncentral chi-square distribution, but identify the appropriate degrees of freedom and the noncentrality parameter.

*10. Suppose $X_1, \dots, X_n$ are i.i.d. $N_p(\mu, I)$.

a) Derive the likelihood ratio test statistic $2\log\lambda$ to test $H_0 : \|\mu\| = r$ versus $H_1 : \|\mu\| \ne r$, where $r$ is a fixed constant.

b) Give a formula for the power of the likelihood ratio test that rejects $H_0$ when $2\log\lambda > c$ in terms of the cumulative distribution function for a noncentral chi-square distribution.

c) If $r = 1$, what sample size will be necessary for the test with $\alpha \approx 5\%$ to have power approximately 90% when $\|\mu\| = 1.1$?

11. Let $W_1, \dots, W_n$, $X_1, \dots, X_n$, and $Y_1, \dots, Y_n$ be independent random samples. The $W_i$ have density $e^{-|x-\theta_1|}/2$, the $Y_i$ have density $e^{-|x-\theta_2|}/2$, and the $X_i$ have density $e^{-|x-\theta_3|}/2$. Derive the approximate power for the likelihood ratio test with $\alpha = 5\%$ of $H_0 : \theta_1 = \theta_2 = \theta_3$ if $n = 200$, $\theta_1 = 1.8$, $\theta_2 = 2.0$, and $\theta_3 = 2.2$. You can express the answer in terms of

a noncentral chi-square distribution, but identify the appropriate degrees of freedom and the noncentrality parameter.

*12. *Errors in variables models.* Consider a regression model in which

$$Y_i = \beta X_i + \epsilon_i, \qquad i = 1, \dots, n,$$

with the $X_i$ a random sample from $N(0,1)$ and the $\epsilon_i$ an independent random sample, also from $N(0,1)$. In some situations, the independent variables $X_i$ may not be observed directly. One possibility is that they are measured with error. For a specific model, let

$$W_i = X_i + \eta_i, \qquad i = 1, \dots, n.$$

The $\eta_i$ are modeled as a random sample from $N(0, \sigma^2)$, independent of the $X_i$ and $\epsilon_i$. The data are $W_1, \dots, W_n$ and $Y_1, \dots, Y_n$, with $\beta$ and $\sigma$ unknown parameters.

a) Determine the joint distribution of $W_i$ and $Y_i$.
b) Describe how to compute the generalized likelihood ratio test statistic to test $H_0 : \beta = 0$ versus $H_1 : \beta \neq 0$. An explicit formula for the maximum likelihood estimators may not be feasible, but you should give equations that can be solved to find the maximum likelihood estimators.
c) Show that the least squares estimate for $\beta$ when $\sigma = 0$ (the estimator that one would use when the model ignores measurement error for the independent variable) is inconsistent if $\sigma > 0$.
d) Derive an approximation for the power of the generalized likelihood ratio test with level $\alpha \approx 5\%$ when $\beta = \Delta/\sqrt{n}$. How does $\sigma$ effect the power of the test?

*13. *Goodness-of-fit test.* Let $X_1, \dots, X_n$ be a random sample from some continuous distribution on $(0, \infty)$, and let $Y_1$ be the number of observations in $(0,1)$, $Y_2$ the number of observations in $[1,2)$, and $Y_3$ the number of observations in $[2, \infty)$. Then $Y$ has a multinomial distribution with $n$ trials and success probabilities $p_1, p_2, p_3$. If the distribution of the $X_i$ is exponential with failure rate $\theta$, then $p_1 = 1 - e^{-\theta}$, $p_2 = e^{-\theta} - e^{-2\theta}$, and $p_3 = e^{-2\theta}$.

a) Derive a generalized likelihood ratio test of the null hypothesis that the $X_i$ come from an exponential distribution. The test should be based on data $Y_1, Y_2, Y_3$.
b) If $\alpha \approx 5\%$ and $Y = (36, 24, 40)$, would the test in part (a) accept or reject the null hypothesis? What is the attained significance with these data (approximately)?
c) How large should the sample size be if we want the test with level $\alpha \approx 5\%$ to have power 90% when $p_1 = 0.36$, $p_2 = 0.24$, and $p_3 = 0.4$?

14. Let $X_1, \dots, X_n$ be i.i.d. from a Fisher distribution on the unit sphere in $\mathbb{R}^3$ with common density

$$f_\theta(x) = \frac{\|\theta\| e^{\theta \cdot x}}{4\pi \sinh(\|\theta\|)},$$

with respect to surface area on the unit sphere. When $\theta = 0$, $f_0(x) = 1/(4\pi)$, so the variables are uniformly distributed. (These distributions are often used to model solid angles.)

a) Describe how you would test $H_0 : \theta_2 = \theta_3 = 0$ versus $H_1 : \theta_2 \neq 0$ or $\theta_3 \neq 0$, giving the normal equations you would use to solve for $\hat{\theta}_n$ and $\tilde{\theta}_n$. If $\alpha = 5\%$, $n = 100$, and $\overline{X}_n = (0.6, 0.1, 0.1)$, would you accept or reject $H_0$?

b) What is the approximate power of the likelihood test if $n = 100$, $\alpha = 5\%$, and $\theta = (1, .2, 0)$? Express the answer using the noncentral chi-square distribution, but identify the degrees of freedom and the noncentrality parameter.

15. Let $X_{ij}$, $i = 1, \dots, p$, $j = 1, \dots, n$, be i.i.d. from a standard exponential distribution, and given $X = x$, let $Y_{ij}$, $i = 1, \dots, p$, $j = 1, \dots, n$, be independent Poisson variables with $EY_{ij} = x_{ij}\theta_i$. Then $(X_{1j}, \dots, X_{pj}, Y_{1j}, \dots, Y_{pj})$, $j = 1, \dots, n$, are i.i.d. random vectors. Consider testing $H_0 : \theta_1 = \dots = \theta_p$ versus $H_1 : \theta_i \neq \theta_k$ for some $i \neq k$.

a) Find the likelihood ratio test statistic to test $H_0$ against $H_1$.

b) What is the approximate power for the likelihood ratio test if $\alpha = 5\%$, $n = 100$, $p = 5$, $\theta_1 = 1.8$, $\theta_2 = 1.9$, $\theta_3 = 2.0$, $\theta_4 = 2.1$, and $\theta_5 = 2.2$? Express your answer using the noncentral chi-square distribution, but give the degrees of freedom and the noncentrality parameter.

c) If $p = 2$ and $\alpha \approx 5\%$, how large should the sample size $n$ be if power 90% is desired when $\theta_1 = 1.9$ and $\theta_2 = 2.1$?

16. Define

$$\Omega_0 = \{x \in (0, \infty)^3 : x_1 x_2 x_3 = 10\},$$

a manifold in $\mathbb{R}^3$.

a) What is the dimension of $\Omega_0$?

b) Let $V$ be the tangent space for this manifold at $x = (1, 2, 5)$. Find an orthonormal basis $e_1, e_2, e_3$ with $e_1, \dots, e_q$ spanning $V$ and $e_{q+1}, \dots, e_3$ spanning $V^\perp$.

c) Find projection matrices $P$ and $Q$ onto $V$ and $V^\perp$, respectively.

17. Define a vector-valued function $\eta : \mathbb{R}^2 \to \mathbb{R}^3$ by $\eta_1(x, y) = x^2$, $\eta_2(x, y) = y^2$, and $\eta_3(x, y) = (x - y)^2$, and let $\Omega_0 = \eta\big((0, \infty)^2\big)$. Let $V$ be the tangent space for $\Omega_0$ at $(4, 1, 1)$. Find orthonormal vectors that span $V$.

18. Let $(N_{11}, \dots, N_{22})$ have a multinomial distribution, as in Example 17.4, and consider testing whether the marginal distributions are the same, that is, testing $H_0 : p_{+1} = p_{1+}$ versus $H_1 : p_{+1} \neq p_{1+}$.

a) Derive a formula for the generalized likelihood ratio test statistic $2 \log \lambda$.

b) How large should the number of trials be if we want the test with level $\alpha \approx 5\%$ to have power 90% when $p_{11} = 30\%$, $p_{12} = 15\%$, $p_{21} = 20\%$, and $p_{22} = 35\%$?

# 18

# Nonparametric Regression

Regression models are used to study how a dependent or response variable depends on an independent variable or variables. The regression function $f(x)$ is defined as the mean for a response variable $Y$ when the independent variable equals $x$. In model form, with $n$ observations, we may write

$$Y_i = f(x_i) + \epsilon_i, \qquad i = 1, \dots, n,$$

with $E\epsilon_i = 0$.

Classically, the regression function $f$ is assumed to lie in a class of functions specified by a finite number of parameters. For instance, in quadratic regression $f$ is a quadratic function, $f(x) = \beta_0 + \beta_1 x + \beta_2 x^2$, specified by three parameters, $\beta_0$, $\beta_1$, and $\beta_2$. This approach feels natural with a small- or moderate-size data set, as the data in this case may not be rich enough to support fitting a more complicated model. But with more data a researcher will often want to consider more involved models, since in most applications there is little reason to believe the regression function lies exactly in some narrow parametric family. Of course one could add complexity by increasing the number of parameters, fitting perhaps a cubic or quartic function, say, instead of a quadratic. But this approach may have limitations, and recently there has been considerable interest in replacing parametric assumptions about $f$ with more qualitative assumptions about the smoothness of $f$.

In this chapter we explore two approaches. We begin with kernel methods, based on Clark (1977), which exploit the assumed smoothness of $f$ in a fairly direct fashion. With this approach the regression function is estimated as a weighted average of the responses with similar values for the independent variable. The other approach, splines, is derived by viewing the regression function $f$ as an unknown parameter taking values in an infinite-dimensional vector space. This approach is developed in Section 18.3, following a section extending results about finite-dimensional vector spaces to Hilbert spaces. The chapter closes with a section showing how similar ideas can be used for density estimation.

R.W. Keener, *Theoretical Statistics: Topics for a Core Course*, Springer Texts in Statistics, DOI 10.1007/978-0-387-93839-4_18, © Springer Science+Business Media, LLC 2010

## 18.1 Kernel Methods

Consider a regression model in which

$$Y_i = f(x_i) + \epsilon_i, \qquad i = 1, \dots, n,$$

where $\epsilon_1, \dots, \epsilon_n$ are mean zero, uncorrelated random variables with a common variance $\sigma^2$. The independent variables $x_1, \dots, x_n$ are viewed as (observed or known) constants, and the response variables $Y_1, \dots, Y_n$ are observed and random. The errors $\epsilon_1, \dots, \epsilon_n$ are not observed, and $\sigma^2$ is an unobserved parameter. The regression function $f$ is unknown and is not assumed to lie in any parametric class. But we do assume it is twice continuously differentiable. Finally, for convenience, assume $x_1 < \cdots < x_n$.

One conceivable estimator for $f$ might be the function $\hat{h}$ obtained from the data by linear interpolation. This function is $Y_i$ when $x = x_i$ and is linear between adjacent values for $x$, so

$$\hat{h}(x) = Y_i \frac{x - x_{i+1}}{x_i - x_{i+1}} + Y_{i+1} \frac{x - x_i}{x_{i+1} - x_i}, \qquad x \in [x_i, x_{i+1}]. \tag{18.1}$$

If the errors are very small then $\hat{h}$ may lie close to $f$, but when the errors are appreciable $\hat{h}$ will jump up and down too much to be a sensible estimator. This can be seen in the plot to the left in Figure 18.1, with the true regression function $f$ shown as a dashed line and $\hat{h}$ given as a solid line.

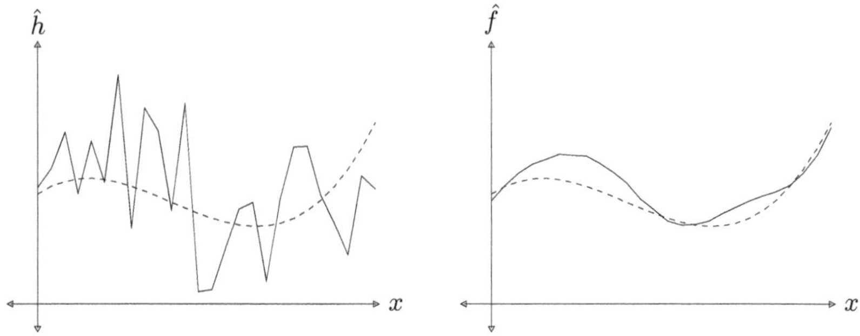


**Fig. 18.1.** Kernel smoothing: left: $\hat{h}$; right: $\hat{f}$.

One way to make a function smoother is through convolution. Doing this with $\hat{h}$ leads to an estimator

$$\hat{f}(x) = \frac{1}{b} \int \hat{h}(t) W\left(\frac{x-t}{b}\right) dt, \tag{18.2}$$

where $W$ is a probability density and $b$, called the *bandwidth*, controls the amount of smoothing. The plot to the right in Figure 18.1 shows[1] $\hat{f}$, and it is

[1] The graphs are based on simulated data with $f(x) = 16x^3 - 19.2x^2 + 4.68x + 2.3$, $\epsilon_i \sim N(0,1)$, $W(x) = (1-|x|)^+$, $b = 0.24$, and $x_{i+1} - x_i = 0.04$. To avoid problems with edge effects, $\hat{f}$ is based on observations (not graphed) with $x \notin [0,1]$.

indeed smoother than $\hat{h}$. Viewing this integral in (18.2) as an expectation,

$$\hat{f}(x) = E_Z \hat{h}(x - bZ),$$

where $Z$ is an absolutely continuous variable with density $W$. From this we can see that $\hat{f}(x)$ is a weighted average of $\hat{h}(y)$ over values $y$ of order $b$ from $x$. With increasing $b$ there is more averaging and, as intuition suggests, this will improve the variance of $\hat{f}(x)$. But if $f$ curves near $x$, averaging will induce bias in the estimator $\hat{f}$ that grows as $b$ increases. We explore this below when the independent variables are equally spaced, trying to choose the bandwidth to balance these concerns.

The estimators $\hat{h}$ and $\hat{f}$ are both linear functions of $Y$. Using (18.1),

$$\hat{h}(x) = \sum_{i=1}^{n} u_i(x) Y_i,$$

where

$$u_i(x) = \begin{cases} \dfrac{x - x_{i-1}}{x_i - x_{i-1}}, & x \in [x_{i-1}, x_i]; \\[2ex] \dfrac{x - x_{i+1}}{x_i - x_{i+1}}, & x \in [x_i, x_{i+1}]; \\[2ex] 0, & \text{otherwise}. \end{cases}$$

Using this in (18.2),

$$\hat{f}(x) = \frac{1}{b} \int \sum_{i=1}^{n} u_i(t) Y_i W\left(\frac{x - t}{b}\right) dt = \sum_{i=1}^{n} v_i(x) Y_i, \tag{18.3}$$

where

$$v_i(x) = \frac{1}{b} \int u_i(t) W\left(\frac{x - t}{b}\right) dt. \tag{18.4}$$

With this linear structure, moments of $\hat{f}(x)$ should be easy to compute.

If we let $h(x) = E\hat{h}(x)$, then by (18.1),

$$h(x) = f(x_i)\frac{x - x_{i+1}}{x_i - x_{i+1}} + f(x_{i+1})\frac{x - x_i}{x_{i+1} - x_i}, \qquad x \in [x_i, x_{i+1}],$$

so $h$ is the linear interpolant of $f$. The difference between $h$ and $f$ can be estimated by Taylor approximation. If $x \in [x_i, x_{i+1}]$, then

$$f(x) = f(x_i) + (x - x_i) f'(x_i) + \int_{x_i}^{x} (x - u) f''(u)\, du.$$

In particular, if $x = x_{i+1}$, we have

$$f(x_{i+1}) = f(x_i) + (x_{i+1} - x_i) f'(x_i) + \int_{x_i}^{x_{i+1}} (x_{i+1} - u) f''(u)\, du,$$

and we can use this to eliminate $f'(x_i)$ from the first equation in favor of $f(x_{i+1})$. The algebra is messy but straightforward and gives

$$\begin{aligned}
f(x) &= f(x_i) + (x - x_i)\frac{f(x_{i+1}) - f(x_i)}{x_{i+1} - x_i} \\
&\quad - \int_{x_i}^{x} (u - x_i)\left(1 - \frac{x - x_i}{x_{i+1} - x_i}\right) f''(u)\,du \\
&\quad - \int_{x}^{x_{i+1}} (x - x_i)\frac{x_{i+1} - u}{x_{i+1} - x_i} f''(u)\,du \\
&= h(x) - \frac{1}{2}(x - x_i)(x_{i+1} - x)\int_{x_i}^{x_{i+1}} p_{i,x}(u) f''(u)\,du,
\end{aligned}$$

where $p_{i,x}$ is a probability density concentrated on $(x_i, x_{i+1})$. If $f''$ is continuous, then by the (first) intermediate value theorem for integrals,

$$f(x) = h(x) - \frac{1}{2}(x - x_i)(x_{i+1} - x) f''(u_{i,x}),$$

where $u_{i,x} \in (x_i, x_{i+1})$. In particular, since $(x - x_i)(x_{i+1} - x) \le \frac{1}{4}(x_{i+1} - x_i)^2$, if $M_2 = \sup|f''(x)|$, then

$$\bigl|f(x) - h(x)\bigr| \le \frac{M_2}{8}(x_{i+1} - x_i)^2. \tag{18.5}$$

For clarity, let us now assume that the $x_i$ are equally spaced between 0 and 1, $x_i = i/n$. Assume also that $W$ is continuous, symmetric, and has support $[-1, 1]$. Two popular choices for the kernel are $W_2(x) = \bigl(1 - |x|\bigr)^+$ and $W_4(x) = \frac{3}{4}(1 - x^2)^+$. Then

$$\begin{aligned}
E\hat{f}(x) &= E\frac{1}{b}\int \hat{h}(t) W\left(\frac{x - t}{b}\right) dt \\
&= \frac{1}{b}\int E\hat{h}(t) W\left(\frac{x - t}{b}\right) dt \\
&= \frac{1}{b}\int f(t) W\left(\frac{x - t}{b}\right) dt + \frac{1}{b}\int \bigl(h(t) - f(t)\bigr) W\left(\frac{x - t}{b}\right) dt.
\end{aligned}$$

By (18.5), the difference between $f$ and $h$ is at most $M_2/(8n^2)$, and since $W$ is a probability density integrating to one, this also bounds the magnitude of the final term.[2] If $Z$ has density $W$, then the other term is

$$E_Z f(x - bZ) = E_Z\bigl[f(x) - bZf'(x) + \tfrac{1}{2}b^2 Z^2 f''(x - bZ^*)\bigr],$$

where $Z^*$ is an intermediate value in $(-1, 1)$. Because $W$ is symmetric, $EZ = 0$, and by dominated convergence

[2] We neglect "edge effects" here, assuming $x_i > b$ and $x_{i+1} < 1 - b$.

$$\text{Bias}\big(\hat{f}(x)\big) = E\hat{f}(x) - f(x) = \frac{1}{2}b^2 f''(x)EZ^2 + o(b^2) + O(1/n^2),$$

as $b \to 0$. With the regularity imposed, the asymptotics here also hold uniformly in $x$ provided we stay away from the endpoints 0 and 1.

From the representation (18.3),

$$\text{Var}\big(\hat{f}(x)\big) = \sigma^2 \sum_{i=1}^{n} v_i^2(x).$$

Note that in this sum, the number of nonzero terms is of order $nb$ because $v_i(x)$ is zero unless $|x - x_i| \le b + 1/n$. To approximate the terms in this sum, note that $nu_i$ is a probability measure concentrated on $(x_{i-1}, x_{i+1})$, and so, by (18.4),

$$v_i(x) = \frac{1}{nb} W\left(\frac{x - t_i^*(x)}{b}\right),$$

with $t_i^*(x)$ some value in $(x_{i-1}, x_{i+1})$. If $M_1 = \sup |W'(t)|$, then

$$\left| v_i(x) - \frac{1}{nb} W\left(\frac{x - x_i}{b}\right)\right| \le \frac{M_1}{n^2 b^2}.$$

Since the points $x_i/b$ are uniformly spaced and separated by an amount $1/(nb)$, in a limit in which $b \to 0$ but $nb \to \infty$, then

$$\frac{1}{nb}\sum_{i=1}^{n} W^2\left(\frac{x - x_i}{b}\right)$$

is a Riemann approximation for $\int W^2(x-t)\,dt$ converging to $\int W^2(t)\,dt$. Thus

$$\begin{aligned}\text{Var}\big(\hat{f}(x)\big) &= \frac{\sigma^2}{nb}\frac{1}{nb}\sum_{i=1}^{n}\left[W^2\left(\frac{x - x_i}{b}\right) + O\left(\frac{1}{nb}\right)\right] \\ &= \frac{\sigma^2}{nb}\left[\int W^2(t)\,dt + o(1)\right] + O\left(\frac{1}{n^2 b^2}\right) \\ &\sim \frac{\sigma^2}{nb}\int W^2(t)\,dt.\end{aligned}$$

Combining our approximations for the bias and variance of $\hat{f}(x)$, we can approximate the mean square error of $\hat{f}(x)$ as

$$\begin{aligned}\text{MSE}(x) = E\big[\hat{f}(x) - f(x)\big]^2 &= \text{Var}\big(\hat{f}(x)\big) + \text{Bias}^2\big(\hat{f}(x)\big) \\ &\approx \frac{\sigma^2}{nb}\int W^2(t)\,dt + b^4\frac{1}{4}\big(f''(x)\big)^2(EZ^2)^2.\end{aligned}$$

The mean square error measures the performance of $\hat{f}$ at individual values for the independent variable $x$. For a more global assessment, the integrated mean square error may be a natural measure:

$$\begin{aligned}\text{IMSE} &= \int_0^1 \text{MSE}(x)\,dx \\ &= \int_0^1 E\big[\hat{f}(x) - f(x)\big]^2\,dx = E\int_0^1 \big[\hat{f}(x) - f(x)\big]^2\,dx.\end{aligned}$$

Using the approximation for the mean square error,

$$\text{IMSE} \approx \frac{c_1}{nb} + b^4 c_2,$$

where

$$c_1 = \sigma^2 \int W^2(t)\,dt \text{ and } c_2 = \frac{1}{4}(EZ^2)^2 \int_0^1 \big(f''(x)\big)^2\,dx.$$

The approximation for the integrated mean square error, viewed as a function of $b$, is minimized at a value where the derivative is zero; that is, at a value solving

$$-\frac{c_1}{nb^2} + 4c_2 b^3 = 0.$$

This gives

$$b_{opt} = \left(\frac{c_1}{4nc_2}\right)^{1/5}$$

as an optimal choice (approximately), and with this choice

$$\text{IMSE} \approx 5c_2^{1/5}\left(\frac{c_1}{4n}\right)^{4/5} = K_f\left(\frac{\sigma^2}{n}\right)^{4/5}\left(\left[\int W^2(t)\,dt\right]^4 [EZ^2]^2\right)^{1/5},$$

where

$$K_f = \frac{5}{4}\left(\int_0^1 \big(f''(x)\big)^2\,dx\right)^{1/5},$$

which depends on $f$ but is independent of the kernel $W$. Using $W_2$, IMSE $\approx 0.353K_f(\sigma^2/n)^{4/5}$, and using $W_4$, IMSE $\approx 0.349K_f(\sigma^2/n)^{4/5}$. Thus $W_4$ has a slight theoretical advantage.

In practice, the choice of the kernel is less important than the choice of the bandwidth. The formula for $b_{opt}$ cannot be used directly, since the constants $c_1$ and $c_2$ depend on $\sigma^2$ and $\int_0^1 \big(f''(x)\big)\,dx$. One natural idea is to estimate these quantities somehow and choose the bandwidth by plugging the estimates into the formula for $b_{opt}$. This is feasible, but a bit tricky since derivatives of $f$ are often harder to estimate than $f$ itself. Another idea would be to use a *cross-validation* approach based on prediction error. Suppose we wanted to predict the outcome at a new location $x$. The expected squared prediction error would be

$$E\big[Y - \hat{f}(x)\big]^2 = E\big[f(x) + \epsilon - \hat{f}(x)\big]^2 = \sigma^2 + \text{MSE}(x).$$

If $x$ were chosen at random from a uniform distribution on $(0,1)$, then the expected squared prediction error would be $\sigma^2 + \text{IMSE}$. Here is a resampling method to estimate this error:

1. Partition the data at random into two samples, an estimation sample with $n_1$ observations and a validation sample with $n_2$ observations. Let $\hat{f}_b^*$ denote the kernel estimate for $f$ based on the estimation sample with bandwidth $b$.
2. Define

$$C(b) = \frac{1}{n_2} \sum_{i=1}^{n_2} \bigl(y_{2,i} - \hat{f}_b^*(x_{2,i})\bigr)^2,$$

where $(x_{2,i}, y_{2,i})$, $i = 1, \dots, n_2$, are the data in the validation sample.
3. Repeat steps 1 and 2 $m$ times and define

$$\overline{C}(b) = \frac{1}{m} \sum_{i=1}^{m} C_i(b),$$

where $C_i(b)$ is the function $C(b)$ for the $i$th partition.

The value $\hat{b}$ minimizing $\overline{C}$ would be the cross-validation choice for the bandwidth.

## 18.2 Hilbert Spaces

If $V$ is a vector space in $\mathbb{R}^n$ with an orthonormal basis $e_1, \dots, e_p$, then any $x \in V$ can be written as

$$x = \sum_{i=1}^{p} c_i e_i,$$

with the constants $c_i$ in the expansion given by

$$c_i = e_i' x, \qquad i = 1, \dots, p.$$

Classes of functions may also form vector spaces over $\mathbb{R}$, but these spaces are rarely spanned by a finite set of functions. However, expansions like those above may be possible with an infinite collection of basis vectors. To deal with infinite sums we need a notion of convergence, and this is based here on the norm or length of a vector. And, for the geometric structure of interest, we also need inner products. Here are formal definitions of norms and inner products.

**Definition 18.1.** *Let $V$ be a vector space over $\mathbb{R}$. A* norm *on $V$ is a real-valued function on $V$, $\|\cdot\| : V \to \mathbb{R}$, satisfying the following conditions:*

1. *For all $x$ in $V$, $\|x\| \ge 0$, and $\|x\| = 0$ only if $x = \mathbb{O}$ (the zero vector in $V$).*
2. *For all $x$ and $y$ in $V$, $\|x + y\| \le \|x\| + \|y\|$.*
3. *For all $x \in V$ and $c \in \mathbb{R}$, $\|cx\| = |c| \times \|x\|$.*

Using a norm we can define convergence $x_n \to x$ to mean that $\|x_n - x\| \to 0$. Then a function $f$ from one normed space to another is *continuous* if $f(x_n) \to f(x)$ whenever $x_n \to x$. For instance the function $f(x) = \|x\|$ is continuous. The second property of norms implies

$$\|x_n\| \le \|x_n - x\| + \|x\| \text{ and } \|x\| \le \|x - x_n\| + \|x_n\|.$$

Together these imply

$$\big|f(x_n) - f(x)\big| = \big|\|x_n\| - \|x\|\big| \le \|x_n - x\|,$$

which tends to zero (by definition) whenever $x_n \to x$.

**Definition 18.2.** *Let $V$ be a vector space over $\mathbb{R}$. An* inner product *is a function $\langle\cdot,\cdot\rangle : V \times V \to \mathbb{R}$ that is* symmetric*:*

$$\langle x, y\rangle = \langle y, x\rangle, \qquad \forall y, x \in V;$$

bilinear*:*

$$\langle x, ay + bz\rangle = a\langle x, y\rangle + b\langle x, z\rangle$$

*and*

$$\langle ax + by, z\rangle = a\langle x, z\rangle + b\langle y, z\rangle,$$

*for all $x$, $y$, $z$ in $V$ and all $a$, $b$ in $\mathbb{R}$; and* positive definite*:*

$$\langle x, x\rangle \ge 0,$$

*with equality only if $x = \mathbb{O}$. The pair $(V, \langle\cdot,\cdot\rangle)$ is called an* inner product space.

**Proposition 18.3.** *In an inner product space, $\|x\| = \sqrt{\langle x, x\rangle}$ defines a norm satisfying the Cauchy–Schwarz inequality,*

$$|\langle x, y\rangle| \le \|x\| \times \|y\|. \tag{18.6}$$

*Proof.* The first property of a norm follows because the inner product is positive definite, the third property follows from the bilinearity which gives $\langle cx, cx\rangle = c^2\langle x, x\rangle$, and, anticipating (18.6), the second property of a norm follows because

$$\begin{aligned}\big|\langle x + y, x + y\rangle\big| &= \big|\langle x, x\rangle + 2\langle x, y\rangle + \langle y, y\rangle\big| \\ &\le \|x\|^2 + 2\|x\| \times \|y\| + \|y\|^2 = \big(\|x\| + \|y\|\big)^2.\end{aligned}$$

To finish we must verify the Cauchy–Schwarz inequality. It is not hard to show that $\langle x, \mathbb{O}\rangle = 0$, and so the inequality is immediate unless $x$ and $y$ are both nonzero. In this case,

$$\langle x - cy, x - cy\rangle = \|x\|^2 - 2c\langle x, y\rangle + c^2\|y\|^2,$$

viewed as a function of $c$, is minimized when $c = \langle x, y\rangle/\|y\|^2$. But the function is nonnegative for all $c$, and so plugging in the minimizing value we have

$$\|x\|^2 - 2\frac{\langle x, y\rangle^2}{\|y\|^2} + \frac{\langle x, y\rangle^2}{\|y\|^2} \geq 0.$$

After a bit of rearrangement this gives (18.6). □

One consequence of the Cauchy–Schwarz inequality is that the inner product $\langle\cdot,\cdot\rangle$ is continuous, because

$$\begin{aligned}\big|\langle \tilde{x}, \tilde{y}\rangle - \langle x, y\rangle\big| &= \big|\langle \tilde{x}, \tilde{y} - y\rangle + \langle \tilde{x} - x, y\rangle\big| \\ &\leq \|\tilde{x}\| \times \|\tilde{y} - y\| + \|\tilde{x} - x\| \times \|y\|,\end{aligned}$$

which tends to zero as $\tilde{x} \to x$ and $\tilde{y} \to y$.

If a norm $\|\cdot\|$ comes from an inner product, then

$$\|x \pm y\|^2 = \|x\|^2 \pm 2\langle x, y\rangle + \|y\|^2.$$

Adding these two relations we have the *parallelogram law*, stating that

$$\|x + y\|^2 + \|x - y\|^2 = 2\|x\|^2 + 2\|y\|^2. \tag{18.7}$$

Elements $x$ and $y$ in an inner product space $V$ are called *orthogonal*, written $x \perp y$, if $\langle x, y\rangle = 0$. Since

$$\langle x + y, x + y\rangle = \langle x, x\rangle + 2\langle x, y\rangle + \langle y, y\rangle,$$

we then have the Pythagorean relation,

$$\|x + y\|^2 = \|x\|^2 + \|y\|^2.$$

If $W$ is a subspace of $V$, then the orthogonal complement of $W$ is

$$W^\perp = \big\{x \in V : \langle x, y\rangle = 0, \forall y \in W\big\}.$$

If $x_n \to x$, then $\|x_n - x\| \to 0$, which implies that

$$\sup_{m:m\geq n} \|x_m - x\| \to 0,$$

as $n \to \infty$. Because

$$\|x_n - x_m\| \leq \|x_n - x\| + \|x - x_m\|,$$

convergence implies

$$\lim_{n\to\infty} \sup_{m\geq n} \|x_n - x_m\| = 0. \tag{18.8}$$

Sequences satisfying this equation are called *Cauchy*, but if the space is not rich enough some Cauchy sequences may not converge. For instance, 3, 3.1, 3.14, ... is a Cauchy sequence in $\mathbb{Q}$ without a limit in $\mathbb{Q}$, because $\pi$ is irrational.

**Definition 18.4.** *A normed vector space $V$ is* complete *if every Cauchy sequence in $V$ has a limit in $V$. A complete inner product space is called a* Hilbert space.

The next result extends our notion of projections in Euclidean spaces to Hilbert spaces.

**Theorem 18.5.** *Let $V$ be a closed subspace of a Hilbert space $\mathcal{H}$. For any $x \in \mathcal{H}$ there is a unique $y \in V$, called the* projection *of $x$ onto $V$, minimizing $\|x - z\|$ over $z \in V$. Then $x - y \in V^{\perp}$, and this characterizes $y$: if $\tilde{y} \in V$ and $x - \tilde{y} \in V^{\perp}$, then $\tilde{y} = y$.*

*Proof.* Let $d = \inf_{z \in V} \|x - z\|$ (the distance from $x$ to $V$), and choose $y_n$ in $V$ so that $\|x - y_n\| \to d$. By the parallelogram law,

$$\|x - y_m + x - y_n\|^2 + \|y_n - y_m\|^2 = 2\|x - y_n\|^2 + 2\|x - y_m\|^2.$$

But

$$\|2x - y_m - y_n\|^2 = 4\left\|x - \tfrac{1}{2}(y_n + y_m)\right\|^2 \geq 4d^2,$$

and so

$$\|y_n - y_m\|^2 \leq 2\|x - y_n\|^2 + 2\|x - y_m\|^2 - 4d^2 \to 0,$$

as $m, n \to \infty$. So $y_n$, $n \geq 1$, is a Cauchy sequence converging to some element $y \in \mathcal{H}$. Since $V$ is closed, $y \in V$, and by continuity, $\|x - y\| = d$. Next, suppose $\tilde{y} \in V$ and $\|x - \tilde{y}\| = d$. If $z \in V$ then $\tilde{y} + cz \in V$ for all $c \in \mathbb{R}$ and

$$0 \leq \|x - \tilde{y} - cz\|^2 - \|x - \tilde{y}\|^2 = -2c\langle x - \tilde{y}, z\rangle + c^2\|z\|^2.$$

This can only hold for all $c \in \mathbb{R}$ if $\langle x - \tilde{y}, z\rangle = 0$, and thus $x - \tilde{y} \in V^{\perp}$. Finally, since $y$ and $\tilde{y}$ both lie in $V$ and $x - y$ and $x - \tilde{y}$ both lie in $V^{\perp}$,

$$\|y - \tilde{y}\|^2 = \langle x - \tilde{y}, y - \tilde{y}\rangle - \langle x - y, y - \tilde{y}\rangle = 0,$$

showing that $y$ is unique. □

If $V$ is a closed subspace of a Hilbert space $\mathcal{H}$, let $P_V x$ denote the projection of $x$ onto $V$. The following result shows that $P_V$ is a linear operator with operator norm one (see Problem 18.4).

**Proposition 18.6.** *If $V$ is a closed subspace of a Hilbert space $\mathcal{H}$, $x \in \mathcal{H}$, $y \in \mathcal{H}$, and $c \in \mathbb{R}$, then*

$$P_V(cx + y) = cP_V x + P_V y$$

*and*

$$\|P_V x\| \leq \|x\|.$$

*Proof.* Since $V$ is a subspace and $P_V x$ and $P_V y$ lie in $V$, $cP_V x + P_V y \in V$; and if $z \in V$,

$$\langle cx + y - cP_V x - P_V y, z\rangle = c\langle x - P_V x, z\rangle + \langle y - P_V y, z\rangle = 0,$$

because $x - P_V x \perp z$ and $y - P_V y \perp z$. Using Theorem 18.5, $cP_V x + P_V y$ must be the projection of $cx + y$ onto $V$. For the second assertion, since $P_V x \in V$ and $x - P_V x \in V^\perp$ are orthogonal, by the Pythagorean relation

$$\|P_V x\|^2 + \|x - P_V x\|^2 = \|x\|^2. \qquad \square$$

**Definition 18.7.** *A collection $e_t$, $t \in T$, is said to be* orthonormal *if $e_s \perp e_t$ for all $s \neq t$ and $\|e_t\| = 1$, for all $t$.*

As in the finite-dimensional case, we would like to represent elements in our Hilbert space as linear combinations of elements in an orthonormal collection, but extra care is necessary because some infinite linear combinations may not make sense.

**Definition 18.8.** *The* linear span *of $S \subset \mathcal{H}$, denoted* span$(S)$, *is the collection of all finite linear combinations $c_1x_1 + \cdots + c_nx_n$ with $c_1, \dots, c_n$ in $\mathbb{R}$ and $x_1, \dots, x_n$ in $S$. The closure of this set is denoted $\overline{\operatorname{span}}(S)$.*

**Definition 18.9.** *An orthonormal collection $e_t$, $t \in T$, is called an* orthonormal basis *for a Hilbert space $\mathcal{H}$ if $\langle e_t, x\rangle \neq 0$ for some $t \in T$, for every nonzero $x \in \mathcal{H}$.*

**Theorem 18.10.** *Every Hilbert space has an orthonormal basis.*

The proof in general relies on the axiom of choice. (The collection of all orthonormal families is inductively ordered, so a maximal element exists by Zorn's lemma, and any maximal element is an orthonormal basis.) When $\mathcal{H}$ is separable, a basis can be found by applying the Gram–Schmidt algorithm to a countable dense set, and in this case the basis will be countable.

**Theorem 18.11.** *If $e_n$, $n \geq 1$, is an orthonormal basis, then each $x \in \mathcal{H}$ may be written as*

$$x = \sum_{k=1}^{\infty} \langle x, e_k\rangle e_k.$$

*Proof.* Let

$$x_n = \sum_{k=1}^{n} \langle x, e_k\rangle e_k.$$

The infinite sum in the theorem is the limit of these partial sums, so we begin by showing that these partial sums form a Cauchy sequence. If $j \leq n$,

$$\langle x - x_n, e_j \rangle = \langle x, e_j \rangle - \sum_{k=1}^{n} \langle x, e_k \rangle \langle e_k, e_j \rangle = 0,$$

since $\langle e_k, e_j \rangle = 0$, unless $k = j$, and in that case it is 1. From this, $x - x_n \in \text{span}\{e_1, \dots, e_n\}^\perp$, and by Theorem 18.5, $x_n$ is the projection of $x$ onto $\text{span}\{e_1, \dots, e_n\}$. By the Pythagorean relation,

$$\|x_n\|^2 = \sum_{k=1}^{n} \langle x, e_k \rangle^2,$$

and by Proposition 18.6, $\|x_n\| \le \|x\|$. From this we have Bessel's inequality,

$$\sum_{k=1}^{n} \langle x, e_k \rangle^2 \le \|x\|^2,$$

and since $n$ here is arbitrary, the coefficients $\langle x, e_k \rangle$, $k \ge 1$, are square summable. By the Pythagorean relation, if $n < m$,

$$\|x_m - x_n\|^2 = \left\| \sum_{k=n+1}^{m} \langle x, e_k \rangle e_k \right\|^2 = \sum_{k=n+1}^{m} \langle x, e_k \rangle^2,$$

which tends to zero as $m$ and $n$ tend to infinity. So $x_n$, $n \ge 1$, is a Cauchy sequence, and since $\mathcal{H}$ is complete the sequence must have a limit $x_\infty$. Because the inner product $\langle \cdot, \cdot \rangle$ is a continuous function of its arguments, for any $j \ge 1$,

$$\langle x - x_\infty, e_j \rangle = \lim_{n \to \infty} \langle x - x_n, e_j \rangle.$$

But if $n \ge j$, $\langle x - x_n, e_j \rangle = 0$ because $x_n$ is the projection of $x$ onto $\text{span}\{e_1, \dots, e_n\}$, and so the limit in this expression must be zero. Therefore

$$\langle x - x_\infty, e_j \rangle = 0, \qquad j \ge 1.$$

Finally, since $e_k$, $k \ge 1$, form an orthonormal basis, $x - x_\infty$ must be zero, proving the theorem. ☐

## 18.3 Splines

Let us consider again our nonparametric regression model

$$Y_i = f(x_i) + \epsilon_i, \qquad i = 1, \dots, n,$$

where $\epsilon_1, \dots, \epsilon_n$ are mean zero, uncorrelated random variables with a common variance $\sigma^2$. As with the kernel approach, there is a presumption that $f$ is smooth. The smoothing spline approach tries to take direct advantage of this smoothness by augmenting the usual least squares criteria with a penalty for roughness. For instance, if the $x_i$ lie in $[0,1]$, the estimator $\hat{f}$ might be chosen to minimize

$$J(f) = \sum_{i=1}^{n} \bigl(Y_i - f(x_i)\bigr)^2 + \lambda \|f^{(m)}\|_2^2, \tag{18.9}$$

where $\|\cdot\|_2$ is the $\mathcal{L}_2$-norm of functions on $[0,1]$ under Lebesgue measure,

$$\|g\|_2^2 = \int_0^1 g^2(x)\,dx.$$

The constant $\lambda$ is called a *smoothing parameter.* Larger values for $\lambda$ will lead to a smoother $\hat{f}$, smaller values will lead to an estimate $\hat{f}$ that follows the observed data more closely, that is, with $\hat{f}(x_i)$ closer to $Y_i$.

For the roughness penalty in our criteria to make sense, $f^{(m-1)}$ will need to be absolutely continuous according to the following definition.

**Definition 18.12.** *A real-valued function $g$ defined on an interval of $\mathbb{R}$ is* absolutely continuous *if there exists a function $g'$ such that*

$$g(b) - g(a) = \int_a^b g'(x)\,dx, \qquad \forall a < b.$$

*If $g$ is differentiable, then $g'$ must be the derivative a.e., so use of a common notation should not cause any confusion. Also, if $f$ has $m-1$ continuous derivatives and $g = f^{(m-1)}$ is absolutely continuous, then we denote $g'$ as $f^{(m)}$.*

**Definition 18.13.** *The* Sobolev space $W_m[0,1]$ *is the collection of all functions $f : [0,1] \to \mathbb{R}$ with $m-1$ continuous derivatives, $f^{(m-1)}$ absolutely continuous, and $\|f^{(m)}\|_2 < \infty$. With an inner product $\langle\cdot,\cdot\rangle$ defined by*

$$\langle f, g\rangle = \sum_{k=0}^{m-1} f^{(k)}(0) g^{(k)}(0) + \int_0^1 f^{(m)}(x) g^{(m)}(x)\,dx, \qquad f, g \in W_m[0,1],$$

*$W_m[0,1]$ is a Hilbert space.*

These Hilbert spaces have an interesting structure. Suppose we define

$$K(x,y) = \sum_{k=0}^{m-1} \frac{1}{k!^2} x^k y^k + \int_0^{x\wedge y} \frac{(x-u)^{m-1}(y-u)^{m-1}}{(m-1)!^2}\,du.$$

Then

$$\frac{\partial^k}{\partial x^k} K(x,y)\Big|_{x=0} = \frac{y^k}{k!}, \qquad k = 0, \dots, m-1,$$

and

$$\frac{\partial^m}{\partial x^m} K(x,y) = \frac{(y-x)^{m-1}}{(m-1)!} I\{x \le y\}. \tag{18.10}$$

Comparing this with the Taylor expansion

$$f(y) = \sum_{k=0}^{m-1} \frac{1}{k!} f^{(k)}(0) y^k + \int_0^y \frac{(y-x)^{m-1} f^{(m)}(x)}{(m-1)!}\, dx,$$

we see that

$$f(y) = \langle f, K(\cdot, y) \rangle.$$

This formula shows that the evaluation functional, $f \rightsquigarrow f(y)$, is a bounded linear operator. Hilbert spaces in which this happens are called *reproducing kernel Hilbert spaces.* The function $K$ here is called the reproducing kernel, reproducing because

$$K(x,y) = \langle K(\cdot, x), K(\cdot, y) \rangle.$$

The kernel $K$ is a positive definite function. To see this, first note that

$$\sum_{i,j} c_i c_j K(x_i, x_j) = \sum_{i,j} c_i c_j \langle K(\cdot, x_i), K(\cdot, x_j) \rangle = \left\| \sum_i c_i K(\cdot, x_i) \right\|^2,$$

which is nonnegative. If this expression is zero, then $h = \sum_i c_i K(\cdot, x_i)$ is zero. But then $\langle h, f \rangle = \sum_i c_i f(x_i)$ will be zero for all $f$, which can only happen if $c_i = 0$ for all $i$.

To minimize $J(f)$ in (18.9) over $f \in W_m[0,1]$, let $\Pi_m$ denote the vector space of all polynomials of degree at most $m-1$, let $\eta_i = K(\cdot, x_i)$, $i = 1, \dots, n$, and define

$$V = \Pi_m \oplus \operatorname{span}\{\eta_1, \dots, \eta_n\}.$$

An arbitrary function $f$ in $W_m[0,1]$ can be written as $g + h$ with $g \in V$ and $h \in V^\perp$. Because $h$ is orthogonal to $\eta_i$, $h(x_i) = \langle h, \eta_i \rangle = 0$. Also, if $k \le m-1$, then the inner product of $h$ with the monomial $x^k$ is $k! h^{(k)}(0)$, and because $h$ is orthogonal to these monomials, $h^{(k)}(0) = 0$, $k = 0, \dots, m-1$. It follows that $\|h\| = \|h^{(m)}\|_2$, and

$$\langle g, h \rangle = \int_0^1 g^{(m)}(x) h^{(m)}(x)\, dx = 0.$$

But then

$$\|g^{(m)} + h^{(m)}\|_2^2 = \int_0^1 \bigl(g^{(m)}(x) + h^{(m)}(x)\bigr)^2\, dx = \|g^{(m)}\|_2^2 + \|h^{(m)}\|_2^2,$$

and so

$$J(f) = J(g+h) = \sum_{i=1}^{n}\bigl(Y_i - g(x_i) - h(x_i)\bigr)^2 + \lambda\|g^{(m)} + h^{(m)}\|_2^2$$
$$= \sum_{i=1}^{n}\bigl(Y_i - g(x_i)\bigr)^2 + \lambda\|g^{(m)}\|_2^2 + \lambda\|h^{(m)}\|_2^2 = J(g) + \lambda\|h\|^2.$$

From this it is evident that a function minimizing $J$ must lie in $V$.

Using (18.10),

$$\eta_i^{(m)}(x) = \frac{(x_i - x)^{m-1}}{(m-1)!} I\{x \le x_i\}.$$

From this, on $[0, x_i]$, $\eta_i$ must be a polynomial of degree $2m-1$, and on $[x_i, 1]$, $\eta_i$ is a polynomial of degree at most $m-1$. Taking more derivatives,

$$\eta_i^{(m+j)}(x) = \frac{(-1)^j (x_i - x)^{m-1-j}}{(m-1-j)!} I\{x \le x_i\}, \qquad j = 1, \dots, m-2,$$

and so the derivatives of $\eta_i$ of order $2m-2$ or less are continuous. Linear combinations of the $\eta_i$ are piecewise polynomials. Functions like these are called *splines.*

**Definition 18.14.** *A function $f : [0,1] \to \mathbb{R}$ is called a* spline *of order $q$ with (simple)* knots $0 < x_1 < \cdots < x_n < 1$ *if, for any $i = 0, \dots, n$, the restriction of $f$ to $[x_i, x_{i+1}]$ (with the convention $x_0 = 0$ and $x_{n+1} = 1$) is a polynomial of degree $q-1$ or less, and if the first $q-2$ derivatives of $f$ are continuous on the whole domain $[0,1]$. The collection of all splines of order $q$ is denoted $\mathcal{S}_q = \mathcal{S}_q(x_0, \dots, x_{n+1})$. The space $\mathcal{S}_q$ is a vector space.*

From the discussion above, any function $f \in V$ must be a spline of order $2m$. In addition, all functions $f \in V$ are polynomials of degree $m-1$ on the last interval $[x_n, 1]$. So if $\hat{f}$ minimizes $J(f)$, it will be a polynomial of degree at most $m-1$ on $[x_n, 1]$, and by time reversal[3] it will also be a polynomial of degree $m-1$ or less on the first interval $[0, x_1]$. It will then be a *natural spline* according to the following definition.

**Definition 18.15.** *A function $f \in \mathcal{S}_{2q}$ is called a* natural spline *of order $2q$ if its restrictions to the first and last intervals, $[0, x_1]$ and $[x_n, 1]$, are polynomials of degree $q-1$ or less. Let $\tilde{\mathcal{S}}_{2q}$ denote the set of all natural splines.*

These spline spaces (with fixed knots) are finite-dimensional vector spaces, and once we know that the function $\hat{f}$ minimizing $J$ lies in a finite-dimensional vector space, $\hat{f}$ can be identified using ordinary linear algebra. To see how, let $e_j$, $j = 1, \dots, k$, be linearly independent functions with

---

[3] Formally, the argument just given shows that the function $\tilde{f}(t) \stackrel{\text{def}}{=} \hat{f}(1-t)$, that minimizes $\sum\bigl(Y_i - f(1-x_i)\bigr)^2 + \lambda\|f^{(m)}\|_2^2$, must be a polynomial of degree at most $m-1$ on $[1-x_1, 1]$.

$$\tilde{S}_{2m} \subset \operatorname{span}\{e_1, \ldots, e_k\}.$$

If

$$f = c_1 e_1 + \cdots + c_k e_k,$$

then

$$f(x_i) = \langle f, \eta_i \rangle = \sum_j c_j \langle e_j, \eta_i \rangle = [Ac]_i,$$

where $A$ is a matrix with entries

$$A_{ij} = \langle e_j, \eta_i \rangle, \qquad i = 1, \ldots, n, \quad j = 1, \ldots, k.$$

Then

$$\sum_{i=1}^{n} \bigl(Y_i - f(x_i)\bigr)^2 = \|Y - Ac\|^2 = Y'Y - 2Y'Ac + c'A'Ac.$$

For the other term in $J$,

$$\|f^{(m)}\|_2^2 = \left\| \sum_{j=1}^{k} c_j e_j^{(m)} \right\|_2^2 = \sum_{i,j} c_i c_j \int_0^1 e_i^{(m)}(x) e_j^{(m)}(x)\, dx = c' B c,$$

where $B$ is a $k \times k$ matrix with

$$B_{ij} = \int_0^1 e_i^{(m)}(x) e_j^{(m)}(x)\, dx, \qquad i = 1, \ldots, k, \quad j = 1, \ldots, k.$$

Using these formulas,

$$J(f) = Y'Y - 2Y'Ac + c'A'Ac + \lambda c'Bc.$$

This is a quadratic function of $c$ with gradient

$$2(A'A + \lambda B)c - 2A'Y.$$

Setting the gradient to zero, if

$$\hat{c} = (A'A + \lambda B)^{-1} A'Y$$

and

$$\hat{f} = \sum_{j=1}^{k} \hat{c}_j e_j,$$

then $\hat{f}$ minimizes $J(f)$ over $W_m[0, 1]$.

One collection of linearly independent functions with span containing[4] $\tilde{S}_{2q}$ is given by

[4] For some linear combinations of these functions, restriction to the final interval $[x_n, 1]$ will give a polynomial of degree greater than $q - 1$. So the linear span of these functions is in fact strictly larger than $\tilde{S}_{2q}$.

$$e_i(x) = (x - x_i)_+^{2q-1}, \qquad i = 1, \dots, n,$$

along with the monomials of degree $q - 1$ or less,

$$e_{n+j}(x) = x^{j-1}, \qquad j = 1, \dots, q.$$

This can be seen recursively. If $f \in \tilde{\mathcal{S}}_{2q}$, let

$$p = \sum_{j=1}^{q} c_j e_{n+j}$$

be a polynomial of degree $q - 1$ equal to $f$ on $[0, x_1]$. Then $f - p \in \tilde{\mathcal{S}}_{2q}$ is zero on $[0, x_1]$. By the enforced smoothness for derivatives at the knots, on $[x_1, x_2]$ $f - p$ will be a polynomial of degree $2q - 1$ with $2q - 2$ derivatives equal to zero at $x_1$. Accordingly, on this interval $f - p = c_1(x - x_1)^{2q-1}$, and it follows that $f - p - c_1 e_1$ is zero on $[0, x_2]$. Next, with a proper choice of $c_2$, $f - p - c_1 e_1 - c_2 e_2$ will be zero on $[0, x_3]$. Further iteration eventually gives $f - p - \sum_{j=1}^{n} c_j e_j = 0$ on $[0, 1]$.

The choice of the spanning functions $e_1, \dots, e_k$ may not seem important from a mathematical perspective. But a careful choice can lead to more efficient numerical algorithms. For instance, $\tilde{\mathcal{S}}_2$ contains the functions

$$e_1(x) = \begin{cases} 1, & x \in [0, x_1]; \\ \dfrac{x_2 - x}{x_2 - x_1}, & x \in [x_1, x_2]; \\ 0, & \text{otherwise}, \end{cases}$$

$$e_n(x) = \begin{cases} \dfrac{x - x_{n-1}}{x_n - x_{n-1}}, & x \in [x_{n-1}, x_n]; \\ 1, & x \in [x_n, 1]; \\ 0, & \text{otherwise}, \end{cases}$$

and

$$e_i(x) = \begin{cases} \dfrac{x - x_{i-1}}{x_i - x_{i-1}}, & x \in [x_{i-1}, x_i]; \\ \dfrac{x_{i+1} - x}{x_{i+1} - x_i}, & x \in [x_i, x_{i+1}]; \\ 0, & \text{otherwise}, \end{cases} \qquad i = 2, \dots, n - 1.$$

These functions are called *B-splines.* The first two, $e_1$ and $e_2$, are plotted in Figure 18.2. The $B$-splines are linearly independent and form what is called a *local basis* for $\tilde{\mathcal{S}}_2$, because each basis function vanishes except on at most two adjacent intervals of the partition induced by $x_1, \dots, x_n$. Functions expressed in this basis can be computed quickly at a point $x$ since all but at most two terms in $\sum c_j e_j(x)$ will be zero. In addition, with this basis, the matrix $A$ is the identity, and $B$ will be "tridiagonal," with $B_{ij} = 0$ if $|i - j| \geq 2$. Matrices

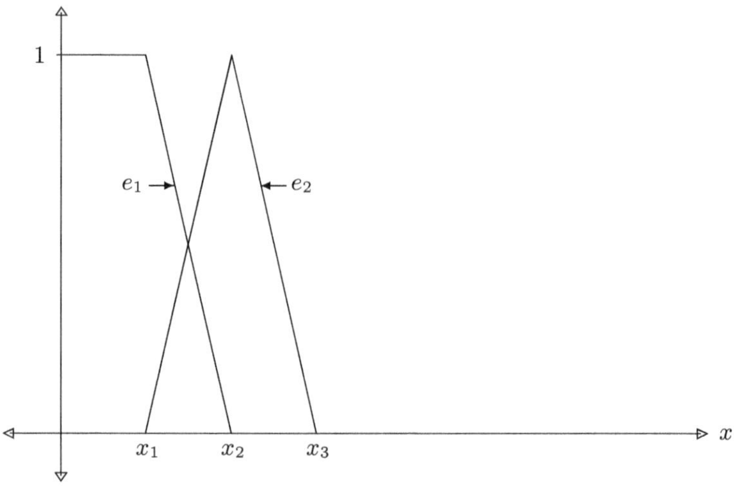


**Fig. 18.2.** $B$-splines $e_1$ and $e_2$.

with a banded structure can be inverted and multiplied much more rapidly than matrices with arbitrary entries.

$B$-spline bases are also available for other spline spaces. The notation needed to define them carefully is a bit involved, but it is not too hard to understand why they should exist. In $\mathcal{S}_q$, the $q+1$ functions

$$(x - x_i)_+^{q-1}, \dots, (x - x_{i+q})_+^{q-1}$$

restricted to $[x_{i+q}, 1]$ are all polynomials of degree $q-1$. Because polynomials of degree $q-1$ form a vector space of dimension $q$, the restrictions cannot be linearly dependent, and some nontrivial linear combination of these functions must be zero on $[x_{i+q}, 1]$. This linear combination gives a function in $\mathcal{S}_q$ that is zero unless its argument lies in $(x_i, x_{i+q})$. With a suitable normalization, functions such as this form a $B$-spline basis for $\mathcal{S}_q$. For further information on splines see Wahba (1990) or De Boor (2001)

## 18.4 Density Estimation

The methods just developed for nonparametric regression can also be applied to nonparametric density estimation. Let $X_1, \dots, X_n$ be i.i.d. from some distribution $Q$. One natural estimator for $Q$ would be the *empirical distribution* $\hat{Q}$ defined by

$$\hat{Q}(A) = \frac{1}{n}\#\{i \le n : X_i \in A\}.$$

This estimator $\hat{Q}(\cdot)$ is a discrete distribution placing atoms with mass $1/n$ at each observation $X_i$. So integrals against $\hat{Q}_n$ are just averages,

$$\int g\,d\hat{Q}_n = \frac{1}{n}\sum_{i=1}^{n} g(X_i).$$

If we believe $Q$ is absolutely continuous with a smooth density $f$, $\hat{Q}$ is not a very sensible estimator; it is too rough, in much the same way the linear interpolant $\hat{h}$ of the data was too rough for estimating a smooth regression function. A kernel approach to estimating $f$ uses convolution, as in Problem 18.10, to smooth $\hat{Q}$. Intuition suggests this may give a reasonable estimate for $f$ if the convolving distribution is concentrated near zero. To accomplish this we incorporate a bandwidth $b$, tending to zero as $n \to \infty$, and consider estimators of the form

$$\hat{f}(x) = \frac{1}{b}\int W\left(\frac{x-t}{b}\right)d\hat{Q}(t) = \frac{1}{nb}\sum_{i=1}^{n} W\left(\frac{x-X_i}{b}\right),$$

with $W$ a fixed symmetric probability density.

With the linear structure, formulas for the mean and variance for $\hat{f}(x)$ are easy to derive and study. If $f''$ is continuous and bounded, then

$$E\hat{f}(x) = f(x) + \frac{1}{2}b^2 f''(x)\int t^2 W(t)\,dt + o(b^2) \tag{18.11}$$

and

$$\operatorname{Var}\bigl(\hat{f}(x)\bigr) = \frac{1}{nb}f(x)\|W\|_2^2 + o\bigl(1/(nb)\bigr) \tag{18.12}$$

uniformly in $n$ as $b \downarrow 0$. Combining these, the mean square error for $\hat{f}(x)$ is

$$\begin{aligned}\operatorname{MSE}(x) &= E\bigl(\hat{f}(x) - f(x)\bigr)^2\\ &= \frac{f(x)}{nb}\|W\|_2^2 + \tfrac{1}{4}b^4\bigl(f''(x)\bigr)^2\left(\int t^2 W(t)\,dt\right)^2 + o\bigl(b^4 + 1/(nb)\bigr)\end{aligned}$$

as $b \downarrow 0$. If $b = b_n$ varies with $n$ so that $b_n \to 0$ and $nb_n \to \infty$, then this mean square error will tend to zero.

With suitable regularity, this approximation can be integrated, giving

$$\begin{aligned}\text{IMSE} &= E\int\bigl(\hat{f}(x) - f(x)\bigr)^2\,dx = \int \operatorname{MSE}(x)\,dx\\ &= \frac{1}{nb}\|W\|_2^2 + \frac{1}{4}b^4\|f''\|_2^2\left(\int t^2 W(t)\,dt\right)^2 + o\bigl(b^4 + 1/(nb)\bigr).\end{aligned}$$

Minimizing this approximation,

$$b \sim \frac{\|W\|_2^{2/5}}{n^{1/5}\|f''\|_2^{2/5}\left(\int t^2 W(t)\,dt\right)^{2/5}}$$

will be asymptotically optimal, and with this choice

$$\text{IMSE} \sim \tfrac{5}{4}\|W\|_2^{8/5}\|f''\|_2^{2/5}\left(\int t^2W(t)\,dt\right)^{2/5} n^{-4/5}.$$

For further discussion, see Chapter 2 of Wand and Jones (1995)

A spline approach to density estimation is more challenging. If we assume $f > 0$ and take $\theta = \log f$, then an estimator $\hat{f} = e^{\hat{\theta}}$ for $f$ will automatically be positive. For regularity, let us assume $\theta \in W_m[0,1]$. In contrast to nonparametric regression, since $f$ must integrate to one, $\theta$ cannot vary freely over $W_m[0,1]$, but must satisfy the constraint

$$\int_0^1 e^{\theta(x)}\,dx = 1.$$

Let $\Omega$ denote the class of all functions in $W_m[0,1]$ satisfying this constraint. The log-likelihood function is given by

$$l(\theta) = \sum_{i=1}^n \theta(X_i).$$

A direct maximum likelihood approach to estimating $\theta$ or $f$ fails because

$$\sup_{\theta\in\Omega} l(\theta) = \infty,$$

with arbitrarily high values for the likelihood achieved by densities with very large spikes at the data values. To mitigate this problem, we incorporate a penalty for smoothness and choose $\hat{\theta}$ to maximize

$$J_0(\theta) = \frac{1}{n}\sum_{i=1}^n \theta(X_i) - \lambda\|\theta^{(m)}\|_2^2.$$

To ameliorate troubles with the constraint, Silverman (1982) introduces another functional,

$$J(\theta) = \frac{1}{n}\sum_{i=1}^n \theta(X_i) - \int_0^1 e^{\theta(x)}\,dx - \lambda\|\theta^{(m)}\|_2^2.$$

**Theorem 18.16.** *The function $\hat{\theta} \in \Omega$ maximizes $J_0$ over $\Omega$ if and only if $\hat{\theta}$ maximizes $J$ over $W_m[0,1]$.*

*Proof.* If $\theta \in W_m[0,1]$ and $c = \int_0^1 e^{\theta(x)}\,dx$, then $\tilde{\theta} = \theta - \log c \in \Omega$ and $\|\theta^{(m)}\|_2 = \|\tilde{\theta}^{(m)}\|_2$. Thus

$$J(\tilde{\theta}) = J(\theta) - \log c - 1 + c.$$

But $c - \log c \ge 1$ for all $c > 0$, with equality only if $c = 1$, and thus $J(\tilde{\theta}) \ge J(\theta)$, with equality only if $c = 1$, that is, only if $\theta \in \Omega$. So any $\theta$ maximizing $J$ over $W_m[0,1]$ must lie in $\Omega$. But on $\Omega$, $J = J_0 - 1$, and the theorem follows. □

The *null family* associated with the smoothness penalty here is defined as

$$\Omega_0 = \{\theta \in \Omega : \|\theta^{(m)}\|_2 = 0\}.$$

Functions in $\Omega_0$ must be polynomials and can be parameterized as

$$\theta_\eta(x) = \eta_1 x + \cdots + \eta_{m-1}x^{m-1} - A(\eta),$$

where

$$A(\eta) = \log \int_0^1 \exp\left[\sum_{i=1}^{m-1} \eta_i x^i\right] dx.$$

Then $\Omega_0 = \{\theta_\eta : \eta \in \mathbb{R}^{m-1}\}$ and the corresponding densities, $f_\eta = e^{\theta_\eta}$, $\eta \in \mathbb{R}^{m-1}$, form an exponential family.

If $\lambda$ is large, then functions $\theta \in \Omega$ that are not close to $\Omega_0$ will incur a substantial smoothness penalty. For this reason, if $\hat{\theta}_\lambda$ is the estimator maximizing $J$, then $\hat{\theta}_\lambda$ should converge as $\lambda \to \infty$ to the maximum likelihood estimator for the null family $\Omega_0$. Let us call this estimator $\hat{\theta}_\infty$. In applications, this observation might be used in a reverse fashion to choose a smoothness penalty. If there is reason to believe that the data come from some particular exponential family, a researcher may want to choose a penalty with these target distributions as its null family. The next result shows that existence is also tied to estimation for the null family.

**Theorem 18.17.** *With a given data set $X_1, \dots, X_n$, $J$ will have a maximizer in $\Omega$ if $\hat{\theta}_\infty$, the maximum likelihood estimator for the null family, exists. This will hold with probability one if $n \ge m$.*

**Proposition 18.18.** *The functional $J$ is strictly concave.*

*Proof.* Given $\theta_1$, $\theta_2$ in $W_m[0,1]$ and $\alpha \in (0,1)$, since the exponential function is convex,

$$\begin{aligned}\int_0^1 e^{\alpha\theta_1(x)+(1-\alpha)\theta_2(x)}\,dx &\le \int_0^1 \left[\alpha e^{\theta_1(x)} + (1-\alpha)e^{\theta_2(x)}\right] dx \\ &= \alpha \int_0^1 e^{\theta_1(x)}\,dx + (1-\alpha)\int_0^1 e^{\theta_2(x)}\,dx,\end{aligned}$$

with equality only if $\theta_1 = \theta_2$. Also,

$$\begin{aligned}\|\alpha\theta_1^{(m)} + (1-\alpha)\theta_2^{(m)}\|_2^2 &\le \left[\alpha\|\theta_1^{(m)}\|_2 + (1-\alpha)\|\theta_2^{(m)}\|_2\right]^2 \\ &\le \alpha\|\theta_1^{(m)}\|_2^2 + (1-\alpha)\|\theta_2^{(m)}\|_2^2.\end{aligned}$$

So

$$J\big(\alpha\theta_1 + (1-\alpha)\theta_2\big) \ge \alpha J(\theta_1) + (1-\alpha)J(\theta_2),$$

with equality if and only if $\theta_1 = \theta_2$. ∎

As a consequence of this result, if an estimator $\hat{\theta}$ maximizing $J$ exists, it must be unique, for if $\hat{\theta}_1$ and $\hat{\theta}_2$ both maximize $J$, then

$$J(\hat{\theta}_1) = \frac{1}{2}J(\hat{\theta}_1) + \frac{1}{2}J(\hat{\theta}_2) \leq J\left(\frac{\hat{\theta}_1 + \hat{\theta}_2}{2}\right) \leq J(\hat{\theta}_1).$$

The inequalities here must be equalities, and strict concavity then implies $\hat{\theta}_1 = \hat{\theta}_2$.

The following result, from Silverman (1982), shows that with a suitable choice for the penalty scale $\lambda$, the estimator $\hat{\theta}$ is consistent.

**Theorem 18.19.** *Suppose* $\theta \in W_{2m}[0,1]$ *and*

$$\theta^{(2m-1)}(0) = \theta^{(2m-1)}(1).$$

*If* $\lambda \to 0$ *and* $n^{m-\delta}\lambda \to \infty$ *for some* $\delta > 0$, *then for every* $\epsilon > 0$,

$$\|\hat{\theta} - \theta\|_\infty^2 = O_p\{\lambda^{-\epsilon}(n^{-1}\lambda^{-1/m} + \lambda^{(4m-1)/(2m)})\}.$$

*In particular, if* $\lambda = n^{-(2m)/(4m+1)}$,

$$\|\hat{\theta} - \theta\|_\infty^2 = O_p\left(n^{-(4m-1)/(4m+1)+\epsilon}\right),$$

*for every* $\epsilon > 0$.

## 18.5 Problems

1. *Estimating* $\sigma^2$. Consider nonparametric regression with the assumptions in Section 18.1 and i.i.d. errors $\epsilon_i$. A natural estimator for $\sigma^2$ might be

$$\hat{\sigma}^2 = \frac{1}{n}\sum_{i=1}^n \left(Y_i - \hat{f}(x_i)\right)^2,$$

   with $\hat{f}$ the kernel estimator for $f$. Suppose $b = cn^{-1/5}$. Is this estimator necessarily consistent? Prove or explain why not. In your argument you can ignore edge effects.
2. *Locally weighted regression.* Like kernel smoothing, locally weighted regression is a linear approach to nonparametric regression. Let $W$ be a continuous, nonnegative, symmetric ($W(x) = W(-x)$) function, decreasing on $(0,\infty)$ with support $[-1,1]$. The estimate for $f$ at a point $x$ is based on weighted least squares, fitting a polynomial to the data with weights emphasizing data with $x_i$ near $x$. This problem considers quadratic models in which $\hat{\beta}$ is chosen to minimize

$$\sum_{i=1}^{n} W\left(\frac{x_i - x}{b}\right)\left(y_i - \beta_1 - \beta_2(x_i - x) - \beta_3(x_i - x)^2\right)^2.$$

The estimate for $f(x)$ is then $\hat{f}(x) = \hat{\beta}_1$. Here the bandwidth $b$ is taken to be a small constant (decreasing as $n$ increases), although in practice $b$ is often chosen using the $x_i$ so that the estimate for $f(x)$ is based on a fixed number of data points. For simplicity, you may assume below that $x_i = i/n$.

a) Derive an explicit formula for $\hat{f}(x_j)$ when $x_j \in (b, 1-b)$.
b) Derive approximations for the bias of $\hat{f}(x_j)$ as $n \to \infty$ and $b \downarrow 0$ with $nb \to \infty$ and $x_j = \lfloor nx \rfloor / n$, $x \in (0, 1)$.
c) Derive an approximation for the variance of $\hat{f}(x_j)$ in the same limit.
d) What choice for bandwidth $b$ minimizes the mean square error of $\hat{f}(x_j)$ (approximately).

3. Show that the stated inner product $\langle \cdot, \cdot \rangle$ in Definition 18.13 for $W_m[0,1]$ satisfies the conditions in Definition 18.2
4. Let $\mathcal{X}$ and $\mathcal{Y}$ be normed vector spaces over $\mathbb{R}$. A function $T : \mathcal{X} \to \mathcal{Y}$ is called a *linear operator* if

$$T(cx_1 + x_2) = cT(x_1) + T(x_2), \qquad \forall x_1, x_2 \in \mathcal{X}, c \in \mathbb{R}.$$

The *operator norm* (or spectral norm) of $T$ is defined as

$$\|T\| = \sup\bigl\{\|T(x)\| : \|x\| \le 1\bigr\},$$

and $T$ is called *bounded* if $\|T\| < \infty$.

a) Show that a bounded operator $T$ is continuous: If $\|x_n - x\| \to 0$, then $\|T(x_n) - T(x)\| \to 0$.
b) Show that a continuous linear operator $T$ is bounded.
c) Let $\mathcal{X} = \mathbb{R}^m$ and $\mathcal{Y} = \mathbb{R}^n$, with the usual Euclidean norms. Let $A$ be an $n \times m$ matrix, and define a linear operator $T$ by $T(x) = Ax$. Relate the operator norm $\|T\|$ to the eigenvalues of $A'A$.

5. Consider the set $\mathcal{C}[0,1]$ of continuous real functions on $[0,1]$ with the $\mathcal{L}_2$ inner product

$$\langle x, y \rangle = \int_0^1 x(t)y(t)\,dt$$

and associated norm

$$\|x\|_2 = \sqrt{\int_0^1 x^2(t)\,dt}.$$

a) Find a Cauchy sequence $x_n$ (for this norm) that does not converge, showing that $\mathcal{C}[0,1]$ is not complete (with this norm). (The usual norm for $\mathcal{C}[0,1]$ is $\|x\| = \sup_{t\in(0,1)} |x(t)|$, and with this norm $\mathcal{C}[0,1]$ is complete.)

b) Let $T_t$ be the evaluation operator, $T_t(x) = x(t)$. Show that $T_t$ is an unbounded linear operator.

6. Show that if $f_n \to f$ in $W_2[0,1]$, then $f_n'(x) \to f'(x)$.
7. Find a nontrivial function $f \in \mathcal{S}_4(0, 0.1, 0.2, \dots, 1)$ which is zero unless its argument lies between 0.2 and 0.6.
8. Suppose $m = 1$, $x = (0.2, 0.4, 0.6, 0.8)$, and we use the $B$-spline basis described for $\tilde{\mathcal{S}}_2$. Calculate the matrices $A$ and $B$ that arise in the formula to compute $\hat{f} = \sum \hat{c}_j e_j$.
9. *Semiparametric models.* Consider a regression model with two explanatory variables $x$ and $w$ in which

$$Y_i = f(x_i) + \beta w_i + \epsilon_i, \qquad i = 1, \dots, n,$$

with $0 < x_1 < \cdots < x_n < 1$, $f \in W_m[0,1]$, $\beta \in \mathbb{R}$, and the $\epsilon_i$ i.i.d. from $N(0, \sigma^2)$. This might be called a semiparametric model because the dependence on $w$ is modeled parametrically, but the dependence on $x$ is nonparametric. Following a penalized least squares approach, consider choosing $\hat{f}$ and $\hat{\beta}$ to minimize

$$J(f, \beta) = \sum_{i=1}^{n} \bigl(Y_i - f(x_i) - \beta w_i\bigr)^2 + \lambda \|f^{(m)}\|_2^2.$$

a) Show that the estimator $\hat{f}$ will still be a natural spline of order $2m$.

b) Derive explicit formulas based on linear algebra to compute $\hat{\beta}$ and $\hat{f}$.

10. *Convolutions.* Suppose $X \sim Q$ and $Y \sim W$ are independent, and that $W$ is absolutely continuous with Lebesgue density $w$. Show that $T = X + Y$ is absolutely continuous with density $h$ given by

$$h(t) = \int w(t - x)\, dQ(x).$$

The distribution of $T$ is called the convolution of $Q$ with $W$, and this shows that if either $Q$ or $W$ is absolutely continuous, their convolution is absolutely continuous.

11. Use dominated convergence to prove (18.11).
12. Prove (18.12).

# 19

# Bootstrap Methods

Bootstrap methods use computer simulation to reveal aspects of the sampling distribution for an estimator $\hat{\theta}$ of interest. With the power of modern computers the approach has broad applicability and is now a practical and useful tool for applied statisticians.

## 19.1 Introduction

To describe the bootstrap approach to inference, let $X_1, \dots, X_n$ be i.i.d. from some unknown distribution $Q$, and let

$$\mathbf{X} = (X_1, \dots, X_n)$$

denote all $n$ observations. For now we proceed nonparametrically with $Q$ an arbitrary distribution. Natural modifications when $Q$ comes from a parametric family are introduced in Section 19.3. With $Q$ arbitrary, a natural estimator for it would be the *empirical distribution*

$$\hat{Q} = \frac{1}{n} \sum_{i=1}^{n} \delta_{X_i}.$$

Here $\delta_x$ represents a "point mass" at $x$, that assigns full probability to the point $x$, $\delta_x(\{x\}) = 1$, and zero probability to all other points, $\delta_x(\{x\}^c) = 0$. Then the estimator $\hat{Q}$ is a discrete distribution that assigns mass $1/n$ to each data point $X_i$, $1 \le i \le n$, and $\hat{Q}(A)$ is just the proportion of these values that lie in $A$:

$$\hat{Q}(A) = \frac{1}{n} \#\{i \le n : X_i \in A\}.$$

Note that by the law of large numbers, $\hat{Q}(A) \xrightarrow{p} Q(A)$ as $n \to \infty$, supporting the notion that $\hat{Q}$ is a reasonable estimator for $Q$.

Suppose next that $\hat{\theta} = \hat{\theta}(\mathbf{X})$ is an estimator for some parameter $\theta = \theta(Q)$. Anyone using $\hat{\theta}$ should have interest in the distribution for the error $\hat{\theta} - \theta$, since

R.W. Keener, *Theoretical Statistics: Topics for a Core Course*, Springer Texts in Statistics, DOI 10.1007/978-0-387-93839-4_19, © Springer Science+Business Media, LLC 2010

this distribution provides information about the bias, variance, and accuracy of $\hat{\theta}$. Unfortunately, this error distribution typically varies with $Q$, and because $Q$ is unknown we cannot hope to know it exactly. Bootstrap methods are based on the hope or intuition that the true error distribution may be similar to the error distribution if the observations were sampled from $\hat{Q}$ instead of $Q$.

In principle, the error distribution with observations drawn from $\hat{Q}$ is a specific function of $\hat{Q}$, but exact calculations are generally intractable. This is where computer simulation plays an important role in practice. Given the original data $\mathbf{X}$, a computer routine draws a *bootstrap sample* $\mathbf{X}^* = (X_1^*, \dots, X_n^*)$, with the variables in this sample conditionally i.i.d. from $\hat{Q}$, so $\mathbf{X}^*|\mathbf{X} \sim \hat{Q}^n$. Note that since $\hat{Q}$ assigns mass $1/n$ to each observation, $X_1^*, \dots, X_n^*$ can be viewed as a random sample drawn with replacement from the set $\{X_1, \dots, X_n\}$. So these variables are very easy to simulate. If $\hat{\theta}^*$ is the estimate from the bootstrap sample,

$$\hat{\theta}^* = \hat{\theta}(\mathbf{X}^*) = \hat{\theta}(X_1^*, \dots, X_n^*),$$

then the distribution of $\hat{\theta}^* - \hat{\theta}$ is used to estimate the unknown distribution of the error $\hat{\theta} - \theta$. To be more precise, the estimate for the error distribution is the conditional distribution for $\hat{\theta}^* - \hat{\theta}$ given the original data $\mathbf{X}$. The following examples show ways this estimate for the error distribution might be used.

*Example 19.1. Bias Reduction.* Let

$$b = b(Q) = E_Q[\hat{\theta} - \theta]$$

denote the bias of an estimator $\hat{\theta} = \hat{\theta}(\mathbf{X})$ for a parameter $\theta = \theta(Q)$. If this bias were known, subtracting it from $\hat{\theta}$ would give an unbiased estimator. The true bias depends only on the error distribution. Substituting the bootstrap estimate for the true error distribution gives

$$\hat{b} = E[\hat{\theta}^* - \hat{\theta}|\mathbf{X}]$$

as the bootstrap estimate for the bias. Subtracting this estimate from $\hat{\theta}$ gives a new estimator $\hat{\theta} - \hat{b}$, generally less biased that the original estimator $\hat{\theta}$. Results detailing improvement are derived in the next section for a special case in which $\hat{\theta} = \overline{X}^3$ and $\theta$ is the cube of the mean of $Q$.

In practice, $\hat{b}$ would typically be computed by numerical simulation, having a computer routine draw multiple random samples, $\mathbf{X}_1^*, \dots, \mathbf{X}_B^*$, each from $\hat{Q}^n$. Letting $\hat{\theta}_i^* = \hat{\theta}(\mathbf{X}_i^*)$ denote the estimate from $i$th bootstrap sample $\mathbf{X}_i^*$, if the number of replications $B$ is large, then by the law of large numbers $\hat{b}$ should be well approximated by the average

$$\frac{1}{B}\sum_{i=1}^{B}[\hat{\theta}_i^* - \hat{\theta}].$$

A natural assumption, relating the unknown parameter $\theta = \theta(Q)$ and the estimator $\hat{\theta}(\mathbf{X})$, is that the estimator has no error whenever proportions in the sample agree with probabilities from $Q$, which happens if $Q = \hat{Q}$. With this assumption,

$$\hat{\theta}(\mathbf{X}) = \theta(\hat{Q}), \tag{19.1}$$

and so $\hat{b} = b(\hat{Q})$. Hence, from a technical viewpoint, the bootstrap estimator here is found by plugging the empirical distribution $\hat{Q}$ into the functional of interest, $b(\cdot)$. This mathematical structure occurs generally and underlies various results in the literature showing that bootstrap methods perform well when functionals of interest are smooth in an appropriate sense.

*Example 19.2. Confidence Intervals.* Quantiles for $|\hat{\theta}-\theta|$ are useful in assessing the accuracy of $\hat{\theta}$, for if $q = q(Q)$ is the upper $\alpha$th quantile[1] for the distribution of $|\hat{\theta} - \theta|$, then

$$P\bigl(\theta \in [\hat{\theta} - q, \hat{\theta} + q]\bigr) = 1 - \alpha.$$

The bootstrap estimator $\hat{q}$ for $q$ would be the upper $\alpha$th quantile for the conditional distribution of $|\hat{\theta}^* - \hat{\theta}|$ given $\mathbf{X}$. If this estimator is reasonably accurate, we expect that

$$P\bigl(\theta \in [\hat{\theta} - \hat{q}, \hat{\theta} + \hat{q}]\bigr) \approx 1 - \alpha,$$

so that $[\hat{\theta} - \hat{q}, \hat{\theta} + \hat{q}]$ is an approximate $1 - \alpha$ interval for $\theta$. As in the bias example, $\hat{q}$ can be approximated numerically by simulation, still with random samples $\mathbf{X}_1^*, \dots, \mathbf{X}_B^*$ from $\hat{Q}^n$, again taking $\hat{\theta}_i^* = \hat{\theta}(\mathbf{X}_i^*)$. Then $\hat{q}$ could be approximated as the upper $\alpha$th quantile for the list of values $|\hat{\theta}_i^* - \hat{\theta}|$, $i = 1, \dots, B$, generated in the simulation, or more formally as

$$\hat{q} \approx \inf\left\{x : \frac{1}{B}\#\{i \le B : |\hat{\theta}_i^* - \hat{\theta}| \le x\} \ge 1 - \alpha\right\}.$$

Mathematically, the structure is much the same as the bias example, with the bootstrap estimator $\hat{q}$ obtained by plugging $\hat{Q}$ into $q(\cdot)$; that is, $\hat{q} = q(\hat{Q})$.

In practice, bootstrap confidence intervals can often be improved by modifying the approach and approximating the distribution of *studentized* errors, obtained by dividing the error $\hat{\theta} - \theta$ by an estimate of the standard deviation of $\hat{\theta}$. If $\tilde{q} = \tilde{q}(Q)$ is the upper $\alpha$th quantile for the distribution of the absolute studentized error $|\hat{\theta} - \theta|/\hat{\tau}$, where $\hat{\tau} = \hat{\tau}(\mathbf{X})$ is an estimate for the standard deviation of $\hat{\theta}$, then

$$P\bigl(\theta \in [\hat{\theta} - \tilde{q}\hat{\tau}, \hat{\theta} + \tilde{q}\hat{\tau}]\bigr) = 1 - \alpha.$$

---

[1] For convenience, we assume that $\hat{\theta} - \theta$ has a continuous distribution. When this is not the case, some of the equations above may not hold exactly, but discrepancies will be quite small if masses for atoms of the error distribution are small.

The bootstrap estimator $\hat{\tilde{q}}$ for $\tilde{q}$ is the upper $\alpha$th quantile for the conditional distribution of $|\hat{\theta}^* - \hat{\theta}|/\hat{\tau}^*$ given $\mathbf{X}$, where $\hat{\tau}^* = \hat{\tau}(\mathbf{X}^*)$. If this estimator is reasonably accurate, then

$$[\hat{\theta} - \hat{\tilde{q}}\hat{\tau}, \hat{\theta} + \hat{\tilde{q}}\hat{\tau}]$$

should be an approximate $1-\alpha$ interval for $\theta$. Again, the bootstrap estimate $\hat{\tilde{q}}$ can be computed numerically by simulation as the upper $\alpha$th quantile for the list $|\hat{\theta}_i^* - \hat{\theta}|/\hat{\tau}_i^*$, $i = 1, \dots, B$, with $\hat{\theta}_i^*$ as before, and $\hat{\tau}_i^* = \hat{\tau}(\mathbf{X}_i^*)$.

To appreciate the value of studentizing, note that in various settings, including those detailed in the large-sample theory developed in Chapter 8, $(\hat{\theta} - \theta)/\hat{\tau}$ is approximately standard normal. If this is the case, the quantile $\tilde{q}$ is nearly independent of $Q$, making it easy to estimate. For instance, if the studentized error distribution happened to be exactly standard normal[2] for any $Q$, $\tilde{q}$ would always equal $z_{\alpha/2}$ and could be "estimated" perfectly, even without data.

## 19.2 Bias Reduction

In this section we explore a simple case of bias reduction where the performance of bootstrap estimators can be determined explicitly. Specifically, in Example 19.1 let

$$\theta = \theta(Q) = \mu^3,$$

where $\mu = EX_i$ is the mean of $Q$,

$$\mu = \mu(Q) = \int x\,dQ(x).$$

The mean of $\hat{Q}$ is the average,

$$\mu(\hat{Q}) = \int x\,d\hat{Q}(x) = \overline{X} = \frac{1}{n}\sum_{i=1}^{n} X_i,$$

and so

$$\hat{\theta} = \theta(\hat{Q}) = \overline{X}^3.$$

To find the bias $b$ we need $E\hat{\theta} = E\overline{X}^3$. Let $\sigma^2 = \operatorname{Var}(X_i)$ and $\gamma = E(X_i - \mu)^3$. Since $\gamma$ is the third cumulant for $X_i$, the third cumulant for $n\overline{X} = X_1 + \cdots + X_n$ is $n\gamma$, and so

$$E(n\overline{X} - n\mu)^3 = n^3 E(\overline{X} - \mu)^3 = n\gamma.$$

Thus $E(\overline{X} - \mu)^3 = \gamma/n^2$. Also, $E(\overline{X} - \mu)^2 = \operatorname{Var}(\overline{X}) = \sigma^2/n$. Using these identities,

---

[2] This could only happen in parametric situations.

$$\begin{aligned} E\overline{X}^3 &= E(\overline{X} - \mu + \mu)^3 \\ &= E\big[\mu^3 + 3\mu^2(\overline{X} - \mu) + 3\mu(\overline{X} - \mu)^2 + (\overline{X} - \mu)^3\big] \\ &= \mu^3 + \frac{3\mu\sigma^2}{n} + \frac{\gamma}{n^2}. \end{aligned} \tag{19.2}$$

So

$$b = b(Q) = E\hat{\theta} - \theta = E\overline{X}^3 - \mu^3 = \frac{3\mu\sigma^2}{n} + \frac{\gamma}{n^2}. \tag{19.3}$$

Because $\hat{b} = b(\hat{Q})$, we find the bootstrap estimate for $b$ with the same calculations but with data drawn from $\hat{Q}$. The mean, variance, and third central moment of $\hat{Q}$ are

$$\overline{X} = \int x \, d\hat{Q}(x), \qquad \hat{\sigma}^2 = \int (x - \overline{X})^2 \, d\hat{Q}(x) = \frac{1}{n}\sum_{i=1}^n (X_i - \overline{X})^2,$$

and

$$\hat{\gamma} = \frac{1}{n}\sum_{i=1}^n (X_i - \overline{X})^3.$$

Using these in (19.3),

$$\hat{b} = \frac{3\overline{X}\hat{\sigma}^2}{n} + \frac{\hat{\gamma}}{n^2}.$$

Subtracting this from $\overline{X}^3$, the "bias-reduced" estimator is

$$\hat{\theta} - \hat{b} = \overline{X}^3 - \frac{3\overline{X}\hat{\sigma}^2}{n} - \frac{\hat{\gamma}}{n^2}.$$

To see whether bootstrapping actually reduces the bias, we need to evaluate the mean of this new estimator. From (19.2) with $n = 1$, $EX_1^3 = \mu^3 + 3\mu\sigma^2 + \gamma$. Next, by symmetry, for any $j$,

$$\begin{aligned} EX_j \sum_{i=1}^n X_i^2 = EX_1 \sum_{i=1}^n X_i^2 &= EX_1^3 + \sum_{i=2}^n EX_1 X_i^2 \\ &= \mu^3 + 3\mu\sigma^2 + \gamma + (n-1)\mu(\mu^2 + \sigma^2). \end{aligned}$$

Averaging over $j$,

$$E\overline{X} \sum_{i=1}^n X_i^2 = \mu^3 + 3\mu\sigma^2 + \gamma + (n-1)\mu(\mu^2 + \sigma^2).$$

Finally, since

$$\hat{\sigma}^2 = \frac{1}{n}\sum_{i=1}^n X_i^2 - \overline{X}^2,$$

we have

$$\begin{aligned} E\overline{X}\hat{\sigma}^2 &= E\left[\frac{1}{n}\overline{X}\sum_{i=1}^n X_i^2 - \overline{X}^3\right] \\ &= \mu\sigma^2 + \frac{1}{n}(\gamma - \mu\sigma^2) - \frac{1}{n^2}\gamma. \end{aligned}$$

To find $E\hat{\gamma}$, note that

$$E(X_1-\mu)^2(\overline{X}-\mu) = \frac{1}{n}E(X_1-\mu)^3 = \frac{\gamma}{n},$$

and by symmetry

$$E(X_1-\mu)(\overline{X}-\mu)^2 = \frac{1}{n}\sum_{i=1}^n E(X_i-\mu)(\overline{X}-\mu)^2 = E(\overline{X}-\mu)^3 = \frac{\gamma}{n^2}.$$

Using these and symmetry,

$$\begin{aligned} E\hat{\gamma} &= E(X_1-\overline{X})^3 \\ &= E\big[(X_1-\mu)^3 - 3(X_1-\mu)^2(\overline{X}-\mu) \\ &\qquad + 3(X_1-\mu)(\overline{X}-\mu)^2 - (\overline{X}-\mu)^3\big] \\ &= \gamma\left(1 - \frac{3}{n} + \frac{2}{n^2}\right). \end{aligned}$$

Using these formulas, the mean of $\hat{\theta} - \hat{b}$ is

$$E[\hat{\theta} - \hat{b}] = \mu^3 + \frac{3}{n^2}(\mu\sigma^2 - \gamma) + \frac{6\gamma}{n^3} - \frac{2\gamma}{n^4}.$$

For large $n$ the bias of this estimator is of order $1/n^2$ compared with a bias of order $1/n$ for $\hat{\theta}$. So bootstrap correction here typically improves the bias.[3]

## 19.3 Parametric Bootstrap Confidence Intervals

In this section we consider parametric models in which our data are i.i.d. from a distribution $Q$ in some parametric family $\{Q_\lambda : \lambda \in \Lambda\}$. Knowing the marginal distribution lies in this family, $\hat{Q} = Q_{\hat{\lambda}}$, with $\hat{\lambda}$ the maximum likelihood estimator of $\lambda$, is a more natural estimator for $Q$ than the empirical distribution used in earlier sections. With this modification, the bootstrap approach is essentially the same as before. If $\hat{\theta}$ is the maximum likelihood

[3] If bias is the sole concern, there are unbiased estimators for $\mu^3$. The most natural one might be the $U$-statistic $\sum_{i<j<k} X_iX_jX_k/\binom{n}{3}$. Other resampling approaches, such as the jackknife, could also be used to reduce bias. See Problem 19.2.

estimator[4] for a parameter $\theta = \theta(\lambda)$, and if $\mathbf{X}^*$ is a bootstrap sample with entries conditionally i.i.d. from $\hat{Q} = Q_{\hat{\lambda}}$, so

$$\mathbf{X}^*|\mathbf{X} \sim \hat{Q}^n,$$

then the error distribution for $\hat{\theta}$ would be estimated as the conditional distribution for $\hat{\theta}^* - \hat{\theta}$ given $\mathbf{X}$, with $\hat{\theta}^* = \hat{\theta}(\mathbf{X}^*)$.

As in Example 19.2, interval estimation for $\theta$ can be approached with an attempt to estimate the upper $\alpha$th quantile $q = q(\lambda)$ for $|\hat{\theta}-\theta|$. The bootstrap estimate $\hat{q}$ for $q$ is the upper $\alpha$th quantile for the conditional distribution of $|\hat{\theta}^* - \hat{\theta}|$ given $\mathbf{X}$, and the approximate confidence interval based on $\hat{q}$ is

$$(\hat{\theta} - \hat{q}, \hat{\theta} + \hat{q}).$$

As before, studentizing the error distribution may give a more accurate confidence interval. If $\hat{\tau} = \hat{\tau}(\mathbf{X})$ is an estimate for the standard deviation of $\hat{\theta}$, and $\tilde{q} = \tilde{q}(\lambda)$ denotes the upper $\alpha$th quantile for the distribution of the absolute studentized error $|\hat{\theta} - \theta|/\hat{\tau}$, then the bootstrap estimate $\hat{\tilde{q}}$ for $\tilde{q}$ is the upper $\alpha$th quantile for the conditional distribution of $|\hat{\theta}^* - \hat{\theta}|/\tau^*$ given $\mathbf{X}$, with $\hat{\tau}^* = \hat{\tau}(\mathbf{X}^*)$, and the associated approximate confidence interval is

$$(\hat{\theta} - \hat{\tilde{q}}\hat{\tau}, \hat{\theta} + \hat{\tilde{q}}\hat{\tau}).$$

If the functions $q(\cdot)$ or $\tilde{q}(\cdot)$ are tractable, quantile estimates $\hat{q}$ and $\hat{\tilde{q}}$ can be found by evaluation at the maximum likelihood estimator: $\hat{q} = q(\hat{\lambda})$ and $\hat{\tilde{q}} = \tilde{q}(\hat{\lambda})$. When this is not feasible or practical, these estimators can be approximated by bootstrap simulation. Specifically $\hat{q}$ or $\hat{\tilde{q}}$ would be approximated numerically as upper $\alpha$th quantiles for the lists $|\hat{\theta}_i^* - \hat{\theta}|$, $i = 1, \dots, B$, or $|\hat{\theta}_i^* - \hat{\theta}|/\hat{\sigma}_i^*$, $i = 1, \dots, B$, with $\hat{\theta}_i^* = \hat{\theta}(\mathbf{X}_i^*)$, $\hat{\sigma}_i^* = \hat{\sigma}(\mathbf{X}_i^*)$, and $\mathbf{X}_i^*$, $i = 1, \dots, B$, conditionally i.i.d. from $\hat{Q}^n = Q_{\hat{\lambda}}^n$, given the original data $\mathbf{X}$.

To illustrate these ideas, consider interval estimation of the mean of a normal distribution. Taking $\lambda = (\theta, \sigma)$, $\hat{\lambda} = (\hat{\theta}, \hat{\sigma})$, with $\hat{\theta} = \overline{X}$ and

$$\hat{\sigma}^2 = \frac{1}{n}\sum_{i=1}^{n}(X_i - \overline{X})^2.$$

Given $\mathbf{X}$, the resampled data $X_1^*, \dots, X_n^*$ are i.i.d. from $N(\overline{X}, \hat{\sigma}^2)$. Since $\hat{\theta}^* = (X_1^* + \cdots + X_n^*)/n$, we have

$$\hat{\theta}^*|\mathbf{X} \sim N(\overline{X}, \hat{\sigma}^2/n) \text{ and } \hat{q} = z_{\alpha/2}\hat{\sigma}/\sqrt{n},$$

and the bootstrap confidence interval is

$$\left(\overline{X} - z_{\alpha/2}\frac{\hat{\sigma}}{\sqrt{n}}, \overline{X} + z_{\alpha/2}\frac{\hat{\sigma}}{\sqrt{n}}\right).$$

[4] The maximum likelihood structure here replaces assumption (19.1).

Exploiting the independence of $\overline{X}$ and $\hat{\sigma}$, the coverage probability for this bootstrap confidence interval can be expressed as

$$\begin{aligned} P\left(\overline{X} - z_{\alpha/2}\frac{\hat{\sigma}}{\sqrt{n}} < \theta < \overline{X} + z_{\alpha/2}\frac{\hat{\sigma}}{\sqrt{n}}\right) &= P\left(-z_{\alpha/2}\frac{\hat{\sigma}}{\sigma} < \frac{\overline{X}-\theta}{\sigma/\sqrt{n}} < z_{\alpha/2}\frac{\hat{\sigma}}{\sigma}\right) \\ &= EP\left(-z_{\alpha/2}\frac{\hat{\sigma}}{\sigma} < \frac{\overline{X}-\theta}{\sigma/\sqrt{n}} < z_{\alpha/2}\frac{\hat{\sigma}}{\sigma} \,\middle|\, \hat{\sigma}\right) \\ &= 1 - 2E\Phi\left(-\sqrt{z_{\alpha/2}^2\frac{\hat{\sigma}^2}{\sigma^2}}\right). \end{aligned}$$

By Taylor expansion about $\hat{\sigma}^2/\sigma^2 = 1$,

$$\Phi\left(-\sqrt{z_{\alpha/2}^2\frac{\hat{\sigma}^2}{\sigma^2}}\right) = \Phi(-z_{\alpha/2}) - \frac{1}{2}z_{\alpha/2}\phi(z_{\alpha/2})\left(\frac{\hat{\sigma}^2}{\sigma^2} - 1\right) + o(\hat{\sigma}^2 - \sigma^2).$$

Taking expectations of this approximation[5] and using

$$E\hat{\sigma}^2 = \frac{n-1}{n}ES^2 = \frac{n-1}{n}\sigma^2,$$

the coverage probability of the bootstrap confidence interval is

$$1 - \alpha - \frac{z_{\alpha/2}\phi(z_{\alpha/2})}{n} + o(1/n).$$

Because $\hat{\theta} = \overline{X}$ has standard deviation $\tau = \sigma/\sqrt{n}$, for the studentized approach we take $\hat{\tau} = \hat{\sigma}/\sqrt{n}$ and define $\tilde{q}(\lambda)$ as the upper $\alpha$th quantile for $|\overline{X} - \theta|/\hat{\tau}$. If $S^2 = n\hat{\sigma}^2/(n-1)$ is the sample variance, then $T \stackrel{\text{def}}{=} \sqrt{n}(\overline{X} - \theta)/S \sim t_{n-1}$. Since

$$\frac{|\hat{\theta} - \theta|}{\hat{\tau}} = \frac{\sqrt{n}|T|}{\sqrt{n-1}z},$$

we have

$$\tilde{q}(\lambda) = \frac{\sqrt{n}t_{\alpha/2,n-1}}{\sqrt{n-1}}.$$

This quantile is independent of $\lambda$, and so $\hat{\tilde{q}} = \tilde{q}$ and the studentized bootstrap confidence interval is

$$(\hat{\theta} - \hat{\tilde{q}}\hat{\tau}, \hat{\theta} + \hat{\tilde{q}}\hat{\tau}) = (\overline{X} - t_{\alpha/2,n-1}S/\sqrt{n}, \overline{X} + t_{\alpha/2,n-1}S/\sqrt{n}).$$

This interval is the same as the usual $t$-confidence interval, so in this case studentizing works perfectly, giving a bootstrap confidence interval with coverage exactly $1 - \alpha$.

[5] This can be justified using dominated convergence.

## 19.4 Nonparametric Accuracy for Averages

In this section we consider in some detail the performance of the nonparametric bootstrap when $\hat{\theta}$ is the sample mean $\overline{X}_n$, estimating $\theta = EX_i$. In this case error distributions can also be approximated using central limit theory, and it is of interest to see if a bootstrap approach does as well or better. Let $\sigma^2 = \operatorname{Var}(X_i)$,

$$\hat{\sigma}_n^2 = \frac{1}{n}\sum_{i=1}^n (X_i - \overline{X}_n)^2,$$

the variance of the empirical distribution $\hat{Q}_n$, and $\gamma = E(X_i - \theta)^3$. If

$$Z_n \stackrel{\text{def}}{=} \frac{\sqrt{n}(\hat{\theta} - \theta)}{\sigma},$$

then by the central limit theorem, $Z_n \Rightarrow N(0,1)$ and $P(Z_n \le x) \to \Phi(x)$. The next result shows that the error of this large-sample approximation is typically of order $1/\sqrt{n}$. Note that if $Q$ is a lattice distribution, assigning probability one to a set of the form $\{a + bj : j \in \mathbb{Z}\}$, then this would have to be the case as the distribution for $\overline{X}_n$ would have to have jumps of order $1/\sqrt{n}$. The theorem assumes that $Q$ is nonlattice.

**Theorem 19.3.** *If $Q$ is nonlattice and $E|X_i|^3 < \infty$, then*

$$P(Z_n \le x) = \Phi(x) - \frac{\gamma(x^2-1)}{6\sigma^3\sqrt{n}}\phi(x) + o\bigl(1/\sqrt{n}\bigr)$$

*as $n \to \infty$, uniformly in $x$.*

A proof of this result is given in Appendix A.7.3. Let $Y_i = (X_i - \theta)/\sigma$, so that $Z_n = \sqrt{n}\overline{Y}_n$, let $\mathfrak{f}$ be the characteristic function[6] of $Y_i$, defined as

$$\mathfrak{f}(t) = Ee^{itY_i}, \qquad t \in \mathbb{R},$$

and let $\mathfrak{f}_n$ be the characteristic function of $Z_n$,

$$\mathfrak{f}_n(t) = Ee^{itZ_n} = \mathfrak{f}^n\bigl(t/\sqrt{n}\bigr).$$

Using a smoothing lemma based on an inversion formula for characteristic functions (Lemma A.14), this theorem follows from two facts: first, that

$$\mathfrak{f}_n(t) = e^{-t^2/2}\left(1 - \frac{i\gamma t^3}{6\sigma^3\sqrt{n}}\right) + o\bigl(1/\sqrt{n}\bigr), \tag{19.4}$$

which follows by Taylor expansion of $\mathfrak{f}$ using the first three moments of $Y$: $EY_i = 0$, $EY_i^2 = 1$, and $EY_i^3 = \gamma/\sigma^3$; and second, that for any $0 < \delta < c < \infty$,

[6] See Appendix A.7.1.

$$\sup_{|t|\in[\delta,c]} \left|\mathfrak{f}(t)\right| < 1, \tag{19.5}$$

which follows from the nonlattice assumption.

A bootstrap approach to inference would approximate the distribution for $Z_n$ by the conditional distribution for

$$Z_n^* \stackrel{\text{def}}{=} \frac{\sqrt{n}(\hat{\theta}^* - \hat{\theta})}{\hat{\sigma}_n}.$$

The following result, due to Singh (1981), shows that this conditional distribution is the same to $o_p(1/\sqrt{n})$, which represents an improvement over normal approximation.

**Theorem 19.4.** *If $Q$ is nonlattice and $E|X_i|^3 < \infty$, then*

$$P(Z_n^* \le x|\mathbf{X}) = \Phi(x) - \frac{\gamma(x^2-1)}{6\sigma^3\sqrt{n}}\phi(x) + o_p(1/\sqrt{n})$$

*as $n \to \infty$, uniformly in $x$.*

*Proof (Sketch).* If we define $Y_i^* = (X_i^* - \overline{X}_n)/\hat{\sigma}_n$, then $Z_n^* = \sqrt{n}\overline{Y}_n^*$, and given $\mathbf{X}$, $Y_1^*, \dots, Y_n^*$ are conditionally i.i.d. So the conditional structure here is identical to that in Theorem 19.3. Also,

$$E(Y_i^*|\mathbf{X}) = 0, \qquad E(Y_i^{*2}|\mathbf{X}) = 1,$$

and

$$E(Y_i^{*3}|\mathbf{X}) = \frac{\hat{\gamma}_n}{\hat{\sigma}_n^3},$$

with

$$\hat{\gamma}_n = \frac{1}{n}\sum_{i=1}^n (X_i - \overline{X}_n)^3.$$

So the first two conditional moments for $Y_i^*$ are exactly the same as the corresponding moments of $Y_i$, and by the law of large numbers,

$$E(Y_i^{*3}|\mathbf{X}) = \hat{\gamma}_n/\hat{\sigma}_n^3 \xrightarrow{p} EY_i^3.$$

The same Taylor expansion argument used to show (19.4) gives

$$\mathfrak{f}_n^*(t) = e^{-t^2/2}\left(1 - \frac{i\gamma t^3}{6\sigma^3\sqrt{n}}\right) + o_p(1/\sqrt{n}),$$

with $\mathfrak{f}_n^*$ the conditional characteristic function for $Z_n^*$, $\mathfrak{f}_n^*(t) = E[e^{itZ_n^*}|\mathbf{X}]$.

The proof is completed, using the same argument used to prove Theorem 19.3, by showing that for any $0 < \delta < c < \infty$, there exists $\epsilon > 0$ such that

$$P\left(\sup_{|t|\in[\delta,c]} \bigl|\mathfrak{f}^*(t)\bigr| > 1-\epsilon\right) \to 0, \tag{19.6}$$

where $\mathfrak{f}^*$ is the conditional characteristic function for $Y_i^*$, given by

$$\mathfrak{f}^*(t) = E\bigl(e^{itY_i^*} \bigm| \mathbf{X}\bigr) = \frac{1}{n}\sum_{j=1}^n \exp\left[\frac{it(X_j - \overline{X}_n)}{\hat\sigma_n}\right], \qquad t \in \mathbb{R}.$$

Using Theorem 9.2, our law of large numbers for random functions, it is not hard to show that

$$\sup_{|t|\in[\delta,c]} \bigl|\mathfrak{f}(t) - \mathfrak{f}^*(t)\bigr| \xrightarrow{p} 0,$$

and (19.6) then follows from (19.5). ☐

Because the approximations in Theorems 19.3 and 19.4 hold uniformly in $x$,

$$P\bigl[\sqrt{n}(\hat\theta - \theta) \le x\bigr] = P(Z_n \le x/\sigma) = \Phi(x/\sigma) - \frac{\gamma(x^2-\sigma^2)}{6\sigma^5\sqrt{n}}\phi(x/\sigma) + o\bigl(1/\sqrt{n}\bigr)$$

and

$$P\bigl[\sqrt{n}(\hat\theta^* - \hat\theta) \le x \bigm| \mathbf{X}\bigr] = \Phi(x/\hat\sigma_n) - \frac{\gamma(x^2-\sigma^2)}{6\sigma^5\sqrt{n}}\phi(x/\sigma) + o_p\bigl(1/\sqrt{n}\bigr).$$

Since $\hat\sigma_n - \sigma = O_p\bigl(1/\sqrt{n}\bigr)$, by the delta method the leading terms in these approximations differ by $O_p\bigl(1/\sqrt{n}\bigr)$. So in this case, bootstrap methods do a better job of approximating the distribution of the standardized variable $Z_n$ than the distribution of the scaled error $\sqrt{n}(\hat\theta - \theta)$. Although the issues are somewhat different, this provides some support for the notion that studentizing generally improves the bootstrap performance.

The results above on the accuracy of bootstrap approximations can be extended in various ways. Perhaps the first thing worth noting is that Edgeworth expansions for distributions can be used to derive corresponding approximations, called *Cornish–Fisher expansions*, for quantiles. These expansions naturally play a central role in studying the performance of bootstrap confidence intervals.

Although the algebra is more involved, Edgeworth expansions can be derived to approximate distributions of averages of random vectors. And in principle these expansions lead directly to expansions for the distributions of smooth functions of averages. In his monograph, Hall (1992) uses this approach to study the performance of the bootstrap when $\hat\theta$ is a smooth function of averages. With suitable regularity, the discrepancy between the true coverage probability and the desired nominal value for the symmetric two-sided bootstrap confidence intervals described in Example 19.2 is of order $O(1/n)$ without studentizing, and of order $O(1/n^2)$ with studentizing.

As mentioned earlier, if $\theta = \theta(Q)$ and $\hat{\theta} = \theta(\hat{Q})$, then bootstrapping should work well if $\theta(\cdot)$ is suitably smooth. One regularity condition, studied in Bickel and Freedman (1981), is that $\theta(\cdot)$ is Gâteaux differentiable with the derivative representable as an integral. Such $\theta(\cdot)$ are often called *von Mises functionals.* Bickel and Freedman (1981) also give examples showing that bootstrapping can fail when $\theta(\cdot)$ is not smooth.

## 19.5 Problems

1. Bootstrap methods can also be used to reduce bias in parametric estimation. As an example, suppose $X_1, \dots, X_n$ are i.i.d. from $N(\mu, 1)$, and consider estimating $\theta = \sin \mu$.
   a) The maximum likelihood estimator, $\hat{\theta} = \sin \overline{X}$, is (for most $\mu$) a biased estimator of $\theta$. Derive an approximation for the bias of $\hat{\theta}$, accurate to $o(1/n^2)$ as $n \to \infty$.
   b) Consider a parametric bootstrap approach to estimating the bias $b(\mu) = E_\mu \hat{\theta} - \theta$, in which, given $\mathbf{X} = (X_1, \dots, X_n)$, $X_1^*, \dots, X_n^*$ are conditionally i.i.d. from $N(\overline{X}, 1)$. Letting $\hat{b} = E[\hat{\theta}^* - \hat{\theta} | \mathbf{X}]$, derive an approximation for the bias of $\hat{\theta} - \hat{b}$, accurate to $o(1/n^2)$ as $n \to \infty$.
2. Another resampling approach to inference is called the jackknife. Let $\hat{\theta}$ be an estimator for $\theta$ based on i.i.d. observations $X_1, \dots, X_n$, let $\hat{\theta}_{-i}$ be the estimator obtained omitting observation $X_i$ from the data set, and define

$$\tilde{\theta}_i = n\hat{\theta} - (n-1)\hat{\theta}_{-i}, \qquad i = 1, \dots, n,$$

   called *pseudo-values* by Tukey.
   a) Let $\tilde{\theta}$ denote the average of the pseudo-values, and assume

$$E\hat{\theta} = \theta + \frac{a_1}{n} + \frac{a_2}{n^2} + o(1/n^2),$$

   as $n \to \infty$. Derive an approximation for the bias of $\tilde{\theta}$ as $n \to \infty$, accurate to $o(1/n)$.
   b) Assume now that the observations $X_i$ are random variables with a finite mean $\mu$ and variance $\sigma^2 \in (0, \infty)$, and that $\theta = A(\mu)$ and $\hat{\theta} = A(\overline{X})$ for some function $A$, with $A'$ and $A''$ bounded and continuous. Show that

$$\frac{\sqrt{n}(\tilde{\theta} - \theta)}{\tilde{S}} \Rightarrow N(0,1),$$

   as $n \to \infty$, where $\tilde{S}$ is the sample standard deviation for the pseudo-values:

$$\tilde{S}^2 = \frac{1}{n-1} \sum_{i=1}^{n} (\tilde{\theta}_i - \tilde{\theta})^2.$$

3. Consider a parametric bootstrap approach to interval estimation with observations from a location family. Let $X_1, \dots, X_n$ be i.i.d. from an absolutely continuous distribution with density $f(x-\theta)$. Let $\hat{\theta}$ be the maximum likelihood estimator of $\theta$, and define $q = q(\theta)$ as the upper $\alpha$th quantile for the distribution of $|\hat{\theta} - \theta|$. The bootstrap estimate of $q$ is $\hat{q} = q(\hat{\theta})$. Show that the associated confidence interval

$$(\hat{\theta} - \hat{q}, \hat{\theta} + \hat{q})$$

has exact coverage $1 - \alpha$.
4. Let $q_\alpha$ denote the upper $\alpha$th quantile for $Z_n = \sqrt{n}(\overline{X}_n - \theta)/\sigma$ in Section 19.4. Use Theorem 19.3 to derive an approximation for $q_\alpha$, accurate to $o\bigl(1/\sqrt{n}\bigr)$ as $n \to \infty$.
5. Let $X_1, X_2, \dots$ be i.i.d. from a nonlattice distribution $Q$, with common mean $\mu$, common variance $\sigma^2$, and $E|X_i|^3 < \infty$; and let $g$ be a twice continuously differentiable function with $g(\mu) \neq 0$. Use Theorem 19.3 to derive an approximation for

$$P\bigl[\sqrt{n}\bigl(g(\overline{X}_n) - g(\mu)\bigr) \le x\bigr],$$

accurate to $o\bigl(1/\sqrt{n}\bigr)$ as $n \to \infty$.
6. Let $X_1, X_2, \dots$ be i.i.d. absolutely continuous variables from a canonical exponential family with marginal density

$$f_\eta(x) = \exp\{\eta x - A(\eta)\}h(x), \qquad x \in \mathbb{R},$$

for $\eta \in \Xi$. The maximum likelihood estimator of $\eta$ based on the first $n$ observations is then $\hat{\eta}_n = \psi(\overline{X}_n)$ with $\psi$ the inverse of $A'$. Consider a parametric bootstrap approach to estimating the error distribution for $\hat{\eta}_n$. Given $\mathbf{X} = (X_1, \dots, X_n)$, let $X_1^*, \dots, X_n^*$ be conditionally i.i.d. with marginal density $f_{\hat{\eta}_n}$. Assume that the approximation derived in Problem 19.5 holds uniformly in some neighborhood of $\eta$, and use it to derive approximations for

$$P\bigl(\sqrt{n}(\hat{\eta}_n - \eta) \le x\bigr)$$

and

$$P\bigl(\sqrt{n}(\hat{\eta}_n^* - \hat{\eta}) \le x \bigm| \mathbf{X}\bigr),$$

both accurate to $o_p\bigl(1/\sqrt{n}\bigr)$. Are these the same to $o_p\bigl(1/\sqrt{n}\bigr)$?

# 20

# Sequential Methods

Sequential experiments, introduced in Chapter 5, call for design decisions as data are collected. Optional stopping, in which the data are observed sequentially and used to decide when to terminate the experiment, would be the simplest example. A sequential approach can lead to increased efficiency, or it may achieve objectives not possible with a classical approach, but there are technical, practical, and philosophical issues that deserve attention.

*Example 20.1. Sampling to a Foregone Conclusion.* Let $X_1$, $X_2$, ... be i.i.d. from $N(\mu, 1)$, and let $S_n$ denote the sum of the first $n$ observations. The standard level $\alpha$ test of $H_0 : \mu = 0$ versus $H_1 : \mu \neq 0$ based on these observations will reject $H_0$ if $|S_n| > z_{\alpha/2}\sqrt{n}$.

Suppose a researcher proceeds sequentially, stopping the first time $n$ that $|S_n|$ exceeds $z_{\alpha/2}\sqrt{n}$, so the sample size is

$$N = \inf\{n \geq 1 : |S_n| > z_{\alpha/2}\sqrt{n}\}.$$

Whenever $N$ is finite, the classical test will reject $H_0$. If $\mu \neq 0$, then $N$ will be finite almost surely by the law of large numbers. In fact, $N$ will also be finite almost surely if $\mu = 0$. To see this, note that for any $k$, $\{N = \infty\}$ implies

$$\{|S_k| \leq \sqrt{k} z_{\alpha/2}, |S_{2k}| \leq \sqrt{2k} z_{\alpha/2}\},$$

which in turn implies

$$\{|S_{2k} - S_k| \leq (\sqrt{k} + \sqrt{2k}) z_{\alpha/2}\}.$$

These events have constant probability

$$p = P\bigl(|Z| \leq (1 + \sqrt{2}) z_{\alpha/2}\bigr) < 1,$$

where $Z \sim N(0, 1)$, and so by independence,

$$P(N = \infty) \leq P\left(\bigcap_{j=1}^{\infty} \{|S_{2^{j+1}} - S_{2^j}| \leq \bigl(\sqrt{2^j} + \sqrt{2^{j+1}}\bigr) z_{\alpha/2}\}\right) = \prod_{j=1}^{\infty} p = 0.$$

R.W. Keener, *Theoretical Statistics: Topics for a Core Course*, Springer Texts in Statistics,
DOI 10.1007/978-0-387-93839-4_20, 

This example highlights one central technical problem with sequential experiments; sampling distributions may change with optional stopping. For any fixed sample size $N$, if $\mu = 0$, then $S_N/\sqrt{N} \sim N(0,1)$, but with the random sample size $N$ in the example, $|S_N/\sqrt{N}|$ exceeds $z_{\alpha/2}$ almost surely. Historically, there was controversy and concern when this was noted. If a researcher conducts an experiment sequentially, a standard frequentist analysis is not appropriate. For a proper frequentist analysis there must be a specific protocol detailing how the sample size will be determined from the data, so that the effects of optional stopping can be taken into account properly when probabilities, distributions, and moments of statistics are computed.

Surprisingly, likelihood functions after a sequential experiment are found in the usual way. Since Bayesian inference is driven by the likelihood, posterior distributions will be computed in the usual fashion, and a sequential design will not affect Bayesian analysis of the data. Due to this, design problems are often more tractable with a Bayesian formulation.

In Section 20.1 a central limit theorem is derived for sequential experiments and used to find stopping rules that allow asymptotic interval estimation with specified fixed accuracy. Section 20.2 studies stopping times in a more formal fashion, explaining why they do not affect likelihood functions. In Section 20.3, the backwards induction method, used to find optimal stopping times, is explored, focusing on a Bayesian approach to hypothesis testing. Section 20.4 introduces Wald's sequential probability ratio test for simple versus simple testing in a sequential context. Finally, Section 20.5 explores design issues beyond optional stopping, specifically stochastic approximation recursions in which independent regression variables are chosen adaptively, and "bandit" allocation problems.

## 20.1 Fixed Width Confidence Intervals

Let $X_1, X_2, \dots$ be i.i.d. from a one-parameter exponential family with marginal density

$$f_\theta(x) = h(x)\exp\{\eta(\theta)T(x) - B(\theta)\}.$$

Let $T_i = T(X_i)$ and $\overline{T}_n = (T_1 + \cdots + T_n)/n$. Then the maximum likelihood estimator $\hat{\lambda}_n$ of a parameter $\lambda = g(\theta)$ based on the first $n$ observations is a function of $\overline{T}_n$,

$$\hat{\lambda}_n = \hat{\lambda}(\overline{T}_n),$$

and by the delta method,

$$\sqrt{n}(\hat{\lambda}_n - \lambda) \Rightarrow N\bigl(0, \nu^2(\theta)\bigr),$$

where

$$\nu^2(\theta) = \bigl[\hat{\lambda}'\bigl(\mu_T(\theta)\bigr)\bigr]^2\sigma_T^2(\theta),$$

with $\mu_T(\theta) = E_\theta T_i$ and $\sigma_T^2(\theta) = \text{Var}_\theta(T_i)$. Using this, if $\nu(\cdot)$ is continuous and $\hat{\theta}_n$ is the maximum likelihood estimator of $\theta$, then

$$\left(\hat{\lambda}_n \pm \frac{z_{\alpha/2}}{nu(\hat{\theta}_n)}/\sqrt{n}\right) \tag{20.1}$$

is an asymptotic $1-\alpha$ confidence interval for $\lambda$.

If a researcher is interested in estimating $\lambda$ with fixed precision, a confidence interval with a fixed width $w$ would be desired. Since $\nu(\theta)$ will generally vary with $\theta$, the interval (20.1) from any fixed sample may fail. But following a sequential strategy, the researcher may choose to continue sampling until the width $2z_{\alpha/2}\nu(\hat{\theta}_n)/\sqrt{n}$ of interval (20.1) is less than $w$. This leads to a sequential experiment with sample size

$$N = N_w = \inf\{n : w^2 n \geq 4z_{\alpha/2}^2 \nu^2(\hat{\theta}_n)\}. \tag{20.2}$$

If $w$ is small, $N$ will be large, and it seems reasonable to hope that the interval

$$\left(\hat{\lambda}_N \pm \frac{z_{\alpha/2}\nu(\hat{\theta}_N)}{\sqrt{N}}\right) \tag{20.3}$$

from a sequential experiment will have coverage approximately $1-\alpha$. And by construction, the width of this interval is at most $w$. This is correct, but a proper demonstration takes a bit of care, because the sample size $N$ is a random variable, whereas sample sizes in our prior results on weak convergence were constant. The main result we need is a central limit theorem due to Anscombe (1952) in which the number of summands is random. Almost sure convergence, introduced in Section 8.7, and the strong law of large numbers play a role here.

The proposed sample size $N_w$ in (20.2) tends to $\infty$ almost surely as $w \downarrow 0$. If $\hat{\theta}_n \to \theta$ almost surely,[1] then $\hat{\theta}_N \to \theta$ almost surely and $\hat{\theta}_{N-1} \to \theta$ almost surely. Since

$$4z_{\alpha/2}^2\nu^2(\hat{\theta}_N) \leq w^2 N < w^2 + 4z_{\alpha/2}^2\nu^2(\hat{\theta}_{N-1}),$$

it follows that

$$w^2 N \to 4z_{\alpha/2}^2\nu^2(\theta)$$

almost surely as $w \downarrow 0$. If we define

$$n_w = \left\lfloor \frac{4z_{\alpha/2}^2\nu^2(\theta)}{w^2} \right\rfloor,$$

then $w^2(N - n_w) \to 0$ almost surely as $w \downarrow 0$. The idea behind Anscombe's central limit theorem is that a shift in the sample size from $N$ to $n_w$ will change the limiting variable by an amount that is $o_p(1)$.

[1] When $\eta$ is continuous, $\hat{\theta}_n$ is a continuous function of $\overline{T}_n$, and this follows from the strong law of large numbers.

**Definition 20.2.** *Random variables* $W_n$, $n \ge 1$, *are* uniformly continuous in probability *(u.c.i.p.) if for all* $\epsilon > 0$ *there exists* $\delta > 0$ *such that*

$$P\left\{\max_{0\le k\le n\delta} |W_{n+k} - W_n| \ge \epsilon\right\} < \epsilon, \qquad \textit{for all } n \ge 1.$$

**Theorem 20.3 (Anscombe).** *If* $N_w$, $w > 0$, *are positive integer-valued random variables with* $w^2 N_w \xrightarrow{p} c \in (0,\infty)$ *as* $w \downarrow 0$, *if* $n_w = \lfloor c/w^2 \rfloor$, *and if* $W_n$, $n \ge 1$ *are u.c.i.p., then*

$$W_{N_w} - W_{n_w} \xrightarrow{p} 0$$

*as* $w \downarrow 0$.

*Proof.* Fix $\epsilon > 0$. For any $\delta > 0$,

$$P\big(|W_{N_w} - W_{n_w}| > \epsilon\big) \le P\big(w^2|N_w - n_w| > \delta\big) + P\big(\max_{w^2|n-n_w|\le\delta} |W_n - W_{n_w}| > \epsilon\big).$$

The first term here tends to zero regardless of the choice of $\delta$. By the triangle inequality, if $m = \lceil n_w - \delta/w^2 \rceil$ (the smallest integer $m$ with $w^2|m - n_w| \le \delta$), then

$$|W_n - W_{n_w}| \le |W_n - W_m| + |W_{n_w} - W_m|,$$

and so

$$\max_{w^2|n-n_w|\le\delta} |W_n - W_{n_w}| \le 2 \max_{w^2|n-n_w|\le\delta} |W_n - W_m|.$$

Therefore

$$P\big(\max_{w^2|n-n_w|\le\delta} |W_n - W_{n_w}| > \epsilon\big) \le P\big(\max_{w^2|n-n_w|\le\delta} |W_n - W_m| > \epsilon/2\big).$$

Since the $W_n$ are u.c.i.p., this probability will be less than $\epsilon/2$ if $\delta$ is sufficiently small, and the theorem follows as $\epsilon$ is arbitrary. ☐

In Theorem 20.3, if $W_n \Rightarrow W$, $W_{n_w} \Rightarrow W$, and so

$$W_{N_w} = W_{n_w} + o_p(1) \Rightarrow W.$$

One example of particular interest would be normalized partial sums, $W_n = \sqrt{n}\overline{Y}_n$ with $Y_1, Y_2, \dots$ i.i.d. mean zero, and $\overline{Y}_n$ the average of the first $n$ of these variables. The following maximal inequality, due to Kolmogorov, is used to show these variables are u.c.i.p. Let $S_k = Y_1 + \cdots + Y_k$.

**Lemma 20.4.** *If* $Y_1, \dots, Y_n$ *are i.i.d. with mean zero and common variance* $\sigma_Y^2 \in (0,\infty)$, *then for any* $c > 0$,

$$P\left(\max_{1\le k\le n} |S_k| \ge c\right) \le \frac{n\sigma_Y^2}{c^2}.$$

*Proof.* Let $A_k$ be the event that $S_k$ is the first partial sum with magnitude at least $c$, that is,

$$A_k = \{|S_1| < c, \dots, |S_{k-1}| < c, |S_k| \ge k\}.$$

Because $A_k$ is determined by $Y_1, \dots, Y_k$, $S_k 1_{A_k}$ is independent of $S_n - S_k = Y_{k+1} + \cdots + Y_n$, and for $k \le n$

$$E\big[S_k(S_n - S_k)1_{A_k}\big] = E[S_k 1_{A_k}] \times E[S_n - S_k] = 0.$$

But on $A_k$, $S_k^2 \ge c^2$, and so for $k \le n$,

$$E[S_n^2 1_{A_k}] = E\big[(S_n - S_k)^2 + 2S_k(S_n - S_k) + S_k^2\big]1_{A_k} \ge c^2 P(A_k).$$

Since $\{\max_{1 \le k \le n} |S_k| \ge c\}$ is the disjoint union of $A_1, \dots, A_n$,

$$c^2 P(\max_{1 \le k \le n} |S_k| \ge c) = \sum_{k=1}^{n} c^2 P(A_k) \le E\left[S_n^2 \sum_{i=1}^{n} A_k\right] \le ES_n^2 = n\sigma_Y^2,$$

proving the lemma. ☐

Considering the normalized partial sums, since

$$|W_{n+k} - W_n| = \left| \frac{1}{\sqrt{n}} \sum_{i=n+1}^{k} Y_i - \left(\sqrt{\frac{n+k}{n}} - 1\right) W_{n+k} \right|,$$

we have

$$P(\max_{0 \le k \le n\delta} |W_{n+k} - W_n| \ge \epsilon) \le P\big((\sqrt{1+\delta} - 1)|W_{n+k}| \ge \epsilon/2\big) + P\left(\max_{0 \le k \le n\delta} \left|\sum_{i=n+1}^{n+k} Y_i\right| \ge \epsilon\sqrt{n}/2\right).$$

By Chebyshev's inequality, the first term here is at most

$$\frac{4(\sqrt{1+\delta} - 1)^2 \sigma_Y^2}{\epsilon^2},$$

and by Lemma 20.4 the second term is at most

$$\frac{4\sigma_Y^2 \delta}{\epsilon^2}.$$

These bounds tend to zero as $\delta \downarrow 0$, uniformly in $n$, and so $W_n$, $n \ge 1$, are u.c.i.p.

Returning to fixed width interval estimation and the coverage probability for interval (20.3), by Theorem 20.3,

$$\sqrt{N}\big(\overline{T}_N - \mu(\theta)\big) \Rightarrow N\big(0, \mathrm{Var}_\theta(T_i)\big)$$

as $w \downarrow 0$, and by the delta method,

$$\sqrt{N_w}(\hat{\lambda}_N - \lambda) \Rightarrow N\big(0, \nu^2(\theta)\big).$$

Since $N \to \infty$ as $w \downarrow 0$ and $\hat{\theta}_n \to \theta$ as $n \to \infty$, both almost surely, $\hat{\theta}_N \to \theta$ almost surely as $w \downarrow 0$. It then follows that

$$\frac{\sqrt{N}(\hat{\lambda}_N - \lambda)}{\nu(\hat{\theta}_N)} \Rightarrow N(0,1)$$

as $w \downarrow 0$. So the coverage probability for the confidence interval (20.3),

$$P_\theta\Big[\lambda \in \big(\hat{\lambda}_N \pm z_{\alpha/2}\nu(\hat{\theta}_N)/\sqrt{N}\big)\Big] = P_\theta\left(\frac{\sqrt{N}|\hat{\lambda}_N - \lambda|}{\nu(\hat{\theta}_N)} < z_{\alpha/2}\right),$$

converges to $1 - \alpha$ as $w \downarrow 0$.

## 20.2 Stopping Times and Likelihoods

In Chapter 5 we had some trouble representing data from a sequential experiment as a random vector, because this kind of experiment's sample size is not a fixed constant. The most elegant and standard way to ameliorate this problem is to use $\sigma$-fields to represent information. To understand how this is done, consider an experiment in which a coin is tossed two times, so the sample space is

$$\mathcal{E} = \{TT, TH, HT, HH\}.$$

Let $\mathcal{F}$ be the $\sigma$-field of all subsets of $\mathcal{E}$, $\mathcal{F} = 2^{\mathcal{E}}$, and let the random variable $X$ give the number of heads. If we observe $X$ we will know if certain events occur. For instance, we will know whether

$$\{X = 1\} = \{HT, TH\}$$

occurs. But other events, such as $\{HH, HT\}$ (the first toss lands heads), will remain in doubt. The collection of all events we can resolve,

$$\sigma(X) = \{\emptyset, \{TT\}, \{HT, TH\}, \{HH\}, \{TT, HT, TH\},$$
$$\{TT, HH\}, \{HT, TH, HH\}, \{TT, TH, HT, HH\}\},$$

is a $\sigma$-field. A means to learn which events in $\sigma(X)$ occur would provide exactly the same information about the outcome $e$ as the value for $X$. Thus $X$ and $\sigma(X)$ in a natural sense provide the same information.

The notions in the coin tossing example generalize easily. If we observe a random vector $X$, then we will know whether $\{X \in B\} = X^{-1}(B)$ occurs.

Here we insist that $B$ is a Borel set to guarantee that $\{X \in B\}$ is an event in $\mathcal{F}$. We can then define

$$\sigma(X) \stackrel{\text{def}}{=} \{X^{-1}(B) : B \text{ Borel}\},$$

and it is easy to show that $\sigma(X)$ is a $\sigma$-field.

Consider now an experiment in which random vectors $X_1, X_2, \ldots$ are observed sequentially. Let

$$\mathcal{F}_n = \sigma\{X_1, \ldots, X_n\}, \tag{20.4}$$

the events we can resolve observing the first $n$ variables, and take $\mathcal{F}_0 = \{\emptyset, \mathcal{E}\}$. These $\sigma$-fields are increasing, $\mathcal{F}_0 \subset \mathcal{F}_1 \subset \mathcal{F}_2 \subset \cdots$. In general, any increasing sequence of $\sigma$-fields $\mathcal{F}_n$, $n \geq 0$, is called a *filtration*. The filtration given by (20.4) would be called the *natural filtration* for $X_i$, $i \geq 1$. We can also define $\mathcal{F}_\infty$ as the smallest $\sigma$-field containing all events in $\bigcup_{n \geq 1} \mathcal{F}_n$. For the natural filtration, $\mathcal{F}_\infty$ would represent the information available from all the $X_i$. This $\sigma$-field may not equal the underlying $\sigma$-field $\mathcal{F}$; it will be strictly smaller if $\mathcal{F}$ contains events that cannot be determined from the $X_i$.

Sample sizes for a sequential experiment cannot depend on the observations in an arbitrary fashion. For instance, a design calling for two observations if and only if $X_5 > 10$ would be absurd. Clairvoyance needs to be prohibited in the mathematical formulation. In particular, the decision to stop or continue after $n$ observations needs to be based on the information $\mathcal{F}_n$ available from those data. Specifically, the event $\{N = n\}$ should lie in $\mathcal{F}_n$. These variables are called *stopping times* according to the following definition.

**Definition 20.5.** *A random variable*[2] *$N$ taking values in $\{0, 1, 2, \ldots, \infty\}$ is called a* stopping time *with respect to a filtration $\mathcal{F}_n$, $n \geq 0$, if*

$$\{N = n\} \in \mathcal{F}_n, \qquad \textit{for all } n \geq 0.$$

Next we would like to find a $\sigma$-field that represents the information available when data are observed until a stopping time $N$. Any event $B \in \mathcal{F}$ can be written as the disjoint union of the sets $B \cap \{N = n\}$, $n = 1, 2, \ldots, \infty$. If we can determine $B$ from the data, it must be the case that that part of $B$ where $N = n$ (i.e., $B \cap \{N = n\}$) must lie in $\mathcal{F}_n$, for any $n$, and we define

$$\mathcal{F}_N = \bigl\{B : B \cap \{N = n\} \in \mathcal{F}_n, \forall n = 0, 1, \ldots, \infty\bigr\}.$$

It is not hard to show that $\mathcal{F}_N$ is a $\sigma$-field, and it represents the information available observing the data until stopping time $N$. We may also want to consider what random variables $Y$ are based on the observed data. Because the event $\{Y \in B\} = Y^{-1}(B)$ can be resolved by observing $Y$, this event

---

[2] Since "$+\infty$" is an allowed value for $N$, it may be slightly more proper to call $N$ an extended random variable.

should lie in $\mathcal{F}_N$. But this requirement is simply that $Y$ is $\mathcal{F}_N$ measurable. For instance, with the natural filtration, the stopping time $N$ is $\mathcal{F}_N$ measurable, and $\overline{X}_N = (X_1 + \cdots + X_N)/N$ is $\mathcal{F}_N$ measurable. But $X_{N+1}$ is not.

If we use $\sigma$-fields to represent information, we will also be interested in conditioning to revise probabilities and expectations in light of the information from a $\sigma$-field. With a random vector $X$ and an integrable random variable $Y$, $E(Y|X)$ should be a measurable function of $X$, and smoothing must work for $YI\{X \in B\}$ with $B$ an arbitrary Borel set:

$$EYI\{X \in B\} = EI\{X \in B\}E(Y|X),$$

for all Borel sets $B$. The requirements for conditioning on a $\sigma$-field $\mathcal{G}$ are similar. First, $E(Y|\mathcal{G})$ should be $\mathcal{G}$ measurable (based only on information available from $\mathcal{G}$); and second, smoothing should work for $Y1_B$ with $B$ any event in $\mathcal{G}$:

$$EY1_B = E1_B E(Y|\mathcal{G}), \qquad B \in \mathcal{G}.$$

The next result, Wald's fundamental identity, is the basis for likelihood calculations. In this result, there are two probability measures $P_0$ and $P_1$. Let $f_{0n}$ and $f_{1n}$ denote joint densities for $(X_1, \dots, X_n)$ under $P_0$ and $P_1$, and let $L_n = L_n(X_1, \dots, X_n)$ denote the likelihood ratio

$$L_n(X_1, \dots, X_n) = \frac{f_{1n}(X_1, \dots, X_n)}{f_{0n}(X_1, \dots, X_n)}.$$

**Theorem 20.6 (Wald's Fundamental Identity).** *If $f_{1n} = 0$ whenever $f_{0n} = 0$, and if $P_0(N < \infty) = P_1(N < \infty) = 1$, then*

$$P_1(B) = E_0 1_B L_N, \qquad \forall B \in \mathcal{F}_N.$$

*Proof.* Because $\{N = n\} \cap B \in \mathcal{F}_n$, by Lemma 12.18,

$$P_1(N = n, B) = E_0 I\{N = n\}1_B L_n = E_0 I\{N = n\}1_B L_N.$$

Because $P_0(N < \infty) = P_1(N < \infty) = 1$,

$$\begin{aligned} P_1(B) = P_1(B, N < \infty) &= \sum_{n=1}^{\infty} P_1(N = n, B) \\ &= \sum_{n=1}^{\infty} E_0 I\{N = n\}1_B L_N = E_0 I\{N < \infty\}1_B L_N = E_0 1_B L_N. \end{aligned}$$ ☐

If $P_0$ and $P_1$ are restricted and only considered as measures on $\mathcal{F}_N$, this result asserts that $P_1$ has density $L_N$ with respect to $P_0$. Theorem 5.4 follows from this. It can be shown that any $\sigma$-finite measure $\mu$ is equivalent to, or has the same null sets as, a probability measure. So in Theorem 5.4 we can assume that the dominating measure $\mu$ is a probability measure. If the $X_i$ are i.i.d. with density $f_\theta$ under $P_\theta$, viewed as $P_1$ in Wald's fundamental identity, and they are i.i.d. from $\mu$ under $P_0$, then the density for the restriction of $P_\theta$ to $\mathcal{F}_N$ is

$$L_N = \prod_{i=1}^{N} f(X_i),$$

with respect to the restriction of $P_0$ to $\mathcal{F}_N$.

## 20.3 Optimal Stopping

This section provides an introduction to the theory of optimal stopping, used to select the best stopping time $N$ for a sequential experiment. The main ideas are developed in the context of Bayesian hypothesis testing.

Given $\Theta = \theta$, let potential observations $X_1, X_2, \dots$ be i.i.d. from $Q_\theta$, and consider testing $H_0 : \Theta \in \Omega_0$ versus $H_1 : \Theta \in \Omega_1$. To be specific in our goals we proceed in a decision-theoretic fashion, assigning costs for the consequences of our inferential actions, with additional costs to perform the experiment and collect data. Inferential actions and stopping times are chosen to minimize expected costs.

After data collection, one of the hypotheses, $H_0$ or $H_1$, will be accepted. Let variable $A$ specify this action, with $A = 0$ if we accept $H_0$ and $A = 1$ if we accept $H_1$. This action depends on the observed data $\mathcal{F}_N$, so $A$ must be $\mathcal{F}_N$ measurable. Let $L(\Theta)$ denote the loss if we make the wrong decision: $A = 0$ when $\Theta \in \Omega_1$, or $A = 1$ when $\Theta \in \Omega_0$. The following result characterizes an optimal action $A$.

**Theorem 20.7.** *The inferential risk associated with action $A$, given by*

$$R(A) = EL(\Theta)\big[I\{A = 0, \Theta \in \Omega_1\} + I\{A = 1, \Theta \in \Omega_0\}\big],$$

*will be minimal if*

$$A = 0 \ \text{ on } \ E\big[L(\Theta)I\{\Theta \in \Omega_1\} \mid \mathcal{F}_N\big] < E\big[L(\Theta)I\{\Theta \in \Omega_0\} \mid \mathcal{F}_N\big]$$

*and*

$$A = 1 \ \text{ on } \ E\big[L(\Theta)I\{\Theta \in \Omega_1\} \mid \mathcal{F}_N\big] > E\big[L(\Theta)I\{\Theta \in \Omega_0\} \mid \mathcal{F}_N\big].$$

*Proof.* Because $A$ is $\mathcal{F}_N$ measurable,

$$E\big[L(\Theta)I\{A = 0\}I\{\Theta \in \Omega_1\} \mid \mathcal{F}_N\big] = I\{A = 0\}E\big[L(\Theta)I\{\Theta \in \Omega_1\} \mid \mathcal{F}_N\big]$$

and

$$E\big[L(\Theta)I\{A = 1\}I\{\Theta \in \Omega_0\} \mid \mathcal{F}_N\big] = I\{A = 1\}E\big[L(\Theta)I\{\Theta \in \Omega_0\} \mid \mathcal{F}_N\big].$$

So by smoothing,

$$\begin{aligned}R(A) &= EE\Big[L(\Theta)\big[I\{A=0,\Theta\in\Omega_1\}+I\{A=1,\Theta\in\Omega_0\}\big]\Bigm|\mathcal{F}_N\Big]\\&= E\Big[I\{A=0\}E\big[L(\Theta)I\{\Theta\in\Omega_1\}\mid\mathcal{F}_N\big]\\&\qquad+I\{A=1\}E\big[L(\Theta)I\{\Theta\in\Omega_0\}\mid\mathcal{F}_N\big]\Big]\\&\geq E\min\Big\{E\big[L(\Theta)I\{\Theta\in\Omega_1\}\mid\mathcal{F}_N\big],E\big[L(\Theta)I\{\Theta\in\Omega_0\}\mid\mathcal{F}_N\big]\Big\}.\end{aligned}$$

This bound is achieved if $A$ has the form indicated in the theorem. □

Using this result, if we define

$$\rho_N=\min\Big\{E\big[L(\Theta)I\{\Theta\in\Omega_1\}\mid\mathcal{F}_N\big],E\big[L(\Theta)I\{\Theta\in\Omega_0\}\mid\mathcal{F}_N\big]\Big\},$$

the inferential risk with an optimal action $A$ is $E\rho_N$, and an optimal stopping rule $N$ should balance this risk against expected costs running the experiment. A simple assumption, natural and fairly appropriate in many cases, is that each observation costs some fixed amount $c>0$. The total cost to run the experiment is then $cN$, and an optimal stopping rule $N$ minimizes

$$E[cN+\rho_N]. \tag{20.5}$$

To illustrate some ideas useful in a broader context, let us now restrict attention to a simple example, testing $H_0:\Theta\leq 1/2$ versus $H_1:\Theta>1/2$ with $\Theta$ the success probability for a sequence of independent Bernoulli trials. We develop a recursive method, called backwards induction, to find an optimal stopping time. For the loss function, let us take

$$L(\Theta)=K|\Theta-1/2|.$$

This function decreases as $\Theta$ tends to $1/2$, which seems natural because incorrect inference when $\Theta$ is near $1/2$ should be less of a concern than incorrect inference when $\Theta$ lies farther from $1/2$. Finally, for a prior distribution assume

$$\Theta\sim\text{Beta}(\alpha,\beta).$$

To begin, let us consider our inferential risk $\rho_0$ if we stop immediately with no data collection. Noting that

$$L(\Theta)I\{\Theta>1/2\}=\frac{1}{2}K|\Theta-1/2|+\frac{1}{4}K-\frac{1}{2}K(1-\Theta)$$

and

$$L(\Theta)I\{\Theta\leq 1/2\}=\frac{1}{2}K|\Theta-1/2|+\frac{1}{4}K-\frac{1}{2}K\Theta,$$

$$\rho_0 = \frac{1}{2}KE|\Theta - 1/2| + \frac{1}{4}K - \frac{1}{2}K\frac{\max\{\alpha,\beta\}}{\alpha+\beta}.$$

In form, $\rho_0$ is a function of $\alpha$ and $\beta$, $\rho_0 = H(\alpha,\beta)$, but the specific representation given is convenient. Calculations to find $\rho_N$ are similar, but involve conditional expectations given $\mathcal{F}_N$. By Wald's identity (Theorem 20.6), the likelihood function for the data is proportional to

$$\theta^{S_N}(1-\theta)^{N-S_N},$$

where $S_N = X_1 + \cdots + X_N$. Calculations identical to those in Example 7.3 show that

$$\Theta|\mathcal{F}_N \sim \text{Beta}(\alpha_N, \beta_N),$$

where

$$\alpha_n = \alpha + S_n \quad\text{and}\quad \beta_n = \beta + n - S_n.$$

The calculations for $\rho_0$ involved expectations for functions of $\Theta$, which can be viewed as integrals against the $\text{Beta}(\alpha,\beta)$ distributions. The calculations for $\rho_N$ are identical, except that the expectations are integrals against $\text{Beta}(\alpha_N,\beta_N)$, the posterior distribution for $\Theta$ given $\mathcal{F}_N$. Using this observation,

$$\rho_N = \frac{1}{2}KE\bigl[|\Theta - 1/2| \bigm| \mathcal{F}_N\bigr] + \frac{1}{4}K - \frac{1}{2}K\frac{\max\{\alpha_N,\beta_N\}}{\alpha_N+\beta_N}.$$

By smoothing,

$$E|\Theta - 1/2| = EE\bigl[|\Theta - 1/2| \bigm| \mathcal{F}_N\bigr],$$

so according to criteria (20.5), our stopping time $N$ should be chosen to minimize

$$\frac{1}{2}KE|\Theta - 1/2| + \frac{1}{4}K - \frac{1}{2}KE\left[\frac{\max\{\alpha_N,\beta_N\}}{\alpha_N+\beta_N}\right] + cEN.$$

Equivalently, since the first two terms here are independent of $N$, the stopping time $N$ should be chosen to maximize

$$\frac{1}{2}KE\left[\frac{\max\{\alpha_N,\beta_N\}}{\alpha_N+\beta_N}\right] - cEN. \tag{20.6}$$

This expression has an interesting interpretation as the expected reward in a "guess the next observation" game. In this game, $c$ is the cost for each observation, and after sampling, the player tries to guess whether $X_{N+1}$ will be 0 or 1, winning $K/2$ for a correct guess.

Recursive algorithms to maximize (20.6) or minimize (20.5) rely on the conditional independence of the $X_i$ given $\Theta$. Since

$$P(X_{k+1} = x_1, \dots, X_{k+n} = x_n | \Theta, \mathcal{F}_k) = \Theta^{s_n}(1-\Theta)^{n-s_n},$$

where $s_n = x_1 + \cdots + x_k$, smoothing gives

$$P(X_{k+1} = x_1, \dots, X_{k+n} = x_n | \mathcal{F}_k) = E\bigl[\Theta^{s_n}(1-\Theta)^{n-s_n} \bigm| \mathcal{F}_k\bigr].$$

Since the variable of interest here is a function of $\Theta$, this could be computed as an integral against Beta$(\alpha_k, \beta_k)$, the posterior distribution of $\Theta$ given $\mathcal{F}_k$. Formally, the answer would be exactly the same as $P(X_1 = x_1, \dots, X_n = x_n)$ if the prior distribution were Beta$(\alpha_k, \beta_k)$. So the joint distribution of future observations given $\mathcal{F}_k$ only depends on the posterior distribution of $\Theta$. The observed data $X_1, \dots, X_k$, once they have been used to compute this posterior distribution, have no other effect on the distribution of future observations. The posterior distributions, characterized by the sequence $(\alpha_n, \beta_n)$, $n \geq 0$, form a Markov chain (see Section 15.3), and given $(\alpha_k, \beta_k)$, the initial observations $X_1, \dots, X_k$ and the future observations $X_{k+1}, X_{k+2}, \dots$ are independent.

To use the Markov structure described, let $V(\alpha, \beta)$ denote the supremum of (20.6),

$$V(\alpha, \beta) = \sup_{\text{stopping times } N} \left[\frac{1}{2} K E\left(\frac{\max\{\alpha_N, \beta_N\}}{\alpha_N + \beta_N}\right) - cEN\right],$$

called the *value* of the game. Suppose the player takes an initial observation and proceeds after this observation stopping optimally at some later stage. How much should he expect to win? Given $X_1$, future observations will evolve as if the prior were Beta$(\alpha_1, \beta_1)$ and we had no data. Since we have to pay $c$ for the first observation, the expected winnings will be

$$EV(\alpha_1, \beta_1) - c.$$

Since $\alpha_1 = \alpha + X_1$, $\beta_1 = \beta + 1 - X_1$, and

$$P(X_1 = 1) = EP(X_1 = 1|\Theta) = E\Theta = \frac{\alpha}{\alpha + \beta},$$

the expected winnings can be written as

$$\frac{\alpha V(\alpha + 1, \beta) + \beta V(\alpha, \beta + 1)}{\alpha + \beta} - c.$$

If instead the player stops immediately, he will win

$$\frac{1}{2} K \frac{\max\{\alpha, \beta\}}{\alpha + \beta}.$$

The optimal expectation must be the larger of these, so

$$V(\alpha, \beta) = \max\left\{\frac{1}{2} K \frac{\max\{\alpha, \beta\}}{\alpha + \beta}, \frac{\alpha V(\alpha + 1, \beta) + \beta V(\alpha, \beta + 1)}{\alpha + \beta} - c\right\}. \quad (20.7)$$

This key equation can be used to calculate $V$ recursively. To get started, if the sum $\alpha_n + \beta_n = \alpha + \beta + n$ is extremely large, the value of information from a single additional observation cannot offset its cost $c$, and so $V(\alpha_n, \beta_n)$ will be

$$V(\alpha_n, \beta_n) = \frac{1}{2} K \frac{\max\{\alpha_n, \beta_n\}}{\alpha + \beta + n}.$$

There are $n+1$ possible values for $(\alpha_n, \beta_n)$: $(\alpha, \beta + n)$, $(\alpha + 1, \beta + n - 1)$, $\dots$, $(\alpha + n, \beta)$ and the values for $V$ on this grid can be saved as a vector or array on a computer. Using these values, $V$ can be computed at $(\alpha, \beta + n - 1)$, $\dots$, $(\alpha + n - 1, \beta)$, all of the possible values for $(\alpha_{n-1}, \beta_{n-1})$, using (20.7). Continuing in this fashion we will eventually find $V$ at all of the possible values for $(\alpha_k, \beta_k)$, $k = 0, \dots, n$. This recursive approach to calculating $V$ is called *backwards induction.*

Once $V$ has been computed, it is easy to characterize an optimal stopping time $N$. From the discussion above, it should be clear that stopping immediately will be optimal if

$$\frac{1}{2} K \frac{\max\{\alpha, \beta\}}{\alpha + \beta} \ge \frac{\alpha V(\alpha + 1, \beta) + \beta V(\alpha, \beta + 1)}{\alpha + \beta} - c.$$

With the Markov structure, the decision to stop or continue at stage $k$ will be the same as the decision to stop or continue initially if the prior were $\text{Beta}(\alpha_k, \beta_k)$. Thus it is optimal to stop the first time $k$ that

$$\frac{1}{2} K \frac{\max\{\alpha_k, \beta_k\}}{\alpha + \beta + k} \ge \frac{\alpha_k V(\alpha_k + 1, \beta_k) + \beta_k V(\alpha_k, \beta_k + 1)}{\alpha + \beta + k} - c.$$

Given the data, the left-hand side here is what you expect to win if you stop at stage $k$, and the right-hand side is what you can expect to win if you take an additional observation and stop optimally at some later time.

If $\alpha = \beta = 1$, so the prior distribution for $\Theta$ is the uniform distribution on $(0, 1)$, and if $K = 200$ and $c = 1$, the optimal stopping time continues until $(\alpha_n, \beta_n)$ leaves the set

$$\big\{(1,1), (2,1), (1,2), (2,2), (3,2), (2,3), (3,3), (4,3), (3,4), (4,4), \\ (5,4), (4,5), (5,5), (6,5), (5,6), (6,6), (7,6), (6,7), (7,7)\big\}.$$

The expected value for this stopping time is 3.12683, the chance of a correct guess for $X_{N+1}$ is 0.71447, and the inferential risk is 3.55256. In contrast, the best fixed sample design will sample three observations, and the inferential risk with this sample size is 5.0. The sequential experiment is more efficient. Although the expected sample size is slightly larger, its inferential risk is 29% smaller.

## 20.4 Sequential Probability Ratio Test

The sequential probability ratio test is suggested in Wald (1947) for simple versus simple testing with i.i.d. observations with optional stopping. Take

$\Omega = \{0, 1\}$, let $X_1$, $X_2$, ... be i.i.d. from $Q_\theta$ with density $q_\theta$, and consider testing $H_0 : \theta = 0$ versus $H_1 : \theta = 1$. Define

$$L_n = L_n(X_1, \dots, X_n) = \frac{\prod_{i=1}^n q_1(X_i)}{\prod_{i=1}^n q_0(X_i)},$$

the likelihood ratio for the first $n$ observations. By convention, for $n = 0$ we take $L_0 = 1$. We know from the Neyman–Pearson theory in Section 12.2, that with a fixed sample size the best test rejects $H_0$ according to the size of $L_n$. The sequential probability ratio test (SPRT) has a similar feel. At each stage, the researcher has three options: stop and accept $H_0$, stop and accept $H_1$, or continue sampling. For the SPRT these options are resolved by comparing the likelihood ratio with two critical values $A < 1 < B$ in the following manner;

$$\begin{array}{lll} \text{if} & L_n \in (A, B), & \text{take another observation;} \\ \text{if} & L_n \geq B, & \text{reject } H_0; \\ \text{if} & L_n \leq A, & \text{accept } H_0. \end{array} \tag{20.8}$$

Formally, the sample size[3] for this SPRT is then

$$N = \inf\{n : L_n \notin (A, B)\}.$$

To understand the optimality properties of the SPRT, let us consider this testing problem from a Bayesian perspective. Let $\Theta$ be a Bernoulli variable with success probability $\pi = P(\Theta = 1)$, and given $\Theta = \theta$, potential observations $X_1$, $X_2$, ... be i.i.d. from, $Q_\theta$. Because

$$P(\Theta = \theta | X_1, \dots, X_n) \propto_\theta P(\Theta = \theta) \prod_{i=1}^n q_\theta(X_i),$$

it is not hard to show that

$$\pi_n \overset{\text{def}}{=} P(\Theta = 1 | X_1, \dots, X_n) = \frac{\pi L_n}{1 - \pi + \pi L_n}, \tag{20.9}$$

an increasing function of $L_n$. For convenience, let $\mathcal{F}_n = \sigma(X_1, \dots, X_n)$. Given $\Theta$ the data are i.i.d., and so, for any Borel set $B \in \mathbb{R}^k$,

$$P\big[(X_{n+1}, \dots, X_{n+k}) \in B \bigm| \Theta, \mathcal{F}_n\big] = Q_\Theta^k(B).$$

So by smoothing,

$$\begin{aligned} P\big[(X_{n+1}, \dots, X_{n+k}) \in B \bigm| \mathcal{F}_n\big] &= E\big[Q_\Theta^n(B) | \mathcal{F}_n\big] \\ &= (1 - \pi_n) Q_0^k(B) + \pi_n Q_1^k(B). \end{aligned}$$

If data $X_1, \dots, X_n$ are observed, the conditional distribution of the remaining observations is a function only of $\pi_n$ and does not depend in any other way

[3] Some authors also consider a procedure that stops immediately ($N = 0$) a SPRT.

on either $n$ or values for the first $n$ observations. This Markov structure is essentially the same as that for the testing problem considered in Section 20.3, and also holds more generally. See Problem 20.5.

As in the last section, let $L(\Theta)$ denote the cost for accepting the wrong hypothesis. If we stop without collecting any data, our minimal inferential risk, taking the smaller of the risks for the two actions, is

$$\rho(\pi) = \min\bigl\{\pi L(1), (1-\pi)L(0)\bigr\}.$$

If we collect data, as in Theorem 7.1 it will be optimal to minimize posterior inferential risk. Since optional stopping does not change the likelihood, $P(\Theta = 1|\mathcal{F}_N) = \pi_N$, and the minimal posterior inferential risk, again minimizing over the two actions, is just $\rho(\pi_N)$.

To have a definite design objective, let us now assume, as in Section 20.3, that each observation costs an amount $c$, so the total sampling costs with a stopping rule $N$ will be $cN$. Then the total risk for a stopping time $N$ and ($\mathcal{F}_N$ measurable) test function $\varphi$ will be

$$\begin{aligned} R(\pi, N, \varphi) &= E\bigl[cN + L(1)I\{\Theta = 1\}(1-\varphi) + L(0)I\{\Theta = 0\}\varphi\bigr] \\ &= \pi c E_1 N + (1-\pi)cE_0 N + \pi L(1)E_1(1-\varphi) + (1-\pi)L(0)E_0\varphi, \end{aligned}$$

where the second equality follows by conditioning on $\Theta$, with $E_0$ and $E_1$ denoting conditional expectation given $\Theta = 0$ and $\Theta = 1$, respectively. Note that $R(\pi, N, \varphi)$ is a linear function of $\pi$. By the argument above, if $\varphi$ is chosen optimally the posterior inferential risk is $\rho(\pi_N)$, and so

$$R(\pi, N) \stackrel{\text{def}}{=} \inf_{\varphi} R(\pi, N, \varphi) = E\bigl[cN + \rho(\pi_N)\bigr].$$

Let $R(\pi)$ be the optimal risk, obtained minimizing over stopping times $N$:

$$R(\pi) = \inf_N R(\pi, N) = \inf_{N,\varphi} R(\pi, N, \varphi).$$

Note that since $R(\pi)$ is the infimum of a collection of linear functions, it must be concave.

Let us next consider whether stopping immediately is optimal. Since there are no sampling costs, the risk for the stopping time $N = 0$ is $\rho(\pi)$. The best possible risk, taking at least one observation, is

$$R_1(\pi) = \inf_{N \geq 1} R(\pi, N),$$

which is concave by the same reasoning as that for $R$. Comparing these risks, stopping immediately is optimal only if

$$\rho(\pi) \leq R_1(\pi).$$

The functions $\rho$ and $R_1$ are graphed in Figure 20.1, and the values $\pi_\pm$ where the two functions agree are indicated on the horizontal axis. As pictured, the function $R_1$ is continuous,[4] and approaches $c$ as $\pi \to 0$ or $\pi \to 1$. In some

[4] Because $R_1$ is concave, continuity is immediate on $(0, 1)$. The argument for continuity at the endpoints 0 and 1 is more delicate.

**Fig. 20.1.** $\rho$ and $R_1$.

cases, the function $\rho$ may lie entirely below $R_1$. Barring this possibility, if the functions ever cross, the values $\pi_- \le \pi_+$ are uniquely determined. By the Markov structure, the decision to stop or continue at stage $n$ should be formed in the same fashion after replacing the prior probability $\pi$ with the posterior probability $\pi_n$. Thus the rule that stops the first time $\pi_n \notin (\pi_-, \pi_+)$ is optimal.

**Theorem 20.8.** *If $R_1(\pi) < \rho(\pi)$, so $N = 0$ is suboptimal, then the SPRT with*

$$A = \frac{1-\pi}{\pi}\frac{\pi_-}{1-\pi_-} \quad \textit{and} \quad B = \frac{1-\pi}{\pi}\frac{\pi_+}{1-\pi_+}$$

*is optimal, minimizing $R(\pi, N)$ over all stopping times $N$.*

*Proof.* This follows from the discussion above and the monotonic relationship between $\pi_n$ and $L_n$ in (20.9). With $A$ and $B$ as defined in this theorem, straightforward algebra shows that $\pi_n \in (\pi_-, \pi_+)$ if and only if $L_n \in (A, B)$. □

To allow comparisons with procedures that may accept or reject $H_0$ in a suboptimal fashion, let $N$ and $\varphi$ denote the sample size and test function for a sequential procedure. As usual, $E_\theta\varphi$ gives the chance of rejecting $H_0$, so the error probabilities for this test are $\alpha_0 \overset{\text{def}}{=} E_0\varphi$ and $\alpha_1 \overset{\text{def}}{=} 1 - E_1\varphi$. The risk for this procedure in the Bayesian model is

$$R(\pi, N, \varphi) = \pi c E_1 N + (1-\pi)cE_0 N + \pi L(1)\alpha_1 + (1-\pi)L(0)\alpha_0,$$

a linear combination of $\alpha_0$, $\alpha_1$, $E_0N$, and $E_1N$. Because SPRTs are optimal, they must minimize the linear combination of the expected sample sizes $E_0N$

and $E_1 N$ in this risk among all procedures with the same or better error probabilities. The following striking result of Wald and Wolfowitz (1948) is stronger, asserting that the SPRT simultaneously minimizes both of these expected sample sizes.

**Theorem 20.9.** *Let $\tilde{\alpha}_0$, $\tilde{\alpha}_1$, and $\tilde{N}$ be the error probabilities and sample size for a SPRT with $0 < \tilde{A} < 1 < \tilde{B} < \infty$. If $\alpha_0$ and $\alpha_1$ are error probabilities for a competing procedure $(N, \varphi)$, and if $\alpha_0 \le \tilde{\alpha}_0$ and $\alpha_1 \le \tilde{\alpha}_1$, then $E_0 N \ge E_0 \tilde{N}$ and $E_1 N \ge E_1 \tilde{N}$.*

The proof of this result is based on showing that the SPRT $\tilde{N}$ is optimal for Bayesian models with different values for the prior probability $\pi$, which may seem plausible because the loss structure for the problem depends on several values, $c$, $L(0)$, and $L(1)$, which can be varied, whereas the SPRT is completely specified by $A$ and $B$.

A rescaling of costs just amounts to measuring them in different monetary units, and has no impact on a procedure's optimality. So let us assume that $L(0) + L(1) = 1$ and define $\omega = L(1)$, so $L(0) = 1 - \omega$. For notation, we write $\pi_\pm = \pi_\pm(c, \omega)$ and $R_1(\pi) = R_1(\pi, c, \omega)$ to indicate how these critical values and risk depend on $\omega$ and $c$.

**Lemma 20.10.** *For any values $0 < \hat{\pi}_- < \hat{\pi}_+ < 1$ there exist values $\omega \in (0, 1)$ and $c > 0$ such that $\pi_-(c, \omega) = \hat{\pi}_-$ and $\pi_+(c, \omega) = \hat{\pi}_+$.*

A careful proof of this lemma takes some work. It is not hard to argue that $R_1(\pi, c, \omega)$ is continuous, strictly increasing in $c$ when $\pi$ and $\omega$ are fixed, and tends to zero as $c \downarrow 0$, again with $\pi$ and $\omega$ fixed. It follows that $\pi_\pm(c, \omega)$ are continuous, and that with $\omega$ fixed, $\pi_+(c, \omega)$ is an increasing function of $c$, $\pi_-(c, \omega)$ is a decreasing function of $c$, and $\pi_-(c, \omega) \to 0$ and $\pi_+(c, \omega) \to 1$ as $c \downarrow 0$. From this, for fixed $\omega$, ratio $A/B$ for the SPRT,

$$\frac{\pi_-(c, \omega)}{1 - \pi_-(c, \omega)} \frac{1 - \pi_+(c, \omega)}{\pi_+(c, \omega)},$$

is continuous in $c$ and increases from 0 to 1 as $c$ varies. The proof of the lemma is completed showing that if $c$ is chosen to keep this ratio $A/B$ fixed, then $\pi_+(c, \omega)$ will increase from 0 to 1 as $\omega$ varies over $(0, 1)$. Intuitively, this occurs because as $\omega$ increases, the risk for accepting $H_0$ increases while the risk for accepting $H_0$ decreases, leading to an increase in critical value $\pi_+(c, \omega)$ necessary to accept $H_1$. A careful proof takes some care; details are available in Lehmann (1959).

*Proof of Theorem 20.9.* Given any value $\pi \in (0, 1)$, define $0 < \hat{\pi}_- < \pi < \hat{\pi}_+ < 1$ by

$$\hat{\pi}_- = \frac{\tilde{A}\pi}{\tilde{A}\pi + 1 - \pi} \quad \text{and} \quad \hat{\pi}_+ = \frac{\tilde{B}\pi}{\tilde{B}\pi + 1 - \pi}. \tag{20.10}$$

Using Lemma 20.10, choose $\omega$ and $c$ so that

$$\pi_-(\omega, c) = \hat{\pi}_- \text{ and } \pi_+(\omega, c) = \hat{\pi}_+. \tag{20.11}$$

By Theorem 20.8, with the loss structure $(c, \omega)$ and $\pi$ as the prior probability, the optimal sequential procedure will be a SPRT with

$$A = \frac{1-\pi}{\pi} \frac{\pi_-(\omega, c)}{1 - \pi_-(\omega, c)} \text{ and } B = \frac{1-\pi}{\pi} \frac{\pi_+(\omega, c)}{1 - \pi_+(\omega, c)}.$$

But by (20.10) and (20.11), $\tilde{A} = A$ and $\tilde{B} = B$, and so this SPRT is the same as the one with stopping time $\tilde{N}$ in the theorem. Because this SPRT minimizes risk,

$$\begin{aligned} &\pi c E_1 \tilde{N} + (1-\pi) c E_0 \tilde{N} + \pi L(1) \tilde{\alpha}_1 + (1-\pi) L(0) \tilde{\alpha}_0 \\ &\qquad \leq \pi c E_1 N + (1-\pi) c E_0 N + \pi L(1) \alpha_1 + (1-\pi) L(0) \alpha_0, \end{aligned}$$

which implies

$$\pi c E_1 \tilde{N} + (1-\pi) c E_0 \tilde{N} \leq \pi c E_1 N + (1-\pi) c E_0 N.$$

But $\pi \in (0,1)$ is arbitrary, and this bound can hold for all $\pi \in (0,1)$ only if $E_0 N \geq E_0 \tilde{N}$ and $E_0 N \geq E_0 \tilde{N}$. ☐

## 20.5 Sequential Design

In other sections of this chapter, the decision concerning when to stop experimentation has been the central design issue. Here we move beyond stopping problems and consider procedures with other design options. The first example, called *stochastic approximation*, concerns adaptive variable selection in regression models. In the other example we consider allocation, or bandit, problems.

### Stochastic Approximation

Let $Q_x$, $x \in \mathbb{R}$, be the distribution for a response variable $Y$ when an input variable $X$, chosen by the researcher, equals $x$. The mean of $Q_x$,

$$f(x) \stackrel{\text{def}}{=} E[Y|X = x] = \int y \, dQ_x(y),$$

is called the *regression function*. Let $\sigma^2(x)$ denote the variance of $Q_x$,

$$\sigma^2(x) = \int \bigl(y - f(x)\bigr)^2 \, dQ_x(y).$$

Stochastic approximation, introduced in Robbins and Monro (1951), considers situations in which the input variables are chosen adaptively. Specifically, $X_1$ is a constant, and for $n \geq 1$, $X_{n+1}$ is a function of the first $n$

observations. So if $\mathcal{F}_n$ is the $\sigma$-field representing information from the first $n$ observations,

$$\mathcal{F}_n = \sigma(X_1, Y_1, \dots, X_n, Y_n),$$

$X_{n+1}$ is $\mathcal{F}_n$ measurable. Conditional distributions for the $Y_n$ are given by

$$Y_n | \mathcal{F}_{n-1} \sim Q_{X_n}.$$

If the mechanism to select independent variables is specified, these conditional distributions determine the joint distribution for the data.

Let $t$ be a fixed constant representing a target value for the regression function, and let $\theta$ denote the value for the independent variable that achieves this target, so

$$f(\theta) = t.$$

The design objective in stochastic approximation is to find an adaptive strategy that drives the independent variables $X_n$ to $\theta$ as quickly as possible. It should be clear that this will be possible only with a sequential approach.

*Example 20.11.* Bioassay experiments are designed to investigate the relationship between a dose level $x$ and the chance of some response. The median[5] effective dose, ED50, is defined as the dose that gives a 50% chance of response. If the variable $Y$ is a response indicator and its conditional distribution given $X = x$ is a Bernoulli distribution with success probability $f(x)$, then ED50 will be $\theta$ if the target $t$ is 1/2.

If the regression function $f$ is assumed to be increasing, then a natural strategy would be to decrease the independent variable $X$ if the response were to lie above the target and increase $X$ if the response were to lie below the target. A specific recursion suggested in Robbins and Monro (1951) takes

$$X_{n+1} = X_n - a_n(Y_n - t), \qquad n \geq 1, \tag{20.12}$$

where $a_n$, $n \geq 1$, is a sequence of positive constants.

**Theorem 20.12.** *If $f$ is continuous and strictly increasing, $\sum a_n^2 < \infty$, $\sum a_n = \infty$, and*

$$\sup_{x \in \mathbb{R}} \frac{\sigma(x) + |f(x)|}{1 + |x|} < \infty,$$

*then $X_n \to \theta$ almost surely as $n \to \infty$.*

Asymptotic normality for $X_n$ can be established under slightly stronger assumptions using results of Fabian (1968). In particular, $a_n = c/n$ gives the best rate of convergence. With this choice and suitable regularity,

---

[5] The language here, though somewhat natural, is a bit unfortunate because this dose is not in any natural sense the median for any list or distribution for doses.

$$\sqrt{n}(X_n - \theta) \Rightarrow N\left(0, \frac{c^2\sigma^2(\theta)}{2cf'(\theta) - 1}\right),$$

provided $2c > 1/f'(\theta)$. The asymptotic variance here is minimized taking $c = 1/f'(\theta)$. For more discussion, see the review articles by Ruppert (1991) and Lai (2003).

In practice, procedures seeking the maximum of a regression function may have even more practical value. Kiefer and Wolfowitz (1952) suggest an approach similar to that of Robbins and Monro. Let $b_n$, $n \geq 1$ be positive constants decreasing to zero. Responses are observed in pairs with the independent variable set to $X_n \pm b_n$. Then the conditional mean of $Z_n = (Y_{n+} - Y_{n-})/(2c_n)$ will be approximately $f'(X_n)$, and their recursion takes

$$X_{n+1} = X_n - a_n Z_n, \qquad n \geq 1,$$

again with $a_n$, $n \geq 1$, a sequence of constants.

In industrial applications, the independent variables $X_n$ are often multivariate, and the response surface methodology suggested in Box and Wilson (1951) has been popular. Experimentation proceeds in stages. For early stages, the independent variables are selected to allow a linear fit to estimate the gradient of the regression function. At successive stages, this information about the gradient is used to shift to a region with higher average response. Then at later stages richer designs are used that allow quadratic models to be fit. The maximum of the fitted quadratic is then taken as the estimate for the maximum of the regression function.

## Bandit Problems

Bandit or allocation problems have a rich history and literature. We only touch on a few ideas here to try to give a feel for the general area. The "bandit" language refers to a slot machine for gambling, operated by pulling a lever (arm) and called informally a *one-armed bandit.* Playing the machine repeatedly, a gambler will receive a sequence of rewards, until he tires or runs out of quarters. If the gaming establishment has several machines, then the gambler may choose to switch among them, playing different arms over time. Our main concern is to find an optimal strategy for the gambler, that identifies the best arm to play at each stage.

Mathematically, bandit problems can be formulated in various ways. A special case of interest in statistics is discussed here, although extensions are mentioned. For more extensive developments, see Berry and Fristedt (1985).

At each stage $n \geq 0$, the researcher chooses among $k$ arms. In a clinical setting, "arms" might correspond to giving one of $k$ treatments for some medical condition. Each time an arm is played, the researcher observes a random variable (or vector) $X$, and the distribution of this variable for arm $a$ is governed by an unknown random parameter $\Theta_a$. These parameters $\Theta_1, \dots, \Theta_k$ are independent with prior marginal distributions $\Theta_a \sim \pi_0(a)$, $a = 1, \dots, k$.

Let $\mathcal{F}_n = \sigma(X_0, \dots, X_{n-1})$, the information from the first $n$ observations, and let $A_n$ denote the arm played at stage $n \geq 0$. Since $A_n$ must be chosen based on past data, $A_n$ is $\mathcal{F}_n$ measurable. Distributions for the observations, given the arm played and the value for the parameter of the arm, are denoted $Q(a, \theta_a)$. If a strategy to select arms to play has been fixed, then the joint distribution for the observations $X_n$, $n \geq 1$, is determined by the conditional distributions

$$X_n | \mathcal{F}_n, \Theta \sim Q(A_n, \Theta_{A_n}).$$

Every time arm $a$ is played, the researcher receives a reward $r(a, \Theta_a)$, but the value of this reward if it is acquired at stage $n$ is discounted geometrically by the factor $\beta^n$ for some constant $\beta \in (0, 1)$, called the *discount factor*. For regularity, we assume the reward function $r(\cdot, \cdot)$ is bounded. The total discounted reward, if arms are played indefinitely, is

$$\sum_{n=0}^{\infty} r(A_n, \Theta_{A_n}) \beta^n,$$

and the design objective is to maximize the expected value of this variable. Using a conditioning argument the expectation $V$ of this variable can be expressed in another way. If $\pi$ is an arbitrary probability measure, define

$$r(a, \pi) = \int r(a, \theta)\, d\pi(\theta),$$

and let $\pi_n(a)$ denote the conditional distribution for $\Theta_a$ given $\mathcal{F}_n$. Then

$$E\bigl[r(A_n, \Theta_{A_n}) \bigm| \mathcal{F}_n\bigr] = r\bigl(A_n, \pi_n(A_n)\bigr).$$

Using Fubini's theorem to justify interchanging expectation and summation,

$$\begin{aligned} V = \sum_{n=0}^{\infty} \beta^n E r(A_n, \Theta_{A_n}) &= \sum_{n=0}^{\infty} \beta^n E E\bigl[r(A_n, \Theta_{A_n}) \bigm| \mathcal{F}_n\bigr] \\ &= E \sum_{n=0}^{\infty} \beta^n r\bigl(A_n, \pi_n(A_n)\bigr). \end{aligned} \tag{20.13}$$

In this form, the Markov structure of this allocation problem is clearer. When arm $a$ is played, the posterior distributions $\pi_n(a)$, $n \geq 0$, evolve as a time homogeneous Markov chain, the same structure noted in Sections 20.3 and 20.4, and this is really the intrinsic structure needed for Theorem 20.13 below. So a Markov formulation for bandit allocation problems, seen in much of the literature, is formally more general, and also makes application to scheduling problems in operations research more evident.

An *allocation index* $\nu$ for arm $a$ is a function of $\pi_n(a)$ (the current state of arm $a$), determined solely by the reward function $r(a, \cdot)$ for arm $a$ and by the family $Q(a, \cdot)$ that implicitly determines the stochastic transition kernel that dictates how posterior distributions for $\Theta_a$ evolve when arm $a$ is played. An *index strategy* is an allocation procedure that always selects the arm with the largest allocation index.

A formula for allocation indices can be derived by considering simple two-armed bandit problems in which the researcher plays either arm $a$ or an arm with a fixed reward $\lambda$. It is natural to calibrate indices so that the allocation index for the fixed arm is simply $\lambda$. These special bandit problems are actually stopping problems, for if it is ever correct to play the arm with a fixed reward, playing it at the current and all future stages is optimal. So we can restrict attention to strategies that play arm $a$ before some stopping time $\tau$, and the maximal expected discounted reward is

$$H\bigl(a, \pi_0(a), \lambda\bigr) = \sup_\tau E\left[\sum_{n=0}^{\tau-1} \beta^n r\bigl(a, \pi_n(a)\bigr) + \lambda \frac{\beta^\tau}{1-\beta}\right].$$

If $\lambda$ is large enough, stopping immediately ($\tau = 0$) is optimal, and we have

$$H\bigl(a, \pi_0(a), \lambda\bigr) = \frac{\lambda}{1-\beta}.$$

But for sufficiently small $\lambda$, $\tau = 0$ is suboptimal, and arm $a$ should be played at least once. In this case,

$$H\bigl(a, \pi_0(a), \lambda\bigr) > \frac{\lambda}{1-\beta}.$$

If index strategies are optimal in these problems, since the index for the fixed arm is $\lambda$, the initial index for arm $a$ will be at most $\lambda$ in the former case, and will exceed $\lambda$ in the latter. So the index for arm $a$ will be the smallest value $\lambda$ where stopping immediately is optimal, the critical value dividing these two regions. Thus

$$\nu\bigl(a, \pi_0(a)\bigr) = \inf\Bigl\{\lambda : (1-\beta)H\bigl((a, \pi_0(a), \lambda\bigr) = \lambda\Bigr\} \tag{20.14}$$

should be the initial allocation index for arm $a$. Allocation indices at later stages are obtained in the same way, replacing $\pi_0(a)$ with posterior distribution $\pi_n(a)$, in effect treating the posterior distribution as the prior distribution in the stopping problems.

**Theorem 20.13 (Gittins).** *An index strategy, with indices given by* (20.14), *is optimal for the $k$-armed bandit problem, maximizing the expected discounted reward $V$ in* (20.13).

This beautiful result first appears in Gittins and Jones (1974). Whittle (1980) gives an elegant proof. Gittins (1979) gives a characterization of allocation indices using a notion of *forwards induction*, and Katehakis and Veinott (1987) relates the index to the value for a game based on a single arm, with the option of "restarting" at any stage.

## 20.6 Problems

1. *Two-stage procedures.* Let $X_1, X_2, \dots$ be i.i.d. from $N(\mu, \sigma^2)$. In a two-stage sequential procedure, the sample size $N_1$ for the first stage is a fixed constant, and the sample size $N_2$ for the second stage is based on observations $X_1, \dots, X_{N_1}$ from the first stage. Let $N = N_1 + N_2$ denote the total number of observations, let $\overline{X}_1$ and $\overline{X}_2$ denote sample averages for the first and second stages,
$$\overline{X}_1 = \frac{X_1 + \cdots + X_{N_1}}{N_1} \text{ and } \overline{X}_2 = \frac{X_{N_1+1} + \cdots + X_N}{N_2},$$
and let $\overline{X}$ denote the average of all $N$ observations. For this problem, assume that $N_2$ is a function of
$$S_1^2 = \frac{1}{N_1 - 1} \sum_{i=1}^{N_1} (X_i - \overline{X}_1)^2,$$
the sample variance for the first stage. For convenience, assume $N_2 \geq 1$ almost surely.
   a) Use smoothing to find the distribution of $\sqrt{N_2}(\overline{X}_2 - \mu)$.
   b) Show that $\sqrt{N_2}(\overline{X}_2 - \mu)$ and $(\overline{X}_1, S_1)$ are independent.
   c) Show that $S_1$ and $\sqrt{N}(\overline{X} - \mu)$ are independent.
   d) Determine the distribution of $T = \sqrt{N}(\overline{X} - \mu)/S_1$ and give a confidence interval for $\mu$ with coverage probability (exactly) $1 - \alpha$.
   e) The second-stage sample size $N_2$ is a function of $S_1$. Suggest a choice for this function if the researcher would like a confidence interval for $\mu$ with width at most some fixed value $w$.
2. If $N_1$ and $N_2$ are stopping times with respect to the same filtration $\mathcal{F}_n$, $n \geq 0$, show that $N_1 \wedge N_2$ and $N_1 \vee N_2$ are also stopping times.
3. Extend Theorem 20.6, giving an identity for $P_1(B)$, $B \in \mathcal{F}_N$, that holds if $P_1(N < \infty) = 1$ but $P_0(N = \infty) > 0$.
4. Given $\Theta_1 = \theta_1$ and $\Theta_2 = \theta_2$, let $X_1, X_2, \dots$ and $Y_1, Y_2, \dots$ be independent with the $X_i$ from a Bernoulli distribution with success probability $\theta_1$ and the $Y_i$ from a Bernoulli distribution with success probability $\theta_2$. Consider a sequential experiment in which pairs of these variables, $(X_1, Y_1), (X_2, Y_2), \dots$, are observed until a stopping time $N$. Assume that $\Theta_1$ and $\Theta_2$ are *a priori* independent, each uniformly distributed on $(0, 1)$. Consider testing $H_0 : \Theta_1 \leq \Theta_2$ versus $H_1 : \Theta_1 > \Theta_2$ with loss $K|\Theta_1 - \Theta_2|$ if we accept the wrong hypothesis and total sampling costs $cN$. Derive a recursive "backwards induction" algorithm to find the optimal stopping time $N$. This algorithm will involve the posterior inferential risk $\rho_n$.
5. Consider a Bayesian model in which $\Theta \sim \Lambda$ and given $\Theta = \theta$, $X_i$, $i \geq 1$, are i.i.d. from $Q_\theta$. Let $\Lambda_n$ denote the posterior distribution for $\Theta$ given $X_1, \dots, X_n$. Use smoothing to show that

$$P\big[(X_n+1,\dots,X_{n+k})\in B \bigm| X_1,\dots,X_n\big] = \int Q_\theta^n(B)\,d\Lambda_n(\theta).$$

This equation shows that the distribution of future observations given the past depends on the conditioning variables only through the posterior distribution $\Lambda_n$.

6. *Secretary problems.* Let $X_1,\dots,X_n$ be i.i.d. from a uniform distribution, let $B_{ij} = \{X_i < X_j\}$, and define

$$\mathcal{F}_k = \sigma(B_{ij} : 1 \le i < j \le k)$$

($\mathcal{F}_1 = \{\emptyset, \mathcal{E}\}$, the trivial $\sigma$-field), so that $\mathcal{F}_k$ provides information about the relative ranks of $X_i$, $i \le k$, but not their values. Also, let

$$\mathcal{G}_k = \sigma(X_1,\dots,X_k),$$

so the filtration $\mathcal{G}$ has information about the actual values of the $X_i$. Let

$$p(N) = P(X_N > X_i, 1 \le i \le n, i \ne N),$$

the chance $X_N$ is maximal.

a) Take $n = 5$. Find a stopping time $N$ with respect to $\mathcal{F}$ that maximizes $p(N)$.

b) Take $n = 2$. Find a stopping time $N$ with respect to $\mathcal{G}$ that maximizes $p(N)$.

7. *Wald's identity.* Let $X_1, X_2, \dots$ be i.i.d. with $E|X_i| < \infty$ and mean $\mu = EX_i$. Define $S_n = X_1 + \cdots + X_n$ and let $N$ be a (positive) stopping time with respect to the filtration $\mathcal{F}$ generated by these variables,

$$\mathcal{F}_n = \sigma(X_1,\dots,X_n), \qquad n \ge 1.$$

Assume $EN < \infty$. Wald's identity asserts that

$$ES_N = \mu EN.$$

a) Use indicator variables and Fubini's theorem to show that

$$EN = \sum_{n=1}^{\infty} P(N \ge n).$$

b) Prove Wald's identity if the $X_i$ are nonnegative, $X_i \ge 0$. Hint: Show that

$$S_N = \sum_{n=1}^{\infty} X_n I\{N \ge n\},$$

and use Fubini's theorem. Independence, related to the condition that $N$ is a stopping time, will play an important role.

c) Prove Wald's identity if the stopping time $N$ is bounded, $N \leq k$, almost surely for some $k \geq 1$. An argument like that for part (a) should suffice.

d) Prove Wald's identity in general. Hint: Take

$$N_k = \min\{N, k\},$$

so that $N_k \uparrow N$ as $k \to \infty$. Using part (b), it should be enough to show that $E[S_N - S_{N_k}]$ tends to zero. But this expectation is $E(S_N - S_k)I\{N > k\}$ (explain why). Use dominated convergence and independence to show this expectation tends to zero.

8. *Power one tests and one-sided SPRTs.* Power one tests arise in a sequential setting if we reject $H_0$ whenever our sample size $N$ is finite and we stop. Since $N = \infty$ is desirable when $H_0$ is correct, sampling costs in this case should be zero. For the simple versus simple model considered for the SPRT in Section 20.4, if $c$ represents the cost per observation when $\Theta = 1$ and $L$ the loss for stopping if $\Theta = 0$, our stopping time $N$ should be chosen to minimize

$$E\big[LI\{N < \infty, \Theta = 0\} + cNI\{\Theta = 1\}\big] = (1 - \pi)LP_0(N < \infty) + c\pi E_1 N.$$

a) Use a convexity argument to show that an optimal stopping will have form

$$N = \inf\{n \geq 0 : \pi_n > \pi_+\},$$

for some constant $\pi_+ \in (0, 1)$.

b) Define $Y_n = \log\big[q_1(X_i)/q_0(X_i)\big]$ and take $S_n = Y_1 + \cdots + Y_n = \log L_n$. Introduce ladder times

$$\mathcal{T}_+ \stackrel{\text{def}}{=} \inf\{n \geq 1 : S_n > 0\}, \qquad \mathcal{T}_- \stackrel{\text{def}}{=} \inf\{n \geq 1 : S_n \leq 0\}.$$

Note that if $\pi_0$ equals $\pi_+$, then $\mathcal{T}_+$ will be the optimal stopping time in part (a). But $N = 0$ should also be optimal. Use this observation and the duality formula

$$P_0(\mathcal{T}_+ = \infty) = \frac{1}{E_0 \mathcal{T}_-}$$

to derive an explicit formula relating $\pi_+/(1 - \pi_+)$ to moments of the ladder variables $\mathcal{T}_\pm$.

9. Consider the mean square performance of the Robbins–Monro recursion with $t = 0$ and $a_n = 1/n$, so that

$$X_{n+1} = X_n - \frac{Y_n}{n}, \qquad n \geq 1.$$

Assume that $f(x) = x$, so $\theta = 0$, and that the variance function is constant, $\sigma^2(x) = \sigma^2$.

a) Let $\mu_n = EX_n$. Use a conditioning argument to derive a recursion relating $\mu_{n+1}$ to $\mu_n$. Solve this recursion, expressing $\mu_n$, $n \geq 2$, as a function of the initial value $\mu_1 = X_1$.

b) Define $m_n = EX_n^2$, the mean square error at stage $n$. Use conditioning to derive a recursion relating $m_{n+1}$ to $m_n$. Solve this recursion and show that $m_n \to 0$ as $n \to \infty$. Hint: Let $w_n = (n-1)^2 m_n$, $n \geq 1$. The recursion for $w_n$ should be easy to solve.

# A

# Appendices

## A.1 Functions

Informally, a function $f$ can be viewed as a rule that associates with every point $x$ in some domain $D$ an image value $f(x)$. More formally, a function can be defined as a collection of ordered pairs with the property that if $(x, y_1)$ and $(x, y_2)$ are both in the collection, then $y_1$ and $y_2$ must be equal. The *domain* of the function can then be defined as $D = \{x : (x, y) \in f\}$, and the *range* of the function is $R = \{y : (x, y) \in f\}$. For $x \in D$, $f(x)$ denotes the unique value $y \in R$ with $(x, y) \in f$, and we say $f$ maps $x$ to this value $f(x)$.

A function $f$ with domain $D$ and range $R$ is said to be *into* a set $S$ if $R \subset S$, and *onto* $S$ if $R = S$. The notation $f : D \to S$ means that $f$ has domain $D$ and maps this domain into $S$.

A function is *one-to-one* if every value $y$ in the range $R$ is the image of a *unique* value $x$ in the domain $D$. In this case the collection formed reversing all of the ordered pairs, $\{(y, x) : (x, y) \in f\}$, is also a function, with domain $R$ and range $D$, called the *inverse function*, denoted $f^{\leftarrow}$. Then $f^{\leftarrow}(y)$ for $y \in R$ is the unique value $x$ with $f(x) = y$. Functions also have an inverse that maps sets to sets, denoted $f^{-1}$, given by

$$f^{-1}(S) = \{x \in D : f(x) \in S\}.$$

This inverse always exists, even if $f$ is not one-to-one.

*Example A.1.* If $D = \{HH, HT, TH, TT\}$, perhaps viewed as all possible outcomes of tossing a coin twice, then

$$f = \{(HH, 2), (HT, 1), (TH, 1), (TT, 0)\}$$

is the function mapping the outcome to the number of heads. The range of this function is $R = \{0, 1, 2\}$. The function is not one-to-one because the value $1 \in R$ is the image of $HT$ and $TH$, $f(HT) = f(TH)$. So the inverse function $f^{\leftarrow}$ does not exist. The other inverse, $f^{-1}$, does exist. For instance, $f^{-1}(1, 2) = \{HT, TH, HH\}$ and $f^{-1}(4) = \emptyset$.

The definitions given above are quite general, covering situations where $D$ and $R$ are very complicated sets. For instance, a measure is a function with domain $D$ a $\sigma$-field of subsets of some sample space. The domain for a function could even be itself a collection of functions. For instance, if $C[0,1]$ is the set of all continuous functions on $[0,1]$, then

$$h \rightsquigarrow f(h) = \int_0^1 h(x)\,dx$$

describes a function $f$ with domain $D = C[0,1]$ and range $R = \mathbb{R}$, $f : C[0,1] \to \mathbb{R}$. A function is called *real-valued* if $R \subset \mathbb{R}$ and *real-valuedvector-valued* if $R \subset \mathbb{R}^n$ for some $n$.

## A.2 Topology and Continuity in $\mathbb{R}^n$

For $x \in \mathbb{R}^n$ and $\epsilon > 0$, let $B_\epsilon(x) = \{y : \|y - x\| < \epsilon\}$, the open ball around $x$ with radius $\epsilon$. A set $O \subset \mathbb{R}^n$ is called *open* if it has the property that for every point $x$ in $O$ there is some open ball $B_\epsilon(x)$ that is contained in $O$ ($B_\epsilon(x) \subset O$). The collection $\mathcal{T}$ of all open sets is called a *topology.* A set $C$ is *closed* if its complement is open.

Topologies can be used to characterize convergence and continuity in general settings. Note that the topology $\mathcal{T}$ for $\mathbb{R}^n$ is closed under finite intersections and arbitrary unions. Also, $\mathbb{R}^n$ itself and the empty set $\emptyset$ are both open sets in $\mathcal{T}$. In general, a topology is any collection of sets with these properties. One example arises when we are only concerned with points in a subset $S$ of $\mathbb{R}^n$. In this case, we use the *relative topology* in which all sets of the form $O \cap S$ with $O \in \mathcal{T}$ are open. For instance, in the relative topology with $S = [0,2]$, $[0,1)$ is an open set, even though this set is not open in $\mathbb{R}$. Sets in the relative topology are called *open relative to* $S$.

A set $N$ is called a *neighborhood* of $x$ if $N$ contains an open set $O$ with $x \in O$. A sequence of vectors $x_n$, $n \geq 1$, in $\mathbb{R}^n$ *converges* to $x \in \mathbb{R}^n$ if for any neighborhood $N$ of $x$, $x_n$ lies in $N$ for all sufficiently large $n$. This definition is equivalent to the usual definition in calculus courses. But because it is based only on the topology $\mathcal{T}$, this definition of convergence can be used to define convergence on any space with a topology of open sets, even if there is no notion of distance between elements in the space.

A point $x$ lies in the boundary $\partial S$ of a set $S \subset \mathbb{R}^n$ if for any $\epsilon > 0$, $B_\epsilon(x)$ contains at least one point in $S$ and at least one point in $S^c$. The *closure* $\overline{S}$ of a set $S$ is the union of $S$ with its boundary $\partial S$, $\overline{S} = S \cup \partial S$. This closure $\overline{S}$ is the smallest closed set that contains $S$. The *interior* $S^\circ$ of $S$ is $S - \partial S$. The interior is the largest open set contained in $S$.

A function $f : D \to \mathbb{R}$ is *continuous* at a point $x \in D$ if $f(x_n) \to f(x)$ whenever $x_n$, $n \geq 1$, is a sequence in $D$ converging to $x$. The function $f$ is called continuous if it is continuous at every point $x$ in $D$. Continuity can also be characterized using open sets. A function $f : D \to \mathbb{R}$ is continuous if $f^{-1}(O)$ is open relative to $D$ for any $O \in \mathcal{T}$.

A collection $\{O_\alpha : \alpha \in A\}$ of open sets is called an *open cover* of $K$ if $K$ is a subset of the union,

$$K \subset \bigcup_{\alpha \in A} O_\alpha.$$

A set $K$ in $\mathbb{R}^n$ (or any other topological space) is *compact* if any open cover $\{O_\alpha : \alpha \in A\} \subset \mathcal{T}$ has a finite subcover, $\{O_{\alpha_1}, \dots, O_{\alpha_m}\}$ with $K \subset \bigcup_{i=1}^m O_{\alpha_i}$. The following result provides a useful characterization of compact sets in $\mathbb{R}^n$.

**Theorem A.2 (Heine–Borel).** *A set $K \subset \mathbb{R}^n$ is compact if and only if it is closed and bounded,* $\sup_{x \in K} \|x\| < \infty$.

If $x_n$, $n \ge 1$, is a sequence of points in some space, and if $n_1 < n_2 < \cdots$ are positive integers, then $x_{n_m}$, $m \ge 1$, is called a *subsequence*. A set $K$ is called *sequentially compact* if any subsequence $x_n$, $n \ge 1$, has a convergent subsequence, $x_{n_m} \to x$ with $x \in K$. In general, compactness implies sequential compactness, but in $\mathbb{R}^n$ (or any metric space), compactness and sequential compactness are the same. Let $C(K)$ denote the collection of continuous real-valued functions on $K$. The next result shows that functions in $C(K)$ achieve their supremum if $K$ is compact.

**Proposition A.3.** *Suppose $K$ is compact and $f \in C(K)$, and let $M = \sup_{x \in K} f(x)$. Then $f(x) = M$ for some $x \in K$.*

*Proof.* There must be values in the range of $f$ arbitrarily close to $M$. Thus for any $n \ge 1$ there exists $x_n$ with $f(x_n) > M - 1/n$. Since the points $x_n$, $n \ge 1$, lie in $K$, by compactness there is a subsequence $x_{n_m} \to x \in K$. Since $f$ is continuous, $f(x_{n_m}) \to f(x)$. But because $M - 1/n_m < f(x_{n_m}) \le M$, $f(x_{n_m}) \to M$. Thus $f(x) = M$. ☐

If $f : D \to \mathbb{R}$ is continuous, then for any $x \in D$ and any $\epsilon > 0$, there exists $\delta > 0$ such that $|f(x) - f(y)| < \epsilon$ whenever $\|x - y\| < \delta$. In general, the value for $\delta$ will need to depend on both $\epsilon$ and $x$. If the choice can be made independently of $x$, then $f$ is called *uniformly continuous*. Equivalently, $f$ is uniformly continuous if

$$\sup_{\|x-y\|<\epsilon} |f(x) - f(y)| \to 0$$

as $\epsilon \downarrow 0$.

**Proposition A.4.** *If $K$ is compact and $f \in C(K)$, then $f$ is uniformly continuous.*

*Proof.* Fix $\epsilon > 0$ and for every $x \in D$ choose $\delta_x$ so that $|f(x) - f(y)| < \epsilon/2$ whenever $y \in B_{\delta_x}(x)$. Since $x \in B_{\delta_x/2}(x)$, these balls $B_{\delta_x/2}(x)$, $x \in D$, form an open cover of $K$. Because $K$ is compact, there must be a finite subcover $B_{\delta_{x_i}/2}(x_i)$, $i = 1, \dots, m$. Define $\delta = \min_{i=1,\dots,m} \delta_{x_i}/2$. Suppose $\|x - y\| < \delta$. Since $x$ is in one of the sets in the finite open cover, $\|x - x_i\| < \delta_{x_i}/2$ for some $i$. By the triangle inequality,

$$\|y - x_i\| \le \|y - x\| + \|x - x_i\| < \tfrac{1}{2}\delta + \tfrac{1}{2}\delta_{x_i} \le \delta_{x_i}.$$

From the definition of $\delta_x$,

$$|f(x) - f(x_i)| < \frac{\epsilon}{2} \text{ and } |f(y) - f(x_i)| < \frac{\epsilon}{2},$$

and so, by the triangle inequality,

$$|f(x) - f(y)| \le |f(x) - f(x_i)| + |f(x_i) - f(y)| < \epsilon. \qquad \square$$

Functions in $C(K)$, $K$ compact, achieve their supremum and are uniformly continuous. A final useful property of $C(K)$ is that monotone sequences converge uniformly.

**Theorem A.5 (Dini).** *If $f_1 \ge f_2 \ge \cdots$ are positive functions in $C(K)$, $K$ compact, and if $f_n(x) \to 0$ as $n \to \infty$ for every $x \in K$, then $\sup_{x\in D} f_n(x) \to 0$ as $n \to \infty$.*

*Proof.* Fix $\epsilon > 0$ and define

$$O_n = f_n^{-1}[(-\infty, \epsilon)] = \{x \in K : f_n(x) < \epsilon\}.$$

Since $f$ is continuous these sets $O_n$, $n \ge 1$, are open relative to $K$, and since the functions are decreasing, $O_1 \subset O_2 \subset \cdots$. Because $f_n(x) \to 0$, the point $x$ will be in $O_n$ once $n$ is large enough, and so these sets cover $K$. By compactness, there is a finite subcover, $O_{n_1}, \dots, O_{n_m}$. If $N = \max n_i$, then the union of these sets is $O_N$, and thus $O_N = K$. So $f_N(x) < \epsilon$ for all $x \in K$, and since the functions are decreasing, $f_n(x) < \epsilon$ for all $x \in K$ and all $n \ge N$. So

$$\limsup_{n\to\infty} \sup_{x\in K} f_n(x) \le \epsilon,$$

and since $\epsilon > 0$ is arbitrary, $\sup_{x\in K} f_n(x) \to 0$. $\square$

## A.3 Vector Spaces and the Geometry of $\mathbb{R}^n$

**Definition A.6.** *A set $V$ is a* vector space *over the real numbers*[1] $\mathbb{R}$ *if elements of $V$ can be added, with the sum again an element of $V$, and multiplied by a constant, with the product an element of $V$, so that:*

[1] Vector spaces can be defined analogously over complex numbers or any other field.

1. *(Commutative and associative laws for addition) If $u \in V$, $v \in V$, and $w \in V$,*
$$u + v = v + u \text{ and } (u + v) + w = u + (v + w).$$
2. *There is a zero element in $V$, denoted $\mathbb{O}$, such that $\mathbb{O} + u = u$, for all $u \in V$.*
3. *Given any $u \in V$ there exists an element $-u \in V$ such that $u + (-u) = \mathbb{O}$.*
4. *(Associative law for multiplication) If $v \in V$, $a \in \mathbb{R}$, and $b \in \mathbb{R}$, then $(ab)v = a(bv)$.*
5. *(Distributive laws) If $u \in V$, $v \in V$, $a \in \mathbb{R}$, and $b \in \mathbb{R}$, then*
$$(a + b)v = av + bv \text{ and } a(u + v) = au + av.$$
6. *If $v \in V$, $1v = v$.*

A subset $W \subset V$ is called a *subspace* if $W$ is closed under addition (if $u \in W$ and $v \in W$, then $u + v \in W$) and multiplication (if $c \in \mathbb{R}$ and $v \in W$, then $cv \in W$). If these hold, then $W$ must also be a vector space.

Most of the vector spaces in this book are $\mathbb{R}^n$ or subspaces of $\mathbb{R}^n$. Other examples of interest include the set of all $n \times p$ matrices with real entries. Collections of functions can also form vector spaces. For instance, the collection of all functions $f$ with form $f(x) = a \sin x + b \cos x$ is a vector space. And the set $C[0, 1]$ of all continuous real-valued functions on $[0, 1]$ is a vector space.

A vector $u$ is a linear combination of vectors $v_1, \dots, v_n$ if $u = c_1v_1 + \cdots + c_nv_n$ for some constants $c_i \in \mathbb{R}$, $i = 1, \dots, n$. The set of all linear combinations of vectors from a set $S$ is called the *linear span* of $S$, and the linear span of any set is then a vector space. For instance, $\mathbb{R}^2$ is the linear span of $S = \{(1, 0), (0, 1)\}$ because an arbitrary vector $(x, y) \in \mathbb{R}^2$ can be expressed as $x(1, 0) + y(0, 1)$. The *dimension* of a vector space $V$ is the smallest number of vectors needed to span $V$. If $V$ is not the span of any finite set, then $V$ is called infinite-dimensional. By convention, the dimension of the trivial vector space $\{\mathbb{O}\}$ is zero.

Vectors $v_1, \dots, v_p$ are *linearly independent* if the linear combination $c_1v_1 + \cdots + c_pv_p = \mathbb{O}$ only if $c_1 = \cdots = c_p = 0$. In this case, the linear span $W$ of $S = \{v_1, \dots, v_p\}$ has dimension $p$. If these vectors are points in $\mathbb{R}^n$, and if $X = (v_1, \dots, v_p)$, an $n \times p$ matrix with columns $v_1, \dots, v_p$, then $X\beta = \beta_1v_1 + \cdots + \beta_pv_p$ for $\beta \in \mathbb{R}^p$. Since this is an arbitrary linear combination of the vectors in $S$,
$$W = \text{span}(S) = \{X\beta : \beta \in \mathbb{R}^p\}.$$

The *rank*[2] of $X$ is the dimension of $W$. Because $X$ has $p$ columns, its rank is at most $p$. If the columns of $X$ are linearly independent, then the rank of $X$ is $p$ and $X$ is called *full rank.*[3]

The *dot* or *inner* product of two vectors $u$ and $v$ in $\mathbb{R}^n$ is $u \cdot v = u_1v_1 + \cdots + u_nv_n$, and these vectors are called *orthogonal*, denoted $u \perp v$, if $u \cdot v = 0$. If $W$ is a subspace of $\mathbb{R}^n$, the set of vectors orthogonal to all points in $W$ is called the orthogonal complement of $W$, denoted $W^\perp$. The sum of the dimensions of $W$ and $W^\perp$ is $n$.

The (Euclidean) *length* $\|u\|$ of a vector $u \in \mathbb{R}^n$ is

$$\|u\| = \sqrt{u \cdot u} = \sqrt{u_1^2 + \cdot + u_n^2},$$

and $u$ is called a *unit vector* if $\|u\| = 1$. If a subspace $W$ of $\mathbb{R}^n$ has dimension $p$, then there exist $p$ unit vectors $e_1, \dots, e_p$ that are mutually orthogonal, $e_i \cdot e_j = 0$, $i \neq j$, and span $W$. This collection $\{e_1, \dots, e_p\}$ is called an *orthonormal basis* for $W$, and any vector $v \in W$ can be expressed uniquely as $v = c_1e_1 + \cdots + c_pe_p$. Then if $e_{p+1}, \dots, e_n$ is an orthonormal basis for $W^\perp$, $e_1, \dots, e_n$ forms an orthonormal basis for $\mathbb{R}^n$. Writing $w \in \mathbb{R}^n$ uniquely as $c_1e_1 + \cdots + c_ne_n$, we see that $w = u + v$ with $u = c_1e_1 + \cdots + c_pe_p \in W$ and $v = c_{p+1}e_{p+1} + \cdots + c_ne_n \in W^\perp$. These vectors $u$ and $v$ are called the (orthogonal) *projections* of $w$ onto $W$ and $W^\perp$. The projection $u$ can be characterized as the vector in $W$ closest to $w$. We can show this using the Pythagorean formula that if $u \perp v$,

$$\|u + v\|^2 = \|u\|^2 + \|v\|^2.$$

If $x$ is an arbitrary point in $W$, then

$$\|w - x\|^2 = \|(u - x) + (w - u)\|^2 = \|u - x\|^2 + \|w - u\|^2 \le \|w - u\|^2,$$

because $w - x \in W$ and $w - u = v \in W^\perp$ are orthogonal.

## A.4 Manifolds and Tangent Spaces

Manifolds are sets in $\mathbb{R}^r$ that look locally like $\mathbb{R}^q$ for some $q$ called the *dimension of the manifold.* For instance, the unit circle in $\mathbb{R}^2$ is a manifold with dimension one. A precise definition involves reparameterizing the manifold locally by a differentiable function. If $U$ is an open subset of $\mathbb{R}^q$, let $\mathcal{C}_1(U)$

[2] This might also be called the column rank of $X$ since it is the dimension of the linear span of the columns of $X$. But the row rank of $X$ can be defined similarly, and these two ranks must agree.

[3] Since the row and column ranks agree, the rank of $X$ is also at most $n$. So it is also natural to call $X$ full rank if $n < p$ and the rank of $X$ is $n$. This happens if the rows of $X$ are linearly independent.

be the collection of all functions $h : U \to \mathbb{R}^r$ with continuous first partial derivatives. For $h \in \mathcal{C}_1(U)$, let $Dh$ be the matrix of partial derivatives,

$$[Dh(x)]_{i,j} = \frac{\partial h_i(x)}{\partial x_j}.$$

When $U$ is not open, a function $h : U \to \mathbb{R}^r$ is in $\mathcal{C}_1(U)$ if there exists an open set $V \supset U$ and a function $H \in \mathcal{C}_1(V)$ that coincides with $h$ on $U$; that is, $h(x) = H(x)$ for $x \in U$. A set $\Omega_0 \subset \mathbb{R}^r$ is a *manifold* of dimension $q < r$ if for every $\theta_0 \in \Omega_0$ there exists an open neighborhood $N_0$ of $\theta_0$, an open set $M_0 \subset \mathbb{R}^q$ and a one-to-one function $h : M_0 \to N_0 \cap \Omega_0$ with $h \in \mathcal{C}_1(M_0)$ and $h^{-1} \in \mathcal{C}_1(N_0 \cap \Omega_0)$. Using the inverse function theorem, the condition that $h^{-1}$ is differentiable can be replaced with the condition that $Dh(x)$ has full rank for all $x \in M_0$. The definition of a manifold is sometimes given in terms of constraints: $\Omega_0 \subset \mathbb{R}^r$ is a manifold of dimension $q < r$ if for every $\theta_0 \in \Omega_0$ there exists an open neighborhood $N_0$ of $\theta_0$ and a function $g : N_0 \to \mathbb{R}^{r-q}$ in $\mathcal{C}_1(N_0)$ such that $Dg(x)$ has full rank for all $x \in N_0$ and

$$\Omega_0 \cap N_0 = \{x \in N_0 : g(x) = 0\}.$$

Because $g$ maps onto $\mathbb{R}^{r-q}$, the last assertion means that in some neighborhood of $\theta_0$, the points in $\Omega_0$ are those that satisfy $r - q$ nonlinear constraints. The assumption on the rank of $Dg$ is needed so that the constraints are not redundant.

Tangent spaces of a manifold $\Omega_0$ are defined so that if points $\theta_0$ and $\theta_1$ in $\Omega_0$ are close to each other, their difference should lie approximately in the tangent space at $\theta_0$. To be specific, let $\theta_0$ be an arbitrary point of the manifold $\Omega_0$ and let $h$ be the local reparameterization given in our first definition of a manifold. Assume that $h(y) = \theta_0$. First-order Taylor expansion of $h_i$ about $y$ gives

$$h_i(x) \approx h_i(y) + \sum_{j=1}^{q} \frac{\partial}{\partial y_j} h_i(y)(x_j - y_j),$$

or, in matrix notation,

$$h(x) \approx \theta_0 + Dh(y)(x - y).$$

The *tangent space* at $\theta_0$ is defined as

$$V_{\theta_0} = \{Dh(y)x : x \in \mathbb{R}^q\}.$$

Equivalently, $V_{\theta_0}$ is the linear span of the columns of $Dh(y)$. It is worth noting that $V_{\theta_0}$ is a vector space that passes through the origin, but typically does not pass through $\theta_0$. Since $h(x)$ lies in $\Omega_0$ for $x$ near $y$, by the Taylor expansion above

$$h(x) - \theta_0 \approx Dh(y)(x - y) \in V_{\theta_0},$$

so if $\theta_1 \in \Omega_0$ is close to $\theta_0$, then $\theta_1 - \theta_0$ is "almost" in $V_{\theta_0}$. The tangent space $V_{\theta_0}$ can also be identified from a local constraint function $g$. By first-order Taylor expansion, if $\theta \in \Omega_0$ is close to $\theta_0$, then

$$0 = g(\theta) - g(\theta_0) \approx Dg(\theta_0)(\theta - \theta_0).$$

Hence $\theta - \theta_0$ (which should lie almost in $V_{\theta_0}$) is approximately orthogonal to each of the rows of $Dg(\theta_0)$. A careful argument along these lines shows that $V_{\theta_0}^{\perp}$ is the vector space spanned by the rows of $Dg(\theta_0)$.

*Example A.7.* Let $\Omega_0$ be the unit circle in $\mathbb{R}^2$. Then $\Omega_0$ is a manifold with dimension $q = 1$. Let

$$\theta_0 = \begin{pmatrix} 1/\sqrt{2} \\ 1/\sqrt{2} \end{pmatrix} \in \Theta_0.$$

The tangent space at $\theta_0$ is just the line $V_{\theta_0} = \{x \in \mathbb{R}^2 : x_1 + x_2 = 0\}$. The graph to the left in Figure A.1 shows the circle $\Omega_0$ with the tangent line, and the graph to the right shows the tangent space $V_{\theta_0}$.

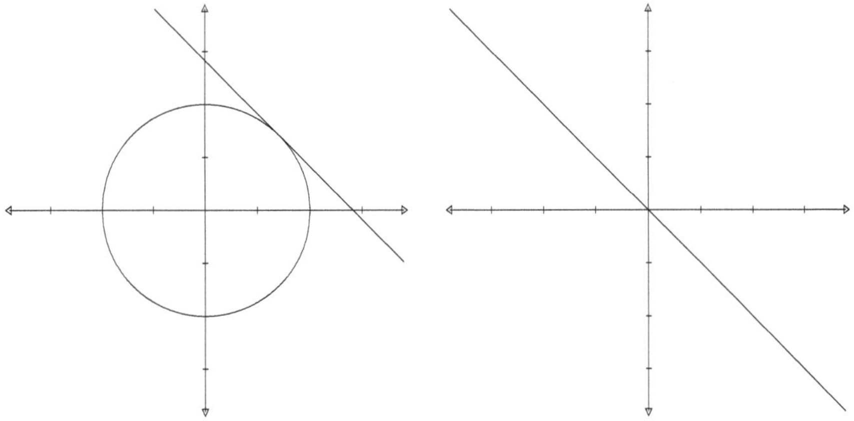

**Fig. A.1.** Tangent lines and spaces.

## A.5 Taylor Expansion for Functions of Several Variables

Let $f : \mathbb{R} \to \mathbb{R}$ have a continuous derivative $f'$. Taylor's theorem with Lagrange's form for the remainder asserts that

$$f(x) = f(x_0) + (x - x_0)f'(x^*), \tag{A.1}$$

where $x^*$ is an intermediate point between $x$ and $x_0$. If $f''$ is continuous, then

$$f(x) = f(x_0) + (x - x_0)f'(x_0) + \frac{1}{2}(x - x_0)^2 f''(x^{**}), \tag{A.2}$$

where $x^{**}$ is an intermediate value between $x$ and $x_0$. The goal of this section is to derive analogous results when $f : \mathbb{R}^n \to \mathbb{R}$. Define the function $\nabla_i f$ by

$$\nabla_i f(x) = \frac{\partial}{\partial x_i} f(x),$$

and let

$$\nabla f(x) = \begin{pmatrix} \nabla_1 f(x) \\ \vdots \\ \nabla_n f(x) \end{pmatrix}.$$

The first step is deriving a chain rule formula for computing $\partial f(hx)/\partial h$. We assume that $\nabla f$ is continuous. Using the definition of derivatives,

$$\begin{aligned} \frac{\partial}{\partial h} f(hx) &= \lim_{\epsilon \to 0} \frac{f\big((h+\epsilon)x_1, \dots, (h+\epsilon)x_n\big) - f\big(hx_1, \dots, hx_n\big)}{\epsilon} \\ &= \lim_{\epsilon \to 0} \frac{f\big((h+\epsilon)x_1, \dots, (h+\epsilon)x_n\big) - f\big(hx_1, (h+\epsilon)x_2, \dots, (h+\epsilon)x_n\big)}{\epsilon} \\ &\quad + \lim_{\epsilon \to 0} \frac{f\big(hx_1, (h+\epsilon)x_2, \dots, (h+\epsilon)x_n\big) - f\big(hx_1, \dots, hx_n\big)}{\epsilon}, \end{aligned}$$

provided both limits exist. By (A.1), the argument of the first limit equals

$$\begin{aligned} &\frac{\big((h+\epsilon)x_1 - hx_1\big)\nabla_1 f\big(z^*, (h+\epsilon)x_2, \dots, (h+\epsilon)x_n\big)}{\epsilon} \\ &\qquad\qquad = x_1 \nabla_1 f\big(z^*, (h+\epsilon)x_2, \dots, (h+\epsilon)x_n\big), \end{aligned}$$

where $z^*$ lies between $(h+\epsilon)x_1$ and $hx_1$. Since $\nabla f$ is continuous, this term approaches $x_1 \nabla_1 f(hx)$ as $\epsilon \to 0$. The other limit is like the original derivative except that $\epsilon$ does not appear in the first argument. Repeating the argument we can remove $\epsilon$ from the second argument, but an additional term arises: $x_2 \nabla_2 f(hx)$. Iteration $n$ times gives

$$\frac{\partial}{\partial h} f(hx) = \sum_{i=1}^{n} x_i \nabla_i f(hx) = x' \nabla f(hx).$$

Also,

$$\begin{aligned} \frac{\partial^2}{\partial h^2} f(hx) &= \sum_{i=1}^{n} x_i \frac{\partial}{\partial h} \nabla_i f(hx) \\ &= \sum_{i=1}^{n} x_i \sum_{j=1}^{n} x_j \nabla_j \nabla_i f(hx) \\ &= x' \nabla^2 f(hx) x, \end{aligned}$$

where $\nabla^2 f$ is the Hessian matrix of partial derivatives of $f$:

$$\nabla^2 f = \begin{pmatrix} \nabla_1 \nabla_1 f & \nabla_1 \nabla_2 f & \cdots \\ \nabla_2 \nabla_1 f & \nabla_2 \nabla_2 f & \cdots \\ \vdots & \vdots & \ddots \end{pmatrix}.$$

Viewing $f(hx)$ as a function of $h$ with $x$ a fixed constant, we can use (A.1) and (A.2) to express the value of the function at $h = 1$ by an expansion about $h = 0$. This gives

$$f(x) = f(0) + x'\nabla f(x^*)$$

and

$$f(x) = f(0) + x'\nabla f(0) + \frac{1}{2}x'\nabla^2 f(x^{**})x,$$

where $x^* = h^* x$, $x^{**} = h^{**}x$, and $h^*$ and $h^{**}$ are intermediate points between 0 and 1; so $x^*$ and $x^{**}$ lie on the chord between 0 and $x$. As $x \to 0$, $\nabla f(x^*) \to \nabla f(0)$ and $\nabla^2 f(x^{**}) \to \nabla^2 f(0)$, which justifies the approximations

$$f(x) \approx f(0) + x'\nabla f(0)$$

and

$$f(x) \approx f(0) + x'\nabla f(0) + \frac{1}{2}x'\nabla^2 f(0)x.$$

The corresponding Taylor approximations expanding about a point $x_0 \neq 0$ are

$$f(x) \approx f(x_0) + (x - x_0)'\nabla f(x_0)$$

and

$$f(x) \approx f(x_0) + (x - x_0)'\nabla f(x_0) + \frac{1}{2}(x - x_0)'\nabla^2 f(x_0)(x - x_0).$$

Both of these approximations hold with equality if the argument of the highest derivative is changed to an intermediate point on the chord between $x$ and $x_0$.

## A.6 Inverting a Partitioned Matrix

Let $A$ be a nonsingular matrix partitioned into blocks,

$$A = \begin{pmatrix} A_{11} & A_{12} \\ A_{21} & A_{22} \end{pmatrix},$$

and let $B = A^{-1}$, also partitioned into blocks,

$$B = \begin{pmatrix} B_{11} & B_{12} \\ B_{21} & B_{22} \end{pmatrix},$$

Then

$$AB = \begin{pmatrix} A_{11}B_{11} + A_{12}B_{21} & A_{11}B_{12} + A_{12}B_{22} \\ A_{21}B_{11} + A_{22}B_{21} & A_{21}B_{12} + A_{22}B_{22} \end{pmatrix} = \begin{pmatrix} I & 0 \\ 0 & I \end{pmatrix}.$$

This leads to the following two equations:

$$A_{11}B_{12} + A_{12}B_{22} = 0 \tag{A.3}$$

and

$$A_{21}B_{12} + A_{22}B_{22} = I. \tag{A.4}$$

Using (A.3),

$$B_{12} = -A_{11}^{-1}A_{12}B_{22}.$$

Using this to eliminate $B_{12}$ in (A.4),

$$-A_{21}A_{11}^{-1}A_{12}B_{22} + A_{22}B_{22} = I$$

and hence

$$B_{22} = (A_{22} - A_{21}A_{11}^{-1}A_{12})^{-1}.$$

Using this in the equation for $B_{12}$,

$$B_{12} = -A_{11}^{-1}A_{12}(A_{22} - A_{21}A_{11}^{-1}A_{12})^{-1}.$$

Similar calculations show that

$$B_{11} = (A_{11} - A_{12}A_{22}^{-1}A_{21})^{-1}$$

and

$$B_{21} = -A_{22}^{-1}A_{21}B_{11} = -A_{22}^{-1}A_{21}(A_{11} - A_{12}A_{22}^{-1}A_{21})^{-1}.$$

So $A^{-1}$ equals

$$\begin{pmatrix} (A_{11} - A_{12}A_{22}^{-1}A_{21})^{-1} & -A_{11}^{-1}A_{12}(A_{22} - A_{21}A_{11}^{-1}A_{12})^{-1} \\ -A_{22}^{-1}A_{21}(A_{11} - A_{12}A_{22}^{-1}A_{21})^{-1} & (A_{22} - A_{21}A_{11}^{-1}A_{12})^{-1} \end{pmatrix}.$$

## A.7 Central Limit Theory

This appendix derives the central limit theorem and a few of the extensions used in the main body of this text. The approach is based on an inversion formula for characteristic functions and smoothing.

### A.7.1 Characteristic Functions

If $U$ and $V$ are random variables, then the function $X = U + iV$, mapping the sample space to the complex numbers $\mathcal{C}$, is called a *complex random variable.* The mean of $X$ is defined as

$$EX \stackrel{\text{def}}{=} EU + iEV.$$

As with ordinary random variables,

$$|EX| \le E|X|, \tag{A.5}$$

which follows from Jensen's inequality. The integral of a complex function $u + iv$ against a measure $\mu$ is defined similarly as

$$\int (u + iv)\, d\mu = \int u\, d\mu + i \int v\, d\mu.$$

The *characteristic function* of a random variable $X \sim F$ is

$$\mathfrak{f}(t) = Ee^{itX} = \int e^{itx}\, dF(x), \qquad t \in \mathbb{R}.$$

Formally, the characteristic function is just the moment generating function for $X$ evaluated at the imaginary argument $it$, so it is natural that derivatives of $\mathfrak{f}$ are related to moments for $X$. By dominated convergence, if $E|Y|^k < \infty$,

$$\frac{d^k}{dt^k}\mathfrak{f}(t) = E\frac{\partial^k}{\partial t^k}e^{itY} = E(iY)^k e^{itY}.$$

In particular, the $k$th derivative at zero is $i^k EX^k$, and Taylor expansion gives

$$\mathfrak{f}(t) = 1 + itEY - \frac{1}{2}t^2EY^2 + \cdots + \frac{1}{k!}(it)^k EY^k + o(t^k) \tag{A.6}$$

as $t \to 0$.

Suppose $X \sim F$ and $Y \sim G$ are independent with characteristic functions $\mathfrak{f}$ and $\mathfrak{g}$. Then

$$e^{-ity}\mathfrak{f}(y) = \int e^{iy(x-t)}\, dF(x).$$

Integrating this against $G$,

$$\int e^{-ity}\mathfrak{f}(y)\, dG(y) = \int \mathfrak{g}(x - t)\, dF(x),$$

an important identity called *Parseval's relation.*

*Example A.8.* If $Z \sim N(0,1)$, then

$$\begin{aligned} Ee^{itZ} &= \int \frac{\exp(itz - z^2/2)}{\sqrt{2\pi}}\,dz \\ &= e^{-t^2/2} \int \frac{\exp\left[-\frac{1}{2}(z-it)^2\right]}{\sqrt{2\pi}}\,dz \\ &= e^{-t^2/2}. \end{aligned}$$

From this, if $X \sim N(\mu, \sigma^2)$, then

$$Ee^{itX} = Ee^{it(\mu+\sigma Z)} = e^{it\mu}Ee^{it\sigma Z} = e^{i\mu t - t^2\sigma^2/2}.$$

Suppose $Y = Z/a$ in Parseval's relation. Then

$$\mathfrak{g}(t) = Ee^{itZ/a} = \exp\left[-\tfrac{1}{2}(t/a)^2\right].$$

The density of $Y$ is $a\phi(ay)$, and so

$$\int e^{-ity}\mathfrak{f}(y)a\phi(ay)\,dy = \int \exp\left[-\frac{1}{2}\left(\frac{x-t}{a}\right)^2\right]dF(x),$$

or

$$\frac{1}{2\pi}\int \mathfrak{f}(y)e^{-ity-a^2y^2/2}\,dy = \int \frac{1}{a}\phi\left(\frac{x-t}{a}\right)dF(x). \tag{A.7}$$

The right-hand side of this equation is the density of $X + aZ$, and by this formula, if $\mathfrak{f}$ is known the density for $X + aZ$ can be computed for any $a > 0$. Because $X + aZ \Rightarrow X$ as $a \downarrow 0$, we have the following result.

**Theorem A.9.** *Distinct probability distributions on $\mathbb{R}$ have distinct characteristic functions.*

The next result is a bit more constructive. It gives an inversion formula for the density when the characteristic function is integrable.

**Theorem A.10.** *Suppose $\int |\mathfrak{f}(t)|\,dt < \infty$. Then $F$ is absolutely continuous with a bounded density given by*

$$f(x) = \frac{1}{2\pi}\int e^{-itx}\mathfrak{f}(t)\,dt.$$

*Proof.* Let $f_a$ be the density of $X + aZ$ in (A.7). Then $f_a(x) \to f(x)$ as $a \downarrow 0$ for every $x \in \mathbb{R}$ by dominated convergence. Also,

$$f_a(x) \le \frac{1}{2\pi}\int |\mathfrak{f}(t)|\,dt, \qquad \forall x \in \mathbb{R},\ a > 0.$$

Because $X + aZ \Rightarrow X$, by the portmanteau theorem (Theorem 9.25) and dominated convergence, for any $b < c$,

$$P\big[X \in (b,c)\big] \le \liminf_{a \downarrow 0} P\big[X + aZ \in (b,c)\big]$$
$$= \liminf_{a \downarrow 0} \int_b^c f_a(x)\,dx = \int_b^c f(x)\,dx,$$

and

$$P\big[X \in [b,c]\big] \ge \limsup_{a \downarrow 0} \int_b^c f_a(x)\,dx = \int_b^c f(x)\,dx.$$

So $P\big[X \in (b,c)\big] = \int_b^c f(x)\,dx$, and $X$ has density $f$. □

*Remark A.11.* When $F$ is absolutely continuous with density $f$, the characteristic function $\mathfrak{f}$ is also called the *Fourier transform* of $f$. Because $\mathfrak{f}(t) = \int e^{itx} f(x)\,dx$, Fourier transforms can be defined for measurable functions $f$ that are not densities, provided $f$ is integrable, $\int |f(x)|\,dx < \infty$. When $\mathfrak{f}$ is integrable, the inversion formula in this theorem remains valid and gives a constructive way to compute $f$ from $\mathfrak{f}$.

*Remark A.12.* The inversion formula for the standard normal distribution gives

$$\phi(x) = \frac{1}{2\pi} \int e^{-itx - t^2/2}\,dt. \tag{A.8}$$

Dominated convergence and repeated differentiation then give

$$\frac{1}{2\pi} \int (-it)^k e^{-itx - t^2/2}\,dt = \phi^{(k)}(x). \tag{A.9}$$

### A.7.2 Central Limit Theorem

Let $X, X_1, X_2, \dots$ be i.i.d. with common mean $\mu$ and finite variance $\sigma^2$, and define

$$Z_n = \frac{\sqrt{n}(\overline{X}_n - \mu)}{\sigma}.$$

Let $\mathfrak{f}$ denote the characteristic function of $Y = (X - \mu)/\sigma$

$$\mathfrak{f}(t) = Ee^{itY},$$

and let $\mathfrak{f}_n$ denote the characteristic function of $Z_n$. Noting that

$$e^{itZ_n} = \prod_{i=1}^n \exp\left[\frac{it}{\sqrt{n}} \frac{X_i - \mu}{\sigma}\right],$$

a product of independent variables, we have

$$\mathfrak{f}_n(t) = \mathfrak{f}^n(t/\sqrt{n}).$$

Since $EY = 0$ and $EY^2 = 1$, by Taylor expansion as in (A.6),

$$\mathfrak{f}(t/\sqrt{n}) = 1 - \frac{t^2}{2n} + o(1/n)$$

as $n \to \infty$ with $t$ fixed. It follows that

$$\log \mathfrak{f}_n(t) = n \log \mathfrak{f}(t/\sqrt{n}) \to -t^2/2,$$

and so

$$\mathfrak{f}_n(t) \to e^{-t^2/2},$$

the characteristic function for the standard normal distribution. Convergence of these characteristic functions certainly suggests that the corresponding distributions should converge, but a careful argument using our inversion formula is a bit delicate, mainly because $\mathfrak{f}_n$ need not be integrable. To circumvent this problem, we use a smoothing approach due to Berry. Let $h$ be a density with support $[-1, 1]$ and bounded derivatives of all orders. One concrete possibility is

$$h(x) = c \exp\bigl[-1/(1-x^2)\bigr] 1_{(-1,1)}(x).$$

Let $\mathfrak{h}$ be the corresponding characteristic function, and let $W$ be a random variable with density $h$ independent of $Z_n$. Repeated integration by parts gives

$$\mathfrak{h}(t) = \int h(x) e^{itx}\, dx = \frac{i}{t} \int h'(x) e^{itx}\, dx = \cdots = \left(\frac{i}{t}\right)^j \int h^{(j)}(x) e^{itx}\, dx.$$

So $\mathfrak{h}(t) = O\bigl(|t|^{-j}\bigr)$ as $t \to \pm\infty$, for any $j = 1, 2, \ldots$.

Instead of approximating the distribution of $Z_n$ directly, we consider the distribution of

$$\tilde{Z}_n = Z_n + \epsilon_n W,$$

with $\epsilon_n$, $n \geq 1$, a sequence of constants tending to zero. By the independence, $\tilde{Z}_n$ has characteristic function

$$\tilde{\mathfrak{f}}_n(t) = \mathfrak{f}_n(t) \mathfrak{h}(\epsilon_n t),$$

and since $\mathfrak{h}$ is integrable, $\tilde{Z}_n$ has a bounded density $\tilde{f}_n$ given by the inversion formula in Theorem A.10. Because $|W| \leq 1$, we have the bounds

$$P(\tilde{Z}_n \leq x - \epsilon_n) \leq P(Z_n \leq x) \leq P(\tilde{Z}_n \leq x + \epsilon_n),$$

or

$$\tilde{F}_n(x - \epsilon_n) \leq F_n(x) \leq \tilde{F}_n(x + \epsilon_n),$$

where $F_n$ and $\tilde{F}_n$ denote the cumulative distribution functions for $Z_n$ and $\tilde{Z}_n$. Since differences $|\tilde{F}_n(x \pm \epsilon_n) - \tilde{F}_n(x)|$ are at most $\epsilon_n \|\tilde{f}_n\|_\infty$, these bounds imply

$$\|F_n - \tilde{F}_n\|_\infty \leq \epsilon_n \|\tilde{f}_n\|_\infty. \tag{A.10}$$

With this bound, the central limit theorem follows easily from the following proposition.

**Proposition A.13.** *Taking* $\epsilon_n = 1/n^{1/4}$, *as* $n \to \infty$,

$$\tilde{f}_n(x) \to \phi(x).$$

*Proof.* The desired result essentially follows by dominated convergence from the inversion formula

$$\tilde{f}_n(x) = \frac{1}{2\pi} \int e^{-itx} \mathfrak{f}_n(t) \mathfrak{h}(\epsilon_n t)\, dx. \tag{A.11}$$

To be specific, since $\log \mathfrak{f}(t) \sim -t^2/2$ as $t \to 0$, then in some neighborhood of zero, $(-\delta, \delta)$ say, we have

$$\Re\bigl[\log \mathfrak{f}(t)\bigr] \le -\frac{t^2}{4},$$

which implies

$$|\mathfrak{f}(t)| \le e^{-t^2/4}, \qquad |t| < \delta,$$

and

$$|\mathfrak{f}_n(t)| \le e^{-t^2/4}, \qquad |t| < \delta\sqrt{n}.$$

Then

$$\tilde{f}_n(x) = \frac{1}{2\pi} \int_{|t|<\delta\sqrt{n}} e^{-itx} \mathfrak{f}_n(t) \mathfrak{h}(\epsilon_n t)\, dt + \frac{1}{2\pi} \int_{|t|\ge\delta\sqrt{n}} e^{-itx} \mathfrak{f}_n(t) \mathfrak{h}\bigl(\epsilon_n t\bigr)\, dt.$$

If we take $\epsilon_n = 1/n^{1/4}$, it is easy to see that the second term here tends to zero, since $|\mathfrak{f}_n| \le 1$ and $\mathfrak{h}(t) = O\bigl(|t|^{-j}\bigr)$ as $t \to \infty$ for any $j$. The integrand in the first term is dominated by $e^{-t^2/4}/(2\pi)$, an integrable function. So by dominated convergence the first term tends to

$$\frac{1}{2\pi} \int e^{-t^2/2-itx}\, dt.$$

This integral equals $\phi(x)$ by (A.8), proving the proposition. ☐

By Proposition A.13, the densities for $\tilde{Z}_n$ converge pointwise to $\phi$. So by Scheffé's theorem, given in Problem 2.19,

$$\tilde{Z}_n \Rightarrow N(0,1).$$

Because $Z_n = \tilde{Z}_n - \epsilon_n W$ and $\epsilon_n W \xrightarrow{p} 0$, it follows that

$$Z_n \Rightarrow N(0,1),$$

proving the central limit theorem (Theorem 8.12).

### A.7.3 Extensions

The central limit theorem in the last subsection was proved using the inversion formula, smoothing, and Taylor expansion to approximate $\mathfrak{f}_n$. The Taylor approximation for $\mathfrak{f}_n$ can be improved keeping an extra term in the expansion. By (A.6), if $E|X|^3 < \infty$,

$$\mathfrak{f}(t/\sqrt{n}) = 1 - \frac{t^2}{2n} - \frac{i\gamma t^3}{6\sigma^3 n\sqrt{n}} + o(n^{-3/2}),$$

as $n \to \infty$, where $\gamma = E(X-\mu)^3$. Because $\log(1+\epsilon) = \epsilon + O(\epsilon^2)$,

$$\log \mathfrak{f}_n(t) = n \log \mathfrak{f}(t/\sqrt{n}) = -\frac{1}{2}t^2 - \frac{i\gamma t^3}{6\sigma^3\sqrt{n}} + o(1/\sqrt{n}).$$

Finally, since $e^\epsilon = 1 + \epsilon + O(\epsilon^2)$ as $\epsilon \to 0$,

$$\mathfrak{f}_n(t) = e^{-t^2/2}\left(1 - \frac{i\gamma t^3}{6\sigma^3\sqrt{n}}\right) + o(1/\sqrt{n}). \tag{A.12}$$

By Fourier inversion, if we define

$$\begin{aligned} g_n(x) &= \frac{1}{2\pi}\int e^{-itx}e^{-t^2/2}\left(1 - \frac{i\gamma t^3}{6\sigma^3\sqrt{n}}\right) dt \\ &= \phi(x) - \frac{\gamma\phi'''(x)}{6\sigma^3\sqrt{n}}, \end{aligned}$$

with the integral in this expression evaluated using (A.9), then the Fourier transform of $g_n$ is the approximation for $\mathfrak{f}_n$ in (A.12). Integrating $g_n$, the function $G_n$, defined as

$$G_n(x) = \int_{-\infty}^x g_n(u)\,du = \Phi(x) - \frac{\gamma\phi''(x)}{6\sigma^3\sqrt{n}} = \Phi(x) - \frac{\gamma(x^2-1)\phi(x)}{6\sigma^3\sqrt{n}},$$

is the natural candidate to approximate $F_n$. There is one technical issue worth noting. Because $g_n/\phi$ is a cubic polynomial, $g_n(x)$ will be negative for some $x$ (unless $\gamma = 0$), and so $G_n$ may not be nondecreasing. We call a function, such as $G_n$, that tends to zero at $-\infty$ and one at $+\infty$ a *pseudo-cdf*, and if a pseudo-cdf is differentiable, we define its characteristic function as the Fourier transform of its derivative.

To use (A.12) to argue that $G_n$ is a good approximation for $F_n$ we need the following technical lemma. Although there are technical differences, the approach used to prove this result is similar to that used above to prove the central limit theorem. For details, see Section 16.3 of Feller (1971).

**Lemma A.14.** *Let $F$ be the cumulative distribution function for a distribution with mean zero and characteristic function $\mathfrak{f}$, and let $G$ be a differentiable pseudo-cdf with density $g = G'$ and characteristic function $\mathfrak{g}$ satisfying $\mathfrak{g}'(0) = 0$. Then for any $T > 0$,*

$$\|F - G\|_\infty \le \frac{1}{\pi}\int_{-T}^{T}\left|\frac{\mathfrak{f}(t) - \mathfrak{g}(t)}{t}\right| dt + \frac{24\|g\|_\infty}{\pi T}.$$

*Proof of Theorem 19.3.* Let us first show that the nonlattice assumption implies $|\mathfrak{f}(t)| < 1$ for $t \ne 0$. To see this, first note that by (A.5),

$$|\mathfrak{f}(t)| = \left|Ee^{itX}\right| \le E\left|e^{itX}\right| = 1,$$

and we only need to rule out $|\mathfrak{f}(t)| = 1$, that is, that $\mathfrak{f}(t) = e^{i\omega}$ for some $\omega \in \mathbb{R}$. But in this case

$$1 = \Re\big(\mathfrak{f}(t)e^{-i\omega}\big) = \Re\big(Ee^{itX - i\omega}\big) = E\cos(tX - \omega),$$

and since the cosine function is at most one, this implies

$$P\big(\cos(tX - \omega) = 1\big) = P\big(X = \omega + 2\pi k/t, \exists k \in \mathbb{Z}\big) = 1,$$

and $Q$ is a lattice distribution.

Next, fix $\epsilon > 0$ and take

$$c = 24 \sup_{n\ge 1} \frac{\|g_n\|_\infty}{\pi\epsilon}.$$

By Lemma A.14 with $T = c\sqrt{n}$,

$$\|F_n - G_n\|_\infty \le \frac{1}{\pi}\int_{-c\sqrt{n}}^{c\sqrt{n}}\left|\frac{\mathfrak{f}_n(t) - \mathfrak{g}_n(t)}{t}\right| dt + \frac{\epsilon}{\sqrt{n}}. \tag{A.13}$$

Because $\epsilon > 0$ is arbitrary, the theorem will follow if the integral in this formula is $o\big(1/\sqrt{n}\big)$. For some $\delta > 0$, the contribution integrating over $|t| < \delta\sqrt{n}$ will be $o\big(1/\sqrt{n}\big)$ by dominated convergence and the expansion (A.12). If instead, $|t| \in \big[\delta\sqrt{n}, c\sqrt{n}\big]$,

$$\big|\mathfrak{f}_n(t)\big| = \big|\mathfrak{f}^n(t/\sqrt{n})\big| \le M^n$$

with

$$M = \sup_{|t|\in[\delta,c]} |\mathfrak{f}(t)|.$$

Since $\mathfrak{f}$ is continuous and $\big|\mathfrak{f}(t)\big| < 1$ for $t \ne 0$, $M < 1$, and $\mathfrak{f}_n$ is exponentially small for $\delta\sqrt{n} \le |t| \le c\sqrt{n}$. Because $\mathfrak{g}_n$ is also exponentially small over this region, the contribution to the integral in (A.13) over $|t| \ge \delta\sqrt{n}$ is also $o\big(1/\sqrt{n}\big)$, and the theorem follows. ☐

The approximation in Theorem 19.3 is called an *Edgeworth expansion* for $F_n$. The same method can be used to obtain higher-order expansions. For regularity, higher-order moments of $X$ must be finite to improve the Taylor approximation for $\mathfrak{f}_n$. In addition, the nonlattice assumption needs to be strengthened. This occurs because $T$ in Lemma A.14 will need to grow at a rate faster than $\sqrt{n}$, and exponential decay for $\mathfrak{f}_n$ over this region can fail. A suitable replacement, due to Cramér, is that

$$\limsup_{|t|\to\infty} \bigl|\mathfrak{f}(t)\bigr| < 1. \tag{A.14}$$

This assumption fails if $Q$ is discrete, but holds when $Q$ is absolutely continuous.[4] Similar expansions are possible if $Q$ is a lattice distribution. For these results and a derivation of the Berry–Esséen theorem (equation (8.2)) based on Taylor expansion and Lemma A.14, see Feller (1971). For Edgeworth expansions in higher dimensions, see Bhattacharya and Rao (1976).

[4] The Riemann–Lebesgue lemma asserts that $\mathfrak{f}(t) \to 0$ as $|t| \to \infty$ when $Q$ is absolutely continuous.

# B

# Solutions

## B.1 Problems of Chapter 1

1. If $j < k$, then $B_j \subseteq B_{k-1}$, which implies $B_{k-1}^c \subseteq B_j^c$. Since $A_j \subset B_k$ and $A_k \subset B_{k-1}^c \subseteq B_k^c$, $A_j$ and $A_k$ are disjoint. By induction, $B_n = \bigcup_{j=1}^n A_j$. Also, $\bigcup_{j=1}^\infty A_j = B$, for if $x \in B$, $x \in B_n$ for some $n$, and then $x \in A_j$ for some $j \leq n$. Conversely, if $x \in \bigcup_{j=1}^\infty A_j$, $x \in A_n \subset B_n \subset B$ for some $n$. By countable additivity,

$$\mu(B) = \sum_{j=1}^{\infty} \mu(A_j) = \lim_{n\to\infty} \sum_{j=1}^{n} \mu(A_j) = \lim_{n\to\infty} \mu(B_n).$$

8. If $B$ is the union of the $B_i$, then $1_B \leq \sum 1_{B_i}$, and by Fubini's theorem, viewing summation as integration against counting measure,

$$\mu(B) = \int 1_B \, d\mu \leq \int \sum_i 1_{B_i} \, d\mu = \sum_i \int 1_{B_i} \, d\mu = \sum_i \mu(B_i).$$

10. a) Let $B$ be an arbitrary set in $\mathcal{B}$. Since $\mu(B) \geq 0$ and $\nu(B) \geq 0$, $\eta(B) = \mu(B) + \nu(B) \geq 0$. Thus $\mu : \mathcal{B} \to [0, \infty]$. Next, if $B_1$, $B_2$, ... are disjoint sets in $\mathcal{B}$, then

$$\begin{aligned}\eta\left(\bigcup_{i=1}^{\infty} B_i\right) &= \mu\left(\bigcup_{i=1}^{\infty} B_i\right) + \nu\left(\bigcup_{i=1}^{\infty} B_i\right) \\ &= \sum_{i=1}^{\infty} \mu(B_i) + \sum_{i=1}^{\infty} \nu(B_i) = \sum_{i=1}^{\infty} \big[\mu(B_i) + \nu(B_i)\big] = \sum_{i=1}^{\infty} \eta(B_i).\end{aligned}$$

Thus $\eta$ is a measure.
b) To establish the integration identity, suppose $f$ is a simple function: $f = \sum_{i=1}^n c_i 1_{A_i}$. Then

$$\int f\,d\eta = \sum_{i=1}^{n} c_i\eta(A_i) = \sum_{i=1}^{n} c_i\left[\mu(A_i)+\nu(A_i)\right]$$
$$= \sum_{i=1}^{n} c_i\mu(A_i) + \sum_{i=1}^{n} c_i\nu(A_i) = \int f\,d\mu + \int f\,d\nu.$$

So the identity holds for simple functions. For the general case, let $f_n$ be simple functions increasing to $f$. Then from our definition of the integral,

$$\int f\,d\eta = \lim_{n\to\infty}\int f_n\,d\eta = \lim_{n\to\infty}\left\{\int f_n\,d\mu + \int f_n\,d\nu\right\}$$
$$= \lim_{n\to\infty}\int f_n\,d\mu + \lim_{n\to\infty}\int f_n\,d\nu = \int f\,d\mu + \int f\,d\nu.$$

11. By finite additivity,

$$\mu\big((0,1/2]\big) + \mu\big((1/2,\pi]\big) = \mu\big((0,\pi]\big).$$

Since $\mu\big((0,1/2]\big) = 1/\sqrt{2}$ and $\mu\big((0,\pi]\big) = \sqrt{\pi}$,

$$\mu\big((1/2,\pi]\big) = \sqrt{\pi} - \frac{1}{\sqrt{2}}.$$

Similarly $\mu\big((1,2]\big) = \sqrt{2} - 1$. Then

$$\int f\,d\mu = \mu\big((1/2,\pi]\big) + 2\mu\big((1,2]\big) = \sqrt{\pi} + 2\sqrt{2} - 2 - \frac{1}{\sqrt{2}}.$$

12. The integral is $\int f\,d\mu = 4 + 21\pi$.
13. For $x \in (0,1]$, let $f_n(x) = \lfloor 2^n x\rfloor/2^n$, where $\lfloor y\rfloor$ is $y$ rounded down to the nearest integer. If $x \notin (0,1]$, let $f_n(x) = 0$. So, for instance, $f_1(x)$ is $1/2$ for $x \in [1/2,1)$, $f_1(1) = 1$, and $f_1(x) = 0$, $x \notin [1/2,1]$. (Draw a picture of $f_2$.) Then

$$\int f_n\,d\mu = \frac{1+2+\cdots+(2^n-1)}{4^n} = \frac{2^n(2^n-1)}{2\times 4^n} \to \frac{1}{2}.$$

16. Define $B_n = \{X \le a - 1/n\}$. Then $B_1 \subset B_2 \subset \cdots$. Also, $\{X < a\} = \bigcup_{n=1}^{\infty} B_n$, for if $X(e) < a$, $e \in \{X \le a - 1/n\}$ for some $n$, and if $e \in \bigcup_{n=1}^{\infty} B_n$, $e \in \{X \le a - 1/n\}$ for some $n$, which implies $e \in \{X < a\}$. Then by Problem 1.1,

$$P(X < a) = \lim_{n\to\infty} P(X \le a - 1/n) = \lim_{n\to\infty} F(a - 1/n) = F(a-).$$

For the second part, $\{X < a\}$ and $\{X = a\}$ are disjoint with union $\{X \le a\}$, and so $P(X < a) + P(X = a) = P(X \le a)$; that is, $F_X(a-) + P(X = a) = F_X(a)$.

17. By countable additivity, the chance $X$ is even is

$$P(X=0)+P(X=2)+\cdots=\theta+\theta(1-\theta)^2+\cdots=\frac{1}{2-\theta}.$$

18. For the first assertion, an outcome $e$ lies in $X^{-1}(A\cap B)$ if and only if $X(e)\in A\cap B$ if and only if $X(e)\in A$ and $X(e)\in B$ if and only if $e\in X^{-1}(A)$ and $e\in X^{-1}(B)$ if and only if $e\in X^{-1}(A)\cap X^{-1}(B)$. For the third assertion, an outcome $e$ lies in $X^{-1}\big(\bigcup_{i=1}^\infty A_i\big)$ if and only if $X(e)\in\bigcup_{i=1}^\infty A_i$ if and only if $X(e)\in A_i$, for some $i$, if and only if $e\in X^{-1}(A_i)$, for some $i$, if and only if $e\in\bigcup_{i=1}^\infty X^{-1}(A_i)$. The second assertion follows in the same way.
19. Since $P_X(B)=P\big(X^{-1}(B)\big)\geq 0$, we only need to establish countable additivity. Let us first show that $X^{-1}\big(\bigcup_i B_i\big)=\bigcup_i X^{-1}(B_i)$. Suppose $e\in X^{-1}\big(\bigcup_i B_i\big)$. Then $X(e)\in\bigcup_i B_i$, which implies $X(e)\in B_j$ for some $j$. But then $e\in X^{-1}(B_j)$, and so $e\in\bigcup_i X^{-1}(B_i)$. Conversely, if $e\in\bigcup_i X^{-1}(B_i)$, then $e\in X^{-1}(B_j)$ for some $j$, which implies $X(e)\in B_j$, and so $X(e)\in\bigcup_i B_i$. Thus $e\in X^{-1}\big(\bigcup_i B_i\big)$. Next, suppose $B_i$ and $B_j$ are disjoint. Then $X^{-1}(B_i)$ and $X^{-1}(B_j)$ are disjoint, for if $e$ lies in both of these sets, $X(e)$ lies in $B_i$ and $B_j$. Finally, if $B_1,B_2,\ldots$ are disjoint Borel sets with union $B=\bigcup_i B_i$, then $X^{-1}(B_1),X^{-1}(B_2),\ldots$ are disjoint sets with union $X^{-1}(B)$, and so

$$\sum_i P_X(B_i)=\sum_i P\big(X^{-1}(B_i)\big)=P\big(X^{-1}(B)\big)=P_X(B).$$

21. The probabilities are all the same, $P(Y_1=0,Y_2=0)=P(Y_1=0,Y_2=1)=P(Y_1=1,Y_2=0)=P(Y_1=1,Y_2=1)=1/4$. For instance, $P(Y_1=0,Y_2=1)=P(1/4\leq X<1/2)=1/4$.
22. First note that

$$\begin{aligned}\{X\in B\}&=\{y\in(0,1):X(y)\in B\}\\&=\begin{cases}B\cap(0,1/2), & 1/2\notin B;\\ [B\cap(0,1/2)]\cup[1/2,1), & 1/2\in B.\end{cases}\end{aligned}$$

Let $\lambda$ be Lebesgue measure, and let $\nu$ be counting measure on $\{1/2\}$. Then $\mu(B)=0$ if and only if $\lambda(B)=0$ and $1/2\notin B$. But then $P_X(B)=P(X\in B)=P\big(B\cap(0,1/2)\big)=\lambda\big(B\cap(0,1/2)\big)=0$, and hence $P_X$ is absolutely continuous w.r.t. $\lambda+\nu$. From the equation above,

$$P(X\in B)=\lambda\big(B\cap(0,1/2)\big)+\frac{1}{2}1_B(1/2).$$

If $f$ is the density, this should equal $\int_B f\,d(\lambda+\nu)$, and $f=1_{(0,1/2)}+\frac12 1_{\{1/2\}}$ works because

$$\begin{aligned}\int f1_B d(\lambda+\nu) &= \int \left(1_{B\cap(0,1/2)} + \frac{1}{2}1_{\{1/2\}\cap B}\right) d(\lambda+\nu)\\ &= \lambda\big(B\cap(0,1/2)\big) + \nu\big(B\cap(0,1/2)\big)\\ &\quad + \frac{1}{2}\lambda(\{1/2\}\cap B) + \frac{1}{2}\nu(\{1/2\}\cap B)\\ &= \lambda\big(B\cap(0,1/2)\big) + \frac{1}{2}1_B(1/2).\end{aligned}$$

23. Define $g(x) = x1_{(-1,1)}(x)$, so that $Y = g(X)$. Integrating against the density of $X$,

$$Ef(Y) = Ef\big(g(X)\big) = \int f\big(g(x)\big)\phi(x)\,dx = \int_{-1}^{1} f(x)\phi(x)\,dx + cf(0),$$

where $c = \int_{|x|>1}\phi(x)\,dx = 2\Phi(-1)$. But integrating against the density $p$ of $Y$,

$$\begin{aligned}Ef(Y) = \int fp\,d(\lambda+\nu) &= \int fp\,d\lambda + \int fp\,d\mu\\ &= \int f(x)p(x)\,dx + f(0)p(0).\end{aligned}$$

These expressions must agree for any integrable function $f$. If $f = 1_{\{0\}}$, this gives $p(0) = 2\Phi(-1)$. And if $f(0) = 0$, we must have $\int f(x)p(x)\,dx = \int f(x)\phi(x)1_{(-1,1)}(x)\,dx$. This will hold if $p(x) = \phi(x)$ when $0 < |x| < 1$. So the density is $p(x) = 2\Phi(-1)1_{\{0\}}(x) + \phi(x)1_{(0,1)}\big(|x|\big)$.

24. The problem is trivial if $\mu$ is finite (just divide $\mu$ by $\mu(X)$). If $\mu$ is infinite but $\sigma$-finite, there exist sets $A_1, A_2, \dots$ in $\mathcal{B}$ with $\bigcup_i A_i = X$ and $0 < \mu(A_i) < \infty$. Define truncated measures $\mu_i$, as suggested, by $\mu_i(B) = \mu(B\cap A_i)$. (Routine calculations show that $\mu_i$ is indeed a measure.) Note that $\mu_i(X) = \mu(A_i)$, so each $\mu_i$ is a finite measure. Let $b_i = 1/2^i$ (or any other sequence of positive constants summing to one) and define $c_i = b_i/\mu(A_i)$. Then $P = \sum_i c_i\mu_i$ is a probability measure since $P(X) = \sum_i c_i\mu_i(X) = \sum_i [b_i/\mu(A_i)]\mu(A_i) = 1$. Suppose $P(N) = \sum_i c_i\mu_i(N) = 0$. Then $\mu_i(N) = \mu(N\cap A_i) = 0$ for all $i$. By Boole's inequality (Problem 1.8), $\mu(N) = \mu\big(\bigcup_i [N\cap A_i]\big) \le \sum_i \mu(N\cap A_i) = 0$. This shows that any null set for $P$ is a null set for $\mu$, and $\mu$ is thus absolutely continuous with respect to $P$.

25. a) Suppose $f = 1_A$. Then $f\big(X(e)\big) = 1$ if and only if $X(e) \in A$, so $f\circ X = 1_B$, where $B = \{e : X(e) \in A\}$. Note that the definition of $P_X$ has $P_X(A) = P(B)$. Now

$$\int f\big(X(e)\big)\,dP(e) = \int 1_B(e)\,dP(e) = P(B),$$

and

$$\int f(x)\,dP_X(x) = \int 1_A(x)\,dP_X(x) = P_X(A).$$

So the equation holds for indicator functions. Next, suppose that $f$ is a simple function: $f = \sum_{i=1}^n c_i 1_{A_i}$. Because integration is linear,

$$\begin{aligned}
\int f(X(e))\,dP(e) &= \int \sum_{i=1}^n c_i 1_{A_i}(X(e))\,dP(e) \\
&= \sum_{i=1}^n c_i \int 1_{A_i}(X(e))\,dP(e) = \sum_{i=1}^n c_i \int 1_{A_i}(x)\,dP_X(x) \\
&= \int \sum_{i=1}^n c_i 1_{A_i}(x)\,dP_X(x) = \int f(x)\,dP_X(x).
\end{aligned}$$

Finally, let $f$ be an arbitrary nonnegative measurable function, and let $f_n$ be nonnegative simple functions increasing to $f$. Then $f_n \circ X$ increase to $f \circ X$, and using the monotone convergence theorem twice,

$$\begin{aligned}
\int f(X(e))\,dP(e) &= \lim_{n\to\infty} \int f_n(X(e))\,dP(e) \\
&= \lim_{n\to\infty} \int f_n(x)\,dP_X(x) \\
&= \int f(x)\,dP_X(x).
\end{aligned}$$

b) Using the same general approach, since $P_X$ has density $p$ with respect to $\mu$,

$$\int 1_A\,dP_X = P_X(A) = \int_A p\,d\mu = \int 1_A p\,d\mu,$$

and the equation holds for indicator functions. Next, if $f$ is a simple function, $f = \sum_{i=1}^n c_i 1_{A_i}$, linearity gives

$$\begin{aligned}
\int f\,dP_X &= \int \sum_{i=1}^n c_i 1_{A_i}\,dP_X \\
&= \sum_{i=1}^n c_i \int 1_{A_i}\,dP_X \\
&= \sum_{i=1}^n c_i \int 1_{A_i} p\,d\mu \\
&= \int \sum_{i=1}^n c_i 1_{A_i} p\,d\mu \\
&= \int f p\,d\mu.
\end{aligned}$$

For the general case, let $f_n$ be nonnegative simple functions increasing to $f$. Then $f_n p$ increase to $fp$, and using the monotone convergence theorem twice,

$$\int f\,dP_X = \lim_{n\to\infty}\int f_n\,dP_X = \lim_{n\to\infty}\int f_n p\,d\mu = \int fp\,d\mu.$$

26. a) Integrating $e^{-x}$ and differentiating $x^\alpha$, integration by parts gives

$$\begin{aligned}\Gamma(\alpha+1) &= \int_0^\infty x^\alpha e^{-x}\,dx\\ &= -x^\alpha e^{-x}\Big|_0^\infty + \int_0^\infty \alpha x^{\alpha-1}e^{-x}\,dx\\ &= \alpha\Gamma(\alpha).\end{aligned}$$

Using this repeatedly, $\Gamma(x+1) = x\Gamma(x) = x(x-1)\Gamma(x-1) = \cdots x(x-1)\cdots 1\Gamma(1)$. But $\Gamma(1) = \int_1^\infty e^{-x}\,dx = 1$, and so $\Gamma(x+1) = x!$, $x = 0, 1, \ldots$.

b) The change of variables $u = x/\beta$ (so $dx = \beta\,du$) gives

$$\int p(x)\,dx = \int_0^\infty \beta p(\beta u)\,du = \int_0^\infty \frac{1}{\Gamma(\alpha)}u^{\alpha-1}e^{-u}\,du = \frac{\Gamma(\alpha)}{\Gamma(\alpha)} = 1.$$

c) The same change of variables gives

$$\begin{aligned}EX^r &= \int x^r p(x)\,dx = \int_0^\infty \beta^{r+1}u^r p(\beta u)\,du\\ &= \frac{\beta^r}{\Gamma(\alpha)}\int_0^\infty u^{\alpha+r}e^{-u}\,du = \frac{\beta^r\Gamma(\alpha+r)}{\Gamma(\alpha)}.\end{aligned}$$

Using this, $EX = \beta\Gamma(\alpha+1)/\Gamma(\alpha) = \beta\alpha$, $EX^2 = \beta^2\Gamma(\alpha+2)/\Gamma(\alpha) = \beta(\alpha+1)\alpha$, and $\mathrm{Var}(X) = EX^2 - (EX)^2 = \alpha\beta^2$.

27. Integrating against the density, $EX^p = \int_0^1 x^p\,dx = 1/(p+1)$. So $EX = 1/2$, $EX^2 = 1/3$, $\mathrm{Var}(X) = EX^2 - (EX)^2 = 1/3 - 1/4 = 1/12$, $\mathrm{Var}(X^2) = EX^4 - (EX^2)^2 = 1/5 - 1/9 = 4/45$ and $\mathrm{Cov}(X, X^2) = EX^3 - (EX)(EX^2) = 1/4 - 1/6 = 1/12$. So the mean and covariance are

$$E\begin{pmatrix}X\\X^2\end{pmatrix} = \begin{pmatrix}1/2\\1/3\end{pmatrix} \text{ and } \mathrm{Cov}\begin{pmatrix}X\\X^2\end{pmatrix} = \begin{pmatrix}1/12 & 1/12\\1/12 & 4/45\end{pmatrix}.$$

28. The mean and covariance are

$$E\begin{pmatrix}X\\I\{X>c\}\end{pmatrix} = \begin{pmatrix}0\\1-\Phi(c)\end{pmatrix}$$

and

$$\mathrm{Cov}\begin{pmatrix}X\\I\{X>c\}\end{pmatrix} = \begin{pmatrix}1 & \phi(c)\\\phi(c) & \Phi(c)(1-\Phi(c))\end{pmatrix}.$$

32. By Fubini's theorem,

$$\begin{aligned}
\int_0^\infty h(t)\,dt &= \int_0^\infty E\frac{1-\cos(tX)}{t^2}\,dt \\
&= \int_0^\infty \int \frac{1-\cos(tx)}{t^2}\,dP_X(x)\,dt \\
&= \iint_0^\infty \frac{1-\cos(tx)}{t^2}\,dt\,dP_X(x) \\
&= \iint_0^\infty |x|\frac{1-\cos u}{u^2}\,du\,dP_X(x) \\
&= \int \frac{\pi}{2}|x|\,dP_X(x) \\
&= \frac{\pi}{2}E|X|.
\end{aligned}$$

33. The sum is $\sum_{n=1}^\infty c_n = 1$.

36. a) By independence of $X$ and $Y$, $P(X+Y \le s|Y=y) = P(X \le s-y) = F_X(s-y)$, and so, $P(X+Y \le s|Y) = F_X(s-Y)$. Then

$$F_S(s) = P(X+Y \le s) = EP(X+Y \le s|Y) = EF_X(s-Y).$$

b) Using the independence, for $y > 0$,

$$P(XY \le w|Y=y) = P(X \le w/y|Y=y) = F_X(w/y).$$

So, $P(XY \le w|Y) = F_X(w/Y)$ almost surely, and

$$F_W(w) = P(XY \le w) = EP(XY \le w|Y) = EF_X(w/Y).$$

37. Note that $p_X(x-Y) = 0$ if $Y \ge x$. The change of variables $u = y/x$ gives

$$\begin{aligned}
p_S(x) &= E\frac{(x-Y)^{\alpha-1}e^{-(x-Y)}}{\Gamma(\alpha)}1_{(0,x)}(Y) = \int_0^x \frac{(x-y)^{\alpha-1}y^{\beta-1}e^{-x}}{\Gamma(\alpha)\Gamma(\beta)}\,dy \\
&= \int_0^1 \frac{x^{\alpha+\beta-1}e^{-x}u^{\beta-1}(1-u)^{\alpha-1}}{\Gamma(\alpha)\Gamma(\beta)}\,du = \frac{x^{\alpha+\beta-1}e^{-x}}{\Gamma(\alpha+\beta)},
\end{aligned}$$

for $x > 0$. So $X+Y \sim \Gamma(\alpha+\beta,1)$.

38. a) The mean of the exponential distribution $Q_\lambda$ is

$$\int x\,dQ_\lambda(x) = \int xq_\lambda(x)\,dx = \int_0^\infty x\lambda e^{-\lambda x}\,dx = \frac{1}{\lambda}.$$

So $E[Y|X=x] = 1/x$ and $E[Y|X] = 1/X$.

b) By smoothing,

$$EY = EE[Y|X] = E(1/X) = \sum_{x=1}^n \frac{1}{x}\frac{2x}{n(n+1)} = \frac{2}{n+1}.$$

39. a) $P[Y > y|X = k] = Q_k\big((y,\infty)\big) = \int_y^\infty ke^{-ku}\,du = e^{-ky}$. So $P(Y > y|X) = e^{-Xy}$.
b) By smoothing,

$$P(Y > y) = EP(Y > y|X) = Ee^{-Xy} = \sum_{k=1}^{n} \frac{e^{-ky}}{n} = \frac{1-e^{-ny}}{n(e^y - 1)}.$$

(The sum here is geometric.)
c) The density is

$$F_Y'(y) = \frac{d}{dy}\big(1 - P(Y > y)\big) = \frac{e^y(1-e^{-ny})}{n(e^y-1)^2} - \frac{e^{-ny}}{e^y - 1}.$$

## B.2 Problems of Chapter 2

1. a) The mass functions can be written as $\exp\bigl[\log(1-p)x + \log(p)\bigr]$, which has exponential family form.
b) The canonical parameter is $\eta = \log(1-p)$. Solving, $p = 1 - e^\eta$, and so the mass function in canonical form is $\exp\bigl[\eta x + \log(1 - e^\eta)\bigr]$, with $A(\eta) = -\log(1 - e^\eta)$.
c) Since $T = X$, by (2.4), $EX = A'(\eta) = e^\eta/(1-e^\eta) = (1-p)/p$.
d) The joint mass functions are

$$\prod_{i+1}^{n} \bigl[p(1-p)_i^x\bigr] = p^n(1-p)^{T(x)},$$

where $T(x) = \sum_i x_i$. With the same definition for $\eta$, this can be rewritten as

$$\exp\bigl[\log(1-p)T(x) + n\log(p)\bigr] = \exp\bigl[\eta T(x) + n\log(1-e^\eta)\bigr].$$

Now $A(\eta) = -n\log(1-e^\eta)$, and so

$$ET = A'(\eta) = \frac{ne^\eta}{1-e^\eta} = \frac{n(1-p)}{p},$$

and

$$\operatorname{Var}(T) = A''(\eta) = \frac{ne^\eta}{(1-e^\eta)^2} = \frac{n(1-p)}{p^2}.$$

2. From the definition,

$$e^{A(\eta)} = \iint h(x,y)e^{\eta xy}dx\,dy = \frac{1}{\sqrt{2\pi}}\int \exp\left[-\frac{1}{2}(1-\eta^2)y^2\right]dy.$$

This integral is finite if and only if $|\eta| < 1$, and so $\Xi = (-1,1)$. Doing the integral, $e^{A(\eta)} = 1/\sqrt{1-\eta^2}$ and the densities are $\exp\bigl[\eta xy + \log(1 - \eta^2)/2\bigr]h(x,y)$.

4. The parameter space is $\Xi = (-1, \infty)$, and the densities are

$$p_\eta(x) = \frac{1}{2}(\eta+1)(\eta+2)(\eta+3)x^\eta(1-x)^2, \qquad x \in (0,1).$$

5. The integral (below) defining $A(\eta)$ is finite if and only if $\eta \geq 0$, and so $\Xi = [0, \infty)$. To evaluate the integral, let $y = \sqrt{x}$. Then

$$\begin{aligned} e^{A(\eta)} &= \int_0^\infty e^{-\eta x - 2\sqrt{x}} \frac{dx}{\sqrt{x}} = 2\int_0^\infty e^{-\eta y^2 - 2y}\,dy \\ &= \frac{2\sqrt{\pi} e^{1/\eta}}{\sqrt{\eta}} \int_0^\infty \frac{\exp\left[-\frac{(y+1/\eta)^2}{2[1/(2\eta)]}\right]}{\sqrt{2\pi[1/(2\eta)]}}\,dy. \end{aligned}$$

If $Y \sim N\bigl(-1/\eta, 1/(2\eta)\bigr)$, then the integral here is $P(Y > 0) = \Phi(-\sqrt{2/\eta})$. So

$$A(\eta) = \log\sqrt{4\pi} + 1/\eta - \log\sqrt{\eta} + \log\Phi(-\sqrt{2/\eta}),$$

and $p_\eta(x) = \exp\bigl[-2\sqrt{x} - \eta x - A(\eta)\bigr]/\sqrt{x}$, $x > 0$. Then

$$E_\eta X = -E_\eta T = -A'(\eta) = \frac{1}{\eta^2} + \frac{1}{2\eta} - \frac{\phi(\sqrt{2/\eta})}{\Phi(-\sqrt{2/\eta})\sqrt{2\eta^3}}.$$

6. The parameter space is $\Xi = \mathbb{R}^2$, and the densities are

$$p_\eta(x) = \frac{e^{\eta_1 x + \eta_2 x^2}}{1 + e^{\eta_1+\eta_2} + e^{2\eta_1 + 4\eta_2}}, \qquad x = 0, 1, 2.$$

7. The joint densities are

$$\prod_{i=1}^n \bigl[1 - e^{\alpha+\beta t_i}\bigr]\bigl[e^{\alpha+\beta t_i}\bigr]^{x_i} = \exp\left[\alpha\sum_{i=1}^n x_i + \beta\sum_{i=1}^n x_i t_i + \sum_{i=1}^n \log\bigl[1 - e^{\alpha+\beta t_i}\bigr]\right],$$

forming an exponential family with $T_1 = \sum_{i=1}^n X_i$ and $T_2 = \sum_{i=1}^n t_i X_i$.

8. The joint densities are

$$\begin{aligned} &\prod_{i=1}^n \frac{1}{\sqrt{2\pi}} \exp\left[-\frac{1}{2}(x_i - \alpha - \beta t_i)^2\right] \\ &\quad = \frac{1}{\sqrt{2\pi}^n} \exp\left[\alpha\sum_{i=1}^n x_i + \beta\sum_{i=1}^n t_i x_i - \frac{1}{2}\sum_{i=1}^n x_i^2 - \frac{1}{2}\sum_{i=1}^n (\alpha + \beta t_i)^2\right], \end{aligned}$$

which is a two-parameter exponential family with $T_1 = \sum_{i=1}^n X_i$ and $T_2 = \sum_{i=1}^n t_i X_i$.

9. Since $P(X_i = x_i) = \exp(\alpha x_i + \beta t_i x_i)/\bigl(1 + \exp(\alpha + \beta t_i)\bigr)$, the joint densities are

$$\prod_{i=1}^{n} \frac{\exp(\alpha x_i + \beta t_i x_i)}{1+\exp(\alpha+\beta t_i)} = \frac{\exp\bigl[\alpha \sum_{i=1}^n x_i + \beta \sum_{i=1}^n t_i x_i\bigr]}{\prod_{i=1}^n \bigl(1+\exp(\alpha+\beta t_i)\bigr)},$$

which form a two-parameter exponential family with $T_1 = \sum_{i=1}^n X_i$ and $T_2 = \sum_{i=1}^n t_i X_i$.

15. Differentiating the identity

$$e^{B(\theta)} = \int \exp\{\eta(\theta)\cdot T(x)\} h(x)\, d\mu(x)$$

with respect to $\theta_i$ under the integral sign, gives

$$e^{B(\theta)} \frac{\partial B(\theta)}{\partial \theta_i} = \int \sum_{j=1}^{s} \frac{\partial \eta_j(\theta)}{\partial \theta_i} T_j(x) \exp\{\eta(\theta)\cdot T(x)\} h(x)\, d\mu(x).$$

Division by $e^{B(\theta)}$ then gives

$$\frac{\partial B(\theta)}{\partial \theta_i} = \sum_{j=1}^{s} \frac{\partial \eta(\theta)_j}{\partial \theta_i} E_\theta T_j, \qquad i = 1,\dots,s.$$

(This also follows from the chain rule because $B(\theta) = A\bigl(\eta(\theta)\bigr)$ and $ET_j = \partial A(\eta)/\partial \eta_j$.) These equations can be written as $\nabla B(\theta) = D\eta(\theta)' E_\theta T$, where $D\eta(\theta)$ denotes an $s \times s$ matrix with $(i,j)$th entry $\partial \eta_i(\theta)/\partial \theta_j$. Solving, $E_\theta T = [D\eta(\theta)']^{-1} \nabla B(\theta)$.

17. a) Let $f_n(k) = f(k)$ for $k \le n$, and $f_n(k) = 0$ for $k > n$. Then $f_n \to f$ pointwise, and $|f_n| \le |f|$. Note that $f_n$ is a simple function and that $\int f_n\, d\mu = \sum_{k=1}^n f(k)$. The dominated convergence theorem gives $\int f_n\, d\mu \to \int f\, d\mu$, which is the desired result.
b) Define $f_n$ as in part (a). Then $f_n \uparrow f$, and so $\sum_{k=1}^n f(k) = \int f_n\, d\mu \to \int f\, d\mu$ by the monotone convergence theorem.
c) Begin the $g$-sequence taking positive terms from the $f$-sequence until the sum exceeds $K$. Then, take negative terms from the $f$ sequence until the sum is below $K$. Then take positive terms again, and so on. Because the summands tend to zero, the partial sums will have limit $K$.

19. a) Since $\int (p-p_n)\, d\mu = \int (p-p_n)^+\, d\mu - \int (p-p_n)^-\, d\mu = 0$, we have $\int (p-p_n)^+\, d\mu = \int (p-p_n)^-\, d\mu$. Using $|p_n - p| = |p-p_n| = (p-p_n)^+ + (p-p_n)^-$,

$$\begin{aligned}\int |p_n - p|\, d\mu &= \int (p-p_n)^+\, d\mu + \int (p-p_n)^-\, d\mu \\ &= 2\int (p-p_n)^+\, d\mu.\end{aligned}$$

But $|(p-p_n)^+| \le p$, which is an integrable function, and $\bigl(p(x)-p_n(x)\bigr)^+ \to 0$. So by dominated convergence

$$\int |p_n - p|\, d\mu = 2\int (p-p_n)^+\, d\mu \to 0.$$

b) Since $P_n(A) = \int 1_A p_n \, d\mu$ and $P(A) = \int 1_A p \, d\mu$,

$$\begin{aligned} |P_n(A) - P(A)| &= \left| \int 1_A (p_n - p) \, d\mu \right| \\ &\leq \int 1_A |p_n - p| \, d\mu \leq \int |p_n - p| \, d\mu. \end{aligned}$$

22. Because $p_\theta(x) = \phi(x) \exp\bigl[\theta x - \log \Phi(\theta) - \theta^2/2\bigr]$, we have a canonical exponential family with $A(\theta) = \log \Phi(\theta) + \theta^2/2$. So $M_X(u) = \exp\bigl[A(\theta+u) - A(\theta)\bigr] = \Phi(\theta+u)\exp(u\theta + u^2/2)/\Phi(\theta)$. Also, $E_\theta X = A'(\theta) = \theta + \phi(\theta)/\Phi(\theta)$ and $\operatorname{Var}_\theta(X) = A''(\theta) = 1 - \theta\phi(\theta)/\Phi(\theta) - \phi^2(\theta)/\Phi^2(\theta)$.
23. The exponential family $N(0, \sigma^2)$ has densities

$$\frac{1}{\sqrt{2\pi\sigma^2}} \exp\left[-\frac{x^2}{2\sigma^2}\right] = \frac{1}{\sqrt{2\pi}} \exp\left[\eta x^2 + \frac{1}{2}\log(-2\eta)\right],$$

where $\eta = -1/(2\sigma^2)$. So $A(\eta) = -\frac{1}{2}\log(-2\eta)$ and $K_T(u) = -\frac{1}{2}\log\bigl[-2(\eta + u)\bigr] + \frac{1}{2}\log(-2\eta)$, which simplifies to $-\frac{1}{2}\log(1-2u)$ when $\eta = -1/2$ (or $\sigma^2 = 1$). Then

$$K_T'(u) = \frac{1}{1-2u}, \qquad K_T''(u) = \frac{2}{(1-2u)^2},$$
$$K_T'''(u) = 8(1-2u)^3, \qquad K_T''''(u) = \frac{48}{(1-2u)^4},$$

and so the first four cumulants of $T \sim Z^2$ are 1, 2, 8, and 48.
24. The first four cumulants of $T = XY$ are 0, 1, 0, and 4.
25. From the last part of Problem 2.1 with $n = 1$, the first two cumulants are $\kappa_1 = A'(\eta) = (1-p)/p$ and $\kappa_2 = A''(\eta) = (1-p)/p^2$. Because $p = p(\eta) = 1 - e^\eta$, $p' = -e^\eta = p - 1$. Using this,

$$\kappa_3 = A'''(\eta) = \frac{-2p'}{p^3} + \frac{p'}{p^2} = \frac{2}{p^3} - \frac{3}{p^2} + \frac{1}{p},$$

and

$$\kappa_4 = A''''(\eta) = \frac{6p'}{p^4} - \frac{6p'}{p^3} + \frac{p'}{p^2} = \frac{6}{p^4} - \frac{12}{p^3} + \frac{7}{p^2} - \frac{1}{p}.$$

26. The third cumulant is

$$\kappa_3 = np(1-p)(1-2p),$$

and the third moment is

$$EX^3 = np(1-p)(1-2p) + 3n^2p^2(1-p) + n^3p^3.$$

27. a) For notation, let $f_{ij}(u) = \partial^{i+j} f(u)/(\partial u_1^i \partial u_2^j)$, and let $M$ and $K$ be the moment generating function and cumulant generating function for $T$, so that $K = \log M$. Taking derivatives,

$$K_{10} = \frac{M_{10}}{M}, \qquad K_{11} = \frac{M_{11}M - M_{10}M_{01}}{M^2},$$

and

$$K_{21} = \frac{M_{21}M^2 - M_{20}M_{01}M - 2M_{11}M_{10}M + 2M_{10}^2 M_{01}}{M^3}.$$

At zero we get

$$\kappa_{2,1} = ET_1^2T_2 - (ET_1^2)(ET_2) - 2(ET_1T_2)(ET_1) + 2(ET_1)^2(ET_2).$$

b) Taking one more derivative,

$$K_{22} = \frac{\begin{array}{c} M_{22}M^3 - 2M_{21}M_{01}M^2 - 2M_{12}M_{10}M^2 \\ -M_{20}M_{02}M^2 + 2M_{20}M_{01}^2M + 2M_{02}M_{10}^2M \\ -2M_{11}M_{11}M^2 + 8M_{11}M_{10}M_{01}M - 6M_{10}^2M_{01}^2 \end{array}}{M^4}.$$

Since $M_{10}(0) = ET_1 = 0$ and $M_{01}(0) = ET_2 = 0$, at zero we get

$$\kappa_{22} = ET_1^2T_2^2 - (ET_1^2)(ET_2^2) - 2(ET_1T_2)^2.$$

28. Taking $\eta = (-\lambda, \alpha)$, $X$ has density

$$\frac{\exp\bigl[\eta_1 T_1(X) + \eta_2 T_2(x) - \log\Gamma(\eta_2) + \eta_2\log(-\eta_1)\bigr]}{x}, \qquad x > 0.$$

These densities form a two-parameter exponential family with cumulant generating function $A(\eta) = \log\Gamma(\eta_2) - \eta_2\log(-\eta_1)$. The cumulants of $T$ are derivatives of $A$:

$$\kappa_{10} = \frac{\partial A(\eta)}{\partial \eta_1} = -\frac{\eta_2}{\eta_1} = \frac{\alpha}{\lambda},$$
$$\kappa_{01} = \frac{\partial A(\eta)}{\partial \eta_2} = \psi(\eta_2) - \log(-\eta_1) = \psi(\alpha) - \log\lambda,$$
$$\kappa_{20} = \frac{\partial^2 A(\eta)}{\partial \eta_1^2} = \frac{\eta_2}{\eta_1^2} = \frac{\alpha}{\lambda^2},$$
$$\kappa_{11} = \frac{\partial^2 A(\eta)}{\partial \eta_1 \partial \eta_2} = -\frac{1}{\eta_1} = \frac{1}{\lambda},$$
$$\kappa_{02} = \frac{\partial^2 A(\eta)}{\partial \eta_2^2} = \psi'(\eta_2) = \psi'(\alpha),$$
$$\kappa_{30} = \frac{\partial^3 A(\eta)}{\partial \eta_1^3} = -\frac{2\eta_2}{\eta_1^3} = \frac{2\alpha}{\lambda^3},$$
$$\kappa_{21} = \frac{\partial^3 A(\eta)}{\partial \eta_1^2 \partial \eta_2} = \frac{1}{\eta_1^2} = \frac{1}{\lambda^2},$$
$$\kappa_{12} = \frac{\partial^3 A(\eta)}{\partial \eta_1 \partial \eta_2^2} = 0,$$
$$\kappa_{03} = \frac{\partial^3 A(\eta)}{\partial \eta_2} = \psi''(\eta_2) = \psi''(\alpha).$$

## B.3 Problems of Chapter 3

2. The marginal density of $X_i$ is $t_i\theta x^{t_i\theta}/x$, $x \in (0,1)$. Multiplying these together, the joint density is
$$\theta^n \left(\prod_{i=1}^n \frac{t_i}{x_i}\right)\left(\prod_{i=1}^n x_i^{t_i}\right)^\theta .$$
By the factorization theorem, $T = \prod_{i=1}^n X_i^{t_i}$ is sufficient. An equivalent sufficient statistic is $\sum_{i=1}^n t_i \log X_i$.
3. The joint densities are
$$\exp\left[\mu \sum_{i=1}^n \frac{x_i}{\sigma_i^2} - \sum_{i=1}^n \frac{x_i^2+\mu^2}{2\sigma_i^2} - \sum_{i=1}^n \log\sigma_i - n\log\sqrt{2\pi}\right],$$
and $T = \sum_{i=1}^n X_i/\sigma_i^2$ is sufficient by the factorization theorem. The weighted average $T/\sum_{i=1}^n \sigma_i^{-2}$ is a natural estimator for $\theta$.
4. By independence, the joint mass functions are
$$P(X_1 = x_1, \dots, X_n = x_n) = p_1^{n_1(x)} p_2^{n_2(x)} p_3^{n_3(x)},$$

where $p_i = P(\{i\})$, $i = 1, 2, 3$, and $n_i(x) = \#\{j : x_j = i\}$. Since $n_1(x) + n_2(x) + n_3(x) = n$, we can write the joint mass functions as $p_1^{n_1(x)} p_2^{n_2(x)} p_3^{n-n_1(x)-n_2(x)}$, and $T = (n_1, n_2)$ is sufficient by the factorization theorem.

6. a) The joint densities are

$$\exp\left[(\alpha - 1)\sum_{i=1}^{n} \log x_i + (\beta - 1)\sum_{i=1}^{n} \log(1 - x_i) + n \log \frac{\Gamma(\alpha + \beta)}{\Gamma(\alpha)\Gamma(\beta)}\right],$$

a full rank exponential family with

$$T = \left(\sum_{i=1}^{n} \log X_i, \sum_{i=1}^{n} \log(1 - X_i)\right)$$

a minimal sufficient statistic.

b) Now the joint densities are

$$\exp\left[(\beta - 1)\sum_{i=1}^{n} \log\left[x_i^2(1 - x_i)\right] + \sum_{i=1}^{n} \log x_i + n \log \frac{\Gamma(3\beta)}{\Gamma(2\beta)\Gamma(\beta)}\right],$$

a full rank exponential family with minimal sufficient statistic

$$\sum_{i=1}^{n} \log\left[x_i^2(1 - x_i)\right] = 2T_1 + T_2.$$

c) The densities, parameterized by $\beta$, are

$$p_\beta(x) = \exp\left[(\beta^2 - 1)T_1(x) + (\beta - 1)T_2(x) + n \log \frac{\Gamma(\beta + \beta^2)}{\Gamma(\beta)\Gamma(\beta^2)}\right].$$

Suppose $p_\beta(x) \propto_\beta p_\beta(y)$. Then

$$\frac{p_2(x)}{p_1(x)} = \frac{p_2(y)}{p_1(y)} \text{ and } \frac{p_3(x)}{p_1(x)} = \frac{p_3(y)}{p_1(y)}.$$

Taking the logarithm of these and using the formula for $p_\beta$,

$$3T_1(x) + T_2(x) + n \log 20 = 3T_1(y) + T_2(y) + n \log 20,$$

and

$$8T_1(x) + 2T_2(x) + n \log 495 = 8T_1(y) + 2T_2(y) + n \log 495.$$

These equations imply $T(x) = T(y)$, and $T$ is minimal sufficient by Theorem 3.11.

7. The statistic $T = \left(\sum_{i=1}^{n} X_i, \sum_{i=1}^{n} t_i X_i\right)$, is minimal sufficient.

8. a) The statistic $(N_{11}, N_{12}, N_{21})$ is minimal sufficient. (The statistic $(N_{11}, N_{12}, N_{21}, N_{22})$ is also minimal sufficient.)
b) With the constraint, $(N_{11} + N_{12}, N_{11} + N_{21})$ is minimal sufficient.
9. a) The joint density is zero unless $x_i > \theta$, $i = 1, \dots, n$, that is, unless $M(x) = \min\{x_1, \dots, x_n\} > \theta$. Using this, the joint density can be written as
$$p_\theta(x) = c^n(\theta) I\{M(x) > \theta\} \prod_{i=1}^{n} f(x_i),$$
and $M(X)$ is sufficient by the factorization theorem.
b) If $p_\theta(x) \propto_\theta p_\theta(y)$, then the region where the two functions are zero must agree, and $M(x)$ must equal $M(y)$. So $M$ is minimal sufficient by Theorem 3.11.
10. The joint densities are $p_\theta(x) = 2^{-n} \prod(1 + \theta x_i) = 2^{-n} \prod(1 + \theta x_{(i)})$, where $x_{(1)} \le \dots \le x_{(n)}$ are the ordered values. Note that $p_\theta$ is a polynomial in $\theta$ with degree $n$, with roots $-1/x_{(i)}$. Suppose $p_\theta(x) \propto_\theta p_\theta(y)$. Then these polynomials must have the same roots, and we must have $x_{(i)} = y_{(i)}$, $i = 1, \dots, n$. So the order statistics are minimal sufficient by Theorem 3.11.
16. a) Let $T(x) = \max\{x_1, \dots, x_n\}$, and $M(x) = \min\{x_1, \dots, x_n\}$. Then the joint density will be positive if and only if $M(x) > 0$ and $T(x) < \theta$. Introducing suitable indicator functions, the joint density can be written $\prod_1^n (2x_i) I\{M(x) > 0\} I\{T(x) < \theta\}/\theta^{2n}$. So $T = T(X)$ is sufficient by the factorization theorem.
b) For $t \in (0, \theta)$, $P(X_i \le t) = \int_0^t 2x\,dx/\theta^2 = t^2/\theta^2$. So $P(T \le t) = P(X_1 \le t, \dots, X_n \le t) = P(X_1 \le t) \times \dots \times P(X_n \le t) = t^{2n}/\theta^{2n}$. Taking the derivative of this with respect to $t$, $T$ has density $2nt^{2n-1}/\theta^{2n}$, $t \in (0, \theta)$.
c) Suppose $E_\theta f(T) = c$, for all $\theta > 0$. Then $\int_0^\theta f(t) 2nt^{2n-1}\,dt/\theta^{2n} = c$, which implies $\int_0^\theta t^{2n-1} f(t)\,dt = c\theta^{2n}/(2n)$, for all $\theta > 0$. Taking a derivative with respect to $\theta$, $\theta^{2n-1} f(\theta) = c\theta^{2n-1}$, for a.e. $\theta$, and hence $f(t) = c$, for a.e. $t$.
17. a) If $y > 0$, then $P(Y \le y) = P(\lambda X \le y) = \int_0^{y/\lambda} \lambda e^{-\lambda x}\,dx = 1 - e^{-y}$. So $Y$ has density $d(1 - e^{-y})/dy = e^{-y}$, $y > 0$, the standard exponential density.
b) The joint densities are $\lambda^n \exp\{-n\lambda\overline{x}\}$, a full rank exponential family with $T = \overline{X}$ a complete sufficient statistic. Let $Y_i = \lambda X_i$, so that regardless of the value of $\lambda$, $Y_1, \dots, Y_n$ are i.i.d. from the standard exponential distribution. Then $(X_1^2 + \dots + X_n^2)/\overline{X}^2 = (Y_1^2 + \dots + Y_n^2)/\overline{Y}^2$ is ancillary, and independence follows by Basu's theorem.
29. $f(x) = 1/(1 + x)$ is bounded and convex on $(0, \infty)$.
30. Because $\eta_0 < \eta < \eta_1$, $\eta = \gamma\eta_0 + (1 - \gamma)\eta_1$ for some $\gamma \in (0, 1)$, and because the exponential function is convex,
$$e^{\eta T(x)} < \gamma e^{\eta_0 T(x)} + (1 - \gamma) e^{\eta_1 T(x)}.$$
Multiplying by $h(x)$ and integrating against $\mu$,

$$\int e^{\eta T(x)} h(x)\,d\mu(x) < \gamma \int e^{\eta_0 T(x)} h(x)\,d\mu(x) + (1-\gamma)\int e^{\eta_1 T(x)} h(x)\,d\mu(x).$$

From the definition of $\Xi$, the upper bound here is finite, and $\eta$ must then also lie in $\Xi$.

31. Suppose $X$ is absolutely continuous with density $f$, and define $Y = g(X)/f(X)$. Then

$$EY = \int \frac{g(x)}{f(x)} f(x)\,dx = \int g(x)\,dx = 1.$$

The function $h(y) = -\log y = \log(1/y)$ is strictly convex on $(0,\infty)$ (its second derivative is $1/y^2$). So by Jensen's inequality,

$$Eh(Y) = E\log\left[\frac{f(X)}{g(X)}\right] = \int \log\left[\frac{f(x)}{g(x)}\right] f(x)\,dx \ge h(EY) = -\log 1 = 0.$$

The inequality is strict unless $Y$ is constant a.e. If $Y$ is constant a.e., then $Y = EY = 1$ and $f(x) = g(x)$ a.e.

## B.4 Problems of Chapter 4

1. a) The joint densities form a two-parameter exponential family with $T = (T_x, T_y) = (\sum_{i=1}^m X_i, \sum_{j=1}^n Y_j)$ as a complete sufficient statistic. Since $T_x$ has a gamma distribution,

$$ET_x^{-1} = \frac{\lambda_x^m}{\Gamma(m)} \int_0^\infty t^{m-2} e^{-\lambda_x t}\,dt = \frac{\lambda_x}{m-1}.$$

Also, $ET_y = n/\lambda_y$. So $(m-1)T_y/(nT_x)$ is unbiased for $\lambda_x/\lambda_y$ and must be UMVU since it is a function of $T$.

b) Integrating against the gamma density,

$$ET_x^{-2} = \frac{\lambda_x^m}{\Gamma(m)} \int_0^\infty t^{m-3} e^{-\lambda_x t}\,dt = \frac{\lambda_x^2}{(m-1)(m-2)}.$$

Also, $ET_y^2 = (ET_y)^2 + \mathrm{Var}(T_y) = n(n+1)/\lambda_x^2$. So

$$\begin{aligned} E\left(d\frac{T_y}{T_x} - \frac{\lambda_x}{\lambda_y}\right)^2 &= d^2 E\frac{T_y^2}{T_x^2} - 2d\frac{\lambda_x}{\lambda_y} E\frac{T_y}{T_x} + \frac{\lambda_x^2}{\lambda_y^2} \\ &= \left(d^2 \frac{n(n+1)}{(m-1)(m-2)} - 2d\frac{n}{m-1} + 1\right)\frac{\lambda_x^2}{\lambda_y^2}, \end{aligned}$$

which is minimized taking $d = (m-2)/(n+1)$. So the best multiple of $\overline{Y}/\overline{X}$ is $(m-2)T_y/\big[(n+1)T_x\big] = \frac{n(m-2)}{(n+1)m}\overline{Y}/\overline{X}$.

c) Since $\delta = I\{X_1 > 1\}$ is evidently unbiased, the UMVU estimator must be $E[\delta|T]$, and by independence this must be $P(X_1 > 1|T_x]$. The joint density of $X_1$ and $S = X_2 + \cdots + X_m$ is $g(x,s) = \lambda_x^m s^{m-2} e^{-\lambda_x(s+x)}/(m-2)!$, $s > 0$ and $x > 0$. From this, the joint density of $X_1$ and $T_x$ is $f(x,t) = \lambda_x^m (t-x)^{m-2} e^{-\lambda_x t}/(m-2)!$, $0 < x < t$. Dividing by the marginal density of $T_x$, $\lambda_x^m t^{m-1} e^{-\lambda_x t}/(m-1)!$, the conditional density of $X$ given $T = t$ is $f(x|t) = (m-1)(1-x/t)^{m-2}/t$, $0 < x < t$. Integrating this conditional density, $P(X_1 > 1|T_x] = I\{T_x \ge 1\}(1 - 1/T_x)^{m-1}$.

2. a) The joint densities are

$$\frac{\exp\left[\dfrac{\mu_x \sum_{i=1}^n x_i}{\sigma^2} + \dfrac{\mu_y \sum_{j=1}^m y_j}{2\sigma^2} - \dfrac{2\sum_{i=1}^n x_i^2 + \sum_{j=1}^m y_j^2}{4\sigma^2} - \dfrac{2n+m}{4\sigma^2}\right]}{(2\pi\sigma^2)^{n/2}(4\pi\sigma^2)^{m/2}},$$

a full rank exponential family with complete sufficient statistic $T = (\sum_{i=1}^n X_i, \sum_{j=1}^m Y_j, 2\sum_{i=1}^n X_i^2 + \sum_{j=1}^m Y_j^2)$.
b) Expanding the squares and simplifying, $2(n-1)S_x^2 + (m-1)S_y^2 = 2\sum_{i=1}^n X_i^2 + \sum_{j=1}^m Y_j^2 - 2n\overline{X}^2 - m\overline{Y}^2$ is a function of $T$ with mean $2(n+m-2)\sigma^2$. So $S_p^2 = \left[2(n-1)S_x^2 + (m-1)S_y^2\right]/(2n+2m-4)$ is UMVU for $\sigma^2$.
c) Because $E(\overline{X} - \overline{Y})^2 = (\mu_x - \mu_y)^2 + \sigma^2/n + 2\sigma^2/m$, $(\overline{X} - \overline{Y})^2 - (1/n + 2/m)S_p^2$ is UMVU for $(\mu_x - \mu_y)^2$.
d) With the additional constraint, $(2\sum_{i=1}^n X_i + 3\sum_{j=1}^m Y_j, 2\sum_{i=1}^n X_i^2 + \sum_{j=1}^m Y_j^2)$ is complete sufficient. The first statistic here has mean $(2n + 9m)\mu_x$, so $(2\sum_{i=1}^n X_i + 3\sum_{j=1}^m Y_j)/(2n+9m)$ is the UMVU estimator of $\mu_x$.

3. The joint mass functions are

$$\frac{\lambda^{x_1+\cdots+x_n} e^{-n\lambda}}{x_1! \times \cdots \times x_n!}.$$

These densities form a full rank exponential family with $T = X_1 + \cdots + X_n$ as a complete sufficient statistic. Since $T$ has a Poisson distribution with mean $n\lambda$, $\delta(T)$ will be an unbiased estimator of $\cos\lambda$ if

$$\sum_{t=0}^\infty \frac{\delta(t)(n\lambda)^t e^{-n\lambda}}{t!} = \cos\lambda$$

or if

$$\begin{aligned}\sum_{t=0}^\infty \frac{\delta(t)n^t}{t!}\lambda^t = e^{n\lambda}\cos\lambda &= \frac{e^{(n+i)\lambda} + e^{(n-i)\lambda}}{2} \\ &= \sum_{t=0}^\infty \frac{(n+i)^t + (n-i)^t}{2t!}\lambda^t.\end{aligned}$$

Equating coefficients of $\lambda^t$ in these expansions,

$$\delta(t) = \frac{1}{2}\left[\left(1+\frac{i}{n}\right)^t + \left(1-\frac{i}{n}\right)^t\right] = \left(1+\frac{1}{n^2}\right)^{t/2}\cos(t\omega),$$

where $\omega = \arctan(1/n)$.

4. The joint densities are

$$\frac{1}{(2\pi)^{n/2}}\exp\left[\alpha\sum_{i=1}^n t_i x_i + \beta\sum_{i=1}^n t_i^2 x_i - \frac{1}{2}\sum_{i=1}^n x_i^2 - \frac{1}{2}\sum_{i=1}^n(\alpha t_i + \beta t_i^2)^2\right].$$

These densities form a full rank exponential family with

$$T = \left(\sum_{i=1}^n t_i X_i, \sum_{i=1}^n t_i^2 X_i\right)$$

a complete sufficient statistic. Now

$$ET_1 = \alpha\sum_{i=1}^n t_i^2 + \beta\sum_{i=1}^n t_i^3 \text{ and } ET_2 = \alpha\sum_{i=1}^n t_i^3 + \beta\sum_{i=1}^n t_i^4.$$

Using these,

$$\frac{T_1\sum_{i=1}^n t_i^4 - T_2\sum_{i=1}^n t_i^3}{\sum_{i=1}^n t_i^2\sum_{i=1}^n t_i^4 - \left(\sum_{i=1}^n t_i^3\right)^2} \text{ and } \frac{T_2\sum_{i=1}^n t_i^2 - T_1\sum_{i=1}^n t_i^3}{\sum_{i=1}^n t_i^2\sum_{i=1}^n t_i^4 - \left(\sum_{i=1}^n t_i^3\right)^2}$$

are unbiased estimators for $\alpha$ and $\beta$. Since they are functions of the complete sufficient statistic $T$, they are UMVU.

5. a) Expanding the quadratic, $S^2 = (\sum_{i=1}^n X_i^2 - n\overline{X}^2)/(n-1)$. If we let $\mu(\theta) = E_\theta X_i$, then $E_\theta X_i^2 = \mu^2(\theta) + \sigma^2(\theta)$, and $E_\theta \overline{X} = \mu^2(\theta) + \sigma^2(\theta)/n$. So

$$E_\theta S^2 = \frac{1}{n-1}\left(\sum_{i=1}^n[\mu^2(\theta)+\sigma^2(\theta)] - n\left[\mu^2(\theta) + \frac{1}{n}\sigma^2(\theta)\right]\right) = \sigma^2(\theta).$$

b) If $X_i$ is Bernoulli, then $X_i = X_i^2$ and

$$S^2 = \frac{\sum_{i=1}^n X_i - n\overline{X}^2}{n-1} = \frac{n\overline{X}(1-\overline{X})}{n-1}.$$

The joint mass functions form a full rank exponential family with $\overline{X}$ as a complete sufficient statistic. Since $\delta$ is unbiased and is a function of $\overline{X}$, $\delta$ is UMVU.

c) Again we have a full rank exponential family with $\overline{X}$ as a complete sufficient statistic. Because $EX_i = 1/\theta$ and $\mathrm{Var}(X_i) = 1/\theta^2$, $E\overline{X}^2 = \theta^{-2} +$

$\theta^{-2}/n$, and $n\overline{X}^2/(n+1)$ is an unbiased estimator of $\sigma^2$. This estimator is UMVU because it is a function of the complete sufficient statistic. Next, by symmetry $E_\theta[X_1^2|\overline{X}=c] = \cdots = E_\theta[X_n^2|\overline{X}=c]$. The UMVU estimator must equal $E_\theta[\delta|\overline{X}]$, and therefore

$$\frac{nc^2}{n+1} = E_\theta\left[\frac{\sum_{i=1}^n X_i^2 - n\overline{X}^2}{n-1}\,\middle|\,\overline{X}=c\right] = \frac{nE_\theta[X_i^2|\overline{X}=c] - nc^2}{n-1}.$$

From this,

$$E_\theta[X_i^2|\overline{X}=c] = \frac{2nc^2}{n+1}.$$

6. Because $\delta + cU$ is unbiased and $\delta$ is UMVU, for any $\theta$,

$$\operatorname{Var}_\theta(\delta + cU) = \operatorname{Var}_\theta(\delta) + 2c\operatorname{Cov}_\theta(\delta, U) + c^2\operatorname{Var}_\theta(U) \ge \operatorname{Var}_\theta(\delta).$$

So $h(c) = c^2\operatorname{Var}_\theta(U) + 2c\operatorname{Cov}_\theta(\delta, U) \ge 0$. Since $h(0) = 0$, this will hold for all $c$ only if $h'(0) = 2\operatorname{Cov}_\theta(\delta, U) = 0$.

7. Suppose $\delta$ is unbiased for $g_1(\theta) + g_2(\theta)$, and that $\operatorname{Var}_\theta(\delta) < \infty$. Then $U = \delta - \delta_1 - \delta_2$ is an unbiased estimator of zero. By Problem 4.6, $\operatorname{Cov}_\theta(U, \delta_1 + \delta_2) = \operatorname{Cov}_\theta(U, \delta_1) + \operatorname{Cov}_\theta(U, \delta_2) = 0$. Since these variables are uncorrelated,

$$\begin{aligned}\operatorname{Var}_\theta(\delta) &= \operatorname{Var}_\theta(U + \delta_1 + \delta_2)\\ &= \operatorname{Var}_\theta(U) + \operatorname{Var}_\theta(\delta_1 + \delta_2) \ge \operatorname{Var}_\theta(\delta_1 + \delta_2).\end{aligned}$$

8. If $M(x) = \min x_i$, then the joint densities are

$$p_\theta(x) = \frac{\theta^n}{\prod_{i=1}^n x_i^2} I\{M(x) > \theta\},$$

and $M$ is sufficient by the factorization theorem. Next, for $x > \theta$,

$$\begin{aligned}P_\theta(M > x) &= P_\theta(X_1 > x, \dots, X_n > x)\\ &= P_\theta(X_1 > x) \times \cdots \times P_\theta(X_n > x) = (\theta/x)^n.\end{aligned}$$

So $M$ has cumulative distribution function $1 - (\theta/x)^n$, $x > \theta$, and density $n\theta^n/x^{n+1}$, $x > \theta$. If $\delta(M)$ is an unbiased estimator of $g(\theta)$, then

$$\int_\theta^\infty \delta(x)\frac{n\theta^n}{x^{n+1}}\,dx = g(\theta)$$

or

$$\int_\theta^\infty \frac{n\delta(x)}{x^{n+1}}\,dx = \frac{g(\theta)}{\theta^n}.$$

Taking a derivative with respect to $\theta$,

$$\frac{n\delta(x)}{x^{n+1}} = \frac{ng(x)}{x^{n+1}} - \frac{g'(x)}{x^n}.$$

In particular, if $g(\theta) = c$ for all $\theta > 0$, this calculation shows that $\delta(x) = c$, and so $M$ is complete. In general, $\delta(M) = g(M) - Mg'(M)/n$ is the UMVU estimator of $g(\theta)$.

10. If we assume $\delta$ is unbiased and can be written as a power series $\delta(x) = c_0 + c_1 x + \cdots$, then by Fubini's theorem we anticipate

$$\begin{aligned} E_\theta \delta(X) &= \int_0^\infty \left( \sum_{n=0}^\infty c_n x^n \right) \theta e^{-\theta x}\, dx \\ &= \sum_{n=0}^\infty \int_0^\infty c_n x^n \theta e^{-\theta x}\, dx = \sum_{n=0}^\infty \frac{n! c_n}{\theta^n}. \end{aligned}$$

The form here is a power series in $1/\theta$. Writing

$$\frac{1}{1+\theta} = \frac{1/\theta}{1+1/\theta} = -\sum_{n=1}^\infty (-1)^n \theta^{-n},$$

by matching coefficients for powers of $1/\theta$, $c_0 = 0$ and $c_n = (-1)^{n+1}/n!$, $n = 1, 2, \ldots$. This gives $\delta = 1 - e^{-X}$. The steps in this derivation only work if $\theta > 1$, but it is easy to show directly that $\delta$ is unbiased. Because the densities form a full rank exponential family, $X$ is complete, and $\delta$ is UMVU.

11. The joint mass functions form a full rank exponential family with $T = X_1 + \cdots + X_3$ complete sufficient. The estimator $\delta = I\{X_1 = X_2 = 0\}$ is unbiased. By Theorem 4.4, $\eta(t) = E[\delta(X)|T = t] = P(X_1 = X_2 = 0|T = t)$ is UMVU. To calculate $\eta$ we need $P_\theta(T = t)$. This event occurs if and only if trial $t + 3$ is a success, and there are exactly two successes in the first $t + 2$ trials. Thus

$$P_\theta(T = t) = \binom{t+2}{2} \theta^3 (1-\theta)^t.$$

Since $P_\theta(X_1 = X_2 = 0, T = t) = P(X_1 = X_2 = 0, X_3 = t) = \theta^3(1-\theta)^t$,

$$\eta(t) = \frac{P_\theta(X_1 = X_2 = 0, T = t)}{P_\theta(T = t)} = \frac{2}{(t+1)(t+2)}.$$

12. a) Since

$$E_\theta X = \int_{-1}^1 \frac{1}{2}(x + \theta x^2)\, dx = \frac{\theta}{3},$$

$3X$ is unbiased for $\theta$.

b) Integrating against the density,

$$b = E_\theta |X| = \int_{-1}^1 \frac{1}{2}(1 + \theta x)|x|\, dx = \frac{1}{2},$$

for all $\theta \in [-1, 1]$.

c) Since

$$E_\theta X^2 = \int_{-1}^{1} \frac{1}{2}(x^2 + \theta x^3)\,dx = \frac{1}{3},$$

$\mathrm{Var}_\theta(3X) = 3 - \theta^2$. Also, $\mathrm{Var}_\theta(|X|) = 1/3 - 1/4 = 1/12$. Finally,

$$E_\theta 3X|X| = \int_{-1}^{1} \frac{3}{2}(x + \theta x^2)|x|\,dx = \frac{3\theta}{4},$$

and so $\mathrm{Cov}_\theta(3X, |X|) = 3\theta/4 - \theta/2 = \theta/4$. So

$$\begin{aligned} \mathrm{Var}_{\theta_0}(3X + c|X|) &= \mathrm{Var}_{\theta_0}(3X) + 2c\mathrm{Cov}_{\theta_0}(3X, |X|) + c^2\mathrm{Var}_{\theta_0}(|X|) \\ &= (3 - \theta_0^2) + \frac{c\theta_0}{2} + \frac{c^2}{12}, \end{aligned}$$

minimized when $c = -3\theta_0$. Since the variance of this estimator is smaller than the variance of $3X$ when $\theta = \theta_0$, $3X$ cannot be UMVU.

24. Since $\overline{X}$ and $S^2$ are independent, using (4.10),

$$Et = E\frac{\sqrt{n}\,\overline{X}}{S} = \sqrt{n}\big(E\overline{X}\big)\big(ES^{-1}\big) = \delta\frac{\sqrt{n-1}\,\Gamma\big((n-2)/2\big)}{\sqrt{2}\,\Gamma\big((n-1)/2\big)},$$

$$Et^2 = n\big(E\overline{X}^2\big)\big(ES^{-2}\big) = (1 + \delta^2)\frac{n-1}{n-3},$$

and

$$\mathrm{Var}(t) = \frac{n-1}{n-3} + \delta^2\left(\frac{n-1}{n-3} - \frac{(n-1)\Gamma^2\big((n-2)/2\big)}{2\Gamma^2\big((n-1)/2\big)}\right).$$

28. a) Since $g(\theta) = \theta$, $g(\theta + \Delta) - g(\theta) = \Delta$, so the lower bound is

$$\frac{\Delta^2}{E_\theta\left(\frac{p_{\theta+\Delta}(X)}{p_\theta(X)} - 1\right)^2}.$$

Because we need $p_{\theta+\Delta} = 0$ whenever $p_\theta = 0$, $\Delta$ must be negative. Also, $\theta + \Delta$ must be positive, so $\Delta \in (-\theta, 0)$. To evaluate the expectation, note that $p_{\theta+\Delta}(X)/p_\theta(X)$ will be $\theta^n/(\theta + \Delta)^n$ if $M = \max\{X_1, \ldots, X_n\} < \theta + \Delta$, which happens with probability $(\theta + \Delta)^n/\theta^n$ under $P_\theta$. Otherwise, $p_{\theta+\Delta}(X)/p_\theta(X)$ will be zero. After a bit of algebra, the lower bound is found to be

$$\frac{\Delta^2}{\big[\theta^n/(\theta + \Delta)^n\big] - 1} = \frac{c^2\theta^2/n^2}{(1 - c/n)^{-n} - 1}$$

b) $g_n(c) = c^2/\big[(1 - c/n)^{-n} - 1\big] \to c^2/(e^c - 1) = g(c)$.

c) Setting derivatives to zero, the value $c_0$ maximizing $g$ over $c \in (0, \infty)$ is the unique positive solution of the equation $e^{-c} = 1 - c/2$. This gives $c_0 = 1.59362$, and an approximate lower bound of $g(c_0)\theta^2/n^2 = 0.64761\theta^2/n^2$.

30. a) We have $\log p_\theta(X) = -\sum_{i=1}^n (X_i - \alpha - \beta t_i)^2/2 - n\log\sqrt{2\pi}$, and so

$$I(\alpha,\beta) = -E\nabla^2 \log p_\theta(X) = \begin{pmatrix} n & \sum_{i=1}^n t_i \\ \sum_{i=1}^n t_i & \sum_{i=1}^n t_i^2 \end{pmatrix}.$$

b) If $g(\theta) = \alpha$, then $\nabla g(\theta) = \binom{1}{0}$ and the lower bound is

$$\begin{aligned}(1,0)I^{-1}(\alpha,\beta)\begin{pmatrix}1\\0\end{pmatrix} &= \frac{(1,0)\begin{pmatrix} \sum_{i=1}^n t_i^2 & -\sum_{i=1}^n t_i \\ -\sum_{i=1}^n t_i & n \end{pmatrix}\begin{pmatrix}1\\0\end{pmatrix}}{n\sum_{i=1}^n t_i^2 - \left(\sum_{i=1}^n t_i\right)^2} \\ &= \frac{\sum_{i=1}^n t_i^2}{n\sum_{i=1}^n t_i^2 - \left(\sum_{i=1}^n t_i\right)^2} \\ &= \frac{1/n}{1 - \left(\sum_{i=1}^n t_i/n\right)^2/\left(\sum_{i=1}^n t_i^2/n\right)}.\end{aligned}$$

c) Now $I(\alpha) = n$, and so the lower bound for the variance is $1/n$.
d) The bound in (b) is larger. This is clear from the final expression, with equality only if $\sum_{i=1}^n t_i = 0$.
e) If $g(\theta) = \alpha\beta$, then $\nabla g(\theta) = \binom{\beta}{\alpha}$ and the lower bound is

$$\frac{(\beta,\alpha)\begin{pmatrix} \sum_{i=1}^n t_i^2 & -\sum_{i=1}^n t_i \\ -\sum_{i=1}^n t_i & n \end{pmatrix}\begin{pmatrix}\beta\\\alpha\end{pmatrix}}{n\sum_{i=1}^n t_i^2 - \left(\sum_{i=1}^n t_i\right)^2} = \frac{\beta^2\sum_{i=1}^n t_i^2 - 2\alpha\beta\sum_{i=1}^n t_i + n\alpha^2}{n\sum_{i=1}^n t_i^2 - \left(\sum_{i=1}^n t_i\right)^2}.$$

31. This is a location model, and so

$$I(\theta) = \int \frac{\left[f'(x)\right]^2}{f(x)}\,dx = \int \frac{4x^2}{\pi(1+x^2)^3}\,dx = \frac{1}{2}.$$

If $\xi = \theta^3$, $\theta = \xi^{1/3} = h(\xi)$, and the information for $\xi$ is $I^*(\xi) = I\left[h(\xi)\right] \times \left[h'(\xi)\right]^2 = 1/(18\xi^{4/3})$.

32. Since $p_\theta(x) = \theta^{2x}e^{-\theta^2}/x!$,

$$I(\theta) = \operatorname{Var}_\theta\left(\frac{\partial \log p_\theta(X)}{\partial\theta}\right) = \operatorname{Var}_\theta\left(\frac{2X}{\theta} - 2\theta\right) = 4.$$

33. Since the exponential distributions form a canonical exponential family with $A(\lambda) = -\log\lambda$, the Fisher information for $\lambda$ is $\tilde I(\lambda) = A''(\lambda) = 1/\lambda^2$. Using (4.18),

$$I(\theta) = \frac{\tilde I(\lambda)}{\left[h'(\lambda)\right]^2} = \frac{1}{\left[\lambda h'(\lambda)\right]^2}.$$

From this, $I(\theta)$ will be constant if $h(\lambda) = \log\lambda$.

34. a) Multiplying the conditional densities,

$$\log p_{\theta,\sigma}(X) = -\frac{(1-\rho^2)(X_1-\theta)^2}{2\sigma^2} - \frac{1}{2\sigma^2}\sum_{j=1}^{n-1}\bigl(X_{j+1}-\rho X_j-(1-\rho)\theta\bigr)^2$$
$$-\frac{n}{2}\log(2\pi) - n\log\sigma + \log\sqrt{1-\rho^2}.$$

Taking derivatives,

$$-\frac{\partial^2}{\partial\theta^2}\log p_{\theta,\sigma}(X) = \frac{1-\rho^2}{\sigma^2} + \frac{(n-1)(1-\rho)^2}{\sigma^2},$$
$$-\frac{\partial^2}{\partial\theta\partial\sigma}\log p_{\theta,\sigma}(X) = \frac{2(1-\rho^2)\epsilon_1}{\sigma^3} + \frac{2(1-\rho)}{\sigma^3}\sum_{j=1}^{n-1}\eta_{j+1},$$

and

$$-\frac{\partial^2}{\partial\sigma^2}\log p_{\theta,\sigma}(X) = \frac{3(1-\rho^2)(X_1-\theta)^2}{\sigma^4} + \frac{3}{\sigma^4}\sum_{j=1}^{n-1}\eta_{j+1}^2 - \frac{n}{\sigma^2},$$

where $\epsilon_j = X_j - \theta$ and $\eta_{j+1} = \epsilon_{j+1} - \rho\epsilon_j$. The conditional distribution of $\eta_{j+1}$ given $X_1 = x_1, \dots, X_j = x_j$, or equivalently, given $\epsilon_1$ and $\eta_2, \dots, \eta_j$, is $N(0,\sigma^2)$. From this, $\eta_2, \dots, \eta_n$ are i.i.d. from $N(0,\sigma^2)$, and these variables are independent of $\epsilon_1 \sim N\bigl(0, \sigma^2/(1-\rho^2)\bigr)$. Using this, it is easy to take expectations of these logarithmic derivatives, giving

$$I(\theta,\sigma) = \begin{pmatrix} \dfrac{1-\rho^2+(n-1)(1-\rho)^2}{\sigma^2} & 0 \\ 0 & \dfrac{2n}{\sigma^2} \end{pmatrix}.$$

b) The lower bound is $\sigma^2/\bigl[1-\rho^2+(n-1)(1-\rho)^2\bigr]$.
c) Since $\epsilon_2 = \rho\epsilon_1 + \eta_2$, $\mathrm{Var}(\epsilon_2) = \rho^2\sigma^2/(1-\rho^2) + \sigma^2 = \sigma^2/(1-\rho^2)$. Further iteration gives $\mathrm{Var}(X_i) = \mathrm{Var}(\epsilon_i) = \sigma^2/(1-\rho^2)$. If $i > j$, then

$$X_i = \rho X_{i-1} + \eta_i = \cdots = \rho^{i-j}X_j + \rho^{i-j+1}\eta_{j+1} + \cdots + \rho\eta_{i-1} + \eta_i,$$

and from this it is easy to see that

$$\mathrm{Cov}(X_i, X_j) = \frac{\rho^{|i-j|}\sigma^2}{1-\rho^2}.$$

Noting that $\sum_{j=1}^{n-1}\rho^j = (\rho-\rho^n)/(1-\rho)$ and

$$\sum_{i=1}^{n-1} j\rho^j = \rho\frac{d}{d\rho}\sum_{j=1}^{n-1}\rho^j = \frac{\rho-\rho^{n+1}}{(1-\rho)^2} - \frac{n\rho^n}{1-\rho},$$

$$\begin{aligned}
\operatorname{Var}(\overline{X}) &= \frac{\sigma^2}{n^2(1-\rho^2)}\sum_{i=1}^n\sum_{j=1}^n \rho^{|i-j|}\\
&= \frac{\sigma^2}{n^2(1-\rho^2)}\left[n+2\sum_{j=1}^{n-1}(n-j)\rho^j\right]\\
&= \frac{\sigma^2}{n(1-\rho)^2} - \frac{2\rho\sigma^2(1-\rho^n)}{n^2(1-\rho)^3(1+\rho)}.
\end{aligned}$$

## B.5 Problems of Chapter 5

1. a) The joint mass functions are

$$\binom{m}{x}\binom{n}{y}\exp\left[x\log\frac{\theta}{1-\theta}+y\log\frac{\theta^2}{1-\theta^2}+n\log\theta+2m\log\theta\right].$$

These mass functions form a curved exponential family with minimal sufficient statistic $(X,Y)$.
b) $E_\theta[X^2+X]=m(m-1)\theta^2$, and $E_\theta Y = n\theta^2$. So $E_\theta[nX^2+nX-m(m-1)Y]=0$ for all $\theta\in(0,1)$.

2. a) The joint densities are

$$ph(p)\exp\left[x\log\left(\frac{p}{1-p}\right)+y\log\left(\frac{h(p)}{1-h(p)}\right)\right].$$

These form a curved exponential family unless the canonical parameters are linearly related, that is, unless

$$\log\left(\frac{h(p)}{1-h(p)}\right)=a+b\log\left(\frac{p}{1-p}\right)$$

for some constants $a$ and $b$. Solving for $h(p)$, this is the same as the equation stated in the problem.
b) If $h(p)=p/2$, we have a curved family with $(X,Y)$ minimal sufficient, but not complete, because $E_p(X-2Y)=0$ for all $p\in(0,1)$. For an example where $(X,Y)$ is complete, note that $E_pg(X,Y)=ph(p)g(1,1)+p\bigl(1-h(p)\bigr)g(1,0)+(1-p)h(p)g(0,1)+(1-p)\bigl(1-h(p)\bigr)g(0,0)$. We need to find a function $h$ with $ph(p)$, $p\bigl(1-h(p)\bigr)$, and $(1-p)h(p)$ linearly independent. One choice that works is $h(p)=p^2$. Then

$$\begin{aligned}
E_pg(X,Y) &= p^3\bigl[g(1,1)-g(1,0)-g(0,1)+g(0,0)\bigr]\\
&\quad + p^2\bigl[g(0,1)-g(0,0)\bigr]+p\bigl[g(1,0)-g(0,0)\bigr]+g(0,0).
\end{aligned}$$

If this is zero for all $p\in(0,1)$, then the coefficients of the various powers of $p$ must vanish, and it is easy to see that this can only happen if $g(0,0)=g(1,0)=g(0,1)=g(1,1)=0$. Thus $g(X,Y)=0$ and $(X,Y)$ is complete.

6. a) Consider the event $X = 4$ and $Y = y$. This happens if and only if A wins exactly 3 of the first $3 + y$ games and then wins the next game. The chance of 3 wins in $3+y$ trials is $\binom{3+y}{3}\theta^3(1-\theta)^y$. The outcome of the next game is independent of this event, and so

$$P(X = 4, Y = y) = \binom{3+y}{3}\theta^4(1-\theta)^y, \qquad y = 0, \dots, 3.$$

Similarly,

$$P(X = x, Y = 4) = \binom{3+x}{3}\theta^x(1-\theta)^4, \qquad x = 0, \dots, 3.$$

The joint mass functions have form $h(x, y)\exp\bigl[x \log\theta + y\log(1-\theta)\bigr]$. Because the relationship between the canonical parameters $\log\theta$ and $\log(1-\theta)$ is nonlinear, this exponential family is curved.
b) Let $f$ be an arbitrary function, and suppose

$$\begin{aligned} h(\theta) &= E_\theta f(X, Y) \\ &= \sum_{x=0}^{3} f(x, 4)\binom{3+x}{3}\theta^x(1-\theta)^4 + \sum_{y=0}^{3} f(4, y)\binom{3+y}{3}\theta^4(1-\theta)^y \\ &= 0, \end{aligned}$$

for all $\theta \in (0, 1)$. This function $h$ is a polynomial in $\theta$. Letting $\theta$ tend to zero, the constant term in this polynomial is $f(0, 4)$, and so $f(0, 4)$ must be zero. If $f(0, 4)$ is zero, then the linear term (dividing by $\theta$ and letting $\theta$ tend to zero) is $4f(1, 4)$, so $f(1, 4)$ must be zero. Similarly $f(2, 4) = f(3, 4) = 0$. Then

$$\frac{h(\theta)}{\theta^4} = \sum_{y=0}^{3} f(4, y)\binom{3+y}{3}(1-\theta)^y = 0.$$

Because this is a polynomial in $1-\theta$, the coefficients of powers of $1-\theta$ must vanish, giving $f(4, 0) = \cdots = f(4, 3) = 0$. Thus $f(X, Y) = 0$ almost surely, demonstrating that $T$ is complete.
c) Let $\delta$ be an indicator that team A wins the first game. Then $\delta$ is unbiased for $\theta$, and the UMVU estimator must be $E(\delta|X, Y) = P(\delta = 1|X, Y)$. Arguments similar to those used deriving the joint mass function give

$$\begin{aligned} P(\delta = 1|X = 4, Y = y) &= \frac{P(\delta = 1, X = 4, Y = y)}{P(X = 4, Y = y)} \\ &= \frac{\theta^2\binom{2+y}{2}\theta^2(1-\theta)^y}{\binom{3+y}{3}\theta^4(1-\theta)^y} = \frac{3}{3+y}, \end{aligned}$$

and

$$P(\delta = 1|X = x, Y = 4) = \frac{\theta(1-\theta)\binom{x+2}{x-1}\theta^{x-1}(1-\theta)^3}{\binom{3+x}{3}\theta^x(1-\theta)^4} = \frac{x}{3+x}.$$

The UMVU is $(X - I\{Y = 4\})/(X + Y - 1)$.

7. a) $P(T = 0) = P(X = 0) = e^{-\lambda}$, and for $k = 0, 1, \ldots,$

$$\begin{aligned} P(T = k+1) &= P(X + Y = k, X > 0) \\ &= P(X + Y = k) - P(X + Y = k, X = 0) \\ &= \frac{(2\lambda)^k e^{-2\lambda}}{k!} - \frac{\lambda^k e^{-2\lambda}}{k!}. \end{aligned}$$

c) Let $N$ denote the sample size, so $N = 1$ if $X = 0$ and $N = 2$ if $X > 0$, and let $W = 0$ if $X = 0$ and $W = X + Y$ if $X > 0$. Using Theorem 5.4, the joint densities form an exponential family with canonical parameter $\eta = (\log \lambda, -\lambda)$ and sufficient statistic $(W, N)$. Because $\eta$ does not satisfy a linear constraint, the exponential family is curved, and $(W, N)$ is minimal sufficient.

b) There is a one-to-one relationship between $T$ and $(W, N)$, so $T$ is minimal sufficient. (This can also be shown directly.)

d) Suppose $E_\lambda g(T) = 0$ for all $\lambda > 0$. Then

$$e^{2\lambda} E_\lambda g(T) = e^\lambda g(0) + \sum_{k=0}^\infty \frac{g(k+1)(2^k - 1)}{k!}\lambda^k = 0.$$

The constant term in this power series is $g(0)$, so $g(0) = 0$. Setting the coefficient of $\lambda^k$ to zero, $g(k+1) = 0$ for $k = 1, 2, \ldots$. Since $T$ is never 1, $g(T)$ must be zero. Thus $T$ is complete.

13. a) Solving, $N_{22} = (n + D - R - C)/2$, $N_{11} = (R + C + D - n)/2$, $N_{12} = (R - C - D + n)/2$ and $N_{21} = (C - R - D + n)/2$. So the joint mass function can be written as

$$\begin{aligned} \binom{n}{n_{11}, \ldots, n_{22}} \exp\Bigg[ r \log \sqrt{\frac{p_{11}p_{12}}{p_{21}p_{22}}} + c \log \sqrt{\frac{p_{11}p_{21}}{p_{12}p_{22}}} \\ + d \log \sqrt{\frac{p_{11}p_{22}}{p_{12}p_{21}}} + n \log \sqrt{\frac{p_{12}p_{21}p_{22}}{p_{11}}} \Bigg], \end{aligned}$$

where $r = R(n_{11}, \ldots, n_{22}) = n_{11} + n_{12}$, and $c$ and $d$ are defined similarly. These densities form a full rank exponential family: $(R, C, D)$ cannot satisfy a linear constraint because there is a one-to-one linear association between it and $(N_{11}, N_{12}, N_{21})$, and the three canonical parameters $\eta_r$, $\eta_c$, and $\eta_d$ can vary freely over $\mathbb{R}^3$.

b) They are related by $\eta_d = \log\sqrt{\alpha}$.
c) Multiplying the marginal mass functions, the joint mass function has form

$$h(x)\exp\left[\sum_{i=1}^{m} r_i\eta_{r,i} + \sum_{i=1}^{m} c_i\eta_{c,i} + \log\sqrt{\alpha}\sum_{i=1}^{m} d_i - \sum_{i=1}^{m} A_i(\eta_i)\right].$$

These mass functions form a full rank $(2m+1)$-parameter exponential family with complete sufficient statistic

$$T = \left(R_1,\dots,R_m,C_1,\dots,C_m,\sum_{i=1}^{m} D_i\right).$$

16. a) Since $N_{i+}$ and $N_{+j}$ are independent with

$$N_{i+} \sim \text{Binomial}(n,p_{i+}) \text{ and } N_{+j} \sim \text{Binomial}(n,p_{+j}),$$

$$\begin{aligned} E(\hat p_{i+}\hat p_{+j})^2 &= \frac{1}{n^4}EN_{i+}^2EN_{+j}^2 \\ &= \frac{1}{n^2}\bigl[np_{i+}^2 + p_{i+}(1-p_{i+})\bigr]\bigl[np_{+j}^2 + p_{+j}(1-p_{+j})\bigr]. \end{aligned}$$

Subtracting $p_{i+}^2p_{+j}^2$,

$$\begin{aligned} \operatorname{Var}(\hat p_{i+}\hat p_{+j}) &= \frac{p_{i+}(1-p_{i+})p_{+j}^2 + p_{+j}(1-p_{+j})p_{i+}^2}{n} \\ &\quad + \frac{p_{i+}(1-p_{i+})p_{+j}(1-p_{+j})}{n^2}. \end{aligned}$$

b) Unbiased estimates of $p_{i+}(1-p_{i+})$, $p_{+j}(1-p_{+j})$, $p_{i+}^2$, and $p_{+j}^2$ are $N_{i+}(n-N_{i+})/(n^2-n)$, $N_{+j}(n-N_{+j})/(n^2-n)$, $N_{i+}(N_{i+}-1)/(n^2-n)$, and $N_{+j}(N_{+j}-1)/(n^2-n)$, respectively. From these, the UMVU estimator for the variance above is

$$\frac{N_{i+}(n-N_{i+})N_{+j}(N_{+j}-1) + N_{+j}(n-N_{+j})N_{i+}(N_{i+}-1)}{n^3(n-1)^2} + \frac{N_{i+}(n-N_{i+})N_{+j}(n-N_{+j})}{n^4(n-1)^2}.$$

## B.6 Problems of Chapter 6

2. a) Conditioning on $X$, for $z \ge 0$,

$$P(X^2Y^2 \le z) = EP(X^2Y^2 \le z|X) = E\bigl[F_Y(\sqrt{z/X^2}) - F_Y(-\sqrt{z/X^2})\bigr].$$

Taking a derivative with respect to $z$, the density is

$$E\frac{f_Y(\sqrt{z/X^2}) + f_Y(-\sqrt{z/X^2})}{2\sqrt{zX^2}}, \qquad z \geq 0.$$

b) Differentiating (1.21), since $-Y$ has density $\lambda e^{\lambda x} I\{x < 0\}$, the density is

$$E\lambda e^{\lambda(z-X)} I\{X > z\} = \int_{z\vee 0}^{\infty} \lambda^2 e^{\lambda(z-2x)}\,dx = \frac{\lambda}{2}e^{-\lambda|z|}.$$

3. Since $X$ and $Y$ are positive and $x \in (0,1)$, $X/(X+Y) \leq x$ if and only if $X \leq xY/(1-x)$. So

$$\begin{aligned} P\left(\frac{X}{X+Y} \leq x \,\middle|\, Y = y\right) &= E\left(I\left\{X \leq \frac{xY}{1-x}\right\} \,\middle|\, Y = y\right) \\ &= F_X\left(\frac{xy}{1-x}\right). \end{aligned}$$

Thus $P\big(X/(X+Y) \leq x \mid Y\big) = F_X\big(xY/(1-x)\big)$, and the desired identity follows by smoothing.

4. The change of variables $u = y/(1-x)$ in the integral against the density of $Y$ gives,

$$\begin{aligned} p_V(x) &= \frac{1}{\Gamma(\alpha)\Gamma(\beta)} \int_0^\infty \frac{y}{(1-x)^2}\left(\frac{xy}{1-x}\right)^{\alpha-1} y^{\beta-1}e^{-y/(1-x)}dy \\ &= \frac{x^{\alpha-1}(1-x)^{\beta-1}}{\Gamma(\alpha)\Gamma(\beta)} \int_0^\infty u^{\alpha+\beta-1}e^{-u}\,du \\ &= \frac{\Gamma(\alpha+\beta)}{\Gamma(\alpha)\Gamma(\beta)} x^{\alpha-1}(1-x)^{\beta-1}. \end{aligned}$$

5. a) For $x \in (0,1)$,

$$p_X(x) = \int p(x,y)\,dy = \int_x^1 2\,dy = 2(1-x).$$

Similarly, for $y \in (0,1)$,

$$p_Y(y) = \int p(x,y)\,dx = \int_0^y 2\,dx = 2y.$$

b) For $y \in (x,1)$,

$$p_{Y|X}(y|x) = \frac{p(x,y)}{p_X(x)} = \frac{2}{2(1-x)} = \frac{1}{1-x}.$$

c) Because

$$E[Y|X=x] = \int_x^1 \frac{y}{1-x}\,dy = \frac{1+x}{2},$$

$E[Y|X] = (1+X)/2$.

d) Integrating against the joint density,

$$EXY = \iint xyp(x,y)\,dx\,dy = \int_0^1 \int_0^y 2xy\,dx\,dy = \int_0^1 y^3\,dy = \frac{1}{4}.$$

e) By smoothing,

$$\begin{aligned} EXY = EE[XY|X] = E\{XE[Y|X]\} &= E\left[X\frac{1}{2}(1+X)\right] \\ &= \int_0^1 2(1-x)\frac{1}{2}x(1+x)\,dx = \int_0^1 [x-x^3]\,dx = \frac{1}{4}. \end{aligned}$$

6. a) For $x \in (0,1)$,

$$p_X(x) = \sum_{y_1=0}^{\infty}\sum_{y_2=0}^{\infty} x^2(1-x)^{y_1+y_2} = 1.$$

b) Integrating the joint density,

$$\begin{aligned} p_Y(y_1,y_2) &= \int_0^1 x^2(1-x)^{y_1+y_2}\,dx \\ &= \frac{\Gamma(3)\Gamma(y_1+y_2+1)}{\Gamma(y_1+y_2+4)} = \frac{2(y_1+y_2)!}{(y_1+y_2+3)!}, \end{aligned}$$

and so

$$p_{X|Y}(x|y) = \frac{p(x,y_1,y_2)}{p_Y(y_1,y_2)} = \frac{(y_1+y_2+3)!}{2(y_1+y_2)!}x^2(1-x)^{y_1+y_2}.$$

c) Using the formula in the hint,

$$E(X|Y=y) = \int_0^1 \frac{(y_1+y_2+3)!}{2(y_1+y_2)!}x^3(1-x)^{y_1+y_2}\,dx = \frac{3}{y_1+y_2+4}$$

and

$$\begin{aligned} E(X^2|Y=y) &= \int_0^1 \frac{(y_1+y_2+3)!}{2(y_1+y_2)!}x^4(1-x)^{y_1+y_2}\,dx \\ &= \frac{12}{(y_1+y_2+4)(y_1+y_2+5)}. \end{aligned}$$

So,

$$E(X|Y) = \frac{3}{Y_1 + Y_2 + 4},$$

and

$$E(X^2|Y) = \frac{12}{(Y_1 + Y_2 + 4)(Y_1 + Y_2 + 5)}.$$

d) Since

$$EX = \frac{1}{2} = EE(X|Y) = E\left[\frac{3}{Y_1 + Y_2 + 4}\right],$$

$$E\left[\frac{1}{Y_1 + Y_2 + 4}\right] = \frac{1}{6}.$$

11. Given $M = m$ and $Z = z$, the conditional distribution for $(X, Y)$ must concentrate on the two points $(m, z)$ and $(z, m)$. By symmetry, a natural guess is

$$\begin{aligned}\frac{1}{2} &= P(X = m, Y = z | M = m, Z = z) \\ &= P(X = z, Y = m | M = m, Z = z).\end{aligned}$$

To see that this is correct we need to check that smoothing works. Let $h$ be an arbitrary function and define $g(x, y) = h(x, y, x \vee y, x \wedge y)$ so that $h(X, Y, M, Z) = g(X, Y)$. Then

$$\begin{aligned}E\bigl[h(X, Y, M, Z) \bigm| M, Z\bigr] &= E\bigl[g(X, Y) \bigm| M, Z\bigr] \\ &= \frac{1}{2} g(M, Z) + \frac{1}{2} g(Z, M).\end{aligned}$$

So smoothing works if $Eg(X, Y) = E\bigl[\frac{1}{2} g(M, Z) + \frac{1}{2} g(Z, M)\bigr]$, that is, if

$$\begin{aligned}&\iint g(x, y) f(x) f(y)\, dx\, dy \\ &\quad = \iint \frac{1}{2}\bigl[g(x \vee y, x \wedge y) + g(x \wedge y, x \vee y)\bigr] f(x) f(y)\, dx\, dy \\ &\quad = \iint \frac{1}{2}\bigl[g(x, y) + g(y, x)\bigr] f(x) f(y)\, dx\, dy.\end{aligned}$$

This holds because $\iint g(x, y)\, dx\, dy = \iint g(y, x)\, dx\, dy$, and the stated conditional distribution is correct.

12. From the example of Section 3.2, $T$ and $U = X/(X + Y)$ are independent with $U \sim \text{Unif}(0, 1)$. Then

$$\begin{aligned}E\bigl[f(X, Y) \bigm| T = t\bigr] &= E\bigl[f\bigl(TU, T(1 - U)\bigr) \bigm| T = t\bigr] \\ &= \int_0^1 f\bigl(tu, t(1 - u)\bigr)\, du.\end{aligned}$$

(It is easy to check that smoothing works by viewing all expectations as integrals against the joint density of $T$ and $U$.)

14. a) Integrating against $y$,

$$p_X(x) = \int p(x,y)\,dy = \int_0^x e^{-x}\,dy = xe^{-x}, \qquad x > 0,$$

and integrating against $x$,

$$p_Y(y) = \int p(x,y)\,dx = \int_y^\infty e^{-x}\,dx = e^{-y}, \qquad y > 0.$$

b) Integration against the marginal density gives

$$EY = \int_0^\infty ye^{-y}\,dy = 1 \text{ and } EY^2 = \int_0^\infty y^2e^{-y}\,dy = 2.$$

c) Dividing the joint density by the marginal density,

$$p_{Y|X}(y|x) = p(x,y)/p_X(x) = \frac{1}{x}, \qquad y \in (0,x).$$

Integrating against this conditional density

$$E[Y|X=x] = \int_0^x \frac{y}{x}\,dy = \frac{x}{2}$$

and

$$E[Y^2|X=x] = \int_0^x \frac{y^2}{x}\,dy = \frac{x^2}{3}.$$

So $E[Y|X] = X/2$, and $E[Y^2|X] = X^2/3$.
d) Integrating against the marginal density of $X$,

$$EE[Y|X] = \frac{1}{2}EX = \frac{1}{2}\int_0^\infty x^2e^{-x}\,dx = 1,$$

and

$$EE[Y^2|X] = \frac{1}{3}EX^2 = \frac{1}{3}\int_0^\infty x^3e^{-x}\,dx = 2.$$

## B.7 Problems of Chapter 7

1. The likelihood is $p_\theta(x) = \theta^{T(x)}e^{-n\theta}/\prod_1^n x_i!$, where $T(x) = \sum_{i=1}^n x_i$. So the Bayes estimator is

$$\delta(x) = \frac{\int_0^\infty \theta^{p+1}p_\theta(x)\lambda(\theta)\,d\theta}{\int_0^\infty \theta^p p_\theta(x)\lambda(\theta)\,d\theta} = \frac{\int_0^\infty \theta^{T(x)+p+1}e^{-(n+\eta)\theta}\,d\theta}{\int_0^\infty \theta^{T(x)+p}e^{-(n+\eta)\theta}\,d\theta} = \frac{T(x)+p+1}{n+\eta}.$$

2. The marginal density of $X$ is

$$
\begin{aligned}
q(x) &= \int p_\theta(x)\lambda(\theta)\,d\theta \\
&= \int_x^\infty \frac{1}{\theta(1+\theta)^2}\,d\theta \\
&= \int_x^\infty \left[\frac{1}{\theta} - \frac{1}{1+\theta} - \frac{1}{(1+\theta)^2}\right] d\theta \\
&= \log\left(\frac{1+x}{x}\right) - \frac{1}{1+x}.
\end{aligned}
$$

So $p(\theta|x) = 1/\bigl[\theta(1+\theta)^2 q(x)\bigr]$, $\theta > x$, and

$$
\begin{aligned}
E\bigl[|\Theta - d| \bigm| X = x\bigr] &= \int_x^\infty \frac{|\theta - d|}{\theta(1+\theta)^2 q(x)}\,d\theta \\
&= \int_x^d \frac{d-\theta}{\theta(1+\theta)^2 q(x)}\,d\theta + \int_d^\infty \frac{\theta - d}{\theta(1+\theta)^2 q(x)}\,d\theta.
\end{aligned}
$$

Note that these integrals are like the integral for $q$ when $d$ is in the numerator and are easy to integrate when $\theta$ is in the numerator. After a bit of algebra,

$$
E\bigl[|\Theta - d| \bigm| X = x\bigr] = d - \frac{2dq(d)}{q(x)} - \frac{1}{(1+x)q(x)} + \frac{2}{(1+d)q(x)}.
$$

Since $q'(d) = -1/\bigl[d(1+d)^2\bigr]$, the derivative of this expression is $1 - 2q(d)/q(x)$. This function is strictly increasing from $-1$ to 1 as $d$ varies from $x$ to infinity. So it will have a unique zero, and this zero determines the Bayes estimator: $q\bigl(\delta_\Lambda(X)\bigr) = q(X)/2$. From this equation,

$$
P[\delta_\Lambda(X) < \Theta | X = x] = \int_{\delta_\Lambda(x)}^\infty \frac{1}{\theta(1+\theta)^2 q(x)}\,d\theta = \frac{q\bigl(\delta_\Lambda(x)\bigr)}{q(x)} = \frac{1}{2}.
$$

3. Completing squares, the conditional density is proportional to

$$
\begin{aligned}
&\lambda(\theta)p_\theta(y) \\
&\quad \propto_\theta \exp\left[-\frac{\theta_1^2}{2\tau_1^2} - \frac{\theta_2^2}{2\tau_2^2} - \frac{n\theta_1^2}{2\sigma^2} - \frac{\theta_2^2}{2\sigma^2}\sum_{i=1}^n x_i^2 + \frac{\theta_1}{\sigma^2}\sum_{i=1}^n y_i + \frac{\theta_2}{\sigma^2}\sum_{i=1}^n x_i y_i\right] \\
&\quad \propto_\theta \exp\left[-\frac{\left[\theta_1 - \dfrac{\sum_{i=1}^n y_i}{n + \sigma^2/\tau_1^2}\right]^2}{2\bigl(n/\sigma^2 + 1/\tau_1^2\bigr)^{-1}} - \frac{\left[\theta_2 - \dfrac{\sum_{i=1}^n x_i y_i}{\sum_{i=1}^n x_i^2 + \sigma^2/\tau_2^2}\right]^2}{2\bigl(\sum_{i=1}^n x_i^2/\sigma^2 + 1/\tau_2^2\bigr)^{-1}}\right].
\end{aligned}
$$

So given the data, $\Theta_1$ and $\Theta_2$ are independent normal variables. The Bayes estimates are the posterior means,

$$E[\Theta_1|X,Y] = \frac{\sum_{i=1}^n Y_i}{n+\sigma^2/\tau_1^2} \text{ and } E[\Theta_2|X,Y] = \frac{\sum_{i=1}^n x_iY_i}{\sum_{i=1}^n x_i^2 + \sigma^2/\tau_2^2}.$$

4. a) Since $\lambda'(\theta) = \big[\alpha - \beta A'(\theta)\big]\lambda(\theta)$,

$$0 = \int_\Omega \lambda'(\theta)\,d\theta = E\big[\alpha - \beta A'(\Theta)\big].$$

So $EA'(\Theta) = \alpha/\beta$.
b) The joint density (or likelihood) is proportional to $e^{\theta n\overline{T} - nA(\theta)}$. Multiplying by $\lambda_{\alpha,\beta}$, the conditional density of $\Theta$ given $X = x$ is proportional to

$$e^{(\alpha+n\overline{T})\theta - (\beta+n)A(\theta)} \propto_\theta \lambda_{\alpha+n\overline{T},\beta+n}.$$

So $\Theta|X = x \sim \Lambda_{\alpha+n\overline{T},\beta+n}$. Using the result from part (a), the Bayes estimator of $A'(\Theta)$ is

$$E\big[A'(\Theta)|X\big] = \frac{\alpha+n\overline{T}}{\beta+n} = \frac{\beta}{\beta+n}\frac{\alpha}{\beta} + \frac{n}{\beta+n}\overline{T},$$

where the last equality expresses this estimator as a weighted average of $EA'(\Theta) = \alpha/\beta$ and $\overline{T}$.
c) Since $p_\theta(x) = \theta e^{-\theta x}$, $x > 0$, we should take $T(x) = -x$, and $A(\theta) = -\log\theta$. Then

$$\lambda_{\alpha,\beta}(\theta) \propto_\theta e^{\alpha\theta + \beta\log\theta} = \theta^\beta e^{\alpha\beta}, \qquad \theta > 0.$$

For convergence, $\alpha$ should be negative. This density is proportional to a gamma density, and so $\Lambda_{\alpha,\beta}$ is the gamma distribution with shape parameter $\beta+1$ and failure rate $-\alpha$. Since $1/\theta = -A'(\theta)$, the Bayes estimator, using results from part (b), is

$$-E\big[A'(\Theta)|X\big] = -\frac{\alpha+n\overline{T}}{\beta+n} = \frac{|\alpha| + n\overline{X}}{\beta+n}.$$

6. a) The joint density is $\lambda(\theta)p_\theta(x) = f_\theta(x)/2$. Integrating (summing) out $\theta$, the marginal density of $X$ is $q(x) = [f_0(x) + f_1(x)]/2$. So the conditional density of $\Theta$ given $X$ is

$$\lambda(\theta|x) = \frac{\lambda(\theta)p_\theta(x)}{q(x)} = \frac{f_\theta(x)}{f_0(x)+f_1(x)}, \qquad \theta = 0, 1.$$

This is the mass function for a Bernoulli distribution with success probability $p = p(x) = f_1(x)/\big(f_0(x) + f_1(x)\big)$. The Bayes estimator under squared error loss is the mean of this conditional distribution,

$$E(\Theta|X) = \frac{f_1(X)}{f_0(X) + f_1(X)}.$$

b) From Theorem 7.1, the Bayes estimator should minimize the posterior risk. For zero-one loss the estimator should be one if $p(X) > 1/2$, and zero if $p(X) < 1/2$. If $p(X) = 1/2$, either is optimal. Equivalently, $\delta$ should be one if $f_1(X) > f_0(X)$, and should be zero if $f_1(X) < f_0(X)$.

7. Using Theorem 7.1, the Bayes estimator $\delta(x)$ should be the value $d$ minimizing

$$E\left[\frac{(d-\Theta)^2}{d}\,\middle|\, X = x\right] = d - 2E[\Theta|X = x] + \frac{E[\Theta^2|X = x]}{d}.$$

Setting the derivative to zero, $\delta = \sqrt{E[\Theta^2|X]}$. If $T(x) = x_1 + \cdots + x_n$, then since $\lambda(\theta|x) \propto \lambda(\theta)p_\theta(x) \propto \theta^n e^{-[1+T(x)]\theta}$,

$$E[\Theta^2|X] = \frac{\int_0^\infty \theta^{n+2}e^{-(1+T)\theta}\,d\theta}{\int_0^\infty \theta^n e^{-(1+T)\theta}\,d\theta} = \frac{(n+1)(n+2)}{(1+T)^2}.$$

So the Bayes estimator is $\sqrt{n^2+3n+2}/(1+T)$.

16. a) Let $f$ and $F$ be the density and cumulative distribution functions for the standard Cauchy distribution, so the respective density and cumulative distribution functions of $X$ (or $Y$) are $f(x-\theta)$ and $F(x-\theta)$. By smoothing,

$$P(A \le x) = EP(X \le 2x - Y|Y) = EF(2x - Y - \theta).$$

Taking $d/dx$, $A$ has density

$$\begin{aligned} 2Ef(2x - Y - \theta) &= \int \frac{2\,dy}{\pi^2\big[(2x-y-\theta)^2+1\big]\big[y^2+1\big]} \\ &= \frac{1}{\pi\big[(x-\theta)^2+1\big]}. \end{aligned}$$

So $A$ and $X$ have the same density.

b) We have $|A-\theta| = |X-\theta|$ along two lines: $Y-\theta = X-\theta$ and $Y-\theta = -3(X-\theta)$. These lines divide the $(X,Y)$ plane into four regions: two where $|A-\theta| < |X-\theta|$ (boundaries for these two regions form obtuse angles), and two where $|A-\theta| > |X-\theta|$. Similarly, the lines $Y-\theta = X-\theta$ and $Y-\theta = -(X-\theta)$ divide the plane into four regions. But these regions are symmetric, and so the chance $(X,Y)$ lies any of them is $1/4$. The region where $|A-\theta| < |X-\theta|$ contains two of the symmetric regions, and so $P(|A-\theta| < |X-\theta|) > 1/2$.

## B.8 Problems of Chapter 8

1. By the covariance inequality (equation (4.11)) $\big|\mathrm{Cov}(X_i, X_j)\big| \le \sigma^2$. Of course, by the independence, this covariance must be zero if $|i-j| \ge m$. So for any $i$, $\sum_{j=1}^n \mathrm{Cov}(X_i, X_j) \le (2m-1)\sigma^2$. Thus

$$E(\overline{X}_n - \xi)^2 = \frac{1}{n^2}\operatorname{Var}\left(\sum_{i=1}^n X_i\right) = \frac{1}{n^2}\sum_{i=1}^n\sum_{j=1}^n \operatorname{Cov}(X_i, X_j)$$
$$\leq \frac{1}{n^2}\sum_{i=1}^n (2m-1)\sigma^2 = \frac{(2m-1)\sigma^2}{n} \to 0.$$

2. Since $\log(1-u)/u \to -1$ as $u \to 0$ and

$$P(M_n \leq x) = P(X_i \leq x, i = 1, \dots, n) = (1 - e^{-\lambda x})^n,$$

if $\epsilon > 0$,

$$\log P\left(\frac{\log n}{M_n} \geq \lambda + \epsilon\right) = n^{\epsilon/(\lambda+\epsilon)} \frac{\log\bigl[1 - e^{-\lambda \log(n)/(\lambda+\epsilon)}\bigr]}{e^{-\lambda\log(n)/(\lambda+\epsilon)}} \to -\infty.$$

So $P(\log n/M_n \geq \lambda + \epsilon) \to 0$. Similarly, if $\epsilon \in (0, \lambda)$,

$$\log P\left(\frac{\log n}{M_n} > \lambda - \epsilon\right) = n^{-\epsilon/(\lambda-\epsilon)} \frac{\log\bigl[1 - e^{-\lambda \log(n)/(\lambda-\epsilon)}\bigr]}{e^{-\lambda\log(n)/(\lambda-\epsilon)}} \to 0.$$

So $P(\log n/M_n > \lambda - \epsilon) \to 1$, and $P(\log n/M_n \leq \lambda - \epsilon) \to 0$. Hence $\log(n)/M_n$ is consistent.

3. Because $M_n$ lies between 0 and $\theta$, $n(\hat{\theta} - \theta)$ lies between $-n\theta$ and $\theta$. So $P_\theta\bigl(n(\hat{\theta}-\theta) \leq y\bigr)$ is one if $y \geq \theta$. If $y \leq \theta$, then for $n$ sufficiently large,

$$P_\theta\bigl(n(\hat{\theta}-\theta) \leq y\bigr) = P_\theta\left(M_n \leq \theta - \frac{\theta - y}{n+1}\right)$$
$$= \left(1 - \frac{1 - y/\theta}{n+1}\right)^n \to e^{y/\theta - 1}.$$

So $Y_n \to Y$ where $Y$ has cumulative distribution function $H_\theta$ given by $H_\theta(y) = \min\{1, e^{y/\theta - 1}\}$.

4. a) By the central limit theorem, $\sqrt{n}(\hat{p} - p) \Rightarrow N\bigl[0, p(1-p)\bigr]$, and so by the delta method, Proposition 8.14, with $f(x) = x^2$, $\sqrt{n}(\hat{p}_n^2 - p^2) \Rightarrow N\bigl[0, 4p^3(1-p)\bigr]$.
b) $T_n = X_1 + \cdots + X_n$ is complete sufficient and $4X_1X_2X_3(1-X_4)$ is unbiased. The UMVU estimator is $\delta_n(T) = E[4X_1X_2X_3(1-X_4)|T]$, given explicitly by

$$\delta_n(t) = \frac{4P(X_1 = X_2 = X_3 = 1, X_4 = 0, X_5 + \cdots + X_n = t-3)}{P(T = t)}$$
$$= \frac{4t(t-1)(t-2)(n-t)}{n(n-1)(n-2)(n-3)}.$$

c) The difference $n\delta_n - n\hat{\sigma}_n^2$ is

$$\frac{4T(T-1)(T-2)(n-T)}{(n-1)(n-2)(n-3)} - \frac{4T^3(n-T)}{n^3}$$
$$= \frac{4T(n-T)\big[T^2(6n^2-11n+6)-3Tn^3+2n^3\big]}{n^3(n-1)(n-2)(n-3)}$$
$$= \frac{4n^3\hat{p}_n(1-\hat{p}_n)}{(n-1)(n-2)(n-3)}\left(\hat{p}_n^2\frac{6n^2-11n+6}{n^2} - 3\hat{p}_n + \frac{2}{n}\right),$$

which converges in probability to $4p(1-p)(6p^2-3p) = 27/32$. By the central limit theorem, $\sqrt{n}(\hat{p}_n - p) \Rightarrow Z \sim N(0, 3/16)$. Now $\hat{\sigma}^2 = f(\hat{p}_n)$ with $f(p) = 4p^3(1-p)$, and since $f'(3/4) = 0$, a two-term Taylor approximation is necessary to derive the limiting distribution using the delta method. With a suitable intermediate value $p_n$, lying between $p$ and $\hat{p}_n$,

$$n(\hat{\sigma}^2 - \sigma^2) = n\big(f(\hat{p}_n) - f(p)\big) = \frac{1}{2}\big[\sqrt{n}(\hat{p}_n - p)\big]^2 f''(p_n).$$

Because $p_n \xrightarrow{p} p$ and $f''$ is continuous, $f''(p_n) \xrightarrow{p} f''(p) = -9$. Using Theorem 8.13, $n(\hat{\sigma}^2 - \sigma^2) \Rightarrow -9Z^2/2$ and $n(\delta_n - \sigma^2) \Rightarrow 27/32 - 9Z^2/2$.

5. If $\epsilon > 0$,

$$P(X_i \geq \theta + \epsilon) = \int_{\theta+\epsilon}^{\infty} (x-\theta)e^{\theta - x}\,dx = (1+\epsilon)e^{-\epsilon} < 1,$$

and so

$$P(M_n \geq \theta + \epsilon) = P(X_i \geq \theta + \epsilon, i = 1, \dots, n) = \big[(1+\epsilon)e^{-\epsilon}\big]^n \to 0.$$

Since $M_n \geq \theta$, $M_n$ is consistent. Next, for $x > 0$,

$$P\big(\sqrt{n}(M_n - \theta) > x\big) = \left[\left(1 + \frac{x}{\sqrt{n}}\right)e^{-x/\sqrt{n}}\right]^n.$$

To evaluate the limit, we use the facts that if $c_n \to c$, then $(1+c_n/n)^n \to e^c$, and that $\big[(1+u)e^{-u} - 1\big]/u^2 \to -1/2$, which follows from Taylor expansion or l'Hôpital's rule. Since

$$n\left[\left(1 + \frac{x}{\sqrt{n}}\right)e^{-x/\sqrt{n}} - 1\right] = x^2\frac{(1+x/\sqrt{n})e^{-x/\sqrt{n}} - 1}{(x/\sqrt{n})^2} \to -x^2/2,$$

$P\big(\sqrt{n}(M_n - \theta) > x\big) \to e^{-x^2/2}$. So, $\sqrt{n}(M_n - \theta) > x\big) \Rightarrow Y$, where $Y$ has cumulative distribution function $P(Y \leq y) = 1 - e^{-y^2/2}$, $y > 0$ (a Weibull distribution).

6. Let $F$ denote the cumulative distribution function of $Y$, and let $y > 0$ be a continuity point of $F$. If $y + \epsilon$ is also a continuity point of $F$, then since

$$\{A_nY_n \leq y\} \subset \{Y_n \leq y + \epsilon\} \cup \{A_n \leq y/(y+\epsilon)\},$$

we have

$$P(A_nY_n \le y) \le P(Y_n \le y+\epsilon) + P\big(A_n \le y/(y+\epsilon)\big) \to F(y+\epsilon).$$

From this, $\limsup P(A_nY_n \le y) \le F(y+\epsilon)$. Because $F$ is continuous at $y$ and $\epsilon$ can be arbitrarily small ($F$ can have at most a countable number of discontinuities), $\limsup P(A_nY_n \le y) \le F(y)$. Similarly, if $y-\epsilon$ is positive and a continuity point of $F$, then

$$P(A_nY_n \le y) \ge P(Y_n \le y-\epsilon) - P\big(A_n < 0 \text{ or } A_n \ge y/(y-\epsilon)\big) \to F(y-\epsilon).$$

From this, $\liminf P(A_nY_n \le y) \ge F(y-\epsilon)$, and since $\epsilon$ can be arbitrarily small, $\liminf P(A_nY_n \le y) \ge F(y)$. Thus

$$\lim P(A_nY_n \le y) = F(y).$$

Similar arguments show that

$$\lim P(A_nY_n \le y) = F(y)$$

when $y$ is negative or zero with $F$ continuous at $y$.

16. The log-likelihood $l(\alpha,\beta,\sigma^2)$ is

$$-\frac{1}{2\sigma^2}\sum_{i=1}^n Y_i^2 + \frac{\alpha}{\sigma^2}\sum_{i=1}^n Y_i + \frac{\beta}{\sigma^2}\sum_{i=1}^n x_iY_i - \frac{n\alpha^2}{2\sigma^2}$$
$$-\frac{\beta^2}{2\sigma^2}\sum_{i=1}^n x_i^2 - \frac{\alpha\beta}{\sigma^2}\sum_{i=1}^n x_i - n\log\sqrt{2\pi\sigma^2}.$$

With any fixed value for $\sigma^2$ this is a quadratic function of $\alpha$ and $\beta$, maximized when both partial derivatives are zero. From the form, the answer is the same regardless of the value for $\sigma^2$. This gives the following equations for $\hat\alpha$ and $\hat\beta$: $-2\sum_{i=1}^n Y_i + 2n\hat\alpha + 2\hat\beta\sum_{i=1}^n x_i = 0$ and $-2\sum_{i=1}^n x_iY_i + 2\hat\beta\sum_{i=1}^n x_i^2 + 2\hat\alpha\sum_{i=1}^n x_i = 0$. Solving, $\hat\beta = \big(\sum_{i=1}^n x_iY_i - n\overline{x}\overline{Y}\big)/\big(\sum_{i=1}^n x_i^2 - n\overline{x}^2\big)$ and $\hat\alpha = \overline{Y} - \hat\beta\overline{x}$. Next, $\hat\sigma^2$ must maximize $l(\hat\alpha,\hat\beta,\sigma^2) = -\sum_{i=1}^n e_i^2/(2\sigma^2) - n\log\sqrt{2\pi\sigma^2}$, where $e_i = Y_i - \hat\alpha - \hat\beta x_i$ (called the $i$th residual). The derivative here with respect to $\sigma^2$ is $\sum_{i=1}^n e_i^2/(2\sigma^4) - n/(2\sigma^2)$ which has a unique zero when $\sigma^2$ is $\hat\sigma^2 = \sum_{i=1}^n e_i^2/n$. Note that the function goes to $-\infty$ as $\sigma^2 \downarrow 0$ or as $\sigma^2 \to \infty$. So this value must give the maximum, and $\hat\sigma^2$ is thus the maximum likelihood estimator of $\sigma^2$.

17. The likelihood is

$$\frac{1}{\sqrt{2\pi}^n}\exp\left[-\frac{X_1^2}{2} - \frac{1}{2}\sum_{i=1}^{n-1}(X_{i+1} - \rho X_i)^2\right]$$
$$\propto \exp\left[\rho\sum_{i=1}^{n-1} X_iX_{i+1} - \frac{\rho^2}{2}\sum_{i=1}^{n-1} X_i^2\right].$$

Maximizing the quadratic function in the exponential, the maximum likelihood estimator is

$$\hat{\rho} = \sum_{i=1}^{n-1} X_i X_{i+1} \Bigg/ \sum_{i=1}^{n-1} X_i^2.$$

19. By the central limit theorem, $\sqrt{n}(\overline{X}_n - 1/\theta) \Rightarrow N(0, 1/\theta^2)$. Using the delta method with $f(x) = 1/x$, $\sqrt{n}(1/\overline{X}_n - \theta) \Rightarrow N(0, \theta^2)$. Next, by the central limit theorem, $\sqrt{n}(\hat{p}_n - e^{-\theta}) \Rightarrow N\bigl[0, e^{-\theta}(1 - e^{-\theta})\bigr]$, and so by the delta method with $f(x) = -\log x$, $\sqrt{n}(-\log \hat{p}_n - \theta) \Rightarrow N(0, e^{\theta} - 1)$. So the asymptotic relative efficiency is $\theta^2/(e^{\theta} - 1)$.
20. By the central limit theorem, $\sqrt{n}(\overline{X}_n - \theta) \Rightarrow N(0, \theta)$, and so by the delta method with $f(x) = x(x+1)$, $\sqrt{n}\bigl[\overline{X}_n(\overline{X}_n + 1) - \theta(\theta + 1)\bigr] \Rightarrow N(0, 4\theta^3 + 4\theta^2 + \theta)$. Next, let $Z = (X_1 - \theta)/\sqrt{\theta} \sim N(0,1)$. Then $\operatorname{Var}(X_1^2) = \operatorname{Var}(\theta Z^2 + 2\theta^{3/2} Z + \theta^2) = \operatorname{Var}(\theta Z^2) + \operatorname{Var}(2\theta^{3/2} Z) + 2\operatorname{Cov}(\theta Z^2, 2\theta^{3/2} Z) = 2\theta^2 + 4\theta^3 + 0$, and by the central limit theorem, $\sqrt{n}\bigl(\delta_n - \theta(\theta+1)\bigr) \Rightarrow N(0, 4\theta^3 + 2\theta^2)$. So the asymptotic relative efficiency of $\overline{X}_n(\overline{X}_n + 1)$ with respect to $\delta_n$ is $(4\theta^3 + 2\theta^2)/(4\theta^3 + 4\theta^2 + \theta)$.
21. Since $\operatorname{Var}(X_i^2) = 2\sigma^4$ and $\hat{\sigma}^2 = \sum_{i=1}^n X_i^2/n$, by the central limit theorem, $\sqrt{n}(\hat{\sigma}^2 - \sigma^2) \Rightarrow N(0, 2\sigma^4)$. By the delta method with $f$ the square root function, $\sqrt{n}(\hat{\sigma} - \sigma) \Rightarrow N(0, \sigma^2/2)$. By Theorem 8.18, $\sqrt{n}(Q_n - c\sigma) \Rightarrow N\bigl[0, 3\sigma^2/\bigl(16\phi^2(c)\bigr)\bigr]$. So $\sqrt{n}(\tilde{\sigma} - \sigma) \Rightarrow N\bigl[0, 3\sigma^2/\bigl(16c^2\phi^2(c)\bigr)\bigr]$. The asymptotic relative efficiency of $\tilde{\sigma}$ with respect to $\hat{\sigma}$ is $8c^2\phi^2(c)/3 = 0.1225$ $(c = 0.6745)$.
24. a) For $x > 0$, $P(|X_i| \le x) = \Phi(x/\sigma) - \Phi(-x/\sigma) = 2\Phi(x/\sigma) - 1$. So $|X_i|$ has density $2\phi(x/\sigma)/\sigma$, $x > 0$, and median $\Phi^{\leftarrow}(3/4)\sigma = 0.6745\sigma$. Hence $\tilde{\sigma} = cM \xrightarrow{p} 0.6745c\sigma$. This estimator will be consistent if $c = 1/0.6745 = 1.4826$.
    b) By (8.5),
    $$\sqrt{n}(M - 0.6745\sigma) \Rightarrow N\bigl[0, \sigma^2/\bigl((16\phi^2(0.6745)\bigr)\bigr] = N(0, 0.6189\sigma^2),$$
    and so
    $$\sqrt{n}(\tilde{\sigma} - \sigma) \Rightarrow N(0, 1.3604\sigma^2).$$
    c) The log-likelihood is
    $$l_n(\sigma) = -\frac{1}{2\sigma^2}\sum_{i=1}^n X_i^2 - n\log(\sqrt{2\pi}\sigma).$$
    Setting $l_n'(\sigma)$ to zero, the maximum likelihood estimator is $\hat{\sigma} = \sum_{i=1}^n X_i^2/n$. The summands in $\hat{\sigma}^2$ have common variance $2\sigma^4$, and by the central limit theorem,
    $$\sqrt{n}(\hat{\sigma}^2 - \sigma^2) \Rightarrow N(0, 2\sigma^4).$$
    Using the delta method,

$$\sqrt{n}(\hat{\sigma} - \sigma) \Rightarrow N(0, \sigma^2/2).$$

d) Dividing the variances, the asymptotic relative efficiency is

$$\frac{\sigma^2/2}{1.3604\sigma^2} = 0.3675.$$

## B.9 Problems of Chapter 9

8. Define $S_x^2 = \sum_{i=1}^m (X_i - \overline{X})^2/(m-1)$ and $S_y^2 = \sum_{j=1}^n (Y_i - \overline{Y})^2/(n-1)$. Then $(m-1)S_x^2/\sigma_x^2 \sim \chi_{m-1}^2$ and $(n-1)S_y^2/\sigma_y^2 \sim \chi_{n-1}^2$ are independent, and $F = \sigma_y^2 S_x^2/(\sigma_x^2 S_y^2)$ has an $F$ distribution with $m-1$ and $n-1$ degrees of freedom. The distribution of $F$ does not depend on unknown parameters. Therefore it is a pivot. If $c_1$ and $c_2$ are the $(\alpha/2)$th and $(1-\alpha/2)$th quantiles for this $F$ distribution, then $P(c_1 < F < c_2) = 1 - \alpha$. This event, $c_1 < F < c_2$, occurs if and only if

$$\frac{\sigma_x}{\sigma_y} \in \left( \sqrt{\frac{S_x^2}{c_2 S_y^2}}, \sqrt{\frac{S_x^2}{c_1 S_y^2}} \right),$$

and this is the desired $1 - \alpha$ confidence interval.
9. a) The likelihood function is $\theta^{-n} I\{X_{(n)} \le \theta\}$, maximized at $\hat{\theta} = X_{(n)}$.
b) For $x \in (0,1)$, $P(\hat{\theta}/\theta \le x) = P(X_{(n)} \le x\theta) = x^n$, and so $\hat{\theta}/\theta$ is a pivot. The $(\alpha/2)$th and $(1-\alpha/2)$th quantiles for this distribution are $(\alpha/2)^{1/n}$ and $(1-\alpha/2)^{1/n}$. So

$$\begin{aligned} 1 - \alpha &= P\big[(\alpha/2)^{1/n} < \hat{\theta}/\theta < (1-\alpha/2)^{1/n}\big] \\ &= P\big[\theta \in \big(\hat{\theta}/(1-\alpha/2)^{1/n},\, \hat{\theta}/(\alpha/2)^{1/n}\big)\big]. \end{aligned}$$

Hence $\big(\hat{\theta}/(1-\alpha/2)^{1/n}, \hat{\theta}/(\alpha/2)^{1/n}\big)$ is the desired $1-\alpha$ confidence interval.
10. For $t > 0$,

$$P(\theta X_i \le t) = P(X_i \le t/\theta) = \int_0^{t/\theta} \theta e^{-\theta x} = 1 - e^{-t},$$

and so $\theta X_i$ has a standard exponential distribution. So $\theta T$ has a gamma distribution with density $x^{n-1}e^{-x}/\Gamma(n)$, $x > 0$. If $\gamma_{\alpha/2}$ and $\gamma_{1-\alpha/2}$ are the upper and lower $(\alpha/2)$th quantiles for this distribution, then

$$P(\gamma_{1-\alpha/2} < \theta T < \gamma_{\alpha/2}) = P\big(\theta \in (\gamma_{1-\alpha/2}/T, \gamma_{\alpha/2}/T)\big) = 1 - \alpha,$$

which shows that $(\gamma_{1-\alpha/2}/T, \gamma_{\alpha/2}/T)$ is a $1 - \alpha$ confidence interval for $\theta$.

11. a) The cumulative distribution function of $Y = (X - \theta)/\sigma$ is

$$F_Y(y) = P\left(\frac{X-\theta}{\sigma} \le y\right) = P(X \le \theta + \sigma y)$$
$$= \int_{-\infty}^{\theta+\sigma y} g\left(\frac{x-\theta}{\sigma}\right)\frac{dx}{\sigma} = \int_{-\infty}^{y} g(u)\,du.$$

From this, $Y$ has density $g$.
b) Let $Y_i = (X_i - \theta)/\sigma$, $i = 1, 2$. Then $X_i = \theta + \sigma Y_i$, and using this, $W = (Y_1 + Y_2)/|Y_1 - Y_2|$. Since $Y_1$ and $Y_2$ are independent, $(Y_1, Y_2)$ has joint density $g(y_1)g(y_2)$, which does not depend on $\theta$ or $\sigma$. Because $W$ is a function of $Y_1$ and $Y_2$, its distribution does not depend on $\theta$ or $\sigma$.
c) Let $q_{\alpha/2}$ and $q_{1-\alpha/2}$ denote the upper and lower $(\alpha/2)$th quantiles for the distribution of $W$. Then

$$1-\alpha = P\left(q_{1-\alpha/2} < \frac{X_1 + X_2 - 2\theta}{|X_1 - X_2|} < q_{\alpha/2}\right)$$
$$= P\left[\theta \in \left(\overline{X} - \tfrac{1}{2}|X_1 - X_2|q_{\alpha/2}, \overline{X} + \tfrac{1}{2}|X_1 - X_2|q_{1-\alpha/2}\right)\right],$$

and the interval in this expression is a $1-\alpha$ confidence interval for $\theta$.
d) The variable $V = |X_1 - X_2|/\sigma = |Y_1 - Y_2|$ is a pivot. If $q_{\alpha/2}$ and $q_{1-\alpha/2}$ are the upper and lower $(\alpha/2)$th quantiles for the distribution of $V$, then

$$1-\alpha = P\left(q_{1-\alpha/2} < \frac{|X_1 - X_2|}{\sigma} < q_{\alpha/2}\right)$$
$$= P\left[\sigma \in \left(\frac{|X_1 - X_2|}{q_{\alpha/2}}, \frac{|X_1 - X_2|}{q_{1-\alpha/2}}\right)\right],$$

and the interval in this expression is a $1-\alpha$ confidence interval for $\sigma$.

12. By the addition law,

$$\begin{aligned} P_\theta\big(g(\theta) \in S_1 \cap S_2\big) &= P_\theta\big(g(\theta) \in S_1\big) + P_\theta\big(g(\theta) \in S_2\big) \\ &\quad - P_\theta\big(g(\theta) \in S_1 \cup S_2\big) \\ &\ge P_\theta\big(g(\theta) \in S_1\big) + P_\theta\big(g(\theta) \in S_2\big) - 1 \\ &\ge 1 - 2\alpha. \end{aligned}$$

13. a) Multiplying the marginal density of $X_i$ times the conditional density of $Y_i$ given $X_i$, the joint density of $(X_i, Y_i)$ is $\exp\{-x^2/2 - (y - x\theta)^2/2\}/(2\pi)$. So the joint density for the entire sample is $\prod_{i=1}^n \big[\exp\{-x_i^2/2 - (y_i - x_i\theta)^2/2\}/(2\pi)\big]$, and the log-likelihood function is

$$l_n(\theta) = -\frac{1}{2}\sum_{i=1}^n X_i^2 - \frac{1}{2}\sum_{i=1}^n (Y_i - X_i\theta)^2 - n\log(2\pi).$$

This is a quadratic function of $\theta$, maximized when

$$l_n'(\theta) = -2\sum_{i=1}^n X_i(Y_i - X_i\theta) = 0.$$

This gives $\hat\theta = \sum_{i=1}^n X_iY_i / \sum_{i=1}^n X_i^2$.
b) Fisher information is

$$I(\theta) = -E_\theta \frac{\partial^2 \log f_\theta(X,Y)}{\partial\theta^2} = E_\theta X^2 = 1.$$

c) As $n \to \infty$, $\sqrt{n}(\hat\theta - \theta) \Rightarrow N(0,1)$.
d) $(\hat\theta \pm z_{\alpha/2}/\sqrt{n})$.
e) The observed Fisher information is $-l_n''(\hat\theta) = \sum_{i=1}^n X_i^2$, and the associated confidence interval is $\bigl(\hat\theta \pm z_{\alpha/2}/\bigl[\sum_{i=1}^n X_i^2\bigr]^{1/2}\bigr)$. The main difference is that now the width of the interval varies according to the observed information.
f) Given $X_1 = x_1, \dots, X_n = x_n$, the variables $Y_1, \dots, Y_n$ are conditionally independent with $N(x_i\theta, 1)$ as the marginal distribution for $Y_i$. So, given $X_1 = x_1, \dots, X_n = x_n$, $\sum_{i=1}^n X_iY_i \sim N(\theta\sum_{i=1}^n x_i^2, \sum_{i=1}^n x_i^2)$. From this, the conditional distribution of $\bigl[\sum_{i=1}^n X_i^2\bigr]^{1/2}(\hat\theta - \theta)$ given $X_1 = x_1, \dots, X_n = x_n$ is $N(0,1)$. By smoothing,

$$\begin{aligned} P\left[\left(\sum_{i=1}^n X_i^2\right)^{1/2}(\hat\theta - \theta) \le x\right] &= EP\left[\left(\sum_{i=1}^n X_i^2\right)^{1/2}(\hat\theta - \theta) \le x \,\middle|\, X_1, \dots, X_n\right] \\ &= E\Phi(x) = \Phi(x). \end{aligned}$$

So $\bigl[\sum_{i=1}^n X_i^2\bigr]^{1/2}(\hat\theta - \theta) \sim N(0,1)$, and using this it is easy to show that the coverage probability for the interval in part (e) is exactly $1-\alpha$.

15. Since the Fisher information for a single Bernoulli observation is $I(p) = 1/[p(1-p)]$, the first two confidence regions/intervals are

$$\begin{aligned} \mathrm{CI}_1 &= \left\{p : \frac{\sqrt{n}|\hat p - p|}{\sqrt{p(1-p)}} < z_{\alpha/2}\right\} \\ &= \left(\frac{\hat p + z_{\alpha/2}^2/(2n)}{1 + z_{\alpha/2}^2/n} \pm \frac{z_{\alpha/2}\sqrt{\hat p(1-\hat p) + z_{\alpha/2}^2/(4n)}}{\sqrt{n}(1 + z_{\alpha/2}^2/n)}\right), \end{aligned}$$

and

$$\mathrm{CI}_2 = \left(\hat p \pm z_{\alpha/2}\sqrt{\frac{\hat p(1-\hat p)}{n}}\right).$$

Since $l_n(p) = \log\binom{n}{X} + n\hat p \log p + n(1-\hat p)\log(1-p)$, the observed Fisher information is $-l_n''(\hat p) = n/[\hat p(1-\hat p)]$. Using this, $\mathrm{CI}_3$, based on the observed

Fisher information, is the same as $\mathrm{CI}_2$. Finally, the profile confidence region is

$$\mathrm{CR}_4 = \left\{p : \hat{p}\log[\hat{p}/p] + (1-\hat{p})\log[(1-\hat{p})/(1-p)] < z^2_{\alpha/2}/(2n)\right\}.$$

This region is an interval, but it can only be found numerically. With the stated data, the confidence intervals are

$$\mathrm{CI}_1 = (0.2189, 0.3959),$$
$$\mathrm{CI}_2 = \mathrm{CI}_3 = (0.2102, 0.3898),$$

and

$$\mathrm{CI}_4 = (0.2160, 0.3941).$$

20. a) The joint densities are $p^{x_1}(1-p)p^{2x_2}(1-p^2)$, an exponential family with sufficient statistic $T = X_1 + 2X_2$. The maximum likelihood estimator solves

$$l'(p) = -\frac{(3+T)p^2 + p - T}{p(1-p^2)} = 0,$$

which gives

$$\hat{p} = \hat{p}(T) = \frac{-1 + \sqrt{1 + 4T(3+T)}}{6 + 2T}.$$

b) Since

$$\begin{aligned} P(Y = y) &= \sum_{x=0}^{y} P(X_1 = y - x, X_2 = x) \\ &= p^y(1-p)(1-p^2)(1 + p + \cdots + p^y) \\ &= p^y(1-p^2)(1-p^{1+y}), \end{aligned}$$

we have

$$P(X_2 = x | Y = y) = \frac{P(X_1 = y - x, X_2 = x)}{P(Y = y)} = \frac{p^x(1-p)}{1 - p^{1+y}}$$

and

$$\begin{aligned} e(y,p) &= E(T|Y = y) = y + E(X_2|Y = y) \\ &= y + \frac{1-p}{1-p^{1+y}}\sum_{x=0}^{y} xp^x = y + \frac{p(1-p^y) - p(1-p)yp^y}{(1-p)(1-p^{1+y})}. \end{aligned}$$

The algorithm then evolves with $\hat{p}_j = \hat{p}(T_j)$ and $T_{j+1} = e(y, \hat{p}_j)$.
c) The iterates are $T_1 = 124/21 = 5.9048$, $\hat{p}_1 = 0.76009$, $T_2 = 6.7348$, and $\hat{p}_2 = 0.78198$.

21. a) The joint densities are $\exp\bigl[\theta T(x) - n\log[(2\sinh\theta)/\theta]\bigr]$, where $T(x) = x_1 + \cdots + x_n$. This is an exponential family, and the maximum likelihood estimator $\hat{\theta}_x$ is the unique solution of the equation

$$\frac{T}{n} = \frac{A'(\theta)}{n} = \coth\theta - \frac{1}{\theta}.$$

b) Data $Y_1, \dots, Y_n$ are i.i.d. Bernoulli variables with success probability $p = P(X_i > 0) = (e^\theta - 1)/(e^\theta - e^{-\theta})$. Noting that $1 - p = (1 - e^{-\theta})/(e^\theta - e^{-\theta}) = p/e^\theta$, $\theta = \log[p/(1-p)]$, and the relation between $\theta$ and $p$ is one-to-one. Naturally, the maximum likelihood estimator of $p$ based on $Y_1, \dots, Y_n$ is $\hat{p} = (Y_1 + \cdots + Y_n)/n$, and so $\hat{\theta}_y = \log[\hat{p}/(1 - \hat{p})]$.
c) From the independence, $E[X_j|Y = y] = E[X_j|Y_j = y_j]$, which is

$$\frac{\int_0^1 \theta x e^{\theta x}\,dx}{\int_0^1 \theta e^{\theta x}\,dx} = 1 - \frac{1}{\theta} + \frac{1}{e^\theta - 1}, \qquad \text{if } y_j = 1,$$

$$\frac{\int_{-1}^0 \theta x e^{\theta x}\,dx}{\int_{-1}^0 \theta e^{\theta x}\,dx} = -\frac{1}{\theta} + \frac{1}{e^\theta - 1}, \qquad \text{if } y_j = 0.$$

So $E[T|Y] = \sum_{i=1}^n Y_i - n/\theta + n/(e^\theta - 1)$.
d) Because

$$\sum_{i=1}^n Y_i = n\hat{p} = \frac{n(e^{\hat{\theta}_y} - 1)}{e^{\hat{\theta}_y} - e^{-\hat{\theta}_y}},$$

if we start the algorithm at $\hat{\theta}_y$, then

$$\frac{T_1}{n} = \frac{e^{\hat{\theta}_y} - 1}{e^{\hat{\theta}_y} - e^{-\hat{\theta}_y}} - \frac{1}{\hat{\theta}_y} + \frac{1}{e^{\hat{\theta}_y} - 1} = \coth(\hat{\theta}_y) - \frac{1}{\hat{\theta}_y}.$$

From the equation in part (a), the next estimate, $\hat{\theta}_1$, will also be $\hat{\theta}_y$.
e) The iterates are $T_1 = 1/2$, $\hat{\theta}_1 = 0.30182$, $T_2 = 0.62557$, and $\hat{\theta}_2 = 0.37892$.

27. a) Since $\sqrt{n}(\hat{\theta}_n - \theta) \Rightarrow Y \sim N(0, I^{-1}(\theta))$ and $\sqrt{n}(\hat{\eta}_n - \eta) = \binom{1}{0} \cdot \sqrt{n}(\hat{\theta}_n - \theta)$, we have $\sqrt{n}(\hat{\eta}_n - \eta) \Rightarrow \binom{1}{0} \cdot Y \sim N(0, \tau^2)$ with

$$\tau^2 = (1, 0) I^{-1}(\theta) \tbinom{1}{0} = [I^{-1}(\theta)]_{11} = \frac{I_{22}}{I_{11}I_{22} - I_{12}^2}.$$

b) The limiting variance is $\nu^2 = 1/E[\partial \log f_\theta(X)/\partial\eta]^2 = 1/I_{11}$. So $\nu^2 \ge \tau^2$ if and only if $I_{11}I_{22}/(I_{11}I_{22} - I_{12}^2) \le 1$, which always holds, and $\nu^2 = \tau^2$ if and only if $I_{12} = 0$.
c) If $I(\cdot)$ is continuous, then

$$\hat{\tau}^2 = \frac{I_{22}(\hat{\theta})}{I_{11}(\hat{\theta})I_{22}(\hat{\theta}) - I_{12}(\hat{\theta})^2}$$

is a consistent estimator of $\tau^2$. Then $\sqrt{n}(\hat{\eta}_n - \eta)/\hat{\tau} \Rightarrow N(0,1)$, and using this asymptotic pivot, $(\hat{\eta}_n \pm z_{\alpha/2}\hat{\tau}/\sqrt{n})$ is an asymptotic $1-\alpha$ confidence interval for $\eta$.

d) The observed Fisher information divided by $n$, $\tilde{I} = -\nabla^2 l(\hat{\theta}_n)/n$, is a consistent estimator of $I(\theta)$, and so

$$\tilde{\tau}^2 = \frac{\tilde{I}_{22}}{\tilde{I}_{11}\tilde{I}_{22} - \tilde{I}_{12}}$$

is a consistent estimator of $\tau^2$. Continuing as in part (c), $(\hat{\eta}_n \pm z_{\alpha/2}\tilde{\tau}/\sqrt{n})$ is an asymptotic $1-\alpha$ confidence interval for $\eta$.

29. a) The joint density of $W, X, Y$ is

$$f_{\alpha,\beta}(w,x,y) = q(w,x)\frac{\exp\left[-\frac{1}{2}(y-\alpha w-\beta x)^2\right]}{\sqrt{2\pi}},$$

and so the Fisher information $I(\alpha,\beta)$ is

$$-E\nabla^2 \log f_{\alpha,\beta}(W,X,Y) = E\begin{pmatrix} W^2 & WX \\ WX & X^2 \end{pmatrix} = \begin{pmatrix} EW^2 & EWX \\ EWX & EX^2 \end{pmatrix}.$$

The gradient of the log-likelihood is

$$\nabla l(\alpha,\beta) = \begin{pmatrix} \sum_{i=1}^n W_i(Y_i - \alpha W_i - \beta X_i) \\ \sum_{i=1}^n X_i(Y_i - \alpha W_i - \beta X_i) \end{pmatrix}.$$

Setting this equal to zero, the maximum likelihood estimators are

$$\hat{\alpha} = \frac{\left(\sum_{i=1}^n X_i^2\right)\left(\sum_{i=1}^n W_iY_i\right) - \left(\sum_{i=1}^n W_iX_i\right)\left(\sum_{i=1}^n X_iY_i\right)}{\left(\sum_{i=1}^n X_i^2\right)\left(\sum_{i=1}^n W_i^2\right) - \left(\sum_{i=1}^n W_iX_i\right)^2}$$

and

$$\hat{\beta} = \frac{\left(\sum_{i=1}^n W_i^2\right)\left(\sum_{i=1}^n X_iY_i\right) - \left(\sum_{i=1}^n W_iX_i\right)\left(\sum_{i=1}^n W_iY_i\right)}{\left(\sum_{i=1}^n X_i^2\right)\left(\sum_{i=1}^n W_i^2\right) - \left(\sum_{i=1}^n W_iX_i\right)^2}.$$

Since $\sqrt{n}(\hat{\theta} - \theta) \Rightarrow N(0, I^{-1})$, $\sqrt{n}(\hat{\alpha} - \alpha) \Rightarrow N\bigl(0, [I(\alpha,\beta)^{-1}]_{1,1}\bigr)$ with $[I(\alpha,\beta)^{-1}]_{1,1} = EX^2/\bigl(EX^2EW^2 - (EXW)^2\bigr)$.

b) If $\beta$ is known, then $\tilde{\alpha}$ solves $l'(\alpha) = \sum_{i=1}^n W_i(Y_i - \alpha W_i - \beta X_i) = 0$, which gives $\tilde{\alpha} = \bigl(\sum_{i=1}^n W_iY_i - \beta\sum_{i=1}^n W_iX_i)\bigr)/\sum_{i=1}^n W_i^2$. The Fisher information is just $EW^2$ in this case, and so $\sqrt{n}(\tilde{\alpha} - \alpha) \Rightarrow N(0, 1/EW^2)$. This is the same as the limiting distribution in part (a) when $EXW = 0$. Otherwise the distribution in part (a) has larger variance.

30. By Taylor expansion about $\theta$,

$$g(\hat{\theta}_n) = g(\theta) + \nabla g(\tilde{\theta}_n)\cdot(\hat{\theta}_n - \theta),$$

where $\tilde{\theta}_n$ is an intermediate value on the line segment between $\hat{\theta}_n$ and $\theta$. Since $\hat{\theta}_n$ is consistent, $\tilde{\theta}_n \xrightarrow{P_\theta} \theta$, and so $\nabla g(\tilde{\theta}_n) \xrightarrow{P_\theta} \nabla g(\theta)$. Because $\sqrt{n}(\hat{\theta}_n - \theta) \Rightarrow Z \sim N(0, I(\theta)^{-1})$, using Theorem 9.30,

$$\sqrt{n}\big(g(\hat{\theta}_n) - g(\theta)\big) = \nabla g(\tilde{\theta}_n) \cdot \sqrt{n}(\hat{\theta}_n - \theta) \Rightarrow \nabla g(\theta) \cdot Z \sim N\big(0, \nu(\theta)\big),$$

where $\nu(\theta) = \nabla g(\theta)' I(\theta)^{-1} \nabla g(\theta)$. This proves Proposition 9.31. To show that (9.13) is a $1-\alpha$ asymptotic confidence interval, if $\hat{\nu} \xrightarrow{P_\theta} \nu(\theta)$, then $1/\sqrt{\hat{\nu}} \xrightarrow{P_\theta} 1/\sqrt{\nu(\theta)}$ by Proposition 8.5. If $Y \sim N\big(0, \nu(\theta)\big)$, then using Theorem 8.13,

$$\frac{1}{\sqrt{\hat{\nu}}}\sqrt{n}\big(g(\hat{\theta}_n) - g(\theta)\big) \Rightarrow \frac{1}{\sqrt{\nu(\theta)}} Y \sim N(0,1).$$

One natural estimate for $\nu$ is $\hat{\nu}_1 = \nu(\hat{\theta}_n)$. Another estimator, based on observed Fisher information, is

$$\hat{\nu}_2 = -n \nabla g(\hat{\theta}_n)' \big(\nabla^2 l(\hat{\theta}_n)\big)^{-1} \nabla g(\hat{\theta}_n).$$

The asymptotic confidence intervals are $\big(g(\hat{\theta}) \pm z_{\alpha/2}\sqrt{\hat{\nu}_i/n}\big)$, $i = 1$ or $2$.

31. The likelihood is

$$L(\theta) = \binom{n}{N_{11}, \dots, N_{22}} \theta_1^{N_{+1}} (1-\theta_1)^{N_{+2}} \theta_2^{N_{1+}} (1-\theta_2)^{N_{2+}},$$

and maximum likelihood estimators for $\theta_1$ and $\theta_2$ are $\hat{\theta}_1 = N_{+1}/n$ and $\hat{\theta}_2 = N_{1+}/n$. The observed Fisher information matrix is

$$-\nabla^2 l(\hat{\theta}) = \begin{pmatrix} \dfrac{N_{+1}}{\hat{\theta}_1^2} + \dfrac{N_{+2}}{(1-\hat{\theta}_1)^2} & 0 \\ 0 & \dfrac{N_{1+}}{\hat{\theta}_2^2} + \dfrac{N_{2+}}{(1-\hat{\theta}_2)^2} \end{pmatrix}$$

$$= \begin{pmatrix} \dfrac{n}{\hat{\theta}_1(1-\hat{\theta}_1)} & 0 \\ 0 & \dfrac{n}{\hat{\theta}_2(1-\hat{\theta}_2)} \end{pmatrix}.$$

Since $-\nabla^2 l(\hat{\theta})/n \xrightarrow{P_\theta} I(\theta)$,

$$I(\theta) = \begin{pmatrix} \dfrac{1}{\theta_1(1-\theta_1)} & 0 \\ 0 & \dfrac{1}{\theta_2(1-\theta_2)} \end{pmatrix}.$$

In this example, $g(\theta) = \theta_1\theta_2$ and the two estimates of $\nu(\theta)$ are the same:

$$(\hat\theta_2, \hat\theta_1)\begin{pmatrix}\hat\theta_1(1-\hat\theta_1) & 0\\ 0 & \hat\theta_2(1-\hat\theta_2)\end{pmatrix}\begin{pmatrix}\hat\theta_2\\ \hat\theta_1\end{pmatrix} = \hat\theta_1\hat\theta_2(\hat\theta_1+\hat\theta_2-2\hat\theta_1\hat\theta_2).$$

The confidence intervals are both

$$\left(\hat\theta_1\hat\theta_2 \pm z_{\alpha/2}\sqrt{\frac{\hat\theta_1\hat\theta_2(\hat\theta_1+\hat\theta_2-2\hat\theta_1\hat\theta_2)}{n}}\right).$$

## B.10 Problems of Chapter 10

1. The joint density can be written as

$$\frac{\exp\Bigl[-\frac12(x_1-\theta)^2-\frac12\sum_{j=1}^{n-1}\bigl(x_{j+1}-\theta-\frac12(x_j-\theta)\bigr)^2\Bigr]}{\sqrt{2\pi}^n},$$

so we have a location family. Ignoring terms that do not depend on $\theta$, the likelihood is proportional to

$$\exp\left[X_1\theta-\frac12\theta^2+\frac12\sum_{j=1}^{n-1}(X_{j+1}-\frac12X_j)\theta-\frac18(n-1)\theta^2\right]$$
$$=\exp\left[\frac14T\theta-\frac18(n+3)\theta^2\right]\propto\exp\left[-\frac{n+3}{8}\left(\theta-\frac{T}{n+3}\right)^2\right],$$

where $T=(3X_1+X_2+\cdots+X_{n-1}+2X_n)$. This likelihood is proportional to a normal density with mean $T/(n+3)$. Because the minimum risk equivariant estimator is the mean of the normalized likelihood, it must be $T/(n+3)$.
2. a) By dominated convergence, if $c$ is a continuity point of $F$,

$$g'(c)=E\frac{\partial}{\partial c}|X-c|=-E\,\mathrm{Sign}(X-c)=P(X<c)-P(X>c),$$

which is zero if $c$ is the median.
b) Dominated convergence (when $c$ is a continuity point of $F$) gives

$$g'(c)=E\bigl[-aI\{X>c\}+bI\{X<c\}\bigr]=bP(X<c)-aP(X>c),$$

which equals zero if $P(X<c)=a/(a+b)$. So $g$ is minimized if $c$ is the $\bigl(a/(a+b)\bigr)$th quantile of $F$.
4. An equivariant estimator must have form $X-c$, with minimal risk if $c$ is chosen to minimize

$$E_\theta L(\theta, X - c) = E_0\big[a(X-c)^+ + b(c-X)\big].$$

From Problem 10.2, $c$ should be the $\big(a/(a+b)\big)$th quantile of the $P_0$-cumulative distribution of $X$, given by $F(t) = \frac{1}{2}\int_{-\infty}^t e^{-|x|}dx$. Solving $F(t) = a/(a+b)$, $c = \log[2a/(a+b)]$ if $a \le b$, and $c = -\log[2b/(a+b)]$ if $a \ge b$.

## B.11 Problems of Chapter 11

1. a) The joint density of $\Lambda_i$ with $X_i$ is $\lambda^\alpha e^{-\lambda(1+x)}/\Gamma(\alpha)$. So the marginal density of $X_i$ is

$$\int_0^\infty \frac{\lambda^\alpha e^{-\lambda(1+x)}}{\Gamma(\alpha)}\,d\lambda = \frac{\alpha}{(1+x)^{1+\alpha}}, \qquad x > 0.$$

b) Dividing the joint density by the marginal density of $X_i$, the conditional density of $\Lambda_i$ given $X_i = x$ is

$$\frac{(1+x)^{1+\alpha}\lambda^\alpha e^{-\lambda(1+x)}}{\Gamma(1+\alpha)},$$

a gamma density with shape parameter $1+\alpha$ and scale $1/(1+x)$. The Bayes estimator is

$$E(\Lambda_i|X_i) = \frac{1+\alpha}{1+X_i}.$$

c) From part (a), the joint density is

$$\prod_{i=1}^p \frac{\alpha}{(1+x_i)^{1+\alpha}} = \exp\left[-(1+\alpha)\sum_{i=1}^p \log(1+x_i) + p\log\alpha\right].$$

So the log-likelihood is $l(\alpha) = -(1+\alpha)\sum_{i=1}^p \log(1+X_i) + p\log\alpha$. Then $l'(\alpha) = -\sum_{i=1}^p \log(1+X_i) + p/\alpha$ which is zero when $\alpha$ is $p/\sum_{i=1}^p \log(1+X_i)$. This is the maximum likelihood estimator.

d) The empirical Bayes estimator for $\Lambda_i$ is

$$\frac{1 + p/\sum_{j=1}^p \log(1+X_j)}{1+X_i}.$$

2. a) Direct calculation shows that

$$\Theta_i|X = x_i, Y = y_i \sim N\left(\frac{x_i y_i \tau^2}{1+x_i^2\tau^2}, \frac{\tau^2}{1+x_i^2\tau^2}\right).$$

The Bayes estimate is $X_iY_i\tau^2/(1+X_i^2\tau^2)$.

b) By smoothing, $EY_i^2 = EE[Y_i^2|X_i,\Theta_i] = E[X_i^2\theta_i^2+1] = 1+\tau^2$. A simple estimator of $\tau^2$ is

$$\hat{\tau}^2 = \frac{1}{p}\sum_{i=1}^p Y_i^2 - 1.$$

c) The empirical Bayes estimator is

$$\hat{\theta}_i = \frac{X_i Y_i \hat{\tau}^2}{1+X_i^2\hat{\tau}^2}.$$

3. a) and b) The joint density of $\Theta_i$ with $X_i$ is $\lambda\theta^x e^{-(1+\lambda)\theta}/x!$, the marginal density of $X_i$ is

$$\int_0^\infty \frac{\lambda\theta^x e^{-(1+\lambda)\theta}}{x!}\,d\theta = \frac{\lambda}{(1+\lambda)^{x+1}},$$

the conditional density of $\Theta_i$ given $X_i = x$ is

$$\frac{(1+\lambda)^{x+1}\theta^x e^{-(1+\lambda)\theta}}{x!},$$

and

$$E[\Theta_i|X_i = x] = \int_0^\infty \frac{(1+\lambda)^{x+1}\theta^{x+1}e^{-(1+\lambda)\theta}}{x!}\,d\theta = \frac{x+1}{1+\lambda}.$$

By independence, $E[\Theta_i|X] = E[\Theta_i|X_i] = (X_i+1)/(1+\lambda)$, which is the Bayes estimate of $\Theta_i$ under compound squared error loss.
c) The joint density of $X_1,\dots,X_p$ in the Bayesian model is

$$\prod_{i=1}^p \frac{\lambda}{(1+\lambda)^{x_i+1}} = \exp\biggl[-T(x)\log(1+\lambda) + p\log\Bigl(\frac{\lambda}{1+\lambda}\Bigr)\biggr],$$

where $T(x) = x_1+\cdots+x_p$. The maximum likelihood estimator $\hat{\lambda}$ of $\lambda$ solves

$$0 = -\frac{T}{1+\lambda} + \frac{p}{\lambda} - \frac{p}{1+\lambda} = \frac{-\lambda T + p}{\lambda(1+\lambda)},$$

giving $\hat{\lambda} = p/T = 1/\overline{X}$.
d) The empirical Bayes estimator for $\theta_i$ is $(X_i+1)/(1+1/\overline{X})$.

## B.12 Problems of Chapter 12

1. By smoothing, $E_\theta\psi = P_\theta\bigl((X,U)\in S\bigr) = E_\theta P_\theta\bigl((X,U)\in S \bigm| X\bigr)$. This will equal $E_\theta\varphi(X)$ if we can choose $S$ so that $P_\theta\bigl((X,U)\in S \bigm| X = x\bigr) = \varphi(x)$. One solution is $S = \bigl\{(x,u) : u < \varphi(x)\bigr\}$.

2. This is like the Neyman–Pearson lemma. If $f_0$ and $f_1$ are densities for $N(0,1)$ and $N(0,4)$, then we want to maximize $\int h(x)f_1(x)\,dx$ with $\int h(x)f_0(x)\,dx = 0$. Adding a Lagrange multiplier, let us try to maximize $\int h(x)[f_1(x) - kf_0(x)]\,dx$. Here there is a bit of a difference. Because $h$ has range $[-M, M]$ instead of $[0,1]$, an optimal function $h^*$ will satisfy

$$h^*(x) = M, \text{ if } \frac{f_1(x)}{f_0(x)} > k, \text{ and } h^*(x) = -M, \text{ if } \frac{f_1(x)}{f_0(x)} < k.$$

The likelihood ratio is $\frac{1}{2}e^{3x^2/8}$, so equivalently,

$$h^*(x) = M, \text{ if } |x| > k', \text{ and } h^*(x) = -M, \text{ if } |x| < k'.$$

To satisfy the constraint, $k' = \Phi^{-1}(3/4) = 0.67449$. Then $Eh^*(2Z) = 0.47186M$. (You can also solve this problem applying the Neyman–Pearson lemma to the test function $\varphi = (h+M)/(2M)$.)

3. By smoothing,

$$P(Z_1/Z_2 \le x) = EP(Z_1/Z_2 \le x|Z_2) = E\Phi(x|Z_2|).$$

(By symmetry, this is true regardless of the sign of $Z_2$.) Taking $d/dx$, the density of $Z_1/Z_2$ is

$$\begin{aligned} E|Z_2|\phi(x|Z_2|) &= \frac{1}{2\pi}\int |z|e^{-(z^2+x^2z^2)/2}\,dz \\ &= \frac{1}{\pi}\int_0^\infty ze^{-(1+x^2)z^2/2}\,dz = -\frac{e^{-(1+x^2)z^2/2}}{\pi(1+x^2)}\Bigg|_0^\infty = \frac{1}{\pi(1+x^2)}. \end{aligned}$$

4. The likelihood ratio is

$$L = \exp\left[(X_1^2 - X_2^2)\left(\frac{1}{2\sigma_1^2} - \frac{1}{2\sigma_2^2}\right)\right].$$

If we assume $\sigma_1^2 < \sigma_2^2$, then Neyman–Pearson likelihood ratio tests will reject $H_0$ if $X_1^2 - X_2^2 \ge k$, and for a symmetric test, $k$ should be zero. Taking $Z_1 = X_1/\sigma_1$ and $Z_2 = X_2/\sigma_2$, the error probability under $H_0$ is

$$\begin{aligned} P(X_1^2 - X_2^2 \ge 0) &= P(\sigma_1^2 Z_1^2 \ge \sigma_2^2 Z_2^2) \\ &= P\big(|Z_2/Z_1| \le \sigma_1/\sigma_2\big) = \frac{2}{\pi}\tan^{-1}(\sigma_1/\sigma_2). \end{aligned}$$

If $\sigma_1^2 > \sigma_2^2$, the error rate is $2\tan^{-1}(\sigma_2/\sigma_1)/\pi$.

6. a) By a change of variables,

$$Eh(X^2/2) = \frac{1}{2}\int_0^2 h(x^2/2)\,dx = \int_0^2 \frac{h(y)}{2\sqrt{2y}}\,dy,$$

and we want to maximize this integral with the constraint $\int_0^2 h(y)\,dy = 0$. Introducing a Lagrange multiplier, consider maximizing

$$\int_0^2 \left[\frac{1}{2\sqrt{2y}} - k\right] h(y)\,dy$$

without constraint. An optimal solution will have $h^*(y) = M$ when $y < 1/(8k^2)$ and $h^*(y) = -M$ when $y > 1/(8k^2)$. This solution will satisfy the constraint if $k = 1/\sqrt{8}$. This gives an upper bound of $(\sqrt{2}-1)M$, and $h^*(y) = M\,\mathrm{Sign}(1-y)$ as a function achieving the bound.
b) Introducing a Lagrange multiplier and proceeding in the same fashion, consider maximizing

$$\int_0^2 \left[\frac{1}{2\sqrt{2y}} - k\right] h(y)\,dy$$

without constraint. Now an optimal solution will be $h^*(y) = My\,\mathrm{Sign}(c-y)$ with $c = 1/(8k^2)$. Then $Eh^*(X) = M(c^2-2)/2$ so $h^*$ will satisfy the constraint if $c = \sqrt{2}$, giving an upper bound of

$$Eh^*(X^2/2) = \int_0^{2^{3/4}} \frac{Mx^2}{4}\,dx - \int_{2^{3/4}}^2 \frac{Mx^2}{4}\,dx = \frac{2}{3}(2^{1/4}-1)M.$$

7. a) If we can interchange differentiation and integration, then

$$\begin{aligned}\beta'(\theta) &= \frac{d}{d\theta}\int_0^\infty \varphi(x)p_\theta(x)\,dx = \int_0^\infty \varphi(x)\frac{\partial}{\partial\theta}p_\theta(x)\,dx \\ &= \int_0^\infty \frac{1-\theta x}{\theta(1+\theta x}\varphi(x)p_\theta(x)\,dx = E_\theta\left[\frac{1-\theta X}{\theta(1+\theta X)}\varphi(X)\right].\end{aligned}$$

Dominated convergence can be used to justify the interchange. Note that $\partial p_\theta(x)/\partial\theta = (1-\theta x)/(1+\theta x)^3$. Let $h = h_n$ be a sequence of positive constants all less than $\theta$ converging to zero, and let $\xi = \xi_n(x)$ be intermediate values in $[\theta, 2\theta]$ chosen so that

$$\frac{\varphi(x)\big(p_{\theta+h}(x) - p_\theta(x)\big)}{h} = \frac{\varphi(x)(1-\xi x)}{(1+\xi x)^3}.$$

These functions converge pointwise to $\varphi(x)\partial p_\theta(x)/\partial\theta$ and are uniformly bounded in magnitude by $(1+2\theta x)/(1+\theta x)^3$, an integrable function.
b) Introducing a Lagrange multiplier $k$, consider unconstrained maximization of

$$\beta'_\varphi(1) - k\beta_\varphi(1) = E_1\left(\frac{1-X}{1+X} - k\right)\varphi.$$

An optimal test function $\varphi*$ should equal one when $(1-X)/(1+X) > k$ and zero, otherwise. Because $(1-X)/(1+X)$ is a decreasing function of $X$, this gives $\varphi^* = I\{X < c\}$, where $k = (1-c)/(1+c)$. This test has level $P_1(X < c) = c/(1+c)$. If $c = \alpha/(1-\alpha)$, or, equivalently, $k = 1-2\alpha$, then $\varphi^*$ has level $\alpha$, satisfying the constraint and maximizing $\beta'_\varphi(1)$ among all tests with $\beta_\varphi(1) = \alpha$.

9. Let $v_\theta(x) = \partial \log p_\theta(x)/\partial\theta$. Then the test $\varphi^*$ should maximize $\int \varphi v_{\theta_0} p_{\theta_0}\, d\mu$ subject to the constraint $\int \varphi p_{\theta_0}\, d\mu = \alpha$. Introducing a Lagrange multiplier and arguing as in the Neyman–Pearson lemma, $\varphi^*$ should have form

$$\varphi^*(x) = \begin{cases} 1, & v_{\theta_0}(x) > k; \\ 0, & v_{\theta_0}(x) < k. \end{cases}$$

10. Since $\log p_\theta(x) = \sum_{i=1}^n \log f_\theta(x_i)$, $v_\theta(x) = \sum_{i=1}^n s_\theta(x_i)$, where $s_\theta(x_i) = \partial \log f_\theta(x_i)/\partial\theta$. Note that $s_\theta(X_1), \dots, s_\theta(X_n)$ are i.i.d., $E_\theta s_\theta(X_i) = 0$, and $\mathrm{Var}_\theta\bigl[s_\theta(X_i)\bigr] = I(\theta)$, the Fisher information from a single observation. By the central limit theorem, under $P_\theta$, $v_\theta(X)/\sqrt{n} \Rightarrow N\bigl(0, I(\theta)\bigr)$ as $n \to \infty$. So the level of the test $\varphi^*$ in Problem 12.9 is approximately

$$P_{\theta_0}\bigl(v_{\theta_0}(X)/\sqrt{n} > k/\sqrt{n}\bigr) \approx 1 - \Phi\bigl(k/\sqrt{nI(\theta_0)}\bigr),$$

which is $\alpha$ if $k = z_\alpha\sqrt{nI(\theta_0)}$.

14. The mass function for $X$ is $p(1-p)^x$, $x \ge 0$, which is an exponential family with canonical parameter $\eta = \log(1-p)$. The two hypotheses can be expressed in terms of $\eta$ as $H_0 : \eta \le -\log 2$ and $H_1 : \eta > -\log 2$. With $\eta$ as the parameter the densities have monotone likelihood ratios in $T = X$, and so there will be a uniformly most powerful test with form $\varphi = 0$ if $X < k$, $\varphi = 1$ if $X > k$, and $\varphi = \gamma$ if $X = k$. For a fair coin ($p = 1/2$), $P(X \le 3) = 93.75\%$ and $P(X = 4) = 3.125\%$. If $k = 4$ and $\gamma = 3/5$ the test will have level $\alpha = 5\%$. The power if $p = 40\%$ is $10.8864\%$.

15. The joint density for the data is

$$\frac{\sqrt{1-\rho^2}}{2\pi} \exp\bigl[\rho xy - (x^2 + y^2)/2\bigr],$$

which is an exponential family with $T = XY$. So, the uniformly most powerful test will reject $H_0$ if $XY > k$. Adjusting $k$ to achieve a given level $\alpha$ is a bit tricky. The null density of $T = XY$ is $K_0(|t|/2)/(2\pi)$, where $K_0$ is a Bessel function. Numerical calculations using this density show that $\alpha = 5\%$ when $k = 3.19$.

16. a) Let $h = \log g$. The family will have monotone likelihood ratios in $x$ if

$$\log\left(\frac{p_{\theta_2}(x)}{p_{\theta_1}(x)}\right) = h(x - \theta_2) - h(x - \theta_1)$$

is nondecreasing in $x$ whenever $\theta_2 > \theta_1$. A sufficient condition for this is that $h'(x - \theta_2) - h'(x - \theta_1) \ge 0$. Since $x - \theta_2 \le x - \theta_1$, this will hold if $h'$ is nonincreasing, which follows if $h''(x) = d^2 \log g(x)/dx^2 \le 0$.

b) As in part (a), a sufficient condition will be that the derivative of $\log\bigl(p_{\theta_2}(x)/p_{\theta_1}(x)\bigr)$ is at least zero whenever $\theta_2 > \theta_1$, and this derivative is

$$\frac{1}{x}\left(\frac{g'(x/\theta_2)x/\theta_2}{g(x/\theta_2)} - \frac{g'(x/\theta_1)x/\theta_1}{g(x/\theta_1)}\right).$$

Since $x/\theta_2 < x/\theta_1$, this will hold if the function $xg'(x)/g(x)$ is nonincreasing, and a sufficient condition for this is

$$\frac{d}{dx}\left(\frac{xg'(x)}{g(x)}\right) = \frac{g(x)g'(x) + xg(x)g''(x) - x\big[g'(x)\big]^2}{g^2(x)} \le 0, \qquad x > 0.$$

17. a) Define $F(t) = P_{\theta_0}(T \le t)$. The uniformly most powerful level $\alpha$ test is $\varphi_\alpha(x) = I\{T(x) > k(\alpha)\}$ with $k(\alpha)$ chosen so that $F\big(k(\alpha)\big) = 1-\alpha$. Suppose $\alpha_0 < \alpha_1$. Then since $F$ is nondecreasing, $k(\alpha_0) > k(\alpha_1)$. So if $T(x) > k(\alpha_0)$, $T(x)$ also exceeds $k(\alpha_1)$, and hence $\varphi_{\alpha_1}(x) = 1$ whenever $\varphi_{\alpha_0}(x) = 1$. Thus $\varphi_{\alpha_1}(x) \ge \varphi_{\alpha_0}(x)$ for all $x$, and since $\alpha_0$ and $\alpha_1$ are arbitrary, $\varphi_\alpha(x)$ is nondecreasing in $\alpha$.
b) Because $F$ is nondecreasing and continuous, if $t > k(\alpha)$ then $F(t) \ge F\big(k(\alpha)\big) = 1-\alpha$, and so $P = \inf\{\alpha : t > k(\alpha)\} \ge \inf\{\alpha : F(t) \ge 1-\alpha\} = 1 - F(t)$. But in addition, if $F(t) > F\big(k(\alpha)\big) = 1-\alpha$, then $t > k(\alpha)$, and so $P = \inf\{\alpha : t > k(\alpha)\} \le \inf\{\alpha : F(t) > 1-\alpha\}$, which is again $1-F(t)$. So the p-value must be $1 - F(t) = P_{\theta_0}(T > t)$.
c) Let $F^{\leftarrow}$ denote the largest inverse function of $F$: $F^{\leftarrow}(c) = \sup\{t : F(t) = c\}$, $c \in (0,1)$. Then $F(T) \le x$ if and only if $T \le F^{\leftarrow}(x)$ and $P_{\theta_0}\big(F(T) \le x\big) = P_{\theta_0}\big(T \le F^{\leftarrow}(x)\big) = F\big(F^{\leftarrow}(x)\big) = x$. So $F(T)$ and the p-value $1 - F(T)$ are both uniformly distributed on $(0,1)$ under $P_{\theta_0}$.
20. Suppose $\theta_2 > \theta_1$. Then the log-likelihood ratio is

$$l(x) = \log L(x,\theta_1,\theta_2) = \log\frac{p_{\theta_2}(x)}{p_{\theta_1}(x)} = \log\left(\frac{1+\theta_2 x}{1+\theta_1 x}\right)$$

and

$$l'(x) = \frac{\theta_2 - \theta_1}{(1+\theta_2 x)(1+\theta_1 x)}.$$

This is positive for $x \in (-1,1)$. So $l(x)$ and $L(x,\theta_1,\theta_2)$ are increasing functions of $x$, and the family has monotone likelihood ratios in $x$.
21. Fix $0 < \theta_1 < \theta_2 < 1$. The likelihood ratio

$$\frac{p_{\theta_2}(x)}{p_{\theta_1}(x)} = \frac{\theta_2 + (1-\theta_2)f(x)}{\theta_1 + (1-\theta_1)f(x)} = \frac{1-\theta_2}{1-\theta_1} + \frac{(\theta_2-\theta_1)/(1-\theta_1)}{\theta_1 + (1-\theta_1)f(x)}$$

is monotone decreasing in $f(x)$, so we can take $T(X) = -f(X)$.
25. a) The joint densities are $f_\lambda(x,u) = \lambda^x e^{-\lambda}/x!$, $u \in (0,1)$, $x = 0,1,\dots$. If $\lfloor t \rfloor$ denotes the greatest integer less than or equal to $t$, then with $T(x,u) = x + u$, these densities can also be written as $\lambda^{\lfloor T(x,u)\rfloor} e^{-\lambda}/x!$. (From this, we see that $T$ is sufficient, but not minimal sufficient, because $X$ is also sufficient, and $T$ is not a function of $X$.) If $\lambda_1 > \lambda_0$, then the likelihood ratio $f_{\lambda_1}(x,u)/f_{\lambda_0}(x,u) = (\lambda_1/\lambda_0)^{\lfloor T(x,u)\rfloor} e^{\lambda_0-\lambda_1}$ is an increasing function of $T$, so the joint densities have monotone likelihood ratios in $T$.
b) Let $F_\lambda(t) = P_\lambda(T \le t)$. For $n = 0,1,2,\dots$, $F_\lambda(n) = P_\lambda(X \le n-1)$, which can be found summing the Poisson mass function. For nonintegral

values $t$, $F_\lambda(t)$ is the linear interpolation of the value of $F_\lambda$ at the two adjacent integers. So, $F_\lambda$ is strictly increasing and continuous on $(0,\infty)$. The UMP test has form $\varphi = 1$ if $T \geq k$, $\varphi = 0$ if $T < k$. (Randomization on the boundary is unnecessary because $T$ is continuous.) The constant $k$ is chosen (uniquely) so that $F_{\lambda_0}(k) = 1-\alpha$. For the particular case, $F_2(5) = 0.947347$ and $F_2(6) = 0.9834364$. By linear interpolation, $F_2(5.073519) = 95\%$, and we should reject $H_0$ if $T \geq 5.073519$.
c) From part (b), if $T = t$, we accept the null hypothesis that the true value of the parameter is $\lambda$ if and only if $F_\lambda(t) < 1-\alpha$. For fixed $t$, $F_\lambda(t)$ is continuous and strictly decreasing on $[0,\infty)$, and so there is a unique value $\lambda_t$ such that $F_{\lambda_t}(t) = 1-\alpha$. The confidence interval is $(\lambda_t,\infty)$. For data $X = 2$ and $U = 0.7$, the observed value of $T$ is 2.7. As in part (a), $F_\lambda(2.7) = (1+\lambda+7\lambda^2/20)e^{-\lambda}$, which is 95% at $\lambda_{2.7} = 0.583407$.

28. As in Example 12.10, the uniformly most powerful test of $\theta = \theta_0$ versus $\theta > \theta_0$ will reject if $T = \max\{X_1,\dots,X_n\} > c$, with $c$ chosen so that

$$P_{\theta_0}(T > c) = 1 - (c/\theta_0)^n = \alpha.$$

Solving, $c = \theta_0(1-\alpha)^{1/n}$, and the acceptance region for this test is

$$A(\theta_0) = \bigl\{x : T(x) < \theta_0(1-\alpha)^{1/n}\bigr\}.$$

The confidence interval $S_1$ dual to these tests is

$$S_1 = \bigl\{\theta : X \in A(\theta)\bigr\} = \bigl\{\theta : T < \theta(1-\alpha)^{1/n}\bigr\} = \bigl(T(1-\alpha)^{-1/n},\infty\bigr).$$

Similarly, the uniformly most powerful test of $\theta = \theta_0$ versus $\theta < \theta_0$ will reject if $T = \max\{X_1,\dots,X_n\} < c$, with $c$ chosen so that

$$P_{\theta_0}(T < c) = (c/\theta_0)^n = \alpha.$$

This gives $c = \theta_0\alpha^{1/n}$, $A(\theta_0) = \bigl\{x : T(x) > \theta_0\alpha^{1/n}\bigr\}$, and

$$S_2 = \{\theta : T > \theta\alpha^{1/n}\} = (0, T\alpha^{-1/n}).$$

By the result in Problem 9.12,

$$S = S_1 \cap S_2 = \bigl(T(1-\alpha)^{-1/n}, T\alpha^{-1/n}\bigr)$$

should have coverage probability at least $1-2\alpha$, which is 95% if we take $\alpha = 2.5\%$. (In fact, it is easy to see that the coverage probability is exactly 95%.)

31. a) By dominated convergence,

$$\begin{aligned}
\beta'(\theta) &= \frac{d}{d\theta}\int \varphi(x)\frac{1}{\sqrt{2\pi}}e^{-(x-\theta)^2/2}\,dx \\
&= \int \varphi(x)\frac{1}{\sqrt{2\pi}}\frac{\partial}{\partial\theta}e^{-(x-\theta)^2/2}\,dx \\
&= \int \varphi(x)\frac{1}{\sqrt{2\pi}}(x-\theta)e^{-(x-\theta)^2/2}\,dx \\
&= E_\theta\bigl[(X-\theta)\varphi(X)\bigr].
\end{aligned}$$

The desired result follows setting $\theta = 0$.
b) Using part (a) and writing expectations as integrals, we wish to maximize

$$\int \varphi(x) f_2(x)\,dx$$

with constraints

$$\int \varphi(x) f_0(x)\,dx = \alpha \text{ and } \int \varphi(x) f_1(x)\,dx = 0,$$

where

$$f_2(x) = \frac{e^{-(x-1)^2/2}}{\sqrt{2\pi}}, \qquad f_0(x) = \frac{e^{-x^2/2}}{\sqrt{2\pi}},$$

and

$$f_1(x) = \frac{xe^{-x^2/2}}{\sqrt{2\pi}}.$$

By the generalization of the Neyman–Pearson lemma, there are constants $k_0$ and $k_1$ such that the optimal test function $\varphi$ has form

$$\varphi(x) = \begin{cases} 1, & f_2(x) \geq k_0 f_0(x) + k_1 f_1(x); \\ 0, & \text{otherwise.} \end{cases}$$

Dividing by $f_0$,

$$\varphi(x) = \begin{cases} 1, & e^{x-1/2} \geq k_0 + k_1 x; \\ 0, & \text{otherwise.} \end{cases}$$

Because $e^{x-1/2}$ is convex, $\varphi(x) = 0$ if and only if $x \in [c_1, c_2]$. To satisfy the second constraint, $c_2$ must be $-c_1$, and then the first constraint gives $c = \Phi^{-1}(1-\alpha/2)$ as the common magnitude. So the optimal test function is $\varphi(x) = I\{|x| \geq c\}$.

32. a) By dominated convergence, we should have

$$\beta'(\theta) = \int \varphi \frac{\partial p_\theta}{\partial \theta}\,d\mu = \int \varphi \frac{\partial \log p_\theta}{\partial \theta} p_\theta\,d\mu = E_\theta \varphi l'(\theta),$$

where $l(\theta)$ is the log-likelihood. Here

$$l'(\theta) = \sum_{i=1}^{2} \left[\frac{1}{\theta} - \frac{2X_i}{1+\theta X_i}\right].$$

b) Reasoning as in the Neyman–Pearson lemma, the locally most powerful test will reject $H_0$ if $l'(\theta_0)$ exceeds some critical value, that is, if

$$\sum_{i=1}^{2} \left[\frac{1}{\theta_0} - \frac{2X_i}{1+\theta_0 X_i}\right] \geq k(\theta_0).$$

To find $k(\theta_0)$ we need the $P_{\theta_0}$-distribution of the sum here. Solving the inequality, for $|x| < 1/\theta_0$,

$$P_{\theta_0}\left(\frac{1}{\theta_0} - \frac{2X_i}{1+\theta_0 X_i} < x\right) = P_{\theta_0}\left(X_i > \frac{1-\theta_0 x}{\theta_0(1+\theta_0 x)}\right).$$

Since $P_\theta(X_i > c) = \int_c^\infty \left[\theta/(1+\theta x)^2\right] dx = 1/(1+\theta c)$, this expression equals

$$\left(1 + \frac{1-\theta_0 x}{1+\theta_0 x}\right)^{-1} = \frac{1+\theta_0 x}{2},$$

which is the cumulative distribution for the uniform distribution on $(-1/\theta_0, 1/\theta_0)$. The density of this distribution is $f(x) = \theta_0/2$ for $|x| < 1/\theta_0$. The density for the sum of two independent variables with this density is "triangular" in shape: $g(s) = \int f(x)f(s-x)\,dx = \left[\frac{1}{2}\theta_0 - \frac{1}{4}|s|\theta_0^2\right]^+$. If $k \in (0, 2/\theta_0)$, $\int_k^{2/\theta_0} g(s)\,ds = (k\theta_0 - 2)^2/8$ Setting this to $\alpha$, $k(\theta_0) = (2 - \sqrt{8\alpha})/\theta_0$ (provided $\alpha < 1/2$), which is $1.6/\theta_0$ when $\alpha = 5\%$.
c) The confidence interval is

$$\left\{\theta : \frac{2}{\theta} - \frac{2X_1}{1+\theta X_1} - \frac{2X_2}{1+\theta X_2} < \frac{1.6}{\theta}\right\}$$
$$= \left(\frac{-0.8(X_1+X_2) + \sqrt{0.64(X_1+X_2)^2 + 0.8X_1X_2}}{2X_1X_2}, \infty\right).$$

33. We want to minimize $\int_0^\infty \varphi(x)2e^{-2x}\,dx$ with

$$\int_0^\infty \varphi(x)e^{-x}\,dx = \int_0^\infty \varphi(x)3e^{-3x}\,dx = 1/2.$$

By the generalized Neyman–Pearson lemma, there should be Lagrange multipliers $k_1$ and $k_2$ so that an optimal test $\varphi^*$ is one if $-2e^{-2x} > k_1 e^{-x} + 3k_2 e^{-3x}$, and zero if the opposite inequality holds. Equivalently, with $c_1 = -k_1/2$ and $c_2 = -3k_2/2$,

$$\varphi^*(x) = \begin{cases} 1, & c_1 e^x + c_2 e^{-x} > 1; \\ 0, & c_1 e^x + c_2 e^{-x} < 1. \end{cases}$$

If $c_1$ and $c_2$ have opposite signs, or one of them equals zero, then the left-hand side of these inequalities will be a monotone function of $x$, and $\varphi^*$ will be a one-sided test. But then its power function will be monotone, and we cannot satisfy the constraints for the power function. So $c_1$ and $c_2$ must both be positive. Then $c_1 e^x + c_2 e^{-x}$ is convex, and $\varphi^*$ will be a two-sided test[1] with form $\varphi^*(x) = 1 - 1_{(b_1,b_2)}(x)$, with $b_1$ and $b_2$ adjusted so that

[1] Reversing the null and alternative hypotheses, an extended argument using similar ideas can show that the test $1-\varphi^*$ is a uniformly most powerful level $\alpha = 1/2$

$$P_1\big(X \in (b_1, b_2)\big) = e^{-b_1} - e^{-b_2} = 1/2$$

and

$$P_3\big(X \in (b_1, b_2)\big) = e^{-3b_1} - e^{-3b_2} = 1/2.$$

Solving,

$$b_1 = \log\left(\frac{4}{\sqrt{5}+1}\right) = 0.2119 \text{ and } b_2 = \log\left(\frac{4}{\sqrt{5}-1}\right) = 1.1744.$$

37. a) Suppose $f$ is integrable, $\int |f|\,d\mu < \infty$. Since $|\varphi_n^2 f| \le |f|$ and $\varphi_n^2 f$ converges pointwise to $\varphi^2 f$, by dominated convergence $\int \varphi_n^2 f\,d\mu \to \int \varphi^2 f\,d\mu$. Because $f$ is an arbitrary integrable function, $\varphi_n^2 \xrightarrow{w} \varphi^2$.
b) A dominated convergence argument now fails because $1/\varphi_n$ can be arbitrarily large. Because of this, it need not be the case that $1/\varphi_n \xrightarrow{w} 1/\varphi$. For instance, suppose $\mu$ is Lebesgue measure on $(0,1)$, and $\varphi_n(x)$ is $1/n$ if $x \in (0, 1/n)$ and is one otherwise. Then $\varphi_n$ converges pointwise to the test function $\varphi$ that is identically one. But if $f$ is one, $\int (1/\varphi_n) f\,d\mu = 2 - 1/n \to 2$, instead of $\int (1/\varphi) f\,d\mu = 1$.
41. The equation $E_{\theta_0}\varphi = \alpha$ gives $\varphi(0) + 4\varphi(1) + 4\varphi(2) = 9\alpha$, and the equation $E_{\theta_0} X\varphi = \alpha E_{\theta_0} X$ gives $\varphi(1) + 2\varphi(2) = 3\alpha$. If $\varphi(1) = 0$, then these equations give $\varphi(0) = 3\alpha$ and $\varphi(2) = 3\alpha/2$. This is the solution if $\alpha \le 1/3$. When $\alpha > 1/3$, then $\varphi(0) = 1$, $\varphi(1) = (3\alpha - 1)/2$, and $\varphi(2) = (1 + 3\alpha)/4$.
42. The joint densities are $\exp\{-T(x)/(2\sigma^2)\}/(4\pi^2\sigma^4)$, where $T(x) = x_1^2 + \cdots + x_4^2$, an exponential family. The UMP unbiased test will reject if and only if $T \le c_1$ or $T \ge c_2$ with $c_1$ and $c_2$ adjusted so that $P_1(T \le c_1) + P_1(T \ge c_2) = 5\%$ and $E_1 T\varphi = 20\%$. The density of $T$ when $\sigma = 1$ is $te^{-t/2}/4$, and after a bit of calculus these equations become

$$1 - (1 + c_1/2)e^{-c_1/2} + (1 + c_2/2)e^{-c_2/2} = 5\%$$

and

$$4 - (4 + 2c_1 + c_1^2/2)e^{-c_1/2} + (4 + 2c_2 + c_2^2/2)e^{-c_2/2} = 20\%.$$

Numerical solution of these equations gives $c_1 = 0.607$ and $c_2 = 12.802$.
44. a) The densities form an exponential family with $T = X$, and by Theorem 12.26 the uniformly most powerful unbiased test will be two-sided with the proper level and uncorrelated with $X$ if $\theta = \theta_0$. Since $X$ is continuous, we do not need to worry about randomization, and can take $\varphi^* = I\{X \notin (c_1, c_2)\}$. The constants $c_1$ and $c_2$ are determined by

---

test of $H_0 : \theta \le 1$ or $\theta \ge 3$ versus $H_1 : \theta \in (1, 3)$. In general, if $\theta_1 < \theta_2$, $\alpha \in (0,1)$, and the data come from a one-parameter exponential family with $\eta(\cdot)$ a monotone function, then there will be a two-sided test $\varphi^*$ with $\beta_{\varphi^*}(\theta_1) = \beta_{\varphi^*}(\theta_1) = 1 - \alpha$, and $1 - \varphi^*$ will be a uniformly most powerful level $\alpha$ test of $H_0 : \theta \le \theta_1$ or $\theta \ge \theta_2$ versus $H_1 : \theta \in (\theta_1, \theta_2)$.

$$1 - \alpha = P_{\theta_0}\big(X \in (c_1, c_2)\big) = \frac{e^{\theta_0 c_2} - e^{\theta_0 c_1}}{2\sinh(\theta_0)},$$

and $\mathrm{Cov}_{\theta_0}(X, \varphi^*) = -\mathrm{Cov}_{\theta_0}(X, 1-\varphi^*) = 0$, which becomes

$$\int_{c_1}^{c_2} \frac{x\theta_0 e^{\theta_0 x}}{2\sinh(\theta_0)}\,dx = (1-\alpha)E_{\theta_0}X,$$

or

$$\frac{c_2 e^{\theta_0 c_2} - c_1 e^{\theta_0 c_1}}{2\sinh(\theta_0)} - \frac{1-\alpha}{\theta_0} = (1-\alpha)\left[\coth(\theta_0) - \frac{1}{\theta_0}\right].$$

b) When $\theta_0 = 0$, $p_{\theta_0}(x) = 1/2$, $x \in (-1, 1)$, the uniform density. The equations for $c_1$ and $c_2$ are

$$1 - \alpha = P_0\big(X \in (c_1, c_2)\big) = \frac{c_2 - c_1}{2}$$

and

$$\int_{c_1}^{c_2} \frac{1}{2} x\,dx = \frac{1}{4}(c_2^2 - c_1^2) = 0.$$

Solving, $c_2 = 1 - \alpha$ and $c_1 = -(1-\alpha)$.

## B.13 Problems of Chapter 13

1. a) Since densities must integrate to one, if $\theta \neq 0$ and $\phi \neq 0$,

$$\begin{aligned} A(\theta,\phi) &= \log \int_0^1 \int_0^1 (x+y)e^{\theta x + \phi y}\,dx\,dy \\ &= \log\left[\frac{\theta(e^\theta - 1)(\phi e^\phi + 1 - e^\phi) + \phi(e^\phi - 1)(\theta e^\theta + 1 - e^\theta)}{\theta^2\phi^2}\right]. \end{aligned}$$

b) The marginal density of $X$ is

$$\int_0^1 p_{\theta,\phi}(x,y)\,dy = \frac{x\phi(e^\phi - 1) + \phi e^\phi + 1 - e^\phi}{\phi^2} e^{\theta x - A(\theta,\phi)}.$$

This has the form in Theorem 13.2, with the dominating measure $\lambda_\phi$ having density $\big[x\phi(e^\phi - 1) + \phi e^\phi + 1 - e^\phi\big]/\phi^2$ with respect to Lebesgue measure on $(0, 1)$.

c) The conditional density is

$$\frac{p_{\theta,\phi}(x,y)}{\int_0^1 p_{\theta,\phi}(u,y)\,du} = \frac{\theta^2(x+y)e^{\theta x}}{\theta e^\theta + 1 - e^\theta + y\theta(e^\theta - 1)}.$$

Again, this has the form from Theorem 13.2, now with a dominating measure $\nu_y$ that has density $x + y$ with respect to Lebesgue measure on $(0, 1)$.

d) The UMP unbiased test will reject $H_0$ if and only if $X \ge c(Y)$ with $c(\cdot)$ chosen so that $P_{\theta=0}\big(X \ge c(y) \mid Y = y\big) = \alpha$. When $\theta = 0$, the conditional density of $X$ given $Y = y$ is $2(x+y)/(1+2y)$. So

$$P_{\theta=0}\big(X \ge c(y) \mid Y = y\big) = \int_{c(y)}^{1} \frac{2(x+y)}{1+2y}\,dx = \frac{1 - c^2(y) + 2y - 2yc(y)}{1+2y}.$$

Solving, this will be $\alpha$ if $c(y) = \sqrt{y^2 + (1+2y)(1-\alpha)} - y$.

e) Now the UMP unbiased test will reject $H_0$ if and only if $X \le c_1(Y)$ or $X \ge c_2(Y)$, with $c_1(\cdot)$ and $c_2(\cdot)$ adjusted so that

$$\int_0^{c_1(y)} \frac{2(x+y)}{1+2y}\,dx + \int_{c_2(y)}^{1} \frac{2(x+y)}{1+2y}\,dx = \frac{c_1^2(y) + 2yc_1(y) + 1 - c_2^2(y) + 2y - 2yc_2(y)}{1+2y} = \alpha,$$

and

$$\int_0^{c_1(y)} \frac{2x(x+y)}{1+2y}\,dx + \int_{c_2(y)}^{1} \frac{2x(x+y)}{1+2y}\,dx = \alpha \int_0^1 \frac{2x(x+y)}{1+2y}\,dx,$$

or

$$\frac{2c_1^3(y) + 3yc_1^2(y) + 2 - 2c_2^3(y) + 3y - 3yc_2^2(y)}{3+6y} = \alpha\frac{2+3y}{3+6y}.$$

Explicit solution of these equations for $c_1(y)$ and $c_2(y)$ does not seem possible.

3. a) Letting $\theta_1 = \lambda_x - \lambda_y$ and $\theta_2 = -\lambda_x$, the joint densities are

$$\frac{\lambda_x^{\alpha_x} x^{\alpha_x - 1} \lambda_y^{\alpha_y} y^{\alpha_y - 1}}{xy\Gamma(\alpha_x)\Gamma(\alpha_y)} e^{-\lambda_x x - \lambda_y y} = \frac{|\theta_2|^{\alpha_x} x^{\alpha_x} |\theta_1 + \theta_2|^{\alpha_y} y^{\alpha_y}}{xy\Gamma(\alpha_x)\Gamma(\alpha_y)} e^{\theta_1 y + \theta_2(x+y)},$$

which is a canonical exponential family with sufficient statistics $T_1 = Y$ and $T_2 = X + Y$. With this parameterization, we are testing $H_0 : \theta_1 \le 0$ versus $H_1 : \theta_1 > 0$, and the UMP unbiased test will reject $H_0$ if and only if $T_1 > z(T_2)$, with $z(\cdot)$ chosen so that $P_{(0,\theta_2)}\big[T_1 > z(t_2) \mid T_2 = t_2\big] = \alpha$. To compute this conditional probability, first note that the joint density of $T_1$ and $T_2$ is (the Jacobian for the transformation is one)

$$\frac{|\theta_2|^{\alpha_x}(t_2 - t_1)^{\alpha_x - 1}|\theta_1 + \theta_2|^{\alpha_y} t_1^{\alpha_y - 1}}{\Gamma(\alpha_x)\Gamma(\alpha_y)} e^{\theta_1 t_1 + \theta_2 t_2}, \qquad 0 < t_1 < t_2.$$

The conditional density when $\theta_1 = 0$ is

$$\frac{(t_2-t_1)^{\alpha_x-1}t_1^{\alpha_y-1}}{\int_0^{t_2}(t_2-\tau)^{\alpha_x-1}\tau^{\alpha_y-1}\,d\tau} = \frac{(t_2-t_1)^{\alpha_x-1}t_1^{\alpha_y-1}}{t_2^{\alpha_x+\alpha_y-1}\int_0^1(1-u)^{\alpha_x-1}u^{\alpha_y-1}\,du}$$
$$= \frac{\Gamma(\alpha_x)\Gamma(\alpha_y)(t_2-t_1)^{\alpha_x-1}t_1^{\alpha_y-1}}{\Gamma(\alpha_x+\alpha_y)t_2^{\alpha_x+\alpha_y-1}}.$$

Here the first equality arises from the change of variables $u = \tau/t_2$. The change of variables $u = t_1/t_2$ now gives

$$P_{(0,\theta_2)}\bigl[T_1 > a \bigm| T_2 = t_2\bigr] = \int_a^{t_2} \frac{\Gamma(\alpha_x)\Gamma(\alpha_y)(t_2-t_1)^{\alpha_x-1}t_1^{\alpha_y-1}}{\Gamma(\alpha_x+\alpha_y)t_2^{\alpha_x+\alpha_y-1}}\,dt_1$$
$$= \int_{a/t_2}^1 \frac{\Gamma(\alpha_x)\Gamma(\alpha_y)}{\Gamma(\alpha_x+\alpha_y)}u^{\alpha_y-1}(1-u)^{\alpha_x-1}\,du$$
$$= 1 - F(a/t_2),$$

where $F$ is the cumulative distribution function for the beta distribution with parameters $\alpha_y$ and $\alpha_x$. If $q$ is the upper $\alpha$th quantile for this distribution, so $F(q) = 1-\alpha$, then the probability will be $\alpha$ if $a = qt_2$. Thus the UMP unbiased test rejects $H_0$ if and only if $T_1 > qT_2$.

b) From the definition, $\sum_{i=1}^n (X_i/\sigma_x)^2 \sim \chi_n^2 = \Gamma(n/2, 1/2)$, and since the reciprocal of the "failure rate" is a scale parameter, $ns_x^2 \sim \Gamma\bigl(n/2, 1/(2\sigma_x^2)\bigr)$. Similarly, $ms_y^2 \sim \Gamma\bigl(m/2, 1/(2\sigma_y^2)\bigr)$. From part (a), the UMP unbiased test will reject $H_0$ if and only if $ns_x^2/(ns_x^2+ms_y^2) > q$, if and only if $F > mq/\bigl(n(1-q)\bigr)$, where $q$ is the upper $\alpha$th quantile for $\beta(m/2, n/2)$. This is the usual $F$-test, and this derivation shows that the upper $\alpha$th quantile of $F_{n,m}$ is $mq/\bigl(n(1-q)\bigr)$.

5. The joint densities are

$$(2\pi)^{-n/2}\exp\left[\beta x'y + \gamma w'y - \frac{1}{2}\|y\|^2 - \frac{1}{2}\|\beta x + \gamma w\|^2\right].$$

Introducing new parameters $\theta = \beta - \gamma$ and $\eta = \gamma$, the joint densities become

$$(2\pi)^{-n/2}\exp\left[\theta x'y + \eta(w+x)'y - \frac{1}{2}\|y\|^2 - \frac{1}{2}\|\theta x + \eta x + \eta w\|^2\right],$$

and we would like to test $H_0 : \theta \le 0$ versus $H_1 : \theta > 0$. These densities form an exponential family with canonical sufficient statistics $U = x'Y$ and $T = (x+w)'Y$. By Theorem 13.6, a uniformly most powerful unbiased test will reject $H_0$ if $U > c(T)$ with $c(\cdot)$ chosen so that

$$P_{\theta=0}\bigl(U > c(t) \bigm| T = t\bigr) = \alpha.$$

When $\theta = 0$, $U$ and $T$ have a bivariate normal distribution with $EU = \eta x'(x+w)$, $ET = \eta\|x+w\|^2$, $\operatorname{Var}(U) = \|x\|^2$, $\operatorname{Var}(T) = \|x+w\|^2$, and $\operatorname{Cov}(U,T) = x'(x+w)$. So, when $\theta = 0$,

$$U|T = t \sim N\left(\frac{x'(x+w)t}{\|x+w\|^2}, \frac{\|x\|^2\|w\|^2 - (x'w)^2}{\|x+w\|^2}\right),$$

and the uniformly most powerful unbiased test will reject $H_0$ if

$$U - \frac{x'(x+w)T}{\|x+w\|^2} > z_\alpha \sqrt{\frac{\|x\|^2\|w\|^2 - (x'w)^2}{\|x+w\|^2}}.$$

6. a) Taking $\theta_1 = \log\lambda_x - \log\lambda_y$ and $\theta_2 = \log\lambda_y$, the likelihood is

$$\frac{\exp\left[\theta_1 T_1 + \theta_2 T_2 - me^{\theta_1+\theta_2} - ne^{\theta_2}\right]}{\prod_{i=1}^m X_i! \prod_{j=1}^n Y_j!},$$

where $T_1 = \sum_{i=1}^m X_i$ and $T_2 = \sum_{i=1}^m X_i + \sum_{j=1}^n Y_j$. The UMP unbiased test has form

$$\varphi = \begin{cases} 1, & T_1 > z(T_2); \\ \gamma(T_2), & T_1 = z(T_2); \\ 0, & T_1 < z(T_2), \end{cases}$$

with $z$, $\gamma$ chosen so that $P\big(T_1 > z(t_2) \bigm| T_2 = t_2\big) + \gamma(t_2)P\big(T_1 = z(t_2) \bigm| T_2 = t_2\big) = \alpha$ when $\lambda_x = \lambda_y$. Note that if $\lambda_x = \lambda_y$

$$\begin{aligned} P(T_1 = t_1 | T_2 = t_2) &= \frac{P\big(\sum_{i=1}^m X_i = t_1, \sum_{j=1}^n Y_j = t_2 - t_1\big)}{P\big(\sum_{i=1}^m X_i + \sum_{j=1}^n Y_j = t_2\big)} \\ &= \frac{(m\lambda)^{t_1}(n\lambda)^{t_2-t_1}e^{-(m+n)\lambda}/\big[t_1!(t_2-t_1)!\big]}{\big((m+n)\lambda\big)^{t_2}e^{-(m+n)\lambda}/t_2!} \\ &= \binom{t_2}{t_1}\left(\frac{m}{m+n}\right)^{t_1}\left(\frac{n}{m+n}\right)^{t_2-t_1}, \end{aligned}$$

and so, when $\lambda_x = \lambda_y$, $T_1|T_2 = t_2 \sim \text{Binomial}\big(t_2, m/(m+n)\big)$.
b) If $\lambda_x = \lambda_y$, $P(T_1 = 9|T_2 = 9) = (2/3)^9 = 2.6\%$ and $P(T_1 = 8|T_2 = 9) = 3(2/3)^8 = 11.7\%$, so in this case $z(9) = 8$ and $\gamma(9) = 20.5\%$. So the chance of rejection is 20.5%.
c) Using normal approximation for the binomial distribution, the approximate test will reject $H_0$ if

$$T_1 > \frac{mT_2 + z_\alpha\sqrt{mnT_2}}{m+n}.$$

## B.14 Problems of Chapter 14

1. a) The matrix $X$ should be

$$\begin{pmatrix} 1 & w_1 & x_1 \\ \vdots & \vdots & \vdots \\ 1 & w_n & x_n \end{pmatrix}.$$

b) Because $w_1 + \cdots + w_n = x_1 + \cdots + x_n = 0$, $w$ and $x$ are both orthogonal to a column of 1s, and $X$ will be of full rank unless $w$ and $x$ are collinear. Algebraically, this would occur if $D = S_{xx}S_{ww} - S_{xw}^2$ is zero.
c) Using the formula $\hat{\beta} = (X'X)^{-1}X'Y$, since

$$X'X = \begin{pmatrix} n & 0 & 0 \\ 0 & S_{ww} & S_{wx} \\ 0 & S_{wx} & S_{xx} \end{pmatrix}, \qquad X'Y = \begin{pmatrix} n\overline{Y} \\ S_{wY} \\ S_{xY} \end{pmatrix},$$

and

$$(X'X)^{-1} = \begin{pmatrix} 1/n & 0 & 0 \\ 0 & S_{xx}/D & -S_{wx}/D \\ 0 & -S_{wx}/D & S_{ww}/D \end{pmatrix},$$

we have

$$\hat{\beta}_1 = \overline{Y}, \quad \hat{\beta}_2 = \frac{S_{xx}S_{wY} - S_{wx}S_{xY}}{S_{xx}S_{ww} - S_{xw}^2}, \text{ and } \hat{\beta}_3 = \frac{S_{ww}S_{xY} - S_{wx}S_{wY}}{S_{xx}S_{ww} - S_{xw}^2}.$$

d) The covariance of $\hat{\beta}$ is $\text{Cov}(\hat{\beta}) = (X'X)^{-1}\sigma^2$, with $(X'X)^{-1}$ given above.
e) The UMVU estimator of $\sigma^2$ is

$$S^2 = \frac{1}{n-3}\sum_{i=1}^{n}(Y_i - \hat{\beta}_1 - w_i\hat{\beta}_2 - x_i\hat{\beta}_3)^2.$$

f) The variance of $\hat{\beta}_1$ is $\sigma^2/n$, estimated by $S^2/n$. So the confidence interval for $\beta_1$ is

$$\left(\hat{\beta}_1 \pm \frac{S}{\sqrt{n}}t_{\alpha/2,n-3}\right).$$

Since $\hat{\beta}_3 - \hat{\beta}_2 = \begin{pmatrix} 0 & -1 & 1 \end{pmatrix}\hat{\beta}$,

$$\begin{aligned} \text{Var}(\hat{\beta}_3 - \hat{\beta}_2) &= \begin{pmatrix} 0 & -1 & 1 \end{pmatrix} \text{Cov}(\hat{\beta}) \begin{pmatrix} 0 \\ -1 \\ 1 \end{pmatrix} \\ &= \begin{pmatrix} 0 & -1 & 1 \end{pmatrix} (X'X)^{-1} \begin{pmatrix} 0 \\ -1 \\ 1 \end{pmatrix} \sigma^2 \\ &= \frac{S_{xx} + S_{ww} + 2S_{xw}}{S_{xx}S_{ww} - S_{xw}^2}\sigma^2 \\ &= \frac{\sum_{i=1}^{n}(x_i + w_i)^2}{S_{xx}S_{ww} - S_{xw}^2}\sigma^2. \end{aligned}$$

Plugging in $S^2$ to estimate $\sigma^2$, the confidence interval for $\beta_3 - \beta_2$ is

$$\left(\hat{\beta}_3 - \hat{\beta}_2 \pm s\sqrt{\frac{\sum_{i=1}^n (x_i + w_i)^2}{S_{xx}S_{ww} - S_{xw}^2}}t_{\alpha/2,n-3}\right).$$

4. a) The dimension $r$ is $2m-1$ because the rows of the design matrix corresponding to $(i,j)$ pairs $(1,1), \dots, (1,m)$ and $(2,1), \dots, (m,1)$ are linearly independent.
b) The least squares estimators $\hat{\alpha}_i$ and $\hat{\gamma}_j$ are not unique in this problem (since $r < p$), but they still minimize

$$L = \sum_{i=1}^m \sum_{j=1}^m (Y_{ij} - \alpha_i - \gamma_j)^2,$$

and must satisfy normal equations, obtained setting $\partial L/\partial \alpha_i$, $i = 1, \dots, m$ and $\partial L/\partial \gamma_j$, $j = 1, \dots, m$ to zero. This gives

$$\hat{\alpha}_i + \hat{\overline{\gamma}} = \overline{Y}_{i\cdot}, \qquad i = 1, \dots, m,$$

and

$$\hat{\gamma}_j + \hat{\overline{\alpha}} = \overline{Y}_{\cdot j}, \qquad j = 1, \dots, m,$$

where

$$\hat{\overline{\alpha}} = \frac{1}{m}\sum_{i=1}^m \hat{\alpha}_i, \qquad \hat{\overline{\gamma}} = \frac{1}{m}\sum_{j=1}^m \hat{\gamma}_j,$$

$$\overline{Y}_{i\cdot} = \frac{1}{m}\sum_{j=1}^m Y_{ij}, \qquad i = 1, \dots, m,$$

and

$$\overline{Y}_{\cdot j} = \frac{1}{m}\sum_{i=1}^m Y_{ij}, \qquad j = 1, \dots, m.$$

Averaging these equations over $i$ or $j$,

$$\hat{\overline{\alpha}} + \hat{\overline{\gamma}} = \overline{Y} = \frac{1}{m^2}\sum_{i=1}^m \sum_{j=1}^m Y_{ij}.$$

So the least squares estimator for $\xi_{ij}$ is

$$\hat{\xi}_{ij} = \hat{\alpha}_i + \hat{\gamma}_j = \overline{Y}_{i\cdot} + \overline{Y}_{\cdot j} - \overline{Y},$$

$i = 1, \dots, m$, $j = 1, \dots, m$.
c) Since $e_{ij} = Y_{ij} - \hat{\xi}_{ij}$, the estimator of $\sigma^2$ is

$$S^2 = \frac{1}{(m-1)^2}\sum_{i=1}^m \sum_{j=1}^m (Y_{ij} - \overline{Y}_{i\cdot} - \overline{Y}_{\cdot j} + \overline{Y})^2.$$

d) Because

$$\overline{\xi}_{i\cdot} = \frac{1}{m}\sum_{j=1}^{m} \xi_{ij} = \alpha_i + \overline{\gamma}, \qquad i = 1, \ldots, m,$$

we have $\alpha_i - \alpha_j = \overline{\xi}_{i\cdot} - \overline{\xi}_{j\cdot}$, expressing this difference as a linear function of $\xi$. The least squares estimator of $\alpha_i - \alpha_j$ is $\hat{\alpha}_i - \hat{\alpha}_j = \overline{Y}_{i\cdot} - \overline{Y}_{j\cdot}$.
e) Let $Z_i = m^{1/2}(\overline{Y}_{i\cdot} - \alpha_i - \overline{\gamma})/\sigma \sim N(0,1)$. These variables depend on different subsets of the $Y_{ij}$, and hence they are independent. Also, averaging $\hat{\xi}_{ij} = \overline{Y}_{i\cdot} + \overline{Y}_{\cdot j} - \overline{Y}$ over $j$ gives $\overline{Y}_{i\cdot} = \hat{\overline{\xi}}_{i\cdot}$, so the variables $Z_i$, $i = 1, \ldots, m$, are functions of $\hat{\xi}$ and are independent of $S^2$. Since $(m-1)^2 S^2/\sigma^2 \sim \chi^2_{(m-1)^2}$, by Definition 14.14,

$$\frac{\max Z_i - \min Z_i}{S/\sigma} = \frac{\max(\overline{Y}_{i\cdot} - \alpha_i) - \min(\overline{Y}_{i\cdot} - \alpha_i)}{S/\sqrt{m}}$$

has the studentized range distribution with parameters $m$ and $(m-1)^2$. If $q$ is the upper $\alpha$th quantile of this distribution, and $I_{ij} = (\hat{\alpha}_i - \hat{\alpha}_j \pm qS/\sqrt{m})$, then, proceeding as in the text,

$$\begin{aligned} P(\alpha_i - \alpha_j \in I_{ij}, \forall i \neq j) &= P\left(\frac{\max(\overline{Y}_{i\cdot} - \alpha_i) - \min(\overline{Y}_{i\cdot} - \alpha_i)}{S/\sqrt{m}} < q\right) \\ &= 1 - \alpha. \end{aligned}$$

So $I_{ij}$, $i \neq j$, are simultaneous confidence intervals for the differences $\alpha_i - \alpha_j$.
f) Under $H_0$, the data are distributed as they are in one-way analysis of variance, so $q = m$, and the least squares estimator for the mean of $Y_{ij}$ is $\overline{Y}_{\cdot j}$. Then

$$\begin{aligned} \|\hat{\xi} - \hat{\xi}_0\|^2 &= \sum_{i=1}^{m}\sum_{j=1}^{m}(\overline{Y}_{i\cdot} - \overline{Y})^2 = m\sum_{i=1}^{m}(\overline{Y}_{i\cdot} - \overline{Y})^2, \\ T &= \frac{m\sum_{i=1}^{m}(\overline{Y}_{i\cdot} - \overline{Y})^2}{(m-1)S^2}, \end{aligned}$$

and we should reject $H_0$ if $T > F_{\alpha, m-1, (m-1)^2}$.
g) The test statistic has a noncentral $F$ distribution with noncentrality parameter

$$\delta^2 = \frac{\|\xi - P_0\xi\|^2}{\sigma^2} = \frac{m\sum_{i=1}^{m}(\alpha_i - \overline{\alpha})^2}{\sigma^2}$$

(found easily from the prior results, since $\xi - P_0\xi$ equals $\hat{\xi} - \hat{\xi}_0$ when $\epsilon = 0$) and degrees of freedom $m-1$ and $(m-1)^2$. The power is the probability this distribution assigns to the interval $(F_{\alpha, m-1, (m-1)^2}, \infty)$.

h) Taking $\psi = (\alpha_1 - \alpha_m, \ldots, \alpha_{m-1} - \alpha_m)'$, we can write any contrast $\sum_{i=1}^m a_i\alpha_i$ as $\sum_{i=1}^{m-1} a_i\psi_i$. From part (d), the least squares estimator of this contrast is

$$\sum_{i=1}^{m-1} a_i(\overline{Y}_{i\cdot} - \overline{Y}_{m\cdot}) = \sum_{i=1}^{m} a_i\overline{Y}_{i\cdot}.$$

The variance of this estimator is $\sum_{i=1}^m a_i^2\sigma^2/m = \|a\|^2\sigma^2/m$, estimated by $\|a\|^2S^2/m$. So the desired simultaneous confidence intervals are

$$\left(\sum_{i=1}^{m} a_i\overline{Y}_{i\cdot} \pm \|a\|S\sqrt{\tfrac{m-1}{m}F_{\alpha,m-1,(m-1)^2}}\right).$$

13. a) Let $\omega_0 = \{v \in \omega : Av = 0\}$, with dimension $r-q$ because $A$ has full rank. Choose $\xi_0 \in \omega$ so that $A\xi_0 = \psi_0$, and introduce $Y^* = Y - \xi_0 \sim N(\xi^*, \sigma^2 I)$, with $\xi^* = \xi - \xi_0 \in \omega$. Since $\psi = \psi_0$ if and only if $A\xi = A\xi_0$ if and only if $A\xi^* = 0$, the null hypothesis is $H_0 : \xi^* \in \omega_0$, tested using the usual statistic

$$T = \frac{(n-r)\|PY^* - P_0Y^*\|^2}{q\|Y^* - PY^*\|^2},$$

where $P$ is the projection onto $\omega$ and $P_0$ is the projection onto $\omega_0$. The level $\alpha$ test rejects $H_0$ if $T \geq F_{\alpha,q,n-r}$.
b) Because $\xi_0 \in \omega$, $P\xi_0 = \xi_0$ and $\|Y^* - PY^*\|^2 = \|Y - PY\|^2$. So $T = \|PY^* - P_0Y^*\|^2/(qS^2)$, where $S^2 = \|Y - PY\|^2/(n-r)$ is the usual estimator of $\sigma^2$. Next, note that $P - P_0$ is the projection onto $\omega \cap \omega_0^\perp$, for if $v = v_0 + v_1 + v_2$ with $v_0 \in \omega_0$, $v_1 \in \omega \cap \omega_0^\perp$, and $v_2 \in \omega^\perp$, then $(P - P_0)v = (v_0 + v_1) - v_0 = v_1$. Because $A = AP$, the rows of $A$ all lie in $\omega$, and so they must span $\omega \cap \omega_0^\perp$. Consequently, $P - P_0 = A'(AA')^{-1}A$, as in the derivation for (14.14). So

$$\begin{aligned}\|(P - P_0)Y^*\|^2 &= (Y - \xi_0)'(P - P_0)(Y - \xi_0)\\ &= (Y - \xi_0)'A'(AA')^{-1}A(Y - \xi_0)\\ &= (\hat\psi - \psi_0)'(AA')^{-1}(\hat\psi - \psi_0),\end{aligned}$$

and

$$T = \frac{(\hat\psi - \psi_0)'(AA')^{-1}(\hat\psi - \psi_0)}{qS^2}.$$

The confidence set consists of all values $\psi_0$ where we would accept the null hypothesis $\psi = \psi_0$, that is, $\{\psi_0 : T(\psi_0) < F_{\alpha,q,n-r}\}$. Using the last formula for $T$, this region is the ellipse given in the problem.

14. a) Because $\sum_{l=1}^c x_{kl} = 0$, the columns of the design matrix $X$ are orthogonal, and $X'X$ is diagonal. The (0,0) entry will be $\sum_{k=1}^p\sum_{l=1}^c x_{kl}^2$, and the other diagonal entries will all equal $c$. Also, the zeroth entry of $X'Y$ will be $\sum_{k=1}^p\sum_{l=1}^c x_{kl}Y_{kl}$, and the $k$th entry will be $\sum_{l=1}^c Y_{kl}$, $1 \leq k \leq p$. Since $\hat\beta = (X'X)^{-1}X'Y$,

$$\hat{\beta}_k = \frac{1}{c}\sum_{l=1}^{c} Y_{kl}, \qquad k = 1, \ldots, p,$$

are the least squares estimators for $\beta_1, \ldots, \beta_p$, and

$$\hat{\beta}_0 = \frac{\sum_{k=1}^{p}\sum_{l=1}^{c} x_{kl}Y_{kl}}{\sum_{k=1}^{p}\sum_{l=1}^{c} x_{kl}^2}$$

is the least squares estimator for $\beta_0$. If $Z_k = \sqrt{p}(\hat{\beta}_k - \beta_k)/\sigma$, $k = 1, \ldots, p$, then $Z_1, \ldots, Z_c$ are i.i.d. $N(0, 1)$ independent of

$$S^2 = \frac{1}{pc - p - 1}\sum_{k=1}^{p}\sum_{l=1}^{c}(Y_{kl} - \hat{\beta}_k - \hat{\beta}_0 x_{kl})^2.$$

Because $(pc - p - 1)S^2/\sigma^2 \sim \chi^2_{pc-p-1}$,

$$M = \frac{\max |Z_k|}{S/\sigma} = \frac{\max |\hat{\beta}_k - \beta_k|}{S/\sqrt{c}}$$

has the studentized maximum modulus distribution with parameters $p$ and $pc - p - 1$. If $q$ is the upper $\alpha$th quantile for this distribution, then

$$P\big[\beta_k \in (\hat{\beta}_k \pm Sq/\sqrt{c}), k = 1, \ldots, p\big] = P(M < q) = 1 - \alpha,$$

and $(\hat{\beta}_k \pm sq/\sqrt{c})$, $k = 1, \ldots, p$, are simultaneous $1-\alpha$ confidence intervals for $\beta_1, \ldots, \beta_p$.
b) Let $\overline{x} = \sum_{l=1}^{c} x_{kl}/c$, and write

$$Y_{kl} = \beta_k + \beta_0\overline{x} + \beta_0(x_{kl} - \overline{x}) + \epsilon_{kl} = \beta_k^* + \beta_0^*(x_{kl} - \overline{x}) + \epsilon_{kl},$$

where $\beta_k^* = \beta_k + \beta_0\overline{x}$ and $\beta_0^* = \beta_0$. Then the design matrix $X^*$ has orthogonal columns, and proceeding as in part (a), $\hat{\beta}_k^* = \overline{Y}_k = \sum_{l=1}^{c} Y_{kl}/c$, $k = 1, \ldots, p$, and

$$\hat{\beta}_0^* = \frac{\sum_{k=1}^{p}\sum_{l=1}^{c} Y_{kl}(x_{kl} - \overline{x})}{\sum_{k=1}^{p}\sum_{l=1}^{c}(x_{kl} - \overline{x})^2}$$

are the least squares estimators of $\beta_k^*$, $k = 1, \ldots, p$, and $\beta_0^*$. Since $Z_k = \sqrt{c}(\overline{Y}_k - \beta_k^*)/\sigma$, $k = 1, \ldots, p$, are i.i.d. $N(0, 1)$, independent of

$$S^2 = \frac{1}{pc - p - 1}\sum_{k=1}^{p}\sum_{l=1}^{c}(Y_{kl} - \hat{\beta}_k^* - \hat{\beta}_0^* x_{kl})^2,$$

and since $(pc - p - 1)S^2/\sigma^2 \sim \chi^2_{pc-p-1}$,

$$R = \frac{\max Z_k - \min Z_k}{S/\sigma} = \frac{\max_{j,k} |\beta_k - \beta_j - \overline{Y}_k + \overline{Y}_j|}{S/\sqrt{c}}$$

has the studentized maximum range distribution with parameters $p$ and $pc - p - 1$. If $q$ is the upper $\alpha$th quantile of this distribution,

$$P\big[\beta_k - \beta_j \in (\overline{Y}_k - \overline{Y}_j \pm qS/\sqrt{c}), \forall k, j\big] = P(R < q) = 1 - \alpha.$$

So $(\overline{Y}_k - \overline{Y}_j \pm qS/\sqrt{c})$, $1 \le j < k \le p$, are simultaneous confidence intervals for the differences $\beta_k - \beta_j$, $1 \le j < k \le p$.

15. a) Since $\hat\beta_i = \sum_{j=1}^{n_i} Y_{ij}/n_i$, the $(i,j)$th entry of $PY$ is $\hat\beta_i$. The $(i,j)$th entry of $P_0 Y$ is $\overline{Y} = \sum_{i=1}^{p}\sum_{j=1}^{n_i} Y_{ij}/n$, where $n = \sum_{i=1}^{p} n_i$. So

$$\|PY - P_0 Y\|^2 = \sum_{i=1}^{p}\sum_{j=1}^{n_i}(\hat\beta_i - \overline{Y})^2 = \sum_{i=1}^{p} n_i(\hat\beta_i - \overline{Y})^2$$

and

$$S^2 = \frac{\|Y - PY\|^2}{n-p} = \frac{1}{n-p}\sum_{i=1}^{p}\sum_{j=1}^{n_i}(Y_{ij} - \hat\beta_i)^2.$$

The $F$-statistic for the test is

$$T = \frac{\sum_{i=1}^{p} n_i(\hat\beta_i - \overline{Y})^2}{(p-1)S^2},$$

and we reject $H_0$ if and only if $T \ge F_{p-1,n-p}(1-\alpha)$.

b) The estimate $a_1\hat\beta_1 + \cdots + a_p\hat\beta_p$ has variance $\sigma^2 \sum_{i=1}^{p} a_i^2/n_i$, estimated by $S^2 \sum_{i=1}^{p} a_i^2/n_i$. So the simultaneous confidence intervals are

$$\sum_{i=1}^{p} a_i\beta_i \in \left(\sum_{i=1}^{p} a_i\hat\beta_i \pm S\sqrt{(p-1)F_{p-1,n-p}(1-\alpha)\sum_{i=1}^{p}\frac{a_i^2}{n_i}}\right).$$

## B.17 Problems of Chapter 17

3. a) As usual, let

$$\overline{Y}_{i\cdot} = \frac{1}{n}\sum_{j=1}^{n} Y_{ij}, \qquad i = 1, \dots, p,$$

and

$$\overline{Y}_{\cdot j} = \frac{1}{p}\sum_{i=1}^{p} Y_{ij}, \qquad j = 1, \dots, n.$$

Introduce

$$SS_e = \sum_{i=1}^{p}\sum_{j=1}^{n}(Y_{ij} - \overline{Y}_{i\cdot} - \overline{Y}_{\cdot j} + \overline{Y}_{\cdot\cdot})^2$$
$$= \sum_{j=1}^{n} \|Y_j - \overline{Y}\|^2 - \frac{1}{p}\sum_{j=1}^{n}[\mathbf{1}'(Y_j - \overline{Y})]^2$$

and

$$SS_\beta = p\sum_{j=1}^{n}(\overline{Y}_{\cdot j} - \overline{Y}_{\cdot\cdot})^2 = \frac{1}{p}\sum_{j=1}^{n}[\mathbf{1}'(Y_j - \overline{Y})]^2,$$

sums of squares that would arise testing null hypotheses if the $\beta_j$ were viewed as fixed constants. Let $P = \mathbf{1}\mathbf{1}'/p$, the projection onto the linear span of $\mathbf{1}$, and let $Q = I - P$. The covariance of $Y_j$ is $\Sigma = \sigma^2 I + \tau^2\mathbf{1}\mathbf{1}' = \sigma^2 Q + (\sigma^2 + p\tau^2)P$, with determinant $\sigma^{2(p-1)}(\sigma^2 + p\tau^2)$ (the eigenvector $\mathbf{1}$ has eigenvalue $\sigma^2 + p\tau^2$, and the other eigenvalues are all $\sigma^2$), and $\Sigma^{-1} = \sigma^{-2}Q + (\sigma^2 + p\tau^2)^{-1}P$. The log-likelihood is

$$l(\alpha, \sigma, \tau) = -\frac{1}{2}\sum_{j=1}^{n}(Y_j - \alpha)'\Sigma^{-1}(Y_j - \alpha) - np\log\sqrt{2\pi} - \frac{n}{2}\log|\Sigma|$$
$$= -\frac{1}{2}\sum_{j=1}^{n}(Y_j - \overline{Y})'\Sigma^{-1}(Y_j - \overline{Y}) - \frac{n}{2}(\overline{Y} - \alpha)'\Sigma^{-1}(\overline{Y} - \alpha)$$
$$- np\log\sqrt{2\pi} - \frac{n(p-1)}{2}\log\sigma^2 - \frac{n}{2}\log(\sigma^2 + p\tau^2).$$

(The cross-product term drops out because $\sum_{j=1}^{n}(Y_j - \overline{Y}) = 0$.) This is maximized over $\alpha$ by $\hat{\alpha} = \overline{Y}$, regardless of the value of $\sigma$ or $\tau$. To maximize over $\sigma$ and $\tau$, introduce $\eta^2 = \sigma^2 + p\tau^2$ and note that

$$(Y_j - \overline{Y})'\Sigma^{-1}(Y_j - \overline{Y}) = \frac{\|Q(Y_j - \overline{Y})\|^2}{\sigma^2} + \frac{\|P(Y_j - \overline{Y})\|^2}{\eta^2}$$
$$= \frac{\|Y_j - \overline{Y}\|^2 - [\mathbf{1}'(Y_j - \overline{Y})]^2/p}{\sigma^2} + \frac{[\mathbf{1}'(Y_j - \overline{Y})]^2}{p\eta^2}.$$

So

$$l(\hat{\alpha}, \sigma, \tau) = -\frac{\sum_1^n \|Y_j - \overline{Y}\|^2 - \sum_1^n[\mathbf{1}'(Y_j - \overline{Y})]^2/p}{2\sigma^2} - \frac{\sum_1^n[\mathbf{1}'(Y_j - \overline{Y})]^2}{2p\eta^2}$$
$$- n(p-1)\log\sigma - n\log\eta$$
$$= -\frac{SS_e}{2\sigma^2} - \frac{SS_\beta}{2\eta^2} - n(p-1)\log\sigma - n\log\eta.$$

Setting derivatives to zero suggests that the maximum likelihood estimators should be

$$\hat{\eta}^2 = \frac{1}{n}SS_\beta \text{ and } \hat{\sigma}^2 = \frac{SS_e}{np - n}.$$

But there is a bit of a problem here because we need $\hat{\eta}^2 \geq \hat{\sigma}^2$. If these formulas give $\hat{\eta}^2 < \hat{\sigma}^2$, then the likelihood is maximized with $\hat{\eta} = \hat{\sigma}$. The common value will maximize $l(\hat{\alpha}, \sigma, \sigma) = \sum_{j=1}^n \|Y_j - \overline{Y}\|^2/\sigma^2 - np\log\sigma$. Setting the derivative to zero, the common value is $\tilde{\sigma}^2 = \sum_{j=1}^n \|Y_j - \overline{Y}\|^2/(np)$, and this is also the maximum likelihood estimator for $\sigma^2$ under $H_0$. Plugging these estimates into the likelihood,

$$2\log\lambda = 2l(\hat{\alpha}, \hat{\sigma}, \hat{\tau}) - 2l(\hat{\alpha}, \tilde{\sigma}, 0) = 2n(p-1)\log(\tilde{\sigma}/\hat{\sigma}) + 2n\log(\tilde{\sigma}/\hat{\eta}).$$

Remark: $H_0$ would not be rejected when $\hat{\eta}^2 = \hat{\sigma}^2$, and when $\hat{\eta}^2 > \hat{\sigma}^2$ there is a monotonic relationship between $\lambda$ and the $F$-statistic that would be used to test $H_0$ if $\beta_j$ were viewed as constants. So the $F$-test and likelihood ratio test here would be the same in practice.

b) Under $H_0$, $\alpha = \alpha_0\mathbf{1}$ and

$$(\overline{Y} - \alpha)'\Sigma^{-1}(\overline{Y} - \alpha) = \frac{\overline{Y}'Q\overline{Y}}{\sigma^2} + \frac{(\mathbf{1}'\overline{Y} - p\alpha_0)^2}{p(\sigma^2 + p\tau^2)},$$

which is minimized when $\alpha_0$ is $\mathbf{1}'\overline{Y}/p$. So

$$(\overline{Y} - \tilde{\alpha})'\Sigma^{-1}(\overline{Y} - \tilde{\alpha}) = \frac{SS_\alpha}{n\sigma^2}$$

and

$$l(\tilde{\alpha}, \sigma, \tau) = -\frac{SS_e + SS_\alpha}{2\sigma^2} - \frac{SS_\beta}{2\eta^2} - n(p-1)\log\sigma - n\log\eta,$$

where $SS_\alpha = \sum_{i=1}^p \sum_{j=1}^n (\overline{Y}_{i\cdot} - \overline{Y}_{\cdot\cdot})^2$. Setting derivatives to zero, and keeping in mind that we must have $\tilde{\eta} \geq \tilde{\sigma}$,

$$\tilde{\sigma}^2 = \frac{SS_e + SS_\alpha}{n(p-1)} \text{ and } \tilde{\eta}^2 = \frac{SS_\beta}{n},$$

when $(p-1)SS_\beta \geq SS_e + SS_\alpha$, and $\tilde{\sigma}^2 = \tilde{\eta}^2 = (SS_e + SS_\alpha + SS_\beta)/(np)$ when $(p-1)SS_\beta < SS_e + SS_\alpha$. Plugging these values into the log-likelihood function, $2\log\lambda = n(p-1)\log(\tilde{\sigma}/\hat{\sigma}) + n\log(\tilde{\sigma}/\hat{\eta})$. The estimators $\tilde{\sigma}$ and $\tilde{\eta}$ depend on $SS_\beta$; thus the $F$-statistic and the likelihood ratio statistic are not equivalent.

4. a) The likelihood function is $L(\theta_x, \theta_y) = \theta_x\theta_y\exp[-\theta_x X - \theta_y Y]$. Setting derivatives to zero, the maximum likelihood estimators are $\hat{\theta}_x = 1/X$ and $\theta_y = 1/Y$, and so $\sup_\Omega L(\theta) = L(\hat{\theta}_x, \hat{\theta}_y) = e^{-2}/(XY)$. The maximum likelihood estimator $\tilde{\theta}_y$ for $\theta_y$ under $H_0$ maximizes $L(2\theta_y, \theta_y) = 2\theta_y^2\exp\bigl[-(2X+Y)\theta_y\bigr]$. A bit of calculus gives $\tilde{\theta}_y = 2/(2X+Y)$, and then $\sup_{\Omega_0} L(\theta) = L(\tilde{\theta}_x, \tilde{\theta}_y) = 8e^{-2}/(2X+Y)^2$. The likelihood ratio test statistic is $\lambda = (2X+Y)^2/(8XY)$.

b) Define $U = \theta_x X$ and $V = \theta_y Y$, so $U$ and $V$ are independent standard exponential variables. Under $H_0$, $\lambda$ can be expressed in terms of these variables as $\lambda = (U+V)^2/(4UV)$. The significance level of the test is

$$\begin{aligned}\alpha = P_{H_0}(\lambda \geq c) &= P\bigl[(U+V)^2 \geq 4cUV\bigr]\\ &= P(U \geq mV) + P(V \geq mU),\end{aligned}$$

where $m = 2c - 1 + 2\sqrt{c^2 - c}$. By smoothing, $P(U \geq mV) = EP(U \geq mV|V) = Ee^{-mV} = \int_0^\infty e^{-(m+1)v}\,dv = 1/(m+1)$. So $\alpha = 2/(m+1)$, and to achieve a specified level $\alpha$, $m$ should be $2/\alpha - 1$. From this, $c = (m+1)^2/(4m) = 1/\bigl[\alpha(2-\alpha)\bigr]$.

8. a) The log-likelihood is $l(\theta_1, \theta_2) = -n\overline{X}/\theta_1 - n\overline{Y}/\theta_2 - n\log\theta_1 - n\log\theta_2$. Setting derivatives to zero, the maximum likelihood estimators are $\hat{\theta}_1 = \overline{X}$ and $\hat{\theta}_2 = \overline{Y}$. The maximum likelihood estimator $\tilde{\theta}_1$ for $\theta_1$ under $H_0$ maximizes $l(\theta_1, \theta_1/c_0)$. Setting the derivative to zero, $\tilde{\theta}_1 = (\overline{X} + c_0\overline{Y})/2$. Then

$$\begin{aligned}\log\lambda &= l(\hat{\theta}_1, \hat{\theta}_2) - l(\tilde{\theta}_1, \tilde{\theta}_1/c_0)\\ &= n\log\left(\frac{\overline{X} + c_0\overline{Y}}{2\overline{X}}\right) + n\log\left(\frac{\overline{X} + c_0\overline{Y}}{2c_0\overline{Y}}\right).\end{aligned}$$

b) The confidence intervals would contain all values $c_0$ for which $2\log\lambda < 3.84$. (Here 3.84 is the 95th percentile of $\chi_1^2$.) For the data and sample size given, if $c_0$ is 2.4, $2\log\lambda = 1.66$. This value is less than 3.84, and so 2.4 is in the confidence interval.

c) The Fisher information matrix is

$$I(\theta) = \begin{pmatrix}\theta_1^{-2} & 0\\ 0 & \theta_2^{-2}\end{pmatrix}.$$

It seems natural to take $\theta_0 = \binom{1}{1}$, and then $\Delta = \binom{1}{-1}\sqrt{n}/10$. Define an orthonormal basis $v_1 = \binom{1}{1}/\sqrt{2}$ and $v_2 = \binom{1}{-1}/\sqrt{2}$. Since $\Omega_0$ is linear, the tangent spaces $V_\theta = V$ at different $\theta \in \Omega$ are all the same, each being the linear span of $v_1$. So $P_0 = v_1v_1'$ and $Q_0 = v_2v_2'$ are the projection matrices onto $V$ and $V^\perp$. Since $\Delta$ lies in $V^\perp$, $Q_0\Delta = \Delta$. Since $I(\theta_0)$ is the identity, $P_0 + Q_0I(\theta_0)^{-1}Q_0 = P_0 + Q_0 = \left(\begin{smallmatrix}1 & 0\\ 0 & 1\end{smallmatrix}\right)$. So the noncentrality parameter is $\delta^2 = \Delta'Q_0\bigl[P_0 + Q_0I(\theta_0)^{-1}Q_0\bigr]^{-1}Q_0\Delta = \Delta'\Delta = n/50$. The asymptotic distribution of $2\log\lambda$ is $\chi_1^2(\delta^2)$. If $Z \sim N(0,1)$, then $(Z+\delta)^2 \sim \chi_1^2(\delta^2)$ has this distribution, and so the power is approximately $P\bigl[(Z+\delta)^2 > 3.84\bigr] = P(Z + \delta > 1.96) + P(Z + \delta < -1.96) = \Phi(\delta - 1.96) + \Phi(-\delta - 1.96)$. The second term here will be negligible, and since $\Phi(1.28) = 0.9$ we will need $\delta - 1.96 = 1.28$, which gives $n = 525$ as the necessary sample size.

9. a) The likelihood $L$ is

$$\frac{\exp\Bigl[-\frac{1}{2\sigma_w^2}\sum_1^n(W_i - \mu_w)^2 - \frac{1}{2\sigma_x^2}\sum_1^n(X_i - \mu_x)^2 - \frac{1}{2\sigma_y^2}\sum_1^n(Y_i - \mu_y)^2\Bigr]}{\sqrt{2\pi}^{3n}\sigma_w^n\sigma_x^n\sigma_y^n},$$

and the Fisher information matrix is

$$I(\mu_w,\mu_x,\mu_y,\sigma_w,\sigma_x,\sigma_y) = \operatorname{diag}\left(\frac{1}{\sigma_w^2},\frac{1}{\sigma_x^2},\frac{1}{\sigma_y^2},\frac{2}{\sigma_w^2},\frac{2}{\sigma_x^2},\frac{2}{\sigma_y^2}\right).$$

Maximum likelihood estimators under the full model and under $H_0$ are given by

$$\hat\mu_w = \tilde\mu_w = \overline{W},\quad \hat\mu_x = \tilde\mu_x = \overline{X},\quad \hat\mu_y = \tilde\mu_y = \overline{Y},$$

$$\hat\sigma_w^2 = \frac{1}{n}\sum_{i=1}^n (W_i-\overline{W})^2,\quad \hat\sigma_x^2 = \frac{1}{n}\sum_{i=1}^n (X_i-\overline{X})^2,\quad \hat\sigma_y^2 = \frac{1}{n}\sum_{i=1}^n (Y_i-\overline{Y})^2,$$

and

$$\tilde\sigma^2 = \frac{\hat\sigma_w^2+\hat\sigma_x^2+\hat\sigma_y^2}{3}.$$

Plugging these in,

$$\lambda = \frac{L(\hat\theta)}{L(\tilde\theta)} = \left[\frac{\tilde\sigma^3}{\hat\sigma_w\hat\sigma_x\hat\sigma_y}\right]^n.$$

b) The natural choice for $\theta_0$ is $(\mu_w,\mu_x,\mu_y,2,2,2)'$, and then

$$I(\theta_0) = \operatorname{diag}\left(\frac14,\frac14,\frac14,\frac12,\frac12,\frac12\right).$$

Introduce the orthonormal basis

$$v_1 = \begin{bmatrix}1\\0\\0\\0\\0\\0\end{bmatrix},\qquad v_2 = \begin{bmatrix}0\\1\\0\\0\\0\\0\end{bmatrix},\qquad v_3 = \begin{bmatrix}0\\0\\1\\0\\0\\0\end{bmatrix},$$

$$v_4 = \frac{1}{\sqrt3}\begin{bmatrix}0\\0\\0\\1\\1\\1\end{bmatrix},\quad v_5 = \frac{1}{\sqrt2}\begin{bmatrix}0\\0\\0\\1\\-1\\0\end{bmatrix},\quad v_6 = \frac{1}{\sqrt6}\begin{bmatrix}0\\0\\0\\1\\2\\-2\end{bmatrix}.$$

As in Example 17.3, all of the tangent spaces $V = V_\theta$ are the same. Here $V$ is the linear span of $v_1$, $v_2$, $v_3$, and $v_4$, and $Q_0 = v_5v_5' + v_6v_6'$ is the orthogonal projection onto $V^\perp$. Noting that $Q_0 I(\theta_0)^{-1}Q_0 = 2Q_0$, that $\Delta = (\theta-\theta_0)\sqrt{n} = (0,0,0,-\sqrt8,\sqrt8,0)'$, and that $Q_0\Delta = \Delta$, the noncentrality parameter is

$$\delta^2 = \Delta' Q_0\bigl[P_0 + Q_0 I(\theta_0)^{-1}Q_0\bigr]^{-1}Q_0\Delta = \Delta'\Delta/2 = 8.$$

The likelihood ratio test with $\alpha \approx 5\%$ rejects $H_0$ if $2\log\lambda \geq 5.99$ (the 95th percentile of $\chi_2^2$). The power of this test is thus $P(2\log\lambda \geq 5.99)$. The distribution of $2\log\lambda$ should be approximately $\chi_2^2(\delta^2 = 8)$. If $F$ is the cumulative distribution function for this distribution, the power is approximately $1 - F(5.99) = 71.77\%$.

10. a) The log-likelihood is

$$\begin{aligned} l(\mu) &= -\frac{1}{2}\sum_{i=1}^{n} \|X_i - \mu\|^2 - np\log\sqrt{2\pi} \\ &= -\frac{n}{2}\|\overline{X} - \mu\|^2 - \frac{1}{2}\sum_{i=1}^{n}\|X_i - \overline{X}\|^2 - np\log\sqrt{2\pi}. \end{aligned}$$

By inspection, $\hat{\mu} = \overline{X}$ is the maximum likelihood estimator under the full model, and the maximum likelihood estimator under $H_0$ minimizes $\|\overline{X}-\mu\|$ over $\mu$ with $\|\mu\| = r$. The natural guess is $\tilde{\mu} = r\overline{X}/\|\overline{X}\|$. That this is indeed correct can be seen using the triangle inequality. Suppose $\|\mu\| = r$. Then if $\|\overline{X}\| > r$, $\|\overline{X}-\mu\|+\|\mu\| \geq \|\overline{X}\| = \|\tilde{\mu}\|+\|\overline{X}-\tilde{\mu}\|$, and so $\|\overline{X}-\mu\| \geq \|\overline{X}-\tilde{\mu}\|$; and if $\|\overline{X}\| < r$, $\|\overline{X}\| + \|\mu - \overline{X}\| \geq \|\mu\| = 1 = \|\overline{X}\| + \|\overline{X} - \tilde{\mu}\|$, and again $\|\overline{X}-\mu\| \geq \|\overline{X}-\tilde{\mu}\|$. So $2\log\lambda = 2l(\hat{\mu})-2l(\tilde{\mu}) = n\|\tilde{\mu}-\overline{X}\|^2 = n\big(\|\overline{X}\|-r\big)^2$.

b) By Lemma 14.8, because $\sqrt{n}\overline{X} \sim N(\sqrt{n}\mu, I)$,

$$n\|\overline{X}\|^2 \sim \chi_p^2(n\|\mu\|^2).$$

If $F$ is the cumulative distribution function for this distribution, then the power of the test, assuming $nr^2 > c$, is

$$\begin{aligned} P\big[n\big(\|\overline{X}\| - r\big)^2 > c\big] &= P\big(\sqrt{n}\|\overline{X}\| < \sqrt{n}r - \sqrt{c}\big) \\ &\quad + P\big(\sqrt{n}\|\overline{X}\| > \sqrt{n}r + \sqrt{c}\big) \\ &= F\big((\sqrt{n}r - \sqrt{c})^2\big) + 1 - F\big((\sqrt{n}r + \sqrt{c})^2\big). \end{aligned}$$

c) Since the Fisher information is the identity, the formula for the noncentrality parameter $\delta^2$ is just $\|Q_0\Delta\|^2$, where $Q_0$ is the projection onto the orthogonal complement of the tangent space at $\mu_0$ and $\Delta = \sqrt{n}(\mu - \mu_0)$. Now $\mu_0$ should be near $\mu$ and in the null parameter space, and the most natural choice is just $\mu_0 = \mu/\|\mu\|$. Then $\mu - \mu_0$ is in the orthogonal complement of the tangent space at $\mu_0$, and the noncentrality parameter is $\delta^2 = \|\Delta\|^2 = n\|\mu - \mu_0\|^2 = n/100$. Arguing as in the Example 17.4, $\delta^2 = 10.51$ will give power 90%, and 1051 observations will be necessary.

12. a) Because $Y_i$ and $W_i$ are linear functions of $X_i$, $\epsilon_i$, and $\eta_i$, they should have a bivariate normal distribution with mean zero and covariances given by

$$\mathrm{Var}(Y_i) = 1 + \beta^2, \qquad \mathrm{Var}(W_i) = 1 + \sigma^2,$$

and

$$\text{Cov}(Y_i, W_i) = \text{Cov}(\beta X_i + \epsilon_i, X_i + \eta_i) = \beta.$$

b) The log-likelihood is

$$l(\beta, \sigma^2) = -\frac{n}{2}\log(2\pi) - \frac{n}{2}\log(1 + \sigma^2 + \sigma^2\beta^2) \\ - n\frac{(1+\sigma^2)T_1 - 2\beta T_2 + (1+\beta^2)T_3}{2(1+\sigma^2+\sigma^2\beta^2)},$$

with

$$T_1 = \frac{1}{n}\sum_{i=1}^n Y_i^2, \quad T_2 = \frac{1}{n}\sum_{i=1}^n W_i Y_1, \text{ and } T_3 = \frac{1}{n}\sum_{i=1}^n W_i^2.$$

Maximum likelihood estimators for $\hat{\beta}$ and $\hat{\sigma^2}$ can be obtained solving the following equations, obtained setting partial derivatives of $l$ to zero:

$$1 + \beta^2 + T_1 = \frac{[(1+\sigma^2)T_1 - 2\beta T_2 + (1+\beta^2)T_3](1+\beta^2)}{1+\sigma^2+\sigma^2\beta^2},$$

$$2\beta\sigma^2 - 2T_2 + 2\beta T_3 = \frac{[(1+\sigma^2)T_1 - 2\beta T_2 + (1+\beta^2)T_3](2\beta\sigma^2)}{1+\sigma^2+\sigma^2\beta^2}.$$

Explicit formulas do not seem possible. Under $H_0 : \beta = 0$, the maximum likelihood estimator for $\sigma^2$ is $\tilde{\sigma^2} = T_3/n - 1$, obtained setting $\partial l(0, \sigma^2)/\partial\sigma^2$ to zero.

c) When $\sigma = 0$, then the least squares estimator for $\beta$ is $T_2/T_1$. By the law of large numbers, $T_2 \xrightarrow{p} EY_1W_1 = \beta$ and $T_1 \xrightarrow{p} EW_1^2 = 1 + \sigma^2$. So the least squares estimator converges in probability to $\beta/(1+\sigma^2)$ and is inconsistent unless $\sigma^2 = 0$ or $\beta = 0$.

d) Let $\theta = (\beta, \sigma^2)'$. The Fisher information is

$$I(\theta) = \frac{1}{D^2}\begin{pmatrix} 2 + 2\sigma^2 + 6\beta^2\sigma^4 & \beta + 2\beta^2 + \beta\sigma^2 + \beta^3\sigma^2 \\ \beta + 2\beta^2 + \beta\sigma^2 + \beta^3\sigma^2 & (1+\beta^2)^2/2 \end{pmatrix},$$

where $D = 1 + \sigma^2 + \beta^2\sigma^2$. Under $H_0$, $\beta = 0$ and $I(\theta)$ is diagonal with inverse

$$I(\theta)^{-1} = \begin{pmatrix} \frac{1}{2}(1+\sigma^2) & 0 \\ 0 & 2(1+\sigma^2)^2 \end{pmatrix}.$$

If $e_1 = \binom{1}{0}$ and $e_2 = \binom{0}{1}$, the standard basis vectors, then the projection matrices in the formula for the noncentrality parameter $\delta^2$ are $P_0 = e_2e_2'$ and $Q_0 = e_1e_1'$. Taking $\theta_0 = \binom{0}{\sigma^2}$, $\Delta = \sqrt{n}\binom{\beta}{0}$ and

$$\begin{aligned}\delta^2 &= n\beta^2 e_1'[P_0 + Q_0 I(\theta_0)^{-1}Q_0]^{-1}e_1 \\ &= n\beta^2 e_1'\left(P_0 + \frac{Q_0}{e_1'I(\theta_0)^{-1}e_1}\right)e_1 \\ &= \frac{2n\beta^2}{1+\sigma^2}.\end{aligned}$$

The power of the test is approximately $P\big((Z+\delta)^2 > 1.96^2\big)$ with $Z \sim N(0,1)$. It decreases as $\sigma^2$ increases.

13. a) Viewing $(p_1, p_2)$ as the parameter with $p_3$ determined as $1-p_1-p_2$, the parameter space $\Omega$ is the triangular-shaped region where $p_1 > 0$, $p_2 > 0$, and $p_1 + p_2 < 1$, an open set in $\mathbb{R}^2$, and the likelihood function is

$$L(p) = \binom{n}{Y_1, Y_2, Y_3} p_1^{Y_1} p_2^{Y_2} p_3^{Y_3}.$$

Maximum likelihood estimators under the full model are $\hat{p}_i = Y_i/n$, $i = 1$, 2, 3. The parameter space under $H_0$ is

$$\Omega_0 = \{(1-e^{-\theta}, e^{-\theta} - e^{-2\theta}) : \theta > 0\}.$$

The maximum likelihood estimator $\tilde{\theta}$ for $\theta$ under $H_0$ maximizes

$$\log L(1-e^{-\theta}, e^{-\theta} - e^{-2\theta}).$$

Setting the $\theta$-derivative to zero, $\tilde{\theta} = -\log[(Y_2 + 2Y_3)/(2n - Y_1)]$. Plugging in the estimators, $2\log\lambda$ is

$$2\left[\sum_{i=1}^{3} Y_i \log\frac{Y_i}{n} - (Y_1+Y_2)\log\frac{Y_1+Y_2}{Y_1+2Y_2+2Y_3} \right.$$
$$\left. - (Y_2+2Y_3)\log\frac{Y_2+2Y_3}{Y_1+2Y_2+2Y_3}\right].$$

b) With $Y = (36, 24, 40)$, $2\log\lambda = 0.0378$ and we accept $H_0$. The $p$-value is $P(\chi^2(1) > 0.0378) = 84.6\%$.

c) Take $\theta_0 = (0.4, 0.24)$, a convenient point in $\Omega_0$ close to $\theta = (0.36, 0.24)$ (any other reasonable choice should give a similar answer). Inverse Fisher information at $\theta$ is

$$I(\theta_0)^{-1} = \begin{pmatrix} 0.240 & -0.0960 \\ -0.096 & 0.1824 \end{pmatrix},$$

and points $\theta$ in $\Omega_0$ satisfy the constraint $g(\theta) = \theta_1^2 + \theta_2 - \theta_1 = 0$. If $V_0$ is the tangent space at $\theta_0$, then $V_0^{\perp}$ is the linear span of

$$v = \nabla g(\theta_0) = \begin{pmatrix} 2\theta_1 - 1 \\ 1 \end{pmatrix}_{\theta=\theta_0} = \begin{pmatrix} -0.2 \\ 1 \end{pmatrix},$$

and $Q_0 = vv'/\|v\|^2$. Also, $\Delta = \sqrt{n}(\theta - \theta_0) = \sqrt{n}\binom{-0.04}{0}$. So

$$\delta^2 = \Delta' Q_0 \big(P_0 + Q_0 I(\theta_0)^{-1} Q_0\big)^{-1} Q_0 \Delta$$
$$= \frac{(\Delta\cdot v)^2}{v' I(\theta_0)^{-1} v} = \frac{n(0.008)^2}{0.2304} = \frac{n}{3600}.$$

For 90% power, $\delta^2$ needs to be 10.51 which leads to $n = 37{,}836$ as the requisite sample size.

# References


Anderson, T. W. (1955). The integral of a symmetric unimodal function over a symmetric convex set and some probability inequalities. *Proc. Amer. Math. Soc. 6*, 170–176.

Anscombe, F. J. (1952). Large sample theory of sequential estimation. *Proc. Cambridge Philos. Soc. 48*, 600–607.

Bahadur, R. R. (1954). Sufficiency and statistical decision functions. *Ann. Math. Statist. 25*, 423–462.

Basu, D. (1955). On statistics independent of a complete sufficient statistic. *Sankhyā 15*, 377–380.

Basu, D. (1958). On statistics independent of a complete sufficient statistic. *Sankhyā 20*, 223–226.

Berger, J. (1985). *Statistical Decision Theory and Bayesian Analysis* (Second ed.). New York: Springer.

Berry, D. A. and B. Fristedt (1985). *Bandit Problems: Sequential Allocation of Experiments.* London: Chapman and Hall.

Bhattacharya, R. N. and R. R. Rao (1976). *Normal Approximation and Asymptotic Expansions.* New York: Wiley.

Bickel, P. J. and K. A. Doksum (2007). *Mathematical Statistics: Basic Ideas and Selected Topics* (Second ed.). Upper Saddle River, NJ: Prentice Hall.

Bickel, P. J. and D. A. Freedman (1981). Some asymptotic theory for the bootstrap. *Ann. Statist. 9*, 1196–1217.

Billingsley, P. (1961). *Statistical Inference for Markov Processes*, Volume 2 of *Statistical Research Monographs.* Chicago: University of Chicago.

Billingsley, P. (1995). *Probability and Measure* (Third ed.). New York: Wiley.

Blackwell, D. A. (1951). Comparison of experiments. In *Proceedings of the Second Berkeley Symposium on Mathematical Statistics and Probability*, pp. 93–102. Berkeley: University of California Press.

Blyth, C. R. (1951). On minimax statistical decision procedures and their admissibility. *Ann. Math. Statist. 22*, 22–42.

Box, G. E. P. and K. B. Wilson (1951). On the experimental attainment of optimum conditions. *J. Roy. Statist. Soc. Ser. B 13*, 1–45. With discussion.

Brouwer, L. (1912). Zur invarianz des $n$-dimensionalen gebiets. *Mathematische Annalen 72*, 55–56.

Brown, B. M. (1971a). Martingale central limit theorems. *Ann. Math. Statist. 42*, 59–66.

Brown, L. D. (1971b). Admissible estimators, recurrent diffusions, and insoluble boundary-value problems. *Ann. Math. Statist. 42*, 855–903.

Brown, L. D. (1986). *Fundamentals of Statistical Exponential Families with Applications in Statistical Decision Theory.* Hayward, CA: Institute of Mathematical Statistics.

Chen, L. (1975). Poisson approximation for dependent trials. *Ann. Prob. 3*, 534–545.

Chernoff, H. and L. E. Moses (1986). *Elementary Decision Theory.* New York: Dover.

Clark, R. M. (1977). Non-parametric estimation of a smooth regression function. *J. Roy. Statist. Soc. Ser. B 39*, 107–113.

Cody, W. J. (1965). Chebyshev approximations for the complete elliptic integrals $K$ and $E$. *Math. of Comput. 19*, 105–112.

DasGupta, A. (2008). *Asymptotic Theory of Statistics and Probability* (First ed.). New York: Springer.

De Boor, C. (2001). *A Practical Guide to Splines.* New York: Springer.

DeGroot, M. H. (1970). *Optimal Statistical Decisions.* New York: McGraw-Hill.

Dempster, A. P., N. M. Laird, and D. B. Rubin (1977). Maximum likelihood from incomplete data via the em algorithm. *J. Roy. Statist. Soc. Ser. B 39*, 1–22.

Draper, N. R. and R. C. van Nostrand (1979). Ridge regression and James–Stein estimation: Review and comments. *Technometrics 21*, 451–466.

Eaton, M. L. (1983). *Multivariate Statistics: A Vector Space Approach.* New York: Wiley.

Eaton, M. L. (1989). *Group Invariance Applications in Statistics*, Volume 1 of *CBMS-NSF Regional Conference Series in Probability and Statistics.* Hayward, CA and Alexandria, VA: IMS and ASA.

Fabian, V. (1968). On asymptotic normality in stochastic approximation. *Ann. Math. Statist. 39*, 1327–1332.

Farrell, R. H. (1964). Estimation of the location parameter in the absolutely continuous case. *Ann. Math. Statist. 35*, 949–998.

Farrell, R. H. (1968a). On a necessary and sufficient condition for admissibility of estimators when strictly convex loss is used. *Ann. Math. Statist. 39*, 23–28.

Farrell, R. H. (1968b). Towards a theory of generalized Bayes tests. *Ann. Math. Statist. 39*, 1–22.

Feller, W. (1971). *An Introduction to Probability Theory and Its Applications*, Volume 2. New York: Wiley.

Ferguson, T. (1967). *Mathematical Statistics: A Decision Theoretic Approach.* New York: Academic Press.

Fieller, E. C. (1954). Some problems in interval estimation. *J. Roy. Statist. Soc. Ser. B 16*, 175–183.

Freedman, D. (2007). How can the score test be inconsistent? *Amer. Statist. 61*, 291–295.

Geman, S. and D. Geman (1984). Stochastic relaxation, Gibbs distributions, and the Bayesian restoration of images. *IEEE Trans. Patt. Anal. Mach. Intell. 6*, 721–741.

Gittins, J. C. (1979). Bandit processes and dynamic allocation indices. *J. Roy. Statist. Soc. Ser. B 41*, 148–177. With discussion.

Gittins, J. C. and D. Jones (1974). A dynamic allocation index for the sequential design of experiments. In J. Gani, K. Sarkadi, and I. Vincze (Eds.), *Progress in Statistics*, pp. 241–266. Amsterdam: North-Holland.

Hájek, J. (1972). Local asymptotic minimax and admissibility in estimation. In *Proc. Sixth Berkeley Symp. on Math. Statist. and Prob.*, Volume 1, pp. 175–194. Berkeley: University of California Press.

Hall, P. (1992). *The Bootstrap and Edgeworth Expansion.* New York: Springer-Verlag.

Hall, P. and C. C. Heyde (1980). *Martingale Limit Theory and Its Application.* New York: Academic.

Hill, B. M. (1963). The three-parameter lognormal distribution and Bayesian analysis of a point-source epidemic. *J. Amer. Statist. Assoc. 58*, 72–84.

Huber, P. (1964). Robust estimation of a location parameter. *Ann. Math. Statist. 35*, 73–101.

James, W. and C. Stein (1961). Estimation with quadratic loss. In *Proc. Fourth Berkeley Symp. on Math. Statist. and Prob.*, Volume 1, pp. 361–379. Berkeley: University of Calififornia Press.

Karlin, S. and H. M. Taylor (1975). *A First Course in Stochastic Processes* (Second ed.). New York: Academic Press.

Katehakis, M. N. and A. F. Veinott, Jr. (1987). The multi-armed bandit problem: Decomposition and composition. *Math. Oper. Res. 12*, 262–268.

Kemeny, J. B. and J. L. Snell (1976). *Finite Markov Chains.* New York: Springer.

Kiefer, J. and J. Wolfowitz (1952). Stochastic estimation of the maximum of a regression function. *Ann. Math. Statist. 23*, 452–466.

Lai, T. L. (2003). Stochastic approximation. *Ann. Statist. 31*, 391–406.

Landers, D. and L. Rogge (1972). Minimal sufficient $\sigma$-fields and minimal sufficient statistics. Two counterexamples. *Ann. Math. Statist. 43*, 2045–2049.

Lange, K. (2004). *Optimization.* New York: Springer.

Le Cam, L. M. (1955). An extension of Wald's theory of statistical decision functions. *Ann. Math. Statist. 26*, 69–81.

Le Cam, L. M. (1986). *Asymptotic Methods in Statistical Decision Theory.* New York: Springer-Verlag.

Le Cam, L. M. and G. L. Yang (2000). *Asymptotics in Statistics: Some Basic Concepts* (Second ed.). New York: Springer-Verlag.

Lehmann, E. L. (1959). *Testing Statistical Hypotheses* (First ed.). New York: Wiley.

Lehmann, E. L. (1981). Interpretation of completeness and Basu's theorem. *J. Amer. Statist. Assoc. 76*, 335–340.

Lehmann, E. L. and J. P. Romano (2005). *Testing Statistical Hypotheses* (Third ed.). New York: Springer.

McLachlan, G. J. and T. Krishnan (2008). *The EM Algorithm and Extensions* (Second ed.). Hoboken, NJ: Wiley.

Meyn, S. P. and R. L. Tweedie (1993). *Markov Chains and Stochastic Stability.* New York: Springer.

Miescke, K.-J. and F. Liese (2008). *Statistical Decision Theory: Estimation, Testing, Selection.* New York: Springer.

Nummelin, E. (1984). *General Irreducible Markov Chains and Non-Negative Operators.* Cambridge, UK: Cambridge University Press.

Rao, C. R. (1948). Large sample tests of statistical hypotheses concerning several parameters with applications to problems of estimation. *Proc. Camb. Phil. Soc. 44*, 50–57.

Rao, M. M. (2005). *Conditional Measures and Applications* (Second ed.). Boca Raton, FL: CRC Press.

Robbins, H. and S. Monro (1951). A stochastic approximation method. *Ann. Math. Statist. 22*, 400–407.

Rockafellar, R. T. (1970). *Convex Analysis.* Princeton, NJ: Princeton University Press.

Roussas, G. G. (1972). *Contiguity of Probability Measures: Some Applications in Statistics*, Volume 63 of *Cambridge Tracts in Mathematics and Mathematical Physics.* Cambridge, UK: Cambridge University Press.

Ruppert, D. (1991). Stochastic approximation. In B. K. Ghosh and P. K. Sen (Eds.), *Handbook of Sequential Analysis*, pp. 503–529. New York: Dekker.

Silverman, B. W. (1982). On the estimation of a probability density function by the maximum penalized likelihood method. *Ann. Statist. 10*, 795–810.

Singh, K. (1981). On the asymptotic accuracy of Efron's bootstrap. *Ann. Statist. 9*, 1187–1195.

Stein, C. (1955). A necessary and sufficient condition for admissibility. *Ann. Math. Statist. 26*, 518–522.

Stein, C. (1956). Inadmissibility of the usual estimator for the mean of a multivariate normal distribution. In *Proc. Third Berkeley Symp. on Math. Statist. and Prob.*, Volume 1, pp. 197–206. Berkeley: University of California Press.

Stein, C. (1986). *Approximate Computation of Expectations*, Volume 7 of *Lecture Notes—Monograph Series.* Hayward, CA: Institute of Mathematical Statistics.

Tierney, L. (1994). Markov chains for exploring posterior distributions. *Ann. Statist. 22*, 1701–1762. With discussion.

van der Vaart, A. W. (1998). *Asymptotic Statistics.* Cambridge, UK: Cambridge University Press.

Wahba, G. (1990). *Spline Models for Observational Data.* Number 59 in CBMS-NSF Regional Conference Series in Applied Mathematics. Philadelphia: Society for Industrial and Applied Mathematics.

Wald, A. (1943). Tests of statistical hypotheses concerning several parameters when the number of observations is large. *Trans. Am. Math. Soc. 54*, 426–482.

Wald, A. (1947). *Sequential Analysis.* New York: John Wiley and Sons.

Wald, A. (1949). Note on the consistency of the maximum likelihood estimate. *Ann. Math. Statist. 20*, 595–601.

Wald, A. and J. Wolfowitz (1948). Optimum character of the sequential probability ratio test. *Ann. Math. Statist. 19*, 326–339.

Wand, M. P. and M. C. Jones (1995). *Kernel Smoothing.* London: Chapman and Hall.

Whittle, P. (1980). Multi-armed bandits and the Gittins index. *J. Roy. Statist. Soc. Ser. B 42*, 143–149.

Woodroofe, M. (1989). Very weak expansions for sequentially designed experiments: Linear models. *Ann. Statist. 17*, 1087–1102.

Woodroofe, M. and G. Simons (1983). The Cramer–Rao inequality holds almost everywhere. In *Recent Advances in Statistics: Papers in Honor of Herman Chernoff on his Sixtieth Birthday*, pp. 69–92. New York: Academic Press.

Wu, C. F. J. (1983). On the convergence properties of the EM algorithm. *Ann. Statist. 11*, 95–103.

# Index

GPSR Compliance
The European Union's (EU) General Product Safety Regulation (GPSR) is a set of rules that requires consumer products to be safe and our obligations to ensure this.

If you have any concerns about our products, you can contact us on

ProductSafety@springernature.com

In case Publisher is established outside the EU, the EU authorized representative is:

Springer Nature Customer Service Center GmbH
Europaplatz 3
69115 Heidelberg, Germany

www.ingramcontent.com/pod-product-compliance
Ingram Content Group UK Ltd.
Pitfield, Milton Keynes, MK11 3LW, UK
UKHW020130290926
15471UKWH00003B/385
*9781461426707*